25 Years
Previous Solved Papers

AF660733

ESE 2025

UPSC ENGINEERING SERVICES EXAMINATION

Preliminary Examination

ELECTRONICS AND TELECOMMUNICATION ENGINEERING

Objective Solved Papers

Volume-II

- Topicwise presentation
- Thoroughly revised & updated

Also useful for

State Engineering Services Examinations, Public Sector Examinations & Other Competitive Examinations

MADE EASY PUBLICATIONS

www.madeeasypublications.org

MADE EASY Publications Pvt. Ltd.

Corporate Office: 44-A/4, Kalu Sarai (Near Hauz Khas Metro Station), New Delhi-110016

Contact: 9021300500

E-mail: infomep@madeeasy.in

Visit us at: www.madeeasypublications.org

ESE-2025 : Preliminary Examination
Electronics and Telecommunication Engineering : Volume-II
Topicwise Objective Solved Questions : (2000-2024)

1st Edition : 2008
2nd Edition : 2009
3rd Edition : 2010
4th Edition : 2011
5th Edition : 2012
6th Edition : 2013
7th Edition : 2014
8th Edition : 2015
9th Edition : 2016
10th Edition : 2017
11th Edition : 2018
12th Edition : 2019
13th Edition: 2020
14th Edition: 2021
15th Edition: 2022
16th Edition: 2023
17th Edition: 2024

Director's Message

B. Singh (Ex. IES)

Engineering is one of the most chosen graduating field. Taking engineering is usually a matter of interest but this eventually develops into "purpose of being an engineer" when you choose engineering services as a career option.

Train goes in tunnel we don't panic but sit still and trust the engineer, even we don't doubt on signalling system, we don't think twice crossing over a bridge reducing our travel time; every engineer has a purpose in his department which when coupled with his unique talent provides service to mankind.

I believe *"the educator must realize in the potential power of his pupil and he must employ all his art, in seeking to bring his pupil to experience this power"*. To support dreams of every engineer and to make efficient use of capabilities of aspirant, MADE EASY team has put sincere efforts in compiling all the previous years' ESE-Pre questions with accurate and detailed explanation. The objective of this book is to facilitate every aspirant in ESE preparation and so, questions are segregated chapterwise and topicwise to enable the student to do topicwise preparation and strengthen the concept as and when they are read.

I would like to acknowledge efforts of entire MADE EASY team who worked hard to solve previous years' papers with accuracy and I hope this book will stand up to the expectations of aspirants and my desire to serve student fraternity by providing best study material and quality guidance will get accomplished.

B. Singh (Ex. IES)
CMD, MADE EASY Group

Volume-II

ELECTRONICS & TELECOM. ENGG.

Objective Solved Questions

of UPSC Engineering Services Examination

Contents

Sl.	Topic	Pages
1.	Control Systems	1 - 114
2.	Signals and Systems + DSP	115 - 193
3.	Electromagnetics	194 - 312
4.	Computer Organization and Architecture	313 - 374
5.	Microprocessors and Microcontrollers	375 - 401
6.	Analog and Digital Communication Systems	402 - 467
7.	Advanced Communication Topics	468 - 500
8.	Advanced Electronics Topics	501 - 511

■■■■

UNIT I

Control Systems

Syllabus

Signal flow graphs, Routh-Hurwitz criteria, root loci, Nyquist/Bode plots; Feedback systems-open &close loop types, stability analysis, steady state, transient and frequency response analysis; Design of control systems, compensators, elements of lead/lag compensation, PID and industrial controllers.

Contents

Sl.	Topic	Page No.
1.	Basics, Block Diagrams and Signal Flow Graphs	2
2.	Time Domain Analysis	8
3.	Routh-Hurwitz Stability Criterion	36
4.	Root Locus	47
5.	Frequency Domain Analysis	57
6.	Compensators and Controllers	80
7.	State Space Analysis	98
8.	Miscellaneous	105

1 Basics, Block Diagrams and Signal Flow Graphs

1.1 **Assertion (A):** Feedback control systems offer more accurate control over open-loop systems.
Reason (R): The feedback path establishes a link for input and output comparison and subsequent error correction.
(a) Both A and R are true and R is the correct explanation of A
(b) Both A and R are true but R is NOT the correct explanation of A
(c) A is true but R is false
(d) A is false but R is true [ESE-2000]

1.2 Consider the following statements:
1. The effect of feedback is to reduce the system error
2. Feedback increases the gain of the system in one frequency range but decreases in another.
3. Feedback can cause a system that is originally stable to become unstable

Which of these statements are correct?
(a) 1, 2 and 3 (b) 1 and 2
(c) 2 and 3 (d) 1 and 3
[ESE-2000]

1.3 For the signal flow diagram shown in the given figure, the transmittance between x_2 and x_1 is

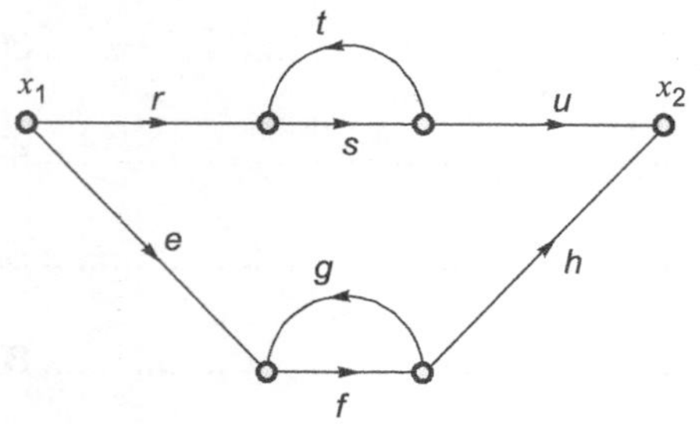

(a) $\frac{rsu}{1-st}+\frac{efh}{1-fg}$ (b) $\frac{rsu}{1-fg}+\frac{efh}{1-st}$

(c) $\frac{efh}{1-ru}+\frac{rsu}{1-eh}$ (d) $\frac{rst}{1-eh}+\frac{rsu}{1-st}$
[ESE-2001]

1.4 Consider the following statements with respect to feedback control systems:
1. Accuracy cannot be obtained by adjusting loop gain.
2. Feedback decreases overall gain.
3. Introduction of noise due to sensor reduces overall accuracy.
4. Introduction of feedback may lead to the possibility of instability of closed loop system.

Which of the statements given above are correct?
(a) 1, 2, 3 and 4 (b) Only 1, 2 and 4
(c) Only 1 and 3 (d) Only 2, 3 and 4
[ESE-2006]

1.5 A control system whose step response is $-0.5(1+e^{-2t})$ is cascaded to another control block whose impulse response is e^{-t}. What is the transfer function of the cascaded combination?

(a) $\frac{1}{(s+1)(s+2)}$ (b) $\frac{1}{s(s+1)}$

(c) $\frac{-1}{s(s+2)}$ (d) $\frac{0.5}{(s+1)(s+2)}$
[ESE-2007]

1.6 If the initial conditions for a system are inherently zero, what does it physically mean?
(a) The system is at rest but stores energy
(b) The system is working but does not store energy
(c) The system is at rest or no energy is stored in any of its parts
(d) The system is working with zero reference input
[ESE-2007]

1.7 The impulse response of a linear time invariant system is given as

$$g(t) = e^{-t}, t > 0.$$

The transfer function of the system is equal to
(a) $1/s$ (b) $1/[s(s+1)]$
(c) $1/(s+1)$ (d) $s/(s+1)$
[ESE-2008]

1.8 In case of DC servo-motor the back-emf is equivalent to an "electric friction" which tends to
(a) improve stability of the motor
(b) slowly decrease stability of the motor
(c) very rapidly decrease stability of the motor
(d) have no effect on stability [ESE-2008]

1.9 The transfer function of a linear-time-invariant system is given as $\frac{1}{(s+1)}$. What is the steady-state value of the unit-impulse response?
(a) Zero (b) One
(c) Two (d) Infinite [ESE-2009]

1.10 In closed loop control system, what is the sensitivity of the gain of the overall system, M to the variation in G?
(a) $\frac{1}{1+G(s)H(s)}$ (b) $\frac{1}{1+G(s)}$
(c) $\frac{G(s)}{1+G(s)H(s)}$ (d) $\frac{G(s)}{1+G(s)}$
[ESE-2009]

1.11 A linear time-invariant system initially at rest, when subjected to a unit-step input, gives a response $y(t) = te^{-t}$, $t > 0$. The transfer function of the system is:
(a) $\frac{1}{(s+1)^2}$ (b) $\frac{1}{s(s+1)^2}$
(c) $\frac{s}{(s+1)^2}$ (d) $\frac{1}{s+1}$ [ESE-2010]

1.12 Consider the function $F(s) = \frac{5}{s(s^2+3s+2)}$, where $F(s)$ is Laplace transform of function $f(t)$. The initial value of $f(t)$ is:
(a) 5 (b) 5/2
(c) 5/3 (d) 0
[ESE-2010]

1.13 The transfer function from $d(s)$ to $y(s)$ is

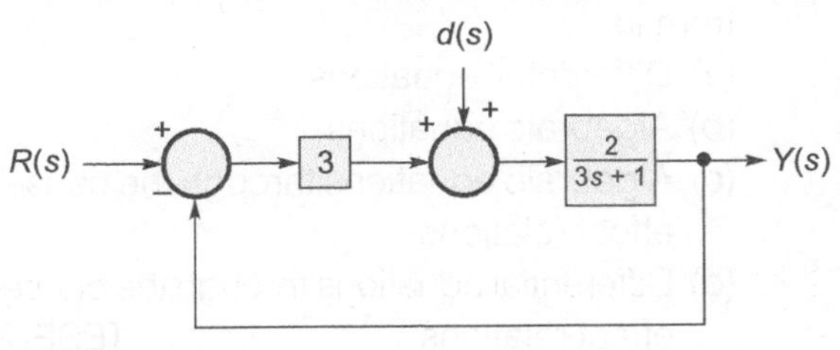

(a) $\frac{2}{3s+7}$ (b) $\frac{2}{3s+1}$
(c) $\frac{6}{3s+7}$ (d) $\frac{2}{3s+6}$ [ESE-2010]

1.14 What is the unit impulse response of the system shown in figure for $t \geq 0$?

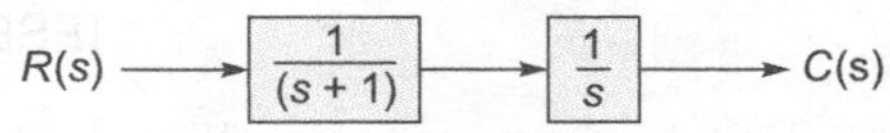

(a) $1 + e^{-t}$ (b) $1 - e^{-t}$
(c) e^{-t} (d) $-e^{-t}$ [ESE-2011]

1.15 An electric motor is developing 10 kW at a speed of 900 rpm. The torque available at the shaft is
(a) 106 N-m (b) 66 N-m
(c) 1600 N-m (d) 90 N-m
[ESE-2013]

1.16 **Statement (I):** Many of the linear control system transfer functions do not have poles or zeros in the right half of s-plane.
Statement (II): These are called minimum-phase transfer functions.
(a) Both Statement (I) and Statement (II) are individually true and Statement (II) is the correct explanation of Statement (I).
(b) Both Statement (I) and Statement (II) are individually true but Statement (II) is not the correct explanation of Statement (I).
(c) Statement (I) is true but Statement (II) is false.
(d) Statement (I) is false but Statement (II) is true.
[ESE-2013]

1.17 A tachometer has a sensitivity of 5 V/1000 rpm. The gain constant of the tachometer is
(a) 0.48 V/rad/sec (b) 0.048 V/rad/sec
(c) 4.8 V/rad/sec (d) 48 V/rad/sec
[ESE-2014]

1.18 The closed loop transfer function of a unity negative feedback system is $\frac{100}{s^2+8s+100}$. Its open loop transfer function is
(a) $\frac{100}{s+8}$ (b) $\frac{1}{s^2+8s}$
(c) $\frac{100}{s^2-8s}$ (d) $\frac{100}{s^2+8s}$
[ESE-2015]

1.19 The Laplace transform of $e^{-2t}\sin(2\omega t)$ is

(a) $\dfrac{2s}{(s+2)^2+2\omega^2}$ (b) $\dfrac{2\omega}{(s-2)^2+4\omega^2}$

(c) $\dfrac{2\omega}{(s+2)^2+4\omega^2}$ (d) $\dfrac{2s}{(s-2)^2+2\omega^2}$

[ESE-2015]

1.20 In a closed loop system for which the output is the speed of a motor, the output rate control can be used to

(a) Limit the speed of the motor

(b) Limit the torque output of the motor

(c) Reduce the damping of the system

(d) Limit the acceleration of the motor

[ESE-2015]

1.21 In a servo-system, the device used for providing derivative feedback is known as

(a) Synchro (b) Servomotor

(c) Potentiometer (d) Techogenerator

[ESE-2015]

1.22 The transfer function of any stable system which has no zeros or poles in the right half of the s-plane is said to be

(a) Minimum phase transfer function

(b) Non-minimum phase transfer function

(c) Minimum frequency response function

(d) Minimum gain transfer function

[ESE-2015]

1.23 A dominant pole is determined as

(a) the highest frequency pole among all poles.

(b) the lowest frequency pole at least two octaves lower than other poles.

(c) the lowest frequency pole among all poles.

(d) the highest frequency pole at least two octaves higher than other poles.

[ESE-2017]

1.24 Consider the following statements for signal flow graph:

1. It represents linear as well as non-linear systems.
2. It is not unique for a given system.

Which of the above statements is /are correct?

(a) 1 only (b) 2 only

(c) Both 1 and 2 (d) Neither 1 nor 2

[ESE-2018]

1.25 The open-loop transfer function of a system is

$$\frac{10K}{1+10s}$$

When the system is converted into a closed-loop with unity feedback, the time constant of the system is reduced by a factor of 20. The value of K is

(a) 1.9 (b) 1.6

(c) 1.3 (d) 1.0 [ESE-2018]

1.26 The closed-loop transfer function $\dfrac{C(s)}{R(s)}$ of the system represented by the block diagram in the figure is

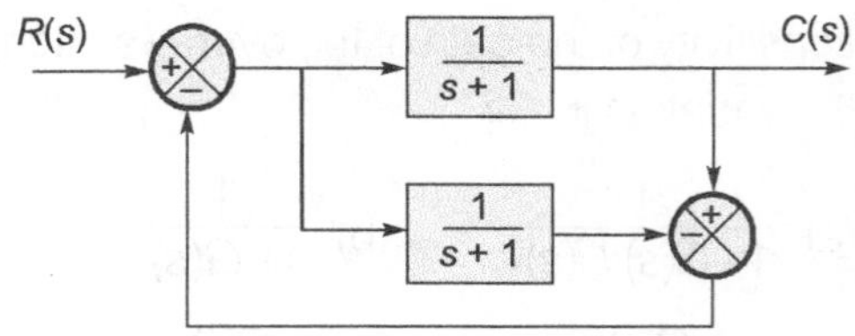

(a) $\dfrac{1}{(s+1)^2}$ (b) $\dfrac{1}{s+1}$

(c) $s + 1$ (d) 1 [ESE-2018]

1.27 The price for improvement in sensitivity by the use of feedback is paid in terms of

(a) loss of system gain

(b) rise of system gain

(c) improvement in transient response, delayed response

(d) poor transient response [ESE-2019]

1.28 Consider the following open-loop transfer function:

$$G = \frac{K(s+2)}{(s+1)(s+4)}$$

The characteristic equation of the unity negative feedback will be

(a) $(s + 1)(s + 4) + K(s + 2) = 0$

(b) $(s + 2)(s + 1) + K(s + 4) = 0$

(c) $(s + 1)(s - 2) + K(s + 4) = 0$

(d) $(s + 2)(s + 4) + K(s + 1) = 0$ [ESE-2019]

1.29 The signal flow graph of a system is constructed from its

(a) Differential equations

(b) Algebraic equations

(c) Algebraic equations through the cause-and-effect relations

(d) Differential equations through the cause-and-effect relations [ESE-2020]

1.30 The important aspects in the study of feedback systems are to control
1. Sensitivity
2. Effect of an internal disturbance
3. Distortion in a nonlinear system

(a) 1 and 2 only (b) 1 and 3 only
(c) 2 and 3 only (d) 1, 2 and 3

[ESE-2020]

1.31 The subsystem that generates the input to the plant or process is known as

(a) controller (b) controlled variable
(c) controllability (d) compensator

[ESE-2022]

1.32 The open-loop DC gain of a unity negative feedback system with closed-loop transfer function $\frac{s+4}{s^2+7s+13}$ is

(a) $\frac{4}{13}$ (b) $\frac{2}{3}$
(c) $\frac{1}{3}$ (d) $\frac{4}{9}$ [ESE-2023]

Answers Basics, Block Diagrams and Signal Flow Graphs

1.1 (a)	1.2 (a)	1.3 (a)	1.4 (d)	1.5 (*)	1.6 (c)	1.7 (c)	1.8 (a)	1.9 (a)
1.10 (a)	1.11 (c)	1.12 (d)	1.13 (a)	1.14 (b)	1.15 (a)	1.16 (a)	1.17 (b)	1.18 (d)
1.19 (c)	1.20 (a)	1.21 (d)	1.22 (a)	1.23 (c)	1.24 (b)	1.25 (a)	1.26 (b)	1.27 (a)
1.28 (a)	1.29 (c)	1.30 (d)	1.31 (a)	1.32 (d)				

Explanations Basics, Block Diagrams and Signal Flow Graphs

1.2 (a)

Feedback is applied to reduce the system error. Consider the example.

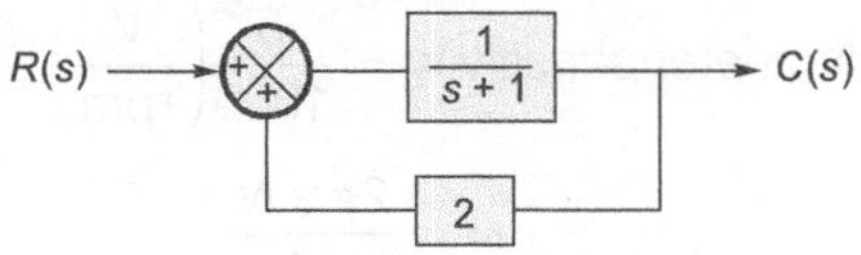

$$\frac{C(s)}{R(s)} = \frac{G(s)}{1-G(s)H(s)} = \frac{\frac{1}{s+1}}{1-\frac{2}{s+1}} = \frac{1}{s-1}$$

Thus, we see that the closed loop system is unstable while the open loop system is stable.

1.3 (a)

$$\frac{X_2}{X_1} = \frac{P_1\Delta_1 + P_2\Delta_2}{\Delta}$$

$P_1 = rsu,\ \Delta_1 = 1 - fg$

$P_2 = efh,\ \Delta_2 = 1 - st,\ \Delta = 1 - fg - st + fgst$

$$\frac{X_2}{X_1} = \frac{rsu(1-fg)+efh(1-st)}{1-fg-st+fgst}$$

$$= \frac{rsu(1-fg)+efh(1-st)}{(1-fg)(1-st)}$$

$$\frac{X_2}{X_1} = \frac{rsu}{1-st}+\frac{efh}{1-fg}$$

1.5 (*)

$$TF_1 = sL[sR]$$

$$= s\left[\frac{-0.5}{5}-\frac{0.5}{s+2}\right] = \frac{-(s+1)}{s+2}$$

$$TF_2 = L[IR] = \frac{1}{s+1}$$

$$\therefore \quad TF = TF_1 \times TF_2 = \frac{-1}{s+2}$$

1.6 (c)

A system with zero initial conditions is said to be at rest since there is no stored energy.

1.7 (c)

Impulse response, $g(t) = e^{-t},\ t > 0$

Transfer function, $G(s) = L\{g(t)\} = \frac{1}{s+1}$

1.9 (a)

Steady state value $= \lim_{s\to 0} s\frac{1}{(s+1)} = 0$

1.10 (a)

$$M(s) = \frac{G(s)}{1+G(s)H(s)}$$

Sensitivity of M to the variation in G is

$$\frac{dM}{dG} \times \frac{G}{M}$$

$$\frac{dM}{dG} = \frac{1+G(s)H(s)-G(s)H(s)}{\{1+G(s)H(s)\}^2}$$

$$\frac{dM}{dG} \times \frac{G}{M} = \frac{1}{\{1+G(s)H(s)\}^2} \times \frac{G(s)}{\frac{G(s)}{1+G(s)}}$$

$$= \frac{1}{1+G(s)H(s)}$$

1.11 (c)

Given, Input $x(t) = u(t)$

and output $y(t) = t \cdot e^{-t}, t > 0$

Taking Laplace transform

$$X(s) = \frac{1}{s} \quad ; \quad Y(s) = \frac{1}{(s+1)^2}$$

Therefore transfer function

$$H(s) = \frac{Y(s)}{X(s)} = \frac{\frac{1}{(s+1)^2}}{1/s} = \frac{s}{(s+1)^2}$$

1.12 (d)

$$F(s) = \frac{5}{s(s^2+3s+2)}$$

The initial value of $f(t)$ is

$$\underset{t\to 0}{f(t)} = \underset{s\to\infty}{sF(s)}$$

$$= \underset{s\to\infty}{s} \cdot \frac{5}{s(s^2+3s+2)} = \frac{5/s^2}{s\to\infty\left(1+\frac{3}{s}+\frac{2}{s^2}\right)}$$

$$\underset{t\to 0}{f(t)} = 0$$

1.13 (a)

To calculate $\frac{y(s)}{d(s)}$, set $R(s) = 0$ and now redraw the given circuit.

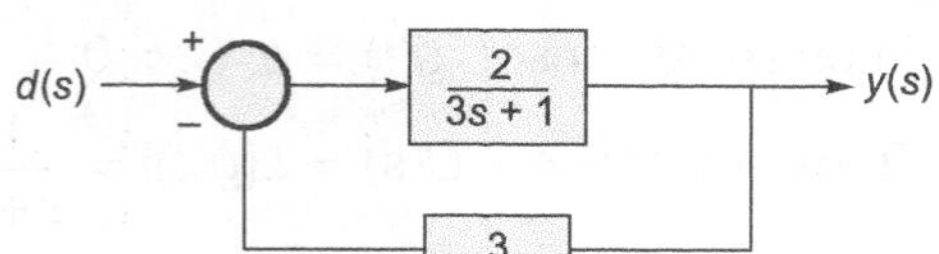

$$\therefore \quad \frac{y(s)}{d(s)} = \frac{\frac{2}{3s+1}}{1+3\times\frac{2}{3s+1}} = \frac{2}{3s+7}$$

1.14 (b)

$$\frac{C(s)}{R(s)} = H(s) = \frac{1}{s(s+1)}$$

$$\Rightarrow \quad H(s) = \frac{1}{s} - \frac{1}{s+1}$$

Taking inverse Laplace transform;

$$h(t) = (1-e^{-t})\, u(t)$$

1.15 (a)

$$\text{Torque} = \frac{9.544\times\text{Power}}{\text{Speed in rpm}} = \frac{9.544\times 10\times 10^3}{900}$$

$$= 106.04 \text{ N-m}$$

1.16 (a)

The transfer function which does not have any pole or zero in the right half of the s-plane is called minimum-phase transfer function.

If all poles lie in L.H.S. of s-plane then the system is linear and always stable.

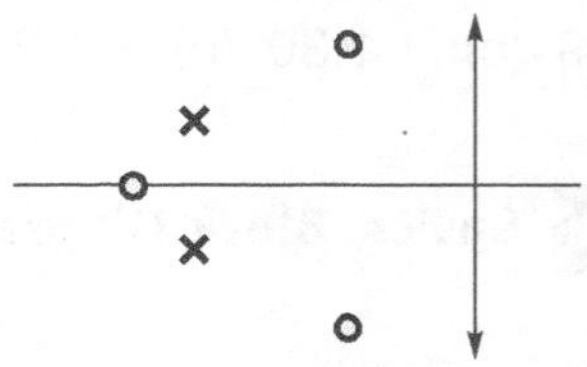

Minimum phase transfer function

1.17 (b)

$$\text{Tachometer sensitivity} = \frac{5}{1000}\left(\frac{V}{\text{rpm}}\right)$$

$$\omega = \frac{2\pi\times N}{60}$$

For $N = 1000$, $\omega = \frac{2\pi\times 1000}{60}$

$$\therefore \quad \text{Gain constant} = \frac{5\times 60}{2\pi\times 1000}\left(\frac{V}{\text{rad/ sec}}\right)$$

$$= 0.048 \text{ V/rad/sec}$$

1.18 (d)

$$G(s)H(s) = \frac{100}{s^2+8s+100} = \frac{\frac{100}{s^2+8s}}{1+\frac{100}{s^2+8s}}$$

These for a unity feedback system with negative feedback

$$G(s) = \frac{100}{s^2+8s} = \frac{100}{s(s+8)}$$

1.19 (c)

$$L\{\sin(at)\} \leftrightarrow \frac{a}{s^2+a^2}$$

$$L\{e^{-bt}\sin(at)\} \leftrightarrow \frac{a}{(s+b)^2+a^2}$$

in the giving question $a = 2\omega$

$b = 2$

$$\therefore L\{e^{-2t}\sin(2\omega t)\} \leftrightarrow \frac{2\omega}{(s+2)^2+4\omega^2}$$

1.21 (d)

Tachogenerator provides a derivative feedback when used in a servo system.

1.22 (a)

For a minimum phase system all the poles and zeroes of a transfer function must lie on the left side of the $(j\omega)$ axis.

1.23 (c)

Dominant pole is the one which has large time constant i.e. small frequency.

1.25 (a)

$$\text{OLTF} = \frac{10K}{1+10s}$$

Open time constant = 10

$$\text{CLTF} = \frac{10K/1+10K}{1+\frac{10}{1+10K}s}$$

$$\text{Closed loop time constant} = \frac{10}{1+10K}$$

$$\text{Given,} \quad \frac{10}{1+10K} = \frac{10}{20}$$

$$\therefore \quad K = 1.9$$

1.26 (b)

Using Mason gain formula

$$\frac{C}{R} = \frac{\frac{1}{s+1}(1-0)}{1-\left[\frac{-1}{s+1}+\frac{1}{s+1}\right]+0} = \frac{1}{s+1}$$

1.27 (a)

Feedback (negative) decreases both gain and sensitivity.

1.28 (a)

$$q(s) = 1 + G(s)H(s) = 0$$

$$q(s) = 1+\frac{K(s+2)}{(s+1)(s+4)} = 0$$

$$q(s) = (s+1)(s+4) + K(s+2) = 0$$

1.29 (c)

SFG is graphical representation and mathematical relation between variables of a system in the form of set of linear algebraic equation in cause-effect form.

1.30 (d)

Feedback in control system is used to control sensitivity, effect of disturbance and non linearities.

1.31 (a)

The subsystem that generates the input to the plant or process is known as controller.

1.32 (d)

The open loop transfer function is given by,

$$\text{OLTF} = \frac{s+4}{s^2+7s+13-s-4}$$

$$\text{OLTF} = \frac{s+4}{s^2+6s+9} \quad \text{...(1)}$$

* For open-loop DC gain of a unity negative feedback can be found by making s = 0 in equation (1).

$$\therefore \quad \text{Open loop DC gain} = \frac{4}{9}$$

■■■

2 Time Domain Analysis

2.1 **Assertion (A):** The largest undershoot corresponding to a unit step input to an underdamped second order system with damping ratio ξ and undamped natural frequency of oscillation ω_n is $e^{-2\xi\pi/\sqrt{1-\xi^2}}$.
Reason (R): The overshoots and undershoots of a second order underdamped system is

$$e^{-\xi n\pi/\sqrt{1-\xi^2}}, n = 1, 2, ...$$

(a) Both A and R are true and R is the correct explanation of A
(b) Both A and R are true but R is NOT the correct explanation of A
(c) A is true but R is false
(d) A is false but R is true

[ESE-2000]

2.2 Which one of the following transfer functions represents the critically damped system?

(a) $H_1(s) = \dfrac{1}{s^2 + 4s + 4}$

(b) $H_2(s) = \dfrac{1}{s^2 + 3s + 4}$

(c) $H_3(s) = \dfrac{1}{s^2 + 2s + 4}$

(d) $H_4(s) = \dfrac{1}{s^2 + s + 4}$

[ESE-2000]

2.3 A second order system has the damping ratio ξ and undamped natural frequency of oscillation ω_n. The settling time at 2% tolerance band of the system is

(a) $2/\xi\omega_n$ (b) $3/\xi\omega_n$
(c) $4/\xi\omega_n$ (d) $\xi\omega_n$

[ESE-2000]

2.4 Two identical first-order systems have been cascaded non-interactively. The unit step response of the systems will be

(a) overdamped (b) underdamped
(c) undamped (d) critically damped

[ESE-2001]

2.5 Which one of the following is the response $y(t)$ of a causal LTI system described by

$$H(s) = \frac{(s+1)}{s^2 + 2s + 2}$$

for a given input $x(t) = e^{-t}\, u(t)$?

(a) $y(t) = e^{-t} \sin t\, u(t)$
(b) $y(t) = e^{-(t-1)} \sin(t-1)\, u(t-1)$
(c) $y(t) = \sin(t-1)\, u(t-1)$
(d) $y(t) = e^{-t} \cos t\, u(t)$

[ESE-2001]

2.6 Which one of the following is the steady-state error for a step input applied to a unity feedback system with the open loop transfer function

$$G(s) = \frac{10}{s^2 + 14s + 50}?$$

(a) $e_{ss} = 0$ (b) $e_{ss} = 0.83$
(c) $e_{ss} = 1$ (d) $e_{ss} = \infty$

[ESE-2001]

2.7 The unit step response of a particular control system is given by $c(t) = 1 - 10e^{-t}$. Then its transfer function is

(a) $\dfrac{10}{s+1}$ (b) $\dfrac{s-9}{s+1}$

(c) $\dfrac{1-9s}{s+1}$ (d) $\dfrac{1-9s}{s(s+1)}$

[ESE-2001]

2.8 A third-order system is approximated to an equivalent second order system. The rise time of this approximated lower order system will be

(a) same as original system for any input
(b) smaller than the original system for any input
(c) larger than the original system for any input
(d) larger or smaller depending on the input

[ESE-2001]

2.9 Match **List-I** (Pole-zero plot of linear control system) with **List-II** (Responses of the system) and select the correct answer:

List-I List-II

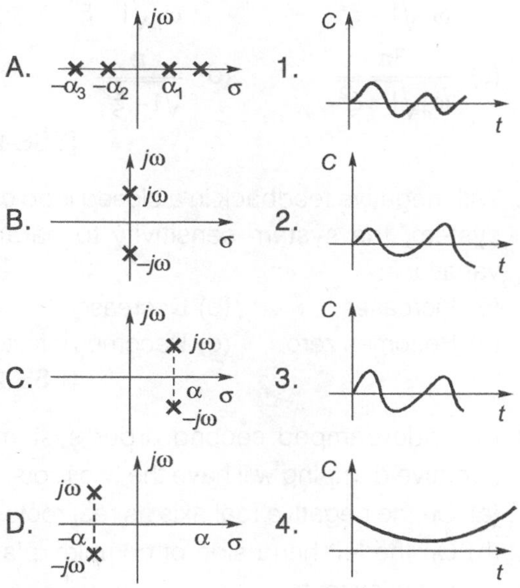

Codes:

	A	B	C	D
(a)	4	3	1	2
(b)	4	3	2	1
(c)	3	4	2	1
(d)	3	4	1	2

[ESE-2002]

2.10 A system has a single pole at origin. Its impulse response will be
(a) constant
(b) ramp
(c) decaying exponential
(d) oscillatory [ESE-2002]

2.11 Consider the unity feedback system as shown below. The sensitivity of the steady state error to change in parameter K and parameter a with ramp inputs are respectively

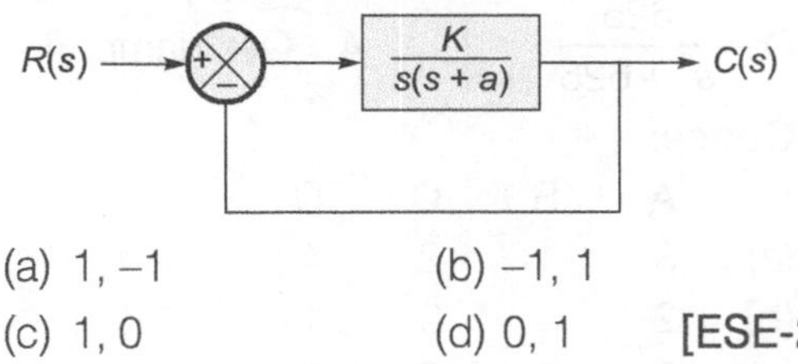

(a) 1, –1 (b) –1, 1
(c) 1, 0 (d) 0, 1 [ESE-2003]

2.12 Which one of the following is the transfer function of a linear system whose output is t^2e^{-t} for a unit step input?

(a) $\dfrac{s}{(s+1)^3}$ (b) $\dfrac{2s}{(s+1)^3}$

(c) $\dfrac{1}{s^2(s+1)}$ (d) $\dfrac{2}{s(s+1)^2}$

[ESE-2003]

2.13 Assuming unit ramp input, match **List-I** (System Type) with **List-II** (Steady State Error) and select the correct answer using the codes given below the lists:

List-I	List-II
A. 0	1. K
B. 1	2. ∞
C. 2	3. 0
D. 3	4. $1/K$

Codes:

	A	B	C	D
(a)	2	4	3	3
(b)	1	2	2	4
(c)	2	1	4	3
(d)	1	2	4	3

[ESE-2003]

2.14 When the time period of observation is large, the type of the error is
(a) Transient error
(b) Steady state error
(c) Half-power error
(d) Position error constant [ESE-2003]

2.15 The unit impulse response of a system having transfer function $K/(s + \alpha)$ is shown below. The value of α is

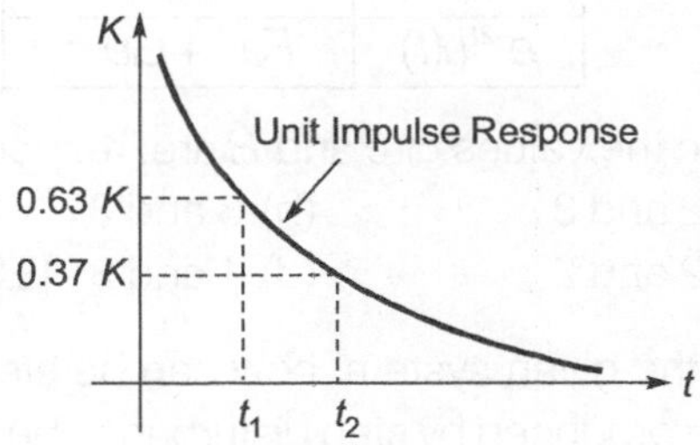

(a) t_1 (b) $1/t_1$
(c) t_2 (d) $1/t_2$

[ESE-2003]

2.16 What is the unit step response of a unity feedback control system having forward path transfer function $G(s) = \dfrac{80}{s(s+18)}$?
(a) Overdamped
(b) Critically damped
(c) Underdamped
(d) Undamped oscillatory

[ESE-2004]

2.17 Consider the following statements:
Feedback in control system can be used
1. to reduce the sensitivity of the system to parameter variations and disturbances
2. to change time constant of the system
3. to increase loop gain of the system

Which of the statements given above are correct?
(a) 1, 2 and 3 (b) 1 and 2
(c) 2 and 3 (d) 1 and 3
[ESE-2004]

2.18 Which one of the following statements is correct?
A second-order system is critically damped when the roots of its characteristic equation are
(a) negative, real and unequal
(b) complex conjugates
(c) negative, real and equal
(d) positive, real and equal
[ESE-2004]

2.19 A linear network has the system function

$$H\frac{(s+c)}{(s+a)(s+b)}$$

The outputs of the network with zero initial conditions for two different inputs are tabled as

Input $x(t)$	Output $y(t)$
$u(t)$	$2+De^{-t}+Ee^{-3t}$
$e^{-2t}u(t)$	$Fe^{-t}+Ge^{-3t}$

Then the values of c and H are, respectively
(a) 2 and 3 (b) 3 and 2
(c) 2 and 2 (d) 1 and 3 [ESE-2005]

2.20 For the given system, how can be steady state error produced by step disturbance be reduced?

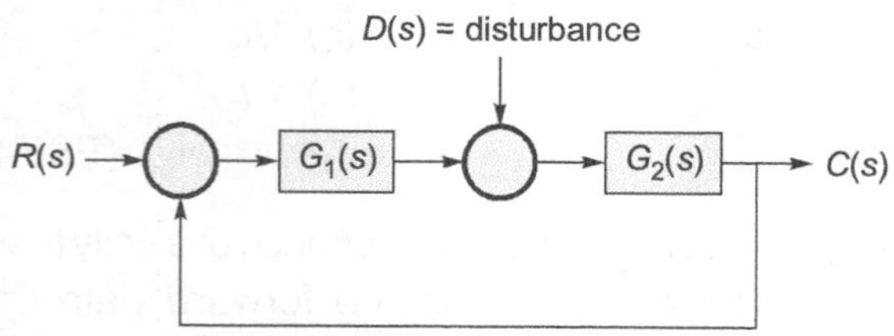

(a) By increasing dc gain of $G_1(s)G_2(s)$
(b) By increasing dc gain of $G_2(s)$
(c) By increasing dc gain of $G_1(s)$
(d) By removing the feedback [ESE-2005]

2.21 Which one of the following expresses the time at which second peak in step response occurs for a second order system?

(a) $\frac{\pi}{\omega_n\sqrt{1-\xi^2}}$ (b) $\frac{2\pi}{\omega_n\sqrt{1-\xi^2}}$
(c) $\frac{3\pi}{\omega_n\sqrt{1-\xi^2}}$ (d) $\frac{\pi}{\sqrt{1-\xi^2}}$
[ESE-2005]

2.22 With negative feedback in a closed loop control system, the system sensitivity to parameter variations:
(a) Increases (b) Decreases
(c) Becomes zero (d) Becomes infinite
[ESE-2005]

2.23 An underdamped second order system with negative damping will have the two roots:
(a) On the negative real axis as real roots
(b) On the left hand side of complex plane as complex roots
(c) On the right hand side of complex plane as complex conjugates
(d) On the positive real axis as real roots
[ESE-2005]

2.24 Match **List-I** (System $G(s)$) with **List-II** (Nature of Response) and select the correct answer using the code given below the Lists:

List-I	List-II
A. $\frac{400}{s^2+12s+400}$	1. Undamped
B. $\frac{900}{s^2+90s+900}$	2. Critically damped
C. $\frac{225}{s^2+30s+225}$	3. Underdamped
D. $\frac{625}{s^2+625}$	4. Overdamped

Codes:

	A	B	C	D
(a)	3	1	2	4
(b)	2	4	3	1
(c)	3	4	2	1
(d)	2	1	3	4

[ESE-2005]

2.25 Given a unity feedback system with $G(s)=\frac{K}{s(s+4)}$, what is the value of K for a damping ratio of 0.5?
(a) 1 (b) 16
(c) 4 (d) 2
[ESE-2005]

2.26 What is the steady state error for a unity feedback control system having $G(s)=\frac{1}{s(s+1)}$, due to unit ramp input?

(a) 1 (b) 0.5
(c) 0.25 (d) $\sqrt{0.5}$ [ESE-2005]

2.27 What is the value of K for a unity feedback system with $G(s)=\frac{K}{s(1+s)}$ to have a peak overshoot of 50%?

(a) 0.53 (b) 5.3
(c) 0.6 (d) 0.047 [ESE-2006]

2.28 Consider the following statements:
For the first order transient systems, the time constant is

1. a specification of transient response
2. reciprocal of real-axis pole location
3. an indication of accuracy of response
4. an indication of speed of the response

Which of the statements given above are correct?

(a) Only 1 and 2 (b) Only 1, 2 and 4
(c) Only 3 and 4 (d) 1, 2, 3 and 4

[ESE-2006]

2.29 The unit step response of a second order system is = $1 - e^{-5t} - 5t\,e^{-5t}$
Consider the following statements:

1. The undamped natural frequency is 5 rad/s.
2. The damping ratio is 1.
3. The impulse response is $25t\,e^{-5t}$.

Which of the statements given above are correct?

(a) Only 1 and 2 (b) Only 2 and 3
(c) Only 1 and 3 (d) 1, 2 and 3

[ESE-2006]

2.30 The unit step response of a system is $1 - e^{-t}(1 + t)$. Which is this system?

(a) Unstable (b) Stable
(c) Critically stable (d) Oscillatory

[ESE-2006]

2.31 **Assertion (A):** The impulse response is only a function of the terms in natural response.
Reason (R): The differentiation and differencing operations eliminate the constant terms associated with the particular solution in the step response and change only the constants associated with exponential terms in the natural response.

(a) Both A and R are true and R is the correct explanation of A
(b) Both A and R are true but R is NOT the correct explanation of A
(c) A is true but R is false
(d) A is false but R is true [ESE-2006]

2.32 The relation between input $x(t)$ and output $y(t)$ of a continuous-time system is given by

$$\frac{dy(t)}{dt}+3y(t)=x(t)$$

What is the forced response of the system when $x(t) = k$ (a constant)?

(a) k (b) $k/3$
(c) $3k$ (d) 0 [ESE-2007]

2.33 How can the steady-state error in a system be reduced?

(a) By decreasing the type of system
(b) By increasing system gain
(c) By decreasing the static error constant
(d) By increasing the input [ESE-2007]

2.34 For second-order system

$$2\frac{d^2y}{dt^2}+4\frac{dy}{dt}+8y=8x$$

what is the damping ratio?

(a) 1 (b) 0.25
(c) 0.333 (d) 0.5 [ESE-2007]

2.35 For a second-order system, ξ is equal to zero in the transfer function given by

$$\frac{\omega_n^2}{s^2+2\xi\omega_n s+\omega_n^2}$$

Which one of the following is correct?

(a) Closed-loop poles are complex conjugate with negative real part
(b) Closed-loop poles are purely imaginary
(c) Closed-loop poles are real, equal and negative
(d) Closed-loop poles are real, unequal and negative

[ESE-2007]

2.36 For the unity feedback system with $G(s)=\frac{10}{s^2(s+4)}$, what is the steady state error resulting from an input $10t$?

(a) 10 (b) 4
(c) Zero (d) 1 [ESE-2007]

2.37 Assertion (A): The system having characteristic equation $4s^2 + 6s + 1 = 0$ gives rise to under-damped oscillations for a step input.

Reason (R): The un-damped natural frequency of oscillation of the system is $\omega_n = 0.5$ rad/s.

(a) Both A and R are individually true and R is the correct explanation of A
(b) Both A and R are individually true but R is not the correct explanation of A
(c) A is true but R is false
(d) A is false but R is true [ESE-2008]

2.38 A diaphragm type pressure sensor has two poles as shown in the figure below. The zeros are at infinity. What is its steady state deformation for a unit step input pressure?

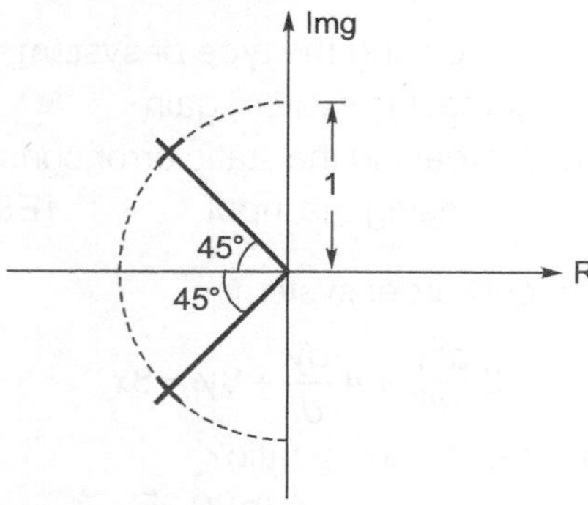

(a) 0.25 (b) 0.5
(c) 0.707 (d) 1 [ESE-2008]

2.39 A second order control system has a transfer function $\frac{16}{s^2+4s+16}$. What is the time for the first overshoot?

(a) $\frac{2\pi}{\sqrt{3}}$ s (b) $\frac{\pi}{\sqrt{3}}$ s
(c) $\frac{\pi}{2\sqrt{3}}$ s (d) $\frac{\pi}{4\sqrt{3}}$ s [ESE-2008]

2.40 The closed loop transfer function of a control system has the following poles and zeros

Poles	Zeros
$p_1 = -0.5$	$z_1 = -6$
$p_2 = -1.0$	$z_2 = -8$
$p_3 = -5$	
$p_4 = -10$	

The closed loop response can be closely approximated by considering which of the following?

(a) p_1 and p_2 (b) p_3 and p_4
(c) p_3 and z_1 (d) p_4 and z_2
[ESE-2008]

2.41 Consider the function $F(s) = \frac{\omega}{s^2+\omega^2}$ where $F(s)$ is the Laplace transform of $f(t)$. What is the steady-state value of $f(t)$?

(a) Zero
(b) One
(c) Two
(d) A value between −1 and +1 [ESE-2009]

2.42 In a unity feedback control system with $G(s) = \frac{4}{s^2+0.4s}$ when subjected to unit step input, it is required that system response should be settled within 2% tolerance band; the system settling time is

(a) 1 sec (b) 2 sec
(c) 10 sec (d) 20 sec [ESE-2010]

2.43 Which of the following is the response of a spring-mass-damper with under-damping?

(a)
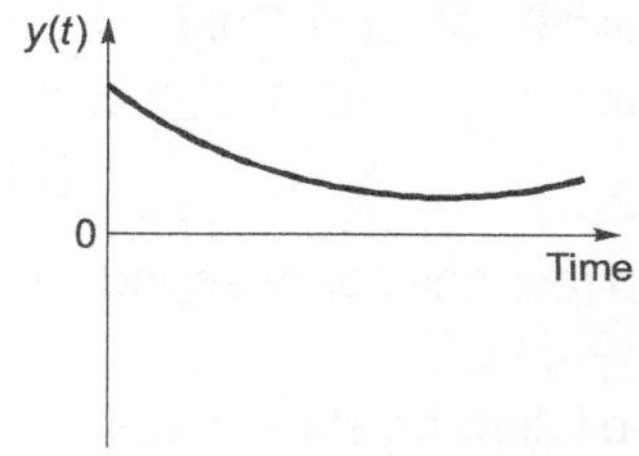

(b)
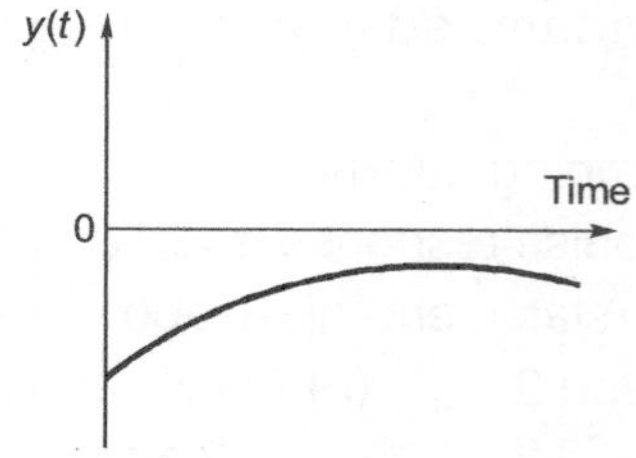

(c)
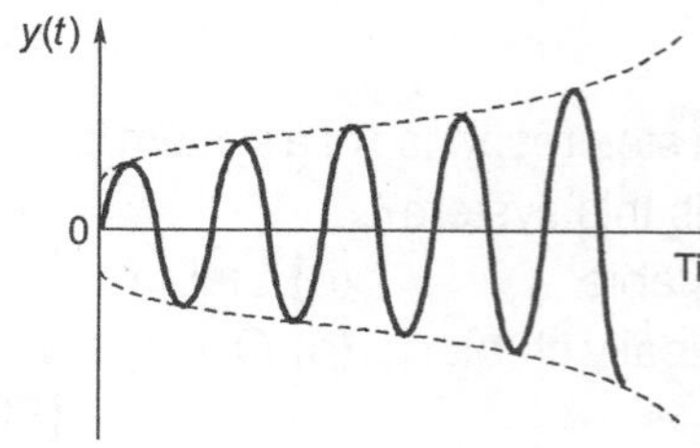

(d)
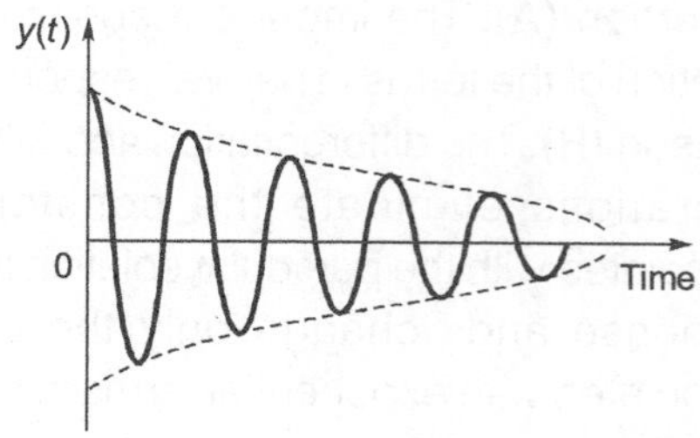

[ESE-2010]

2.44 From the point of view of stability and response speed of a closed loop system, the appropriate range for the value of damping ratio lies between

(a) 0 to 0.2 (b) 0.4 to 0.7
(c) 0.8 to 1.0 (d) 1.1 to 1.5

[ESE-2010]

2.45 **Assertion (A):** Steady state error can be reduced by increasing integral gain.

Reason (R): Overshoot can be reduced by increasing derivative gain.

(a) Both A and R are true and R is the correct explanation of A
(b) Both A and R are true but R is not a correct explanation of A
(c) A is true but R is false
(d) A is false but R is true [ESE-2010]

2.46 **Assertion (A):** A linear system gives a bounded output if the input is bounded.

Reason (R): The roots of the characteristic equation have all negative real parts and response due to initial conditions decay to zero as time t tends to infinity.

(a) Both A and R are true and R is the correct explanation of A
(b) Both A and R are true but R is NOT the correct explanation of A
(c) A is true but R is false
(d) A is false but R is true [ESE-2010]

2.47 What is the steady-state value of the unit-step response of a closed-loop control system shown in figure?

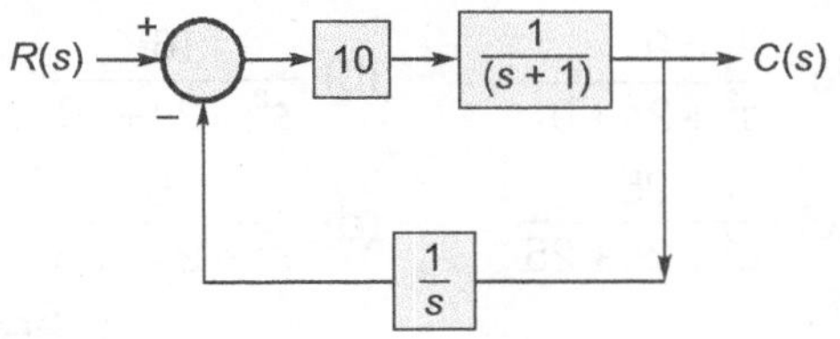

(a) −0.5 (b) 0
(c) 2 (d) ∞ [ESE-2011]

2.48 The block diagram of a closed-loop control system is given in figure. What is the type of this system?

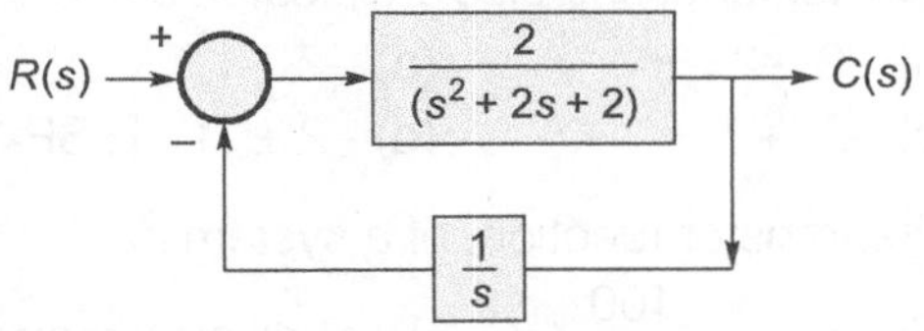

(a) Zero (b) One
(c) Two (d) Three [ESE-2011]

2.49 Given the differential equation model of a physical system, determine the time constant of the system

$$40\frac{dx}{dt} + 2x = f(t)$$

(a) 10 (b) 20
(c) 1/10 (d) 4 [ESE-2011]

2.50 Consider a second order all-pole transfer function model, if the desired settling time (5%) is 0.60 sec and the desired damping ratio 0.707, where should the poles be located in s-plane?

(a) $-5 \pm j4\sqrt{2}$ (b) $-5 \pm j5$
(c) $-4 \pm j5\sqrt{2}$ (d) $-4 \pm j7$ [ESE-2011]

2.51 Given a unity feedback system with

$G(s) = \frac{K}{s(s+6)}$, the value of K for damping ratio of 0.75 is

(a) 1 (b) 4
(c) 16 (d) 64 [ESE-2011]

2.52 **Statement (I):** A second order system subjected to a unit impulse oscillates at its natural frequency.

Statement (II): Impulse input contains frequencies from $-\infty$ to $+\infty$.

(a) Both Statement (I) and Statement (II) are individually true and Statement (II) is the correct explanation of Statement (I).
(b) Both Statement (I) and Statement (II) are individually true but Statement (II) is not the correct explanation of Statement (I).
(c) Statement (I) is true but Statement (II) is false.
(d) Statement (I) is false but Statement (II) is true.

[ESE-2012]

2.53 **Statement (I):** In a prototype second order system the rise time t_r and bandwidth are inversely proportional.

Statement (II): Increasing ω_n increases bandwidth while t_r reduces.

(a) Both Statement (I) and Statement (II) are individually true and Statement (II) is the correct explanation of Statement (I).
(b) Both Statement (I) and Statement (II) are individually true but Statement (II) is not the correct explanation of Statement (I).

(c) Statement (I) is true but Statement (II) is false.
(d) Statement (I) is false but Statement (II) is true.
[ESE-2012]

2.54 The type of system which is used for determination of static error constants is determined from the number of
(a) Zeros at origin for open loop transfer function.
(b) Poles at origin for open loop transfer function.
(c) Zeros at origin for closed loop transfer function.
(d) Poles at origin for closed loop transfer function.
[ESE-2012]

2.55 The time taken for the output to settle within ±2% of step input for the control system represented by $\frac{25}{s^2+5s+25}$ is given by
(a) 1.2 *s* (b) 1.6 *s*
(c) 2.0 *s* (d) 0.4 *s* [ESE-2012]

2.56 The following quantities give a measure of the transient characteristics of a control system, when subjected to unit step excitation:
1. Maximum overshoot.
2. Maximum undershoot.
3. Overall gain.
4. Delay time.
5. Rise time.
6. Fall time.
(a) 1, 3 and 5 (b) 2, 4 and 5
(c) 2, 4 and 6 (d) 1, 4 and 5
[ESE-2012]

2.57 In a feedback control system, if $G(s) = \frac{4}{s(s+3)}$ and $H(s) = \frac{1}{s}$, then the closed-loop system will be of type
(a) 3 (b) 2
(c) 1 (d) 0 [ESE-2012]

2.58 For a second order dynamic system, if the damping ratio is 1 then the poles are
(a) imaginary and complex conjugate
(b) in the right-half of s-plane
(c) equal, negative and real
(d) negative and real [ESE-2012]

2.59 A third order system is approximated to an equivalent second order system. The rise time of this approximated system will be
(a) Same as the original system for any input
(b) Smaller than the original system for any input
(c) Larger than the original system for any input
(d) Smaller or larger depending on the type of input
[ESE-2012]

2.60 A system is described by the transfer function $G(s) = \frac{2s+5}{(s+5)(s+4)}$. The dc gain of the system is
(a) 0.25 (b) 0.5
(c) 1 (d) ∞ [ESE-2012]

2.61 The sensitivity $S_T(K)$ of transfer function $T = \frac{(1+2K)}{(3+4K)}$ with respect to the parameter K is given by
(a) $\frac{K}{3+K^2}$ (b) $\frac{3K}{2+4K+K^2}$
(c) $\frac{2K}{3+10K+8K^2}$ (d) $\frac{4K}{2+5K+7K^2}$
[ESE-2012]

2.62 If the overshoot of the unit-step response of a second order system is 30%, then the time at which peak overshoot occurs (assuming ω_n = 10 rad/sec):
(a) 0.36 sec (b) 0.363 sec
(c) 0.336 sec (d) 0.633 sec
[ESE-2013]

2.63 Which has one of the following transfer functions have the greatest overshoot?
(a) $\frac{9}{s^2+2s+9}$ (b) $\frac{16}{s^2+2s+16}$
(c) $\frac{25}{s^2+2s+25}$ (d) $\frac{36}{s^2+2s+36}$
[ESE-2013]

2.64 The open-loop transfer function of a unity feedback control system is $G(s) = \frac{1}{(s+2)^2}$. The closed loop transfer function poles are located at
(a) –2, –2 (b) –2, –1
(c) –2, +2 (d) $-2 \pm j1$ [ESE-2013]

2.65 The transfer function, of a system is $G(s) = \frac{100}{s^2+10s+100}$. The unit step response of the system will settle in approximately

(a) 2 sec (b) 1 sec
(c) 0.8 sec (d) 1.5 sec
[ESE-2013]

2.66 The DC gain and steady state error for step input for $G(s) = \dfrac{s+1}{s^2+s+1}$ are
(a) 1 and 1 (b) 0 and 1
(c) 1 and 0 (d) 0 and 0
[ESE-2013]

2.67 A second order system is described by the equation

$$\frac{d^2x}{dt^2} + 5\frac{dx}{dt} + 7x = 7y$$

The frequency and damping ratio respectively are
(a) 1 rad/sec and 5
(b) 5 rad/sec and 7
(c) 1 rad/sec and $\sqrt{7}$
(d) $\sqrt{7}$ rad/sec and 0.94 [ESE-2013]

2.68 **Statement (I):** Zero input response is the natural response with zero initial conditions.
Statement (II): Zero state response is the response with given input with zero initial conditions.
(a) Both Statement (I) and Statement (II) are individually true and Statement (II) is the correct explanation of Statement (I).
(b) Both Statement (I) and Statement (II) are individually true but Statement (II) is not the correct explanation of Statement (I).
(c) Statement (I) is true but Statement (II) is false.
(d) Statement (I) is false but Statement (II) is true.
[ESE-2013]

2.69 When damping ratio is equal to zero, the damping frequency of a system is
(a) equal to natural frequency
(b) zero
(c) more than natural frequency
(d) less than natural frequency [ESE-2014]

2.70 A unity feedback system has

$$G(s) = \frac{K(s+12)}{(s+14)(s+18)}.$$

What is the value of K to yield 10% error in steady state?
(a) 672 (b) 189
(c) 100 (d) 21 [ESE-2014]

2.71 A unity feedback system has an open-loop transfer function $G(s) = \dfrac{K}{s(s+10)}$. If the damping ratio is 0.5, then what is the value of K?
(a) 150 (b) 100
(c) 50 (d) 10 [ESE-2014]

2.72 When the unit impulse response of a second order system is $\frac{1}{6}e^{-0.8t}\sin 0.6t$, the natural frequency and damping ratio of the system are respectively
(a) 1 rad/s and 0.8
(b) 0.64 rad/s and 0.8
(c) 1 rad/s and 1
(d) 0.64 rad/s and 1 [ESE-2015]

2.73 The second-order system defined by $\dfrac{25}{s^2+5s+25}$ is given a step input. The time taken for the output to settle within ± 2% is
(a) 1.2 s (b) 2 s
(c) 1.6 s (d) 0.4 s
[ESE-2015]

2.74 The time required for the step response to decrease and stay within a specified percentage of its final value is called
(a) Delay time (b) Rise time
(c) Lag time (d) Settling time
[ESE-2015]

2.75 The rise time of the output response of a low pass filter circuit when a step input is applied will be
(a) Proportional to the band width
(b) Inversely proportional to the band width
(c) Half the value of the band width
(d) $\dfrac{1}{\sqrt{2}}$ of the band width [ESE-2015]

2.76 The transfer function $\dfrac{1}{2s+1}$ will have
(a) dc gain 1 and high frequency gain 1
(b) dc gain 0 and high frequency gain ∞
(c) dc gain 1 and high frequency gain 0
(d) dc gain 0 and high frequency gain 1
[ESE-2016]

2.77 The closed-loop transfer function of a certain control system is given by

$$\frac{C(s)}{R(s)} = \frac{100}{s^2 + 10s + 100}$$

Then the settling time for a 2% tolerance band is given by

(a) 0.8 s (b) 1.2 s
(c) 1.5 s (d) 2.1 s [ESE-2016]

2.78 The unit step input response of a certain control system is given by $c(t) = 1 + 0.2\,e^{-60t} - 1.2\,e^{-10t}$. The undamped natural frequency ω_n and damping ratio ξ are, respectively

(a) 24.5 and 1.27 (b) 33.5 and 1.27
(c) 24.5 and 1.43 (d) 33.5 and 1.43
[ESE-2016]

2.79 For a unity feedback control system having an open-loop transfer function $G(s) = \dfrac{25}{s(s+6)}$, what is the time t_p at which the peak of the step input response occurs?

(a) 0.52 s (b) 2.75 s
(c) 0.79 s (d) 1.57 s
[ESE-2016]

2.80 The closed-loop transfer function of a unity feedback control system is,

$\dfrac{C(s)}{R(s)} = \dfrac{\omega_n^2}{s^2 + 2\xi\omega_n + \omega_n^2}$. The velocity error constant of the system is

(a) $\dfrac{\omega_n}{2\xi}$ (b) $\dfrac{\omega_n}{\xi}$

(c) $\dfrac{2\omega_n}{\xi}$ (d) $\dfrac{3\omega_n}{2\xi}$

[ESE-2016]

2.81 For a second-order differential equation, if the damping ratio ξ, is unity, then

(a) The poles are imaginary and complex conjugate
(b) The poles are in the right half of s-plane
(c) The poles are equal, negative and real
(d) Both the poles are unequal, negative and real
[ESE-2016]

2.82 The largest error between reference input and output during the transient period is called

(a) peak error (b) transient overshoot
(c) peak overshoot (d) transient deviation
[ESE-2017]

2.83 Consider the following statements:
For a type-1 and a unity feedback system, having unity gain in the forward path.

1. positional error constant K_p is equal to zero
2. acceleration error constant K_a is equal to zero
3. steady-state error e_{ss} per unit-step displacement input is equal to 1.

Which of the above statements are correct?

(a) 1, 2 and 3 (b) 1 and 2 only
(c) 2 and 3 only (d) 1 and 3 only
[ESE-2017]

2.84 A control system has $G(s) = \dfrac{10}{s(s+5)}$ and $H(s) = K$.

What is the value of K for which the steady-state error for unit-step input is less than 5%?

(a) 0.913 (b) 0.927
(c) 0.953 (d) 1.050 [ESE-2017]

2.85 What is the time required to reach 2% of steady-state value, for the closed-loop transfer function

$\dfrac{2}{(s+10)(s+100)}$, when the input is $u(t)$?

(a) 20 s (b) 2 s
(c) 0.2 s (d) 0.02 s
[ESE-2017]

2.86 If the characteristic equation of a closed-loop system is $2s^2 + 6s + 6 = 0$, then the system is

(a) overdamped (b) critically damped
(c) underdamped (d) undamped
[ESE-2017]

2.87 In a unity feedback control system, the open-loop transfer function is

$$G(s) = \frac{K(s+2)}{s^2(s^2 + 7s + 12)}$$

Then the error constants K_p, K_v and K_a, respectively, are

(a) ∞, ∞ and $\dfrac{K}{6}$ (b) $0, 0$ and $\dfrac{K}{6}$

(c) $\dfrac{K}{6}$, 0 and 0 (d) $\dfrac{K}{6}$, ∞ and ∞

[ESE-2018]

2.88 Settling time is the time required for the system response to settle within a certain percentage of
(a) maximum value
(b) final value
(c) input amplitude value
(d) transient error value

[ESE-2018]

2.89 A unity feedback system is shown in the figure. What is the magnitude of K so that the system is under-damped?

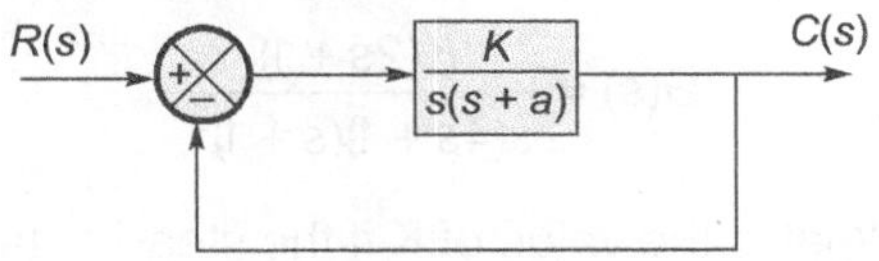

(a) $K = 0$ (b) $K = \frac{a^2}{4}$
(c) $K < \frac{a^2}{4}$ (d) $K > \frac{a^2}{4}$

[ESE-2018]

2.90 For a closed-loop system shown in the figure, what is the settling-time for ±2% settling of the steady-state condition, assuming unit-step input?

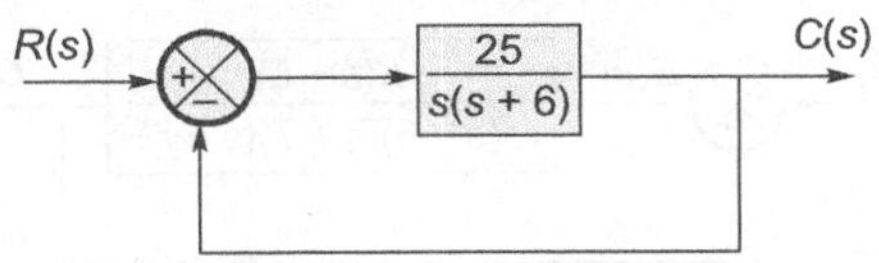

(a) 0.33 s (b) 1.33 s
(c) 2.33 s (d) 3.33 s [ESE-2018]

2.91 A unit-step input to a first-order system $G(s)$ yields a response as shown in the figure. This can happen when the values of K and a, respectively, are

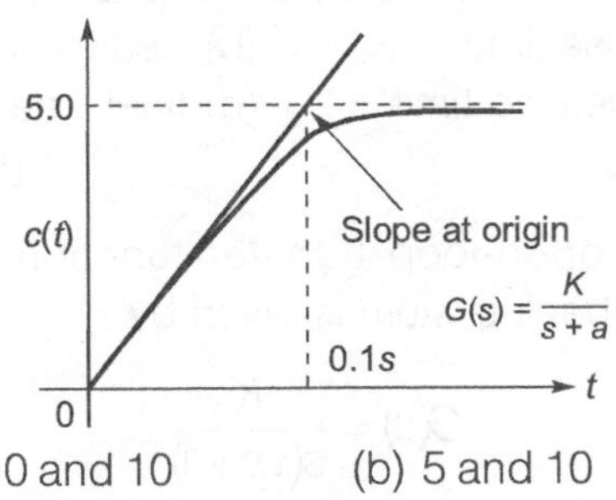

(a) 10 and 10 (b) 5 and 10
(c) 10 and 5 (d) 5 and 5

[ESE-2018]

2.92 The steady-state error for a Type 0 system for unit-step input is 0.2. In a certain instance, this error possibility was removed by insertion of a unity gain block. Thereafter, a unit ramp was applied. The nature of the block and new steady-state error in this changed configuration will, respectively, be
(a) integrator ; 0.25 (b) differentiator ; 0.25
(c) integrator ; 0.20 (d) differentiator ; 0.20

[ESE-2018]

2.93 The unit-impulse response of a system is $16e^{-2t} - 8e^{-t}$. Its unit-step response is
(a) $8 + e^{-t} - 4e^{-2t}$ (b) $8 + e^{-t} + 4e^{-2t}$
(c) $8e^{-t} - 8e^{-2t}$ (d) $e^{-t} - 4e^{-2t}$

[ESE-2018]

2.94 For the given transfer function

$$G(s) = \frac{Y(s)}{R(s)} = \frac{1}{s^2 + 3s + 2}$$

the response $y(t)$ for a step input $r(t) = 5u(t)$ will be

(a) $\left[\frac{5}{2} - 5e^{-t} + \frac{5}{2}e^{-2t}\right]u(t)$

(b) $\left[\frac{5}{2} - 5e^{-t}\right]u(t)$

(c) $\left[\frac{5}{2} + \frac{5}{2}e^{-2t}\right]u(t)$

(d) $\left[-5e^{-t} + \frac{5}{2}e^{-2t}\right]u(t)$

where $u(t)$ is a unit step input. [ESE-2019]

2.95 For a Type-2 system, the steady-state errors for unit step and unit ramp input are
(a) 0 and ∞ (b) ∞ and 0
(c) 0 and 0 (d) ∞ and ∞

[ESE-2019]

2.96 Consider the following statements regarding a parabolic function:
1. A parabolic function is one degree faster than the ramp function.
2. A unit parabolic function is defined as

$$f(t) = \begin{cases} \frac{t^2}{2}, & \text{for } t > 0 \\ 0 & \text{otherwise} \end{cases}$$

3. Laplace transform of unit parabolic function is $\frac{1}{s^3}$.

Which of the above statements are correct?
(a) 1 and 2 only (b) 1 and 3 only
(c) 2 and 3 only (d) 1, 2 and 3

[ESE-2019]

2.97 Unit step response of first order system with transfer function

$$G(s) = \frac{1}{1+\tau s}$$

(a) $1 - e^{-t/\tau}$ (b) $1 + e^{-t/\tau}$
(c) $1 + e^{t/\tau}$ (d) $1 - e^{t/\tau}$ [ESE-2020]

2.98 Step response of the system is defined as
1. The output due to a unit step input signal.
2. The running sum of impulse response.
3. The running integral of impulse response for a continuous-time system.

(a) 1 and 2 only (b) 1 and 3 only
(c) 2 and 3 only (d) 1, 2 and 3

[ESE-2020]

2.99 A unity feedback system is characterized by the open loop transfer function

$$G(s) = \frac{1}{s(0.5s+1)(0.2s+1)}$$

The steady state errors for unit-step and unit-ramp inputs are respectively
(a) 0 and 0 (b) 0 and 1
(c) 1 and 0 (d) 1 and 1 [ESE-2020]

2.100 In a type-1, second-order system, the first undershoot occurs at a time t (with standard notations) is

(a) $\frac{\pi}{\omega_d}$ (b) $\frac{2\pi}{\omega_d}$
(c) $\frac{\pi}{2\omega_d}$ (d) $\frac{2\omega_d}{\pi}$ [ESE-2020]

2.101 The steady-state error of type 1 system with input $r(t) = \frac{t^2}{2}, t \geq 0$; is

(a) 0 (b) $\frac{1}{2}$
(c) 1 (d) ∞ [ESE-2021]

2.102 Which one of the following conditions is justifying a second-order approximation?
(a) Closed-loop zeros near the closed-loop second-order pole pair are nearly cancelled by the close proximity of higher-order closed-loop poles.
(b) Closed-loop zeros cancelled by the close proximity of higher-order closed-loop poles are far removed from the closed-loop second-order pole pair.
(c) Closed-loop zeros near the closed-loop second-order pole pair are not cancelled by the close proximity of higher-order closed-loop poles.
(d) Closed-loop zeros cancelled by the close proximity of higher-order closed-loop poles are far removed from the closed-loop second-order zero pair.

[ESE-2021]

2.103 A unity feedback system has

$$G(s) = \frac{K(2s+1)}{s(4s+1)(s+1)^2}$$

What is the value of K if the steady-state value of error is to be less than 0.1 when an input $r(t) = 1 + 5t$ is applied?
(a) $K = 5$ (b) $6 < K < 10$
(c) $11 < K < 40$ (d) $K > 50$

[ESE-2022]

2.104 For the given control system, what is the value of K so that there is 10% error in the steady state?

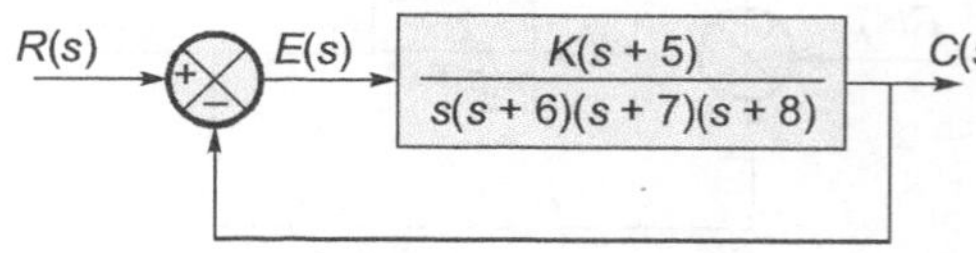

(a) 172 (b) 272
(c) 572 (d) 672

[ESE-2022]

2.105 In an underdamped second-order system the time required for the waveform to go from 0.1 of the final value to 0.9 of the final value is called
(a) rise time (b) peak time
(c) settling time (d) final time

[ESE-2022]

2.106 The open-loop transfer function of a unity feedback system is given by

$$G(s) = \frac{K}{s(\tau s+1)}$$

By what factor should the amplifier gain K be multiplied so that the damping ratio is increased from 0.25 to 0.75?
(a) 0.1111 (b) 1.1111
(c) 0.3333 (d) 3.3333

[ESE-2022]

2.107 Consider the speed control system shown in the figure wherein the inner loop corresponds to motor back e.m.f, the controller is an integrator with gain K observes that the load is inertia only. What is the value of K for which steady-state error to unit ramp input $\left(V_r(s) = \frac{1}{s^2}\right)$ is less than 0.01 rad/sec?

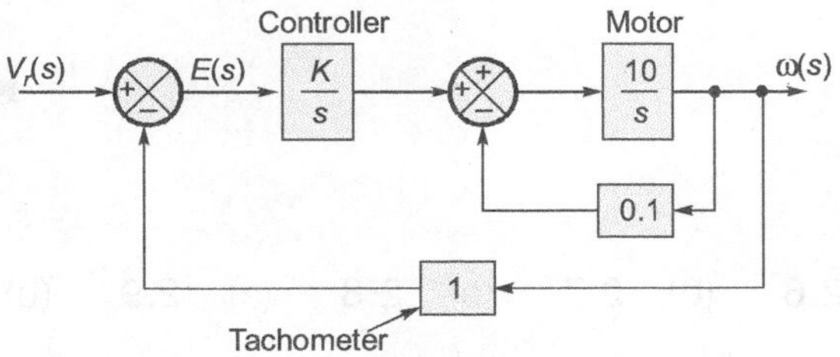

(a) 5 (b) 7
(c) 10 (d) 14 [ESE-2022]

2.108 Consider the following statements regarding time response specifications :
1. Delay time is the time required for the response to reach 10% to 90% or 5% to 95% or 0% to 100% of its final value.
2. Peak time is the time required for the response to reach the first peak overshoot.
3. Settling time is the time required for the response to reach and maintain beyond a specified tolerance band, i.e., either 3% or 5% of the initial value.

Which of the above statements are **not** correct?
(a) 1 and 2 only (b) 1 and 3 only
(c) 2 and 3 only (d) 1, 2 and 3
[ESE-2022]

2.109 A second-order system has a closed loop transfer function given by $G(s) = \frac{25}{s^2 + 8s + 25}$. The settling time for 5 percentage band in tolerance error is
(a) $\frac{1}{3}$ sec (b) $\frac{3}{4}$ sec
(c) 2 sec (d) 4 sec
[ESE-2023]

2.110 The output of a standard second-order system for a unit-step input is given as $y(t) = 1 - \frac{2}{\sqrt{3}}e^{-t}\cos\left(\sqrt{3}t - \frac{\pi}{6}\right)$. What is the transfer function of the system?
(a) $\frac{2}{(s+2)(s+\sqrt{3})}$ (b) $\frac{1}{s^2+2s+1}$
(c) $\frac{3}{s^2+2s+3}$ (d) $\frac{4}{s^2+2s+4}$
[ESE-2023]

2.111 Consider a causal second-order system with the transfer function $G(s) = \frac{1}{s^2+2s+1}$ with a unit-step $R(s) = \frac{1}{s}$ as an input. Let $c(s)$ be the corresponding output. The time taken by the system output $c(t)$ to reach 94% of its steady-state value $\lim_{t\to\infty} c(t)$, rounded off to two decimal places, is
(a) 5.25 (b) 2.81
(c) 4.50 (d) 3.89 [ESE-2023]

2.112 A second-order system has a transfer function given by $G(s) = \frac{25}{s^2+8s+25}$. If the system, initially at rest, is subjected to a unit-step input at $t = 0$, the second peak in the response will occur at
(a) $\frac{\pi}{3}$ sec (b) $\frac{2\pi}{3}$ sec
(c) $\frac{\pi}{2}$ sec (d) π sec [ESE-2023]

2.113 If the open loop transfer function of a servo system with unity feedback is $G(s) = \frac{10}{s(1+0.1s)}$, then the acceleration, velocity and position error constants are respectively
(a) ∞, 0, 1 (b) 0, 1, ∞
(c) 0, 10, ∞ (d) ∞, 1, 10
[ESE-2024]

2.114 A unity feedback system is characterized by an open-loop transfer function $G(s) = \frac{K}{s(s+10)}$.
If the damping ratio is 0.5, then the time to peak overshoot for the unit step input is
(a) 0.125 sec (b) 0.175 sec
(c) 0.254 sec (d) 0.363 sec
[ESE-2024]

2.115 Consider the following statements:

1. Rise time for the underdamped system is the time required for the response to rise from 5% to 95% of its final value.
2. The amount of maximum overshoot directly indicates the relative stability of the system.
3. Settling time is the time required for the response to reach and maintain within a specified tolerance band, i.e. either 3% or 6% of the final value.

Which of the above statements are not correct?

(a) 1 and 2 only (b) 1 and 3 only
(c) 2 and 3 only (d) 1, 2 and 3

[ESE-2024]

■■■

Answers Time Domain Analysis

2.1	(a)	2.2	(a)	2.3	(c)	2.4	(d)	2.5	(a)	2.6	(b)	2.7	(c)	2.8	(a)	2.9	(b)
2.10	(a)	2.11	(b)	2.12	(b)	2.13	(a)	2.14	(b)	2.15	(d)	2.16	(a)	2.17	(b)	2.18	(c)
2.19	(a)	2.20	(c)	2.21	(c)	2.22	(b)	2.23	(c)	2.24	(c)	2.25	(b)	2.26	(a)	2.27	(b)
2.28	(b)	2.29	(d)	2.30	(b)	2.31	(a)	2.32	(b)	2.33	(b)	2.34	(d)	2.35	(b)	2.36	(c)
2.37	(d)	2.38	(d)	2.39	(c)	2.40	(a)	2.41	(d)	2.42	(d)	2.43	(d)	2.44	(b)	2.45	(b)
2.46	(d)	2.47	(b)	2.48	(b)	2.49	(b)	2.50	(b)	2.51	(c)	2.52	(d)	2.53	(a)	2.54	(b)
2.55	(b)	2.56	(d)	2.57	(b)	2.58	(c)	2.59	(a)	2.60	(a)	2.61	(c)	2.62	(c)	2.63	(d)
2.64	(d)	2.65	(c)	2.66	(c)	2.67	(d)	2.68	(d)	2.69	(a)	2.70	(b)	2.71	(b)	2.72	(a)
2.73	(c)	2.74	(d)	2.75	(b)	2.76	(c)	2.77	(a)	2.78	(c)	2.79	(c)	2.80	(a)	2.81	(c)
2.82	(c)	2.83	(*)	2.84	(d)	2.85	(*)	2.86	(c)	2.87	(a)	2.88	(b)	2.89	(d)	2.90	(b)
2.91	(*)	2.92	(a)	2.93	(c)	2.94	(a)	2.95	(c)	2.96	(d)	2.97	(a)	2.98	(d)	2.99	(b)
2.100	(b)	2.101	(d)	2.102	(a)	2.103	(d)	2.104	(d)	2.105	(a)	2.106	(a)	2.107	(d)	2.108	(b)
2.109	(b)	2.110	(d)	2.111	(c)	2.112	(d)	2.113	(c)	2.114	(d)	2.115	(b)				

Explanations Time Domain Analysis

2.1 (a)

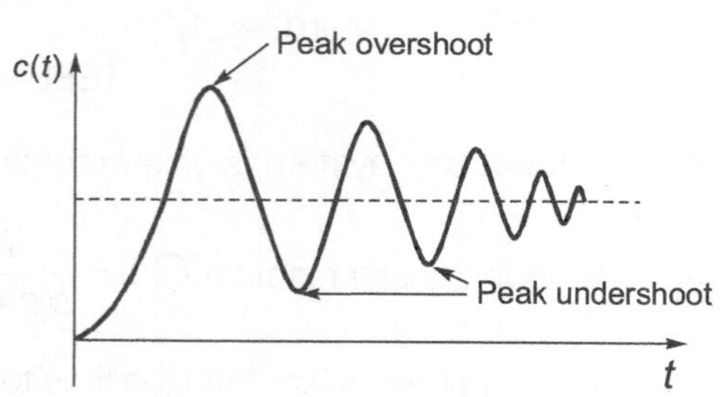

$$M_P = e^{-\xi n\pi/\sqrt{1-\xi^2}} \text{ for } n = 1, 2, 3, ...$$

2.2 (a)

$$H_1(s) = \frac{1}{s^2+4s+4} = \frac{1}{(s+2)^2}$$

since both roots are negative real and equal, it is a critically damped system.

2.3 (c)

Settling time at 2% of tolerance band of the system,

$$t_s = \frac{4}{\xi\omega_n}$$

Settling time at 5% of tolerance band of the system,

$$t_s = \frac{3}{\xi\omega_n}$$

2.4 (d)

$$\left(\frac{1}{s+\tau}\right)\cdot\left(\frac{1}{s+\tau}\right) = \frac{1}{(s+\tau)^2}$$

Since both are cascaded non-interactively, the overall unit step response will be as shown above. It is clear that the above response is critically damped.

Alternate Solution:

Transfer function of first order system is $\frac{1}{1+sT}$

when two such are cascaded overall transfer function is (TF_1)

$$TF_1 = \frac{1}{1+sT}\cdot\frac{1}{1+sT}$$

$$= \frac{1}{s^2T^2+2sT+1} = \frac{1/T^2}{s^2+\frac{2s}{T}+\frac{1}{T^2}}$$

Comparing with standard second order transfer function, we get

$$\omega_n = \frac{1}{T}$$

and $2\xi\omega_n = \frac{2}{T}$

$$\xi = 1$$

∴ Critically underdamped.

2.5 (a)

$$X(s) = \frac{1}{s+1}$$

$$Y(s) = X(s)H(s)$$

$$= \frac{1}{(s+1)}\cdot\frac{(s+1)}{\{(s+1)^2+1\}} = \frac{1}{(s+1)^2+1}$$

$\Rightarrow y(t) = e^{-t}\sin t\, u(t)$

2.6 (b)

$$G(s) = \frac{10}{s^2+14s+50}$$

It is type 0 system. Input is step input.

$$e_{ss} = \frac{1}{1+K_p}$$

where $K_p = \lim_{s\to 0} G(s)H(s)$

$$K_p = \lim_{s\to 0}\frac{10}{s^2+14s+50}$$

$$= \frac{10}{50} = 0.2$$

$$e_{ss} = \frac{1}{1+0.2} = \frac{1}{1.3} = 0.83$$

2.7 (c)

$$\text{TF} = s\,L[\text{Step response}] = \frac{1-9s}{s+1}$$

2.10 (a)

$$G(s) = \frac{1}{s} \Rightarrow g(t) = 1$$

The impulse response of the system is constant.

2.11 (b)

$$R(s) = \frac{1}{s^2}$$

Steady state error

$$e_{ss} = \lim_{s\to 0}\frac{sR(s)}{1+G(s)H(s)}$$

$$\Rightarrow \quad e_{ss} = \lim_{s\to 0}\frac{s.1/s^2}{1+\frac{K}{s(s+a)}.1}$$

$$\Rightarrow \quad e_{ss} = \lim_{s\to 0}\frac{s+a}{s(s+a)+K}$$

$$\Rightarrow \quad e_{ss} = \frac{a}{K}$$

Sensitivity of e_{ss} to change in K is

$$S_K^{e_{ss}} = \frac{de_{ss}}{dK}\times\frac{K}{e_{ss}} = \frac{-a}{K^2}\times\frac{K}{a/K}$$

$$\Rightarrow \quad S_K^{e_{ss}} = -1$$

Now, $S_a^{e_{ss}} = \frac{de_{ss}}{da}\times\frac{a}{e_{ss}} = \frac{1}{K}\times\frac{a}{a/K}$

$$\Rightarrow \quad S_a^{e_{ss}} = 1$$

2.12 (b)

$$c(t) = t^2 e^{-t}$$

$$C(s) = \frac{2}{(s+1)^3}$$

$$R(s) = \frac{1}{s}$$

Transfer function

$$G(s) = \frac{C(s)}{R(s)} = \frac{2/(s+1)^3}{1/s}$$

$$\Rightarrow \quad G(s) = \frac{2s}{(s+1)^3}$$

2.13 (a)

Table for steady state error

Type \ Input	Unit step	Unit Ramp	Unit Parabola
Type 0	$\frac{1}{1+K_p}$	∞	∞
Type 1	0	$\frac{1}{K_v}$	∞
Type 2	0	0	$\frac{1}{K_a}$

Where $K_p = \lim_{s\to 0} G(s)H(s)$

$K_v = \lim_{s\to 0} sG(s)H(s)$

$K_a = \lim_{s\to 0} s^2G(s)H(s)$

2.14 (b)

Steady state error is the error at $t \to \infty$.

2.15 (d)

$$G(s) = \frac{C(s)}{R(s)} = \frac{K}{s+\alpha}$$

$\Rightarrow \quad C(s) = \frac{K}{s+\alpha}$ since $R(s) = 1$

$\Rightarrow \quad c(t) = Ke^{-\alpha t}$

Time constant $\tau = 1/\alpha$

Time constant is the time at which

$c(t) = Ke^{-1} = 0.37$ K

So, $\quad \tau = t_2 = \frac{1}{\alpha} \Rightarrow \alpha = \frac{1}{t_2}$

2.16 (a)

$$\frac{G(s)}{1+G(s)} = \frac{80}{s^2+18s+80} \qquad \omega_n = \sqrt{80}$$

$$\xi = \frac{18}{2\sqrt{80}} = 1.00623$$

So, the system is overdamped.

2.17 (b)

(i) In open-loop system, transfer function $T = G$

Sensitivity of open-loop system is

$$S_G^T = \frac{\partial T}{\partial G}\times\frac{G}{T} = 1 \qquad [\because T = G]$$

In closed-loop system, transfer function

$$T = \frac{G}{1+GH}$$

$$S_G^T = \frac{\partial T}{\partial G}\times\frac{G}{T}$$

$$= \frac{1+GH-GH}{(1+GH)^2}\times\frac{G}{G/(1+GH)}$$

$$S_G^T = \frac{1}{1+GH}$$

Thus feedback is used to reduce the sensitivity of the system.

(ii) Feedback is the fraction of output. It reduces the loop gain.

2.18 (c)

(i) Roots of 2nd order underdamped system are complex conjugates.

(ii) Roots of 2nd order critically damped system are negative, real and equal.

(iii) Roots of 2nd order overdamped system are negative, real and unequal.

2.19 (a)

$$T(s) = H\frac{(s+c)}{(s+a)(s+b)} \qquad \text{...(i)}$$

When input is $u(t)$ output is

$= 2 + De^{-t} + Ee^{-3t}$

When input is $e^{-2t}u(t)$ output is

$= Fe^{-t} + Ge^{-3t}$

Using equation (i) when input is $u(t)$ output is

$$\frac{H(s+c)}{s(s+a)(s+b)} = \frac{K_1}{s}+\frac{D}{s+a}+\frac{E}{s+b}$$

Taking inverse Laplace transform

$= 2 + De^{-t} + Ee^{-3t}$

So, $\quad a = 1$ and $b = 3$

Using final value theorem

$$\lim_{s\to 0}\frac{s\cdot H(s+c)}{s(s+a)(s+b)} = \lim_{t\to\infty} 2+De^{-t}+Ee^{-3t}$$

$$\frac{Hc}{ab} = 2 \text{ and } Hc = 6$$

Using equation (i) when input is $e^{-2t}u(t)$ output is

$$\frac{H(s+c)}{(s+2)(s+a)(s+b)}$$

Only two terms are present in the response.

Hence $s + c = s + 2$

$\Rightarrow \quad c = 2$

$H = 3 \qquad (\because HC = 6)$

2.20 (c)

Output due to disturbance $D(s)$ is

$$C_D(s) = \frac{G_2}{1+G_1G_2}\cdot D(s)$$

$$C_D(s) \equiv \frac{G_2}{G_1G_2}\cdot D(s) \qquad [\because G_1G_2 >> 1]$$

$$C_D(s) \approx \frac{1}{G_1(s)}\cdot D(s)$$

Thus effect of disturbance can be reduced by increasing $G_1(s)$.

2.21 (c)

Time for peak overshoots are

$$t_p = \frac{n\pi}{\omega_n\sqrt{1-\xi^2}} \quad n = 1, 3, 5, ...$$

For first peak overshoot, $n = 1$

$$t_{p1} = \frac{n\pi}{\omega_n\sqrt{1-\xi^2}}$$

For second peak overshoot, $n = 3$

$$t_{p2} = \frac{3\pi}{\omega_n\sqrt{1-\xi^2}}$$

2.22 (b)

With negative feedback, stability of the system increases and as stability is inversely proportional to sensitivity, therefore sensitivity decreases.

2.23 (c)

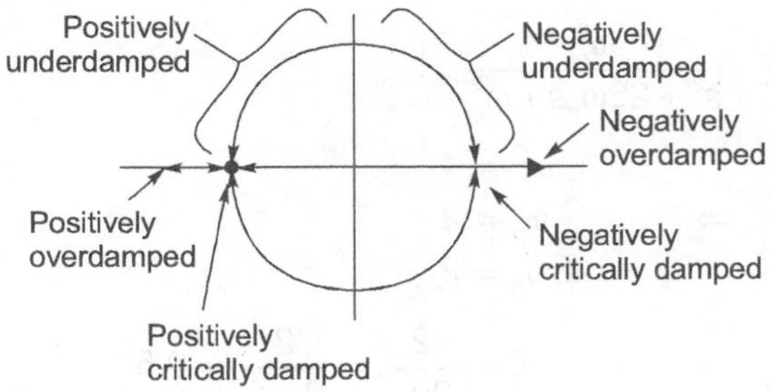

2.24 (c)

$s^2 + 12s + 400 = 0$

$$\Rightarrow \xi = \frac{12}{2\sqrt{400}} = \frac{12}{40} < 1$$

$\Rightarrow$ underdamped $s^2 + 90s + 900 = 0$

$$\Rightarrow \xi = \frac{90}{2\sqrt{900}} = \frac{90}{2\times 30} > 1$$

$\Rightarrow$ overdamped $s^2 + 30s + 225 = 0$

$$\Rightarrow \xi = \frac{30}{2\sqrt{225}} = \frac{30}{2\times 15} = 1$$

$\Rightarrow$ critically damped

$s^2 + 625 = 0$

$\Rightarrow \xi = 0 \Rightarrow$ undamped.

2.25 (b)

$$\frac{G(s)}{1+G(s)} = \frac{K}{s^2+4s+K}$$

$$\xi = \frac{4}{2\sqrt{K}} = 0.5$$

$$\Rightarrow \quad \sqrt{K} = \frac{4}{2\times 0.5} = 4$$

$$\Rightarrow \quad K = 16$$

2.26 (a)

For type 1, ramp input

$$e_{ss} = \frac{1}{K_v}$$

where $\quad K_v = \lim_{s\to 0} sG(s)$

$$= \lim_{s\to 0} s\cdot\frac{1}{s(s+1)} = 1$$

So, $\quad e_{ss} = \dfrac{1}{K_v} = 1$

2.27 (b)

$$\frac{G(s)}{1+G(s)} = \frac{K}{s^2+s+K}$$

$$\xi = \frac{1}{2\sqrt{K}}$$

$$\frac{-\frac{1}{2\sqrt{K}}.\pi}{\sqrt{1-\frac{1}{4K}}} = ln(0.5) = -0.693$$

$$\Rightarrow \quad \frac{\pi^2}{4K} = 0.48\left(1-\frac{1}{4K}\right)$$

$$\Rightarrow \quad 4K - 1 = \frac{\pi^2}{0.48}$$

$$\Rightarrow \quad 4K = 21.56$$

$$\Rightarrow \quad K = 5.39$$

2.28 (b)

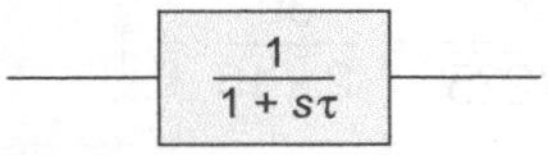

(i) Time constant is a specification of transient response.

(ii) $s = -1/\tau$

(iii) The time constant τ also affect the steady state value of the system. Hence the accuracy is also governed by it.

(iv) Time constant is an indication of speed of the response.

2.29 (d)

$$C(s) = \frac{1}{s} - \frac{1}{s+5} - \frac{5}{(s+5)^2}$$

$$= \frac{(s+5)^2 - (s+5)s - 5s}{s(s+5)^2}$$

$$= \frac{25}{s(s+5)^2}$$

$$C(s) = \frac{25}{s(s^2+10s+25)}$$

$$R(s) = \frac{1}{s}$$

$$G(s) = \frac{C(s)}{R(s)} = \frac{25}{s^2+10s+25}$$

$$\omega_n = \sqrt{25}$$

$$\Rightarrow \quad \omega_n = 5 \text{ rad/s}$$

$$\xi = \frac{10}{2\times 5} = 1$$

$$\text{Impulse response} = \frac{d}{dt}(1 - e^{-5t} - 5te^{-5t})$$

$$= 5e^{-5t} - 5e^{-5t} + 25te^{-5t} = 25te^{-5t}$$

2.30 (b)

$$C(s) = \frac{1}{s} - \frac{1}{s+1} - \frac{5}{(s+1)^2}$$

$$= \frac{(s+1)^2 - s(s+1) - s}{s(s+1)^2} = \frac{1}{s(s+1)^2}$$

$$R(s) = \frac{1}{s}$$

$$G(s) = \frac{C(s)}{R(s)} = \frac{1}{(s+1)^2} = \frac{1}{s^2+2s+1}$$

$$\therefore \quad \xi = 1$$

So given system is critically stable.

2.33 (b)

Steady state error,

$$e_{ss} = \lim_{s\to 0} \frac{sR(s)}{1+G(s)H(s)}$$

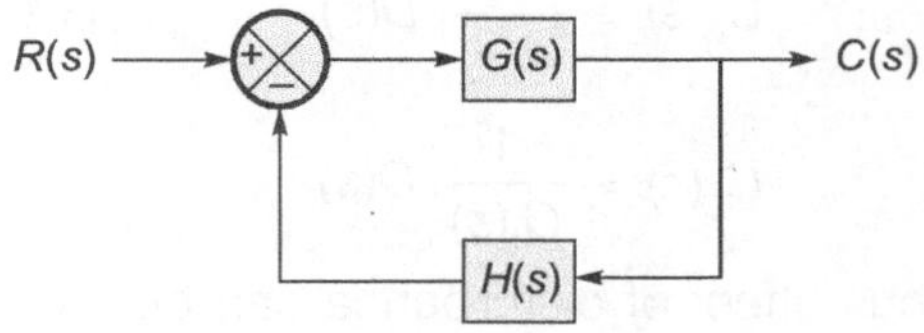

(i) By increasing the input $R(s)$, e_{ss} increases.

(ii) By decreasing the type of system, e_{ss} increases.

(iii) $e_{ss} \propto \frac{1}{\text{Static error constant}}$

Therefore, by decreasing the static error constant (K_p, K_v or K_a), e_{ss} increases.

2.34 (d)

$$2\frac{d^2y}{dt^2} + 4\frac{dy}{dt} + 8y = 8x$$

$$\text{or} \quad \frac{d^2y}{dt^2} + 2\frac{dy}{dt} + 4y = 4x$$

Taking Laplace transform,

$$(s^2 + 2s + 4)\, y(s) = 4X(s)$$

$$\text{or} \quad \frac{Y(s)}{X(s)} = \frac{4}{s^2+2s+4}$$

Comparing the transfer function with

$$\frac{\omega_n^2}{s^2 + 2\xi\omega_n s + \omega_n^2}$$

$$\omega_n^2 = 4$$

$$\Rightarrow \quad \omega_n = 2$$

$$2\xi\omega_n = 2$$

$$\Rightarrow \quad \xi = \frac{2}{2\omega_n} = \frac{2}{2\times 2} = 0.5$$

Therefore, the damping ratio, $\xi = 0.5$.

2.35 (b)

Characteristic equation

$s^2 + 2\xi\omega_n s + \omega_n^2 = 0$

If $\xi = 0$, then $s_2 + \omega_n^2 = 0$

$\Rightarrow \quad s = \pm j\omega_n$

It is clear that the closed-loop poles are purely imaginary.

2.36 (c)

$$e_{ss} = \lim_{s\to 0} \frac{sR(s)}{1+G(s)H(s)}$$

Given that,

input $r(t) = 10t$

$\Rightarrow \quad R(s) = 10/s^2$

$$G(s) = \frac{10}{s^2(s+4)},\ H(s) = 1$$

$$\because \quad e_{ss} = \lim_{s\to 0} \frac{s\cdot 10/s^2}{1+\frac{10}{s^2(s+4)}}$$

$$= \lim_{s\to 0} \frac{10s(s+4)}{s^2(s+4)+10} = 0$$

2.37 (d)

Characteristic equation $= 4s^2 + 6s + 1 = 0$

$$s^2 + \frac{6}{4}s + \frac{1}{4} = 0$$

Comparing with standard characteristic equation

$$s^2 + 2\xi\omega_n s + \omega_n^2 = 0$$

$$\omega_n = \frac{1}{2} = 0.5 \text{ rad/sec}$$

$$2\xi\omega_n = \frac{3}{2}$$

$$\xi = \frac{3}{4}\times 2 = 1.5$$

$\therefore$ System is overdamped.

2.38 (d)

$$C(\infty) = \lim_{s\to 0} s\times\frac{\omega_n^2}{s^2+2\xi\omega_n s+\omega_n^2}\times\frac{1}{s} = 1$$

2.39 (c)

Comparing the transfer function

$$\frac{16}{s^2+4s+16} \text{ with } \frac{\omega_n^2}{s^2+2\xi\omega_n s+\omega_n^2}$$

$\omega_n^2 = 16 \quad \Rightarrow \omega_n = 4$ rad/s

$2\xi\omega_n = 4 \quad \Rightarrow = \dfrac{4}{2\omega_n}$

$$\Rightarrow \quad \xi = \frac{4}{2\times 4}$$

Time for first overshoot

$$t_p = \frac{\pi}{\omega_n\sqrt{1-\xi^2}} = \frac{\pi}{4\sqrt{1-\frac{1}{4}}} = \frac{\pi}{2\sqrt{3}}s$$

2.40 (a)

The response of an amplifier with three (or more) poles is determined approximately by the two lowest poles, p_1 and p_2, provided that $|p_3/p_2| \geq 4$.

2.41 (d)

This is the Laplace transform of sin *t*.

So, $f(t) = \sin t$

Steady-state value of $f(t)$ is undetermined because poles of $F(s)$ are not in LHS of *s*-plane. Therefore, steady-state value will vary between -1 and $+1$.

2.42 (d)

$$G(s) = \frac{4}{s^2+0.4s};\ H(s) = 1$$

Characteristic equation

$$1 + G(s)H(s) = 0$$

$$\Rightarrow \quad 1+\frac{4}{s^2+0.4s}\cdot 1 = 0$$

$$\Rightarrow \quad s^2 + 0.4s + 4 = 0 \quad \text{...(1)}$$

Comparing equation (1) with standard equation of second order system i.e.

$$s^2 + 2\xi\omega_n s + \omega_n^2 = 0$$

we have

$$2\xi\omega_n = 0.4 \Rightarrow \boxed{\xi\omega_n = 0.2}$$

settling time within 2% tolerance band

$$t_s = \frac{4}{\xi\omega_n} = \frac{4}{0.2}$$

$$\Rightarrow \quad t_s = 20 \text{ sec}$$

2.43 (d)

Underdamp system is which oscillation are damped.

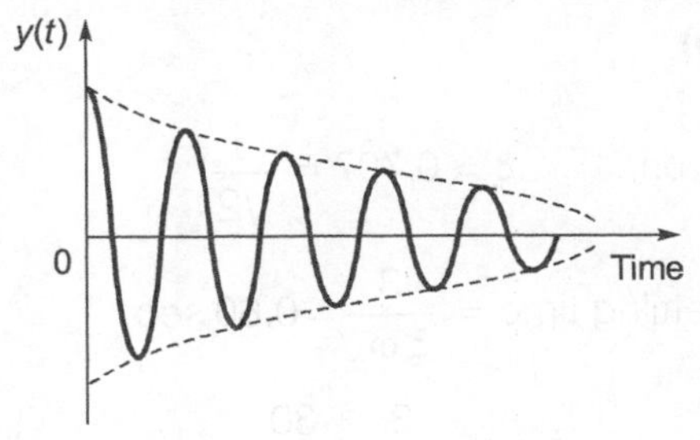

2.45 (b)

Integral controller improves steady state performance while derivative controller improves transient state response.

2.47 (b)

$$\frac{C(s)}{R(s)} = \frac{10\left(\frac{1}{s+1}\right)}{1+\frac{1}{s}\cdot\frac{10}{(s+1)}}$$

$$\Rightarrow \quad C(s) = \frac{10s}{s(s+1)+10}R(s)$$

Given $r(t) = u(t)$

So $R(s) = \frac{1}{s}$

$$\therefore \quad C(s) = \frac{10s}{s(s+1)+10}\cdot\frac{1}{s}$$

$$\Rightarrow \quad C(s) = \frac{10}{s(s+1)+10}$$

Steady state value of response

$$= \lim_{s\to 0} sC(s) = \lim_{s\to 0} s\cdot\frac{10}{s(s+1)+10} = 0$$

2.48 (b)

$$G(s)H(s) = \frac{2}{s(s^2+2s+2)}$$

Since $G(s)H(s)$ has one pole at origin, so given system is type-1 system.

2.49 (b)

$$40\frac{dx}{dt}+2x = f(t)$$

$$\Rightarrow \quad X(s)\,(40s+2) = F(s)$$

$$\therefore \quad \frac{X(s)}{F(s)} = \frac{1}{40s+2}$$

Pole will be at $s = -\frac{1}{20}$

Time constant is reciprocal of location of pole for a first order system.

2.50 (b)

Given, $\xi = 0.707 = \frac{1}{\sqrt{2}}$

$$\text{Settling time} = \frac{3}{\xi\,\omega_n} = 0.60 \text{ sec.}$$

$$\xi\omega_n = \frac{3}{0.6} = \frac{30}{6} = 5$$

Poles are given as

$$s = -\xi\,\omega_n \pm j\omega_n\sqrt{1-\xi^2}$$

$$s = -5 \pm j5\sqrt{2}\sqrt{1-\frac{1}{2}}$$

$$= -5 \pm j\,5\sqrt{2}\cdot\frac{1}{\sqrt{2}} = -5 \pm j5$$

2.51 (c)

$$G(s) = \frac{K}{s(s+6)}$$

Characteristics equation

$$1 + G(s)\,H(s) = 0$$

$$\Rightarrow \quad 1+\frac{K}{s(s+6)} = 0$$

$$\Rightarrow \quad s(s+6)+K = 0$$

$$\Rightarrow \quad s^2+6s+K = 0$$

Comparing above equation with standard equation i.e.

$$s^2+2\xi\,\omega_n\,s+\omega_n^2 = 0$$

We have, $\omega_n = \sqrt{K}$ and $2\xi\,\omega_n = 6$

It is given that $\xi = 0.75$; so

$$2\times 0.75\times\sqrt{K} = 6$$

$$\therefore \quad \sqrt{K} = \frac{6}{1.5} = 4$$

$$K = 16$$

2.52 (d)

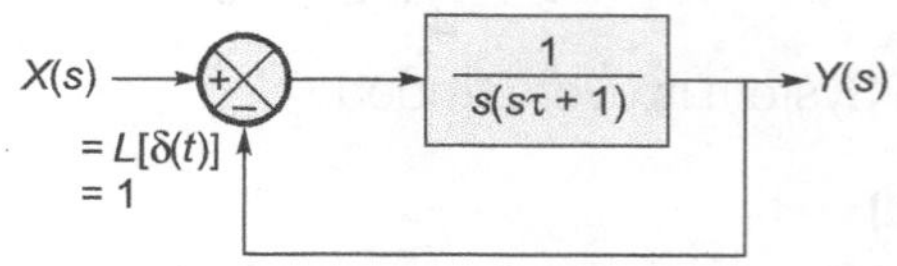

$$\frac{Y(s)}{X(s)} = H(s) = \frac{1}{s^2\tau+s+1}$$

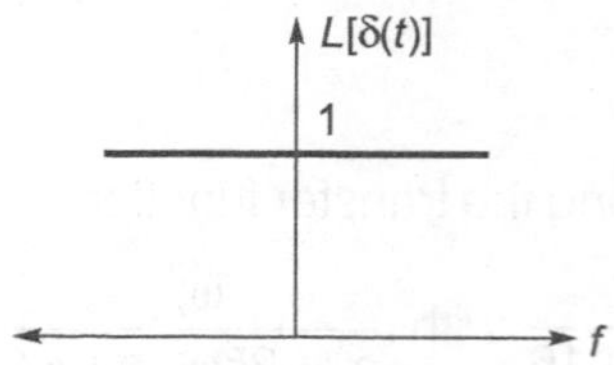

2.53 (a)

- As we know from the formulae

  Rise time, $t_r = \frac{0.35}{\text{Bandwidth}}$

  Thus it can be seen that rise time is inversely proportional to bandwidth.

- Also $\omega_d = \omega_n\sqrt{1-\xi^2}$

  Increasing ω_n causes increase in ω_d and thus bandwidth increases and rise time reduces.

2.54 (b)

The type of system is determined from the number of poles at origin for open loop transfer function.

2.55 (b)

$$G(s)\,H(s) = \frac{25}{s^2+5s+25}$$

Comparing the above transfer function with the standard second order transfer function:

$$\frac{\omega_n^2}{s^2+2\xi\omega_n s+\omega_n^2}$$

So $\omega_n = 5$

$2\xi\omega_n = 5$

$\Rightarrow \xi = 0.5$

Settling time for 2% tolerance band $= \dfrac{4}{\xi\omega_n}$

$= \dfrac{4}{2.5} = 1.6$ sec.

2.56 (d)

Maximum overshoot, rise time and delay time of the system determines the transient characteristics.

2.57 (b)

$$G(s)\,H(s) = \frac{4}{s^2(s+3)}$$

Type of system is found from open loop poles at origin.

Hence type-2 system.

2.58 (c)

The roots of the second order control system is given as

$$s_1, s_2 = -\xi\omega_n \pm \omega_n\sqrt{\xi^2-1}$$

if $\xi = 1$, then

$s_1, s_2 = -\omega_n$

Thus the poles are equal, negative and real.

2.60 (a)

$$G(s) = \frac{(2s+5)}{(s+5)(s+4)}$$

The dc gain is always calculated in $(1 \pm T_s)$ form i.e. time constant form.

So, $$G(s) = \frac{5\left(1+\frac{2}{5}s\right)}{4\times 5(1+0.2s)(1+0.25s)}$$

$$G(s) = 0.25\frac{(1+0.4s)}{(1+0.2s)(1+0.25s)}$$

So dc gain is calculated at $s = 0$

$$\boxed{G(s) = 0.25}$$

2.61 (c)

$$T = \frac{(1+2K)}{(3+4K)}$$

Sensitivity, $$S_K^T = \frac{\partial T/T}{\partial K/K} = \frac{\partial T}{\partial K}\cdot\frac{K}{T}$$

$$\frac{\partial T}{\partial K} = \frac{2(3+4K)-1(1+2K)\cdot 4}{(3+4K)^2}$$

$$= \frac{2}{(3+4K)}$$

$$S_K^T = \frac{2}{(3+4K)^2}\cdot\frac{K}{(1+2K)}\cdot(3+4K)$$

$$= \frac{2K}{3+10K+8K^2}$$

2.62 (c)

$\%M_p = 30\%$

$M_p = 0.3 = e^{-\xi\pi/\sqrt{1-\xi^2}}$

$$\ln 0.3 = \frac{-\xi\pi}{\sqrt{1-\xi^2}}$$

$\Rightarrow \xi = 0.3578$

Peak overshoot occurs at peak time

$$t_p = \frac{\pi}{\omega_d} = \frac{\pi}{\omega_n\sqrt{1-\xi^2}} = 0.336 \text{ sec.}$$

2.63 (d)

Maximum peak overshoot is given as

$$M_p = e^{-\frac{\xi n\pi}{\sqrt{1-\xi^2}}}$$

We conclude $M_p \propto \dfrac{1}{\xi}$

So, lesser the value of ξ, greater will be the value of overshoot. So, go through the options separately.

(a) $T(s) = \dfrac{9}{s^2+2s+9}$; $\omega_n = 2$

$$\xi = \frac{1}{3} = 0.33$$

(b) $T(s) = \dfrac{16}{s^2+2s+16}$; $\omega_n = 4$

$$\xi = \frac{1}{4} = 0.25$$

(c) $T(s) = \dfrac{25}{s^2+2s+25}$; $\omega_n = 5$

$\xi = 0.2$

(d) $T(s) = \dfrac{36}{s^2+2s+36}$; $\omega_n = 6$

$\xi = 0.16$

Thus, option (d) has lowest ξ, so maximum overshoot.

2.64 (d)

$$G(s) = \frac{1}{(s+2)^2}$$

$$\text{Closed loop transfer function} = \frac{G(s)}{1+G(s)}$$

$$= \frac{\frac{1}{(s+2)^2}}{1+\frac{1}{(s+2)^2}} = \frac{1}{s^2+4s+5}$$

$$\therefore \text{ Closed loop poles } s = \frac{-4 \pm \sqrt{16-20}}{2}$$

$$= -2 \pm j1$$

2.65 (c)

$$G(s) = \frac{100}{s^2+10s+100}$$

On comparing it with standard equation, we get

$$\omega_n = 10 \text{ rad/sec.}$$

$$2\xi\omega_n = 10$$

$$\Rightarrow \quad \xi = 0.5$$

Settling time, when unit step is given as an input

$$= \frac{4}{\xi\omega_n} = \frac{4}{5} = 0.8 \text{ sec.}$$

2.66 (c)

The transfer function $G(s) = \dfrac{(s+1)}{(s^2+s+1)}$

The d.c. gain is always found in the time constant form i.e. $(1 \pm T_s)$ at $s = 0$

$$\therefore \quad G(s)\big|_{s=0} = \frac{(1+0)}{(1+0)} = 1$$

Steady state error,

$$e_{ss} = \lim_{s\to 0} \frac{sR(s)}{1+G(s)H(s)}$$

$$K_p = \lim_{s\to 0} G(s)H(s) = \lim_{s\to 0} \frac{s+1}{s^2} = \infty$$

So, $\quad e_{ss} = 0$

2.67 (d)

$$\frac{d^2x}{dt^2} + 5\frac{dx}{dt} + 7x = 7y$$

Taking the Laplace transform of both sides

$$s^2X(s) + 5sX(s) + 7X(s) = 7Y(s)$$

$$H(s) = \frac{X(s)}{Y(s)} = \frac{7}{s^2+5s+7}$$

Comparing the above equation with the standard second-order characteristics equation, i.e.

$$H(s) = \frac{\omega_n^2}{s^2+2\xi\omega_n s+\omega_n^2}$$

Thus, $\quad \omega_n^2 = 7$

$\Rightarrow \quad \omega_n = \sqrt{7}$ rad/sec.

and $\quad 2\xi\omega_n = 5$

$\Rightarrow \quad \xi = \dfrac{5}{2\sqrt{7}} = 0.94$ rad/sec.

2.68 (d)

Statement (I) is false as for zero input response initial conditions should not be equal to zero.

2.69 (a)

At $\quad \xi = 0$

The system becomes undamped system and

$\because \quad \omega_d = \omega_d\sqrt{1-\xi^2}$

$\omega_d = \omega_n =$ Natural frequency damping frequency

2.70 (b)

$$G(s) = \frac{K(s+12)}{(s+14)(s+18)}$$

$$\text{Steady state error} = e_{ss} = \frac{A}{1+K_P}$$

where $\quad K_P = \lim\limits_{s\to 0} G(s) = \dfrac{12K}{18\times 14} = \dfrac{K}{21}$

$\therefore \quad e_{ss} = \dfrac{1}{1+K/21} = 0.10$

or $\quad 10 = 1 + \dfrac{K}{21}$

$\dfrac{K}{21} = 9$ or $K = 189$

2.71 (b)

$$G(s) = \frac{K}{s(s+10)} \quad \text{...(i)}$$

The closed loop system is given by

$$T(s) = \frac{G(s)}{1+G(s)}$$

$$= \frac{K}{s(s+10)+K}$$

$$= \frac{K}{s^2+10s+K} \quad \text{...(ii)}$$

Comparing the above equation with standard transfer function as

$$T(s) = \frac{\omega_n^2}{s^2+2\xi\omega_n s+\omega_n^2} \quad \text{...(iii)}$$

$$\omega_n^2 = K$$

and $\quad 2\xi\omega_n = 10$

or $\quad \omega_n = \frac{10}{2\xi} = \frac{10}{1}$

$\therefore \quad K = 10^2 = 100$

2.72 (a)

Comparing the equation with the standard equation

i.e. $\quad y(t) = Ae^{-\xi\omega_n t}\sin(\omega_d t+\delta)$

thus $\quad \xi\omega_n = 0.8$

Now, only option (a) satisfies this condition.

2.73 (c)

Given, $G(s) = \dfrac{25}{s^2+5s+25}$

Now, $\quad t_s = \dfrac{4}{\xi\omega_n}$

Comparing the above equation with standard equation.

$$G(s) = \frac{\omega_n^2}{s^2+2\xi\omega_n s+\omega_n^2}$$

$\therefore \quad 2\xi\omega_n = 5$

$\xi\omega_n = 2.5$

Thus, $\quad t_s = \dfrac{4}{2.5} = 1.6$ sec

2.74 (d)

Settling time : It is the time required for a system at a given input to increase and stay within a specified percentage of its final value of the input.

2.75 (b)

Rise time $= \dfrac{0.35}{\text{bandwidth}}$

$\therefore$ t_r is inversely proportional to bandwidth.

2.76 (c)

Transfer function $= \dfrac{1}{2s+1}$

$$M = \frac{1}{\sqrt{4\omega^2+1}}$$

Magnitude (M) at $\omega = 0$ is 1 and at $\omega = \infty$ is 0.

2.77 (a)

$$\frac{C(s)}{R(s)} = \frac{100}{s^2+10s+100}$$

$$2\xi\omega_n = 10 \Rightarrow \xi\omega_n = 5$$

$$t_s(2\%) = \frac{4}{\xi\omega_n} = \frac{4}{5} = 0.8\,s$$

2.78 (c)

Step response $= 1 + 0.2\,e^{-60t} - 1.2\,e^{-10t}$ ...(i)

From equation (i) the poles are at –10, –60

Hence, the characteristic equation

$$q(s) = (s+10)(s+60) = 0$$

$$q(s) = s^2 + 70s + 600 = 0$$

$\therefore \quad \omega_n = \sqrt{600} \approx 24.5$ rad/sec

$2\xi\omega_n = 70$

$\therefore \quad \xi = 1.43$

2.79 (c)

$$GH(s) = \frac{25}{s(s+6)}$$

Characteristic equation

$$q(s) = 1 + GH(s) = s^2 + 6s + 25 = 0$$

$$\omega_n = 5\,;\ \xi = 0.6 \therefore \omega_d = 4$$

$\therefore \quad t_p = \dfrac{\pi}{\omega_d} = 0.79$ s

2.80 (a)

$$\text{CLTF} = \frac{C(s)}{R(s)} = \frac{\omega_n^2}{s^2+2\xi\omega_n+\omega_n^2}$$

$$\text{OLTF} = GH(s) = \frac{\omega_n^2}{s(s+2\xi\omega_n)}$$

Velocity error constant $= K_v = \lim_{s\to 0} sGH(s)$

$\therefore \quad K_v = \lim_{s\to 0} s\dfrac{\omega_n^2}{s(s+2\xi\omega_n)} = \dfrac{\omega_n}{2\xi}$

2.81 (c)

$\xi = 1$ means critically damped.

Hence, roots are real, equal, negative.

2.82 (c)

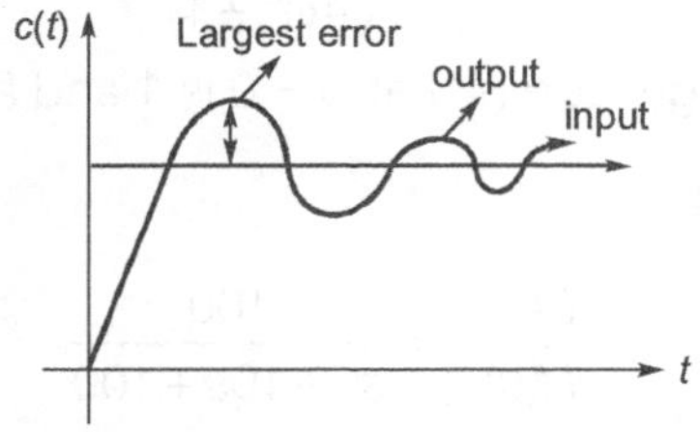

Largest error during transient period is called peak overshoot.

2.83 (*)

$$K_p = \lim_{s\to 0} G(s)H(s)$$

$$K_v = \lim_{s\to 0} sG(s)H(s)$$

$$K_a = \lim_{s\to 0} s^2G(s)H(s)$$

For type-I system, loop transfer function has one pole at origin

Let us take,

$$G(s)H(s) = \frac{1}{s(s+p)}$$

$\therefore$ $K_p = \infty$, $K_v = 1$, $K_a = 0$ and e_{ss} for unit step input $= \frac{1}{1+K_p} = 0$

According to this only statement-2 is correct which is not there in the given options.

2.84 (d)

$$G = \frac{10}{s(s+5)};\ H = K$$

Given is non unity feedback system, first convert to unity feedback system

$$G_{eff} = \frac{G}{1+G(H-1)}$$

$$= \frac{10}{s^2+5s+10(K-1)};\ \text{type} = 0$$

e_{ss} for step input

$$= \frac{1}{1+K_p} = \frac{K-1}{K} < 0.05$$

$$K_p = \frac{1}{K-1}$$

$$\therefore \quad K < 1.05$$

2.85 (*)

$$\frac{C(s)}{R(s)} = \frac{2}{(s+10)(s+100)} \approx \frac{2}{100(s+10)}$$

$\because$ using dominant pole approximation

For unit step excitation,

$$C(s) = \frac{1}{50s(s+10)} = \frac{1}{500}\left[\frac{1}{s} - \frac{1}{s+10}\right]$$

$$c(t) = \frac{1}{500}(1-e^{-10t})u(t)$$

Time required to reach 2% of steady-state value is,

$$\frac{1}{500}(1-e^{-10t}) = \frac{2}{100}\times\frac{1}{500}$$

$$t = \frac{1}{10}ln\left(\frac{50}{49}\right)s \approx 0.002\, s$$

No option is matching with this answer.

2.86 (c)

$$q(s) = 2s^2 + 6s + 6 = 0$$

$$q(s) = s^2 + 3s + 3 = 0$$

$$\omega_n = \sqrt{3};\ 2\xi\omega_n = 3$$

$$\therefore \quad \xi = \frac{3}{2\sqrt{3}} = \frac{\sqrt{3}}{2} < 1$$

Underdamped system.

2.87 (a)

$$G(s) = \frac{K(s+2)}{s^2(s^2+7s+12)}$$

$$K_p = \lim_{s\to 0} G(s) = \infty$$

$$K_v = \lim_{s\to 0} sG(s) = \infty$$

$$K_a = \lim_{s\to 0} s^2G(s) = \frac{K}{6}$$

2.88 (b)

Settling time is defined as the time taken by the response of a system to reach and to stay with in a (specified) certain percentage of final value.

2.89 (d)

$$q(s) = s^2 + as + K = 0$$

$$\omega_n = \sqrt{K}$$

$$2\xi\omega_n = 2$$

$$\xi = \frac{a}{2\sqrt{K}}$$

For underdamped system $\xi < 1$

$$\therefore \quad \frac{a}{2\sqrt{K}} < 1 \Rightarrow K > \frac{a^2}{4}$$

2.90 (b)

$$G(s) = \frac{25}{s(s+6)}$$

$$q(s) = s^2 + 6s + 25 = 0$$

$$2\xi\omega_n = 6$$

$$\Rightarrow \quad \xi\omega_n = 3$$

$$t_s = \frac{4}{\xi\omega_n} = 1.33$$

2.91 (*)

$$G(s) = \frac{K}{s+a}$$

$$a = \frac{1}{0.1} = 10$$

$$\frac{K}{a} = 5$$

$$K = 5a = 50$$

2.92 (a)

Type = 0

$$e_{ss} \text{ for step input} = \frac{1}{1+K_p} = 0.2$$

$$\therefore \quad K_p = 4$$

To remove this e_{ss} type of the system has to be increased.

Hence the block is integrator.

Now type = 1

$$e_{ss} \text{ for ramp input} = \frac{1}{K_v} = \frac{1}{K_p} = 0.25$$

2.93 (c)

$$h(t) = 16 \cdot e^{-2t} - 8e^{-t}$$

$$H(s) = \frac{16}{s+2} - \frac{8}{s+1}$$

$$= \frac{16s + 16 - 8s - 16}{(s+1)(s+2)}$$

$$= \frac{8s}{(s+1)(s+2)}$$

$$x(t) = u(t) \xleftrightarrow{L.T} \frac{1}{s}$$

$$Y(s) = \frac{8}{(s+1)(s+2)}$$

$$y(t) = 8e^{-t} - 8e^{-2t}$$

2.94 (a)

$$\frac{Y(s)}{R(s)} = \frac{1}{s^2 + 3s + 2}$$

$$Y(s) = \frac{1}{(s+1)(s+2)} \times \frac{5}{s}$$

$$= \frac{A}{s} + \frac{B}{s+1} + \frac{C}{s+2}$$

$$A = \frac{5}{2}\ ;\ B = -5\ ;\ C = \frac{5}{2}$$

$$\therefore \quad y(t) = \left[\frac{5}{2} - 5e^{-t} + \frac{5}{2}e^{-2t}\right]u(t)$$

2.95 (c)

$$\text{Type-2}: G(s)\,H(s) = \frac{1}{s^2}$$

$$R(s) = \frac{1}{s}$$

$$\Rightarrow \quad e_{ss} = \lim_{s \to 0} \frac{s \times \frac{1}{s}}{1 + \frac{1}{s^2}} = 0$$

$$R(s) = \frac{1}{s^2}$$

$$\Rightarrow \quad e_{ss} = \lim_{s \to 0} \frac{s \times \frac{1}{s^2}}{1 + \frac{1}{s^2}} = 0$$

2.96 (d)

$$P(t) = \frac{t^2}{2}u(t)$$

$$\frac{d}{dt}p(t) = tu(t)$$

$\therefore$ Parabolic function is one degree faster than ramp function

$$L\left[\frac{t^2}{2}u(t)\right] = \frac{1}{s^3}$$

2.97 (a)

$$y(t) = L^{-1}\left[\frac{1}{s(1+\tau s)}\right] = L^{-1}\left[\frac{A}{s} + \frac{B}{1+\tau s}\right]$$

$$= L^{-1}\left[\frac{1}{s} - \frac{\tau}{1+\tau s}\right]$$

$$y(t) = 1 - e^{-t/\tau}$$

2.98 (d)

$$\text{Step response} = \int_{-\infty}^{t} (\text{Impulse response}) \cdot dt$$

$$= \sum_{-\infty}^{n} \text{Impulse response}$$

2.99 (b)

For a type-1 system, steady state error for

(i)Step input = 0

(ii)Ramp input $= \frac{1}{K_v} = \frac{1}{1} = 1$

2.100 (b)

$$t_p = \frac{n\pi}{\omega_d}$$

For first undershoot $n = 2$.

2.101 (d)

For the parabolic input $r(t) = \frac{t^2}{2}u(t)$, steady-state error

$$e_{ss} = \frac{1}{K_a}$$

where K_a = acceleration error coefficient

$$K_a = \lim_{s\to 0} s^2 G(s) \cdot H(s)$$

$$K_a = \lim_{s\to 0} s^2 \times \frac{(1+sT_1)(1+sT_2)\ldots}{s(1+s\tau_1)(1+s\tau_2)\ldots}$$

[∵ for a type-1 system, one pole is present at origin]

$$K_a = 0$$

Therefore,

$$e_{ss} = \frac{1}{K_a} = \frac{1}{0} = \infty$$

2.102 (a)

The following conditions justify a second-order approximation:

(i) Higher-order poles are much farther into the left half s-plane than the dominant second order pair of poles.

(ii) Closed-loop zeros near the closed-loop second order pole pair are nearly cancelled by the proximity of higher order closed-loop poles.

(iii) Closed-loop zeros not cancelled by the close proximity of higher order closed-loop poles are far removed from the closed-loop second order pole pair.

2.103 (d)

$$G(s) = \frac{K(2s+1)}{s(4s+1)(s+1)^2}$$

The given system is a Type-1 system. Hence, the steady state error is zero for step input and the steady state error for ramp input $Atu(t)$ is given as

$$e_{ss} = \frac{A}{K_V}$$

where, K_V = Velocity error coefficient

$$K_V = \lim_{s\to 0} sG(s)H(s)$$

$$K_V = \lim_{s\to 0} s \times \frac{K(2s+1)}{s(4s+1)(s+1)^2}$$

{∵ $H(s) = 1$ for unity feedback}

$$K_V = K$$

For input $r(t) = 1 + 5t$,

$$e_{ss} = 0 + \frac{5}{K_V}$$

$$e_{ss} = \frac{5}{K} < 0.1 \quad \text{(Given)}$$

$$\Rightarrow \quad K > 50$$

2.104 (d)

The given system is a unity feedback system with open-loop transfer function

$$G(s)H(s) = \frac{K(s+5)}{s(s+6)(s+7)(s+8)}$$

It is a Type-1 system. Hence, the steady-state error is finite for unit ramp input given as

$$e_{ss} = \frac{1}{K_V} = 0.1 \quad \text{(Given)}$$

where, $K_V = \lim_{s\to 0} sG(s)H(s)$

$$K_V = \lim_{s\to 0} s \times \frac{K(s+5)}{(s+6)(s+7)(s+8)}$$

$$K_V = \frac{5K}{336}$$

Hence, $e_{ss} = \frac{1}{K_V} = 0.1$

$$\Rightarrow \quad \frac{336}{5K} = 0.1$$

$$\Rightarrow \quad K = \frac{3360}{5} = 672$$

2.105 (a)

Rise Time : The time required for response to rise from 10% to 90% of the final value is called the rise time.

Peak Time : The time required for the response to reach the first peak or overshoot of the time response is called the peak time.

Settling Time : The time required for the response to reach and stay within the specified range (2% to 5%) of its final value is called the settling time.

2.106 (a)

$$G(s) = \frac{K}{s(\tau s+1)} = \frac{K/\tau}{s(s+1/\tau)}$$

Comparing the given transfer function with standard open loop second order transfer function.

$$G(s) = \frac{\omega_n^2}{s(s+2\xi\omega_n)}$$

We get, $\omega_n^2 = \frac{K}{\tau} \Rightarrow \omega_n = \sqrt{\frac{K}{\tau}}$

and $2\xi\omega_n = \frac{1}{\tau}$

$$\Rightarrow \quad \xi = \frac{1}{2\tau} \times \sqrt{\frac{\tau}{K}}$$

$$\Rightarrow \quad \xi = \frac{1}{2}\sqrt{\frac{1}{\tau K}}$$

$$\Rightarrow \quad \xi \propto \frac{1}{\sqrt{K}}$$

Let K_1 be the amplifier gain corresponding to $\xi_1 = 0.25$ and K_2 be the amplifier gain corresponding to $\xi_2 = 0.75$. Hence,

$$\frac{\xi_1}{\xi_2} = \sqrt{\frac{K_2}{K_1}}$$

$$\Rightarrow \quad \frac{0.25}{0.75} = \sqrt{\frac{K_2}{K_1}}$$

$$\Rightarrow \quad \frac{K_2}{K_1} = \left(\frac{1}{3}\right)^2$$

$$\Rightarrow \quad K_2 = (0.1111)K_1$$

Hence, the amplifier gain must be multiplied by 0.1111 to increase damping ratio from 0.25 to 0.75.

2.107 (d)

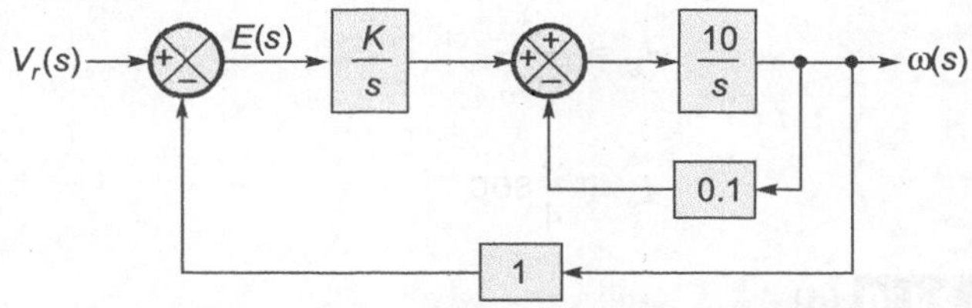

The simplified block diagram can be drawn as below:

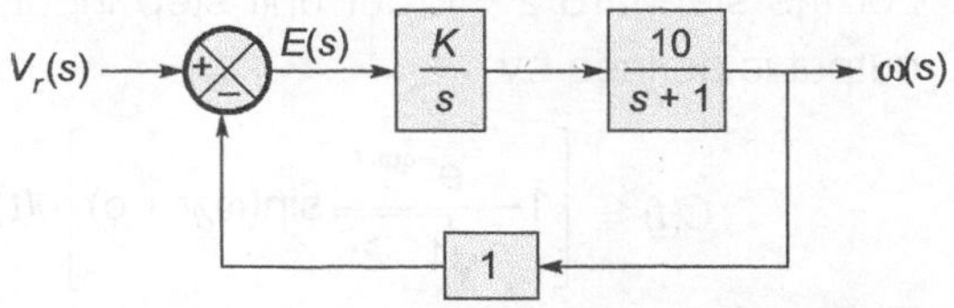

The above block diagram represents a unity feedback control system with open-loop transfer function

$$G(s)H(s) = \frac{10K}{s(s+1)}$$

It is a Type-1 system. The steady state error to unit ramp input is given as

$$e_{ss} = \frac{1}{K_V}$$

where, $K_V = \lim_{s\to 0} sG(s)H(s)$

$$K_V = \lim_{s\to 0} s \times \frac{10K}{s(s+1)}$$

$$K_V = 10\,K$$

For the steady state error to be less than 0.01 rad/sec,

$$e_{ss} = \frac{1}{K_V} = \frac{1}{10K} < 0.01$$

$$\Rightarrow \quad K > 10$$

2.108 (b)

- Delay time (t_d) is the time required to reach at 50% of its final value by a time response signal during its first cycle of oscillation.
- Settling time is the time required for the response to reach and stay within the specified range i.e. either 3% or 5% of the final value.

Therefore, statements 1 and 3 are not correct.

2.109 (b)

Comparing with standard 2nd order system transfer function,

we have $\omega_n = 5$ rad/sec

and $\xi = 0.8$ ..... Underdamped system

$\therefore$ The settling time for 5% band in tolerance error is given as,

$$t_s = \frac{3}{\xi\omega_n}$$

$$t_s = \frac{3}{4}\text{sec}$$

2.110 (d)

For the standard 2nd order unit step input, the output is defined by

$$C(t) = \left[1 - \frac{e^{-\xi\omega_n t}}{\sqrt{1-\xi^2}}\sin(\omega_d t + \phi)\right]u(t)$$

On comparison we get,

$$\xi\omega_n = 1$$

and
$$\frac{1}{\sqrt{1-\xi^2}} = \frac{2}{\sqrt{3}}$$

$$\frac{3}{4} = 1 - \xi^2$$

$$\xi^2 = 1 - \frac{3}{4}$$

$$\xi = \sqrt{\frac{1}{4}} = \frac{1}{2}$$

and $\omega_n = 2$ rad/sec

$\therefore$ Transfer function of the system is given by

$$\text{TF} = \frac{\omega_n^2}{s^2 + 2\xi\omega_n s + \omega_n^2} = \frac{4}{s^2 + 2s + 4}$$

2.111 (c)

On comparing with standard 2nd order system we get,

$$\omega_n = 1 \text{ rad/sec}$$
$$\xi = 1$$

$\therefore$ The given system is critically damped.

$\therefore$ The step response for the critically damped system is given by,

$\therefore$ $C(t) = 1 - e^{-\xi\omega_n t}[1 + \omega_n t]$

$\therefore$ $C(t) = 1 - e^{-t}[1 + t]$

for $t \to \infty$

$$C(\infty) = 1$$

Now, we have to find 't' for which $C(t)$ to reach 94% of its steady-state value,

$1 - e^{-t}[1 + t] = 0.94$

On solving we get, $t = 4.50$ sec

2.112 (d)

On comparing with standard 2nd order system

$$\omega_n = 5 \text{ rad/sec}$$
$$\xi = 0.8$$

$$\therefore \omega_d = \omega_n\sqrt{1-\xi^2} = 5\sqrt{1-0.64} = 5 * 0.6$$

$$\omega_d = 3 \text{ rad/sec}$$

The time at which second peak in the response is given by

$$t_p = \frac{n\pi}{\omega d}$$

where $n = 3$ for second peak

$$\therefore \quad t_p = \frac{3 \times \pi}{3}$$

$$t_p = \pi \text{ sec}$$

2.113 (c)

We have,

OLTF as $G(s)\,H(s) = \dfrac{10}{s(1+0.1s)}$

Position error constant,

$$K_p = \lim_{s\to 0} G(s)H(s) = \lim_{s\to 0} \frac{10}{s(1+0.1s)} = \infty$$

Velocity error constant,

$$K_v = \lim_{s\to 0} sG(s)H(s) = \lim_{s\to 0} \frac{s\times 10}{s(1+0.1s)} = 10$$

Acceleration error constant,

$$K_a = \lim_{s\to 0} s^2G(s)H(s) = \lim_{s\to 0} \frac{s^2\times 10}{s(1+0.1s)} = 0$$

Hence, option (c) is correct.

2.114 (d)

We have, $G(s) = \dfrac{K}{s(s+10)}$; $\xi = 0.5$

Characteristic equation of the feedback system,

$s^2 + 10s + K = 0$

Comparing with the standard second order characteristic equation $s^2 + 2\xi\omega_n s + \omega_n^2 = 0$, we get

$$2\xi\omega_n = 10$$

$$\omega_n = 10 \text{ rad/sec}$$

Peak overshoot time

$$= \frac{n\pi}{\omega_d} = \frac{\pi}{10\sqrt{1-(0.5)^2}}$$

(n = 1 for first peak overshoot)

$$t_p = 0.363 \text{ sec}$$

Hence, option (d) is correct.

2.115 (b)

1. Rise time is defined as the time required for the response to rise from 0% to 100% of the final value for underdamped, 5% to 95% for critically damped and 10% to 90% for overdamped systems. Hence, statement 1 is not correct.
2. The amount of overshoot depends on the damping ratio and directly indicates the relative stability of the system. Hence, statement 2 is correct.
3. Settling time is the time required for the response curve to reach and stay within a range of particular tolerance band (usually 5% or 2% of its final value). Hence, statement 3 is not correct.

Therefore, statements 1 and 3 are not correct.

■■■

3 Routh-Hurwitz Stability Criterion

3.1 The open-loop transfer function of unity feedback control system if

$$G(s) = \frac{K}{s(s+a)(s+b)},\ 0 < a \le b$$

The system is stable if

(a) $0 < K < \dfrac{(a+b)}{ab}$

(b) $0 < K < \dfrac{ab}{(a+b)}$

(c) $0 < K < ab(a + b)$

(d) $0 < K < a/b(a + b)$ **[ESE-2000]**

3.2 The Routh-Hurwitz criterion cannot be applied when the characteristic equation of the system contains any coefficients which is

(a) negative real and exponential functions of s

(b) negative real, both exponential and sinusoidal function of s

(c) both exponential and sinusoidal functions of s

(d) complex, both exponential and sinusoidal functions of s

[ESE-2000]

3.3 Which one of the following characteristic equations of result in the stable operation of the feedback system?

(a) $s^3 + 4s^2 + s - 6 = 0$

(b) $s^3 + s^2 + 5s + 6 = 0$

(c) $s^3 + 4s^2 + 10s + 11 = 0$

(d) $s^4 + s^3 + 2s^2 + 4s + 6 = 0$ **[ESE-2000]**

3.4 Consider the following statements:

Routh-Hurwitz criterion gives

1. absolute stability
2. the number of roots lying on the right half of the s-plane
3. the gain margin and phase margin

Which of these statements are correct?

(a) 1, 2 and 3 (b) 1 and 2

(c) 2 and 3 (d) 1 and 3 **[ESE-2000]**

3.5 The given characteristic polynomial $s^4 + s^3 + 2s^2 + 2s + 3 = 0$ has

(a) zero root in RHS of s-plane

(b) one root in RHS of s-plane

(c) two roots in RHS of s-plane

(d) three roots in RHS of s-plane

[ESE-2001]

3.6 The characteristic equation of a control system is given by $s^6 + 2s^5 + 8s^4 + 12s^3 + 20s^2 + 16s + 16 = 0$. The number of the roots of the equation which lie on the imaginary axis of s-plane is

(a) zero (b) 2

(c) 4 (d) 6 **[ESE-2003]**

3.7 **Assertion (A):** A linear, negative feedback control system is invariable stable if its open loop configuration is stable.

Reason (R): The negative feedback reduces the overall gain of the system.

(a) Both A and R are true and R is the correct explanation of A

(b) Both A and R are true but R is NOT the correct explanation of A

(c) A is true but R is false

(d) A is false but R is true **[ESE-2003]**

3.8 An open loop system has a transfer function $\dfrac{1}{s^3 + 1.5s^2 + s - 1}$. It is converted into a closed loop system by providing a negative feedback having transfer function 20 (s + 1). Which one of the following is correct?

The open loop and closed loop systems are, respectively

(a) stable and stable

(b) stable and unstable

(c) unstable and stable

(d) unstable and unstable **[ESE-2004]**

3.9 The open loop transfer function of a unity negative feedback control system is given by

$$G(s) = \frac{K}{(s+2)(s+4)(s^2+6s+25)}$$

Which is the value of K which causes sustained oscillations in the closed loop system?

(a) 590 (b) 790
(c) 990 (d) 1190

[ESE-2006]

3.10 The characteristic equation of a control system is $s^5 + 15s^4 + 85s^3 + 225s^2 + 274s + 120 = 0$. What are the number of roots of the equation which lie to the left of the line $s + 1 = 0$?

(a) 2 (b) 3
(c) 4 (d) 5 [ESE-2006]

3.11 For the system given below, the feedback does not reduce the closed-loop sensitivity due to variation of which one of the following?

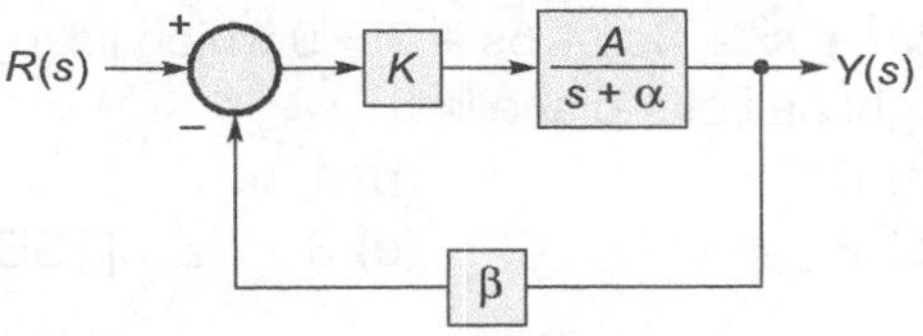

(a) K (b) A
(c) $K\alpha$ (d) β [ESE-2007]

3.12 The characteristic polynomial of a system is

$$q(s) = 2s^5 + s^4 + 4s^3 + 2s^2 + 2s + 1$$

Which one of the following is correct?
The system is

(a) stable (b) marginally stable
(c) unstable (d) oscillatory

[ESE-2007]

3.13 The open loop transfer function of a unity feedback control system is

$$G(s) = \frac{K}{s(s+1)(s+5)}$$

What is the value of K for its stable operation?

(a) $0 < K < 5$ only (b) $0 < K < 6$ only
(c) $1 < K < 5$ only (d) $0 < K < 30$

[ESE-2008]

3.14 Consider the unity feedback system with

$$G(s) = \frac{K}{(s^2+2s+2)(s+2)}$$

The system is marginally stable. What is the radian frequency of oscillation?

(a) $\sqrt{2}$ (b) $\sqrt{3}$
(c) $\sqrt{5}$ (d) $\sqrt{6}$ [ESE-2008]

3.15 The characteristic equation of a control system is given by

$$s^5 + s^4 + 2s^3 + 2s^2 + 4s + 6 = 0$$

What is the number of roots of the equation which lie in the right half of s-plane?

(a) Zero (b) 1
(c) 2 (d) 3 [ESE-2008]

3.16 A closed loop system is shown in the below figure.

What is the ratio of output frequencies $\dfrac{\omega(\text{for } K = 32)}{\omega(\text{for } K = 16)}$?

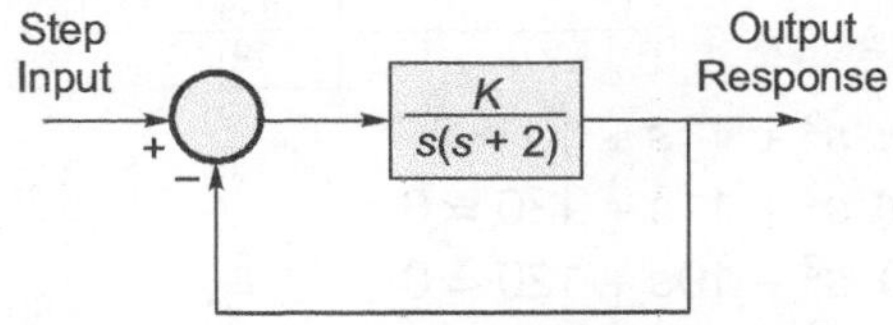

(a) 1.40 (b) 1.42
(c) 1.44 (d) 1.46 [ESE-2008]

3.17 Which of the following may result in instability problem?

(a) Large error (b) High selectivity
(c) High gain (d) Noise [ESE-2009]

3.18 For what positive value of K does the polynomial

$$s^4 + 8s^3 + 24s^2 + 32s + K$$

have roots with zero real parts?

(a) 10 (b) 20
(c) 40 (d) 80 [ESE-2009]

3.19 How many roots with positive real parts do the equation $s^3 + s^2 - s + 1 = 0$ have?

(a) Zero (b) One
(c) Two (d) Three [ESE-2009]

3.20 The characteristic equation of a control system is given as

$$s^4 + 8s^3 + 24s^2 + 32s + K = 0$$

What is the range of values of K for this system to be stable?

(a) $0 < K < 80$ (b) $0 \le K < 100$
(c) $0 \le K < 300$ (d) $0 \le K < 600$

[ESE-2009]

3.21 The characteristic equation of a control system is given as

$$s^4 + 4s^3 + 4s^2 + 3s + K = 0.$$

What is the value of K for which this system is marginally stable?

(a) $\frac{9}{16}$ (b) $\frac{19}{16}$

(c) $\frac{29}{16}$ (d) $\frac{39}{16}$

[ESE-2009]

3.22 The characteristic equation of the below closed-loop system is

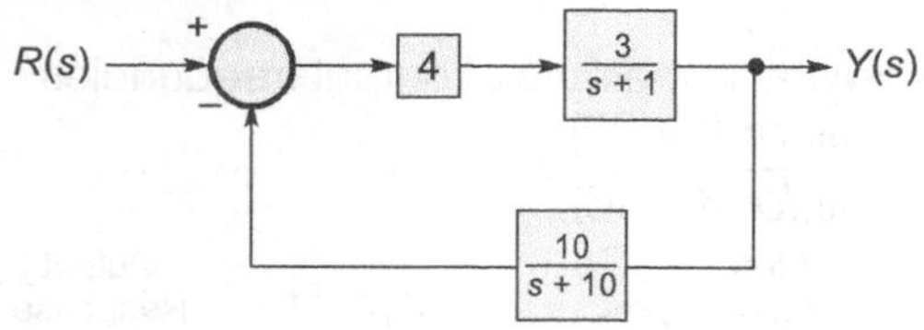

(a) $s^2 + 11s + 10 = 0$

(b) $s^2 + 11s + 130 = 0$

(c) $s^2 + 10s + 120 = 0$

(d) $s^2 + 10s + 12 = 0$ [ESE-2010]

3.23 The characteristic equation of a control system is given as

$$s^4 + 8s^3 + 24s^2 + 32s + K = 0$$

What is the value of K for which the system is unstable?

(a) 10 (b) 20

(c) 60 (d) 100 [ESE-2011]

3.24 The characteristic equation of a particular system is given by $s^3 + 2s^2 + 6s + 12 = 0$. The damping ratio δ will be

(a) $\delta = 0$ (b) $0 < \delta < 1$

(c) $\delta = 1$ (d) $\delta > 1$ [ESE-2012]

3.25 The range of K for stability of a feedback system whose open-loop transfer function is

$$G(s) = \frac{K}{s(s+1)(s+2)} \text{ is}$$

(a) $0 < K < 3$ (b) $0 < K < 6$

(c) $K > 6$ (d) $0 > K > 3$

[ESE-2012]

3.26 The critical value of gain for a system is 40. The system is operating at a gain of 20. The gain margin of the system is

(a) 2 dB (b) 3 dB

(c) 6 dB (d) 4 dB [ESE-2013]

3.27 A feedback system with characteristic equation $s^4 + 20Ks^3 + 5s^2 + 10s + 15 = 0$ is

(a) stable for all values of K

(b) stable only for $K \geq 0$

(c) stable for $\infty > K > 70$

(d) Unstable for all values of K [ESE-2013]

3.28 The system having the characteristic equation

$$s^3 + 4s^2 + s - 6 + K = 0$$

will be stable for

(a) $K > 6$ (b) $0 < K < 6$

(c) $6 < K < 10$ (d) $0 < K < 10$

[ESE-2013]

3.29 The loop transfer function of a system is $\frac{K}{s(s+1)(s+5)}$. The loop gain K is adjusted for inducing sustained oscillations. What is the value of K for this objective?

(a) 15 (b) 25

(c) 30 (d) 45 [ESE-2014]

3.30 The number of roots of the equation $2s^4 + s^3 + 3s^2 + 5s + 7 = 0$ which lie in the right half of s-plane is

(a) 0 (b) 1

(c) 2 (d) 3 [ESE-2014]

3.31 Which one of the following statements is correct?

(a) Phase margin is always positive for stable feedback system.

(b) Phase margin is always negative for stable feedback system.

(c) Phase margin can be negative or positive for stable feedback system.

(d) None of the above [ESE-2014]

3.32 The effects of feedback on stability and sensitivity are

(a) Negative feedback improves stability and system response is less sensitive to external inputs and parameter variations

(b) Feedback does not affect stability but system response is insensitive to disturbances and parameter variations

(c) Feedback does not affect stability and system response is sensitive to disturbances and parameter variations

(d) Negative feedback affects stability and system response is more sensitive to disturbances and parameter variations

[ESE-2015]

3.33 Given that the transfer function $G(s) = \frac{K}{s^2(1+sT)}$, the type and order of this system are respectively
(a) 5 and 2 (b) 2 and 2
(c) 2 and 3 (d) 3 and 3 **[ESE-2015]**

3.34 The oscillation frequency of the system with the characteristic equation
$s^6 + 2s^5 + 3s^4 + 4s^3 + 3s^2 + 2s + 1 = 0$ is
(a) +1 radian/sec (b) –1 radian/sec
(c) $j1$ radian/sec (d) $-j1$ radian/sec
[ESE-2015]

3.35 The feedback system with characteristic equation
$s^4 + 20Ks^3 + 5s^2 + 10s + 15 = 0$ is
(a) Stable for all values of K
(b) Stable for positive values of K
(c) Stable for $7.0 < K < \infty$
(d) Unstable for any value of K **[ESE-2015]**

3.36 The characteristic equation of a certain feedback control system is given by
$s^4 + 4s^3 + 13s^2 + 36s + K = 0$. The range of values of K for which the feedback system is stable, is given by
(a) $0 < K < 4$ (b) $4 < K < 36$
(c) $0 < K < 36$ (d) $13 < K < 36$
[ESE-2016]

3.37 A unity feedback system has open-loop transfer function with two of its poles located at –0.1, 1; and two zeros located at –2 and –1 with a variable gain K. For what value(s) of K would the closed-loop system have one pole in the right half of the s-plane.
(a) $K > 0.3$ (b) $K < 0.05$
(c) $0.05 < K < 0.3$ (d) $K > 0$ **[ESE-2017]**

3.38 The open-loop transfer function of a unity feedback control system is

$$G(s)H(s) = \frac{10}{s(s+2)(s+K)}$$

Here K is a variable parameter. The system will be stable for all vales of
(a) $K > -2$ (b) $K > 0$
(c) $K > 1$ (d) $K > 1.45$
[ESE-2017]

3.39 Consider the stability of the system shown in the figure when analyzed with a positive real value of gain K in
1. open-loop configuration
2. closed-loop configuration

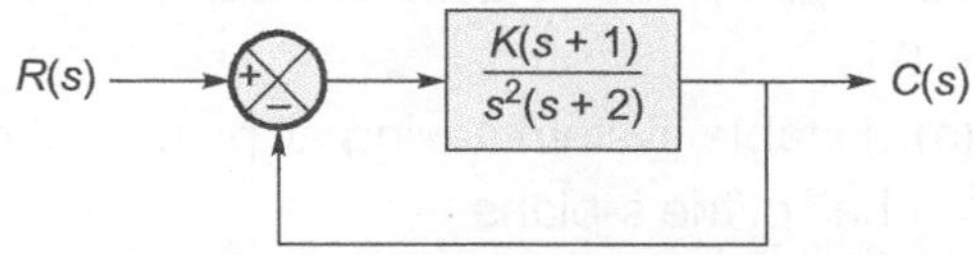

Which of the following statements is correct?
(a) Both 1 and 2 are stable
(b) 1 is stable and 2 is unstable
(c) 1 is unstable and 2 is stable
(d) Both 1 and 2 are unstable **[ESE-2018]**

3.40 If the characteristic equation of a feedback control system is given by,
$s^4 + 20s^3 + 15s^2 + 2s + K = 0$
then the range of values of K for the system to be stable will be
(a) $1 < K < 2.49$ (b) $0 < K < 1.49$
(c) $1 < K < 4.49$ (d) $0 < K < 3.49$
[ESE-2019]

3.41 If all the roots of the characteristic equation have negative real parts, the system is
(a) Stable (b) Unstable
(c) Marginally stable (d) Conditionally stable
[ESE-2020]

3.42 If any root of the characteristic equation has a positive real part or if there is a repeated root on the $j\omega$-axis, then the system is
(a) Limitedly stable (b) Conditionally stable
(c) Stable (d) Unstable
[ESE-2020]

3.43 A feedback system has an open-loop transfer function of

$$G(s)\,H(s) = \frac{K(1-s)}{s(s^2+5s+9)}$$

By using the Routh criterion, the maximum value of K for the closed-loop system to be stable is
(a) 2.5 (b) 5
(c) 7.5 (d) 9
[ESE-2021]

3.44 A feedback system has an open loop transfer function of $G(s)H(s) = \frac{K(1-s)}{s(s^2+5s+9)}$. The maximum value of K for the closed loop system to be stable is
(a) 3.5 (b) 4.8
(c) 6.5 (d) 7.5 **[ESE-2024]**

3.45 The system represented by
$5s^6 + 8s^5 + 12s^4 + 20s^3 + 100s^2 + 150s + 200 = 0$ is
(a) a stable system having all poles in the right half of the s-plane.
(b) an unstable system having one pole in the right half of the s-plane.
(c) a stable system having one pole in the right half of the s-plane.
(d) an unstable system having two poles in the right half of the s-plane.
[ESE-2024]

3.46 For the unity feedback system given by $(s) = \dfrac{K}{s(s+1)(s+2)}$, the range of K for stability is
(a) 0 < K < 2 (b) 0 < K < 3
(c) 0 < K < 6 (d) 1 < K < 6
[ESE-2024]

■■■

Answers Routh-Hurwitz Stability Criterion

3.1 (c)	3.2 (d)	3.3 (c)	3.4 (b)	3.5 (c)	3.6 (c)	3.7 (d)	3.8 (c)	3.9 (*)
3.10 (c)	3.11 (d)	3.12 (c)	3.13 (d)	3.14 (d)	3.15 (c)	3.16 (c)	3.17 (c)	3.18 (d)
3.19 (c)	3.20 (a)	3.21 (d)	3.22 (b)	3.23 (d)	3.24 (a)	3.25 (b)	3.26 (c)	3.27 (d)
3.28 (c)	3.29 (c)	3.30 (c)	3.31 (a)	3.32 (a)	3.33 (c)	3.34 (a)	3.35 (d)	3.36 (c)
3.37 (b)	3.38 (d)	3.39 (c)	3.40 (b)	3.41 (a)	3.42 (d)	3.43 (c)	3.44 (d)	3.45 (d)
3.46 (c)								

Explanations Routh-Hurwitz Stability Criterion

3.1 (c)

$$\frac{G(s)}{1+G(s)} = \frac{K}{s(s+a)(s+b)+K}$$

Characteristic equation is
$s^3 + (a+b)s^2 + abs + K$
Routh array is

s^3	1	$a+b$
s^2	ab	K
s	$\dfrac{ab(a+b)-K}{ab}$	0
s^0	K	

For the system to be stable,
$K > 0$
$ab(a+b) - K > 0 \Rightarrow K < ab(a+b)$
So $0 < K < ab(a+b)$

3.2 (d)

A necessary (but not sufficient) condition for stability of a linear system is that all the coefficients of its characteristic equation be real and have the same sign. Furthermore, none of the coefficients should be zero.

3.3 (c)

Using RH criterion.

3.4 (b)

Routh-Hurwitz criterion gives absolute stability and the number of roots lying on the right half of the s-plane but it does not tell about the gain margin and phase margin.

3.5 (c)

Routh array is

s^4	1	2	3
s^3	1	2	
s^2	$\in$	3	
s^1	$\dfrac{2\in-3}{\in}$	0	
s^0	3		

since the sign changes twice, so there are two roots in RHS of *s*-plane.

3.6 (c)

Characteristic equation is

$s^6 + 2s^5 + 8s^4 + 12s^3 + 20s^2 + 16s + 16 = 0$

The Routh array is

s^6	1	8	20	16
s^5	2	12	16	
s^4	2	12	16	
s^3	0	0	0	

Auxiliary polynomial $A(s) = s^4 + 6s^2 + 8$

Solving for the roots of auxiliary polynomial

$s^4 + 6s^2 + 8 = 0$

$\Rightarrow (s^2 + 2)(s^2 + 4) = 0$

$\Rightarrow s = \pm j\sqrt{2}$ and $\pm j2$

These two pairs of roots are also the roots of the original characteristic equation. Thus the characteristic equation has 4 roots on the imaginary axis of s-plane.

3.7 (d)

Even though there is negative feedback and open loop control system stable we can have closed loop system as unstable if high gain.

3.8 (c)

$$G(s) = \frac{1}{s^3 + 1.5s^2 + s - 1}$$

The coefficient of s^0 is negative. So the open-loop systems is unstable.

$$\frac{G(s)}{1 + G(s)H(s)} = \frac{\dfrac{1}{s^3 + 1.5s^2 + s - 1}}{1 + \dfrac{20s + 20}{s^3 + 1.5s^2 + s - 1}}$$

$$= \frac{1}{s^3 + 1.5s^2 + 21s + 19}$$

Since all the coefficient of the denominator are positive, hence the closed loop system is stable.

3.9 (*)

$$\frac{G(s)}{1 + G(s)} = \frac{K}{(s + 2)(s + 4)(s^2 + 6s + 25) + K}$$

Characteristic equation is

$s^4 + 12s^3 + 69s^2 + 198s + 200 + K = 0$

Routh array is

s^4	1	69	$K + 200$
s^3	2	33	
s^2	52.5	$K + 200$	
s^1	$\dfrac{1332.5 - 2K}{52.5}$	0	

For oscillations,

$1332.5 - 2K = 0$

$\Rightarrow \quad K = 666.25$

3.10 (c)

$s^5 + 15s^4 + 85s^3 + 225s^2 + 274s + 120 = 0$

Put $s = z - 1$

$(z - 1)^5 + 15(z - 1)^4 + 85(z - 1)^3 + 225(z - 1)^2 + 274(z - 1) + 120 = 0$

$\Rightarrow z^5 + 10z^4 + 35z^3 + 50z^2 + 24z = 0$

Routh array is

z^5	1	35	24
z^4	10	50	
z^4	1	5	
z^3	30	24	
z^2	4.2	0	
z^1	24	0	
z^0	0		

There are 4 roots which lie to the left of the line $s + 1 = 0$ and one root lies on $s + 1 = 0$.

3.12 (c)

$$\begin{aligned} q(s) &= 2s^5 + s^4 + 4s^3 + 2s^2 + 2s + 1 = 0 \\ &= (2s + 1)s^4 + (2s + 1)2s^2 + (2s + 1) = 0 \\ &= (2s + 1)(s^4 + 2s^2 + 1) = 0 \\ &= (2s + 1)(s^2 + 1)^2 = 0 \end{aligned}$$

Therefore, the roots of the characteristic equation is $s = -1/2$, $s = \pm j$, $s = \pm j$. Since the poles of the system are repeated on $j\omega$-axis, therefore, the system is unstable.

3.13 (d)

$$\frac{G(s)}{1 + G(s)} = \frac{\dfrac{K}{s(s + 1)(s + 5)}}{1 + \dfrac{K}{s(s + 1)(s + 5)}}$$

$$= \frac{K}{s(s + 1)(s + 5) + K}$$

Characteristic equation is

$s^3 + 6s^2 + 5s + K = 0$

Routh array:

$$\begin{array}{c|cc} s^3 & 1 & 5 \\ s^2 & 6 & K \\ s^1 & \dfrac{30-K}{6} & \\ s^0 & K & \end{array}$$

For stable operation, each element of the first column of the Routh array should have the same sign.

Therefore, $0 < K < 30$

3.14 (d)

$$\frac{G(s)}{1+G(s)} = \frac{K}{(s^2+2s+2)(s+2)+K}$$

Characteristic equation is

$s^3 + 4s^2 + 6s + 4 + K = 0$

Routh array:

$$\begin{array}{c|cc} s^3 & 1 & 6 \\ s^2 & 4 & 4+K \\ s^1 & \dfrac{24-4-K}{4} & 0 \\ s^0 & 4+K & \end{array}$$

For the system to be marginally stable, all elements of the row of s should be zero.

$\therefore \quad 24 - 4 - k = 0$

$\Rightarrow \quad K = 20$

$4s^2 + 4 + K = 0$

$4s^2 + 4 + 20 = 0$

$s^2 + 6 = 0$

$(j\omega)^2 + 6 = 0$

$\Rightarrow \quad \omega^2 = 6$

$\Rightarrow \quad \omega = \sqrt{6}$ rad/s

3.15 (c)

Characteristic equation is

$s^5 + s^4 + 2s^3 + 2s^2 + 4s + 6 = 0$

Putting $s = 1/z$,

$6z^5 + 4z^4 + 2z^3 + 2z^2 + z + 1 = 0$

Routh array:

$$\begin{array}{c|ccc} z^5 & 6 & 2 & 1 \\ z^4 & 4 & 2 & 1 \\ z^3 & \dfrac{8-12}{4} = -1 & \dfrac{4-6}{4} = \dfrac{-1}{2} & \\ z^2 & 0 (\text{Let us take } \epsilon) & 1 & \\ z^0 & 1 & & \end{array}$$

Since only two sign changes in the first column of Routh array, therefore, two roots of the equation lie in the right half of s-plane.

3.16 (c)

$$G(s) = \frac{K}{s(s+2)}$$

$$\text{T.F.} = \frac{Y(s)}{X(s)} = \frac{G}{1+GH} = \frac{K}{s^2+2s+K}$$

$$X(s) = \frac{1}{s}$$

$$Y(s) = \frac{K}{s(s^2+2s+K)} = \frac{A}{s} + \frac{Bs+C}{s^2+2s+K}$$

$$= \frac{(A+B)s^2 + (2A+C)s + AK}{s(s^2+2s+K)}$$

Comparing numerator coefficient

$A = 1$; $B = -1$; $C = -2$

$$Y(s) = \frac{1}{s} - \frac{s+2}{s^2+2s+K} = \frac{1}{s} - \frac{(s+1)+1}{(s+1)^2+K-1}$$

$$= \frac{1}{s} - \frac{s+1}{(s+1)^2+(\sqrt{K-1})^2} - \frac{1}{\sqrt{K-1}} \frac{(\sqrt{K-1})}{(s+1)^2+(\sqrt{K-1})^2}$$

$$y(t) = u(t) - e^{-t}\cos(\sqrt{K-1})t - \frac{1}{\sqrt{K-1}} e^{-t}\sin(\sqrt{K-1})t$$

Output frequency $= \sqrt{K-1}$

$$\therefore \quad \frac{\omega(\text{for } K = 32)}{\omega(\text{for } K = 16)} = \sqrt{\frac{31}{15}} = 1.437 \approx 1.44$$

3.17 (c)

High gain feedback can lead to instability problem.

3.18 (d)

Routh array:

$$\begin{array}{c|ccc} s^4 & 1 & 24 & K \\ s^3 & 8 & 32 & \\ s^3 & 1 & 4 & \\ s^2 & 20 & K & \\ s^1 & \dfrac{80-K}{20} & 0 & \\ s^0 & K & & \end{array}$$

For roots with zero real parts,

$$\frac{80-K}{20} = 0$$

$\Rightarrow \quad K = 80$

3.19 (c)

Routh array:

$$\begin{array}{c|cc} s^3 & 1 & -1 \\ s^2 & 1 & 1 \\ s^1 & -2 & 0 \\ s^0 & 1 & \end{array}$$

Since the sign changes two times in the first column, therefore, two roots have positive parts.

3.20 (a)

Routh array:

$$\begin{array}{c|ccc} s^4 & 1 & 24 & K \\ s^3 & 8 & 32 & \\ s^3 & 1 & 4 & \\ s^2 & 20 & K & \\ s^1 & \dfrac{80-K}{20} & 0 & \\ s^0 & K & & \end{array}$$

For the system to be stable,

$$\frac{80-K}{20} > 0 \quad \text{and } K \geq 0$$

$$\Rightarrow \quad 0 \leq K < 80$$

3.21 (d)

Routh array:

$$\begin{array}{c|ccc} s^4 & 1 & 4 & K \\ s^3 & 4 & 3 & \\ s^2 & \dfrac{16-3}{4} = \dfrac{13}{4} & K & \\ s^1 & \dfrac{\dfrac{39}{4} - 4K}{\dfrac{13}{4}} & 0 & \\ s^0 & K & & \end{array}$$

For the system to be marginally stable,

$$\frac{39}{4} - 4K = 0$$

$$\Rightarrow \quad K = \frac{39}{16}$$

3.22 (b)

Rearranging the given block diagram

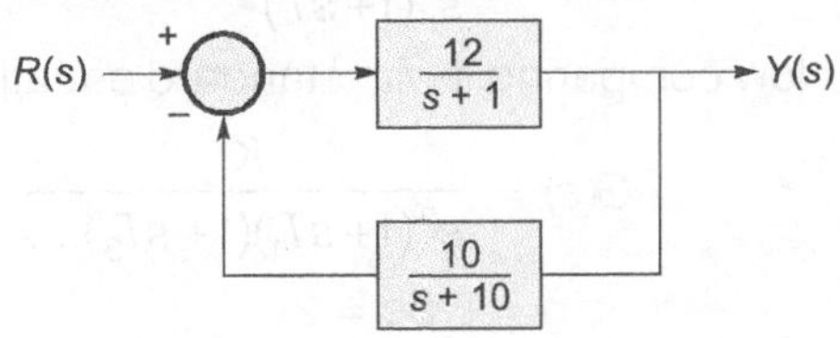

Hence, $G(s) = \dfrac{12}{s+1}$

and $H(s) = \dfrac{10}{s+10}$

The characteristic equation is given by

$$1 + G(s)\,H(s) = 0$$

$$1 + \left(\frac{12}{s+1}\right)\left(\frac{10}{s+10}\right) = 0$$

$$\Rightarrow \quad s^2 + 11s + 130 = 0$$

3.23 (d)

$s^4 + 8s^3 + 24s^2 + 32s + K = 0$

Routh array

$$\begin{array}{c|ccc} s^4 & 1 & 24 & K \\ s^3 & 8 & 32 & \\ s^2 & 20 & K & \\ s^1 & \dfrac{20 \times 32 - 8K}{20} & & \\ s^0 & K & & \end{array}$$

For marginal stability

$$\frac{20 \times 32 - 8K_{mar}}{20} = 0$$

$$\Rightarrow \quad 8K_{mar} = 20 \times 32$$

$$\Rightarrow \quad K_{mar} = 80$$

Hence system will be unstable for $K > K_{mar}$ so option (d) is the correct answer.

3.24 (a)

The characteristic equation is given as

$s^3 + 2s^2 + 6s + 12 = 0$

Using the Routh array, we get

$$\begin{array}{c|cc} s^3 & 1 & 6 \\ s^2 & 2 & 12 \\ s^1 & 0 & 0 \\ s^0 & & \end{array}$$

When one full row becomes zero, then the system will be marginally stable (oscillatory) or unstable. To calculate this stability, we need to check the roots of auxiliary equation i.e. equation of s^2 terms.

so $2s^2 + 12 = 0$

$$\Rightarrow \quad s = \pm j\sqrt{6} \text{ rad/sec}$$

$$\omega = \pm\sqrt{6} \text{ rad/sec}$$

As we can see that system is undamped oscillatory, so $\delta = 0$.

3.25 (b)

$$G(s) = \frac{K}{s(s+1)(s+2)}$$

The range of K is calculated through Routh array using the characteristic equation $1 + G(s)H(s) = 0$.

$$1+\frac{K}{s(s+1)(s+2)} = 0$$

$$s^3 + 3s^2 + 2s + K = 0$$

s^3	1	2
s^2	3	K
s^1	$\frac{3\times 2-K}{3}$	
s^0	$\frac{\left(\frac{6-K}{3}\right)\times K}{\left(\frac{6-K}{3}\right)}$	

For the stable system first column values of the Routh array should always be greater than zero, thus,

$$6-K>0 \Rightarrow K<6$$

also $K > 0$

Now the range is $0 < K < 6$

3.26 (c)

$$\text{G.M.} = \frac{K_{marginal}}{K_{desire}}$$

$$K_{marginal} = 40$$

$$K_{desired} = 20$$

$$\text{G.M.} = 2$$

$$\text{G.M. (dB)} = 20\,[\log 2] = 6 \text{ dB}$$

3.28 (c)

The characteristic equation is given as

$$s^3 + 4s^2 + s - 6 + K = 0$$

The stability can be found out using Routh Array in which 1st column should always be greater than zero.

Now, applying Routh criterion as:

s^3	1	1
s^2	4	$(K-6)$
s^1	$\frac{4-K+6}{4}$	0
s^0	$K-6$	

$10 - K > 0$, $K - 6 > 0$

$\boxed{K<10}$ $\boxed{K>10}$

$\boxed{6<K<10}$

3.29 (c)

$$G(s) = \frac{K}{s(s+1)(s+5)} = \frac{K}{s^3+6s^2+5s}$$

∴ Characteristic equation

$$1 + G(s) = s^3 + 6s^2 + 5s + K = 0$$

Using Routh array,

s^3	1	5
s^2	6	K
s^1	$\frac{30-K}{6}$	0
s^0	K	

For sustained oscillation, $\Rightarrow \frac{30-K}{6} = 0$

$\Rightarrow K = 30$

3.30 (c)

Using Routh Array

s^4	2	3	7
s^3	1	5	
s^2	$(3-10)$	7	
s^1	$\left(\frac{35+7}{7}\right)$	0	
s^0	7		

The total number of sign changes in the first column of Routh array = 2

∴ RHS poles = 2

3.31 (a)

Both phase margin and gain margin are positive for stable system.

3.32 (a)

The negative feedback improves stability and the sensitivity to the system is decreased. The sensitivity to the gain of the system reduces by a factor of $(1 + G(j\omega)H(j\omega))$

3.33 (c)

$$G(s) = \frac{K}{s^2(1+sT)}$$

Now comparing it with standard equation

$$G(s) = \frac{K}{s^n(1+sT_1)(1+sT_2)....}$$

$$n = \text{type} = 2$$

Order is the highest power in the denominator = 3.

3.34 (a)

s^6	1	3	3	1
s^5	2	4	2	
s^4	1	2	1	
s^3	0	0		
s^2	–			
s^1	–			
s^0	–			

Thus the auxiliary equation

$s^4 + 2s^2 + 1 = 0$

now putting $s^2 = x$ we have

$x^2 + 2x + 1 = 0$

$\therefore \quad x = -1$ rad/sec.

$\therefore \quad s^2 = -1$

$(j\omega)^2 = -1$

$-(\omega)^2 = -1$

$\omega = \pm 1$ rad/sec

But only $\omega = 1$ will satisfy the given characteristic equation.

3.36 (c)

Characteristics equation:

$$q(s) = s^4 + 4s^3 + 13s^2 + 36s + K = 0$$

Using Routh's tabular form

s^4	1	13 K
s^3	4	36
s^2	4	K
s^1	$\frac{36\times4-4K}{4}$	
s^0	K	

For stability $K > 0$ and $K < 36$

$\therefore 0 < K < 36$

3.37 (b)

$$GH(s) = \frac{K(s+1)(s+2)}{(s+0.1)(s-1)}$$

$$q(s) = s^2(1 + K) + s(3K - 0.9) + (2K - 0.1) = 0$$

s^2	$1+K$	$2K-0.1$
s^1	$3K-0.9$	
s^0	$2K-0.1$	

For one closed loop pole to be in right side there has to be one sign change.

$\therefore K > -1;\ K < 0.3;\ K < 0.05$

$\therefore -1 < K < 0.05$

3.38 (d)

$$q(s) = s^3 + (2 + K)s^2 + 2Ks + 10 = 0$$

Necessary condition : $K > 0$

s^3	1	$2K$
s^2	$2+K$	10
s^1	$\frac{2K^2+4K-10}{2+K}$	
s^0	10	

Sufficient condition : $K > 1.45$

3.39 (c)

$$\text{OLTF} = \frac{K(s+1)}{s^2(s+2)}$$

$$\text{CLTF} = \frac{K(s+1)}{s^3+2s^2+Ks+K}$$

1. OLTF has two poles at origin. Hence unstable
2. $q(s) = s^3 + 2s^2 + Ks + K = 0$

   $2K > K$

Hence stable.

3.40 (b)

$$q(s) = s^4 + 20s^3 + 15s^2 + 2s + K = 0$$

Necessary condition: $K > 0$

Sufficient condition:

s^4	1	15 K
s^3	20	2
s^2	$\frac{300-2}{20}$	K
s^1	$\frac{2\left(\frac{298}{20}\right)-20K}{\left(\frac{300-2}{20}\right)}$	
s^0	K	

$$\therefore \quad 2\left(\frac{298}{20}\right) - 20K > 0$$

$$\therefore \quad 0 < K < 1.49$$

3.41 (a)

For a stable system all closed loop poles (CLP) must be in left side of s-plane.

3.42 (d)

If any closed loop poles (CLP) is in right side of s-plane (or) if multiple CLP lies on $j\omega$-axis then the system is unstable.

3.43 (c)

The characteristic equation of the system is

$1 + G(s) \cdot H(s) = 0$

$$\Rightarrow 1 + \frac{K(1-s)}{s(s^2+5s+9)} = 0$$

$\Rightarrow s^3 + 5s^2 + s(9 - K) + K = 0$

Applying Routh-Hurwitz criterion,

s^3	1	$9-K$
s^2	5	K
s^1	$\frac{5(9-K)-K}{5}$	
s^0	K	

For the system to be stable,

$5(9 - K) - K > 0$ and $K > 0$

$\Rightarrow \quad 45 > 6K$

$\Rightarrow \quad K < 7.5$

Hence, $0 < K < 7.5$ for the system to be stable. Therefore, the maximum value of K for which the closed-loop system is stable is equal to 7.5.

3.44 (d)

The characteristic equation of the feedback system is given by

$1 + G(s)H(s) = 0$

$s^3 + 5s^2 + 9s + K - Ks = 0$

$s^3 + 5s^2 + s(9 - K) + K = 0$

Applying Routh-Hurwitz criterion,

s^3	1	$9-K$
s^2	5	K
s^1	$\frac{45-5K-K}{5}$	
s^0	K	

For the system to be stable, there should be no sign changes in the first column of the Routh array. Hence,

$45 - 6K > 0 \Rightarrow K < 7.5$

$K > 0$

Hence, $\quad 0 < K < 7.5$

For $K = 7.5$, the system has poles on the imaginary axis. Hence, the system is marginally stable for $K = 7.5$. Hence, the maximum value of K for the closed loop system to be stable is 7.5.

3.45 (d)

The characteristic equation of the system is given as

$5s^6 + 8s^5 + 12s^4 + 20s^3 + 100s^2 + 150s + 200 = 0$

Now, by using R-H criterion,

s^6	5	12	100	200
s^5	8	20	150	
s^4	–0.5	6.25	200	
s^3	120	3350		
s^2	20.2	200		
s^1	2162.35			
s^0	200			

There are two sign changes in the first column of the Routh array. Hence, the system is unstable and having two poles in the right half of s-plane. Hence, option (d) is correct.

3.46 (c)

We have,

$$\text{OLTF} = G(s)H(s) = \frac{K}{s(s+1)(s+2)}$$

($\because$ $H(s) = 1$ for unity feedback)

Characteristic equation,

$1 + G(s)H(s) = 0$

$(s^2 + s)(s + 2) + K = 0$

$s^3 + 2s^2 + s^2 + 2s + K = 0$

$s^3 + 3s^2 + 2s + K = 0$

Now, using R-H criterion,

s^3	1	2
s^2	3	K
s^1	$\frac{6-K}{3}$	
s^0	K	

For the system to be stable, there should be no sign changes in the first column of the Routh Array. Hence,

$6 - K > 0$

$6 > K \; ; \; K > 0$

Hence, for stable system $0 < K < 6$.

■■■

4 Root Locus

4.1 The intersection of asymptotes of root-loci of a system with open-loop transfer function

$G(s)\,H(s) = \dfrac{K}{s(s+1)(s+3)}$ is

(a) 1.44 (b) 1.33
(c) –1.44 (d) –1.33

[ESE-2000]

4.2 The root locus plot of the system having the loop transfer function

$$G(s)H(s) = \frac{K}{s(s+4)\,(s^2+4s+5)} \text{ has}$$

(a) no breakaway point
(b) three real breakaway points
(c) only one breakaway point
(d) one real and two complex breakaway points

[ESE-2001]

4.3 An open loop transfer function is given by

$$G(s)H(s) = \frac{K(s+1)}{s(s+2)\,(s^2+2s+2)}.$$

It has
(a) one zero at infinity
(b) two zeros at infinity
(c) three zeros at infinity
(d) four zeros at infinity

[ESE-2001]

4.4 Which of the following is the open loop transfer function of the root loci shown in figure?

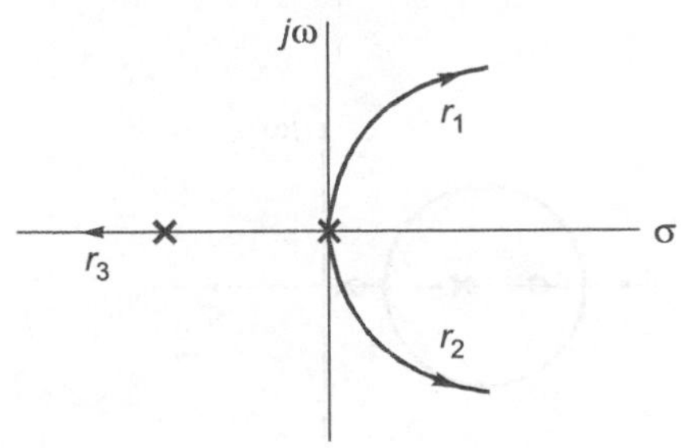

(a) $\dfrac{K}{s(s+T_1)^2}$ (b) $\dfrac{K}{(s+T_1)(s+T_2)^2}$

(c) $\dfrac{K}{(s+T_1)^3}$ (d) $\dfrac{K}{s^2(sT_1+1)}$

[ESE-2002]

4.5 The instrument used for plotting the root locus is called
(a) Slide rule (b) Spirule
(c) Synchro (d) Selsyn

[ESE-2002]

4.6 A control system has

$$G(s)H(s) = \frac{K(s+1)}{s(s+3)(s+4)}$$

Root locus of the system can lie on the real axis
(a) between $s = -1$ and $s = -3$
(b) between $s = 0$ and $s = -4$
(c) between $s = -3$ and $s = -4$
(d) towards left of $s = -4$

[ESE-2002]

4.7 The below figure shows the root locus of a unity feedback system. The open loop transfer function of the system is

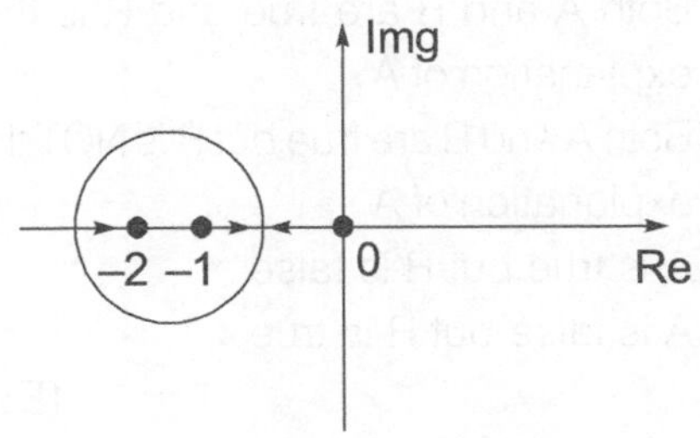

(a) $\dfrac{K}{s(s+1)(s+2)}$ (b) $\dfrac{Ks}{(s+1)(s+2)}$

(c) $\dfrac{K(s+1)}{s(s+2)}$ (d) $\dfrac{K(s+2)}{s(s+1)}$

[ESE-2003]

4.8 Which one of the following application softwares is used to obtain an accurate root locus plot?
(a) LISP (b) MATLAB
(c) dBase (d) Oracle
[ESE-2003]

4.9 Which one of the following techniques is utilized to determine the actual point at which the root locus crosses the imaginary axis?
(a) Nyquist technique
(b) Routh-Hurwitz criterion
(c) Nichol's criterion
(d) Bode technique
[ESE-2003]

4.10 The root loci of a feedback control system for large values of s are asymptotic to the straight lines with angles θ to the real axis given by which one of the following?
(a) $\frac{(p-z)\pi}{2k+1}$ (b) $\frac{(2k+1)\pi}{p-z}$
(c) $2k(p-z)$ (d) $\frac{2k}{p}z$
where p = Number of finite poles of $G(s)\,H(s)$, z = Number of finite zeros of $G(s)\,H(s)$ and $k = 0, 1, 2, ...$
[ESE-2004]

4.11 **Assertion (A):** The number of branches of root locus terminating on infinity is equal to the number of open loop poles minus the number of zeros.
Reason (R): Segment of the real axis having an odd number of real axis open loop poles plus zeros to their right are parts of the root locus.
(a) Both A and R are true and R is the correct explanation of A
(b) Both A and R are true but R is NOT the correct explanation of A
(c) A is true but R is false
(d) A is false but R is true
[ESE-2004]

4.12 Which one of the following is not a property of root loci?
(a) The root locus is symmetrical about jω axis.
(b) They start from the open loop poles and terminate at the open loop zeros.
(c) The breakaway points are determined from $dK/ds = 0$.
(d) Segments of the real axis are part of the root locus, if and only if, the total number of real poles and zeros to their right is odd.
[ESE-2005]

4.13 In root locus, what is the number of separate loci?
(a) The number of zeros of the open loop transfer function
(b) The number of poles of $G(s)\,H(s)$
(c) The number of roots of the characteristic equation with positive real part
(d) The number of zeros of the characteristic equation with the negative real parts
[ESE-2006]

4.14 For a given unity feedback system with $G(s) = \frac{K(s+3)}{s(s+1)(s+2)(s+5)}$, what is the real axis intercept for root locus asymptotes?
(a) 2/3 (b) 1/4
(c) –5/3 (d) –3/2
[ESE-2006]

4.15 Which one of the following open-loop transfer functions has root locus parallel to imaginary axis?
(a) $K/(s+1)$ (b) $K(s+1)/(s+2)^2$
(c) $K/(s+2)^2$ (d) $K(s+2)/(s+1)^2$
[ESE-2007]

4.16 What is the number of root-locus segments which do not terminate on zeros?
(a) The number of poles
(b) The number of zeros
(c) The difference between the number of poles and the number of zeros
(d) The sum of the number of poles and the number of zeros
[ESE-2007]

4.17 Consider the root loci plots:

1. 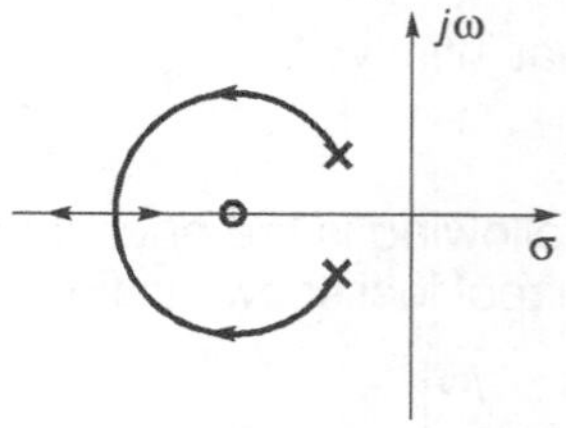

2. 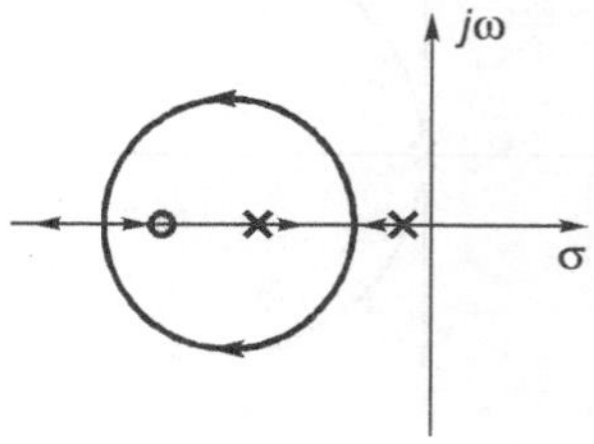

3. 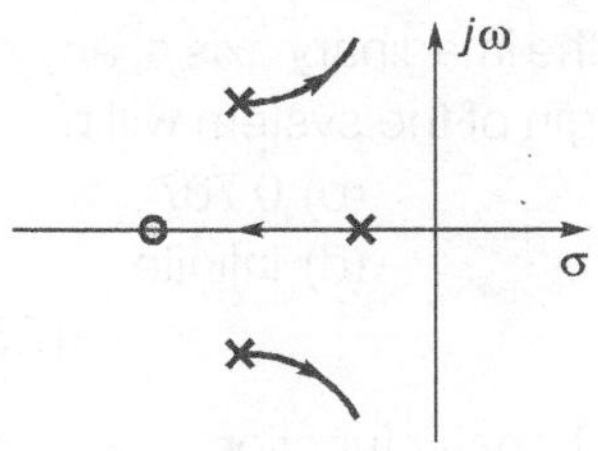

4. 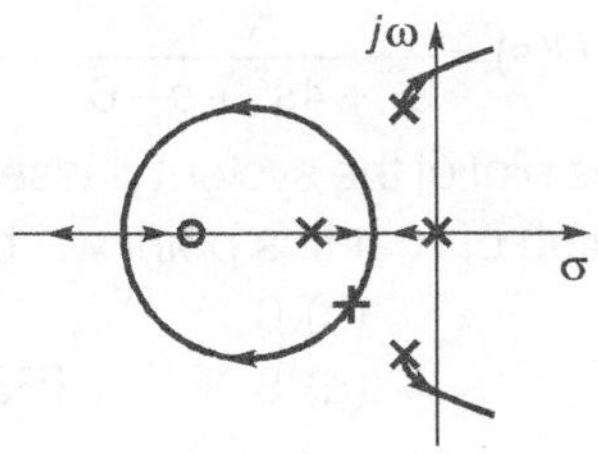

Which one of the above plots is not correct?
(a) 1 (b) 2
(c) 3 (d) 4
[ESE-2008]

4.18 Which one of the following statements is correct for root loci?
(a) The root loci are asymmetrical with respect to the real axis
(b) The root loci are symmetrical with respect to the real axis
(c) The root locus lies in a section of the real axis to the left of which an even number of poles and zeros of open loop system are present
(d) The root locus lies in a section of the real axis to the right of which an even number of poles and zeros of open loop system are present
[ESE-2008]

4.19 Which one of the following is correct?
The root locus is the path of the roots of the characteristic equation traced out in the s-plane
(a) as the input of the system is changed
(b) as the output of the system is changed
(c) as a system parameter is changed
(d) as the sensitivity is changed
[ESE-2008]

4.20 Consider the equation
$s^2 + 2s + 2 + K(s + 2) = 0$.
Where do the roots of this equation break on the root loci plot ?
(a) – 3·414 (b) –2·414
(c) –1·414 (d) –0·414
[ESE-2009]

4.21 How many number of branches the root loci of the equation $s(s + 2)(s + 3) + K(s + 1) = 0$ have?
(a) Zero (b) One
(c) Two (d) Three [ESE-2009]

4.22 If the gain of the system is reduced to zero value, the roots of the system in the s-plane
(a) Coincide with zeros
(b) Move away from the zeros
(c) Move away from the poles
(d) Coincide with the poles [ESE-2010]

4.23 The addition of open loop zero pulls the root-loci towards
(a) The left and therefore system becomes more stable
(b) The right and therefore system becomes unstable
(c) Imaginary axis and therefore system becomes marginally stable
(d) The left and therefore system becomes unstable
[ESE-2010]

4.24 Where are the $K = \pm\infty$ points on the root loci of the characteristic equation of the closed loop control system located at?
(a) Poles of $G(s)\,H(s)$
(b) Zeros of $G(s)\,H(s)$
(c) Both Zeros and Poles of $G(s)\,H(s)$
(d) Neither at Zeros nor at Poles of $G(s)\,H(s)$
[ESE-2011]

4.25 The characteristic equation of a control system is given as

$$1+\frac{K(s+1)}{s(s+4)(s^2+2s+2)}=0$$

For large values of s, the root loci for $K \geq 0$ are asymptotic to asymptotes, where do the asymptotes intersect on the real axis?
(a) $\frac{5}{3}$ (b) $-\frac{2}{3}$
(c) $-\frac{5}{3}$ (d) $\frac{4}{3}$ [ESE-2011]

4.26 Where are the $K = 0$ points on the root loci of the characteristic equation of the closed loop control system located at?
(a) Zeros of $G(s)\,H(s)$
(b) Poles of $G(s)\,H(s)$
(c) Both Zeros and Poles of $G(s)\,H(s)$
(d) Neither at Zeros nor at Poles of $G(s)\,H(s)$
[ESE-2011]

4.27 Given the root locus of a system

$$G(s) = \frac{4K}{(s+1)(s+3)}$$

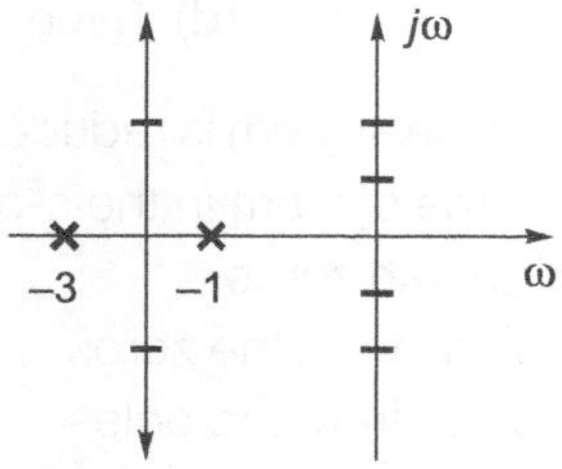

What will be the gain for obtaining the damping ratio 0.707?

(a) 1/4 (b) 5/4
(c) −3/4 (d) 11/4 [ESE-2011]

4.28 The number of individual loci in a root locus plot is equal to

(a) The number of open loop poles
(b) The number of open loop zeros
(c) The difference of the number of open loop poles and the number of open loop zeros
(d) The number of open loop poles or zeros whichever is greater

[ESE-2011]

4.29 **Statement (I):** Root loci are symmetrical with respect to real axis of the s-plane.

Statement (II): Root loci are normally symmetrical with respect to the perpendicular axis of symmetry of the pole-zero combination of the loop transfer function.

(a) Both Statement (I) and Statement (II) are individually true and Statement (II) is the correct explanation of Statement (I).
(b) Both Statement (I) and Statement (II) are individually true but Statement (II) is not the correct explanation of Statement (I).
(c) Statement (I) is true but Statement (II) is false.
(d) Statement (I) is false but Statement (II) is true.

[ESE-2012]

4.30 If a feedback control system has its open-loop transfer function

$$G(s)\,H(s) = \frac{K}{[s(s+2)(s^2+2s+5)]}$$

the coordinates of the centroid of the asymptotes of its root-locus are

(a) −1 and 0 (b) 1 and 0
(c) 0 and −1 (d) 0 and 1 [ESE-2012]

4.31 If root loci plots of a particular control system do not intersect the imaginary axis at any point, then the gain margin of the system will be

(a) zero (b) 0.707
(c) 1.0 (d) infinite

[ESE-2012]

4.32 Consider the transfer function

$$K\,G(s)H(s) = \frac{K}{s^3+4s^2+s-6}$$

The root-locus plot of the system passes through $s = 0$. The value of K at this point will be

(a) 10 (b) 0
(c) 6 (d) 8 [ESE-2013]

4.33 A system has its open-loop transfer function of $\frac{K}{s(s^2+6s+10)}$. The break-points are at $s = -1.18$ and $s = -2.82$, the centroid is at $s = -2$, while the asymptotic angles are ±60° and ±180°. The value of K for the closed loop system to be oscillatory and the frequency of oscillations are respectively:

(a) 600 and 10 rad/sec
(b) 120 and 5 rad/sec
(c) 60 and 3.16 rad/sec
(d) 30 and 3.16 rad/sec [ESE-2013]

4.34 When the number of poles is equal to the number of zeros, how many branches of root locus tend towards infinity?

(a) 1
(b) 2
(c) 0
(d) equal to number of zeros

[ESE-2014]

4.35 The open-loop transfer function of a unity feedback control system is $G(s) = \frac{1}{(s+2)^2}$. The closed-loop transfer function will have poles at

(a) −2, −2 (b) −2, −1
(c) $-2 \pm j$ (d) −2, 2

[ESE-2014]

4.36 The number of branches of the root loci of the equation $s(s + 4)(s + 5) + K(s + 2) = 0$ is

(a) 2 (b) 3
(c) 4 (d) 5

(when K varies from $-\infty$ to $+\infty$)

[ESE-2015]

4.37 Consider the system with $G(s) = \dfrac{K(s+2)}{s^2+2s+3}$ and $H(s) = 1$. The breakaway point(s) of the root loci is/are at
(a) –0.265 only
(b) –3.735 only
(c) –3.735 and –0.265
(d) There is no breakaway point [ESE-2016]

4.38 The main objectives of drawing the root-locus plot are
1. To obtain a clear picture of the open-loop poles and zeros of the system.
2. To obtain a clear picture of the transient response of the system for varying gain, K.
3. To find the range of K to make the system stable.

Which of the above statements are correct?
(a) 1, 2 and 3 (b) 1 and 2 only
(c) 1 and 3 only (d) 2 and 3 only
[ESE-2016]

4.39 A unity feedback system has open-loop poles at $s = -2 \pm j2$, $s = -1$ and $s = 0$ and a zero at $s = -3$. What are the angles made by the root-loci asymptotes with the real axis?
(a) 60°, 180° and – 60° (b) 30°, 90° and 60°
(c) 60°, 120° and – 30° (d) 30°, 60° and 180°
[ESE-2016]

4.40 **Statement (I) :** The complex conjugate poles and zeros of the open-loop transfer function have no effect on the root-loci on the real axis.

Statement (II): Angle contribution of such a pair of conjugate poles or zeros is 2π radians on the real axis.
(a) Both Statement (I) and Statement (II) are individually true and Statement (II) is the correct explanation of Statement (I).
(b) Both Statement (I) and Statement (II) are individually true but Statement (II) is not the correct explanation of Statement (I).
(c) Statement (I) is true but Statement (II) is false.
(d) Statement (I) is false but Statement (II) is true.
[ESE-2016]

4.41 Consider that in a system loop transfer function, addition of a pole results in the following:
1. Root locus gets pulled to the right-hand side.
2. Steady-state error is increased.
3. System response gets slower.

Which of the above are correct?
(a) 1, 2 and 3 (b) 1 and 2 only
(c) 1 and 3 only (d) 2 and 3 only
[ESE-2017]

4.42 For the open-loop system

$$G(s)H(s) = \frac{K}{s(s+1)(s+2)}$$

the breakaway point is
(a) –0.23 (b) –0.42
(c) –1.47 (d) –3.47
[ESE-2018]

4.43 Consider a feedback system with the characteristic equation

$$1 + K\frac{1}{s(s+1)(s+2)} = 0$$

The asymptotes of the three branches of root locus plot of this system will form the following angles with the real axis
(a) 60°, 120° and 300° (b) 60°, 120° and 180°
(c) 60°, 180° and 300° (d) 40°, 120° and 200°
[ESE-2019]

4.44 The angle of departure from a real open-loop pole and the angle of arrival at a real open-loop zero is always equal to
(a) 0° only (b) 90° only
(c) 180° only (d) 0° or 180°
[ESE-2020]

4.45 Which one of the following techniques can be used to analyze and design the effect of loop gain upon the system's transient response and stability?
(a) Open-loop transfer function technique.
(b) Closed-loop transfer function technique.
(c) Root locus technique
(d) Compensation technique
[ESE-2022]

4.46 What are the breakaway points of the root locus of a unity feedback system with open loop transfer function $G(s) = \dfrac{K(s+2)}{(s+1)^2}$?
(a) –1, 2 (b) –1, 3
(c) –1, –2 (d) –1, –3 [ESE-2024]

■■■

Answers Root Locus

4.1 (d) 4.2 (b) 4.3 (c) 4.4 (d) 4.5 (b) 4.6 (c) 4.7 (d) 4.8 (b) 4.9 (b)

4.10 (b) 4.11 (b) 4.12 (a) 4.13 (b) 4.14 (c) 4.15 (c) 4.16 (c) 4.17 (*) 4.18 (b)

4.19 (c) 4.20 (a) 4.21 (d) 4.22 (d) 4.23 (a) 4.24 (b) 4.25 (c) 4.26 (b) 4.27 (b)

4.28 (d) 4.29 (c) 4.30 (a) 4.31 (d) 4.32 (c) 4.33 (c) 4.34 (c) 4.35 (c) 4.36 (b)

4.37 (b) 4.38 (d) 4.39 (a) 4.40 (a) 4.41 (c) 4.42 (b) 4.43 (c) 4.44 (d) 4.45 (c)

4.46 (d)

Explanations Root Locus

4.1 (d)

Intersection of asymptotes, i.e. centroid

$$= \frac{\Sigma \text{real parts of poles} - \Sigma \text{ real parts of zeros}}{\text{Number of poles} - \text{Number of zeros}}$$

$$= \frac{-1-3}{3} = \frac{-4}{3} = -1.33$$

4.2 (b)

$1 + G(s)H(s) = 0$

$\Rightarrow s(s+4)(s^2+4s+5) + K = 0$

$\Rightarrow K = -s(s+4)(s^2+4s+5)$

For breakaway points,

$$\frac{dK}{ds} = 0$$

$\Rightarrow (s+4)(s^2+4s+5) + s(s^2+4s+5) + s(s+4)(2s+4) = 0$

$\Rightarrow (2s+4)(s^2+4s+5) + (s^2+4s)(2s+4)$

$\Rightarrow (2s+4)(2s^2+8s+5) = 0$

$\Rightarrow (2s+4)(s+0.775)(s+3.225) = 0$

$\Rightarrow s = -2, -0.775, -3.225$

Thus the system has three real breakaway points.

4.3 (c)

Number of poles, $n = 4$

Number of zeros, $m = 1$

Since $m < n$, so number of zeros at infinity

$= n - m = 4 - 1 = 3$

4.4 (d)

A point on the real axis lies on the locus if the number of open loop poles plus zeros on the real axis to the right of this point is odd.

4.6 (c)

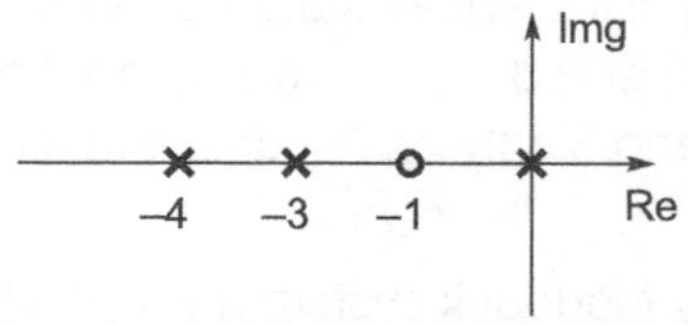

Root locus can lie on the real axis between $s = -3$ and $s = -4$ because, the number of poles + number of zeros to the right half of any point between –4 to –3 are odd.

4.7 (d)

Root locus shows that transfer function has poles at $s = 0, -1$ and zero at $s = -2$.

So, $G(s) = \dfrac{K(s+2)}{s(s+1)}$

4.9 (b)

The corresponding value of ω is found out from auxiliary equation in Routh array.

4.10 (b)

The $(p - z)$ branches of the root loci which tend to infinity, do so along straight line asymptotes whose angles are given by

$$\phi_A = \frac{(2K+1)\pi}{p-z}, K = 0, 1, 2, ..., p-z-1$$

4.12 (a)

The root locus is symmetrical about real axis but it is not symmetrical about $j\omega$-axis.

4.13 (b)

If P = number of open loop poles

Z = number of open loop zeros

The number of individual root locus branches

$= P\,;\, P > Z$

$= Z\,;\, Z > P$

Best possible answer is option (b).

4.14 (c)

Intersection of asymptotes, i.e. centroid

$$= \frac{\Sigma(\text{Real part of poles}) - \Sigma(\text{Real parts of zeros})}{\text{Number of poles} - \text{Number of zeros}}$$

$$= \frac{-1-2-5-(-3)}{4-1} = \frac{-5}{3}$$

The real axis intercept is $\frac{-5}{3}$.

4.15 (c)

For the root locus to be parallel to the imaginary axis, the angle of asymptotes should be ±90°.

Angle of asymptotes,

$$\phi_A = \frac{(2q+1)180^\circ}{n-m}\,;\ q = 0, 1, 2, \ldots$$

where n = number of open-loop poles

m = number of open-loop zeros

ϕ_A for $G(s) = K/(s+2)^2$ is

$$\phi_A = \frac{(2q+1)180^\circ}{2}$$

$= 90^\circ$ for $q = 0$

$= 270^\circ$ or -90° for $q = 1$

Therefore, $\frac{K}{(s+2)^2}$ has root locus parallel to imaginary axis.

4.16 (c)

The number of root-locus segments ending at infinity are equal to n-m, where

n = number of open-loop poles

and m = number of open-loop zeros

4.18 (b)

The root locus is symmetrical about the real axis. segments of the real axis having an odd number of real axis open-loop poles plus zeros to their right are parts of the root locus.

4.19 (c)

The root locus is the locus of closed-loop poles of the system (i.e., the roots of characteristic equation) when the parameter is varied from 0 to ∞.

4.20 (a)

$$s^2 + 2s + 2 + K(s+2) = 0$$

$$K = \frac{-(s^2+2s+2)}{s+2}$$

$$\frac{dK}{ds} = -\left[\frac{(s+2)(2s+2)-(s^2+2s+2)}{(s+2)^2}\right]$$

$$\frac{dK}{ds} = 0$$

$\Rightarrow 2s^2 + 2s + 4s + 4 - s^2 - 2s - 2 = 0$

$\Rightarrow s^2 + 4s + 2 = 0$

$$\Rightarrow \quad s = \frac{-4 \pm \sqrt{16-8}}{2}$$

$= -0.586, -3.414$

Therefore, break-away point is $s = -3.414$

4.21 (d)

$$s(s+2)(s+3) + K(s+1) = 0$$

$$\Rightarrow \quad G(s)H(s) = \frac{K(s+1)}{s(s+2)(s+3)}$$

Since there are 3 poles and 1 zero, therefore, in the root loci one branch will be from a pole to zero and two more branches will be from rest of the poles towards infinity.

4.23 (a)

Consider, $G(s)\,H(s) = \frac{K}{s(s+2)}$

1. $P = 2,\ Z = 0,\ P - Z = 2$

2. 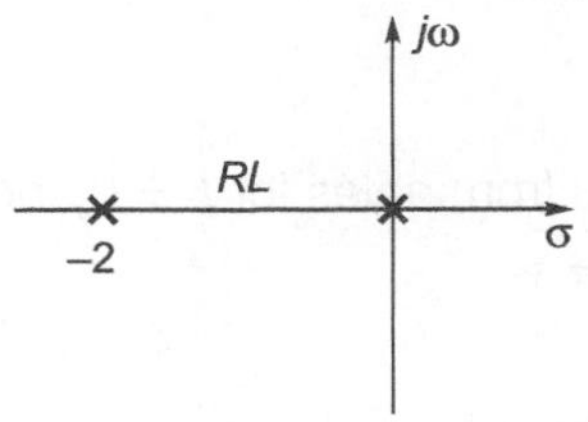

3. $\theta = \frac{(2q+1)180}{P-Z}\,;$

$\theta_1 = 90;\ \theta_2 = 270$

4. Centroid $= \frac{0+(-2)}{2} = -1$

5. B.A. point

$s^2 + 2s + K = 0$

$$\Rightarrow \quad \frac{dK}{ds} = -4s - 2 = 0$$

$\Rightarrow \quad s = -0.5$

Hence, Root locus diagram

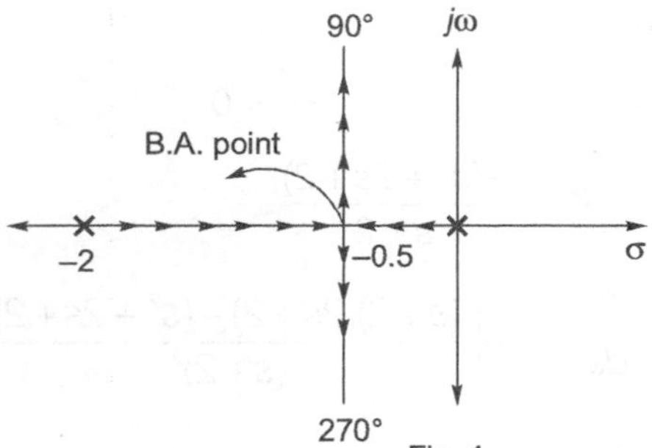

Fig. 1

⇒ Now adding a open zero in the system

i.e. $G(s)\,H(s) = \dfrac{K(s+1)}{s(s+2)}$

1. $P = 2;\ Z = 1;\ P - Z = 1$

2.
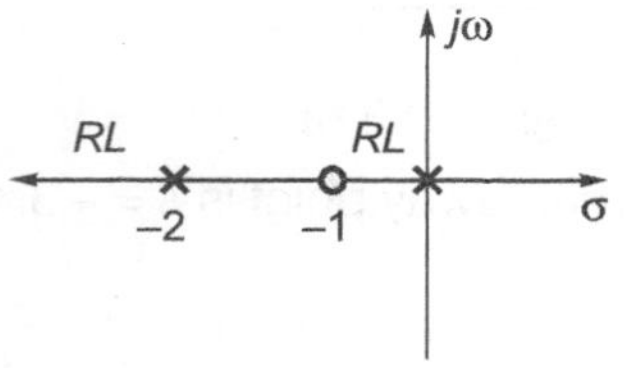

3. $\theta = \dfrac{(2q+1)180}{P-Z} = 180$

Now drawing the root locus

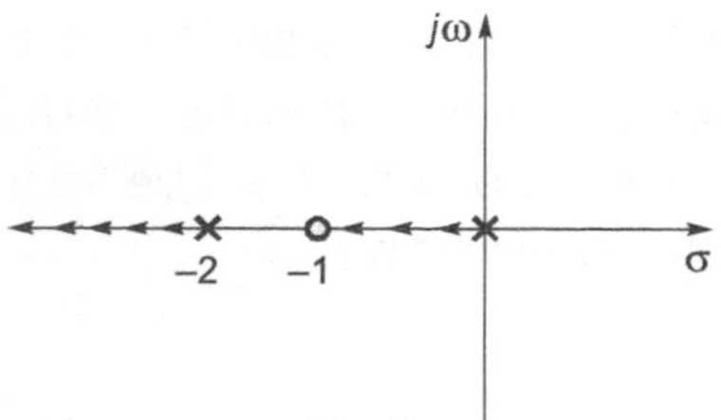

Fig. 2

From figure (1) and (2) it is clear that by adding a zero root locus is shifting towards left making system more stable.

4.24 (b)

Root loci starts from poles for $K = 0$ and ends at zeros for $K = \pm\infty$.

4.25 (c)

Characteristics equation

$$1 + \frac{K(s+1)}{s(s+4)(s^2+2s+2)} = 0$$

Comparing with standard equation $1 + G(s)\,H(s) = 0$, we have;

$$G(s)\,H(s) = \frac{K(s+1)}{s(s+4)(s^2+2s+2)}$$

So open loop poles = 0 ; −4 ; $-1 \pm j$

and open loop zeros = −1

$$\text{Centroid} = \frac{\Sigma\text{real parts of poles} - \Sigma\text{ real parts of zeros}}{\text{Number of poles} - \text{Number of zeros}}$$

$$\text{Centroid} = \frac{(0-4-1-1)-(-1)}{4-1} = -\frac{5}{3}$$

4.26 (b)

Root loci starts from poles of $G(s)\,H(s)$ for $K = 0$.

4.27 (b)

Characteristic equation of the given system is

$$1 + G(s)\,H(s) = 0$$

$$\Rightarrow\quad 1 + \frac{4K}{(s+1)(s+3)}\cdot 1 = 0$$

$$\Rightarrow\quad s^2 + 4s + 4K + 3 = 0$$

Comparing with standard equation

$$2\xi\omega_n = 4$$

$$\Rightarrow\quad \omega_n = \frac{4}{2\xi} = \frac{4}{2\times 1/\sqrt{2}} = 2\sqrt{2}$$

$$\omega_n^2 = 4K + 3$$

or $\quad 4K + 3 = 8$

$\Rightarrow\quad K = 5/4$

4.28 (d)

If $\quad P$ = number of open loop poles

Z = number of open loop zeros

The number of individual root locus branches

$= P;\ P > Z$

$= Z;\ Z > P$

4.29 (c)

Root loci are symmetrical about the real axis (σ-axis).

4.30 (a)

$$\text{Centroid} = \frac{\Sigma\text{real parts of poles} - \Sigma\text{ real parts of zeros}}{\text{Number of poles} - \text{Number of zeros}}$$

$P \rightarrow$ number of open loop poles.

$Z \rightarrow$ number of open loop zeros.

$$G(s)H(s) = \frac{K}{s(s+2)(s^2+2s+5)}$$

Real part of open loop poles = 0, −2, −1, −1

$$\text{Centroid} = \frac{-2-1-1}{4} = -1$$

Thus the centroid lies on the real axis, so centroid is (−1, 0).

4.32 (c)

$$KG(s)H(s) = \frac{K}{s^3+4s^2+s-6}$$

The root-locus will intersect the jω-axis at $s = 0$ and can be found using Routh Array i.e.
$1 + G(s)\,H(s) = s^3 + 4s^2 + s - 6 + K$

s^3	1	1
s^2	4	$K-6$
s^1	$\frac{4-K+6}{4}$	0
s^0	$K-6=0 \Rightarrow K=6$	

$\Rightarrow\ 4 - K + 6 = 0$
$\Rightarrow\ K = 10$

4.33 (c)

Open loop transfer function,

$$G(s)H(s) = \frac{K}{s(s^2+6s+10)}$$

Break away points,
$s = -1.18$ and -2.82
Centroid, $s = -2$

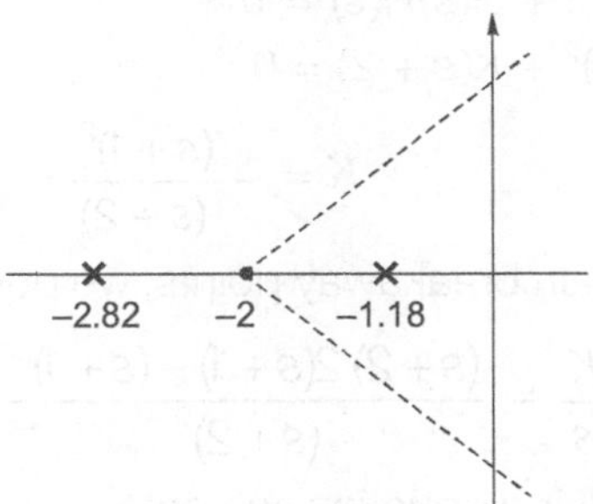

The gain, K can be found out using the Routh criterion i.e. characteristic equation is
$1 + G(s)\,H(s) = 0$
$s^3 + 6s^2 + 10s + K = 0$

s^3	1	10
s^2	6	K
s^1	$\frac{60-K}{6}$	0
s^0	K	

When one full row of routh array becomes zero, then system may be marginally stable or oscillatory

i.e. $\frac{60-K}{6} = 0$

$K_{marginal} = 60$

Auxiliary equation is given as

$6s^2 + K_{marginal} = 0$

$6s^2 + 60 = 0$
$-\omega^2 + 10 = 0$
$\omega = \sqrt{10}$ rad/sec $= 3.16$ rad/sec

4.34 (c)

The total number of root locus branches which tends to infinity is $P - Z$.

4.35 (c)

$$G(s) = \frac{1}{(s+2)^2}$$

Closed loop transfer function

$$T(s) = \frac{G(s)}{1+G(s)} = \frac{1}{(s+2)^2+1}$$

$$= \frac{1}{s^2+4s+5}$$

$\therefore$ Closed loop poles $= \frac{-4 \pm \sqrt{16-20}}{2} = -2 \pm j$

4.36 (b)

$s(s + 4)\,(s + 5) + K(s + 2) = 0$
thus $1 + G(s)\,H(s) = 0$

$$G(s)H(s) = \frac{K(s+2)}{s(s+4)(s+5)}$$

The number of poles = 3
number of zeroes = 1
Thus number of root locus branches = 3
The root will originate from the three poles, one pole will terminate at finite zero and the rest two will terminate at infinity.

4.37 (b)

$$GH(s) = \frac{K(s+2)}{s^2+2s+3}$$

Breakaway point is solution of $\frac{dK}{ds}$ that lies on root locus.
Characteristic equation
$1 + GH(s) = 0$

$$1+\frac{K(s+2)}{s^2+2s+3} = 0$$

or $$K = -\frac{(s^2+2s+3)}{(s+2)}$$

$\therefore$ $$\frac{dK}{ds} = -\frac{d}{ds}\left[\frac{(s^2+2s+3)}{(s+2)}\right] = 0$$

$$\Rightarrow \quad s = -2 \pm \sqrt{3} = -3.735, \ -0.265$$

Only –3.735 lies on the root locus.

4.38 (d)

Root locus obtained mainly to obtain response and stability of system.

4.39 (a)

Total number of poles = 4
Total number of zeros = 1

$$\text{Angles of asymptotes} = \frac{(2K+1)180°}{P-Z}$$

where, $K = 0, 1, 2$
Hence, Angles of asymptotes

$$= 60°, 180°, 300° \text{ (or)} -60°$$

4.40 (a)

Complex poles and zeros exists in pair. Hence they don't have any effect on root locus on real axis. 2π radian angle means 0° angle. Hence it cannot change the phase of the system.

4.41 (c)

Adding a pole causes speed and stability of a system to decrease.

4.42 (b)

$$G(s) = \frac{K}{s(s+1)(s+2)}$$

Breaking point is obtained from solutions of $\frac{dK}{ds}$

$$V(s) = s^3 + 3s^2 + 6s + K = 0$$
$$K = -(s^3 + 3s^2 + 6s)$$
$$\frac{dK}{ds} = -(3s^2 + 6s + 6) = 0$$
$$s = -0.42, -1.58$$

$s = -0.42$ lies on root locus.

4.43 (c)

$$G(s)\,H(s) = \frac{K}{s(s+1)(s+2)}$$

$$\text{Angle of asymptotes, } \phi_A = \frac{(2K+1)\,180°}{P-z}$$
$$K = 0, 1, \ldots, P - z - 1$$
$$\therefore \quad \phi_A = 60°, 180°, 300°$$

4.45 (c)

In root locus technique, the gain of the system K is varied to plot the locus of roots of characteristics equation of the system i.e. the closed-loop poles which can be used to study or analyze the system's transient response and stability.

4.46 (d)

We have,

$$\text{OLTF, } G(s)H(s) = \frac{K(s+2)}{(s+1)^2}$$

The characteristic equation is given by

$$1 + G(s)H(s) = 0$$
$$(s+1)^2 + K(s+2) = 0$$
$$K = -\frac{(s+1)^2}{(s+2)}$$

To obtain breakaway points, we put

$$-\frac{dK}{ds} = \frac{(s+2)\,2(s+1)-(s+1)^2}{(s+2)^2} = 0$$

On solving equation, we get,

$$(s+1)(2s+4-s-1) = 0$$
$$(s+1)(s+3) = 0$$
$$s = -1, -3$$

Hence, option (d) is correct.

■■■

5 Frequency Domain Analysis

5.1 Match **List-I** (Scientist) with **List-II** (Contribution in the area of) and select the correct answer:

List-I	**List-II**
A. Bode	1. Asymptotic plots
B. Evans	2. Polar plots
C. Nyquist	3. Root-locus technique
	4. Constant M and N plots

Codes:

	A	B	C
(a)	1	4	2
(b)	2	3	4
(c)	3	1	4
(d)	1	3	2

[ESE-2000]

5.2 The polar plot (for positive frequencies) for the open-loop transfer function of a unity feedback control system is shown in the given figure

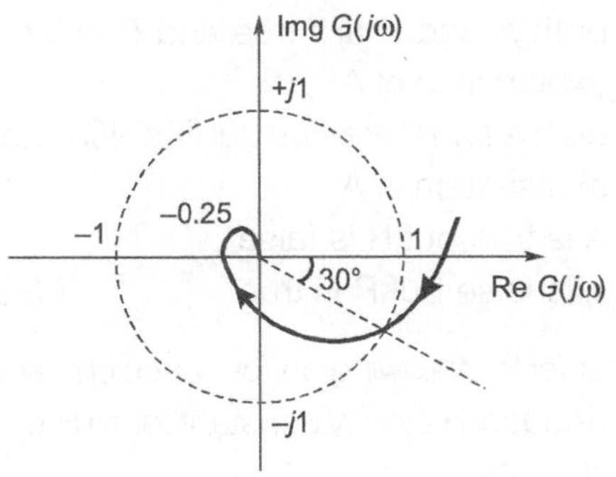

The phase margin and the gain margin of the system are respectively

(a) 150° and 4 (b) 150° and 3/4
(c) 30° and 4 (d) 30° and 3/4

[ESE-2000]

5.3 The open loop transfer function of a system is

$$G(s)H(s) = \frac{K}{(1+s)(1+2s)(1+3s)}$$

The phase crossover frequency ω_{pc} is

(a) $\sqrt{2}$ (b) 1
(c) zero (d) $\sqrt{3}$

[ESE-2001]

5.4 Nyquist plot shown in the given figure is for a type

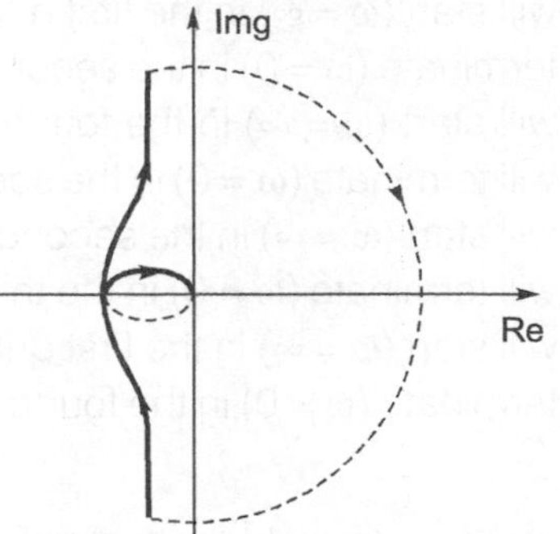

(a) zero system (b) one system
(c) two system (d) three system

[ESE-2001]

5.5 The open-loop transfer function of a unity feedback control system is given as

$$G(s) = \frac{1}{s(1+sT_1)(1+sT_2)}.$$

The phase crossover frequency and the gain margin are, respectively

(a) $\frac{1}{\sqrt{T_1T_2}}$ and $\frac{T_1+T_2}{T_1T_2}$

(b) $\sqrt{T_1T_2}$ and $\frac{T_1+T_2}{T_1T_2}$

(c) $\frac{1}{\sqrt{T_1T_2}}$ and $\frac{T_1T_2}{T_1+T_2}$

(d) $\sqrt{T_1T_2}$ and $\frac{T_1T_2}{T_1+T_2}$

[ESE-2001]

5.6 An open loop transfer function of a unity feedback control system has two finite zeros, two poles at origin and two pairs of complex conjugate poles. The slope of high frequency asymptote in Bode magnitude plot will be

(a) +40 dB/decade (b) 0 dB/decade
(c) –40 dB/decade (d) –80 dB/decade

[ESE-2001]

5.7 Open loop transfer function of a system having one zero with a positive real value is called
(a) zero phase function
(b) negative phase function
(c) positive phase function
(d) non-minimum phase function [ESE-2001]

5.8 The Nyquist plot of

$$G(s)\cdot H(s) = \frac{10}{s^2(1+0.5s).(1+s)}$$

(a) will start ($\omega = \infty$) in the first quadrant and will terminate ($\omega = 0$) in the second quadrant
(b) will start ($\omega = \infty$) in the fourth quadrant and will terminate ($\omega = 0$) in the second quadrant
(c) will start ($\omega = \infty$) in the second quadrant and will terminate ($\omega = 0$) in the third quadrant
(d) will start ($\omega = \infty$) in the first quadrant and will terminate ($\omega = 0$) in the fourth quadrant
[ESE-2002]

5.9 Which of the following is the Nyquist diagram for the open loop function

$$G(s)\cdot H(s) = \frac{5}{s(1+0.1s)(1+0.01s)}?$$

(a)
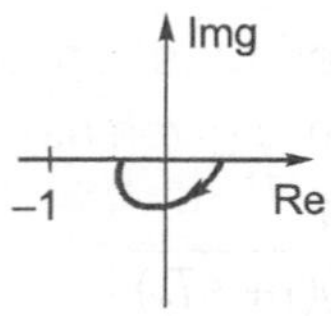

(b)
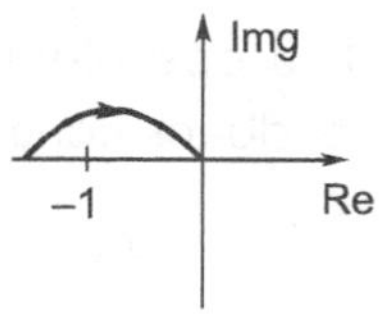

(c)
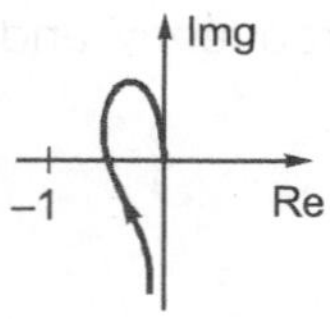

(d)
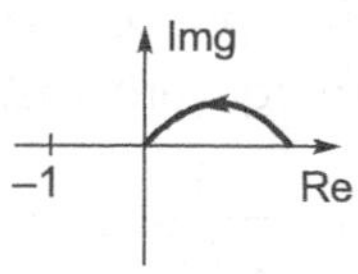

[ESE-2002]

5.10 Consider the following statements associated with phase and gain margins:
1. They are a measure of closeness of the polar plot to the $-1 + j0$ point
2. For a non-minimum phase to be stable it must have positive phase and gain margins
3. For a minimum phase system to be stable, both the margins must be positive

Which of the above statements is/are correct?
(a) 2 and 3 (b) 1 and 3
(c) 1 and 2 (d) 1 alone [ESE-2002]

5.11 Assertion (A): $G(s) = \dfrac{10(s-25)}{s(s+1)(s+5)}$ represents a non-minimum phase transfer function.
Reason (R): A minimum phase transfer function has the property that its magnitude and phase are uniquely related.
(a) Both A and R are true and R is the correct explanation of A
(b) Both A and R are true but R is NOT the correct explanation of A
(c) A is true but R is false
(d) A is false but R is true [ESE-2002]

5.12 Assertion (A): The bandwidth of a control system indicates the noise filtering characteristic of the system.
Reason (R): The bandwidth is a measure of ability of a control system to reproduce the input signal.
(a) Both A and R are true and R is the correct explanation of A
(b) Both A and R are true but R is NOT the correct explanation of A
(c) A is true but R is false
(d) A is false but R is true [ESE-2002]

5.13 Assertion (A): Addition of a pole to the open loop transfer function of a system increases the rise time of the closed loop step-response.
Reason (R): Additional pole has the effect of reducing the bandwidth of the system.
(a) Both A and R are true and R is the correct explanation of A
(b) Both A and R are true but R is NOT the correct explanation of A
(c) A is true but R is false
(d) A is false but R is true [ESE-2002]

5.14 Consider the following performance characteristics:
1. Reduced velocity constant for a given relative stability.
2. Reduced gain cross-over frequency
3. Reduced bandwidth
4. Reduced resonance peak of the system

Which of these performance characteristics are achieved with the phase-lag compensation?
(a) 1 and 2 (b) 1 and 3
(c) 2, 3 and 4 (d) 1, 2, 3 and 4
[ESE-2002]

5.15 Consider the following open loop frequency response of a unity feedback system:

ω, rad/s → :	2	3	4	5	6	8	10
$\lvert G(j\omega)\rvert$ → :	7.5	4.8	3.15	2.25	1.70	1.00	0.64
$\angle G(j\omega)$ → :	–118°	–130°	–140°	–150°	–157°	–170°	–180°

The gain and phase margin of the system are respectively
(a) 0.00 dB, –180° (b) 3.86 dB, –180°
(c) 0.00 dB, –10° (d) 3.86 dB, 10°
[ESE-2003]

5.16 The Nyquist plot shown below, matches with the transfer function

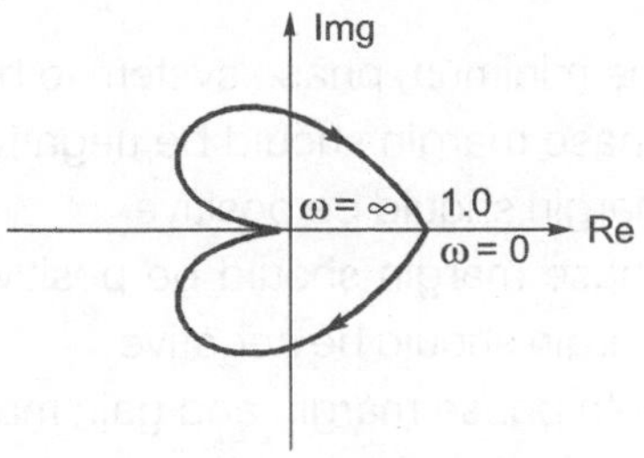

(a) $\dfrac{1}{(s+1)^3}$ (b) $\dfrac{1}{(s+1)^2}$

(c) $\dfrac{1}{(s^2+2s+2)}$ (d) $\dfrac{1}{(s+1)}$

[ESE-2003]

5.17 The phase margin (PM) and the damping ratio (ξ) are related by

(a) $PM = 90° - \tan^{-1}\left\{\dfrac{\sqrt{-2\xi^2+\sqrt{1+4\xi^4}}}{2}\right\}$

(b) $PM = \tan^{-1}\left\{\dfrac{2\xi}{\sqrt{-2\xi^2+\sqrt{1+4\xi^4}}}\right\}$

(c) $PM = 90° + \tan^{-1}\left\{\dfrac{\sqrt{2\xi^2+\sqrt{1+4\xi^4}}}{2\xi}\right\}$

(d) $PM = 180° - \tan^{-1}\left\{\dfrac{\sqrt{2\xi^2-\sqrt{1+4\xi^4}}}{2}\right\}$

[ESE-2003]

5.18 Consider the following statements regarding the frequency response of a system as shown below:

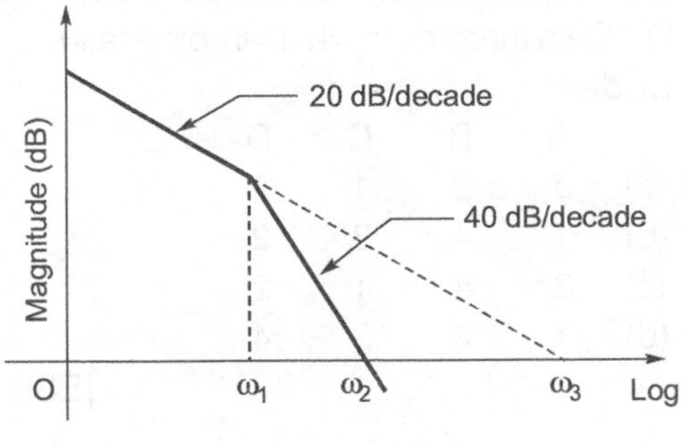

1. The type of the system is one
2. ω_3 = static error coefficient (numerically)
3. $\omega_2 = \dfrac{\omega_1+\omega_3}{2}$

Select the correct answer using the code given below:
(a) 1, 2 and 3 (b) 1 and 2
(c) 2 and 3 (d) 1 and 3 [ESE-2003]

5.19 A closed loop system, employing lag-lead compensator $G_c(s)$ is shown in the figure given below:

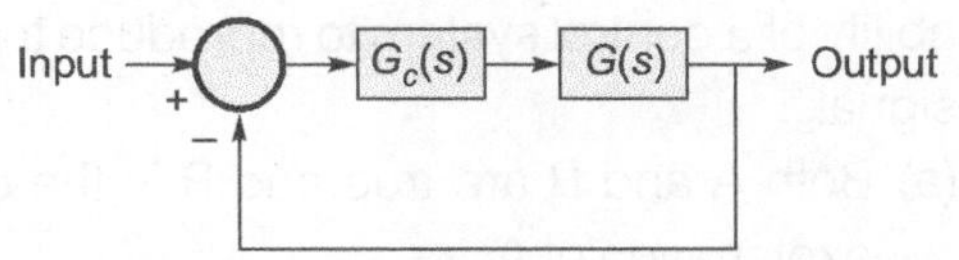

$$G_c(s) = \left[\frac{1+\tau_1 s}{1+\dfrac{\tau_1 s}{\beta}}\right]\left[\frac{1+\tau_2 s}{1+\beta\tau_2 s}\right], \beta < 1$$

If $G(s)$ has 3 poles in the left half of a s-plane, then the shape of the Bode plot for $|G(s)\,G_c(s)|$ in the highest frequency range will be
(a) –20 dB/decade (b) –40 dB/decade
(c) – 60 dB/decade (d) –80 dB/decade
[ESE-2003]

5.20 The forward path transfer function of a unity feedback system is given by

$$G(s) = \frac{100}{s^2+10s+100}$$

The frequency response of this system will exhibit the resonance peak at
(a) 10 rad/s (b) 8.66 rad/s
(c) 7.07 rad/s (d) 5 rad/s [ESE-2004]

5.21 Consider the following Nyquist plot:

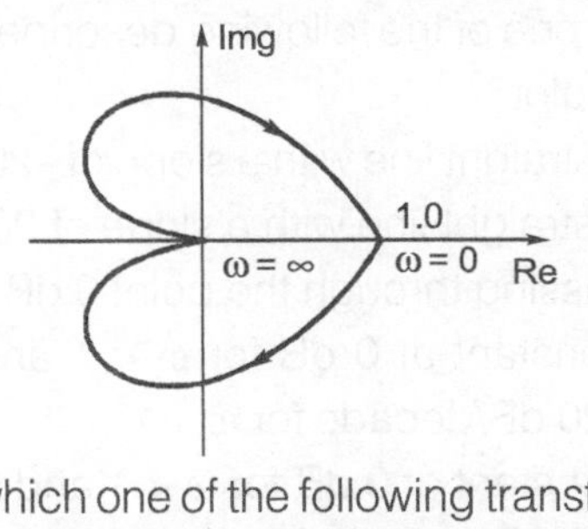

With which one of the following transfer functions, does the above Nyquist plot match?

(a) $\dfrac{1}{(s+1)^3}$ (b) $\dfrac{1}{(s+1)^2}$

(c) $\dfrac{1}{(s^2+2s+2)}$ (d) $\dfrac{1}{(s+1)}$ [ESE-2004]

5.22 The forward path transfer function of a unity feedback system is given by $G(s) = \dfrac{1}{(1+s)^2}$.
What is the phase margin for this system?
(a) $-\pi$ rad (b) 0 rad
(c) $\pi/2$ rad (d) π rad [ESE-2004]

5.23 **Assertion (A):** The bandwidth of a control system indicates the noise filtering characteristic of the system.
Reason (R): The bandwidth is a measure of ability of a control system to reproduce the input signal.
(a) Both A and R are true and R is the correct explanation of A
(b) Both A and R are true but R is NOT the correct explanation of A
(c) A is true but R is false
(d) A is false but R is true [ESE-2004]

5.24 Which one of the following statements is correct in respect of the theory of stability?
(a) Phase margin is the phase angle lagging, in short of 180°, at the frequency corresponding to a gain of 10.
(b) Gain margin is the value by which the gain falls short of unity, at a frequency corresponding to 90° phase lag.
(c) Routh-Hurwitz criterion can determine the degree of stability
(d) Gain margin and phase margin are the measure of the degree of stability
[ESE-2004]

5.25 If $X(s) = 1/(1 + s)$ is the given network function, which one of the following describes the correct Bode plot?
(a) A straight line with a slope of –20 dB/decade
(b) A straight line with a slope of 20 dB/decade passing through the point 0 dB and 1 rad/s
(c) Constant at 0 dB for $\omega < 1$ and a slope of +20 dB/decade for $\omega > 1$
(d) Constant at 0 dB for $\omega < 1$ and a slope of –20 dB/decade for $\omega > 1$
[ESE-2004]

5.26 Consider the following statements:
The frequency response of a control system has very sharp cut off characteristics. This implies that:
1. it has large peak resonance
2. it has large bandwidth.
3. it is a less stable system.
Which of the statements given above is/are correct?
(a) 1 only (b) 2 and 3
(c) 1 and 3 (d) 1, 2 and 3
[ESE-2005]

5.27 For the minimum phase system to be stable:
(a) Phase margin should be negative and gain margin should be positive
(b) Phase margin should be positive and gain margin should be negative
(c) Both phase margin and gain margin should be positive
(d) Both phase margin and gain margin should be negative
[ESE-2005]

5.28 For a stable closed loop system, the gain at phase cross-over frequency should always be:
(a) > 20 dB (b) > 6 dB
(c) < 6 dB (d) < 0 dB
[ESE-2005]

5.29 If the gain of the open loop system is doubled, the gain margin of the system is
(a) not affected
(b) doubled
(c) halved
(d) one fourth of original value
[ESE-2005]

5.30 Match **List-I** (Frequency Response) with **List-II** (Time Response) and select the correct answer using the code given below the Lists:

List-I	List-II
A. Bandwidth	1. Overshoot
B. Phase margin	2. Stability
C. Response peak	3. Speed of time response
D. Gain margin	4. Damping ratio

Codes:

	A	B	C	D
(a)	3	2	1	4
(b)	1	4	3	2
(c)	3	4	1	2
(d)	1	2	3	4

[ESE-2005]

5.31 For a unity feedback control system the damping ratio is 0.421. What is the resonance magnitude?
(a) $M_r = 1$ (b) $M_r = 0.707$
(c) $M_r = 1.30$ (d) $M_r = 1.95$
[ESE-2006]

5.32 Consider the following statements for a minimum phase system:
1. All the poles of the transfer function should lie in the left of *s*-plane.
2. The zeros of the transfer function can lie anywhere in the s-plane
3. Given the magnitude characteristic over the entire frequency range, the phase angle characteristic can be uniquely determined.

Which of the statements given above are correct?
(a) 1, 2 and 3 (b) Only 1 and 2
(c) Only 2 and 3 (d) Only 1 and 3
[ESE-2006]

5.33 Consider the following statements regarding the asymptotic Bode plots used for frequency response analysis:
1. The deviation of the actual magnitude response for a zero on real axis is 3 dB at the corner frequency.
2. The phase angle for a pair of complex conjugate poles at undamped frequency depends upon the value of damping ratio.

What of the statements given above is/are correct?
(a) Only 1 (b) Only 2
(c) Both 1 and 2 (d) Neither 1 and 2
[ESE-2006]

5.34 Consider the unity feedback system with $G(s) = \dfrac{2}{s(s+1)(2s+1)}$. What is the gain margin of the system?

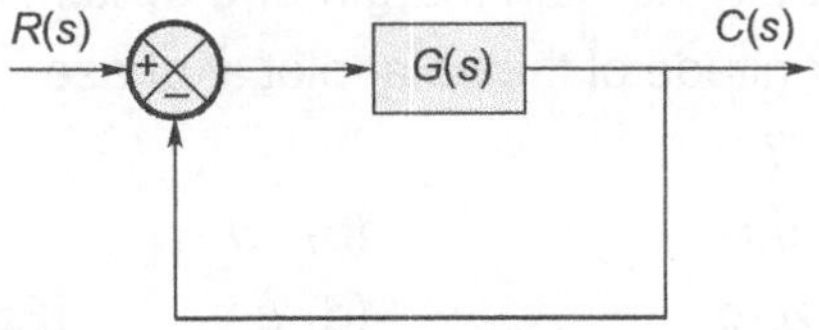

(a) 3/4 (b) 4/3
(c) 1/2 (d) 3/5 [ESE-2006]

5.35 The system having characteristic equation:
$$s^4 + 2s^3 + 3s^2 + 2s + K = 0$$
is to be used as an oscillator. What are the values of *K* and the frequency of oscillation ω?
(a) $K = 1$ and $\omega = 1$ *r/s*
(b) $K = 1$ and $\omega = 2$ *r/s*
(c) $K = 2$ and $\omega = 1$ *r/s*
(d) $K = 2$ and $\omega = 2$ *r/s* [ESE-2006]

5.36 For a stable system, what are the restrictions on the gain margin and phase margin?
(a) Both gain margin and phase margin are negative
(b) Gain margin is negative and phase margin is positive
(c) Gain margin is positive and phase margin is negative
(d) Both gain margin and phase margin are positive
[ESE-2006]

5.37 **Assertion (A):** The closed loop stability can be determined from the poles of an open loop system and the polar plot of frequency response.
Reason (R): Unstable system has right half-poles.
(a) Both A and R are true and R is the correct explanation of A
(b) Both A and R are true but R is NOT the correct explanation of A
(c) A is true but R is false
(d) A is false but R is true [ESE-2006]

5.38 Which one of the following statements is correct for gain margin and phase margin of two closed-loop systems having loop functions $G(s)H(s)$ and $\exp(-s)\, G(s)\, H(s)$?
(a) Both gain and phase margins of the two systems will be identical.
(b) Both gain and phase margins of $G(s)\, H(s)$ will be more.
(c) Gain margins of the two systems are the same but phase margin of $G(s)\, H(s)$ will be more.
(d) Phase margins of the two systems are the same but gain margin of $G(s)\, H(s)$ will be less.
[ESE-2007]

5.39 Consider the following statements with reference to relative stability of a system:
1. Phase margin is related to effective damping of the system.
2. Gain margin gives better estimate of damping ratio than phase margin.
3. When expressed in dB, gain margin is negative for a stable system.

Which of the statements given above are correct?
(a) 1 and 2 only (b) 1 and 3 only
(c) 2 and 3 only (d) 1, 2 and 3
[ESE-2007]

5.40 Which one of the following polar plots corresponds to

$$G(j\omega) = \frac{1}{(j\omega)^2(1+j\omega T)}?$$

(a)
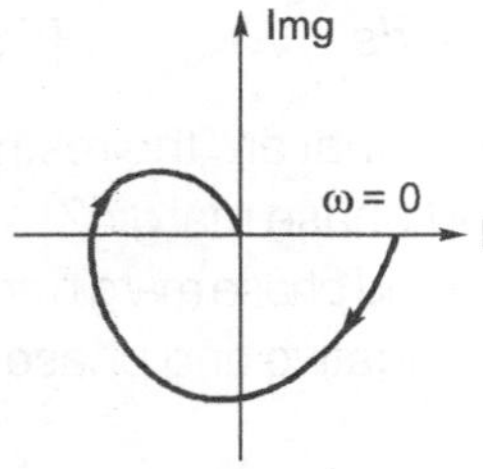

(b)
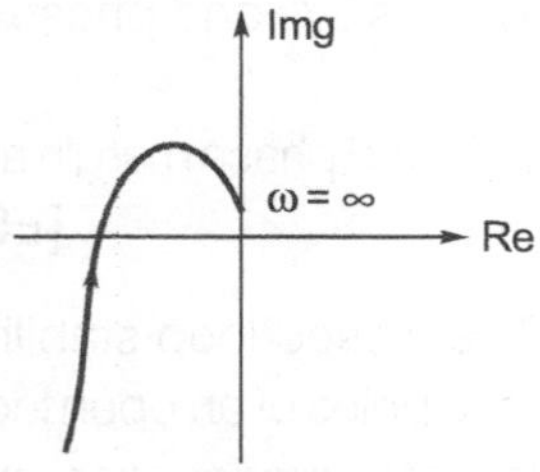

(c)
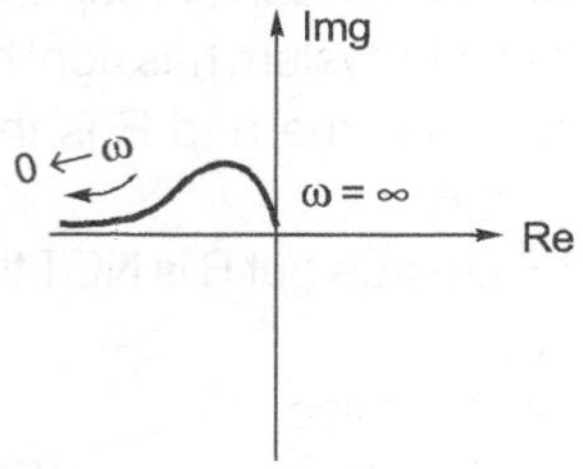

(d)
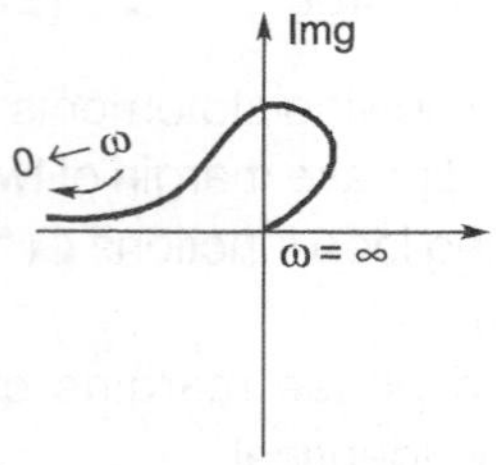

[ESE-2007]

5.41 A system with gain margin close to unity or a phase margin close to zero is

(a) relatively stable (b) oscillatory
(c) stable (d) highly stable

[ESE-2008]

5.42 The open-loop transfer function of a unity feedback control system is given by

$$G(s) = Ke^{-Ts}$$

where K and T are variables and are greater than zero. The stability of the closed-loop system depends on

(a) K only (b) Both K and T
(c) T only (d) Neither K nor T

[ESE-2008]

5.43 Match **List-I** (Shape of Nyquist Plot) with **List-II** (Gain Margin) and select the correct answer using the codes given below the lists:

List-I
A. The plot does not intersect negative real axis
B. The plot intersects negative real axis between 0 and (–1, j0)
C. The plot passes through the point (–1, j0)
D. The plot encloses the point (–1, j0)

List-II
1. < 0 dB
2. 0 dB
3. > 0 dB
4. ∞ dB

Codes:

	A	B	C	D
(a)	3	4	1	2
(b)	4	3	2	1
(c)	3	4	2	1
(d)	4	3	1	2

[ESE-2008]

5.44 The gain-phase plot of a linear control system is shown in the below figure.
What are the gain-margin (GM) and the phase-margin (PM) of the system?

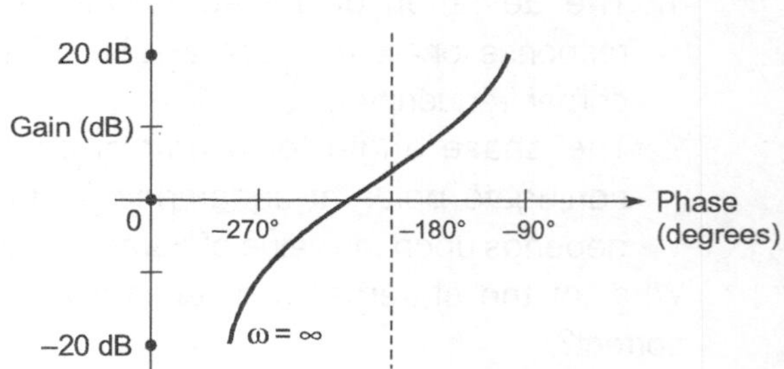

(a) GM > 0 dB and PM > 0 degree
(b) GM > 0 dB and PM < 0 degree
(c) GM < 0 dB and PM > 0 degree
(d) GM < 0 dB and PM < 0 degree

[ESE-2008]

5.45 What is the gain margin of a system when the magnitude of the polar plot at phase cross over is 'a'?

(a) $1/a$ (b) $-a$
(c) zero (d) a [ESE-2008]

5.46 What is the value of the damping ratio of a second order system when the value of the resonant peak is unity?

(a) $\sqrt{2}$ (b) Unity
(c) $1/\sqrt{2}$ (d) Zero [ESE-2008]

5.47 The closed loop transfer function of a control system is

$$\frac{K}{s(s+1)(s+5)+K}$$

What is the frequency of the sustained oscillations for marginally stable condition?

(a) $\sqrt{5}$ rad/s (b) $\sqrt{6}$ rad/s
(c) 5 rad/s (d) 6 rad/s [ESE-2008]

5.48 In the Bode plot of a unity feedback control system, the value of phase angle of $G(j\omega)$ is –90° at the gain cross over frequency of the Bode plot, the phase margin of the system is

(a) –180° (b) +180°
(c) –90° (d) +90° [ESE-2010]

5.49 The Nyquist plot of loop transfer function $G(s)\,H(s)$ of a closed loop control system passes through the point $(-1, j0)$ in the $G(s)\,H(s)$ plane. The phase margin of the system is

(a) 0° (b) 45°
(c) 90° (d) 180° [ESE-2010]

5.50 The magnitude and phase of the transfer function

$G(s) = \dfrac{1}{s+1}$ at $\omega = 1$ is

(a) 0.707 and 45°
(b) –3 dB and 0.78 rad
(c) 0.707 and –45°
(d) 3 dB and –90° [ESE-2010]

5.51 The transfer function $G(s) = \dfrac{1}{(3s+1)}$ has a corner frequency

(a) 3 rad/s (b) 0.33 rad/s
(c) 1 rad/s (d) 30 rad/s [ESE-2010]

5.52 What are the gain and phase angle of a system having the transfer function $G(s) = (s + 1)$ at a frequency of 1 rad/sec?

(a) 0.41 and 0° (b) 1.41 and 45°
(c) 1.41 and –45° (d) 2.41 and 90°
[ESE-2011]

5.53 For the Bode plot of the system

$G(s) = \dfrac{10}{0.66s^2+2.33s+1}$ the corner frequencies are

(a) 0.66 and 0.33 (b) 0.22 and 2.00
(c) 0.30 and 2.33 (d) 0.50 and 3.00
[ESE-2011]

5.54 If the gain margin of a system in decibels is negative, the system is

(a) Stable
(b) Marginally stable
(c) Unstable
(d) Could be stable or unstable or marginally stable
[ESE-2011]

5.55 System is said to be marginally stable, if

(a) Gain crossover frequency > Phase crossover frequency
(b) Gain crossover frequency = Phase crossover frequency
(c) Gain crossover frequency < Phase crossover frequency
(d) Gain crossover frequency ≠ Phase crossover frequency [ESE-2011]

5.56 **Statement (I):** Nyquist criterion is a powerful tool to determine stability of a closed loop system using open loop transfer function.

Statement (II): Nyquist criterion relates the locations of poles and zeros of the closed loop transfer function.

(a) Both Statement (I) and Statement (II) are individually true and Statement (II) is the correct explanation of Statement (I).
(b) Both Statement (I) and Statement (II) are individually true but Statement (II) is not the correct explanation of Statement (I).
(c) Statement (I) is true but Statement (II) is false.
(d) Statement (I) is false but Statement (II) is true.
[ESE-2012]

5.57 **Statement (I):** All the systems which exhibit overshoot in transient response will also exhibit resonance peak in frequency response.

Statement (II): A large resonance peak in frequency response corresponds to a large overshoot in transient response.

(a) Both Statement (I) and Statement (II) are individually true and Statement (II) is the correct explanation of Statement (I).
(b) Both Statement (I) and Statement (II) are individually true but Statement (II) is not the correct explanation of Statement (I).
(c) Statement (I) is true but Statement (II) is false.
(d) Statement (I) is false but Statement (II) is true.
[ESE-2012]

5.58 Statement (I): The phase angle plot in Bode diagram is not affected by the variation in open loop gain of the system.

Statement (II): The variation in gain of the system has no effect on the phase margin.

(a) Both Statement (I) and Statement (II) are individually true and Statement (II) is the correct explanation of Statement (I).

(b) Both Statement (I) and Statement (II) are individually true but Statement (II) is not the correct explanation of Statement (I).

(c) Statement (I) is true but Statement (II) is false.

(d) Statement (I) is false but Statement (II) is true.

[ESE-2012]

5.59 For a unity feedback control system, if its open-loop transfer function is given by

$G(s)H(s) = \frac{10}{(s+5)^3}$, then its gain margin will

(a) 20 dB (b) 40 dB
(c) 60 dB (d) 80 dB [ESE-2012]

5.60 The gain cross-over frequency and phase margin of the transfer function $\frac{1}{s(s+1)}$ are

(a) 1 rad/s and 45° (b) 2 rad/s and 45°
(c) 2 rad/s and 135° (d) 1 rad/s and 135°

[ESE-2012]

5.61 For a type 1 system, the low frequency asymptote of its Bode plot will have a slope of

(a) 0 dB/decade (b) 6 dB/decade
(c) 20 dB/decade (d) –20 dB/decade

[ESE-2012]

5.62 The unit step response y(t) of a linear system is

$y(t) = (1 - 3e^{-t} + 3e^{-2t})\, u(t)$

For the system function, the frequency at which the forced response becomes zero is

(a) $\frac{1}{\sqrt{2}}$ rad/s (b) $\frac{1}{2}$ rad/s
(c) $\sqrt{2}$ rad/s (d) 2 rad/s [ESE-2012]

5.63 In a feedback control system, phase margin (PM) is

1. Directly proportional to ξ
2. Inversely proportional to ξ
3. Independent of ξ
4. Zero when $\xi = 0$

Which of the above statements are correct?

(a) 1 and 2 (b) 2 and 3
(c) 3 and 4 (d) 1 and 4

[ESE-2013]

5.64 The gain margin in dBs of a unity feedback control system whose open-loop transfer function,

$G(s)H(s) = \frac{1}{s(s+1)}$ is

(a) 0 (b) 1
(c) –1 (d) ∞ [ESE-2013]

5.65 A second order system has

$$\frac{C(s)}{R(s)} = \frac{\omega_n^2}{s^2 + 2\xi\omega_n s + \omega_n^2}$$

Its frequency response will have maximum value at the frequency:

(a) $\omega_n\sqrt{1-\xi^2}$ (b) $\omega_n\xi$
(c) $\omega_n\sqrt{1-2\xi^2}$ (d) zero [ESE-2013]

5.66 The transfer function of a system is $\frac{10}{10+s}$. At a frequency of 0.1 rad/sec, the straight line Bode plot will have a magnitude of

(a) 10 dB (b) 20 dB
(c) 0 dB (d) 40 dB [ESE-2013]

5.67 In control systems, excessive bandwidth is NOT employed because

(a) noise is proportional to bandwidth
(b) it leads to low relative stability
(c) it leads to slower response
(d) noise is proportional to the square of the bandwidth

[ESE-2013]

5.68 For a 3rd order system given below, what is the phase crossover frequency?

$$G(s)\,H(s) = \frac{K}{s^3 + 6s^2 + 11s + 6}$$

(a) $\sqrt{6}$ (b) $\sqrt{11}$
(c) $\pm\sqrt{11}$ (d) $\pm\sqrt{6}$ [ESE-2014]

5.69 The Bode plot of the transfer function $G(s) = s$ is

1. Constant magnitude
2. 20 dB/decade
3. Constant phase shift angle
4. Constant phase shift of $\pi/2$

Which of these are correct?

(a) 1 and 3 (b) 1 and 4
(c) 2 and 3 (d) 2 and 4 [ESE-2015]

5.70 Which of the following statements are correct for Bode plots?
1. Gain margin and phase margin can be easily determined.
2. Absolute and relative stability of only minimum-phase systems can be determined.
3. Gain margin is positive and system is stable, if magnitude of $L(j\omega)$ at phase crossover is positive.
(a) 1, 2 and 3 (b) 1 and 3 only
(c) 1 and 2 only (d) 2 and 3 only
[ESE-2015]

5.71 The roots of the characteristic equation $1 + G(s) H(s) = 0$ are the same as the
(a) Poles of the closed loop transfer function
(b) Poles of the open loop transfer function
(c) Zeros of the closed loop transfer function
(d) Zeros of the open loop transfer function
[ESE-2015]

5.72 The transfer function $G(s) = \dfrac{10(s-1)}{(s+10)}$ represents
(a) Unstable system
(b) Minimum phase system
(c) Non-minimum phase system
(d) PID controller system [ESE-2016]

5.73 For a unity feedback system with open-loop transfer function $\dfrac{25}{s(s+6)}$, the resonant peak at output M_m and the corresponding resonant frequency ω_m, are respectively
(a) 2.6 and 2.67 r/s (b) 1.04 and 2.67 r/s
(c) 2.6 and 4.8 r/s (d) 1.04 and 4.8 r/s
[ESE-2016]

5.74 Consider the transfer function

$$G(s) = \frac{5(s^2+10s+100)}{s^2(s^2+15s+1)}$$

The corner frequencies in Bode's plot for this transfer function are as
(a) 10 r/s and 10 r/s (b) 100 r/s and 10 r/s
(c) 10 r/s and 1 r/s (d) 100 r/s and 1 r/s
[ESE-2016]

5.75 Consider the transfer function (0.1 + 0.01s) for a PD controller. What is the frequency at which the magnitude is 20 dB (by using asymptotic Bode's plot)?
(a) 2000 r/s (b) 1000 r/s
(c) 200 r/s (d) 100 r/s
[ESE-2016]

5.76 The open-loop transfer function of a unity feedback system is $G(s) = \dfrac{K}{s(s+5)}$. The gain K that results in a phase margin of 45° is
(a) 35 (b) 30
(c) 25 (d) 20 [ESE-2016]

5.77 Consider the following statements :
The Gain margin and Phase margin of an unstable system may respectively be
1. Positive, negative
2. Negative, positive
3. Negative, negative
Which of the above statements is/are correct?
(a) 3 only (b) 1 and 2 only
(c) 2 and 3 only (d) 1, 2 and 3
[ESE-2016]

5.78 The open-loop transfer function of a unity feedback system is $G(s)\dfrac{10(1+0.2s)}{(1+0.5s)}$. The phase shift at $\omega = 0$ and $\omega = \infty$, will be, respectively
(a) –90° and 180° (b) 0° and 180°
(c) –90° and 90° (d) 0° and 0°
[ESE-2016]

5.79 The Bode plot of the open-loop transfer function of a system is described as follows:
- Slope – 40 dB/decade ; $\omega < 0.1$ rad/s
- Slope – 20 dB/decade ; $0.1 < \omega < 10$ rad/s
- Slope 0 ; $\omega > 10$ rad/s

The system described will have
(a) 1 pole and 2 zeros
(b) 2 poles and 2 zeros
(c) 2 poles and 1 zero
(d) 1 pole and 1 zero [ESE-2016]

5.80 The transfer function of a control system is said to be 'All Pass System', if it has
(a) Unit magnitude at all frequencies with anti-symmetric pole-zero pattern

(b) Unit magnitude at all frequencies with symmetric pole-zero pattern
(c) Magnitude varying with frequency and with anti-symmetric pole-zero pattern
(d) Unit magnitude at some frequencies with symmetric pole-zero pattern

[ESE-2016]

5.81 Consider the following statements regarding 'relative stability':

It is defined

1. in terms of gain margin only
2. in terms of phase margin and certain other parameters
3. in terms of gain margin, phase margin and location of poles in s-plane
4. in relation to another identified system

Which of the above statements are correct?

(a) 1 and 2 (b) 2 and 3
(c) 3 and 4 (d) 1 and 4

[ESE-2017]

5.82 The magnitude plot for the open-loop transfer function of a control system is shown in the figure given below:

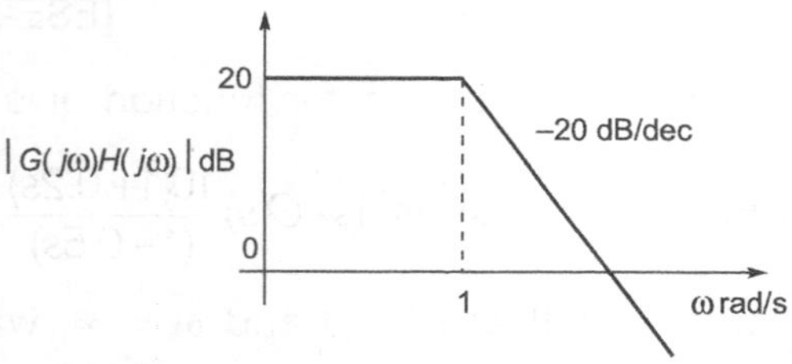

Its open-loop transfer function, $G(s)H(s)$, is

(a) $10(s+1)$ (b) $\frac{1}{s+1}$
(c) $\frac{10}{s+1}$ (d) $20(s+1)$

[ESE-2017]

5.83 **Statement (I):** Elements with non-minimum phase transfer functions introduce large phase lags with increasing frequency resulting in complex compensation problems.

Statement (II): Transportation lag commonly encountered in process control systems is a non-minimum phase element.

(a) Both Statement (I) and Statement (II) are individually true and Statement (II) is the correct explanation of Statement (I).
(b) Both Statement (I) and Statement (II) are individually true but Statement (II) is not the correct explanation of Statement (I).
(c) Statement (I) is true but Statement (II) is false.
(d) Statement (I) is false but Statement (II) is true.

[ESE-2017]

5.84 The low-frequency asymptote in the Bode plot of

$$G(s) = \frac{6(s^2+10s+100)}{s^2(50s^2+15s+1)}$$

has a slope of

(a) –10 dB/dec (b) –20 dB/dec
(c) –40 dB/dec (d) –60 dB/dec

[ESE-2018]

5.85 The open-loop transfer function $G(s)H(s)$ of the Bode plot as shown in the figure is

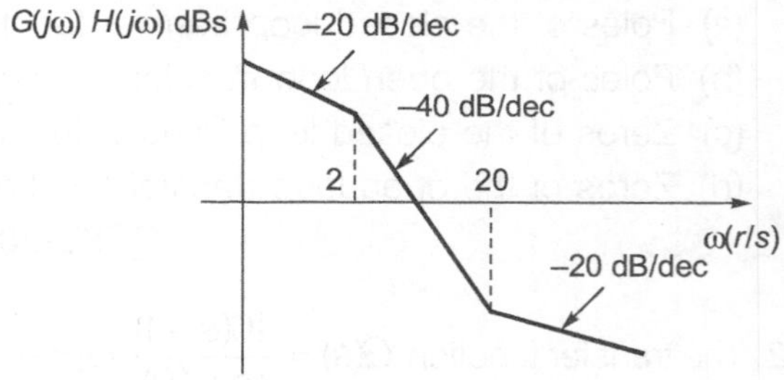

(a) $\frac{Ks(s+2)}{s+20}$ (b) $\frac{K(s+20)}{s(s+2)}$
(c) $\frac{K(s+2)}{s(s+20)}$ (d) $\frac{Ks(s+20)}{s+2}$

[ESE-2018]

5.86 Which one of the following transfer functions represents the Bode plot as shown in the figure (where K is constant)?

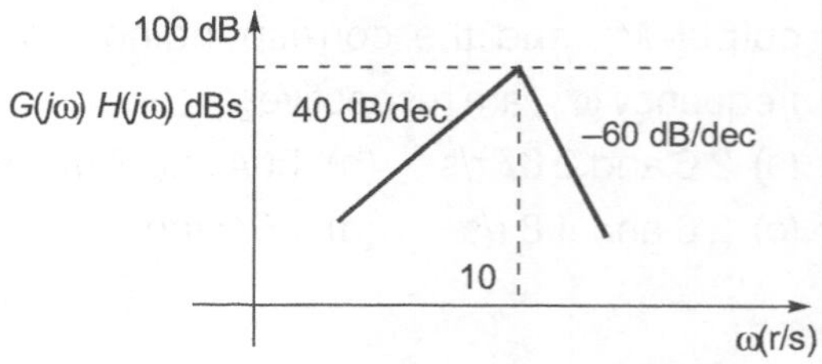

(a) $\frac{Ks^2}{\left(1+\frac{s}{10}\right)^3}$ (b) $\frac{Ks^2}{\left(1+\frac{s}{10}\right)^4}$
(c) $\frac{Ks^2}{\left(1+\frac{s}{10}\right)^5}$ (d) $\frac{Ks^2}{\left(1+\frac{s}{10}\right)^2}$

[ESE-2018]

5.87 The open-loop transfer function of a system has two poles on the imaginary axis, one in the left-half and the other in the right-half, together with a zero at the origin of coordinates and also two zeros in the left-half of the s-plane. The closed-loop response for unity feedback will be stable if the encirclement of the critical point $(-1, j0)$ is

(a) –1 (b) +1
(c) –2 (d) +2

[ESE-2018]

5.88 **Statement (I):** The transportation lag in a system can be easily handled by using Bode plot.

Statement (II): The magnitude plot is unaffected, and only the phase plot shifts by $-\omega T$ rad due to the presence of e^{-st}.

(a) Both Statement I and Statement II are individually true and Statement II is the correct explanation of Statement I
(b) Both Statement I and Statement II are individually true but Statement II is not the correct explanation of Statement I
(c) Statement I is true but Statement II is false
(d) Statement I is false but Statement II is true

[ESE-2018]

5.89 In case of high-pass filter, the transfer function should be with

1. 1 pole and 1 zero
2. 2 poles and 2 zeros
3. 2 poles and 1 zero

Which of the above are correct?

(a) 1, 2 and 3 (b) 1 and 3 only
(c) 2 and 3 only (d) 1 and 2 only

[ESE-2018]

5.90 The magnitude and phase relationship between the sinusoidal input and the steady-state output of a system is called as

(a) magnitude response
(b) transient response
(c) steady-state response
(d) frequency response

[ESE-2019]

5.91 A transfer function having all its poles and zeros only in the left-half of the s-plane is called

(a) a minimum-phase function
(b) a complex transfer function
(c) an all-pass transfer function
(d) a maximum-phase transfer function

[ESE-2019]

5.92 The frequency where magnitude M has a peak value in frequency response is known as

(a) normalized frequency
(b) resonant frequency
(c) peak frequency
(d) tuned frequency

[ESE-2019]

5.93 **Statement (I):** For an unstable feedback system, the gain margin is negative or the phase margin is positive.

Statement (II): For a stable feedback system, both gain margin and phase margin must be positive.

(a) Both Statement (I) and Statement (II) are individually true and Statement (II) is the correct explanation of Statement (I)
(b) Both Statement (I) and Statement (II) are individually true but Statement (II) is **NOT** the correct explanation of Statement (I)
(c) Statement (I) is true but Statement (II) is false
(d) Statement (I) is false but Statement (II) is true

[ESE-2020]

5.94 Which one of the following statements is correct regarding constant N circles?

(a) The locus of constant, closed-loop magnitude frequency response for unity feedback systems.
(b) The locus of constant, closed-loop phase frequency response for unity feedback systems.
(c) A subsystem inserted into the forward or feedback path for the purpose of improving the transient response or steady-state error.
(d) A system that monitors its output and corrects for disturbances. It is characterized by feedback paths from the output.

[ESE-2021]

5.95 A system has poles at 0.01 Hz, 1 Hz and 80 Hz; zeros at 5 Hz, 100 Hz and 200 Hz. The approximate phase of the system response at 20 Hz is

(a) –90° (b) 0°
(c) 90° (d) –180°

[ESE-2023]

5.96 The magnitude of frequency response of an underdamped second-order system is 5 at 0 rad/sec and peaks at $\frac{10}{\sqrt{3}}$ at $5\sqrt{2}$ rad/sec. The transfer function of the system is

(a) $\dfrac{100}{s^2+10s+100}$ (b) $\dfrac{375}{s^2+5s+75}$

(c) $\dfrac{500}{s^2+12s+100}$ (d) $\dfrac{1125}{s^2+25s+225}$

[ESE-2023]

5.97 By considering standard notations, the peak value of the magnitude in the resonant peak M_r is

(a) $\dfrac{2}{\zeta\sqrt{1-\zeta^2}}$ (b) $\dfrac{1}{\zeta\sqrt{2-\zeta^2}}$

(c) $\dfrac{1}{2\zeta\sqrt{1-\zeta^2}}$ (d) $\dfrac{1}{\zeta\sqrt{1-\zeta^2}}$

[ESE-2023]

5.98 The phase margin of a system having the loop transfer function $G(s)H(s)=\dfrac{2\sqrt{3}}{s(s+1)}$ is

(a) 45° (b) 90°
(c) 30° (d) 60°

[ESE-2023]

5.99 The phase margin of a system with the open-loop transfer function $G(s)H(s)=\dfrac{1-s}{(s+1)(s+2)}$ is

(a) 0° (b) 63.4°
(c) 90° (d) ∞°

[ESE-2023]

5.100 The Nyquist criterion for stability states that an amplifier is unstable if the Nyquist curve encloses the $-1+j0$ point, and the amplifier is stable if the curve does not enclose this point. If $A\beta$

(a) extends outside this circle, the feedback is negative.
(b) lies within this circle, then $|1+A\beta|<1$, and the feedback is negative.
(c) does not enclose the point $-1+j0$, i.e., $|1+A\beta|>1$, then the amplifier is unstable and the feedback is negative for all frequencies.
(d) extends inside this circle, the feedback is negative.

[ESE-2023]

5.101 Consider the following statements regarding the Bode plot for the type-2 system having transfer function of the form:

$$G(j\omega)=\frac{K}{(j\omega)^2\,(1+j\omega T)}$$

1. The intercept of the low frequency slope of –40 dB/decade (or its extension) with the 0 dB axis occurs at a frequency ω_X, where $\omega_X = K$
2. The value on the low frequency slope of –40 dB/decade (or its extension) at the frequency $\omega = 1$, is equal to 40 log (K).
3. The gain K is the static ramp error coefficient.

Which of the above statements is/are not correct?

(a) 1 only (b) 1 and 2
(c) 1 and 3 (d) 2 and 3

[ESE-2024]

5.102 Consider the following statements:

1. Gain margin is used to indicate the proximity of the intersection of the negative real axis made by the polar plot of $G(j\omega)H(j\omega)$ to the $(-1+j)$ point.
2. Phase crossover point is the point in the $G(j\omega)$ plane at which the polar $G(j\omega)H(j\omega)$ plot intersects the positive real axis.
3. The phase crossover frequency is the frequency at the phase crossover point or where angle $\{G(j\omega)H(j\omega)\} = 180°$.

Which of the above statements is/are not correct?

(a) 2 only (b) 1 and 2 only
(c) 1 and 3 only (d) 2 and 3 only

[ESE-2024]

■■■

Answers Frequency Domain Analysis

5.1	(d)	5.2	(a)	5.3	(b)	5.4	(b)	5.5	(a)	5.6	(d)	5.7	(d)	5.8	(a)	5.9	(c)
5.10	(b)	5.11	(b)	5.12	(a)	5.13	(a)	5.14	(*)	5.15	(d)	5.16	(b)	5.17	(b)	5.18	(b)
5.19	(c)	5.20	(c)	5.21	(b)	5.22	(d)	5.23	(a)	5.24	(d)	5.25	(d)	5.26	(c)	5.27	(c)
5.28	(d)	5.29	(c)	5.30	(c)	5.31	(c)	5.32	(d)	5.33	(a)	5.34	(a)	5.35	(c)	5.36	(d)
5.37	(b)	5.38	(b)	5.39	(a)	5.40	(c)	5.41	(b)	5.42	(a)	5.43	(b)	5.44	(d)	5.45	(a)
5.46	(c)	5.47	(a)	5.48	(d)	5.49	(a)	5.50	(c)	5.51	(b)	5.52	(b)	5.53	(d)	5.54	(c)
5.55	(b)	5.56	(c)	5.57	(d)	5.58	(c)	5.59	(b)	5.60	(a)	5.61	(d)	5.62	(c)	5.63	(d)
5.64	(d)	5.65	(c)	5.66	(b)	5.67	(a)	5.68	(b)	5.69	(d)	5.70	(c)	5.71	(a)	5.72	(c)
5.73	(b)	5.74	(c)	5.75	(b)	5.76	(a)	5.77	(d)	5.78	(d)	5.79	(b)	5.80	(b)	5.81	(c)
5.82	(c)	5.83	(a)	5.84	(c)	5.85	(b)	5.86	(c)	5.87	(b)	5.88	(a)	5.89	(d)	5.90	(d)
5.91	(a)	5.92	(b)	5.93	(d)	5.94	(b)	5.95	(a)	5.96	(c)	5.97	(c)	5.98	(c)	5.99	(d)
5.100	(b)	5.101	(*)	5.102	(b)												

Explanations Frequency Domain Analysis

5.1 (d)

Bode → Asymptotic plots
Evans → Root - locus technique
Nyquist → Polar plots

5.2 (a)

Phase margin = 180° – 30° = 150°

$$\text{Gain margin} = \frac{1}{0.25} = 4$$

5.3 (b)

At ω_{pc}, $\angle G(j\omega) = -180°$

$$\Rightarrow -\tan^{-1}\omega_{pc} - \tan^{-1}2\omega_{pc} - \tan^{-1}3\omega_{pc} = -180°$$

$$\Rightarrow \tan^{-1}\omega_{pc} + \tan^{-1}2\omega_{pc} = 180 - \tan^{-1}3\omega_{pc}$$

$$\Rightarrow \tan^{-1}\left(\frac{3\omega_{pc}}{1-2\omega_{pc}^2}\right) = 180 - \tan^{-1}3\omega_{pc}$$

$$\Rightarrow \frac{3\omega_{pc}}{1-2\omega_{pc}^2} = -3\omega_{pc}$$

$$\Rightarrow 1 - 2\omega_{pc}^2 = -1$$

$$\Rightarrow 2\omega_{pc}^2 = 2$$

$$\Rightarrow \omega_{pc}^2 = 1$$

$$\Rightarrow \omega_{pc} = 1 \text{ rad/s}$$

5.4 (b)

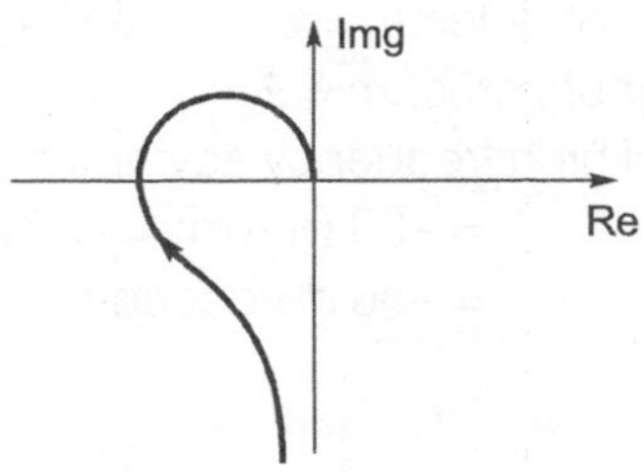

$$G(s)H(s) = \frac{1}{sT_1(1+sT_2)(1+sT_3)}$$

Note:

(i) When a pole is added at negative real axis, the tail of the plot remains at the same position whereas head of plot shifts by 90° in clockwise direction.

(ii) When a pole is added at origin, both the tail and head of the plot shift by 90° in clockwise direction.

5.5 (a)

$$G(j\omega) = \frac{1}{j\omega(1+j\omega T_1)(1+j\omega T_2)}$$

At ω_{pc}, $\angle G(j\omega) = -180°$

$$-90° - \tan^{-1}\omega T_1 - \tan^{-1}\omega T_2 = -180°$$

$$\Rightarrow \tan^{-1}\frac{\omega T_1 + \omega T_2}{1-\omega^2 T_1 T_2} = 90°$$

$$\Rightarrow \frac{\omega T_1 + \omega T_2}{1-\omega^2 T_1 T_2} = \tan 90° = \infty$$

$$\Rightarrow 1-\omega^2 T_1 T_2 = 0$$

$$\Rightarrow \omega_{pc} = \frac{1}{\sqrt{T_1 T_2}}$$

$$|G(j\omega)|_{\omega=\omega_{pc}} = \frac{1}{\omega_{pc}\sqrt{1+\omega_{pc}^2 T_1^2}\sqrt{1+\omega_{pc}^2 T_2^2}}$$

$$= \frac{1}{\frac{1}{\sqrt{T_1 T_2}}\cdot\sqrt{1+\frac{T_1}{T_2}}\cdot\sqrt{1+\frac{T_2}{T_1}}}$$

$$= \frac{T_1 T_2}{T_1 + T_2}$$

$$GM = \left.\frac{1}{|G(j\omega)|}\right|_{\omega=\omega_{pc}} = \frac{T_1+T_2}{T_1 T_2}$$

5.6 (d)

Number of poles, $n = 6$
Number of zeros, $m = 2$
Slope of high frequency asymptote
= –20 (n – m) = –20 (6 – 2)
= –80 dB/decade

5.7 (d)

(i) When transfer function has at least one pole or zero in the RHS of s-plane, it is called **non-minimum phase transfer function.**

(ii) When transfer function has no pole or zero in the RHS of s-plane. It is called **minimum phase transfer function.**

5.8 (a)

Nyquist plot approximated is shown below:

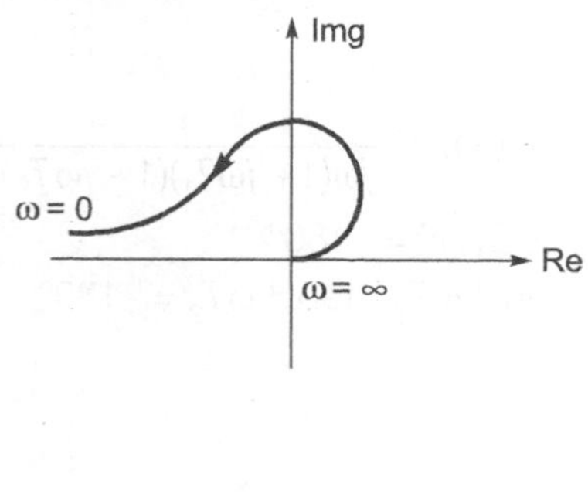

5.9 (c)

$$G(s)H(s) = \frac{5}{s(1+0.1s)(1+0.01s)}$$

Nyquist diagram is

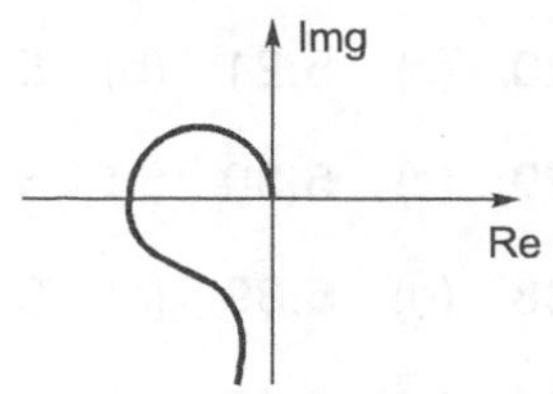

5.14 (*)

Phase lag compensation has the following features:
(i) Pole is nearer to origin
(ii) Bandwidth reduces
(iii) Gain crossover frequency reduces
(iv) Phase crossover frequency reduces
(v) Resonance peak reduces
(vi) Steady state error reduces
(vii) Rise time increases

5.15 (d)

$$GM = \left.\frac{1}{|G(j\omega)|}\right|_{\omega=\omega_{pc}} = \frac{1}{0.64}$$

$$GM = 20\log\left(\frac{1}{0.64}\right) = 3.86 \text{ dB}$$

$$PM = 180° + \angle G(j\omega)|_{\omega=\omega_{gc}}$$

$$= 180° - 170°$$

$$\Rightarrow PM = 10°$$

5.16 (b)

When a pole is added at negative real axis, the tail of the plot remains at same position whereas the head of plot shifts by 90° in clockwise direction.

5.17 (b)

The open-loop transfer function for a unity feedback second-order system is

$$G(s)H(s) = \frac{\omega_n^2}{s(s+2\zeta\omega_n)}$$

$$G(j\omega)H(j\omega) = \frac{\omega_n^2}{j\omega(j\omega+2\zeta\omega_n)}$$

At gain crossover frequency $\omega = \omega_{gc}$,
$|G(j\omega)H(j\omega)| = 1$

$$\Rightarrow \frac{\omega_n^2}{\omega_{gc}\sqrt{\omega_{gc}^2 + 4\zeta^2\omega_n^2}} = 1$$

$$\Rightarrow \quad \omega_n^4 = \omega_{gc}^2(\omega_{gc}^2 + 4\zeta^2\omega_n^2)$$

$$\Rightarrow \omega_{gc}^4 + 4\zeta^2\omega_n^2\omega_{gc}^2 - \omega_n^4 = 0$$

Dividing by ω_n^4:

$$\left(\frac{\omega_{gc}}{\omega_n}\right)^4 + 4\zeta^2\left(\frac{\omega_{gc}}{\omega_n}\right)^2 - 1 = 0$$

$$\Rightarrow \quad \left(\frac{\omega_{gc}}{\omega_n}\right)^2 = \frac{-4\zeta^2 + \sqrt{16\zeta^4 + 4}}{2}$$

$$\Rightarrow \quad \left(\frac{\omega_{gc}}{\omega_n}\right)^2 = -2\zeta^2 + \sqrt{4\zeta^4 + 1}$$

Negative sign has been discarded as square cannot be negative.

$$PM = 180° + \angle G(j\omega)H(j\omega)\big|_{\omega = \omega_{gc}}$$

$$= 180° - 90° - \tan^{-1}\left(\frac{\omega}{2\zeta\omega_n}\right)$$

$$= 90° - \tan^{-1}\left\{\frac{\sqrt{4\zeta^4 + 1} - 2\zeta^2}{2\zeta}\right\}$$

$$= \cot^{-1}\left\{\frac{-2\zeta^2 + \sqrt{1 + 4\zeta^4}}{2\zeta}\right\}$$

$$\left[\because \tan^{-1}x + \cot^{-1}x = \frac{\pi}{2}\right]$$

$$PM = \tan^{-1}\left\{\frac{2\zeta}{\sqrt{-2\zeta^2 + \sqrt{1 + 4\zeta^4}}}\right\}$$

$$\left[\because \tan^{-1}x = \cot^{-1}\left(\frac{1}{x}\right)\right]$$

5.18 (b)

(i) The type of the system can be found out by the initial slope. The type of the system is n where the initial slope is $-20n$ dB/decade.

(ii) $\omega_2 \neq \frac{\omega_1 + \omega_3}{2}$

5.19 (c)

The slope at high frequency range is $-20(n - m)$ dB/decade.

where n = number of poles

m = number of zeros

$\because \quad n = 3, m = 0$

$\therefore \quad$ slope $= -20 \times 3 = -60$ dB/decade

5.20 (c)

$$G(s) = \frac{100}{s^2 + 10s + 100}$$

$$\omega_n = \sqrt{100} \Rightarrow \omega_n = 10 \text{ rad/s}$$

$$\xi = \frac{10}{2\omega_n} \Rightarrow \xi = \frac{10}{2 \times 10} = 0.5$$

$$\omega_r = \omega_n\sqrt{1 - 2\xi^2} = 10\sqrt{1 - 2(0.5)^2}$$

$$\omega_r = 7.07 \text{ rad/s}$$

5.22 (d)

Polar plot of the given transfer function is shown below:

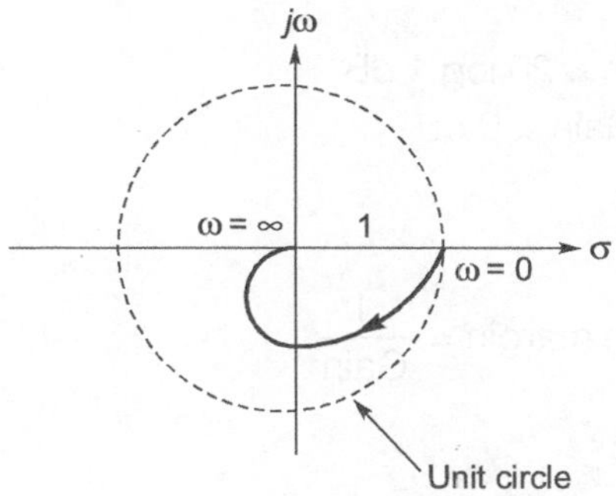

$$\angle G(j\omega)\big|_{\omega = \omega_{gc}} = 0°$$

$$PM = 180° + \angle G(j\omega)\big|_{\omega = \omega_{gc}}$$

$$PM = 180°$$

5.25 (d)

$$X(s) = \frac{1}{(1 + s)}$$

The corner frequency is $\omega = 1$ rad/sec.

$$X(j\omega) = \frac{1}{1 + j\omega} = \frac{1}{(1 + \omega^2)^{1/2}}\angle -\tan^{-1}\omega$$

The log magnitude is

$$20 \log|G(j\omega)| = 20 \log(1 + \omega^2)^{-1/2}$$

$$= -10 \log(1 + \omega^2)$$

For low frequencies ($\omega << 1$), the log magnitude is approximated as

$$20 \log|G(j\omega)| = -10 \log 1 = 0 \text{dB}$$

For high frequencies ($\omega >> 1$), the log magnitude is approximated as $20 \log|G(j\omega)| = -20 \log \omega$. So, the plot is a straight line with a slope of -20 dB/decade for $\omega > 1$.

5.26 (c)

The sharper the cutoff characteristics, the larger the peak resonance and the lesser stable the system.

5.27 (c)

For a minimum phase system to be stable, both phase margin and gain margin should be positive.

5.28 (d)

For a stable closed loop system, the gain at phase crossover frequency should be less than 1.

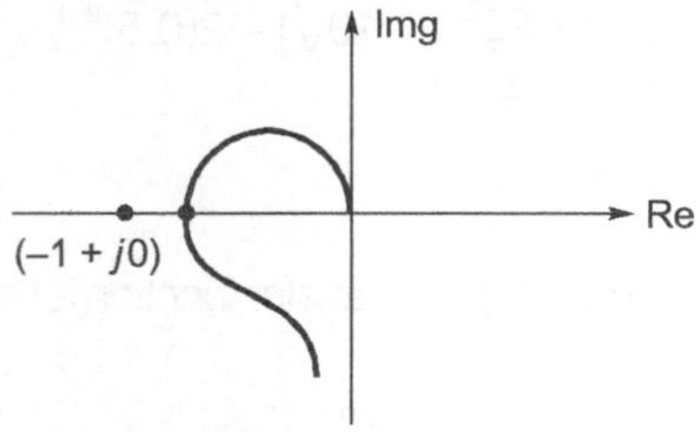

Gain < 20 log 1 dB

⇒ Gain < 0 dB

5.29 (c)

$$\text{Gain margin} = \frac{1}{\text{Gain}}$$

5.30 (c)

(i) Rise time $\propto \dfrac{1}{BW}$

Speed of time response ∝ Bandwidth

(ii) Phase margin = $180° + \phi_r$

$$= 180° - \tan^{-1}\frac{\sqrt{1-2\xi^2}}{\xi}$$

(iii) Response peak is called overshoot.

(iv) Gain margin tells about the stability of the system.

5.31 (c)

$$M_r = \frac{1}{2\xi\sqrt{1-\xi^2}} = \frac{1}{2\times 0.421\sqrt{1-(0.421)^2}}$$

$M_r = 1.30$

5.32 (d)

When transfer function has no pole and zero in RHS of *s*-plane, it is called Minimum phase transfer function.

5.33 (a)

(i) Error in log-magnitude for $0 < \omega \le (1/T)$ is given by $-10 \log (1 + \omega^2T^2) + 10 \log 1$. Therefore, error at corner frequency $\omega = 1/T$ is $-10 \log (1 + 1) + 10 \log 1 = -3$ dB. So, the deviation of the actual magnitude response for a zero on real axis is 3dB at the corner frequency.

(ii) Quadratic factor for a complex conjugate poles is

$$\frac{1}{1+j2\xi u - u^2} \quad \text{where } u = \frac{\omega}{\omega_n}$$

Phase angle of quadratic factor at undamped. frequency, i.e.

$$\omega = \omega_n \left(\Rightarrow u = \frac{\omega}{\omega_n} = 1\right)$$

$$\phi = -\tan^{-1}\left(\frac{2\xi}{1-1}\right) = -\tan^{-1}\infty = -90°.$$

So phase angle is independent of ξ.

5.34 (a)

$$\angle G(j\omega) = -90° - \tan^{-1}\omega - \tan^{-1}2\omega = -180°$$

$$\Rightarrow \tan^{-1}\left(\frac{\omega + 2\omega}{1-2\omega^2}\right) = 90°$$

$$\Rightarrow \quad 1 - 2\omega^2 = 0$$

$$\Rightarrow \quad \omega = \frac{1}{\sqrt{2}}\text{rad/s}$$

$$|G(j\omega)|_{\omega=\frac{1}{\sqrt{2}}} = \frac{2}{\frac{1}{\sqrt{2}}\cdot\sqrt{\frac{1}{2}+1}\cdot\sqrt{\frac{4}{2}+1}}$$

$$= \frac{2\sqrt{2}}{\frac{\sqrt{3}}{\sqrt{2}}\cdot\frac{\sqrt{6}}{\sqrt{2}}} = \frac{4}{3}$$

$$\text{Gain margin} = \left.\frac{1}{|G(j\omega)|}\right|_{\omega=\omega_{pc}} = \frac{3}{4}$$

5.35 (c)

Routh array is

s^4	1	3
s^3	2	2
s^2	2	K
s^1	$2-K$	0
s^0	K	

For oscillations,

$$2 - K = 0$$

$$\Rightarrow \quad K = 2$$

For oscillations,

$$2s^2 + K = 0$$

Putting $s = j\omega$ and $K = 2$,

$$-2\omega^2 + 2 = 0$$

$$\Rightarrow \quad \omega^2 = 1$$

$$\Rightarrow \quad \omega = 1 \text{ rad/s}$$

5.36 (d)

For a stable system, both GM and PM should be positive.

5.38 (b)

e^{-sT} changes phase cross-over frequency hence gain margin also changes.

The factor exp $(-st)$ is the cause of the term transportation lag (time delay). The effect of e^{-st} term is simply to rotate each point of the $G(s)$ $H(s)$ plot by an angle of ωT rad in the clockwise direction. So the phase margin of the system reduces as T increases.

5.39 (a)

(i) Phase Margin,

$$PM = 180° - \tan^{-1}\frac{\sqrt{1-2\xi^2}}{\xi}$$

Thus, the phase margin is related to effective damping of the system.

(ii) Gain Margin: It is the factor by which the system gain can be increased to drive it to the verge of instability.

$$GM = 1/a$$

where $a = |G(j\omega)H(j\omega)|_{\omega=\omega_2}$

$$GM|_{dB} = -20\log_{10} a$$

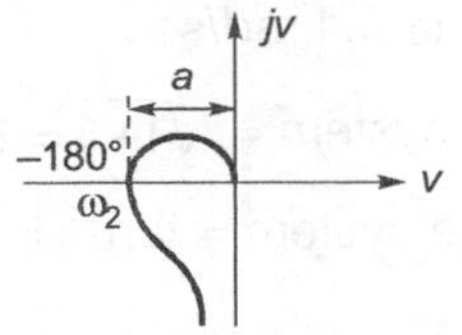

For a stable system, $a < 1$. Therefore, for a stable system, GM(dB) should always be positive.

5.40 (c)

(i) When a pole is added at origin the tail and head of the plot shift by 90° in clockwise direction.

(ii) When a pole is added at negative real axis, the tail of the pole remains at same position whereas head of plot is shifted by 90° in clockwise direction.

Polar plot of $\frac{1}{1+j\omega T}$ is shown below:

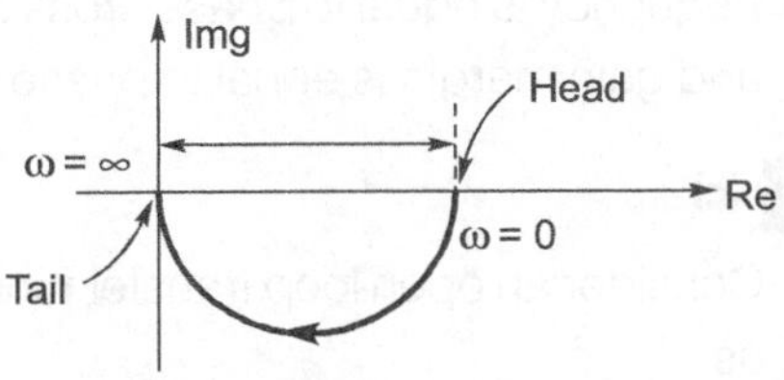

When two poles are added at origin, head and tail both will shift by 90° × 2 = 180° in the clockwise direction.

Therefore, polar plot of

$$G(j\omega) = \frac{1}{(j\omega)^2(1+j\omega T)}$$ is shown below:

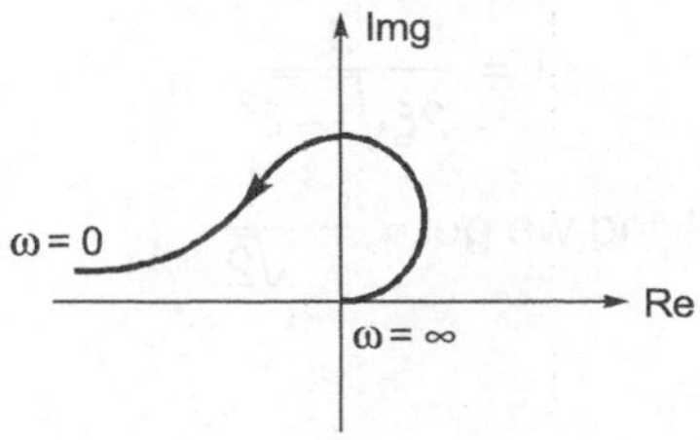

5.41 (b)

System with G.M. ≈ 1
P.M. ≈ 0
is oscillatory.

5.43 (b)

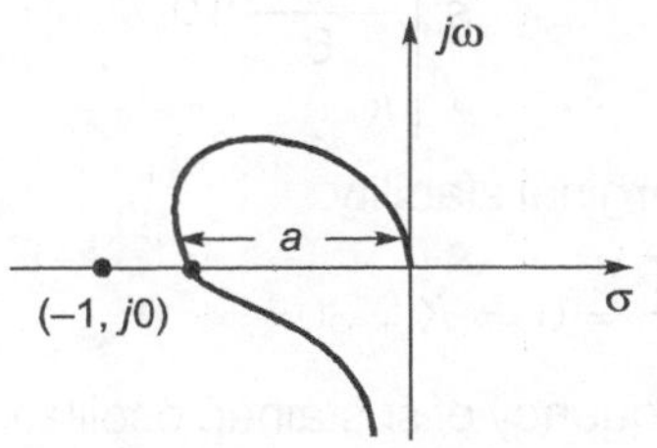

$GM = 20\log(1/a)$

for $a < 1$, GM > 0 dB

for $a = 1$, GM = 0 dB

for $a > 1$, GM < 0 dB

5.44 (d)

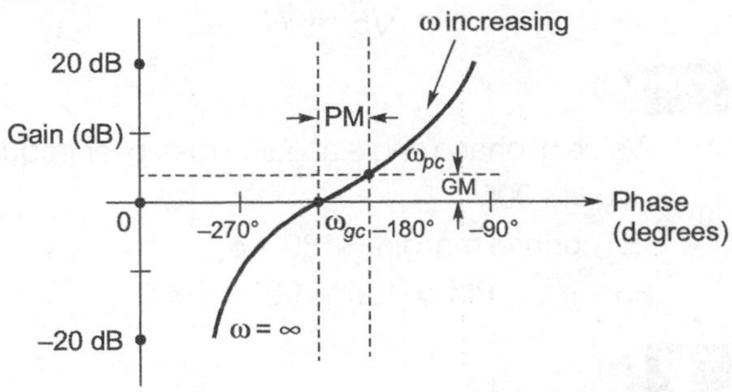

Since $\omega_{pc} < \omega_{gc}$ ... system is unstable

At ω_{pc}, GM < 0

At ω_{gc}, $PM = 180 + \angle GH|_{\omega_{gc}}$

$\therefore \angle GH|_{\omega_{gc}} < -180°$

PM < 0

5.45 (a)

$$GM = \frac{1}{a} \quad \text{(where } a = \text{gain at phase crossover)}$$

5.46 (c)

$$M_r = \frac{1}{2\xi\sqrt{1-\xi^2}}$$

$$1 = \frac{1}{2\xi\sqrt{1-\xi^2}}$$

On solving we get, $\xi = \frac{1}{\sqrt{2}}$

5.47 (a)

Characteristic equation is

$s(s + 1)(s + 5) + K = 0$

i.e. $s^3 + 6s^2 + 5s + K = 0$

Routh array

s^3	1	5
s^2	6	K
s^1	$\frac{30-K}{6}$	0
s^0	K	

For marginal stability

$$\frac{30-K}{6} = 0 \Rightarrow K = 30$$

For frequency of sustained oscillation

$$6s^2 + K = 0$$
$$\Rightarrow 6s^2 + 30 = 0$$
$$\Rightarrow s^2 + 5 = 0$$
$$\Rightarrow (j\omega)^2 + 5 = 0$$
$$\Rightarrow -\omega^2 + 5 = 0$$
$$\Rightarrow \omega^2 = 5$$
$$\Rightarrow \omega = \sqrt{5} \text{ rad/s}$$

5.48 (d)

Value of phase angle at gain cross-over frequency $= \phi = -90°$

$\therefore$ phase margin $= 180 + \phi$

$\Rightarrow$ PM $= 180° - 90° = +90°$

5.49 (a)

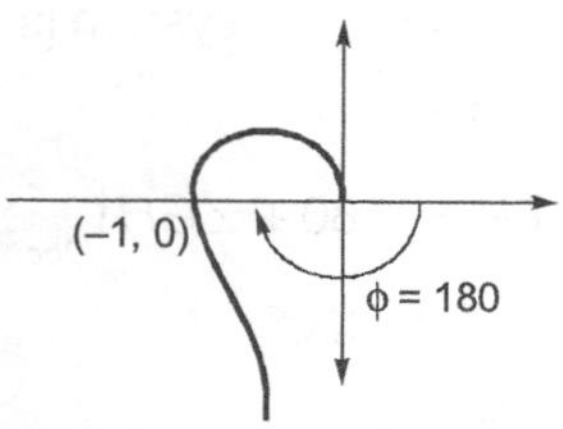

PM $= 180 + \phi = 180 - 180 = 0$

5.50 (c)

$$G(s) = \frac{1}{s+1}$$

$$G(j\omega) = \frac{1}{1+j\omega}$$

$$|G(j\omega)| = \frac{1}{\sqrt{1+\omega^2}}$$

and $\angle G(j\omega) = -\tan^{-1}\omega$

$$|G(j\omega)|_{\omega=1} = \frac{1}{\sqrt{2}}$$

and $\angle G(j\omega)|_{\omega=1} = -\tan^{-1} 1 = -45°$

5.51 (b)

$$G(s) = \frac{1}{(3s+1)} = \frac{1}{(1+3s)}$$

$$T = 3$$

$\Rightarrow$ Corner frequency $= \frac{1}{3} = 0.33$ rad/sec

5.52 (b)

$$G(s) = s + 1$$

Put $s = j\omega$

$$G(j\omega) = j\omega + 1$$

$$G(j\omega) = \sqrt{\omega^2+1}\angle\tan^{-1}\omega$$

At $\omega = 1$ rad/sec

Gain of the system $= \sqrt{1+1} = 1.41$

Phase of the system $= \tan^{-1} 1 = 45°$.

5.53 (d)

$$G(s) = \frac{10}{0.66s^2 + 2.33s + 1} = \frac{10}{(s+0.5)(s+3.03)}$$

Hence corner frequencies are 0.5 and 3.03 rad/sec.

5.54 (c)

System will be stable, only when gain margin in dB and phase margin in degrees both are positive.

5.55 (b)

System will be marginally stable if gain cross over frequency is equal to phase cross over frequency and gain margin is equal to phase margin also.

5.56 (c)

Consider an open loop transfer function $G(s)H(s)$ as

$$G(s)H(s) = \frac{K(s+z_1)(s+z_2)\cdots(s+z_m)}{(s+p_1)(s+p_2)\cdots(s+p_n)}, m < n$$

The characteristic equation of the transfer function is given as:

$1 + G(s)\,H(s) = 0 = q(s)$

$$q(s) = 1 + \frac{k(s+z_1)(s+z_2)\cdots(s+z_m)}{(s+p_1)(s+p_2)\cdots(s+p_n)}$$

$$= \frac{(s+z_1')(s+z_2')\cdots(s+z_n')}{(s+p_1)(s+p_2)\cdots(s+p_n)}$$

- Numerator of above equation determines closed loop poles because characteristic equation determine the closed loop poles. Denominator of above equation determines the open loop poles.
- Observing the encirclement about origin for $1 + G(s)\,H(s) = 0$ is same as the observing encirclement about –1 for $G(s)\,H(s)$. (i.e. open loop transfer function) to determine stability of a closed loop system.

5.58 (c)

Let the open-loop transfer function is

$$G(s)\,H(s) = \frac{K}{s(s+\omega_c)}$$

$$\angle G(s)H(s) = -90° - \tan^{-1}\frac{\omega}{\omega_c} \quad \text{...(i)}$$

Thus it can be seen from equation (i) that phase angle does not depend on the gain of the system.

Statement 2: Phase margin depends on the gain cross-over frequency ω_{gc} and ω_{gc} can be calculated as

$$|G(j\omega)\,H(j\omega)|_{\omega=\omega_{gc}} = 1$$

$$\frac{K}{\omega\sqrt{\omega^2+\omega_c^2}} = 1$$

and $\quad PM = 180° + \phi$

$$\phi = -90° - \tan^{-1}\frac{\omega_{gc}}{\omega_c}$$

Thus ϕ depends on the ω_{gc} and ω_{gc} depends on the gain K. So variation in gain affects the phase margin.

5.59 (b)

To calculate the gain, we need to first calculate the phase cross-over frequency.

$$G(s)H(s) = \frac{10}{(s+5)^3} = \frac{10}{(j\omega+5)^3}$$

So $\quad |\angle G(s)H(s)|_{\omega=\omega_{pc}} = -180°$

$$\Rightarrow \quad -3\tan^{-1}\frac{\omega}{5} = -180°$$

$$\therefore \quad \omega_{pc} = 5\sqrt{3}\text{ rad/sec.}$$

$$X = |G(j\omega)H(j\omega)|_{\omega=5\sqrt{3}}$$

$$= \frac{1}{(\sqrt{\omega^2+5^2})^3} = 0.01$$

$$\therefore \text{ Gain margin} = 20\log\frac{1}{X} = 40\ dB.$$

5.60 (a)

$$G(s)H(s) = \frac{1}{s(s+1)}$$

Condition to calculate gain cross-over frequency is

$$|G(j\omega)H(j\omega)|_{\omega=\omega_{gc}} = 1$$

$$\left|\frac{1}{\omega\sqrt{\omega^2+1}}\right| = 1$$

$$\therefore \quad \omega = 0.768 \approx 1\text{ rad/sec}$$

$$\therefore \text{ angle,} \quad \phi = -90° - \tan^{-1}\omega = -135°$$

Phase margin is calculated as $180 + \phi$

$$= 45°$$

5.61 (d)

The type-1 system is given as

$$G(s)H(s) = \frac{K}{s(s+\omega_1)}$$

Since this represents a slope of –20 dB/decade at the low frequency.

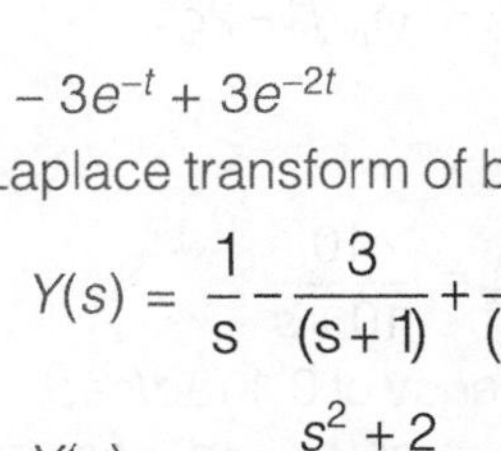

5.62 (c)

$y(t) = 1 - 3e^{-t} + 3e^{-2t}$

Taking Laplace transform of both sides

$$Y(s) = \frac{1}{s} - \frac{3}{(s+1)} + \frac{3}{(s+2)}$$

$$Y(s) = \frac{s^2+2}{s(s+1)(s+2)}$$

So, the forced response become zero when

$$s^2 + 2 = 0$$

$$\therefore \quad s = \pm j\sqrt{2}$$

i.e. $\quad \omega = \sqrt{2}\text{rad/sec.}$

5.63 (d)

Phase margin is given as

$$\text{P.M.} = \tan^{-1}\left\{2\xi\left(\frac{1}{\sqrt{\sqrt{1+4\xi^4}-2\xi^2}}\right)\right\}$$

For small values of ξ, PM is directly to ξ, and at $\xi = 0$

$$\text{P.M.} = \tan^{-1}0 = 0°$$

5.64 (d)

$$G(s)H(s) = \frac{1}{s(s+1)}$$

$$G(j\omega)H(j\omega) = \frac{1}{j\omega(j\omega+1)}$$

Gain margin is always found at phase cross over frequency

$$\angle G(j\omega)H(j\omega)\Big|_{\omega=\omega_{pc}} = \pm 180°$$

$$\Rightarrow \quad -90° - \tan^{-1}\omega = 180°$$

$$\tan^{-1}\omega_{pc} = 90°$$

$$\omega_{pc} = \infty$$

$$|G(j\omega)H(j\omega)| = \frac{1}{\omega\sqrt{\omega^2+1}} = 0$$

Thus, $\text{Gain margin} = \frac{1}{X} = \infty$

$$\text{GM(dB)} = 20\log\infty = \infty \text{ dB}$$

5.65 (c)

$$\frac{C(s)}{R(s)} = \frac{\omega_n^2}{s^2 + 2s\xi\omega_n + \omega_n^2}$$

Resonant frequency of second order system is given as

$$\Rightarrow \quad \omega_r = \omega_n\sqrt{1-2\xi^2}$$

5.66 (b)

$$T(s) = \frac{10}{10+s}$$

Gain at frequency of 0.1 rad/sec.

$$= 20\log_{10}10 = 20 \text{ dB}$$

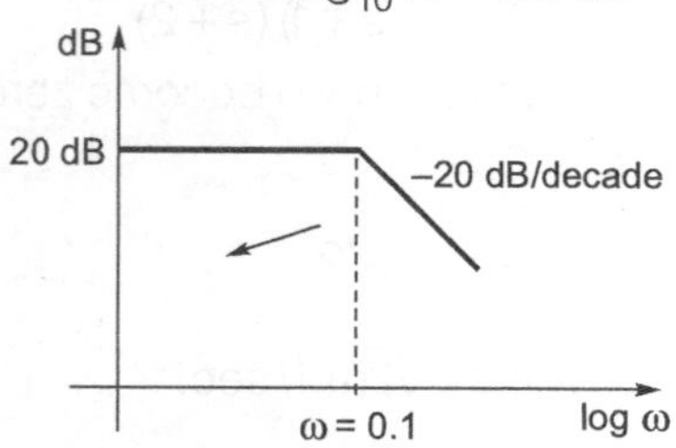

5.67 (a)

Noise is proportional to bandwidth as

$$P_n = kTB$$

When the bandwidth is increased, then system becomes fast as it reduces rise time, increases stability as well as reduces oscillations in the response but according to the above equation, the noise also increases.

5.68 (b)

$$G(s)H(s) = \frac{K}{s^3 + 6s^2 + 11s + 6}$$

Equating the imaginary part of the above equation to zero, we get

$$s^3 + 11s = 0$$

$$s(s^2 + 11) = 0$$

or $\quad s = j\sqrt{11}, \text{ or } \omega_{pc} = \sqrt{11}$

5.69 (d)

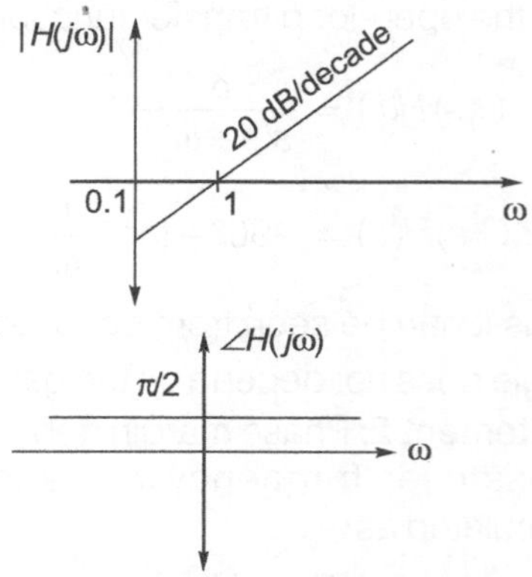

Thus from the two plots it is clear that option (d) is correct.

5.70 (c)

$$(\text{GM})_{\text{dB}} = -M_{\text{dB}}\Big|_{\omega=\omega_{pc}}$$

If $(\text{GM})_{\text{dB}}$ is the system is stable, for this $M_{\text{dB}}\Big|_{\omega=\omega_{pc}}$ must be negative. So the statement 3 is wrong and the correct choice is (c).

5.71 (a)

A closed loop transfer function for a system with feedback $H(s)$ and gain block $G(s)$ is given as

$$H(s) = \frac{G(s)}{1+G(s)H(s)}$$

$$\therefore \quad \text{Poles} = 1 + G(s)H(s) = 0$$

Which is equal to the poles of the characteristic equation.

5.72 (c)

$$GH(s) = \frac{10(s-1)}{s+10}$$

If a system has at least a zero in the right hand side of s-plane then it is called Non-minimum phase system.

5.73 (b)

$$GH(s) = \frac{25}{s(s+6)}$$

Characteristic equation

$$q(s) = 1 + GH(s) = s^2 + 6s + 25 = 0$$

$$\omega_n = 5 : \xi = 0.6$$

$$\therefore \quad \omega_r = \omega_n\sqrt{1-2\xi^2} = 2.67 \text{ rad/sec}$$

$$\therefore \quad M_r = \frac{1}{2\xi\sqrt{1-\xi^2}} = 1.04$$

5.74 (c)

$$G(s) = \frac{5(s^2+10s+100)}{s^2(s^2+15s+1)}$$

Corner frequency in Bode plot is defined for finite poles and zeros, which are complex in given system. Hence, the corner frequency are 10 rad/sec and 1 rad/sec.

5.75 (b)

$$GH(s) = [0.1 + 0.01s] = 0.1\,[1 + 0.1s]$$

Bode plot is

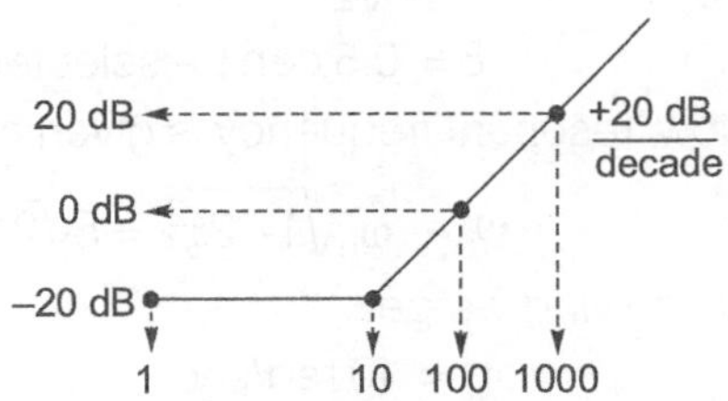

5.76 (a)

$$GH(s) = \frac{K}{s(s+5)}$$

$$q(s) = 1 + GH(s) = s^2 + 5s + K = 0$$

$$\omega_n = \sqrt{K}\,;\ \xi = \frac{5}{2\sqrt{K}}$$

$$\therefore \quad \text{PM} = 100\,\xi = 100\times\frac{5}{2\sqrt{K}} = 45^\circ$$

$$\therefore \quad K \approx 35$$

5.78 (d)

$$GH(s) = \frac{10[1+0.25]}{1+0.5S}$$

type = 0 ; $P - Z = 0$

$\therefore$ Phase at $\omega = 0$ is $-90^\circ \times$ type $= 0^\circ$

$\therefore$ Phase at $\omega = \infty$ is $-90^\circ \times (P - Z) = 0^\circ$

5.79 (b)

As per given details Bode plot is

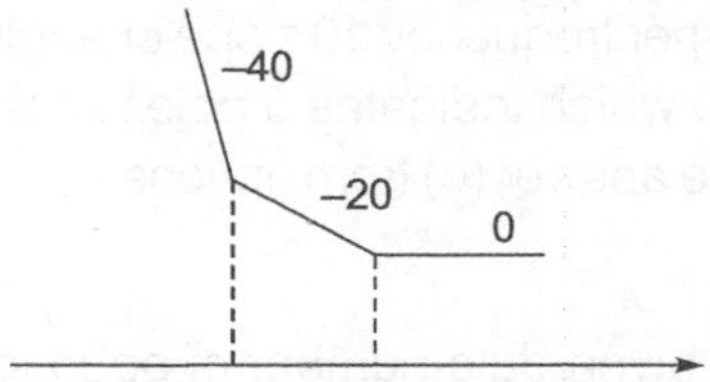

Initial slope indicates 2 poles at origin.

Final slope indicates

$$P = Z$$

$$\therefore \quad P = Z = 2$$

5.80 (b)

All pass system has constant magnitude response for all frequencies. This can happen only if poles and zeros are anti-symmetric with respect to $j\omega$-axis.

5.81 (c)

Relative stability means comparing stability of one system with another system, which is possible with the help of gain margin, phase margin and also location of poles in s-plane.

5.82 (c)

$$GH(s) = \frac{K}{s\left[\frac{s}{\omega}+1\right]}$$

$$20\log_{10}K = 20\ ;\ \omega = 1$$

$$\because \quad K = 10$$

Therefore, $GH(s) = \dfrac{10}{s+1}$

5.83 (a)

Transportation delay (or) lag phase increases with frequency due to which stability decreases due to which compensation become difficult.

5.84 (c)

$$G(s) = \frac{6(s^2+10s+100)}{s^2(50s^2+15s+1)}$$

Type = 2

Low frequency slope = –20 dB/dec × type = –40 dB/dec

5.85 (b)

Initial slope = –20 dB/dec

Hence the transfer function has a pole at origin.

At corner frequency '2' the slope changes to –40 dB/dec which indicates a pole at $S = -2$.

Hence answer (b) from options.

5.86 (c)

Initial slope = +40 dB/dec

Hence two zeros at origin.

At corner frequency 10 slope changing from +40 to –60 which indicates 5 poles at $S = -10$.

Hence answer (c) from options.

5.87 (b)

From given data number of open loop poles in right side $P = 1$.

From closed loop system to be stable $N = P = 1$.

5.88 (a)

Statement (I) : Bode plot means magnitude and phase plots.

Statement (II) : Transportation lag doesn't change magnitude, it only changes phase angle.

5.90 (d)

Magnitude and phase relation between sinusoidal input and the steady state output such that input amplitude and phase kept constant and input frequency is varied is called frequency response of LTI system.

5.91 (a)

A system is said to be minimum phase if it has all finite zeros and poles in left side of s-plane.

5.92 (b)

Resonant frequency is that at which the magnitude of frequency response is maximum.

5.93 (d)

For unstable system both GM and PM are negative and for stable system both GM and PM are positive.

5.94 (b)

For a unity feedback system, the closed-loop frequency response,

$$T(j\omega) = \frac{G(j\omega)}{1+G(j\omega)} = \frac{x+jy}{(1+x)+jy}$$

The closed-loop phase of the system is given by

$$\alpha = \tan^{-1}\left(\frac{y}{x}\right) - \tan^{-1}\left(\frac{y}{1+x}\right)$$

$$\alpha = \tan^{-1}\left(\frac{y}{x^2+x+y^2}\right)$$

We have, $N = \tan\alpha = \dfrac{y}{x^2+x+y^2}$

For a constant value of closed-loop phase α, $N = \tan\alpha$ is constant. Therefore, N-circle is the locus of constant closed-loop phase frequency response for unity feedback systems.

5.95 (a)

There are two poles and one zero are enforces before 20 Hz, hence phase is given by

$$\phi = (P - Z) * (-90°)$$
$$\phi = (2-1) * (-90°) = -90°$$

5.96 (c)

From the given data the graph can be drawn as,

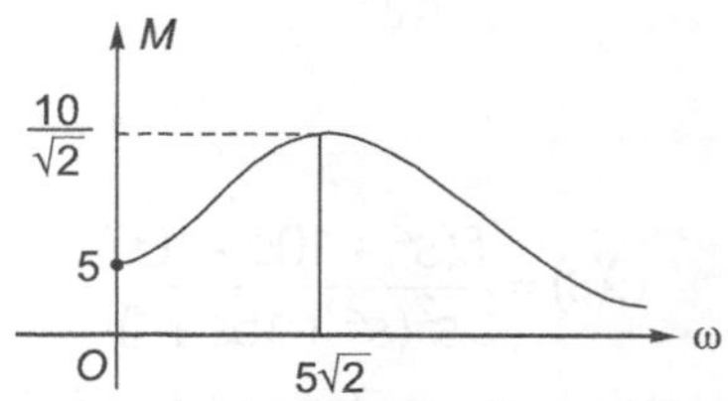

The resonant peak from graph is given as,

$$M_r = \frac{5}{2\xi\sqrt{1-\xi^2}} = \frac{10}{\sqrt{3}}$$

On solving we get, $\xi = 0.5$ and $\xi = 0.86$

For underdamped system

$$\xi < \frac{1}{\sqrt{2}} = 0.707$$

$\therefore$ $\xi = 0.5$ can be selected

Now, resonant frequency is given as,

$$\omega_r = \omega_n\sqrt{1-2\xi^2} = 5\sqrt{2}$$

On solving we get,

$$\omega_n = 10 \text{ rad/sec}$$

$\therefore$ The transfer function is given as

$$TF = \frac{5\times 100}{s^2+10s+100} = \frac{500}{s^2+10s+100}$$

Note: In option (c) there need to be correction in denominator in place of 12s it should be 10s.

5.97 (c)

$\therefore$ The resonant peak for standard second order system is given by

$$M_r = \frac{1}{2\zeta\sqrt{1-\zeta^2}}$$

5.98 (c)

Given open loop transfer function,

$$G(s)H(s) = \frac{2\sqrt{3}}{s(s+1)}$$

Phase margin is given by,

$$PM = 180° + \left[-90° - \tan^{-1}(\omega)\right]\Big|_{\omega=\omega_{gc}}$$

At ω_{gc} magnitude $|M| = 1$

$$\therefore \quad \frac{2\sqrt{3}}{\omega\sqrt{\omega^2+1}} = 1 \text{ at } \omega = \omega_{gc}$$

$$\therefore \quad \omega_{gc} = \sqrt{3}$$

$$\therefore \quad PM = 180° + [-90° - \tan^{-1}[\sqrt{3}]]$$

$$PM = 30°$$

5.99 (d)

For the given open-loop transfer function, the magnitude varies from 1/2 to 0, hence no gain crossover frequency can be defined.

$$\therefore \quad PM = \infty$$

Note : Gain cross over frequency at which magnitude of open loop system is unity.

5.101 (*)

We have,

$$\text{OLTF,} \quad G(j\omega) = \frac{K}{(j\omega)^2(1+j\omega T)}$$

- In the low frequency region, for a system of type 'r',

$$G(j\omega)|_{dB} = -20r\log_{10}\omega + 20\log_{10}K$$

$$M_{dB} = -20r\log_{10}\omega + 20\log_{10}K \quad ...(i)$$

At intersection with the 0-dB axis, $M = 0$. Hence, for $r = 2$,

$$0 = -(20 \times 2)\log_{10}\omega_x + 20\log_{10}K$$

$$\omega_x^2 = K$$ ...Hence, statement 1 is false.

- From equation (i), the magnitude at low frequency slope of –40 dB/decade at the frequency $\omega = 1$, is equal to $20\log K$. Hence, statement 2 is also not correct.
- The static ramp error coefficient for the given system is

$$K_v = \lim_{s\to 0} sG(s)H(s) = \infty$$

Hence, statement 3 is also not correct.

Therefore, none of the options are matched.

5.102 (b)

1. Gain Margin is used to indicate the closeness of the intersection of the negative real axis made by the polar plot of $G(j\omega)H(j\omega)$ to the $(-1 + j0)$ point.
2. Phase crossover frequency is the frequency at which the polar $G(j\omega)H(j\omega)$ plot intersects the negative real-axis such that

$$\angle G(j\omega)H(j\omega) = 180°$$

Hence, statements 1 and 2 are not correct.

■■■

6 Compensators and Controllers

6.1 PID control for a plant is shown in Figures I and II.

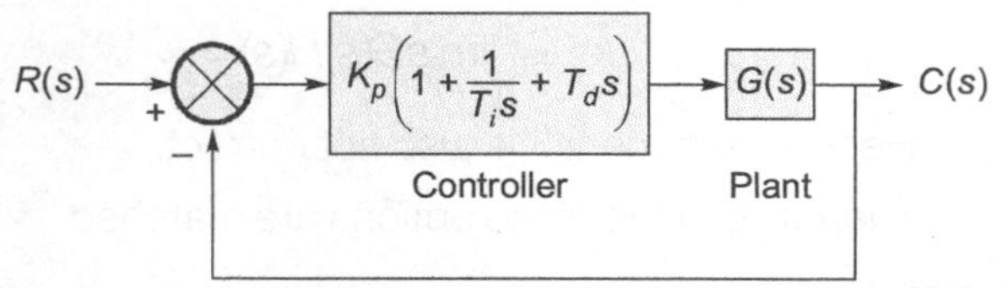

Figure-I

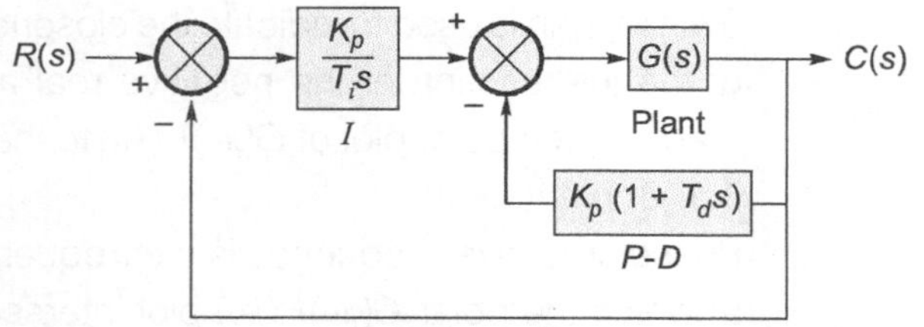

Figure-II

Assertion (A): Figure-II is preferred over Figure-I as it avoids large changes in control signal for a sudden change in reference input.

Reason (R): Placement of P-D action in the feedback path and larger values of K_p and T_d can be chosen in Figure-II.

(a) Both A and R are true and R is the correct explanation of A
(b) Both A and R are true but R is NOT the correct explanation of A
(c) A is true but R is false
(d) A is false but R is true

[ESE-2000]

6.2 **Assertion (A):** An on-off controller gives rise to self-sustained oscillation in the output.

Reason (R): Location of a pair of poles on the imaginary axis gives rise to self-sustained oscillation in the output.

(a) Both A and R are true and R is the correct explanation of A
(b) Both A and R are true but R is NOT the correct explanation of A
(c) A is true but R is false
(d) A is false but R is true

[ESE-2000]

6.3 For the given network, the maximum phase lead ϕ_m of V_o with respect to V_i is

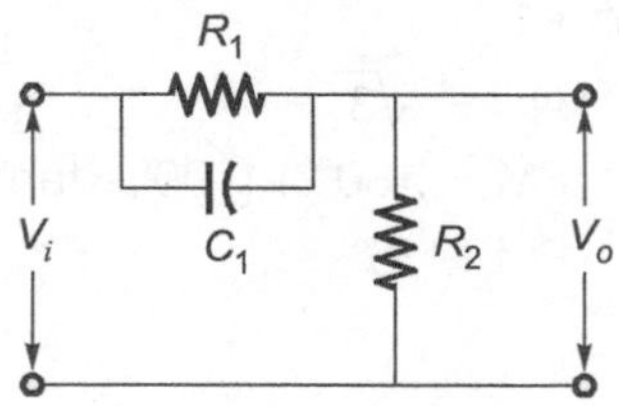

(a) $\sin^{-1}\left(\frac{R_1}{2R_2}\right)$ (b) $\sin^{-1}\left(\frac{R_1}{R_1+2R_2}\right)$
(c) $\sin^{-1}\left(\frac{R_1}{R_1+3R_2}\right)$ (d) $\sin^{-1}\left(\frac{R_1}{2R_2C_1}\right)$

[ESE-2000]

6.4 Consider the following statements:
A proportional plus derivative controller
1. has high sensitivity
2. increases the stability of the system
3. improves the steady-state accuracy

Which of these statements are correct?
(a) 1, 2 and 3 (b) 1 and 2
(c) 2 and 3 (d) 1 and 3 [ESE-2000]

6.5 Match **List-I** with **List-II** and select the correct answer:

List-I	List-II
A. e^{-as}	1. All-pass filter
B. $\frac{1-s}{1+s}$	2. Transport delay
C. $\frac{1+as}{1+bs}, a<b$	3. Lag network
D. $\frac{K}{s(1+as)}$	4. Servomotor

Codes:

	A	B	C	D
(a)	4	3	1	2
(b)	2	1	3	4
(c)	2	3	1	4
(d)	4	1	3	2

[ESE-2001]

6.6 The transfer function of phase-lead compensator is given by $G(s) = \frac{1+aTs}{1+Ts}$, where $T > 0$, $a > 1$. What is the maximum phase shift provided by this compensator?

(a) $\tan^{-1}\left(\frac{a+1}{a-1}\right)$ (b) $\tan^{-1}\left(\frac{a-1}{a+1}\right)$

(c) $\cos^{-1}\left(\frac{a-1}{a+1}\right)$ (d) $\sin^{-1}\left(\frac{a-1}{a+1}\right)$

[ESE-2001]

6.7 The transfer function of a phase lead network can be written as

(a) $\frac{1+sT}{1+s\beta T}$; $\beta > 1$ (b) $\frac{\alpha(1+sT)}{1+s\alpha T}$; $\alpha < 1$

(c) $\frac{\beta(1+sT)}{1+s\beta T+T}$; $\beta < 1$ (d) $\frac{(1+sT)}{\alpha(1+sT)}$; $\alpha > 1$

[ESE-2001]

6.8 Which one of the following compensations is adopted for improving transient response of a negative unity feedback system?
(a) Phase lead compensation
(b) Phase lag compensation
(c) Gain compensation
(d) Both phase lag compensation and gain compensation

[ESE-2001]

6.9 Consider the following statements regarding a phase-lead compensator:
1. It increases the bandwidth of the system
2. It helps in reducing the steady state error due to ramp input
3. It reduces the overshoot due to step input

Which of the above statements is/are correct?
(a) 1 and 2 (b) 1 and 3
(c) 2 and 3 (d) 1 alone [ESE-2002]

6.10 The transfer function of a phase lead network, as shown in the figure below is $\frac{K(1+0.3s)}{(1+0.17s)}$

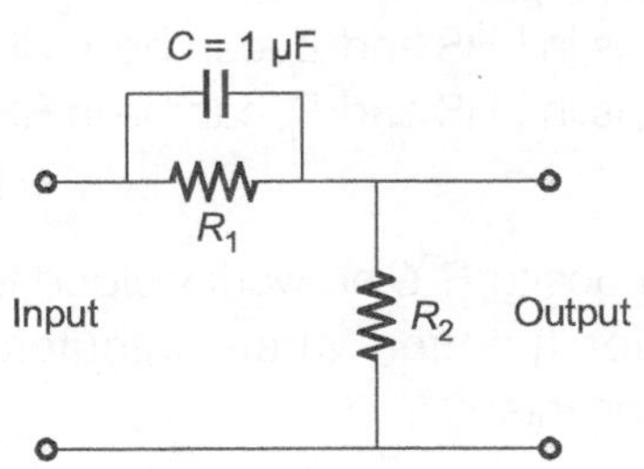

The values of R_1 and R_2 are respectively
(a) 300 kΩ and 300 kΩ
(b) 300 kΩ and 400 kΩ
(c) 400 kΩ and 300 kΩ
(d) 400 kΩ and 400 kΩ [ESE-2003]

6.11 The maximum value of a controller output is 100 V and is obtained when the input error is 1 V. If the controller is working at 20% proportional band, the error and output will be respectively
(a) 0.2 V and 100 V (b) 1 V and 20 V
(c) 1 V and 120 V (d) 0.2 V and 120 V

[ESE-2003]

6.12 Which one of the following statements is correct? A plant is controlled by a proportional controller. If a time delay element is introduced in the loop, its
(a) phase margin remains the same
(b) phase margin increases
(c) phase margin decreases
(d) gain margin increases [ESE-2004]

6.13 Which one of the following statements is correct? The effects of phase lead compensator on gain cross-over frequency (ω_{cg}) and the bandwidth (BW) are
(a) that both are decreased
(b) that ω_{cg} is decreased but BW is increased
(c) that ω_{cg} is increased but BW is decreased
(d) that both are increased

[ESE-2004]

6.14 How does cascading an integral controller in the forward path of a control system affect the relative stability (RS) and the steady-state error (SSE) of that system?
(a) Both are increased
(b) RS is reduced but SSE is increased
(c) RS is increased but SSE is reduced
(d) Both are reduced

[ESE-2004]

6.15 Consider the following statements for phase-lead compensation:
1. Phase-lead compensation shifts the gain cross-over frequency to the right.
2. The maximum phase-lead angle occurs at the arithmetic mean of the corner frequencies of the phase lead network.

3. Phase-lead compensation is effective when the slope of the uncompensated system near the gain cross-over is low.

Which of the statements given above are correct?

(a) 1, 2 and 3 (b) 1 and 2
(c) 2 and 3 (d) 1 and 3 [ESE-2004]

6.16 A tachometer feedback is used as an inner loop in a position control servo-system. What is the effect of feedback on the gain of the sub-loop incorporating tachometer and on the effective time constant of the system?

(a) Both are reduced
(b) Gain is reduced but the time constant is increased
(c) Gain is increased but the time constant is reduced
(d) Both are increased [ESE-2004]

6.17 With regard to the filtering property, the lead compensator and the lag compensator are, respectively:

(a) low pass and high pass filters
(b) high pass and low pass filters
(c) both high pass filters
(d) both low pass filters [ESE-2005]

6.18 The transfer function of phase-lead compensator is given by $G(s) = \frac{1+aTs}{1+Ts}$, where $T > 0$, $a > 1$. What is the maximum phase shift provided by this compensator?

(a) $\tan^{-1}\left(\frac{a+1}{a-1}\right)$ (b) $\tan^{-1}\left(\frac{a-1}{a+1}\right)$
(c) $\cos^{-1}\left(\frac{a-1}{a+1}\right)$ (d) $\sin^{-1}\left(\frac{a-1}{a+1}\right)$

[ESE-2005]

6.19 Match **List-I** (Compensation) with **List-II** (Characteristic) and select the correct answer using the code given below the Lists:

List-I	List-II
A. Lag	**1.** Increases bandwidth
B. Lead	**2.** Attenuation
C. Lag-Lead	**3.** Increases damping factor
D. Rate	**4.** Second order

Codes:

	A	B	C	D
(a)	3	1	4	2
(b)	2	4	1	3
(c)	3	4	1	2
(d)	2	1	4	3

[ESE-2005]

6.20 What is the effect of phase-lag compensation on the performance of a servo system?

(a) For a given relative stability, the velocity constant is increased.
(b) For a given relative stability, the velocity constant is decreased.
(c) The bandwidth of the system is increased
(d) The time response is made faster.

[ESE-2005]

6.21 For which one of the following, given physical realization corresponds to *PD* controller?

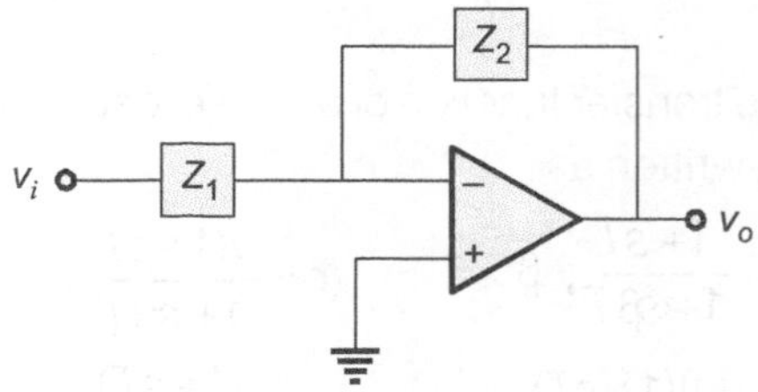

(a) Z_1 = 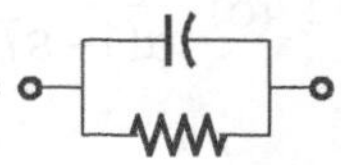 , Z_2 = 

(b) Z_1 = 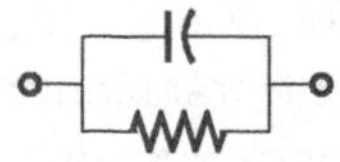 , Z_2 = 

(c) Z_1 =  , Z_2 = 

(d) Z_1 = 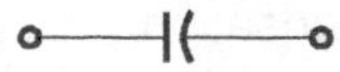 , Z_2 = 

[ESE-2005]

6.22 Which one of the following is an advantage of a *PD* controller in terms of damping (ξ) and natural frequency (ω_n)?

(a) ξ remains fixed but ω_n increases
(b) ξ remains fixed but ω_n decreases
(c) ω_n remains fixed but ξ increases
(d) ω_n remains fixed but ξ decreases

[ESE-2005]

6.23 A phase lead compensating network consists of only capacitors and resistors. The locations of its pole and zero in s-plane are at P_c and Z_c respectively. Which of the following conditions must be satisfied?

(a) Both P_c and Z_c in RHS and $P_c < Z_c$
(b) Both P_c and Z_c in LHS and $P_c > Z_c$
(c) P_c is in LHS and Z_c can be in RHS
(d) Z_c is in LHS and P_c can be in RHS

[ESE-2006]

6.24 A composite *R-C* network yielded the following transfer function when calculated from its components:

$$T(s) = \frac{C(s)}{R(s)} = \frac{1+21s+20s^2}{1+11s+10s^2}$$

This network can be used as which one of the following?

(a) Phase lead compensator
(b) Phase lag compensator
(c) Lag lead compensator
(d) None of the above **[ESE-2006]**

6.25 Which one of the following is a disadvantage of proportional controller?

(a) It destabilises the system
(b) It produces offset
(c) It makes response faster
(d) It has very simple implementation

[ESE-2006]

6.26 A process is controlled by PID controller. The sensor has high measurement noise. How can this effect be reduced?

(a) By use of a bandwidth limited derivative term
(b) By use of proportional and derivative terms in the forward path
(c) By use of high proportional band
(d) By use of low integral gain **[ESE-2007]**

6.27 Consider the following statements for a PI compensator for a control system:

1. It is equivalent to adding a zero at origin.
2. It reduces overshoot.
3. It improves order of the system by 1.
4. It improves steady-state error of the system.

Which of the statements given above are correct?

(a) 1, 2, 3 and 4 (b) 1, 2 and 3 only
(c) 2, 3 and 4 only (d) 1 and 4 only

[ESE-2007]

6.28 A controller transfer function is given by $C(s) = (2s + 1)/(0.2s + 1)$. What is its nature and parameter?

(a) Lag controller, $\alpha = 10$
(b) Lag controller, $\alpha = 2$
(c) Lead controller, $\beta = 0.1$
(d) Lead controller, $\beta = 0.2$ **[ESE-2007]**

6.29 The input to a controller is

(a) sensed signal
(b) error signal
(c) desired variable value
(d) signal of fixed amplitude not dependent on desired variable value

[ESE-2008]

6.30 What is the characteristic of a good control system?

(a) Sensitive to parameter variation
(b) Insensitive to input command
(c) Neither sensitive to parameter variation nor sensitive to input commands
(d) Insensitive to parameter variation but sensitive to input commands

[ESE-2009]

6.31 How can the bandwidth of a control system be increased?

(a) By the use of phase lead network
(b) By the use of phase lag network
(c) By the use of both phase-lag and phase-lead network
(d) By the use of cascaded amplifiers in the system

[ESE-2009]

6.32 The transfer function of a controller is given as $K_p + K_d.s$ where K_p and K_d are constant. What type of controller is this?

(a) Proportional
(b) Proportional plus integral
(c) Proportional plus derivative
(d) Integral plus derivative **[ESE-2009]**

6.33 The transfer function of a controller is given as $K_p + K_d.s + \frac{K_i}{s}$ where K_p, K_d and K_i are constants. What type of controller is this?

(a) Proportional
(b) Proportional plus derivative
(c) Proportional plus integral
(d) Proportional plus integral plus derivative

[ESE-2009]

6.34 Which one of the following represents the pole-zero location in the s-plane for lead-compensator?

(a)
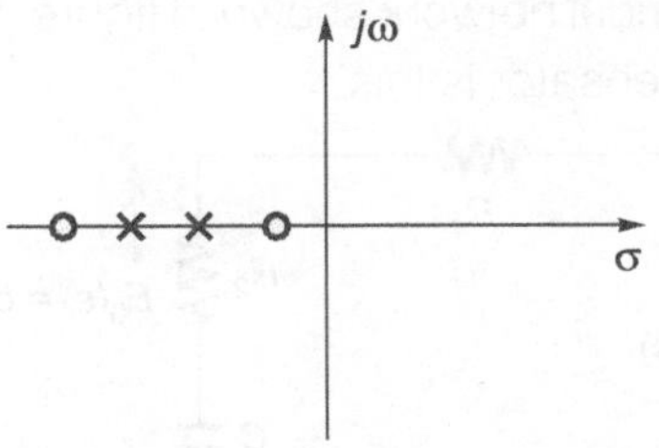

(b)
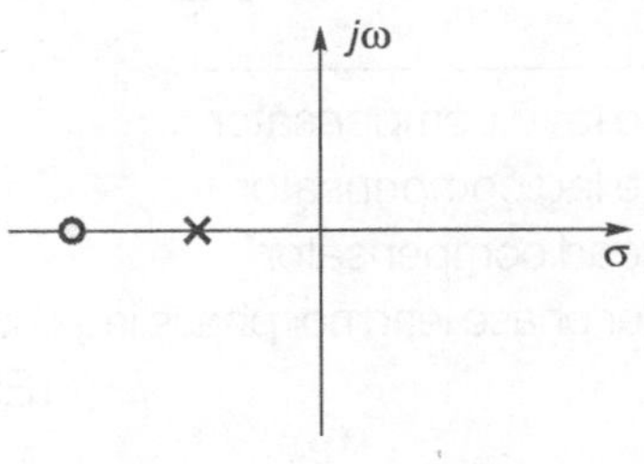

(c) 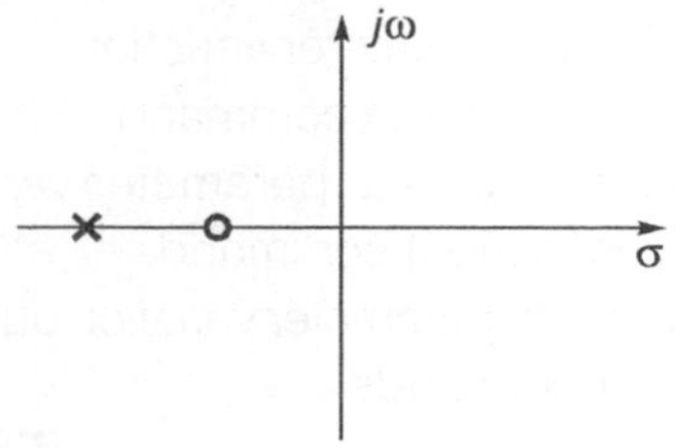

(d) 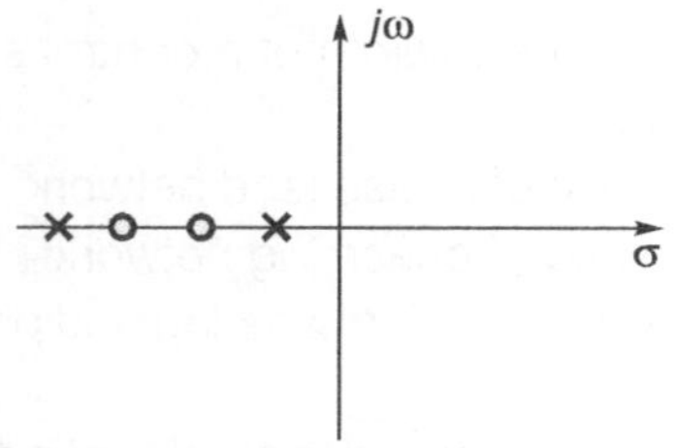

[ESE-2010]

6.35 **Assertion (A):** Integral windup effect in controller causes excessive overshoot.

Reason (R): Presence of saturation in controller and actuator deteriorates the PID control.

(a) Both A and R are true and R is the correct explanation of A
(b) Both A and R are true but R is not a correct explanation of A
(c) A is true but R is false
(d) A is false but R is true [ESE-2010]

6.36 An electrical system transfer function has a pole at $s = -2$ and a zero at $s = -1$ with system gain 10. For sinusoidal current excitation voltage response of the system

(a) is zero
(b) is in phase with the current
(c) leads the current
(d) lags behind the current [ESE-2011]

6.37 An electrical network shown in figure. What type of compensator is this?

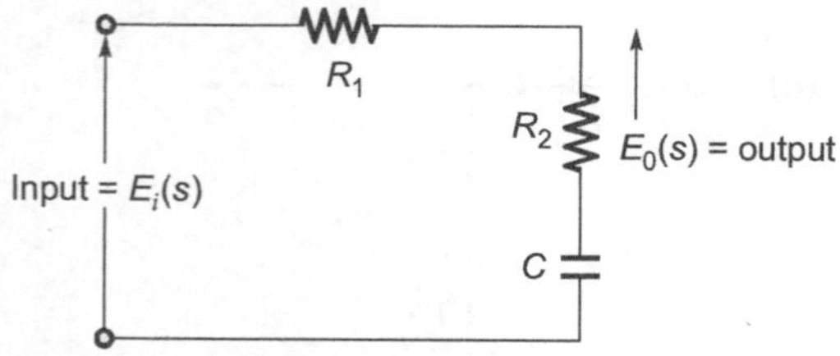

(a) Phase lead compensator
(b) Phase lag compensator
(c) Lag-lead compensator
(d) Neither phase lead nor phase lag compensator [ESE-2011]

6.38 The circuit diagram of an electrical network is given in figure. What type of compensator is this?

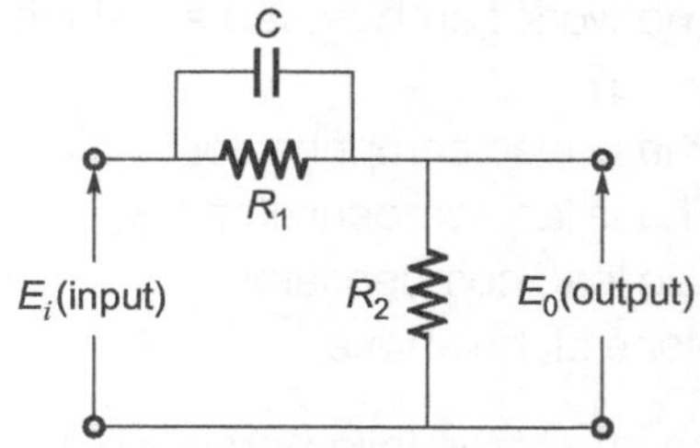

(a) Phase lag compensator
(b) Phase lead compensator
(c) Lag-lead compensator
(d) Neither phase lag nor phase lead compensator

[ESE-2011]

6.39 What is the transfer function of a phase lag compensator? The values of α and τ are given as $\alpha > 1$ and $\tau > 0$

(a) $\dfrac{1}{\alpha}\dfrac{\left(s+\dfrac{1}{\tau}\right)}{\left(s+\dfrac{1}{\alpha\tau}\right)}$ (b) $\dfrac{1}{\alpha}\dfrac{\left(s-\dfrac{1}{\tau}\right)}{\left(s-\dfrac{1}{\alpha\tau}\right)}$

(c) $\dfrac{1}{\alpha}\dfrac{\left(s+\dfrac{1}{\tau}\right)}{\left(s-\dfrac{1}{\alpha\tau}\right)}$ (d) $\dfrac{1}{\alpha}\dfrac{\left(s-\dfrac{1}{\tau}\right)}{\left(s+\dfrac{1}{\alpha\tau}\right)}$

[ESE-2011]

6.40 What is the transfer function of a phase lead compensator? The values of β and τ are given as $\beta < 1$ and $\tau > 0$

(a) $\dfrac{\beta(\tau s+1)}{(\beta\tau s+1)}$ (b) $\dfrac{\beta(\beta\tau s+1)}{(\tau s+1)}$

(c) $\dfrac{\beta(\beta\tau s-1)}{(\tau s+1)}$ (d) $\dfrac{\beta(\beta\tau s-1)}{(\tau s-1)}$

[ESE-2011]

6.41 The circuit diagram of a controller is given in figure. What type of controller is this?

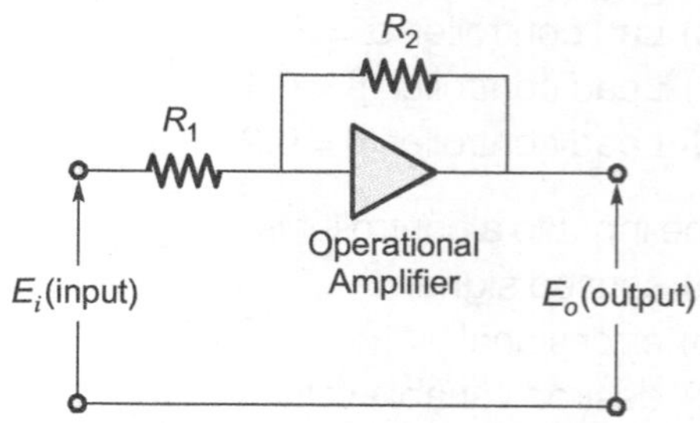

(a) Derivative (b) Integral
(c) Proportional (d) Proportional + Integral

[ESE-2011]

6.42 The circuit diagram of a controller is given in figure. What type of controller is this?

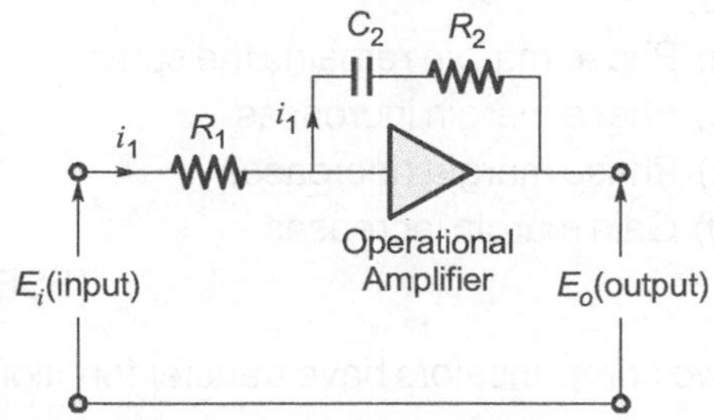

(a) Proportional
(b) Proportional + Derivative
(c) Integral
(d) Proportional + Integral [ESE-2011]

6.43 The effect of integral controller on the steady state error (e_{ss}) and the relative stability (R_s) of the system are
(a) Both are increased
(b) e_{ss} is increased but R_s is reduced
(c) e_{ss} is reduced but R_s is increased
(d) Both are reduced
[ESE-2012]

6.44 A proportional integral (PI) controller results in which of the following?
(a) Improves the transient response without affecting steady state response
(b) Improves the steady state response without affecting transient response
(c) Improves both transient response and steady state response
(d) Improves the steady state response while marginally affecting transient response, for well designed control parameters
[ESE-2012]

6.45 A liquid level controller linearly converts a displacement of 2 m to 3 m into 4 – 20 mA control signal. A relay serves as two position controller to open and close an inlet valve. Relay closes at 12 mA and opens at 10 mA. The hysteresis zone is
(a) 0.1 m (b) 0.125 m
(c) 0.15 m (d) 0.2 m
[ESE-2012]

6.46 Considering the filtering property, the lead compensators and lag compensators are categorized respectively as
(a) Low pass and high pass filters
(b) High pass and low pass filters
(c) High pass and high pass filters
(d) Low pass and low pass filters [ESE-2012]

6.47 Match **List-I** with **List-II** and select the correct answer using the code given below the lists:

List-I	List-II
A. PI control	1. Relay controller
B. PD control	2. Lead lag compensator
C. PID control	3. Lead compensator
D. On-off control	4. Lag compensator

Codes:

	A	B	C	D
(a)	4	2	3	1
(b)	1	2	3	4
(c)	4	3	2	1
(d)	1	3	2	4

[ESE-2012]

6.48 A phase lead compensating network has its transfer function $G_c(s) = \dfrac{10(1+0.04s)}{(1+0.01s)}$. The maximum phase lead occurs at a frequency of
(a) 50 rad/s (b) 25 rad/s
(c) 10 rad/s (d) 4 rad/s
[ESE-2012]

6.49 By adding zero to the system transfer function the improvement to transient response is called as
(a) phase lead compensation
(b) phase lag compensation
(c) phase lag and phase lead compensation
(d) phase lead and phase lag compensation
[ESE-2013]

6.50 The network having transfer function, $G(s) = \dfrac{1+\frac{s}{4}}{1+\frac{s}{25}}$ will provide maximum phase lead at a frequency of
(a) 4 rad/sec (b) 25 rad/sec
(c) 10 rad/sec (d) 100 rad/sec
[ESE-2013]

6.51 Derivative feedback is employed in the control system shown in the figure, to improve its damping. If the required damping factor of the system is 0.5, the value of K_d must be adjusted to

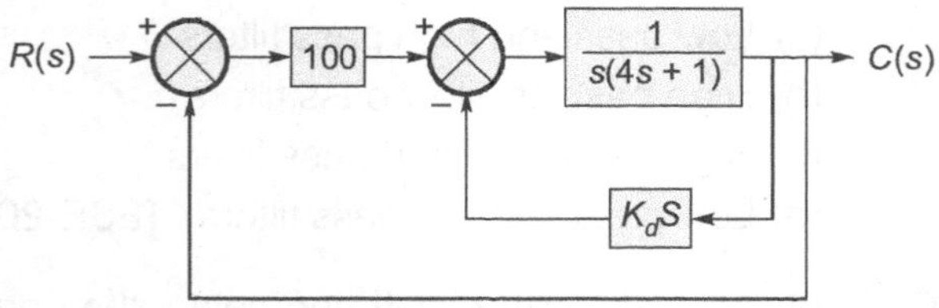

(a) 4 (b) 19
(c) 0.25 (d) 6 [ESE-2013]

6.52 **Statement (I):** Feedback compensation is preferred to series compensation.
Statement (II): No amplification is required as the energy flows from higher level to lower level.
(a) Both Statement (I) and Statement (II) are individually true and Statement (II) is the correct explanation of Statement (I).
(b) Both Statement (I) and Statement (II) are individually true but Statement (II) is not the correct explanation of Statement (I).
(c) Statement (I) is true but Statement (II) is false.
(d) Statement (I) is false but Statement (II) is true.
[ESE-2013]

6.53 **Statement (I):** A PI controller increases the order of a system by units but reduces the steady state error.
Statement (II): A PI controller introduces a pole at either the origin or at a desired points on negative real axis.
(a) Both Statement (I) and Statement (II) are individually true and Statement (II) is the correct explanation of Statement (I).
(b) Both Statement (I) and Statement (II) are individually true but Statement (II) is not the correct explanation of Statement (I).
(c) Statement (I) is true but Statement (II) is false.
(d) Statement (I) is false but Statement (II) is true.
[ESE-2013]

6.54 **Statement (I):** A derivative controller produces a control action for constant error only.
Statement (II): The PD controller increases the damping ratio and reduces the peak overshoot.
(a) Both Statement (I) and Statement (II) are individually true and Statement (II) is the correct explanation of Statement (I).
(b) Both Statement (I) and Statement (II) are individually true but Statement (II) is not the correct explanation of Statement (I).
(c) Statement (I) is true but Statement (II) is false.
(d) Statement (I) is false but Statement (II) is true.
[ESE-2013]

6.55 A plant is controlled by a proportional controller. If a time delay element is introduced in the loop, its
(a) Phase margin remains the same
(b) Phase margin increases
(c) Phase margin decreases
(d) Gain margin increases
[ESE-2014]

6.56 Two compensators have transfer functions
$$G_1(s) = \frac{5(s+10)}{(s+50)} \text{ and } G_2(s) = \frac{(s+50)}{5(s+10)}$$
respectively.
(a) Both are lag
(b) Both are lead
(c) G_1 is lead and G_2 is lag
(d) G_1 is lag and G_2 is lead
[ESE-2014]

6.57 A proportional plus derivative controller
1. has high sensitivity
2. increases the stability of the system
3. improves the steady-state accuracy

Which of the above statements are correct?
(a) 1, 2 and 3 only (b) 1 and 2 only
(c) 1 and 3 only (d) 2 and 3 only
[ESE-2014]

6.58 Which one of the following is the transfer function of the PI-controller?
(a) $G(s) = \dfrac{(K_1 s + K_2)}{K_3}$
(b) $G(s) = \dfrac{(K_1 s + K_2 s + K_3)}{K_4 s}$
(c) $G(s) = \dfrac{(K_1 s + K_2)}{K_3 s}$
(d) $G(s) = \dfrac{K_1 s}{K_2 s}$
[ESE-2014]

6.59 The transfer function of a controller is $G_C(s) = \dfrac{1+3s}{1+s}$. The maximum phase control provided by this controller is
(a) 30° lead (b) 30° lag
(c) 45° lead (d) 45° lag
[ESE-2015]

6.60 Consider the following statements:
The effect of phase lead network is given as
1. Increased velocity constant.
2. Increased phase margin.
3. Increased bandwidth.
4. Slower response.

Which of the above statements are correct?
(a) 1, 2 and 3 only (b) 1, 2 and 4 only
(c) 2, 3 and 4 only (d) 1, 2, 3 and 4

[ESE-2015]

6.61 **Statement (I):** Lead compensation is used to improve system stability margins.
Statement (II): Lead compensation achieves the desired result through the merits of its phase lead contribution.
(a) Both Statement (I) and Statement (II) are individually true and Statement (II) is the correct explanation of Statement (I)
(b) Both Statement (I) and Statement (II) are individually true but Statement (II) is **NOT** the correct explanation of Statement (I)
(c) Statement (I) is true but Statement (II) is false
(d) Statement (I) is false but Statement (II) is true

[ESE-2015]

6.62 In order to improve the system response transient behaviour, the type of controller used is
(a) Phase lead controller
(b) Phase lag controller
(c) *PI* controller
(d) *P* controller

[ESE-2015]

6.63 A phase lead compensator has its transfer function, $G_c(s) = \dfrac{1+0.5s}{1+0.05s}$. The maximum phase lead and the corresponding frequency, respectively are nearly
(a) $\sin^{-1}(0.9)$ and 6 r/s
(b) $\sin^{-1}(0.82)$ and 4 r/s
(c) $\sin^{-1}(0.9)$ and 4 r/s
(d) $\sin^{-1}(0.82)$ and 6 r/s

[ESE-2016]

6.64 Consider the following statements :
1. Lead compensation decreases the bandwidth of the system.
2. Lag compensation increases the bandwidth of the system.

Which of the above statements is/are correct?
(a) 1 only (b) 2 only
(c) Both 1 and 2 (d) Neither 1 nor 2

[ESE-2016]

6.65 A proportional controller with transfer function, K_p is used with a first-order system having its transfer function as $G_c(s) = \dfrac{K}{(1+s\tau)}$, in unity feedback structure. For step inputs, an increase in K_p will
(a) Increase the time constant and decrease the steady state error
(b) Decrease the time constant and decrease the steady state error
(c) Decrease the time constant and increase the steady state error
(d) Increase the time constant and increase the steady state error

[ESE-2016]

6.66 In position control systems, the Tacho-generator feedback is used to
(a) Increase the effective damping in the system
(b) Decrease the effective damping in the system
(c) Decrease the steady state error
(d) Increase the steady state error

[ESE-2016]

6.67 The controller which is highly sensitive to noise is
(a) *PI* (b) *PD*
(c) Both *PI* and *PD* (d) Neither *PI* nor *PD*

[ESE-2016]

6.68 For derivative control action, the actuating signal consists of proportional error signal with addition of
(a) derivative of the error signals
(b) integral of the error signals
(c) steady-state error
(d) a constant which is a function of the system type

[ESE-2017]

6.69 Consider the following statements regarding a PID controller:
1. The error is multiplied by a negative (for reverse action) proportional constant *P*, and added to the current output.
2. The error is integrated (averaged) over a period of time, and then divided by a constant *I*, and added to the current control output.

3. The rate of change of the error is calculated with respect to time, multiplied by another constant D, and added to the output.

Which of the above statements are correct?

(a) 1, 2 and 3 (b) 1 and 3 only
(c) 1 and 2 only (d) 2 and 3 only

[ESE-2017]

6.70 **Statement (I):** PID controller is an essential part of any control loop in process industry.

Statement (II): PID control system performs better than most predictive control methods in the context of measured disturbances.

(a) Both Statement (I) and Statement (II) are individually true and Statement (II) is the correct explanation of Statement (I).
(b) Both Statement (I) and Statement (II) are individually true but Statement (II) is not the correct explanation of Statement (I).
(c) Statement (I) is true but Statement (II) is false.
(d) Statement (I) is false but Statement (II) is true.

[ESE-2017]

6.71 For a lead compensator, whose transfer function is given by $K\dfrac{s+a}{s+b}$; $a, b \geq 0$.

(a) $a < b$ (b) $a > b$
(c) $a \geq Kb$ (d) $a = 0$

[ESE-2017]

6.72 The pole-zero configuration of the transfer function of a compensator and the corresponding Bode plot are shown in the figure. The configuration is indicative of

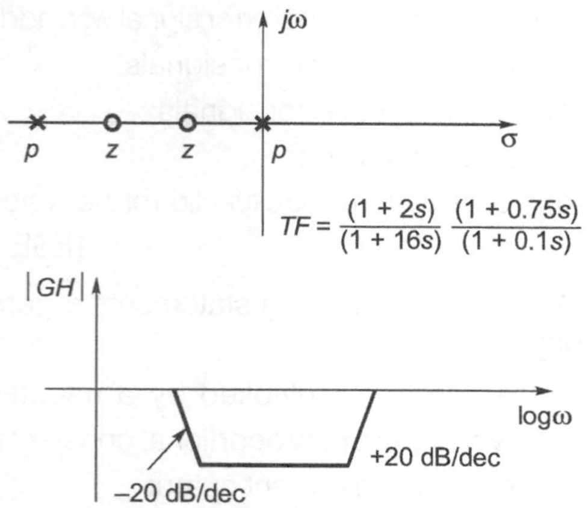

(a) lag compensator (b) lag-lead compensator
(c) P-D compensator (d) lead compensator

[ESE-2018]

6.73 The transfer function $G(s)$ of a PID controller is

(a) $K_1 + K_2s + K_3s^2$ (b) $K_1 + \dfrac{K_2}{s} + K_3s$
(c) $K_1 + \dfrac{K_2}{s}$ (d) $K_1s + K_2s^2 + K_3s^3$

[ESE-2018]

6.74 For a lead compensator having transfer function

$$G_c(s) = \frac{(s+z_c)}{(s+p_c)} = \frac{\left(s+\frac{1}{\tau}\right)}{\left(s+\frac{1}{\alpha\tau}\right)}$$

1. $\alpha = \dfrac{z_c}{p_c} < 1$ 2. $\alpha = \dfrac{z_c}{p_c} > 1$
3. $\tau > 0$ 4. $\tau < 0$

Which of the above are correct?

(a) 1 and 4 (b) 1 and 3
(c) 2 and 4 (d) 2 and 3

[ESE-2019]

6.75 The attenuation (magnitude) produced by a lead compensator at the frequency of maximum phase lead $\omega_m = \sqrt{ab}$ is

(a) $\sqrt{\dfrac{b}{a}}$ (b) $\sqrt{a+b}$
(c) $\sqrt{b-a}$ (d) $\sqrt{\dfrac{a}{b}}$ [ESE-2019]

6.76 The compensator required to improve the steady state response of a system is

(a) Lag (b) Lead
(c) Lag-lead (d) Zero [ESE-2020]

6.77 Which one of the following types of controller is sometimes called automatic reset?

(a) Proportional (b) Integral
(c) Derivative (d) PID [ESE-2020]

6.78 Consider the following statements regarding the disadvantages of a passive lead network over an active PD controller:

1. No additional power supplies are required.
2. Noise due to differentiation is reduced.
3. Additional pole does not reduce the number of branches of the root locus that cross the imaginary axis into the right half-plane.
4. Addition of the single zero of the PD controller tends to reduce the number of branches of the root locus that cross into the right half-plane.

Which of the above statements is/are correct ?
(a) 1 only (b) 2 only
(c) 3 and 4 only (d) 1, 2 and 4 only
[ESE-2021]

6.79 Consider the following statements regarding feedback compensation of control system:
1. A faster response can be achieved by the use of parallel compensation.
2. The environmental conditions in which the feedback control system is to be utilized affect the stability of the controlled quantity.
3. The degree of accuracy and the stability of a control system can be improved by the use of a cascade compensator.

Which of the above statements are correct?
(a) 1 and 2 only (b) 1 and 3 only
(c) 2 and 3 only (d) 1, 2 and 3
[ESE-2022]

6.80 Consider the following statements regarding compensator :
1. The lead compensator reduces the noise signal level relative to the control signal.
2. Introduction of cascade lead compensator increases the gain crossover frequency and, consequently, the speed of the response of resulting system.
3. The lead compensation increases the system bandwidth.

Which of the above statements are correct?
(a) 1 and 2 only (b) 1 and 3 only
(c) 2 and 3 only (d) 1, 2 and 3
[ESE-2022]

6.81 Which of the following is/are used, when the cascade compensator does not employ pure integration?
(a) Lead compensator only
(b) Lag compensator only
(c) Neither lead nor lag compensator
(d) Both lead and lag compensators
[ESE-2022]

6.82 By considering standard notations, the transfer function of a tachometer is of the form
(a) $K_t s$ (b) $\frac{K_t}{s}$
(c) $\frac{K_t}{s+1}$ (d) $\frac{K_t}{s(s+1)}$
[ESE-2023]

6.83 Consider the following statements :
1. In PD controllers, derivative action provides the counterbalancing effect of rapidly changing disturbances, which does not take care of the offset.
2. In PI controllers, proportionality constant provides simplicity, reliability, directness, etc.
3. PI systems have some oscillatory effect.

Which of the above statements are correct?
(a) 1 and 2 only (b) 1 and 3 only
(c) 2 and 3 only (d) 1, 2 and 3
[ESE-2024]

■■■

Answers Compensators and Controllers

6.1 (a)	6.2 (b)	6.3 (b)	6.4 (a)	6.5 (b)	6.6 (d)	6.7 (b)	6.8 (a)	6.9 (b)
6.10 (b)	6.11 (b)	6.12 (c)	6.13 (d)	6.14 (d)	6.15 (d)	6.16 (a)	6.17 (b)	6.18 (d)
6.19 (d)	6.20 (a)	6.21 (a)	6.22 (c)	6.23 (b)	6.24 (a)	6.25 (b)	6.26 (d)	6.27 (c)
6.28 (c)	6.29 (b)	6.30 (d)	6.31 (a)	6.32 (c)	6.33 (d)	6.34 (c)	6.35 (b)	6.36 (c)
6.37 (b)	6.38 (b)	6.39 (a)	6.40 (a)	6.41 (c)	6.42 (d)	6.43 (d)	6.44 (d)	6.45 (b)
6.46 (b)	6.47 (c)	6.48 (a)	6.49 (a)	6.50 (c)	6.51 (b)	6.52 (b)	6.53 (a)	6.54 (d)
6.55 (c)	6.56 (c)	6.57 (a)	6.58 (c)	6.59 (a)	6.60 (*)	6.61 (a)	6.62 (a)	6.63 (d)
6.64 (d)	6.65 (b)	6.66 (a)	6.67 (b)	6.68 (a)	6.69 (a)	6.70 (a)	6.71 (a)	6.72 (b)
6.73 (b)	6.74 (b)	6.75 (d)	6.76 (a)	6.77 (b)	6.78 (c)	6.79 (d)	6.80 (c)	6.81 (b)
6.82 (a)	6.83 (d)							

Explanations Compensators and Controllers

6.3 (b)

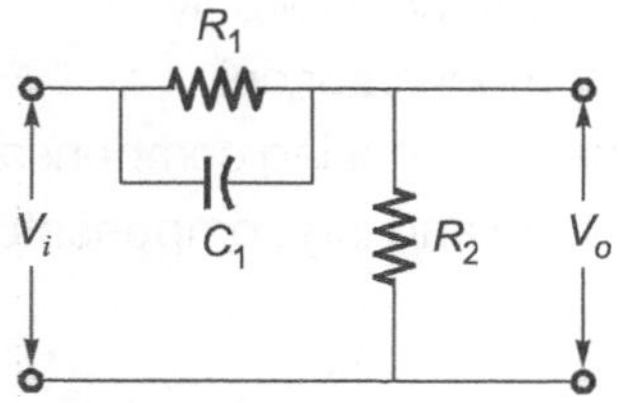

$$\frac{V_o(s)}{V_i(s)} = \frac{R_2}{R_2 + \dfrac{R_1 \dfrac{1}{sC_1}}{R_1 + \dfrac{1}{sC_1}}} = \frac{R_2}{R_2 + \dfrac{R_1}{1 + sR_1C_1}}$$

$$= \frac{R_2(1 + sR_1C_1)}{(R_1 + R_2) + sR_1R_2C_1}$$

Dividing by $R_1R_2C_1$,

$$\frac{V_o(s)}{V_i(s)} = \frac{s + \dfrac{1}{R_1C_1}}{s + \dfrac{(R_1 + R_2)}{R_1R_2C_1}}$$

$$= \frac{s + \dfrac{1}{R_1C_1}}{s + \dfrac{1}{\left(\dfrac{R_2}{R_1 + R_2}\right)R_1C_1}} = \frac{s + \dfrac{1}{\tau}}{s + \dfrac{1}{\alpha\tau}}$$

where, $\tau = \dfrac{1}{R_1C_1}$, $\alpha = \dfrac{R_2}{R_1 + R_2}$

Maximum phase lead

$$\phi_m = \sin^{-1}\left(\frac{1-\alpha}{1+\alpha}\right) = \sin^{-1}\left(\frac{1 - \dfrac{R_2}{R_1 + R_2}}{1 + \dfrac{R_2}{R_1 + R_2}}\right)$$

$$\phi_m = \sin^{-1}\left(\frac{R_1}{R_1 + 2R_2}\right)$$

6.4 (a)

A proportional plus derivative controller has the following features:

(i) It adds an open loop zero on negative real axis.
(ii) Undamped natural frequency remains same and damping ratio increases.
(iii) Peak overshoot decreases.
(iv) Bandwidth increases.
(v) Rise time decreases.
(vi) Effect of external noise increases.
(vii) Settling time decreases, i.e. response becomes faster.
(viii) Stability improves.
(ix) e_{ss} decreases.

6.5 (b)

$e^{-as} \rightarrow$ Transport delay

$\dfrac{1-s}{1+s} \rightarrow$ All – pass filter

$\dfrac{1+as}{1+bs}$, $a < b \rightarrow$ Lag network

$\dfrac{K}{s(1+as)} \rightarrow$ Servomotor

6.6 (d)

$$G(j\omega) = \frac{1 + ja\omega T}{1 + j\omega T} \text{ where } T > 0,\ a > 1$$

Phase angle $\phi = \tan^{-1} a\omega T - \tan^{-1} \omega T$

For maximum phase lead, $\dfrac{d\phi}{d\omega} = 0$

$$\Rightarrow \frac{1}{1 + a^2\omega_m^2 T^2} \cdot aT - \frac{1}{1 + \omega_m^2 T^2} \cdot T = 0$$

$$\Rightarrow T(a + \omega_m^2 T^2 a - 1 - a^2 \omega_m^2 T^2) = 0$$

$$\Rightarrow -\omega_m^2 T^2 a(a-1) + (a-1) = 0$$

$$\Rightarrow (a-1)(1 - \omega_m^2 T^2 a) = 0$$

$$\Rightarrow \omega_m^2 = \frac{1}{aT^2} \Rightarrow \omega_m = \frac{1}{T\sqrt{a}}$$

$$\phi_m = \tan^{-1} a\omega_m T - \tan^{-1} \omega_m T$$

$$= \tan^{-1}\sqrt{a} - \tan^{-1}\frac{1}{\sqrt{a}}$$

$$\Rightarrow \quad \phi_m = \tan^{-1}\left(\frac{a-1}{(1+1)\sqrt{a}}\right) = \tan^{-1}\left(\frac{a-1}{2\sqrt{a}}\right)$$

$$\Rightarrow \quad \phi_m = \sin^{-1}\left(\frac{a-1}{a+1}\right)$$

6.7 (b)

Phase lead network has

$$G(s) = \frac{\alpha(1 + sT)}{1 + s\alpha T}; \alpha < 1$$

6.8 (a)

Phase lead compensation improves transient response. Phase lag compensation improves steady state response.

6.10 (b)

$$G_c(s) = \frac{K(1+0.3s)}{(1+0.17s)} = \frac{\alpha(1+\tau s)}{(1+\alpha\tau s)}$$

Where $\tau = 0.3$, $\alpha\tau = 0.17$

From the given network

$$\tau = R_1C \qquad \alpha = \frac{R_2}{R_1+R_2}$$

So, $\quad 0.3 = 10^{-6} R_1$

$\Rightarrow \quad R_1 = 300\ \Omega$

$$\alpha = \frac{0.17}{0.3} = \frac{R_2}{300+R_2}$$

$\Rightarrow \quad R_2 = 392.3\ \text{k}\Omega$

$\Rightarrow \quad R_2 \approx 400\ \text{k}\Omega$

6.12 (c)

The introduction of a time delay element decreases both phase margin and gain margin.

6.13 (d)

Phase lead compensator acts like a high pass filter. So gain crossover frequency and bandwidth both increase.

6.14 (d)

Integral controller acts like a low pass filter. It reduces the stability as well as steady state error.

6.15 (d)

The maximum phase lead angle occurs at the geometric mean of the corner frequencies of the phase lead network.

6.16 (a)

Techometer feedback reduces both gain and effective time constant.

6.17 (b)

Lead compensator is a high pass filter.
Lag compensator is a low pass filter.

6.18 (d)

$$G(s) = \frac{1+aTs}{1+Ts}$$

Comparing with standard transfer function of phase lead compensator.

$$G(s) = \frac{1+T's}{1+\alpha T's}$$

$\therefore \quad T' = aT$

$\alpha T' = T$

$$\alpha = \frac{1}{a} < 0 \qquad (\because a > 0)$$

$$\therefore \quad \sin\theta = \frac{1-\alpha}{1+\alpha} = \frac{a-1}{a+1}$$

where θ is maximum phase shift

$$\theta = \sin^{-1}\left(\frac{a-1}{a+1}\right)$$

6.19 (d)

Lead compensator is a HPF. So it increases the bandwidth.
In lag compensator,

$$G_c(s) = \frac{1}{\beta}\left(\frac{s+z_c}{s+p_c}\right), \ \beta = \frac{z_c}{p_c} > 1$$

$\because \beta > 1 \Rightarrow 1/\beta < 1$

$\Rightarrow$ Lag compensator attenuates the signal.

6.20 (a)

Phase lag compensation is an integrator. It reduces the steady state error.

$$\text{Velocity constant} = \frac{1}{\text{steady state error}}$$

So, the velocity constant is increased.

6.21 (a)

PD controller behaves like differentiator.
So, Z_1 should be capacitive and Z_2 should be resistive.
Physical realization is possible with

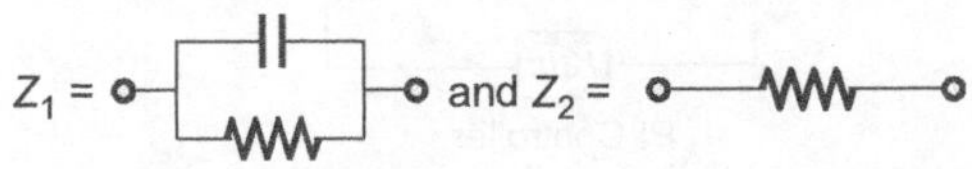

because capacitor alone is not physically realizable.

6.22 (c)

PD Controller:

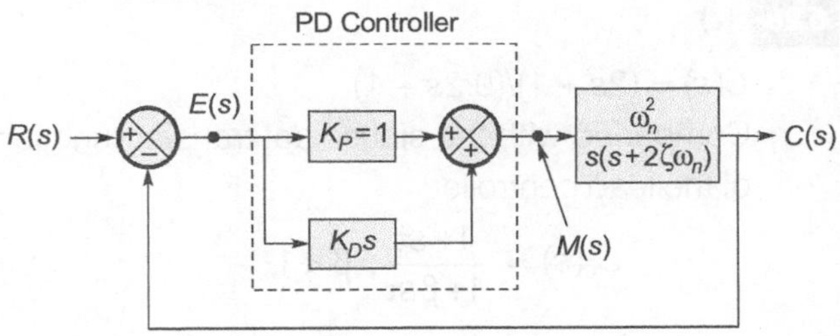

$$\frac{M(s)}{E(s)} = K_P + K_D s$$

$$\frac{C(s)}{R(s)} = \frac{(K_P + K_D s)\omega_n^2}{s^2 + (2\xi\omega_n + K_D\omega_n^2)s + K_P\omega_n^2}$$

Characteristic equation is

$s^2 + (2\xi\omega_n + K_D\omega_n^2)s + \omega_n^2 = 0$

$\because K_P = 1$

Comparing with $s^2 + 2\xi'\omega_n's + \omega_n'^2 = 0$

$\omega_n' = \omega_n$

$$\xi' = \xi + \frac{K_D\omega_n}{2}$$

Thus ω_n remains fixed but ξ increases.

6.23 (b)

A phase lead compensating network has zero nearer to origin than pole.

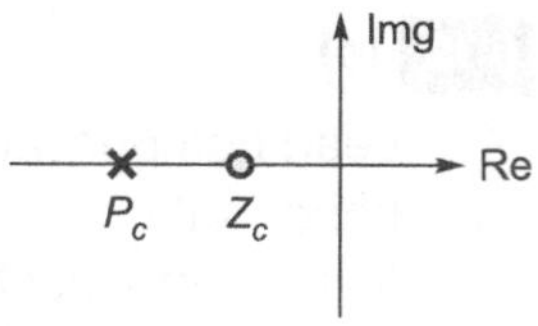

6.24 (a)

$$T(s) = \frac{C(s)}{R(s)} = \frac{(1+s)(1+20s)}{(1+s)(1+10s)} = \frac{1+20s}{1+10s}$$

Since zero $\left(s = -\frac{1}{20}\right)$ is nearer to origin than pole, it is a phase lead compensator.

6.25 (b)

Main disadvantage of proportional controller is, it produces a permanent error called offset error.

6.27 (c)

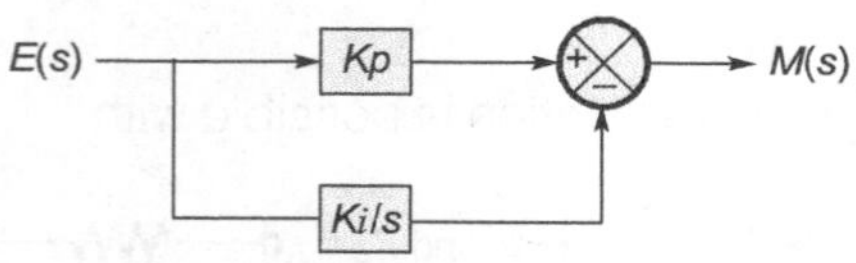

PI Controller

$$\frac{M(s)}{E(s)} = K_p + \frac{K_i}{s} = \frac{sK_P + K_i}{s}$$

PI compensator adds one open-loop pole at origin and one open-loop zero at negative real axis.

6.28 (c)

$C(s) = (2s + 1)/(0.2s + 1)$

Comparing with the sinusoidal transfer function of the lead controller.

$$G_c(s) = \frac{1+s\tau}{1+\beta s\tau};\ \beta < 1$$

$$\tau = 2$$

$$\beta\tau = 0.2$$

$$\Rightarrow \quad \beta = \frac{0.2}{2} = 0.1$$

6.30 (d)

In a good control system, output is sensitive to input variations but insensitive to parameter variations.

6.32 (c)

K_p is for proportional controller and $K_d s$ is for derivative controller.

Therefore, it is proportional plus derivative controller.

6.33 (d)

K_p is for proportional controller, $K_d s$ is for derivative controller and $\frac{K_i}{s}$ is for integral controller.

Therefore, it is proportional plus derivative plus integral controller.

6.34 (c)

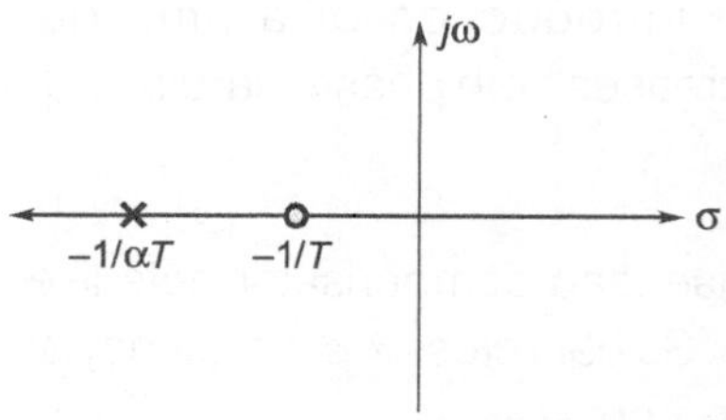

Transfer function of lead compensator

$$= \frac{\alpha(1+Ts)}{(1+\alpha Ts)}$$

where, $\alpha < 1$

6.36 (c)

$$\text{Transfer function} = \frac{10(s+1)}{(s+2)}$$

$$\Rightarrow \quad T(s) = \frac{10(s+1)}{(s+2)}$$

$$\Rightarrow \quad T(j\omega) = \frac{10(1+j\omega)}{(2+j\omega)}$$

Phase of the $T(j\omega)$ is given by

$$\angle T(j\omega) = \tan^{-1}\omega - \tan^{-1}\left(\frac{\omega}{2}\right)$$

Consider the input is $x(t) = \sin t$ then

$$\angle T(j\omega)\big|_{\omega=1} = \tan^{-1}1 - \tan^{-1}\frac{1}{2} > 0$$

Hence voltage response of the system leads the current.

6.37 (b)

For the given circuit

$$\frac{E_0(s)}{E_i(s)} = \frac{1+R_2Cs}{1+(R_1+R_2)Cs}$$

$$\Rightarrow \quad \frac{E_0(s)}{E_i(s)} = \frac{1+Ts}{1+\beta Ts}$$

Where $\quad T = R_2 C$ and $\beta = \dfrac{R_1+R_2}{R_2} > 1$

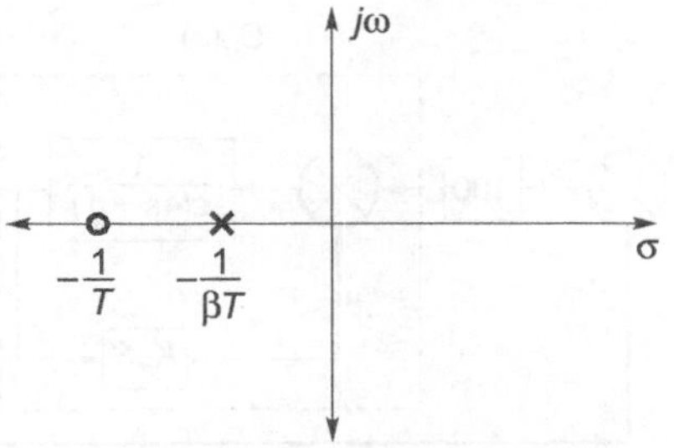

Hence given circuit represents a phase lag compensator.

6.38 (b)

For the given circuit

$$\frac{E_0(s)}{E_i(s)} = \frac{R_2(1+R_1Cs)}{R_1+R_2\left[1+\dfrac{R_1R_2Cs}{R_1+R_2}\right]}$$

$$\Rightarrow \quad \frac{E_0(s)}{E_i(s)} = \frac{\alpha(1+Ts)}{(1+\alpha Ts)}$$

Where $\quad \alpha = \dfrac{R_2}{R_1+R_2} < 1$

and $\quad T = R_1C$

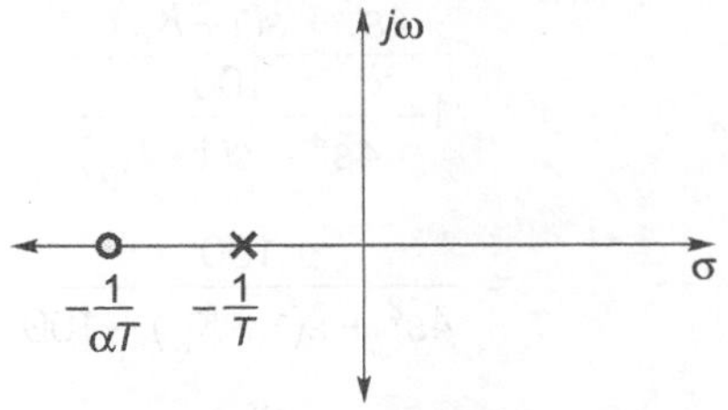

Hence given circuit represents a phase lead compensator.

6.39 (a)

Transfer function for a phase lag compensator is

$$H(s) = \frac{1+\tau s}{1+\alpha\tau s}$$

$$\Rightarrow \quad H(s) = \frac{\tau\left(s+\dfrac{1}{\tau}\right)}{\alpha\tau\left(s+\dfrac{1}{\alpha\tau}\right)} = \frac{\left(s+\dfrac{1}{\tau}\right)}{\alpha\left(s+\dfrac{1}{\alpha\tau}\right)}$$

6.40 (a)

Transfer function for a phase lead compensator

$$H(s) = \frac{\beta(1+\tau s)}{(1+\beta\tau s)}$$

6.41 (c)

$$E_0 = -\frac{R_2}{R_1}E_i$$

$$\Rightarrow \quad E_0 \propto E_i$$

Hence proportional controller.

6.42 (d)

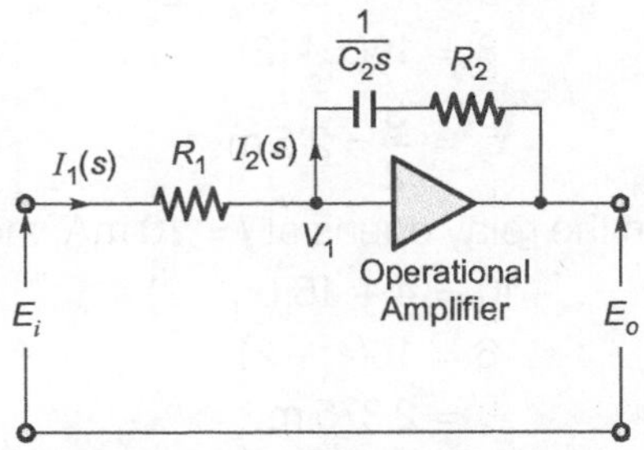

Applying KCL at node V_1 we have

$$\frac{E_i(s)-V_1}{R_1} = \frac{V_1-E_0(s)}{\dfrac{R_2C_2s+1}{C_2s}}$$

But $\quad V_1 = 0$; so

$$\frac{E_i(s)}{R_1} = \frac{-E_0(s)\,C_2s}{R_2C_2s+1}$$

$$\Rightarrow \quad -E_0(s) = E_i(s)\left(\frac{R_2C_2s+1}{C_2s}\right)\frac{1}{R_1}$$

$$= \left(\frac{R_2C_2s}{C_2sR_1}+\frac{1}{R_1C_2s}\right)E_i(s)$$

$$\Rightarrow \quad -E_0(s) = \frac{R_2}{R_1}E_i(s)+\frac{1}{R_1C_1s}E_i(s)$$

Taking inverse Laplace transform ; we have

$$-E_0(t) = \frac{R_2}{R_i}E_i(t)+\frac{1}{R_1C_1}\int E_i(t)$$

Hence proportional + Integral controller.

6.43 (d)

Integral controller reduces both the steady state error and the relative stability (because it adds one pole to the system).

6.44 (d)

$(P+I)$ controller improves the steady state error due to integral action but proportional action improves the transient response marginally by speeding up the transients.

6.45 (b)

Since the liquid level controller linearly converts a displacement of 2 m to 3 m into 4-20 mA control signal. It can be represented as

$$I = 4 + K(x-2)\text{ mA}$$

i.e. when $x = 2$ m $I = 4$ mA

So when $x = 3$ m, then $I = 20$ mA

$$20 = 4 + K(3-2)$$

$$\Rightarrow \quad 16 = K$$

$$\therefore \quad I = 4 + 16\,(x-2)$$

When the relay closed at $I = 12$ mA, then

$$12 = 4 + 16\,(x_2 - 2)$$

$$\Rightarrow \quad 8 = 16(x_2 - 2)$$

$$x_2 = \frac{5}{2} = 2.5\text{ m}$$

When the relay opens at $I = 10$ mA, then

$$10 = 4 + 16\,(x_1 - 2)$$

$$\Rightarrow \quad 6 = 16(x_1 - 2)$$

$$x_1 = 2.375\text{ m}$$

So hysteresis zone $= x_2 - x_1$

$= 0.125$ m

6.48 (a)

$$G(s) = \frac{10(1+0.04s)}{(1+0.01s)}$$

Comparing with the standard phase lead compensating network

$$= \frac{\alpha(1+T_1 s)}{(1+\alpha T_1 s)}$$

$$T_1 = 0.04$$

$$\alpha T_1 = 0.01$$

So, maximum phase lead occurs at frequency ω_m

i.e. $$\omega_m = \frac{1}{\sqrt{\alpha T}\cdot\sqrt{T}} = \frac{1}{\sqrt{0.04}\times\sqrt{0.01}}$$

$$= 50\text{ rad/sec}$$

6.49 (a)

The effect of addition of a zero to a transfer function is providing a phase lead.

6.50 (c)

$$G(s) = \frac{\left(1+\frac{s}{4}\right)}{\left(1+\frac{s}{25}\right)}$$

Comparing it with the standard transfer function of phase lead compensator

$$G(s) = \frac{\alpha(1+Ts)}{(1+\alpha Ts)}$$

$$T = \frac{1}{4}, \quad \alpha T = \frac{1}{25}$$

Now, frequency ω_m occurs at $= \sqrt{\frac{1}{\alpha T}\cdot\frac{1}{T}}$

$$= \sqrt{25\times 4} = 10\text{ rad/sec.}$$

6.51 (b)

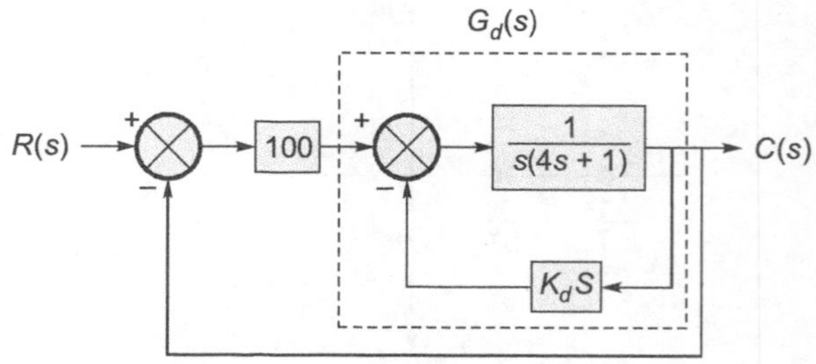

Converting the blocked portion into simplified form as

$$G_d(s) = \frac{\frac{1}{s(4s+1)}}{1+\frac{sK_d}{s(4s+1)}} = \frac{1}{4s^2 + s(1+K_d)}$$

Now,

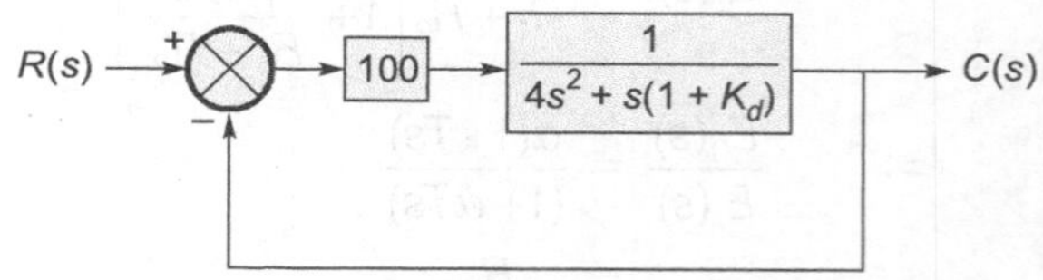

Now, simplifying the above block diagram as

$$G(s) = \frac{\frac{100}{4s^2 + s(1+K_d)}}{1+\frac{100}{4s^2 + s(1+K_d)}}$$

$$= \frac{100}{4s^2 + s(1+K_d) + 100}$$

$$= \frac{25}{s^2 + \frac{s(1+K_d)}{4} + 25}$$

Comparing it with standard equation as

$$\omega_n = 5\text{ rad/sec.}$$

$$2\xi\omega_n = \left(\frac{1+K_d}{4}\right) \qquad \text{Given } \xi = 0.5$$

$$5 = \frac{1+K_d}{4}$$

$$\Rightarrow \quad \boxed{K_d = 19}$$

6.52 (b)

Feedback compensation is preferred to series compensation or cascade compensation because:

(i) In a feedback compensator energy transfer is from higher energy level to a lower one thereby obviating the need for feedback amplification.

(ii) In non-electrical systems, suitable cascade devices may not be available.

(iii) Feedback compensation often provides greater stiffness against load disturbances.

6.53 (a)

PI is proportional + integral controller. Thus integral controller add a pole either at origin or at the desired point on the negative real axis, so order of the system gets increased but it reduces the steady state error or the offset error.

6.54 (d)

- The integral controller produces a control action for constant error not derivative controller.
- A derivative controller improves the transient state of the control system apart from increasing bandwidth, less overshoot, low rise time.
- Since derivative controller increases ξ, so peak overshoot reduces. Bandwidth also increases which causes reduction in rise time, peak-time.

6.55 (c)

When delay element ($e^{-T_d s}$) is introduces in a control system, it only affects the phase margin of the system and the new phase margin has decreased value as

$$PM = 180° + \angle GH = \omega T_d$$

6.56 (c)

For, $G_1(s) = \dfrac{5(s+10)}{(s+50)}$ ...(i)

Comparing equation (i) with standard equation of a compensator as $\dfrac{(1+s\tau)}{(1+\alpha s\tau)}$ ...(ii)

We get,

$$\tau = \frac{1}{10} \text{ and } \alpha\tau = \frac{1}{50}$$

or $$\alpha = \frac{1}{50} \times 10 = \frac{1}{5}$$

$\because$ $\alpha < 1$; lead compensator

Similarly for $G_2(s)$

We get, $\tau = \dfrac{1}{50}$ and $\alpha\tau = \dfrac{1}{10}$

$\therefore$ $\alpha = 5$

$\because$ $\alpha > 1$; lag compensator

6.57 (a)

The PD controller is used for the improvement of transient response of the system where as for study state error a PI controller is used.

6.58 (c)

The proportional plus Integral controller is

$$K_P + \frac{K_I}{s} \quad ...(i)$$

From the given options only, option (c) satisfies the above equation. Hence (c) is the correct explanation for *PI* controller.

6.59 (a)

Transfer function of controller

$$G_e(s) = \frac{1+3s}{1+s}$$

$$\text{zero} = \frac{-1}{3}$$

$$\text{pole} = -1$$

Thus the given controller is a phase lead controller.

total phase $\delta = \tan^{-1}(3\omega) - \tan^{-1}(\omega)$

$$\frac{d\delta}{d\omega} = \frac{3}{1+9\omega^2} - \frac{1}{1+\omega^2} = 0$$

$3 + 3\omega^2 - 1\,9\omega^2 = 0$

$$6\omega^2 = 2$$

$$\omega = \frac{1}{\sqrt{3}}$$

$$\phi = \tan^{-1}\left(\frac{3}{\sqrt{3}}\right) - \tan^{-1}\left(\frac{1}{\sqrt{3}}\right)$$

$$= 60° - 30° = 30°$$

6.60 (*)

Effects of phase lead network

1. Phase margin increases
2. Bandwidth increases
3. Response becomes faster
4. Steady state error is not affected. Hence velocity constant remain same.

Hence only (2) and (3) are correct.

6.61 (a)

A lead compensator provides a better gain margin by introducing a positive shift in the phase plot and thus increasing the phase crossover frequency. Thus the system becomes comparatively more stable by using a lead compensator.

6.62 (a)

A phase lead controller improves the system transient response.

6.63 (d)

$$G_c(s) = \frac{1+0.5s}{1+0.05s}$$

Zero : $S = -2$; pole : $S = -20$; $\alpha = \frac{Z}{P} = 0.1$

$$\therefore \quad \phi_M = \sin^{-1}\left[\frac{1-\alpha}{1+\alpha}\right] = \sin^{-1}[0.82]$$

$$\therefore \quad \omega_M = \sqrt{Z \cdot P} = \sqrt{40} \approx 6 \text{ rad/sec}$$

6.64 (d)

Lead compensator is high pass filter hence it increases bandwidth Lag compensator is low pass filter hence it decreases bandwidth.

6.65 (b)

$$GH(s) = \frac{K}{1+sT}; G_c(s) = K_P$$

$$\text{CLTF} = \frac{KK_p}{1+sT+KK_p} = \frac{\frac{KK_p}{1+KK_p}}{\frac{T}{KK_P}s+1}$$

$$\Rightarrow \text{Time constant} = \frac{T}{1+KK_P}$$

$$\Rightarrow \quad e_{ss} = \frac{1}{KK_P}$$

$\therefore$ If K_p is increased then both time constant and e_{ss} decreases.

6.66 (a)

Tachogenerator (or) derivative controller mainly used to increase system damping

$$\xi_{new} = \xi_{old} + \frac{K_D\omega_n}{2}$$

K_D = tachometer constant

6.67 (b)

PD controller increases system bandwidth since it is analogous to high pass filter. Hence it is highly sensitive to noise because Bandwidth ∝ Noise.

6.68 (a)

In derivative control action i.e. PD controller

$$a(t) \propto \left[e(t) + K_D\frac{d}{dt}e(t)\right]$$

6.69 (a)

In PID controller,

$$a(t) = \pm K_P e(t) + K_D\frac{d}{dt}e(t) + K_I\int_0^t e(t)dt$$

$$K_I = \frac{1}{T_r}$$

6.70 (a)

PID controller can fix the error due to any measured distance hence it is preferred in industry.

6.71 (a)

Lead compensator means zero dominates pole.

$$\frac{a}{b} < 1$$

$$\therefore \quad a < b$$

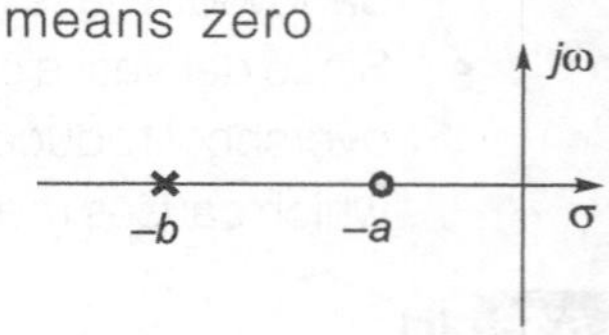

6.72 (b)

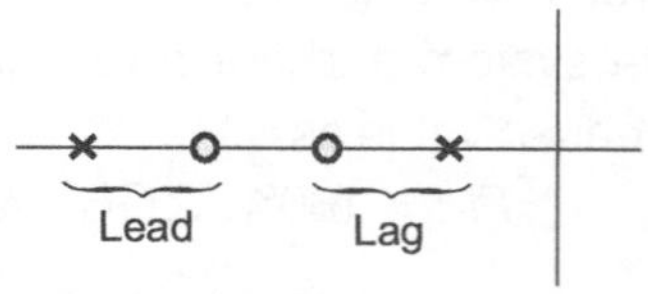

$\therefore$ Lag lead compensator

6.73 (b)

In PID controller,

$$Y(t) = K_1x(t) + K_2\int_0^t x(t)\cdot dt + K_3\frac{d}{dt}x(t)$$

$$TF\frac{Y(s)}{X(s)} = K_1 + \frac{K_2}{s} + K_3s$$

6.74 (b)

$$G_c(s) = \frac{s+z_c}{s+p_c} < 1$$

1. For lead compensator $\frac{Z_c}{P_c} < 1$ i.e. zero should be dominant.
2. T(time constant) is a positive value.

6.75 (d)

$$G_c = \frac{s+a}{s+b}$$

$$M \text{ at } \omega_m = \frac{1}{\sqrt{\alpha}} \text{ (amplification)}$$

$$\text{Attenuation} = \sqrt{\alpha} = \sqrt{\frac{a}{b}}$$

$$\therefore \quad \alpha = \frac{a}{b}$$

6.76 (a)

Lag compensator is analogous to PI controller, which improves steady state response.

6.77 (b)

Integral controller is also called reset controller.

6.78 (c)

The advantages of a passive lead network over an active PD controller is that no additional power supplies are required and noise due to differentiation is reduced.

With the addition of zero in PD controller, root locus shift towards left half of the s-plane. However, in passive lead network, a single zero cannot be produced; rather a compensator zero and a pole result. The addition of pole does not reduce the number of branches of the root locus that cross the imaginary axis into the right-half plane. On the other hand, the addition of the single zero of the PD controller tends to reduce the number of branches of root locus that cross into the right-half s-plane.

Hence, statements 3 and 4 are correct.

6.79 (d)

- Parallel compensation also known as feedback compensation improves the bandwidth and hence, the faster speed of response.
- The environmental conditions in which the feedback control system is to be utilized affect the stability of the controlled quantity.
- The cascaded lead-lag compensator improves the stability and the steady-state error. Hence, the degree of accuracy and the stability of a control system can be improved by the use of a cascade compensator.

Therefore, all the given statements are correct.

6.80 (c)

Introduction of a lag compensator (high pass filter) results in attenuation of the high frequency signals, thereby reducing the noise signal level relative to the control signal. Hence, statement 1 is incorrect.

6.81 (b)

Compensators that use pure integration for improving steady state error is known as PI controller. The lag compensator is used when the cascade compensator does not employ pure integration.

6.82 (a)

Derivative controller is functionally a tachometer.

$\therefore$ Transfer function = $K_t S$

6.83 (d)

- The derivative action in the PD controller helps to react quickly to sudden changes or disturbances, improving the system's response. However, it doesn't eliminate offset errors which requires integral action.
- In a PI controller, proportionality constant provides simplicity, reliability, directness, etc. Disadvantage of offset is eliminated by integral action. But this system will have some oscillatory effect.

Hence, all the statements are correct.

■■■

7 State Space Analysis

7.1 Consider the following equations for the state transition matrix of the linear time-invariant continuous time system '$\phi(t)$':

1. $\phi(-t) = [\phi(t)]^{-1}$.
2. $[\phi(t)]^k = \phi(t^k)$ for any positive integer 'k'.
3. $\phi(t - t_0) = \phi(t)\,\phi(-t_0)$ for any constant t_0.

Which of these equations correctly define the properties of the given system?

(a) 1, 2, and 3 (b) 1 and 2
(c) 1 and 3 (d) 2 and 3 **[ESE-2000]**

7.2 For the system described by $\dot{X} = AX$ match **List-I** (Matrix A) with **List-II** (Position of eigenvalues) and select the correct answer:

List-I	List-II
A. $\begin{bmatrix} -1 & 2 \\ 0 & -2 \end{bmatrix}$	**1.** One eigenvalue at the origin
B. $\begin{bmatrix} -1 & -2 \\ -2 & -4 \end{bmatrix}$	**2.** Both the eigenvalues in the LHP
C. $\begin{bmatrix} 0 & -1 \\ 1 & 0 \end{bmatrix}$	**3.** Both the eigenvalues in RHP
D. $\begin{bmatrix} 1 & 0 \\ 2 & 4 \end{bmatrix}$	**4.** Both the eigenvalues on the imaginary axis

Codes:

	A	B	C	D
(a)	2	1	3	4
(b)	2	1	4	3
(c)	1	2	4	3
(d)	1	2	3	4

[ESE-2001]

7.3 For the system dynamics described by differential equation $\ddot{y} + 3\dot{y} + 2y = u(t)$ the transfer function of the system represented in controllable canonical form is $C[sI - A]^{-1}B$. The matrix A would be

(a) $\begin{bmatrix} 0 & 1 \\ -3 & -2 \end{bmatrix}$ (b) $\begin{bmatrix} 0 & 1 \\ 2 & 3 \end{bmatrix}$

(c) $\begin{bmatrix} 0 & 1 \\ -2 & -3 \end{bmatrix}$ (d) $\begin{bmatrix} 0 & 2 \\ 0 & 3 \end{bmatrix}$ **[ESE-2001]**

7.4 A system can be represented in the form of state equations as

$S(n+1) = AS(n) + Bx(n)$

$y(n) = CS(n) + Dx(n)$

Where A, B, C and D are matrices, $S(n)$ is the state vector. $x(n)$ is the input and $y(n)$ is the output. The transfer function of the system $H(z) = Y(z)/X(z)$ is given by

(a) $A(zI - B)^{-1}\,C + D$
(b) $B(zI - C)^{-1}\,D + A$
(c) $C(zI - A)^{-1}\,B + D$
(d) $D(zI - A)^{-1}\,C + B$ **[ESE-2002]**

7.5 Which one of the following is the state-space model of the circuit shown below?

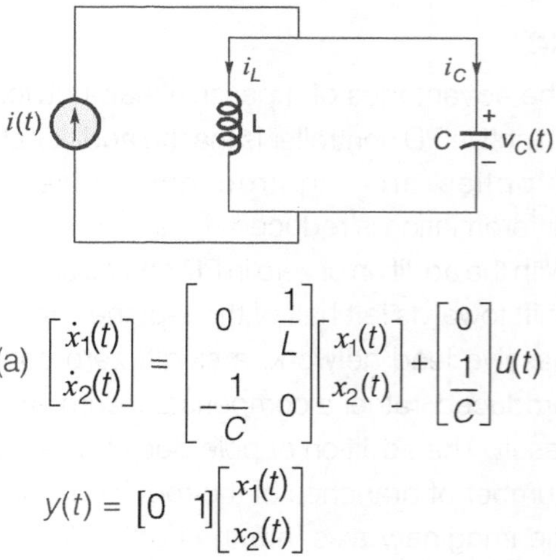

(a) $\begin{bmatrix} \dot{x}_1(t) \\ \dot{x}_2(t) \end{bmatrix} = \begin{bmatrix} 0 & \frac{1}{L} \\ -\frac{1}{C} & 0 \end{bmatrix} \begin{bmatrix} x_1(t) \\ x_2(t) \end{bmatrix} + \begin{bmatrix} 0 \\ \frac{1}{C} \end{bmatrix} u(t)$

$y(t) = [0 \quad 1] \begin{bmatrix} x_1(t) \\ x_2(t) \end{bmatrix}$

(b) $\begin{bmatrix} \dot{x}_1(t) \\ \dot{x}_2(t) \end{bmatrix} = \begin{bmatrix} 0 & 1 \\ -\frac{1}{C} & -\frac{1}{L} \end{bmatrix} \begin{bmatrix} x_1(t) \\ x_2(t) \end{bmatrix} + \begin{bmatrix} 0 \\ 1 \end{bmatrix} u(t)$

$y(t) = [0 \quad 1] \begin{bmatrix} x_1(t) \\ x_2(t) \end{bmatrix}$

(c) $\begin{bmatrix} \dot{x}_1(t) \\ \dot{x}_2(t) \end{bmatrix} = \begin{bmatrix} 0 & \frac{1}{L} \\ -\frac{1}{C} & 0 \end{bmatrix} \begin{bmatrix} x_1(t) \\ x_2(t) \end{bmatrix} + \begin{bmatrix} 1 \\ 0 \end{bmatrix} u(t)$

$y(t) = [0 \quad 1] \begin{bmatrix} x_1(t) \\ x_2(t) \end{bmatrix}$

(d) $\begin{bmatrix} \dot{x}_1(t) \\ \dot{x}_2(t) \end{bmatrix} = \begin{bmatrix} 0 & \frac{1}{L} \\ -\frac{1}{C} & 0 \end{bmatrix} \begin{bmatrix} x_1(t) \\ x_2(t) \end{bmatrix} + \begin{bmatrix} 0 \\ 1 \end{bmatrix} u(t)$

$y(t) = [0 \quad 1] \begin{bmatrix} x_1(t) \\ x_2(t) \end{bmatrix}$ [ESE-2003]

7.6 With the conventional notation $\dot{X} = AX + Bu$ for the state description of a linear time-invariant network, examine the validity of the following statements relating to the matrix A:
1. A is symmetrical if the network is reciprocal.
2. The sum of the natural frequencies of the network is equal to the determinant of A.

Which of these statements is/are true?
(a) Both 1 and 2 (b) 1 only
(c) 2 only (d) Neither 1 nor 2
[ESE-2003]

7.7 Which one of the following is NOT a correct statement about the state-space model of a physical system?
(a) State-space model can be obtained only for a linear system
(b) Eigenvalues of the system represent the roots of the characteristic equation
(c) $\dot{X} = AX + Bu$ represents linear state-space model of a physical system
(d) $X(t)$ represents the state vector of the system
[ESE-2003]

7.8 A linear constant parameter continuous-time system has the following A matrix in its state representation $\dot{X} = AX + Bu$.

$$A = \begin{bmatrix} 0 & \alpha \\ \alpha & 0 \end{bmatrix}$$

Which one of the following is the natural response of the system following the application of any input?
(a) $Me^{\alpha t} + Ne^{-\alpha t}$ (b) $(M + Nt)e^{\alpha t}$
(c) $(M + Nt)e^{-\alpha t}$ (d) $M\cos\alpha t + N\sin\alpha t$
[ESE-2004]

7.9 If the state space equation of a system is

$$\dot{x} = \begin{bmatrix} -3 & 1 \\ 0 & -2 \end{bmatrix} x$$

and the initial state value $x(0) = [10 \ -10]^T$, the steady-state value is $[A \ B]^T$, what are the respective values of A and B?
(a) 0, 0 (b) –3, –2
(c) 0, 1 (d) ∞, ∞
[ESE-2008]

7.10 Transfer function of a certain system is

$$\frac{Y(s)}{U(s)} = \frac{1}{s^4 + 5s^3 + 8s^2 + 6s + 3}$$

Which one of the following will be the A, B matrix pair of state variable representation of this system?

(a) $\begin{bmatrix} 0 & 1 & 0 & 0 \\ 0 & 0 & 1 & 0 \\ 0 & 0 & 0 & 1 \\ -3 & -6 & -8 & -5 \end{bmatrix} \begin{bmatrix} 0 \\ 0 \\ 0 \\ 1 \end{bmatrix}$

(b) $\begin{bmatrix} 0 & 1 & 0 & 0 \\ 0 & 0 & 1 & 0 \\ 0 & 0 & 0 & 1 \\ -5 & -8 & -6 & -3 \end{bmatrix} \begin{bmatrix} 0 \\ 0 \\ 0 \\ 1 \end{bmatrix}$

(c) $\begin{bmatrix} 1 & 0 & 0 & 0 \\ 0 & 1 & 0 & 0 \\ 0 & 0 & 1 & 0 \\ -3 & -6 & -8 & -5 \end{bmatrix} \begin{bmatrix} 1 \\ 0 \\ 0 \\ 0 \end{bmatrix}$

(d) $\begin{bmatrix} 0 & 1 & 0 & 0 \\ 0 & 0 & 1 & 0 \\ 0 & 0 & 0 & 1 \\ -3 & -6 & -8 & -5 \end{bmatrix} \begin{bmatrix} 1 \\ 0 \\ 0 \\ 0 \end{bmatrix}$ [ESE-2009]

7.11 Separation property of state-transition matrix is
(a) $\phi(t - t_0) = \phi(t)\,\phi(t_0)$
(b) $\phi(t - t_0) = \phi^{-1}(t)\,\phi(t_0)$
(c) $\phi(t - t_0) = \phi(t)\,\phi^{-1}(t_0)$
(d) $\phi(t - t_0) = \phi^{-1}(t)\,\phi^{-1}(t_0)$
[ESE-2009]

7.12 Number of state variables of discrete time system, described by

$y[n] - \frac{3}{4}y[n-1] + \frac{1}{8}y[n-2] = x[n]$ is
(a) 2 (b) 3
(c) 4 (d) 1
[ESE-2010]

7.13 The transfer functions for the state representation of continuous time LTI system:

$$\dot{q}(t) = Aq(t) + Bx(t)$$
$$y(t) = Cq(t) + Dx(t)$$

is given by
(a) $[C(sI-A)^{-1}]B+D$
(b) $B(sI-A)^{-1}B+D$
(c) $C[(sI-A)^{-1}B]+D$
(d) $D(sI-A)^{-1}B+C$ [ESE-2010]

7.14 Consider a system described by the state model

$$\dot{X} = \begin{bmatrix} 2 & 1 \\ -1 & 2 \end{bmatrix} X + \begin{bmatrix} 1 \\ 1 \end{bmatrix} U$$

$$Y = [1 \;\; 1]X$$

The system is
(a) controllable but not observable
(b) uncontrollable and observable
(c) both controllable and observable
(d) neither controllable nor observable

[ESE-2012]

7.15 The system represented by the state-variable model

$$\dot{X} = \begin{bmatrix} 0 & -1 \\ 1 & -2 \end{bmatrix} X + \begin{bmatrix} 1 \\ 2 \end{bmatrix} U \text{ is}$$

(a) oscillatory (b) critically damped
(c) over-damped (d) under-damped

[ESE-2012]

7.16 A second-order system represented by state variables has

$$A = \begin{bmatrix} -2 & -4 \\ 1 & 0 \end{bmatrix}$$

The value of natural frequency and damping factor are respectively
(a) 2 and 0.5 (b) 2 and 1
(c) 1 and 2 (d) 0.5 and 2

[ESE-2012]

7.17 The system described by the following state equations

$$\dot{X} = \begin{bmatrix} 0 & 1 \\ 2 & -3 \end{bmatrix} X + \begin{bmatrix} 0 \\ 1 \end{bmatrix} u, \; Y = [1, \;\; 1]X$$

1. Completely controllable
2. Completely observable

Which of the above statements is/are correct?
(a) 1 only (b) 2 only
(c) Both 1 and 2 (d) Neither 1 nor 2

[ESE-2016]

7.18 The state space representation of a linear time invariant system is

$$X(t) = A\,X(t) + Bu(t) \quad ; \quad Y(t) = CX(t)$$

What is the, transfer function H(s) of the system?
(a) $C(sI-A)^{-1}B$ (b) $B(sI-A)^{-1}C$
(c) $C(sI-A)B$ (d) $B(sI-A)C$

[ESE-2016]

7.19 The dynamics of an nth order single-input single-output system can be written by the vector-matrix differential equation $\dot{x} = Ax + Bu$ and output $y = Cx$. What is the size of the output coupling matrix C?
(a) $n \times n$ (b) $n \times 1$
(c) $1 \times n$ (d) 1×1

[ESE-2022]

7.20 For the state variable formulation of R-L-C network, which one of the following statements is correct?
(a) The voltage across the inductor and the current through the capacitor are chosen as the state variables.
(b) Only the current through the inductor is chosen as the state variable.
(c) Only the voltage across the capacitor is chosen as the state variable.
(d) The current through the inductor and the voltage across the capacitor are chosen as the state variables. [ESE-2022]

7.21 If $i(t) = I_0 \sin\omega t$, then the state variable formulation of the *R-L-C* circuit shown in the figure is

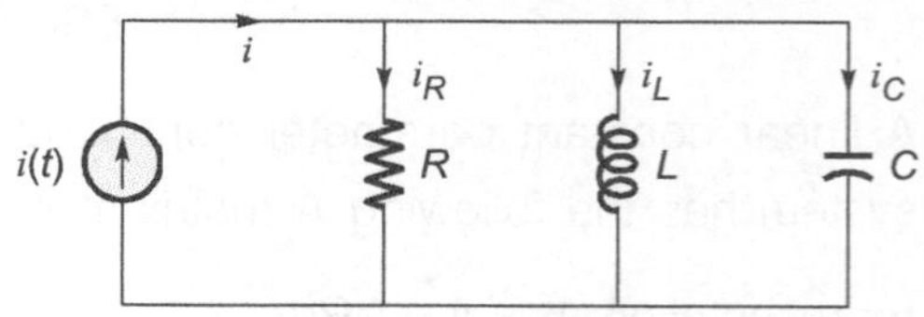

(a) $$\begin{bmatrix} \dot{x}_1 \\ \dot{x}_2 \end{bmatrix} = \begin{bmatrix} 0 & 1 \\ -\dfrac{1}{LC} & -\dfrac{1}{RC} \end{bmatrix} \begin{bmatrix} x_1 \\ x_2 \end{bmatrix} + \begin{bmatrix} 0 \\ I_0(\omega/C)\cos\omega t \end{bmatrix}$$

(b) $$\begin{bmatrix} \dot{x}_1 \\ \dot{x}_2 \end{bmatrix} = \begin{bmatrix} 0 & 1 \\ \dfrac{1}{LC} & \dfrac{1}{RC} \end{bmatrix} \begin{bmatrix} x_1 \\ x_2 \end{bmatrix} + \begin{bmatrix} 0 \\ I_0(\omega/C)\sin\omega t \end{bmatrix}$$

(c) $$\begin{bmatrix} \dot{x}_1 \\ \dot{x}_2 \end{bmatrix} = \begin{bmatrix} 0 & 1 \\ \dfrac{1}{LC} & \dfrac{1}{RC} \end{bmatrix} \begin{bmatrix} x_1 \\ x_2 \end{bmatrix} + \begin{bmatrix} 0 \\ I_0(\omega/C)\cos\omega t \end{bmatrix}$$

(d) $$\begin{bmatrix} \dot{x}_1 \\ \dot{x}_2 \end{bmatrix} = \begin{bmatrix} 0 & 1 \\ -\dfrac{1}{LC} & -\dfrac{1}{RC} \end{bmatrix} \begin{bmatrix} x_1 \\ x_2 \end{bmatrix} + \begin{bmatrix} 0 \\ I_0(\omega/C)\sin\omega t \end{bmatrix}$$

[ESE-2022]

■■■

Answers State Space Analysis

7.1 (c) 7.2 (b) 7.3 (c) 7.4 (c) 7.5 (a) 7.6 (a) 7.7 (a) 7.8 (a) 7.9 (a)
7.10 (a) 7.11 (c) 7.12 (a) 7.13 (c) 7.14 (c) 7.15 (b) 7.16 (b) 7.17 (c) 7.18 (a)
7.19 (c) 7.20 (d) 7.21 (d)

Explanations State Space Analysis

7.1 (c)

Properties of state transition matrix:

(i) $\phi(0) = e^{A0} = I$

(ii) $\phi^{-1}(t) = \phi(-t)$

(iii) $\phi(t_1 + t_2) = \phi(t_1)\,\phi(t_2) = \phi(t_2)\,\phi(t_1)$

7.2 (b)

Eigen values are the roots of $|sI - A| = 0$

Let $A = \begin{bmatrix} -1 & 2 \\ 0 & -2 \end{bmatrix}$

$$[sI - A] = \begin{bmatrix} s+1 & -2 \\ 0 & s+2 \end{bmatrix}$$

$$|sI - A| = (s+1)(s+2)$$

$$|sI - A| = 0$$

$\Rightarrow (s+1)(s+2) = 0$

$\Rightarrow \quad s = -1, -2$

Thus both the eigen values are in the LHP.

Let $A = \begin{bmatrix} -1 & -2 \\ -2 & -4 \end{bmatrix}$

$$[sI - A] = \begin{bmatrix} s+1 & 2 \\ 2 & s+2 \end{bmatrix}$$

$$|sI - A| = 0$$

$\Rightarrow (s+1)(s+4) - 4 = 0$

$\Rightarrow s^2 + 5s + 4 - 4 = 0$

$\Rightarrow \quad s(s+5) = 0$

$\Rightarrow \quad s = 0, -5$

Thus one eigen value is at the origin.

Let $A = \begin{bmatrix} 0 & -1 \\ 1 & 0 \end{bmatrix}$

$$[sI - A] = \begin{bmatrix} s & 1 \\ -1 & s \end{bmatrix}$$

$$|sI - A| = 0$$

$\Rightarrow \quad s^2 + 1 = 0$

$\Rightarrow \quad s = \pm j1$

Thus both the eigen values are on the imaginary axis.

Let $A = \begin{bmatrix} 1 & 0 \\ 2 & 4 \end{bmatrix}$

$$[sI - A] = \begin{bmatrix} s-1 & 0 \\ -2 & s-4 \end{bmatrix}$$

$$|sI - A| = 0$$

$\Rightarrow (s-1)(s-4) = 0$

$\Rightarrow \quad s = 1, 4$

Thus both the eigen values are in the RHP.

7.3 (c)

General representation of phase variable representation:

$\dot{X} = AX + BU \qquad Y = CX + DU$

where

$$A = \begin{bmatrix} 0 & 1 & 0 & 0 & ....0 \\ 0 & 0 & 1 & 0 & ....0 \\ 0 & 0 & 0 & 1 & ....0 \\ \vdots & \vdots & \vdots & & \\ -a_n & -a_{n-1} & -a_{n-2} & .... & ....0 \end{bmatrix}$$

$$B = \begin{bmatrix} 0 \\ 0 \\ \vdots \\ b \end{bmatrix}$$

$$C = [1\ 0\ 0\ ....\ 0]$$

$$D = [0]$$

Differential equation is

$$\frac{d^n y}{dt^n} + a_1 \frac{d^{n-1} y}{dt^{n-1}} + a_2 \frac{d^{n-2} y}{dt^{n-2}} + ... + a_n y = bx$$

Comparing the given equation

$$\ddot{y} + 3\dot{y} + 2y = u(t)$$

with $a_1 \dot{y} + a_1 \dot{y} + a_2 y = bx$

$$A = \begin{bmatrix} 0 & 1 \\ -a_2 & -a_1 \end{bmatrix} \Rightarrow A = \begin{bmatrix} 0 & 1 \\ -2 & -3 \end{bmatrix}$$

7.5 (a)

$$i_c(t) = \frac{Cdv_c(t)}{dt}$$

$$\Rightarrow \quad \frac{dv_c(t)}{dt} = \frac{1}{C}i_c(t) = \frac{1}{C}\{i(t) - i_L(t)\}$$

$$v_c(t) = \frac{Ldi_L(t)}{dt}$$

$$\Rightarrow \quad \frac{di_L(t)}{dt} = \frac{1}{L}v_c(t)$$

Let

$$i_L(t) = x_1(t),$$

$$v_c(t) = x_2(t),\ y(t) = v_c(t),$$

$$i(t) = u(t)$$

Arranging the equations,

$$\dot{x}_1(t) = 0 \cdot x_1(t) + \frac{1}{L} \cdot x_2(t)$$

$$\dot{x}_2(t) = -\frac{1}{C} \cdot x_1(t) + 0 \cdot x_2(t) + \frac{1}{C} \cdot u(t)$$

$$y(t) = 0.x_1(t) + 1.x_2(t)$$

Arranging the above equations in matrix form,

$$\begin{bmatrix} \dot{x}_1(t) \\ \dot{x}_2(t) \end{bmatrix} = \begin{bmatrix} 0 & 1/L \\ -1/C & 0 \end{bmatrix}\begin{bmatrix} x_1(t) \\ x_2(t) \end{bmatrix} + \begin{bmatrix} 0 \\ 1/C \end{bmatrix}u(t)$$

$$y(t) = [0 \quad 1]\begin{bmatrix} x_1(t) \\ x_2(t) \end{bmatrix}$$

7.7 (a)

State space model can also be obtained for a non-linear, time-variant system.

7.8 (a)

$$A = \begin{bmatrix} 0 & \alpha \\ \alpha & 0 \end{bmatrix}$$

$$sI - A = \begin{bmatrix} s & -\alpha \\ -\alpha & s \end{bmatrix}$$

$$sI - A = 0$$

$$\Rightarrow \quad s^2 - \alpha^2 = 0$$

$$\Rightarrow \quad s = \alpha, -\alpha$$

Complementary function = $M e^{\alpha t} + N e^{-\alpha t}$

Therefore, the natural response of the given system is $M e^{\alpha t} + N e^{-\alpha t}$.

7.9 (a)

$$A = \begin{bmatrix} -3 & 1 \\ 0 & -2 \end{bmatrix}$$

$$[sI - A] = \begin{bmatrix} s+3 & -1 \\ 0 & s+2 \end{bmatrix}$$

$$[sI - A]^{-1} = \frac{1}{(s+3)(s+2)}\begin{bmatrix} s+2 & 1 \\ 0 & s+3 \end{bmatrix}$$

$$[sI - A]^{-1} = \begin{bmatrix} \dfrac{1}{s+3} & \dfrac{1}{(s+3)(s+2)} \\ 0 & \dfrac{1}{(s+2)} \end{bmatrix}$$

The steady state value is

$$= \lim_{s\to 0} s \cdot [sI - A]^{-1} \times (0)$$

$$= \lim_{s\to 0} s \cdot \begin{bmatrix} \dfrac{1}{s+3} & \dfrac{1}{(s+2)(s+3)} \\ 0 & \dfrac{1}{(s+2)} \end{bmatrix}\begin{bmatrix} 10 \\ -10 \end{bmatrix}$$

$$= \lim_{s\to 0} s \cdot \begin{bmatrix} \dfrac{10}{s+3} - \dfrac{10}{(s+2)(s+3)} \\ -\dfrac{10}{s+2} \end{bmatrix}$$

$$= \begin{bmatrix} 0 \\ 0 \end{bmatrix}$$

Hence, $A = 0$; $B = 0$

7.10 (a)

$(s^4 + 5s^3 + 8s^2 + 6s + 3)\ Y(s) = U(s)$ ...(i)

Let $x_1 = y$ ...(ii)

$x_2 = \dot{x}_1$ ...(iii)

$x_3 = \dot{x}_2$ ...(iv)

$x_4 = \dot{x}_3$ ...(v)

So, transfer function equation can be written as

$\dot{x}_4 + 5x_4 + 8x_3 + 6x_2 + 3x_1 = u$

$\dot{x}_4 = -3x_1 - 6x_2 - 8x_3 - 5x_4 + u$ ...(vi)

Writing above equations in matrix from,

$$\begin{bmatrix} \dot{x}_1 \\ \dot{x}_2 \\ \dot{x}_3 \\ \dot{x}_4 \end{bmatrix} = \begin{bmatrix} 0 & 1 & 0 & 0 \\ 0 & 0 & 1 & 0 \\ 0 & 0 & 0 & 1 \\ -3 & -6 & -8 & -5 \end{bmatrix}\begin{bmatrix} x_1 \\ x_2 \\ x_3 \\ x_4 \end{bmatrix} + \begin{bmatrix} 0 \\ 0 \\ 0 \\ 1 \end{bmatrix}u$$

7.11 (c)

State transition matrix $[\phi(t)] = e^{At}$

where A is system matrix.

Now, $\phi(t - t_0) = e^{A(t-t_0)} = e^{At}\, e^{-At_0}$

$= \phi(t) \cdot \phi(-t_0)$ or $= \phi(t) \cdot \phi^{-1}(t_0)$

7.12 (a)

Order of difference equation = 2.

$\therefore$ Number of state variables = 2.

7.13 (c)

$$\dot{q}(t) = Aq(t) + Bx(t)$$
$$y(t) = Cq(t) + Dx(t)$$

Taking Laplace transform of above equations

$$s \cdot q(s) = Aq(s) + BX(s)$$
$$Y(s) = Cq(s) + DX(s)$$
$$q(s)[sI - A] = BX(s)$$
$$q(s) = [sI - A]^{-1} \cdot BX(s)$$
$$\therefore \quad Y(s) = C[sI - A]^{-1} \cdot BX(s) + DX(s)$$
$$\Rightarrow \quad Y(s) = \left[C[sI - A]^{-1} \cdot B + D\right]X(s)$$
$$\Rightarrow \quad \boxed{\frac{Y(s)}{X(s)} = C[(sI - A)^{-1}B] + D}$$

7.14 (c)

$$\dot{X} = \begin{bmatrix} 2 & 1 \\ -1 & 2 \end{bmatrix} X + \begin{bmatrix} 1 \\ 1 \end{bmatrix} U$$
$$y = [1 \quad 1]X$$
$$A = \begin{bmatrix} 2 & 1 \\ -1 & 2 \end{bmatrix}, \; B = \begin{bmatrix} 1 \\ 1 \end{bmatrix}, \; C = [1 \quad 1]$$

To check controllability

$$Q_c = [B \quad AB] = \begin{bmatrix} 1 & 3 \\ 1 & 1 \end{bmatrix}$$

For the system to be controllable $|Q_C| \neq 0$

Here $|Q_C| = -3 \neq 0$,

Therefore system is controllable

To check observability

$$Q_0 = [C^T \quad A^T C^T]$$
$$Q_0 = \begin{bmatrix} 1 & 1 \\ 1 & 3 \end{bmatrix}$$
$$|Q_0| = -2 \neq 0,$$

so the system is observable.

7.15 (b)

$$A = \begin{bmatrix} 0 & -1 \\ 1 & -2 \end{bmatrix}, \; B = \begin{bmatrix} 1 \\ 2 \end{bmatrix}$$

Roots can be determined from the characteristic equation.

i.e. $|sI - A| = 0$

$$\begin{bmatrix} s & 0 \\ 0 & s \end{bmatrix} - \begin{bmatrix} 0 & -1 \\ 1 & -2 \end{bmatrix} = 0$$

$$\begin{vmatrix} s & 1 \\ -1 & s+2 \end{vmatrix} = 0 \Rightarrow s^2 + 2s + 1 = 0$$
$$(s + 1)^2 = 0$$

Thus it can be determined that the system is critically damped.

7.16 (b)

$$A = \begin{bmatrix} -2 & -4 \\ 1 & 0 \end{bmatrix}$$

The characteristic equation is

$|sI - A| = 0, \quad I$: identity matrix.

$$\begin{bmatrix} s+2 & 4 \\ -1 & s \end{bmatrix} = 0 \Rightarrow s^2 + 2s + 4 = 0$$

$\Rightarrow$ Therefore $\omega_n = 2$ rad/sec.

$$2\xi\omega_n = 2$$
$$\xi\omega_n = 1$$

7.17 (c)

$$\dot{X} = \begin{bmatrix} 0 & 1 \\ 2 & -3 \end{bmatrix} X + \begin{bmatrix} 0 \\ 1 \end{bmatrix} U$$
$$Y = [1 \quad 1]X$$

Check for controllability:

$$Q_C = [B \quad AB] = \begin{bmatrix} 0 & 1 \\ 1 & -3 \end{bmatrix}$$
$$|Q_C| = -1$$

For controllability

$$|Q_C| \neq 0$$

$\therefore$ Controllable

Check for observability:

$$Q_0 = \begin{bmatrix} C \\ CA \end{bmatrix} = \begin{bmatrix} 1 & 1 \\ 2 & -2 \end{bmatrix}$$
$$|Q_0| = -4$$

For observability:

$$|Q_0| \neq 0$$

$\therefore$ Observable.

7.18 (a)

$$Y(s) = CX(s)$$
$$sX(s) = AX(s) + BU(s)$$
$$\Rightarrow \quad X(s)[s - A] = BU(s)$$
$$X(s) = (sI - A)^{-1} BU(s)$$
$$\frac{Y(s)}{U(s)} = C(sI - A)^{-1} B$$

7.19 (c)

The state space representation of the nth order single input single output system is given as

$$\dot{x} = Ax + Bu$$
$$y = Cx$$

For n^{th} order system, size of matrix A is $(n \times n)$ and size of x is $(n \times 1)$.

Since, it is a single output system.

Hence, $[y]_{1\times 1} = [C]_{1\times n}[x]_{n\times 1}$

$\therefore$ Dimensions of output coupling matrix,

$$C = 1 \times n$$

7.20 (d)

The current through inductor, $i_L(t)$ and voltage across the capacitor, $v_c(t)$ cannot change instantaneously.

Hence, $i_L(t)$ and $v_c(t)$ are chosen as the state variables for the formulation f RLC network.

7.21 (d)

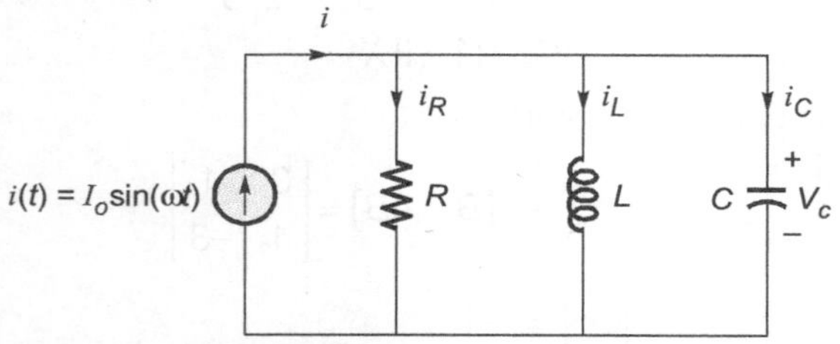

For the state variable formulation, current through inductor, i_L and voltage across capacitor are chosen as state variables i.e.

$$x_1 = Li_L \quad \text{and} \quad x_2 = V_C$$

From the circuit,

$$V_C = L\cdot\frac{di_L}{dt}$$

$$\Rightarrow \quad x_2 = \dot{x}_1$$

$$\Rightarrow \quad \dot{x}_1 = x_2 \quad \text{...(i)}$$

Applying KCL, we get,

$$i = i_R + i_L + i_C$$

$$I_o\sin(\omega t) = \frac{V_C}{R} + i_L + C\cdot\frac{dV_c}{dt}$$

$$\Rightarrow \quad I_o\sin(\omega t) = \frac{x_2}{R} + \frac{x_L}{L} + C\cdot\dot{x}_2$$

$$\Rightarrow \quad \dot{x}_2 = -\frac{x_1}{LC} - \frac{x_2}{RC} + \frac{I_o}{C}\sin(\omega t) \quad \text{...(ii)}$$

Using equations (i) and (ii), the state space model of the given RLC circuit can be written as

$$\begin{bmatrix}\dot{x}_1\\ \dot{x}_2\end{bmatrix} = \begin{bmatrix}0 & 1\\ -\frac{1}{LC} & -\frac{1}{RC}\end{bmatrix}\begin{bmatrix}x_1\\ x_2\end{bmatrix} + \begin{bmatrix}0\\ \frac{I_o\sin(\omega t)}{C}\end{bmatrix}$$

■■■

8 Miscellaneous

8.1 The open-loop transfer function $G(s)$ of a unity feedback control system is $\frac{1}{s(s+1)}$. The system is subjected to an input $r(t) = \sin t$. The steady state error will be

(a) zero (b) 1

(c) $\sqrt{2}\sin\left(t-\frac{\pi}{4}\right)$ (d) $\sqrt{2}\sin\left(t+\frac{\pi}{4}\right)$

[ESE-2000]

8.2 Which one of the following equations represents the constant magnitude locus in G-plane for $M = 1$? {x-axis is Re $G(j\omega)$ and y-axis is Im $G(j\omega)$}

(a) $x = -0.5$ (b) $x = 0$

(c) $x + y^2 = 1$ (d) $(x + 1)^2 + y^2 = 1$

[ESE-2000]

8.3 Which one of the following features is NOT associated with Nichols chart?

(a) (0 dB, –180°) point on Nichols charts represents the critical point $(-1 + j0)$

(b) It is symmetric about –180°

(c) The M loci are centred about (0 dB, –180°) point

(d) The frequency at the intersection of the $G(j\omega)$ locus and M = +3 dB locus gives bandwidth of the closed-loop system [ESE-2000]

8.4 Match **List-I** (Functional components) with **List-II** (Devices) and select the correct answer:

List-I

A. Error detector
B. Servometer
C. Amplifier
D. Feedback

List-II

1. Three-phase FHP induction motor
2. A pair of synchronous transmitter and control transformer
3. Tachogenerator
4. Armature controlled FHP DC motor
5. Amplidyne

Codes:

	A	B	C	D
(a)	2	4	1	5
(b)	4	2	5	3
(c)	2	4	5	3
(d)	1	2	3	5

[ESE-2000]

8.5 Consider the following servomotors:

1. AC two-phase servomotor
2. DC servomotor
3. Hydraulic servomotor
4. Pneumatic servomotor

The correct sequence of these servomotors in increasing order of power handling capacity is

(a) 2, 4, 3, 1 (b) 4, 2, 3, 1

(c) 2, 4, 1, 3 (d) 4, 2, 1, 3

[ESE-2000]

8.6 A constant N-circle having centre at $(-1/2 + j0)$ in the G-plane, represents the phase angle equal to

(a) 180° (b) 90°

(c) 45° (d) 0°

[ESE-2001]

8.7 In industrial control systems, which one of the following methods is most commonly used in designing a system for meeting performance specifications?

(a) The transfer function is first determined and then either a lead compensation or lag compensation is implemented

(b) The transfer function is first determined and PID controllers are implemented by mathematically determining PID constants.

(c) PID controllers are implemented without the knowledge of the system parameters using Ziegler Nichols method

(d) PID controllers are implemented using Ziegler Nichols method after determining the system transfer function

[ESE-2001]

8.8 **Assertion (A):** The stator winding of a control transformer has higher impedance per phase.
Reason (R): The rotor of control transformer is cylindrical in shape.
(a) Both A and R are true and R is the correct explanation of A
(b) Both A and R are true but R is NOT the correct explanation of A
(c) A is true but R is false
(d) A is false but R is true [ESE-2001]

8.9 Consider the mechanical system shown in the given figure. If the system is set into motion by unit impulse force, the equation of the resulting oscillation will be

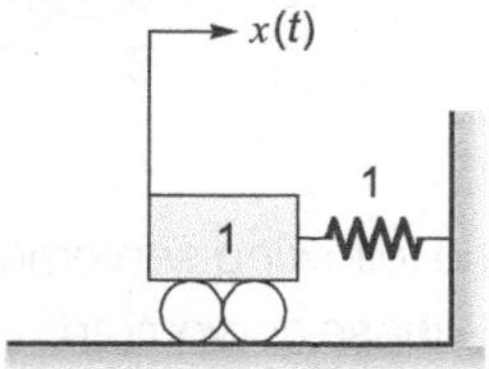

(a) $x(t) = \sin t$ (b) $x(t) = \sqrt{2}\sin t$
(c) $x(t) = 1/2 \sin 2t$ (d) $x(t) = \sin\sqrt{2t}$
[ESE-2001]

8.10 Which one of the following relations holds good for the tachometer shown in the given figure?

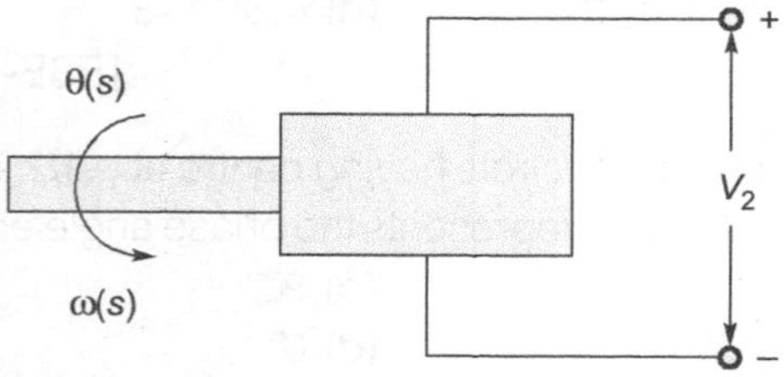

(a) $V_2(s) = sk_t\omega(s)$ (b) $V_2(s) = k_t s^2\theta(s)$
(c) $V_2(s) = k_t s^2\omega(s)$ (d) $V_2(s) = k_t s\,\theta(s)$
[ESE-2001]

8.11 The constant M-circle represented by the equation $x^2 + 2.25x + y^2 = -1.125$ where $x = \text{Re}[G(j\omega)]$ and $y = \text{Im}[G(j\omega)]$ has the value of M equal to
(a) 1 (b) 2
(c) 3 (d) 4 [ESE-2001]

8.12 The mechanical system shown below has its pole(s) at

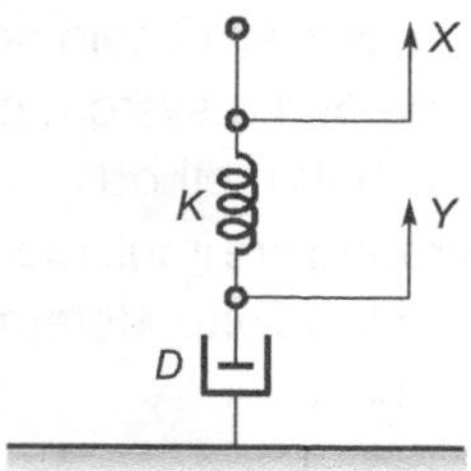

(a) $-K/D$ (b) $-D/K$
(c) $-DK$ (d) 0, $-K/D$
[ESE-2002]

8.13 Consider the following signal-loop feedback structure illustrating the return difference:

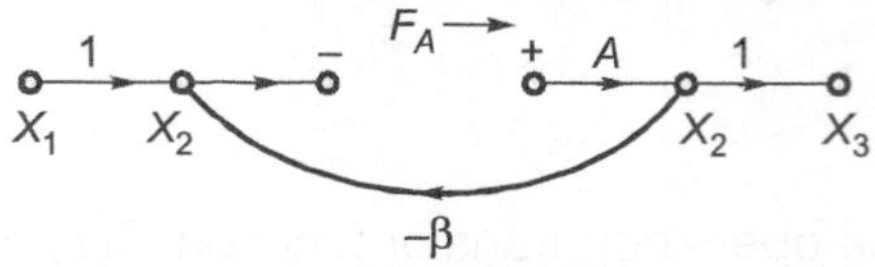

The return difference for A is
(a) $A\beta$ (b) $1 + A\beta$
(c) $\dfrac{A\beta}{1+A\beta}$ (d) $\dfrac{A\beta}{1-A\beta}$
[ESE-2002]

8.14 Match **List-I** (Type of controller) with **List-II** (Operation) and select the correct answer using the codes given below the lists:

List-I	List-II
A. Pneumatic controller	1. Flexible operation
B. Hydraulic controller	2. High torque high-speed operation
C. Electronic controller	3. Fire and explosion proof operation

Codes:

	A	B	C
(a)	1	3	2
(b)	1	2	3
(c)	3	1	2
(d)	3	2	1

[ESE-2002]

8.15 Consider the following statements:
Nichol's chart gives information about
1. closed loop frequency response
2. the value of the peak magnitude of the closed loop frequency response M_p
3. the frequency at which M_p occurs

Which of the above statements are correct?
(a) 2 and 3 (b) 1 and 2
(c) 1 and 3 (d) 1, 2 and 3
[ESE-2002]

8.16 The constant N loci represented by the equation $x^2 + x + y^2 = 0$ where $x = \text{Re}\,|G(j\omega)|$ and $y = \text{Img}\,|G(j\omega)|$ is for the value of phase angle equal to
(a) –45° (b) 0°
(c) + 45° (d) +90° [ESE-2002]

8.17 Consider the following techniques:
1. Bode plot
2. Nyquist plot
3. Nichol's chart
4. Routh-Hurwitz criterion

Which of these techniques are used to determine relative stability of a closed loop linear system?
(a) 1 and 2 (b) 1 and 4
(c) 1, 2 and 3 (d) 2, 3 and 4
[ESE-2003]

8.18 The frequency ω in terms of frequencies α and β in the below figure is

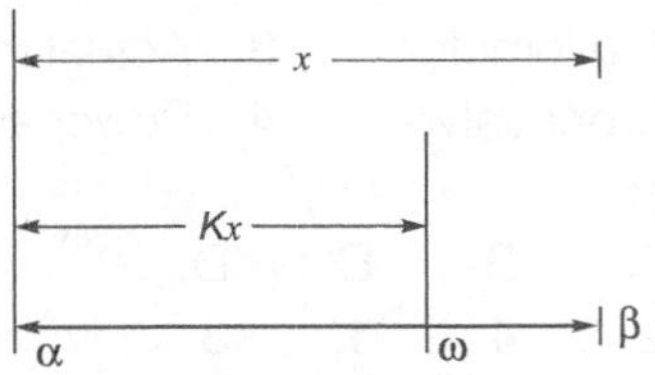

Frequency in log scale

(a) $K(\log\alpha - \log\beta)$ (b) $(\beta/\alpha)^K$
(c) $\alpha^K . \beta^{K-1}$ (d) $\alpha^{1-K} . \beta^K$ [ESE-2003]

8.19 Constant M circles have their centre and radius as

(a) $\left(\frac{-M^2}{M^2-1}, 0\right)$ and $\left(\frac{M^2}{M^2-1}\right)$

(b) $\left(\frac{-M^2}{M^2-1}, 0\right)$ and $\left(\frac{M}{M^2-1}\right)$

(c) $\left(0, \frac{M^2}{M^2-1}\right)$ and $\left(\frac{M^2}{M-1}\right)$

(d) $\left(0, \frac{M^2}{M^2-1}\right)$ and $\left(\frac{M}{M^2-1}\right)$ [ESE-2003]

8.20 **Assertion (A):** In a shunt regulator, the control element is connected in shunt with the load to achieve constant output voltage.
Reason (R): The impedance of the control element varies to keep the total current flowing through the load and the control element constant.
(a) Both A and R are true and R is the correct explanation of A
(b) Both A and R are true but R is NOT the correct explanation of A
(c) A is true but R is false
(d) A is false but R is true [ESE-2003]

8.21 **Assertion (A):** In the error detector configuration using a synchro transmitter and synchro control transformer, the latter is connected to the error amplifier.
Reason (R): Synchro control transformer has almost a uniform reluctance path between the rotor and the stator.
(a) Both A and R are true and R is the correct explanation of A
(b) Both A and R are true but R is NOT the correct explanation of A
(c) A is true but R is false
(d) A is false but R is true [ESE-2004]

8.22 **Assertion (A):** The error detector in a position control system using synchro pairs employs synchro transmitter for reference signal and synchro control transformer for the feedback signal.
Reason (R): Synchro control transformer rotor has a uniform magnetic reluctance.
(a) Both A and R are true and R is the correct explanation of A
(b) Both A and R are true but R is NOT the correct explanation of A
(c) A is true but R is false
(d) A is false but R is true [ESE-2004]

8.23 Consider the following statements for a.c. series motors:
1. The rotor is designed so that its R/X ratio is small.
2. $dT/d\omega < 0$ where T and ω are torque and speed respectively.
3. The reference and control voltages should be in phase quadrature, but their magnitudes need not be equal.

Which of the statements given above are correct?
(a) 1, 2 and 3 (b) 1 and 2
(c) 2 and 3 (d) 1 and 3
[ESE-2004]

8.24 All the constant N-circles in G-planes cross the real axis at the fixed points. Which are these points?
(a) –1 and origin (b) origin and +1
(c) –0.5 and +0.5 (d) –1 and +1
[ESE-2004]

8.25 What is the value of M for the constant M circle represented by the equation
$8x^2 + 18x + 8y^2 + 9 = 0$, where
$x = Re|G(j\omega)|$ and $y = \text{Im}|G(j\omega)|$?
(a) 0.5 (b) 2
(c) 3 (d) 8 [ESE-2004]

8.26 A filter at the input to a processing system is shown in the diagram given below:

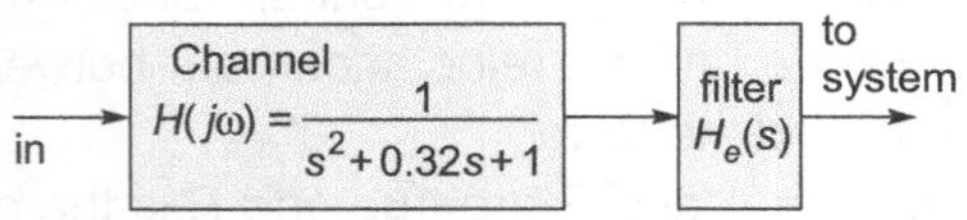

The channel works for toll quality telephone use. If the filter $H_e(s)$ is to be designed so that linear distortion is minimized, then $H_e(s)$ should have which one of the following?

(a) Constant delay
(b) Constant phase
(c) Inverse relationship with $H(s)$
(d) Inverse relationship with $H(s)$ and constant delay

[ESE-2004]

8.27 Which one of the following methods can determine the closed loop system resonance frequency of operation?

(a) Root locus method
(b) Nyquist method
(c) Bode plot
(d) *M* and *N* circle method

[ESE-2005]

8.28 Which one of the following statements is correct? Nichol's chart is useful for detailed study and analysis of

(a) closed loop frequency response
(b) open loop frequency response
(c) close loop and open loop frequency responses
(d) None of the above

[ESE-2006]

8.29 Match **List-I** (Plot/Diagram/Chart) with **List-II** (Characteristic) and select the correct answer using the code given below the lists:

List-I

A. Constant M loci
B. Constant N loci
C. Nichol's chart
D. Nyquist plot

List-II

1. Constant gain and phase shift loci of the closed-loop system
2. Plot of loop gain with variation of ω
3. Circles of constant gain for closed-loop transfer function
4. Circles of constant phase shift of closed-loop transfer function

Codes:

	A	B	C	D
(a)	3	4	2	1
(b)	3	4	1	2
(c)	4	3	2	1
(d)	4	3	1	2

[ESE-2007]

8.30 Match **List-I** (Components) with **List-II** (Functions) and select the correct answer using the code given below the lists:

List-I	**List-II**
A. Servomotor	1. Error detector
B. Amplidyne	2. Transducer
C. Potentiometer	3. Actuator
D. Flapper valve	4. Power amplifier

Codes:

	A	B	C	D
(a)	2	4	1	3
(b)	3	1	4	2
(c)	2	1	4	3
(d)	3	4	1	2

[ESE-2007]

8.31 Consider the following statements for pneumatic and hydraulic control systems:

1. The normal operating pressure of pneumatic control is very much higher than that of hydraulic control.
2. In pneumatic control, external leakage is permissible to a certain extent, but there should be no leakage in a hydraulic control.

Which of the statements given above is/are correct?

(a) 1 only
(b) 2 only
(c) Both 1 and 2
(d) Neither 1 nor 2

[ESE-2008]

8.32 Which one of the following represents the linear mathematical model of the physical system shown in the below figure?

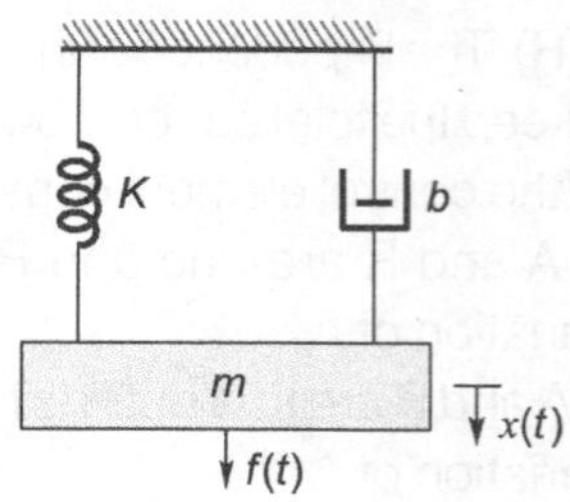

(a) $m\frac{d^2x(t)}{dt^2} + b\frac{dx(t)}{dt} + kx(t) = f(t)$

(b) $m\frac{d^2x(t)}{dt^2} + b\frac{dx(t)}{dt} + kx(t) = 0$

(c) $m\frac{d^2x(t)}{dt^2} + b\frac{dx(t)}{dt} + kx(t) + f(t) = 0$

(d) $m\frac{d^2x(t)}{dt^2} + b\frac{dx(t)}{dt} - kx(t) - f(t) = 0$

[ESE-2008]

8.33 Which of the following can be used as a tachogenerator in control system?
(a) Microsyn (b) DC servomotor
(c) AC servomotor (d) Magnetic amplifier
[ESE-2009]

8.34 For the Nichols plot shown, the system is

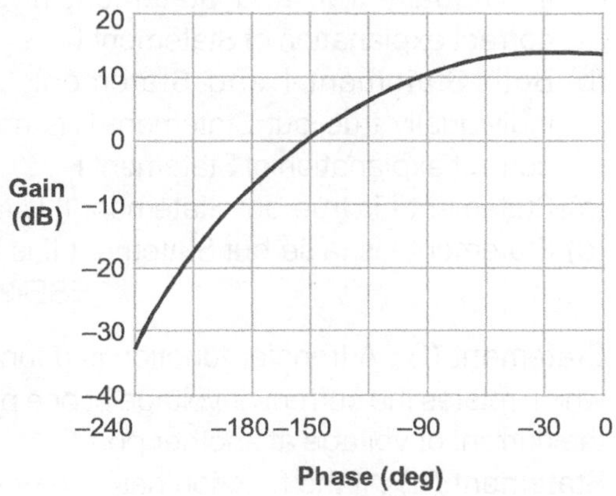

(a) Unstable (b) Stable
(c) Overdamped (d) Critically stable
[ESE-2010]

8.35 A family of constant N circles has the centre as
(a) $X = 1$ and $Y = 2$ N
(b) $X = -\frac{1}{4}$ and $Y = 4$ N
(c) $X = -\frac{1}{2}$ and $Y = \frac{1}{4N}$
(d) $X = -\frac{1}{2}$ and $Y = \frac{1}{2N}$
[ESE-2011]

8.36 System transformation function $H(z)$ for a discrete time LTI system expressed in state variable form with zero initial conditions is
(a) $C(zI - A)^{-1} B + D$
(b) $C(zI - A)^{-1}$
(c) $(zI - A)^{-1} z$
(d) $(zI - A)^{-1}$
[ESE-2011]

8.37 The constant magnitude locus for $M = 1$, in G-plane is given by the following equation where $x = \text{Re}[G(j\omega)]$ and $y = \text{Img}[G(j\omega)]$
(a) $x = -0.5$ (b) $x = 0$
(c) $x^2 + y^2 = 0.25$ (d) $x^2 + y^2 = 1$
[ESE-2012]

8.38 All the constant – N loci in G-plane intersect the real axis in points
(a) –1 and origin (b) –0.5 and +0.5
(c) –1 and +1 (d) Origin and +1
[ESE-2012]

8.39 The necessary conditions for poles and zeros of the transfer function of a bridge-T network containing only resistors and capacitors and used as a compensator are
(a) All the poles and zeros must be imaginary
(b) Poles and zeros both can be complex
(c) Poles can be complex but zeros must be real
(d) Zeros can be complex but poles must be real
[ESE-2012]

8.40 **Statement (I):** Nyquist plot is the locus of $GH(j\omega)$ indicating the magnitude and phase angle on the $GH(j\omega)$ plane.
Statement (II): Given the values of $|GH(j\omega)|$ and $\angle GH(j\omega)$ using the Nichols chart M_m, ω_m and bandwidth can be determined.
(a) Both Statement (I) and Statement (II) are individually true and Statement (II) is the correct explanation of Statement (I).
(b) Both Statement (I) and Statement (II) are individually true but Statement (II) is not the correct explanation of Statement (I).
(c) Statement (I) is true but Statement (II) is false.
(d) Statement (I) is false but Statement (II) is true.
[ESE-2013]

8.41 In industrial control system, which one of the following methods is most commonly used in designing a system for meeting performance specification?
(a) The transfer function is first determined and then either a lead compensation or lag compensation is implemented.
(b) The transfer function is first determined and PID controllers are implemented by mathematically determining PID constants.
(c) PID controllers are implemented without the knowledge of the system parameters using Ziegler–Nichols method.

(d) PID controllers are implemented using Ziegler-Nichols method after determining the system transfer function.

[ESE-2014]

8.42 Which of the following techniques are used to determine relative stability of a closed loop linear system?

1. Bode plot
2. Nyquist plot
3. Nichol's chart
4. Routh-Hurwitz criterion

(a) 1, 2 and 4 only (b) 1, 3 and 4 only
(c) 1, 2 and 3 only (d) 1, 2, 3 and 4

[ESE-2015]

8.43 In Force-Voltage Analogy
(a) Force is analogous to current
(b) Mass is analogous to capacitance
(c) Velocity is analogous to current
(d) Displacement is analogous to magnetic flux linkage [ESE-2016]

8.44 Consider the following:
1. Bode plot **2.** Nyquist plot **3.** Nichols chart
Which of the above frequency response plots are commonly employed in the analysis of control systems?
(a) 1 and 2 only (b) 1 and 3 only
(c) 2 and 3 only (d) 1, 2 and 3

[ESE-2016]

8.45 From the Nichols chart, one can determine the following quantities pertaining to a closed-loop system:
(a) Magnitude, bandwidth and phase
(b) Bandwidth and phase only
(c) Magnitude and phase only
(d) Bandwidth only [ESE-2016]

8.46 **Statement (I):** Constant *M* and *N* circles, as also Nichol's charts, are graphical techniques to assess closed-loop performance in the frequency domain.
Statement (II): While constant *M* and *N* circles use Nyquist polar plots data, Nichol's chart use Bode plots data.
(a) Both Statement (I) and Statement (II) are individually true and Statement (II) is the correct explanation of Statement (I).
(b) Both Statement (I) and Statement (II) are individually true but Statement (II) is not the correct explanation of Statement (I).
(c) Statement (I) is true but Statement (II) is false.
(d) Statement (I) is false but Statement (II) is true.

[ESE-2017]

8.47 **Statement (I):** The stator winding of a control transformer in a synchro pair has a high impedance per phase.
Statement (II): The rotor of the control transformer is cylindrical in shape.
(a) Both Statement I and Statement II are individually true and Statement II is the correct explanation of Statement I
(b) Both Statement I and Statement II are individually true but Statement II is not the correct explanation of Statement I
(c) Statement I is true but Statement II is false
(d) Statement I is false but Statement II is true

[ESE-2018]

8.48 **Statement (I) :** A transfer function is a function which relates the current or voltage at one port to the current or voltage at another port.
Statement (II) : If the function has one or more poles in the right-half plane, then the function is non-minimum phase.
(a) Both Statement (I) and Statement (II) are individually true and Statement (II) is the correct explanation of Statement (I).
(b) Both Statement (I) and Statement (II) are individually true, but Statement (II) is not the correct explanation of Statement (I).
(c) Statement (I) is true, but Statement (II) is false.
(d) Statement (I) is false, but Statement (II) is true.

[ESE-2022]

8.49 Non-minimum phase transfer function is defined as the transfer function
(a) which has zeros in the right-half s-plane.
(b) which has poles in the left-half s-plane.
(c) which has poles in the negative right-half s-plane.
(d) which has zeros only in the left-half s-plane.

[ESE-2023]

■■■

Answers Miscellaneous

8.1 (*)	8.2 (a)	8.3 (d)	8.4 (c)	8.5 (c)	8.6 (b)	8.7 (c)	8.8 (b)	8.9 (a)
8.10 (d)	8.11 (c)	8.12 (a)	8.13 (b)	8.14 (d)	8.15 (d)	8.16 (d)	8.17 (c)	8.18 (d)
8.19 (b)	8.20 (c)	8.21 (c)	8.22 (d)	8.23 (c)	8.24 (a)	8.25 (c)	8.26 (d)	8.27 (d)
8.28 (a)	8.29 (b)	8.30 (d)	8.31 (c)	8.32 (a)	8.33 (c)	8.34 (b)	8.35 (d)	8.36 (a)
8.37 (a)	8.38 (a)	8.39 (d)	8.40 (b)	8.41 (c)	8.42 (c)	8.43 (c)	8.44 (d)	8.45 (a)
8.46 (a)	8.47 (c)	8.48 (b)	8.49 (a)					

Explanations Miscellaneous

8.1 (*)

The steady state calculation is not applicable for sine and cosine function.

8.2 (a)

Closed loop frequency response,

$$T(j\omega) = \frac{G(j\omega)}{1+G(j\omega)} = \frac{x+jy}{1+x+jy}$$

Magnitude, $M = \left[\frac{x^2+y^2}{(1+x)^2+y^2}\right]^{-1/2}$

$$\Rightarrow \quad M^2 = \frac{x^2+y^2}{x^2+2x+1+y^2}$$

Putting $M = 1$,

$$1 = \frac{x^2+y^2}{x^2+2x+1+y^2}$$

$\Rightarrow x^2 + 2x + 1 + y^2 = x^2 + y^2$

$\Rightarrow \quad 2x = -1$

$\Rightarrow \quad x = -0.5$

8.6 (b)

Centre of N-circle is $\left(-\frac{1}{2}, \frac{1}{2N}\right)$

so $\quad \frac{+1}{2N} = 0$

$N = \infty$

$\tan\alpha = \infty$

$\alpha = 90°$

8.9 (a)

Force equation is

$$\frac{Md^2x}{dt^2} + f\frac{dx}{dt} + Kx = F$$

Taking Laplace transform –

$(Ms^2 + fs + k)\,X(s) = F(s)$

Given:

$M = 1,\ k = 1,\ F(s) = 1$

Let $f = 0$ (Assume)

$\Rightarrow (s^2 + 1)\,X(s) = 1$

$\Rightarrow \quad X(s) = \frac{1}{s^2+1}$

$\Rightarrow \quad x(t) = \sin t$

8.10 (d)

$$v_2(t) = K_t\frac{d\theta}{dt}$$

$\Rightarrow \quad V_2(s) = K_t s\theta(s)$

8.11 (c)

$$x^2 + 2.25x + y^2 = -1.125$$

$\Rightarrow \quad (x + 1.125)^2 + y^2 = 0.140625$

M-circle equation is

$$\left(x + \frac{M^2}{M^2-1}\right)^2 + y^2 = \left(\frac{M}{M^2-1}\right)^2$$

Comparing, $\quad \frac{M^2}{M^2-1} = 1.125$

$\Rightarrow \quad M^2 = \frac{1.125}{0.125} = 9$

$\Rightarrow \quad M = 3$

8.12 (a)

F.B.D. of point A:

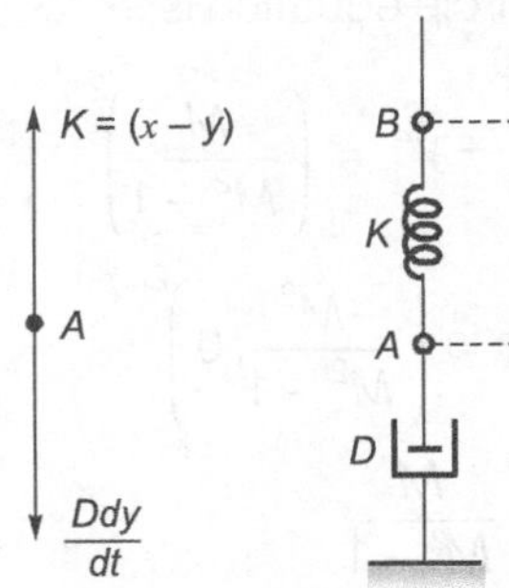

so, $K(x-y) = \dfrac{Ddy}{dt}$

Taking Laplace transform:

$K\{X(s) - Y(s)\} = DsY(s)$

$\Rightarrow (Ds + K)\, Y(s) = KX(s)$

$$\Rightarrow \frac{Y(s)}{X(s)} = \frac{K}{D\left(s+\frac{K}{D}\right)}$$

Thus the pole of the system is at $s = -\dfrac{K}{D}$.

8.16 (d)

$$x^2 + x + y^2 = 0$$

$$\left(x+\frac{1}{2}\right)^2 + y^2 = \frac{1}{4}$$

$$\text{Centre} = \left(-\frac{1}{2}, 0\right)$$

Centre of N-circle is $\left(-\dfrac{1}{2}, \dfrac{1}{2N}\right)$

On comparing we have $\dfrac{1}{2N} = 0$

$$N = \infty$$

$\therefore \quad N = \tan\alpha$

So $\quad \alpha = 90°$

8.18 (d)

$x = C\log\dfrac{\beta}{\alpha}$ where C is a constant.

$$Kx = C\log\frac{\omega}{\alpha} \Rightarrow K = \frac{\log(\omega/\alpha)}{\log(\beta/\alpha)}$$

$$\Rightarrow \quad \log\frac{\omega}{\alpha} = K\log\frac{\beta}{\alpha}$$

$$\Rightarrow \quad \log\frac{\omega}{\alpha} = \log\left(\frac{\beta}{\alpha}\right)^K$$

$$\Rightarrow \quad \frac{\omega}{\alpha} = \left(\frac{\beta}{\alpha}\right)^K$$

$$\Rightarrow \quad \omega = \alpha^{1-K}\cdot\beta^K$$

8.19 (b)

Constant-M circle equation is

$$\left(x+\frac{M^2}{M^2-1}\right)^2 + y^2 = \left(\frac{M}{M^2-1}\right)^2$$

So, the centre is $\left(\dfrac{-M^2}{M^2-1}, 0\right)^2$

and radius is $\dfrac{M}{M^2-1}$

8.24 (a)

Constant $-N$ circle equation is

$$\left(x+\frac{1}{2}\right)^2 + \left(y-\frac{1}{2N}\right)^2 = \frac{N^2+1}{4N^2}$$

where $N = \tan\alpha$

Centre of circles is at $\left(-\dfrac{1}{2}, \dfrac{1}{2N}\right)$

Radius is $\dfrac{\sqrt{N^2+1}}{2N}$

Constant $-N$ circles always pass through $(-1, 0)$ and $(0, 0)$.

8.25 (c)

$$8x^2 + 18x + 8y^2 + 9 = 0$$

$$\Rightarrow x^2 + \frac{9}{4}x + y^2 + \frac{9}{8} = 0$$

$$\Rightarrow \left(x+\frac{9}{8}\right)^2 + y^2 = \frac{81}{64} - \frac{9}{8}$$

Centre of constant $-M$ circles is $\left(\dfrac{-M^2}{M^2-1}, 0\right)$

So $\quad \dfrac{M^2}{M^2-1} = \dfrac{9}{8}$

$$\Rightarrow \quad 8M^2 = 9M^2 - 9$$

$$\Rightarrow \quad M^2 = 9$$

$$\Rightarrow \quad M = 3$$

8.26 (d)

To minimize the distortion $H_e(s)$ should have the inverse relationship with $H(s)$ and constant delay.

8.32 (a)

Free body diagram of mass m is shown below:

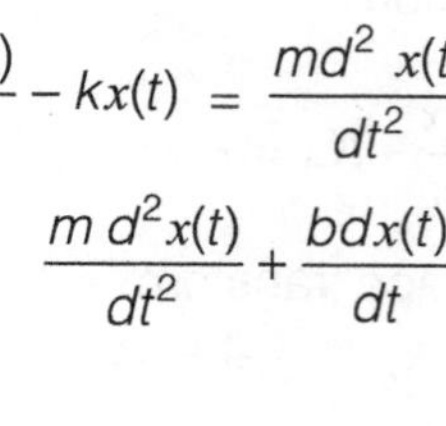

At balance,

$$f(t) - \frac{b\,dx(t)}{dt} - kx(t) = \frac{md^2x(t)}{dt^2}$$

or $$\frac{m\,d^2x(t)}{dt^2} + \frac{bdx(t)}{dt} + kx(t) = f(t)$$

8.34 (b)

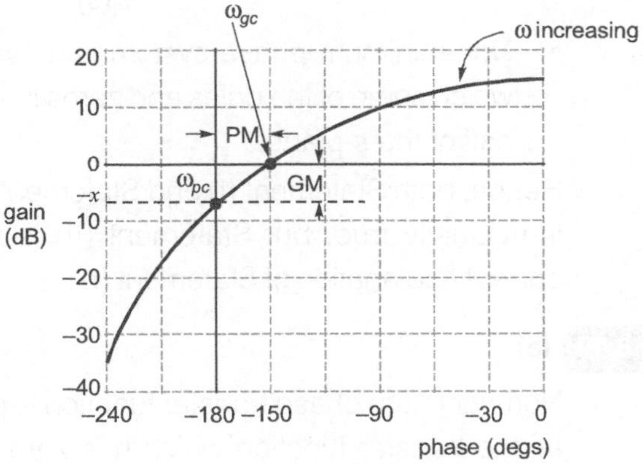

Since $\omega_{gc} < \omega_{pc}$ ... system is stable

From the graph

$$GM = 0 - (-X) = X(+ve)$$

$$PM = 180 - 150 = 30\ (+ve)$$

8.35 (d)

Equation for N-circles is

$$\left[x+\frac{1}{2}\right]^2 + \left[y-\frac{1}{2N}\right]^2 = \frac{1}{4}+\left[\frac{1}{2N}\right]^2$$

$$\text{Hence, Centre} = \left(-\frac{1}{2}, \frac{1}{2N}\right)$$

$$\text{and, radius} = \sqrt{\frac{1}{4}+\left(\frac{1}{2N}\right)^2}$$

8.36 (a)

If state variable equations are as follows

$\dot{X} = AX + BU \qquad Y = CX + DU$

System transfer function $H(z)$ for a discrete LTI system is $= C(zI - A)^{-1} B + D$

8.37 (a)

When $M = 1$ is put in the magnitude equation i.e.

$x^2(M^2 - 1) + 2x M^2 + Y^2 (M^2 - 1) + M^2 = 0$

$$2x + 1 = 0$$

Thus, it is a straight line with $x = 0.5$.

8.38 (a)

The N circles are always drawn between −1 and origin for different values of N.

8.39 (d)

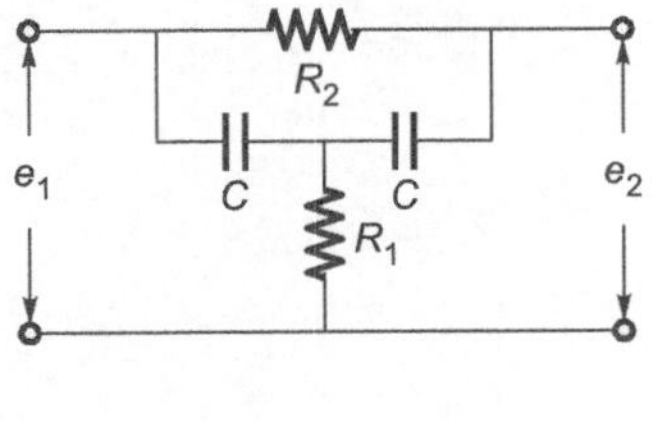

Bridge-*T* network is used for the measurement of resistance at radio frequency.

$$\frac{E_2(s)}{E_i(s)} = \frac{s^2 + \dfrac{2}{R_2 C}s + \dfrac{1}{R_1 R_2 C^2}}{s^2 + \dfrac{2R_1 + R_2 s}{R_1 R_2 C} + \dfrac{1}{R_1 R_2 C^2}}$$

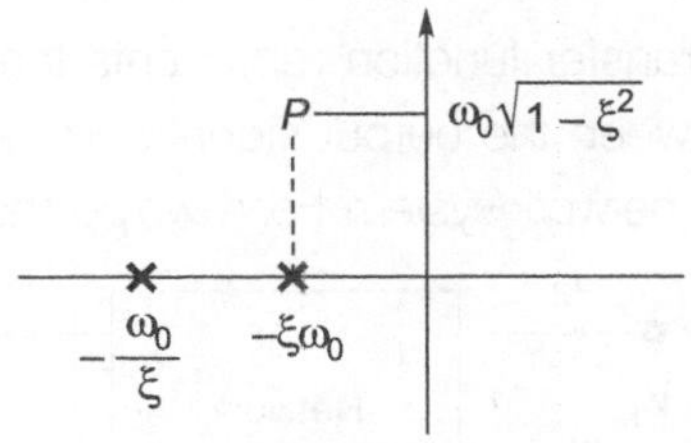

$$\text{for} \qquad \xi = \sqrt{\frac{R_1}{R_2}}, \quad \omega_0 = \frac{1}{C\sqrt{R_1 R_2}}$$

8.40 (b)

- Nyquist plot is a parametric plot of transfer function and is a locus of $GH(j\omega)$ indicating the magnitude and phase angle on the $GH(j\omega)$ plane.
- In using Nichol's chart, graph is obtained by plotting the magnitude of $G(j\omega)\,H(j\omega)$ in dB against the phase of $G(j\omega)\,H(j\omega)$ in degrees on the Nichols chart with ω changing from 0 to ∞. By using open-loop system, one can determine the information which relates the performance of an open-loop system in frequency domain with closed loop system.

8.42 (c)

Routh-Hurwitz criteria can only be used to find the absolute stability of the system.

8.43 (c)

As per theory of analogy,

$$\begin{array}{ccccccc} F & \to & T & \to & V & \to & I \\ M & \to & J & \to & L & \to & C \\ B & \to & K & \to & R & \to & 1/R \\ k & \to & k & \to & 1/C & \to & 1/L \\ v & \to & \omega & \to & i & \to & v \\ x & \to & \theta & \to & q & \to & \phi \end{array}$$

8.44 (d)

All the mentioned plots are popular and commonly used in control analysis.

8.45 (a)

Nichols chart provides complete frequency response of a system.

8.46 (a)

Constant M and N circles uses polar (Nyquist) plot and Nichols chart uses Bode plot concept which is used to assess system performs in frequency domain.

8.48 (b)

- A transfer function represents the relationship between the output signal and input signal of any network/system. For a two port network below,

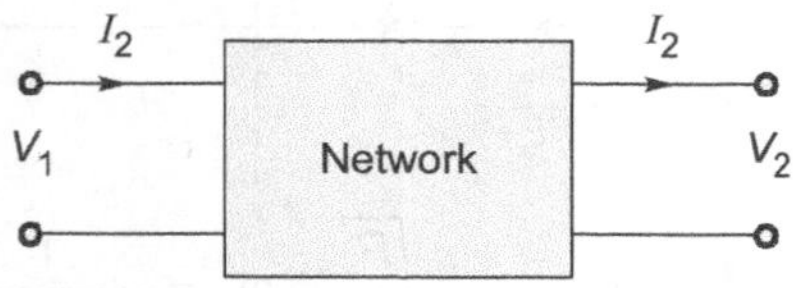

Voltage transfer function = $\dfrac{V_2(s)}{V_1(s)}$

Current transfer function = $\dfrac{I_2(s)}{I_1(s)}$

- Non-minimum phase system is a system in which some of the poles and zeros lie on right half of the s-plane.

Hence, both Statement (I) and Statement (II) are individually true, but Statement (II) is not the correct explanation of Statement (I).

8.49 (a)

Non-minimum phase transfer function is defined as the transfer function which has zeros in the right half of s-plane.

■■■

UNIT

II

Signals and Systems + DSP

Syllabus

Classification of signals and systems; Application of signal and system theory; System realization; Transforms& their applications; DSP: Discrete time signals/systems, uses; Digital filters: FIR/IIR types, design, speech/audio/radar signal processing uses.

Contents

Sl.	Topic	Page No.
1.	Basics of Signals and Systems	116
2.	Continuous and Discrete Time LTI Systems	132
3.	Fourier Series	144
4.	Continuous Time Fourier Transform	150
5.	Laplace Transform	157
6.	Sampling	169
7.	DTFS and DTFT	171
8.	Z-Transform	175
9.	Digital Filters: Design and Applications	189

OOOO

1 Basics of Signals and Systems

1.1 A continuous-time system is governed by the equation $3y^3(t) + 2y^2(t) + y(t) = x^2(t) + x(t)$. {$y(t)$ and $x(t)$ respectively are output and input}. The system is
(a) linear and dynamic
(b) linear and non-dynamic
(c) non-linear and dynamic
(d) non-linear and non-dynamic [ESE-2000]

1.2 Which one of the following systems is a causal system?
[$y(t)$ is output and $u(t)$ is a input step function]
(a) $y(t) = \sin(u(t+3))$
(b) $y(t) = 5u(t)+3u(t-1)$
(c) $y(t) = 5u(t) + 3u(t+1)$
(d) $y(t) = \sin(u(t-3)) + \sin(u(t+3))$
[ESE-2000]

1.3 If $y(t) + \int_0^\infty y(\tau)\, x(t-\tau)\, d\tau = \delta(t) + x(t)$, then $y(t)$ is
(a) $u(t)$ (b) $\delta(t)$
(c) $r(t)$ (d) 1 [ESE-2001]

1.4 If a function $f(t) \cdot u(t)$ is shifted to right side by t_0, then the function can be expressed as
(a) $f(t-t_0)\, u(t)$ (b) $f(t)\, u(t-t_0)$
(c) $f(t-t_0)\, u(t-t_0)$ (d) $f(t+t_0)\, u(t+t_0)$
[ESE-2001]

1.5 The discrete time system described by $y(n) = x(n^2)$ is
(a) causal, linear and time-varying
(b) causal, non-linear and time-varying
(c) non-causal, linear and time-invariant
(d) non-causal, non-linear and time-variant
[ESE-2001]

1.6 The signal $x(t) = A\cos(\omega t + \phi)$ is
(a) an energy signal
(b) a power signal
(c) an energy as well as a power signal
(d) neither an energy nor a power signal
[ESE-2001]

1.7 If $x_1(t) = 2\sin \pi t + \cos 4\pi t$ and $x_2(t) = \sin 5\pi t + 3\sin 13\pi t$, then
(a) x_1 and x_2 both are periodic
(b) x_1 and x_2 both are not periodic
(c) x_1 is not periodic, but x_2 is not periodic
(d) x_1 is not periodic, but x_2 is periodic
[ESE-2001]

1.8 Which one of the systems described by the following input-output relations is time invariant?
(a) $y(n) = nx(n)$
(b) $y(n) = x(n) - x(n-1)$
(c) $y(n) = x(-n)$
(d) $y(n) = x(n)\cos 2\pi f_0 n$ [ESE-2002]

1.9 Match **List-I** (Input-output relation) with **List-II** (Property of the system) and select the correct answer using the codes given below the lists:

List-I	List-II
A. $y(n) = x(n)$	1. Nonlinear, non-causal
B. $y(n) = x(n^2)$	2. Linear, non-causal
C. $y(n) = x^2(-n)$	3. Linear, causal
D. $y(n) = x^2(n)$	4. Nonlinear, causal

Codes:

	A	B	C	D
(a)	1	4	3	2
(b)	3	2	1	4
(c)	1	2	3	4
(d)	3	4	1	2

[ESE-2002]

1.10 The color T.V. picture signal is a
(a) single-channel, one-dimensional signal
(b) single-channel, three-dimensional signal
(c) three-channel, one-dimensional signal
(d) three-channel, three-dimensional signal
[ESE-2002]

1.11 The sum of two or more arbitrary sinusoids is
(a) Always periodic
(b) Periodic under certain conditions
(c) Never periodic
(d) Periodic only if all the sinusoids are identical in frequency and phase [ESE-2003]

1.12 Consider the following waveform diagram

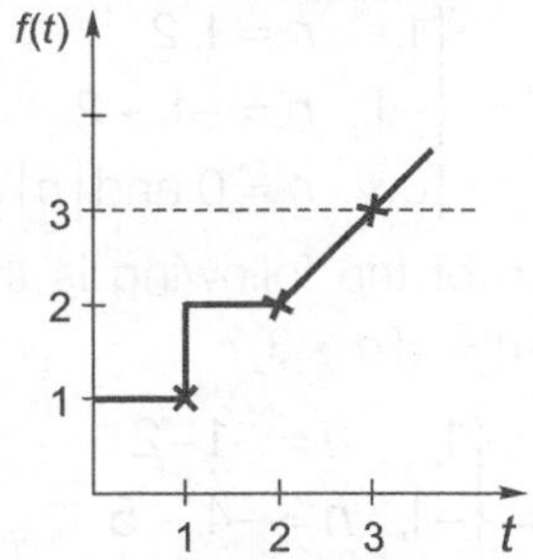

Which one of the following gives the correct description of the waveform shown in the above diagram?

(a) $u(t) + u(t-1)$
(b) $u(t) + (t-1)u(t-1)$
(c) $u(t) + u(t-1) + (t-2)\,u(t-2)$
(d) $u(t) + (t-2)\,u(t-2)$

[ESE-2004]

1.13 Which one of the following must be satisfied if a signal is to be periodic for $-\infty < t < \infty$?

(a) $x(t+T_0) = x(t)$
(b) $x(t+T_0) = dx(t)/dt$
(c) $x(t+T_0) = \int_t^{T_0} x(t)dt$
(d) $x(t+T_0) = x(t) + kT_0$ [ESE-2004]

1.14 If the response of a system to an input does not depend on the future values of the input. Then which one of the following is true for the system?

(a) It is aperiodic (b) It is causal
(c) It is anticipator (d) It is discrete

[ESE-2004]

1.15 What is the phase angle of the composite sinusoidal signal resulting from the addition of $v_1(n) = \sin[5\pi n]$ and $v_2(n) = \sqrt{2}\cos[5\pi n]$?

(a) $+5\pi$ (b) -5π
(c) $-\pi$ (d) $-\pi/3$ [ESE-2004]

1.16 Match **List-I** (Equation Connecting Input $x(n)$ and Output $y(n)$) with **List-II** (System Category) and select the correct answer using the codes given below the lists:

List-I

A. $y(n+2) + y(n+1) + y(n) = 2x(n+1) + x(n)$
B. $n^2y^2(n) + y(n) = x^2(n)$
C. $y(n+1) + ny(n) = 4nx(n)$
D. $y(n+1)\,y(n) = 4x(n)$

List-II

1. Linear, time-variable, dynamic
2. Linear, time-invariant, dynamic
3. Non-linear, time-variable, dynamic
4. Non-linear, time invariant, dynamic
5. Non-linear, time-variable, memoryless

Codes:

	A	B	C	D
(a)	3	5	2	1
(b)	3	2	5	4
(c)	2	3	5	1
(d)	2	5	1	4

[ESE-2004]

1.17 A signal $v(n)$ is defined by

$$v[n] = \begin{cases} 1 & ; \quad n = 1 \\ -1 & ; \quad n = -1 \\ 0 & ; \quad n = 0 \text{ and } |n| > 1 \end{cases}$$

Which is the value of the composite signal defined as $v[n] + v[-n]$?

(a) 0 for all integer values of n
(b) 2 for all integer values of n
(c) 1 for all integer values of n
(d) -1 for all integer values of n [ESE-2005]

1.18 The governing differential equations connecting the output $y(t)$ and the input $x(t)$ of four continuous time systems are given in the **List-I** and **List-II** respectively. Match **List-I** (Equation) with **List-II** (System Category) and select the correct answer using the codes given below the lists:

List-I

A. $2t\dfrac{dy}{dt} + 4y = 2tx$

B. $y\dfrac{dy}{dt} + 4y = 2x$

C. $4\dfrac{d^2y}{dt^2} + 2\dfrac{dy}{dt} + y = 3\dfrac{dx}{dt}$

D. $\left(\dfrac{dy}{dt}\right)^2 + 2ty = 4\dfrac{dx}{dt}$

List-II

1. Linear, time invariant and dynamic
2. Non-linear, time-invariant and dynamic
3. Linear, time-variable and dynamic
4. Non-linear, time-variable and dynamic
5. Non-linear, time-invariant and non-dynamic

Codes:

	A	B	C	D
(a)	3	2	1	4
(b)	4	1	5	3
(c)	3	1	5	4
(d)	4	2	1	3

[ESE-2005]

1.19 In the graph shown below, which one of the following express $v(t)$?

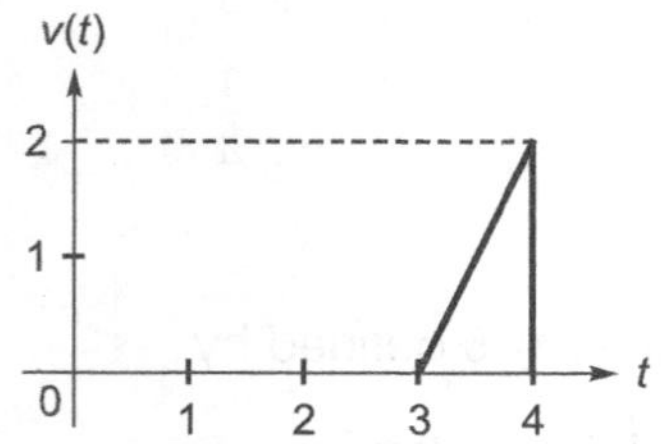

(a) $(2t + 6)[u(t - 3) + 2u(t - 4)]$
(b) $(-2t - 6)[u(t - 3) + u(t - 4)]$
(c) $(-2t + 6)[u(t - 3) + u(t - 4)]$
(d) $(2t - 6)[u(t - 3) - u(t - 4)]$

[ESE-2005]

1.20 Which one of the following is the correct statement?
The system characterized by the equation $y(t) = ax(t) + b$ is
(a) linear for any value of b
(b) linear if $b > 0$
(c) linear if $b < 0$
(d) non-linear

[ESE-2006]

1.21 $x[n]$ is defined as

$$x[n] = \begin{cases} 0 & \text{for } n < -2 \text{ and } n > 4 \\ 1, & \text{otherwise} \end{cases}$$

Determine the value of n for which $x[-n - 2]$ is guaranteed to be zero.
(a) $n < 1$ and $n > 7$ (b) $n < -4$ and $n > 2$
(c) $n < -6$ and $n > 0$ (d) $n < -2$ and $n > 4$

[ESE-2006]

1.22 Which one of the following is the correct statement?
The continuous time system described by $y(t) = x(t^2)$ is
(a) causal, linear and time-varying
(b) causal, non-linear and time-varying
(c) non-causal, non-linear and time-invariant
(d) non-causal and time-variant [ESE-2006]

1.23 The discrete-time signal $x[n]$ is given as

$$x[n] = \begin{cases} 1, & n = 1, 2 \\ -1, & n = -1, -2 \\ 0, & n = 0 \text{ and } |n| > 2 \end{cases}$$

which one of the following is the time-shifted signal $y[n] = x[n + 3]$?

(a) $y[n] = \begin{cases} 1, & n = -1, -2 \\ -1, & n = -4, -5 \\ 0, & n = -3, n < -5 \text{ and } n > -1 \end{cases}$

(b) $y[n] = \begin{cases} 0, & n = -1, -2 \\ -1, & n = -4, -5 \\ 1, & n = -3, n < -5 \text{ and } n > -1 \end{cases}$

(c) $y[n] = \begin{cases} 1, & n = -1, -2 \\ 0, & n = -4, -5 \\ -1, & n = -3, n < -5 \text{ and } n > -1 \end{cases}$

(d) $y[n] = \begin{cases} -1, & n = 1, 2 \\ 1, & n = -4, -5 \\ -1, & n = -3, n < -5 \text{ and } n > -1 \end{cases}$

[ESE-2006]

1.24 Consider two signals $x_1(t) = e^{j20t}$ and $x_2(t) = e^{(-2 + j)t}$.
Which one of the following statements is correct?
(a) Both $x_1(t)$ and $x_2(t)$ are periodic
(b) $x_1(t)$ is periodic but $x_2(t)$ is not periodic
(c) $x_2(t)$ is periodic but $x_1(t)$ is not periodic
(d) Neither $x_1(t)$ nor $x_2(t)$ is periodic

[ESE-2007]

1.25 Which one of the following systems described by the following input-output relations is non-linear?
(a) $y(n) = nx(2n)$ (b) $y(n) = x(n^2)$
(c) $y(n) = n^2x(n)$ (d) $y(n) = x^2(n)$

[ESE-2007]

1.26 Which one of the following is the mathematical representation for the average power of the signal $x(t)$?

(a) $\frac{1}{T}\int_0^T x(t)dt$ (b) $\frac{1}{T}\int_0^T x^2(t)dt$

(c) $\frac{1}{T}\int_{-T/2}^{T/2} x(t)dt$ (d) $\lim_{T \to \infty} \frac{1}{T}\int_{-T/2}^{T/2} x^2(t)dt$

[ESE-2007]

1.27 The discrete time signal $x(n)$ is defined by

$$x(n) = \begin{cases} 1 & n = 1 \\ -1 & n = -1 \\ 0 & n = 0 \text{ and } |n| > 1 \end{cases}$$

Which one of the following is the composite signal $y(n) = x(n) + x(-n)$ for all integer values of n?
(a) 0 (b) 2
(c) ∞ (d) $-\infty$
[ESE-2007]

1.28 The outputs of two systems S_1 and S_2 for the same input $x[n] = e^{j\pi n}$ are 1 and $(-1)^n$, respectively. Which one of the following statements is correct?
(a) Both S_1 and S_2 are linear time invariant (LTI) systems
(b) S_1 is LTI but S_2 is not LTI
(c) S_1 is not LTI but S_2 is LTI
(d) Neither S_1 nor S_2 is LTI [ESE-2007]

1.29 Which one of the following is correct? Energy of a power signal is
(a) finite (b) zero
(c) infinite (d) between 1 and 2
[ESE-2007]

1.30 If v-i characteristic of a circuit is given by $v(t) = t \cdot i(t) + 2$, the circuit is of which type?
(a) Linear and time invariant
(b) Linear and time variant
(c) Non-linear and time invariant
(d) Non-linear and time variant [ESE-2008]

1.31 Which one of the following function is a periodic one?
(a) $\sin(10\pi t) + \sin(20\pi t)$
(b) $\sin(10t) + \sin(20\pi t)$
(c) $\sin(10\pi t) + \sin(20t)$
(d) $\sin(10t) + \sin(25t)$ [ESE-2008]

1.32 A system defined by

$$y[n] = \sum_{k=-\infty}^{n} x[k]$$

is an example of
(a) invertible system
(b) memoryless system
(c) non-invertible system
(d) averaging system
[ESE-2008]

1.33 A function of one or more variables which conveys information on the nature of physical phenomenon is called
(a) Noise (b) Interference
(c) System (d) Signal [ESE-2009]

1.34 The output $y(t)$ of a continuous-time system S for the input $x(t)$ is given by :

$$y(t) = \int_{-\infty}^{t} x(\lambda)\, d\lambda$$

Which one of the following is correct?
(a) S is linear and time-invariant
(b) S is linear and time-varying
(c) S is non-linear and time-invariant
(d) S is non-linear and time-varying
[ESE-2009]

1.35 What is the period of the sinusoidal signal $x(n) = 5\cos[0{\cdot}2\pi n]$?
(a) 10 (b) 5
(c) 1 (d) 0 [ESE-2009]

1.36 The mathematical model of the below shown signal is

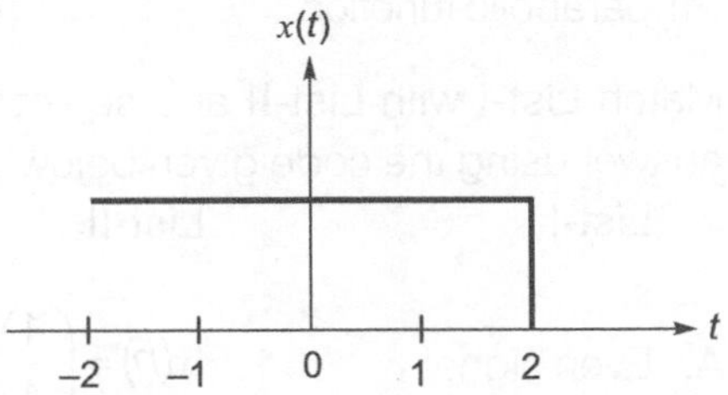

(a) $x(t) = u(2 + t)$ (b) $x(t) = u(t - 2)$
(c) $x(t) = u(2 - t)$ (d) $x(t) = u(t - 1)$
[ESE-2010]

1.37 **Assertion (A):** The system described by $y^2(t) + 2y(t) = x^2(t) + x(t) + c$ is a linear and static system.
Reason (R): The dynamic system is characterized by differential equation.
(a) Both A and R are true and R is the correct explanation of A
(b) Both A and R are true but R is NOT the correct explanation of A
(c) A is true but R is false
(d) A is false but R is true [ESE-2010]

1.38 Which one of the following relations is **not** correct?
(a) $f(t)\,\delta(t) = f(0)\,\delta(t)$

(b) $\int_{-\infty}^{\infty} f(t)\,\delta(\tau)\,d\tau = 1$

(c) $\int_{-\infty}^{\infty} \delta(\tau)\,d(\tau) = 1$

(d) $f(t)\,\delta(t-\tau) = f(\tau)\,\delta(t-\tau)$ [ESE-2011]

1.39 The following equation describes a linear time-varying discrete time system

(a) $y(k+2) + ky(k+1) + y(k) = u(k)$

(b) $y(k+2) + ky^2(k+1) + y(k) = u(k)$

(c) $y(k+2) + 3y(k+1) + 2y(k) = u(k)$

(d) $y(k+2) + y^2(k+1) + ky(k) = u(k)$

[ESE-2012]

1.40 A signal $f(t)$ is described as

$$f(t) = [1 - |t|] \quad \text{when } |t| \le 1$$
$$= 0 \quad \text{when } |t| > 1$$

This represents the unit

(a) sinc function

(b) area triangular function

(c) signum function

(d) parabolic function [ESE-2012]

1.41 Match **List-I** with **List-II** and select the correct answer using the code given below the Lists:

List-I	List-II
A. Even signal	1. $x(n) = \left(\frac{1}{4}\right)^n u(n)$
B. Causal signal	2. $x(-n) = x(n)$
C. Periodic signal	3. $x(t) = u(t)$
D. Energy signal	4. $x(n) = x(n+N)$

Codes:

	A	B	C	D
(a)	2	3	4	1
(b)	1	3	4	2
(c)	2	4	3	1
(d)	1	4	3	2

[ESE-2012]

1.42 The period of the signal

$x(t) = 10\sin 12\pi t + 4\cos 18\pi t$ is

(a) $\frac{\pi}{4}$ (b) $\frac{1}{6}$

(c) $\frac{1}{9}$ (d) $\frac{1}{3}$ [ESE-2012]

1.43 The discrete time system described by $y(n) = x(n)^2$ is

(a) causal and linear

(b) causal and non-linear

(c) non-causal and linear

(d) non-causal and non linear [ESE-2012]

1.44 If a continuous time signal $x(t)$ can take on any value in the continuous interval $(-\infty, \infty)$, it is called

(a) Deterministic signal

(b) Random signal

(c) Analog signal

(d) Digital signal

[ESE-2013]

1.45 The value of the integral

$I = \int_1^2 (5t^2 + 1)\delta(t)dt$ is

(a) 0 (b) 1

(c) $\frac{42}{3}$ (d) $\frac{125}{3}$ [ESE-2013]

1.46 A continuous time system will be BIBO stable if all the Eigen values are

(a) one

(b) distinct and their real parts negative

(c) negative

(d) zero [ESE-2013]

1.47 The value of $\int_{-\infty}^{\infty} \sin t \cdot \delta\left(t - \frac{\pi}{4}\right) dt$ is

(a) $\sqrt{2}$ (b) $\frac{1}{\sqrt{3}}$

(c) $\frac{1}{\sqrt{2}}$ (d) $\sqrt{3}$ [ESE-2013]

1.48 The ramp function can be obtained from the unit impulse at $t = 0$ by

(a) differentiating unit impulse function once

(b) differentiating unit impulse function twice

(c) integrating unit impulse function once

(d) integrating unit impulse function twice

[ESE-2013]

1.49 The signal $x(n)$ shown in the below figure is a

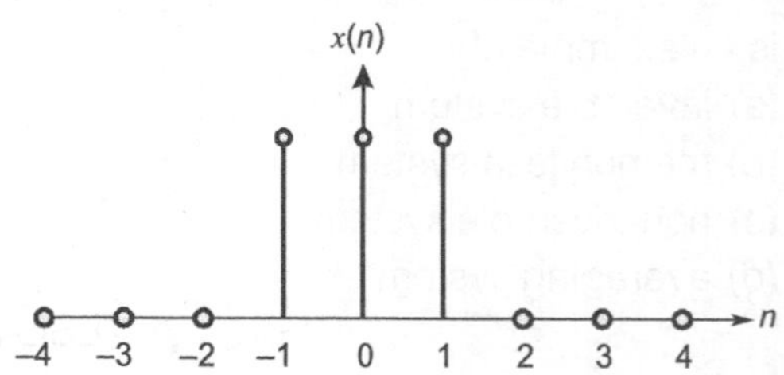

(a) periodic discrete time signal
(b) periodic signal
(c) non-periodic signal
(d) periodic discrete time signal consisting of 3 non-zero samples [ESE-2013]

1.50 A system is characterized by the input-output relation

$$y(t) = x(2t) + x(3t)$$

for all t, where $y(t)$ is the output and $x(t)$ is the input. It is
(a) linear and causal
(b) linear and non-causal
(c) non-linear and causal
(d) non-linear and non-causal [ESE-2014]

1.51 A discrete-time system has input $x|\cdot|$ and output $y|\cdot|$ satisfying

$$y[m] = \sum_{j=-\infty}^{m} x|f|$$

The system is
(a) linear and unstable
(b) linear and stable
(c) non-linear and stable
(d) non-linear and unstable [ESE-2014]

1.52 Consider a complex exponential sequence $e^{j\omega_0 n}$ with frequency ω_0. Suppose $\omega_0 = 1$, then
(a) Such a sequence is periodic
(b) Such a sequence is not periodic at all
(c) Periodic for some value of period 'N'
(d) Some definite range $N_0 < n < N_1$ exists for a periodic sequence [ESE-2015]

1.53 Statement (I): The total energy of an energy signal falls between the limits 0 and ∞.
Statement (II): The average power of an energy signal is zero.
(a) Both Statement (I) and Statement (II) are individually true and Statement (II) is the correct explanation of Statement (I)
(b) Both Statement (I) and Statement (II) are individually true but Statement (II) is **NOT** the correct explanation of Statement (I)
(c) Statement (I) is true but Statement (II) is false
(d) Statement (I) is false but Statement (II) is true [ESE-2015]

1.54 Statement (I): Sinusoidal signals are used as basic function in electrical systems.
Statement (II): The response of a linear system to a sinusoidal input function remains sinusoidal.
(a) Both Statement (I) and Statement (II) are individually true and Statement (II) is the correct explanation of Statement (I)
(b) Both Statement (I) and Statement (II) are individually true but Statement (II) is **NOT** the correct explanation of Statement (I)
(c) Statement (I) is true but Statement (II) is false
(d) Statement (I) is false but Statement (II) is true [ESE-2015]

1.55 In a discrete-time complex exponential sequence of frequency $\omega_0 = 1$, the sequence is
1. Periodic with period $\dfrac{2\pi}{\omega_0}$
2. Non periodic
3. Periodic for some value of period N

Which of the above statements is/are correct?
(a) 1 only (b) 2 only
(c) 3 only (d) 1 and 3 [ESE-2016]

1.56 When is a function $f(n)$ said to be leftsided?
(a) $f(n) = 0$ for $n < 0$
(b) $f(n) < 0$ for $n > 0$
(c) $f(n) = 0$ for $n > n_0$
(d) $f(n) = \infty$ for $n < n_0$
($n_0 \rightarrow$ Positive or negative integer) [ESE-2016]

1.57 The signal $x(t) = u(t + 2) - 2u(t) + u(t - 2)$ is represented by

(a)
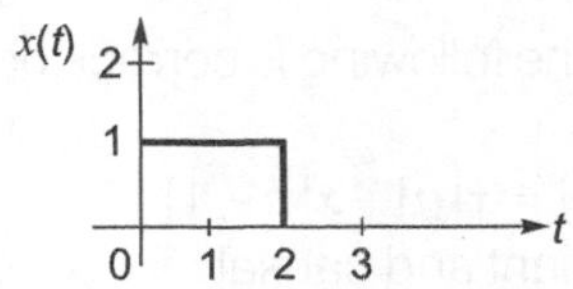

(b)
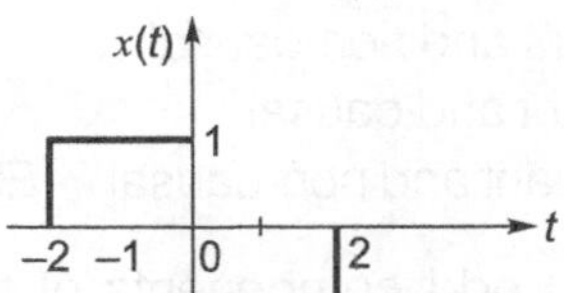

(c)
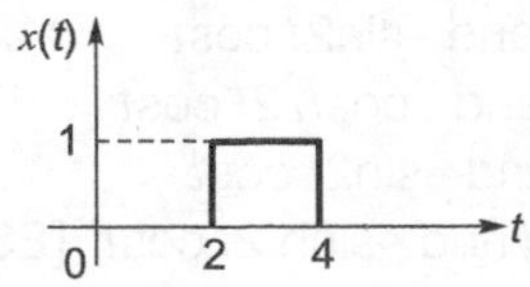

(d)
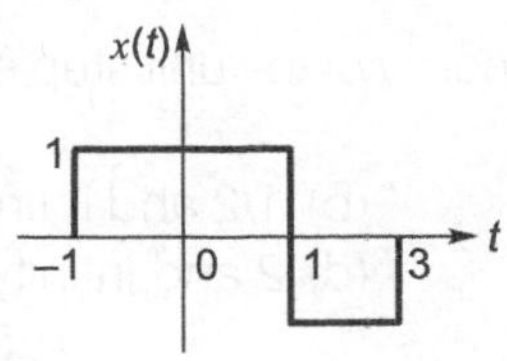

[ESE-2016]

1.58 A discrete time signal is given as

$$x[n] = \cos\frac{\pi n}{9} + \sin\left[\frac{\pi n}{7} + \frac{1}{2}\right]$$

The period N for the periodic signal is

(a) 126 (b) 32
(c) 252 (d) 64 [ESE-2018]

1.59 The special case of a finite-duration sequence is given as

$$x(n) = \{2, \underset{\uparrow}{4}, 0, 3\}$$

The sequence $x(n)$ into a sum of weighted impulse sequences will be

(a) $2\delta(n+1) + 4\delta(n) + 3\delta(n-2)$
(b) $2\delta(n) + 4\delta(n-1) + 3\delta(n-3)$
(c) $2\delta(n) + 4\delta(n-1) + 3\delta(n-2)$
(d) $2\delta(n+1) + 4\delta(n) + 3\delta(n-1)$ [ESE-2019]

1.60 The energy of the signal $x(n) = (-0.4)^n u(n)$ is

(a) $\frac{1}{16}$ (b) $\frac{1}{36}$
(c) $\frac{5}{3}$ (d) $\frac{25}{4}$ [ESE-2021]

1.61 Which one of the following operations is ***not*** commutative?

(a) Scaling and reversal of a signal $x[n]$
(b) Scaling and folding of a signal $x[n]$
(c) Folding and time reversal of a signal $x[n]$
(d) Folding and time delaying of a signal $x[n]$
[ESE-2021]

1.62 Which one of the following is correct for the given system?

$$y[n] = x[n] - x[n-1]$$

(a) Time invariant and causal
(b) Time variant and non-causal
(c) Time variant and causal
(d) Time invariant and non-causal [ESE-2021]

1.63 The even and odd components of the signal $x(t) = e^{-2t} \cos t$ are respectively,

(a) cos $2t$ cos$t$ and –sin$2t$ cos$t$
(b) sin$h$ $2t$ sin$t$ and –cos$h$ $2t$ cos$t$
(c) cos $2t$ sin$t$ and –sin$2t$ cos$t$
(d) cos$h$ $2t$ cos$t$ and –sinh $2t$ cos$t$ [ESE-2021]

1.64 The power and energy of the unit step sequence are respectively,

(a) 1/2 and 0 (b) 1/2 and infinity
(c) 2 and 0 (d) 2 and infinity
[ESE-2021]

1.65 **Statement (I):** Linear system may have multiple equilibrium states.
Statement (II): If a system is BIBO stable, it must also be zero-input or asymptotically stable.

(a) Both Statement (I) and Statement (II) are individually true and Statement (II) is the correct explanation of Statement (I).
(b) Both Statement (I) and Statement (II) are individually true but Statement (II) is **NOT** the correct explanation of Statement (I).
(c) Statement (I) is true, but Statement (II) is false.
(d) Statement (I) is false, but Statement (II) is true.
[ESE-2021]

1.66 Which one of the following statements is correct for the given system?

$$y(n) = x^2(n) + \frac{1}{x^2(n-1)}$$

(a) The given system is linear, non-causal and shift-variant.
(b) The given system is non-linear, causal and shift-invariant.
(c) The given system is non-linear, causal and shift-variant.
(d) The given system is linear, non-causal and shift-invariant.
[ESE-2022]

1.67 Consider the following relations of continuous time unit impulse function:

1. $\delta(t) = \delta(-t)$
2. $\int_{-\infty}^{+\infty} x(t)\,\delta(t)\,dt = x(0) = 0$
3. $\delta(at) = \frac{1}{|a|}\delta(t)$

Which of the above relations is/are correct

(a) 2 only (b) 1 and 2
(c) 1 and 3 (d) 2 and 3
[ESE-2024]

1.68 If $x(n)$ is a periodic signal with fundamental period N and takes infinite values, then it is

(a) an energy signal only
(b) a power signal only
(c) neither energy signal nor power signal
(d) both energy and power signals
[ESE-2024]

■■■

Answers Basics of Signals and Systems

1.1 (d)	1.2 (b)	1.3 (b)	1.4 (c)	1.5 (*)	1.6 (b)	1.7 (a)	1.8 (b)	1.9 (b)
1.10 (d)	1.11 (b)	1.12 (c)	1.13 (a)	1.14 (b)	1.15 (*)	1.16 (d)	1.17 (a)	1.18 (a)
1.19 (d)	1.20 (d)	1.21 (c)	1.22 (d)	1.23 (a)	1.24 (b)	1.25 (d)	1.26 (d)	1.27 (a)
1.28 (c)	1.29 (c)	1.30 (d)	1.31 (a, d)	1.32 (a)	1.33 (d)	1.34 (a)	1.35 (a)	1.36 (c)
1.37 (d)	1.38 (b)	1.39 (c)	1.40 (b)	1.41 (a)	1.42 (d)	1.43 (b)	1.44 (c)	1.45 (a)
1.46 (c)	1.47 (c)	1.48 (d)	1.49 (c)	1.50 (b)	1.51 (a)	1.52 (b)	1.53 (b)	1.54 (a)
1.55 (b)	1.56 (c)	1.57 (b)	1.58 (a)	1.59 (a)	1.60 (*)	1.61 (d)	1.62 (a)	1.63 (d)
1.64 (b)	1.65 (c)	1.66 (b)	1.67 (c)	1.68 (c)				

Explanations Basics of Signals and Systems

1.1 (d)

$$3y^3(t) + 2y^2(t) + y(t) = x^2(t) + x(t)$$

Since the output for each value of the independent variable at a given time is dependent only on the input at the same time, the system is memoryless or static or non-dynamic.

For a system to be linear, if $x(t) = ax_1(t) + bx_2(t)$, then $y(t) = ay_1(t) + by_2(t)$. Since the given system does not satisfy the condition of linearity, it is a non-linear system.

1.2 (b)

A system is causal if the output at any time depends only on values of the input at the present time and in the past.

1.3 (b)

Let $y(t) = \delta(t)$. Then

$$\Rightarrow \quad \delta(t) + \int_0^{\infty} \delta(\tau)x(t-\tau)d\tau$$

$$\Rightarrow \quad \delta(t) + x(t)$$

So, $y(t) = \delta(t)$ satisfies the given equation.

1.4 (c)

Let $g(t) = f(t) \cdot u(t)$

For right shift $t \Rightarrow t - t_0$

$$g(t-t_0) = f(t-t_0) \cdot u(t-t_0)$$

1.5 (*)

$$y(n) = x(n^2)$$

(i) for $n = -2$, $y(-2) = x(4)$

Since, $y(n)$ depends on future states also, it is a non-causal system.

(ii) Let $x_3(n) = ax_1(n) + bx_2(n)$

$$y_3(n) = x_3(n^2) = ax_1(n^2) + bx_2(n^2) = ay_1(n) + by_2(n)$$

$\therefore$ The system is linear

(iii) $x_1(n) \Rightarrow y_1(n) = x_1(n^2)$

Let $x_2(n) = x_1(n - n_0)$

$$\Rightarrow \quad y_2(n) = x_2(n^2) = x_1(n^2 - n_0)$$

$$y_1(n-n_0) = x_1(n-n_0)^2$$

$$y_2(n) \neq y_1(n-n_0)$$

$\therefore$ The system is time-variant.

No option is correct.

1.6 (b)

Energy signal: A signal is said to be an energy signal if the energy of the signal is finite (i.e, $E < \infty$) and power is zero.

Power signal: A signal is said to be a power signal if its power contained is finite. So, energy of power signal is infinite.

$$x(t) = A\cos(\omega t + \phi)$$

Energy of $x(t)$ is

$$E = \int_{-\infty}^{\infty} |x(t)|^2 dt = \int_{-\infty}^{\infty} A^2 dt = \infty$$

Power of $x(t)$ is

$$P = \lim_{T \to \infty} \frac{1}{T} \int_{-T/2}^{T/2} |x(t)|^2 dt$$

$$= \lim_{T\to\infty}\frac{1}{T}\int_{-T/2}^{T/2} A^2 dt = \lim_{T\to\infty}\frac{1}{T}\cdot A^2 T = A^2$$

Since the power of the signal is finite, it is a power signal.

Note: Signals having infinite duration with their value tending to a non zero constant as $t\to\infty$ are in general a power signal.

1.7 (a)

If $f_1(t)$ has the time period $T_1 = p_1/q_1$ and $f_2(t)$ has the time period $T_2 = p_2/q_2$, then $f(t) = f_1(t) + f_2(t)$ has the time period T, given as

$$T = \frac{\text{LCM}\{p_1, p_2\}}{\text{HCF}\{q_1, q_2\}}$$

If the signal $f_1(t)$ is periodic with time period T_1 and $f_2(t)$ is periodic with time period T_2, then $f(t) = f_1(t) + f_2(t)$ will be periodic if and only if T_1/T_2 is a rational number.

$$x_1(t) = 2\sin \pi t + \cos 4\pi t$$

$$\Rightarrow \quad f_1(t) = 2\sin \pi t$$

$$\Rightarrow \quad T_1 = \frac{\pi}{2\pi} = \frac{1}{2}$$

$$\text{and} \quad f_2(t) = \cos 4\pi t$$

$$\Rightarrow \quad T_2 = \frac{4\pi}{2\pi} = 2$$

$$\frac{T_1}{T_2} = \frac{1/2}{2} = \frac{1}{4}$$

Since T_1/T_2 is a rational number, hence $x_1(t)$ is periodic.

$$x_2(t) = \sin 5\pi t + 3\sin 13\pi t$$

$$\Rightarrow \quad T_1 = \frac{5\pi}{2\pi} = \frac{5}{2}$$

$$\text{and} \quad T_2 = \frac{13\pi}{2\pi} = \frac{13}{2}$$

$$\Rightarrow \quad \frac{T_1}{T_2} = \frac{5}{13}$$

$\Rightarrow x_2(t)$ is periodic.

1.8 (b)

Let $\quad y(n) = x(n) - x(n-1)$

$\quad x(n) = x_1(n)$

$\Rightarrow \quad y_1(n) = x_1(n) - x_1(n-1) \quad$ ...(i)

Let $\quad x_2(n) = x_1(n-1)$

$\Rightarrow \quad y_2(n) = x_2(n) - x_2(n-1)$

$\quad = x_1(n-1) - x_1(n-2) \quad$ ...(ii)

Replacing n by $(n-1)$ in equation (i),

$y_1(n-1) = x_1(n-1) - x_1(n-2) \quad$ ...(iii)

Comparing equation (ii) and (iii),

$$y_2(n) = y_1(n-1)$$

Thus, if $\quad x_2(n) = x_1(n-1)$,

then $\quad y_2(n) = y_1(n-1)$

Hence the system is time invariant

1.9 (b)

(i) $\quad y(n) = x(n)$

$\Rightarrow$ Linear, causal system

If $\quad x(n) = ax_1(n) + bx_2(n)$,

then $y(n) = ax_1(n) + bx_2(n)$

$\quad = ay_1(n) + by_2(n)$

Hence system is linear

Since value of $y(n)$ depends only on present value of the $x(n)$, hence the system is causal.

(ii) $\quad y(n) = x(n^2)$

$\Rightarrow$ Linear, non-causal system

It satisfies the condition of linearity

For example, $y(-2) = x(4)$

Since output depends on future value of input, hence the system is non-causal.

1.11 (b)

The sum of two sinusoids is periodic if the ratio of the time periods of the sinusoids is a rational number.

1.12 (c)

$$f(t) = u(t) + u(t-1) + r(t-2)$$

$$f(t) = u(t) + u(t-1) + (t-2)\,u(t-2)$$

1.14 (b)

A system is causal if the output at any time depends only on values of the input at the present time and in the past. In other words, the output at any time does not depend on future values of input.

1.15 (*)

$$v_1(n) = \sin[5\pi n]$$

$$v_2(n) = \sqrt{2}\cos[5\pi n]$$

$$v(n) = v_1(n) + v_2(n)$$

$$= \sqrt{3}\left\{\frac{1}{\sqrt{3}}\sin[5\pi n] + \frac{\sqrt{2}}{\sqrt{3}}\cos[5\pi n]\right\}$$

$$= \sqrt{3}\sin[5\pi n + \theta]$$

where, $\theta = \cos^{-1}\dfrac{1}{\sqrt{3}} = 54.7°$

1.17 (a)

$v[n] + v[-n] = 0$

for all integer values of n.

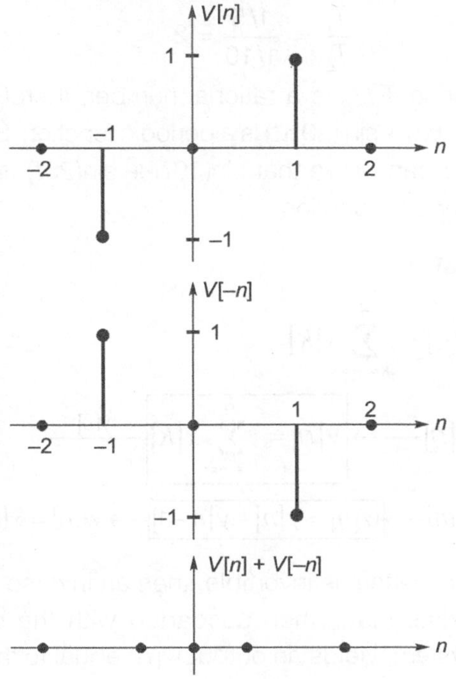

1.19 (d)

$$v(t) = (2t-6)\,[u(t-3) - u(t-4)]$$

1.20 (d)

$$y(t) = ax(t) + b$$

Let $x(t) = x_1(t) + x_2(t)$

Then $y(t) = ax_1(t) + ax_2(t) + b$

$$y(t) \neq \{ax_1(t) + b\} + (ax_2(t) + b\}$$

$$y(t) \neq y_1(t) + y_2(t)$$

Hence the system is non-linear.

1.21 (c)

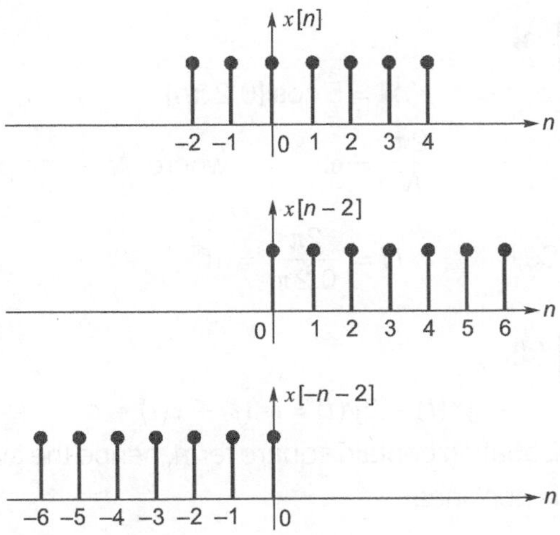

1.22 (d)

$$y(t) = x(t^2)$$

For example, $y(-2) = x(4)$

Since output depends on the future value of input, system is non-causal.

Let $x(t) = x_1(t)$

$\Rightarrow \quad y_1(t) = x_1(t^2)$

Let $x_2(t) = x_1(t-t_0)$

$\Rightarrow \quad y_2(t) = x_2(t^2)$

$\qquad = x_1(t^2 - t_0)$

But $y_1(t-t_0) = x_1\{(t-t_0)^2\}$

So, $y_2(t) \neq y_1(t-t_0)$

Hence, the system is time variant.

1.23 (a)

$$x[n] = \begin{cases} 1, & n = 1,\ 2 \\ -1, & n = -1,\ -2 \\ 0, & n = 0, n > 2 \text{ and } n < -2 \end{cases}$$

$$\Rightarrow x[n+3] = \begin{cases} 1, & n = 1-3,\ 2-3 \\ -1, & n = -1-3,\ -2-3 \\ 0, & n = 0-3, n > 2-3 \\ & \text{and } n < -2-3 \end{cases}$$

$$\Rightarrow x[n+3] = \begin{cases} 1, & n = -2,\ -1 \\ -1, & n = -4, -5 \\ 0, & n = -3, n < -5 \text{ and } n > -1 \end{cases}$$

1.24 (b)

$$x_1(t) = e^{j20t}$$

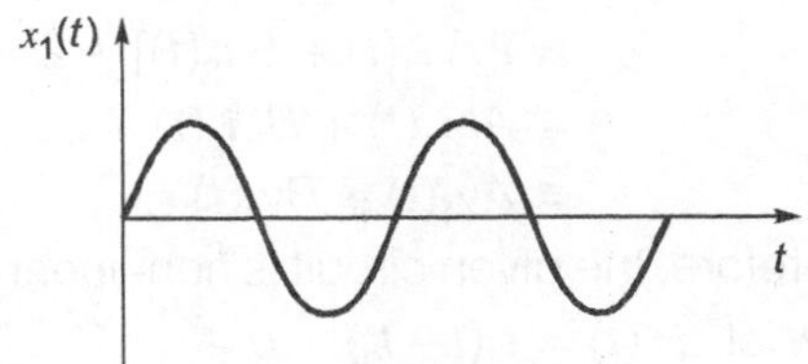

Periodic

$$x_2(t) = e^{(-2+j)t} = e^{-2t} \cdot e^{jt}$$

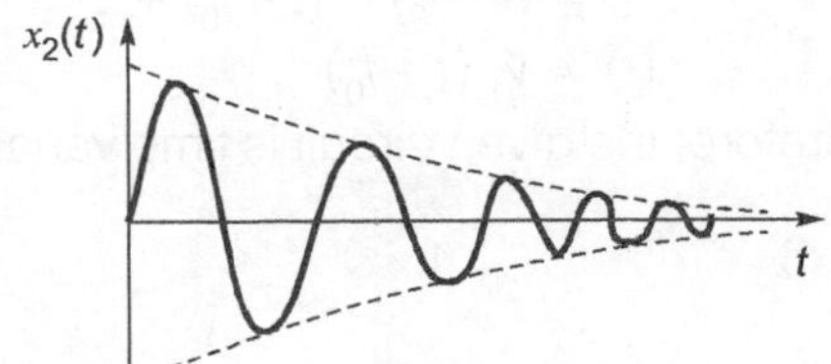

Non-Periodic

Note: A periodic signal on multiplication with a exponential decaying function becomes aperiodic.

1.25 (d)

$$y(n) = x^2(n)$$

If the output $y(n)$ of the system is dependent on the $x(n)$ input having degree other than 1 are most likely to be non linear.

Looking at the option only option (d) satisfies it.

1.27 (a)

$$y(n) = x(n) + x(-n) = 0$$

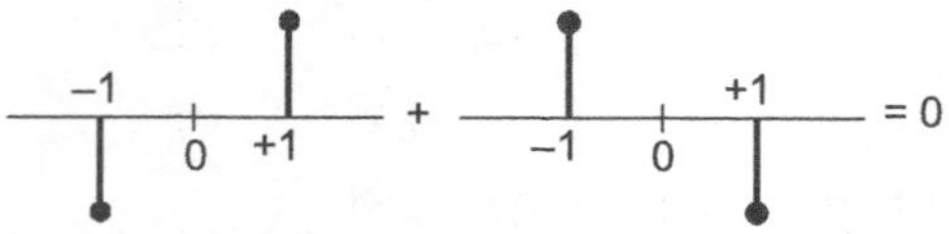

1.28 (c)

$$x(n) = (e^{j\pi})^n$$
$$= (\cos\pi + j\sin\pi)^n$$
$$x(n) = (-1)^n$$

$e^{j\pi n} = (-1)^n \longrightarrow \boxed{S_1} \longrightarrow 1$

$e^{j\pi n} = (-1)^n \longrightarrow \boxed{S_2} \longrightarrow (-1)^n$

In case of system S_2, output is replica of input, so system S_2 is LTI but S_1 is not.

1.29 (c)

Energy of power signal is always infinite.

1.30 (d)

$$v(t) = t\,i(t) + 2$$
$$i_1(t) \rightarrow v_1(t) = t\,i_1(t) + 2$$
$$i_2(t) \rightarrow v_2(t) = t\,i_2(t) + 2$$

Let, $i_3(t) = A\,i_1(t) + B\,i_2(t)$

Then, $v_3(t) = t\,i_3(t) + 2$

$$= t[A\,i_1(t) + B\,i_2(t)] + 2$$
$$= At\,i_1(t) + Bt\,i_2(t) + 2$$
$$\neq Av_1(t) + Bv_2(t)$$

Therefore, the given circuit is non-linear.

Now let $i_2(t) = i_1(t - t_0)$

So $v_2(t) = t\,i_2(t) + 2$

$$= t\,i_1(t - t_0) + 2$$
$$\neq (t - t_0)\,i_1(t - t_0) + 2$$

$\Rightarrow$ $v_2(t) \neq v_1(t - t_0)$

Therefore, the given circuit is time variant.

1.31 (a, d)

Fundamental period of sin ($10\pi t$),

$$T_1 = \frac{2\pi}{10\pi} = \frac{1}{5}$$

Fundamental period of sin ($20\pi t$),

$$T_2 = \frac{2\pi}{20\pi} = \frac{1}{10}$$

$$\frac{T_1}{T_2} = \frac{1/5}{1/10} = 2$$

Since T_1/T_2 is a rational number, therefore, sin ($10\pi t$) + sin ($20\,\pi t$) is a periodic function. Similarly we can prove that sin($10t$) + sin($25t$) is also a periodic function.

1.32 (a)

$$y[n] = \sum_{k=-\infty}^{n} x[k]$$

$x[n] \longrightarrow \boxed{y[n] = \sum_{k=-\infty}^{n} x[k]} \xrightarrow{y[n]}$

$y[n] \longrightarrow \boxed{w[n] = y[n] - y[n-1]} \longrightarrow w[n] = x[n]$

If a system is invertible, then an inverse system exists that, when cascaded with the original system, yields an output $w[n]$ equal to the input $x[n]$ to the first system

1.34 (a)

For a continuous-time system,

$$y(t) = s(t) * x(t)$$
$$= \int_{-\infty}^{\infty} s(t-\lambda)\,x(\lambda)\,d\lambda$$

Taking $s(t) = u(t)$,

$$y(t) = \int_{-\infty}^{\infty} u(t-\lambda)\,x(\lambda)\,d\lambda$$
$$y(t) = \int_{-\infty}^{t} x(\lambda)\,d\lambda$$

Therefore, $s(t) = u(t)$

Which is linear and time-invariant.

1.35 (a)

$$x(n) = 5\cos\{0.2\,\pi n\}$$

$$\frac{2\pi}{N} = \omega \qquad \text{where, } N = \text{time period}$$

So, $$N = \frac{2\pi}{0.2\pi} = 10$$

1.37 (d)

$$y^2(t) + 2y(t) = x^2(t) + x(t) + c$$

Equation contain square term, hence the system is non linear.

1.38 (b)

Properties of delta function:

1. $\int_{-\infty}^{\infty} \delta(t)\, dt = 1$
2. $x(t)\,\delta(t) = x(0)\,\delta(t)$
3. $x(t)\,\delta(t-t_0) = x(t_0)\,\delta(t-t_0)$
4. $\delta(at) = \dfrac{1}{|a|}\delta(t)$
5. $\int_{-\infty}^{\infty} x(t)\,\delta(t)\, d\tau = x(0)$

So option (b) does not represent a property of delta function.

1.39 (c)

The linearity property follows both the superposition and homogeneity property.

i.e. $ax_1(t) \rightarrow ay_1(t)$

$bx_2(t) \rightarrow by_2(t)$

$$x_3(t) = ax_1(t) + bx_2(t) \leftrightarrow ay_1(t) + by_2(t) = ay_3(t)$$

For a system to be time-invariant delayed response should be equal to the response of the system to the delayed input.

Thus option (c) is linear and time-varying system.

1.40 (b)

$$f(t) = \begin{cases} 1-|t| & \text{when } |t| \le 1 \\ 0 & \text{when } |t| > 1 \end{cases}$$

when, $t < 0$; $|t| = -t$

$f(t) = 1 + t$

when, $t > 0$ $|t| = t$

$f(t) = 1 - t$

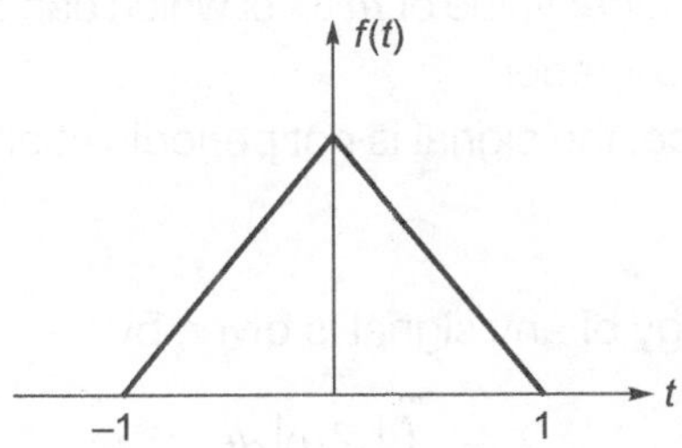

1.41 (a)

- Even signal $x(n) = x(-n)$
- Causal system is one in which output at any time depends only on present and/or past values of input.
- Periodic signal is one which satisfies $x(n) = x(n + N)$; $N \rightarrow$ Fundamental period.
- Energy signal is absolutely summable i.e. $x(n)$ $= \left|\left(\frac{1}{4}\right)^n u(n)\right| < \infty$

1.42 (d)

$x(t) = 10 \sin 12\pi t + 4 \cos 18\,\pi t$

This is combination of two signals having fundamental time period T_1 and T_2.

$$T_1 = \frac{2\pi}{12\pi} = \frac{1}{6}$$

$$T_2 = \frac{2\pi}{18\pi} = \frac{1}{9}$$

Total time period

$$= \frac{\text{L.C.M of numerator of } (T_1, T_2)}{\text{H.C.F. of denominators of } (T_1, T_2)} = \frac{1}{3}$$

Alternatively:

By GCD method

$12\pi = 2(6\pi)$

$18\pi = 3(6\pi)$

So $\omega_0 = 6\pi = \dfrac{2\pi}{T}$

$T = \dfrac{1}{3}$

1.43 (b)

$$y(n) = x(n)^2$$

Linearity check

$x_1(n) \xrightarrow{P} x_1(n)^2$

$x_2(n) \xrightarrow{P} x_2(n)^2$

$$x_3(n) = (x_1(n) + x_2(n)) \xrightarrow{P} (x_1(n) + x_2(n))^2$$
$$= x_1(n)^2 + x_2(n)^2 + 2x_1(n)\,x_2(n)$$
$$\neq x_1(n)^2 + x_2(n)^2$$

So, the system is non-linear.

Causality check

$y(0) = x(0)^2$

$y(1) = x(1)^2$

$y(-1) = x(-1)^2$

Thus output depends only on the present values, so the system is causal and non-linear.

Note: As per RTI filed and reply of **UPSC** answer is option (d).

1.44 (c)

If a continuous time signal can take on any value in the continuous interval $(-\infty, \infty)$ then this signal is known as analog signal.

1.45 (a)

$$\therefore \int_{t}^{t_2} x(t)\cdot\delta(t-t_o) = \begin{cases} x(t_o); & t_1 < t_o < t_2 \\ 0 & ; \text{ otherwise} \end{cases}$$

$$I = \int_1^2 (5t^2+1)\delta(t)dt = 0$$

1.46 (c)

All the roots of characteristic equation should lie in the left half of, s-plane, hence they should have negative real parts.

1.47 (c)

$$\int_{-\infty}^{\infty} \sin t\cdot\delta\left(t-\frac{\pi}{4}\right)dt$$

$$= \int_{-\infty}^{\infty} \sin\frac{\pi}{4}\delta\left(t-\frac{\pi}{4}\right)dt$$

$$= \sin\frac{\pi}{4}\int_{-\infty}^{\infty} \delta\left(t-\frac{\pi}{4}\right)dt$$

$$= \sin\frac{\pi}{4}\times 1 = \frac{1}{\sqrt{2}}$$

1.48 (d)

$$y = \iint \delta(t)dt$$

$$= t\,u(t) = \text{ramp function}$$

1.49 (c)

The given signal is non-periodic discrete time signal because it is a finite duration signal.

1.50 (b)

Given,

$$y(t) = x(2t) + x(3t)$$

Scaling operation is always a non causal operation.

Now check for linearity:

$$y_1(t) = x_1(2t) + x_2(3t)$$

Also $y_2(t) = x_2(2t) + x_2(3t)$

$$y_3(t) = x_3(2t) + x_3(3t) \quad \text{...(i)}$$

where

$$x_3(t) = x_1(t) + x_2(t) \quad \text{...(ii)}$$

from (i) and (ii)

$$y_3(t) = x_1(2t) + x_2(2t) + x_1(3t) + x_2(3t)$$

$$y_3(t) = y_1(t) + y_2(t)$$

Thus, the given system is a linear and non causal system.

1.51 (a)

For the BIBO stability test

Consider an input $x|f| = M < \infty$

thus, $y[m] = \sum_{j=-\infty}^{m} M$

Since we are adding the signal from $-\infty$ thus the output becomes unbounded

thus $y[m] \to \infty$

even if $x|f|$ is bounded

for linearly

$$y_1[m] = \sum_{j=-\infty}^{m} x_1|f|$$

$$y_2[m] = \sum_{j=-\infty}^{m} x_2|f|$$

Now,

$$y_3[m] = a_1 y_1[m] + a_2 y_2[m]$$

$$= a_1\sum_{j=-\infty}^{\infty} x_1|f| + a_2\sum_{j=-\infty}^{m} x_2|f|$$

1.52 (b)

$$h(n) = e^{j\omega_0 n}$$

$$\omega_0 = \frac{2\pi}{N}\cdot m$$

or $N = \frac{2\pi}{\omega_0}\cdot m$

as $\omega_0 = 1$

$\therefore$ $N = 2\pi\cdot m$

No integer value of m exist which can convert the N into integer.

Hence, the signal is not periodic at all.

1.53 (b)

Energy of any signal is given by

$$E = \int_{-\infty}^{\infty} \left|x^2(t)\right| dt$$

and power of a signal is given by

$$P = \lim_{T\to\infty}\int_{-T/2}^{T/2} \frac{1}{T}\left|x^2(t)dt\right|$$

For energy signal, Energy is finite

$$\therefore \quad P = \lim_{T\to\infty}\frac{E}{T}$$

$$P = \frac{E}{\infty} = 0$$

→ All the finite duration and bounded signals are energy signals.
Hence statements (I) and (II) are correct but statement (II) is not correct explanation of statement (I).

1.54 (a)

If input is sinusoidal, output is also sinusoidal having some different phase and amplitude depending on transfer function of system.

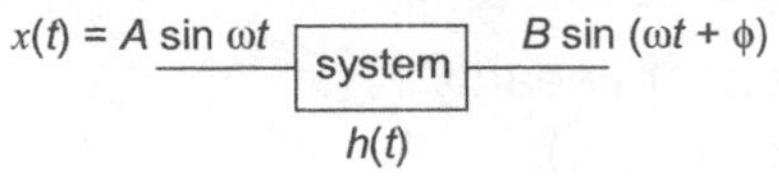

$$B = |X(\omega)|\,|H(\omega)|$$
$$\phi = \angle X(\omega) + \angle H(\omega)$$

and this is the reason we prefer sinusoidal signals. Hence statements (I) and (II) are correct and statement (II) is correct explanation of statement (I).

1.55 (b)

Given that, $\omega_o = 1$
For discrete time exponential to be periodic

$$C = \frac{2\pi}{\omega_o} \text{ (should be rational)}$$

In the present case $= \frac{2\pi}{1} \Rightarrow$ Non-periodic.

1.56 (c)

A signal having a non-zero value towards left of a finite value of time till $t = -\infty$ are called left sided signal.
i.e., for example

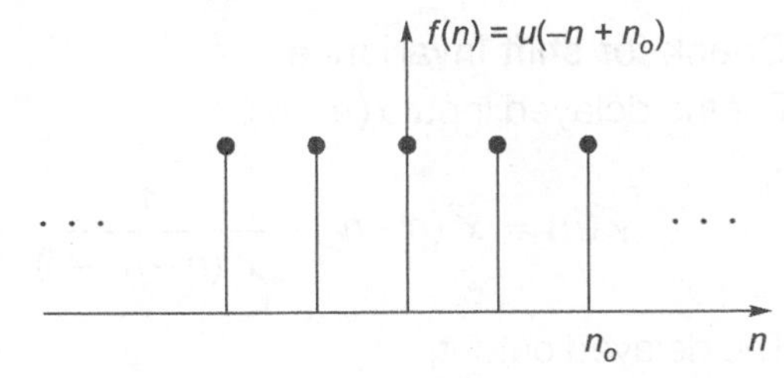

Hence, $f(n) = 0$, for $n > n_o$

1.57 (b)

Shifts represents instants where change in step will occur and coefficients represent the amount of step change at the shifts given to $u(t)$.

i.e. $u(t-(-2)) - 2u(t-0) + u(t-2)$
hence,

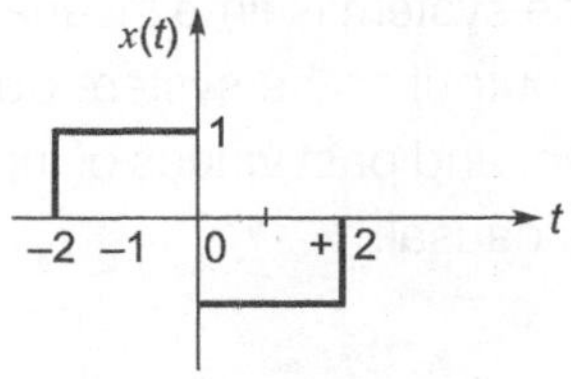

1.58 (a)

$$N_1 = \frac{2\pi}{\pi/9} = 18$$

$$N_2 = \frac{2\pi}{\pi/7} = 14$$

$$N = \text{LCM}\{18, 14\} = 126$$

1.59 (a)

$$x[n] = \{2, \underset{\uparrow}{4}, 0, 3\}$$

So, $x[n] = 2\delta[n+1] + 4\delta[n] + 3\delta[n-2]$

1.60 (*)

$$x(n) = (-0.4)^n u(n)$$

The energy of the signal is given by

$$E = \sum_{n=-\infty}^{\infty} |x(n)|^2$$

$$E = \sum_{n=0}^{\infty} (-0.4)^{2n}$$

$$E = \sum_{n=0}^{\infty} (0.16)^n$$

$$E = \frac{1}{1-0.16} = \frac{1}{0.84} = \frac{25}{21}$$

Hence, all the given options are wrong.

1.61 (d)

Folding and time delaying of a signal $x[n]$ is not commutative as shown below:

$$x[n] \xrightarrow{\text{Folding}} x[-n] \xrightarrow[\text{by } n_0]{\text{Time-delay}} x[-(n-n_0)]$$
$$= x[-n+n_0]$$

$$x[n] \xrightarrow[\text{by } n_0]{\text{Time-delay}} x[n-n_0] \xrightarrow{\text{Folding}} x[-n-n_0]$$

1.62 (a)

Given, $y[n] = x[n] - x[n-1]$
If we delay the input $x[n]$ by n_0, the output,

$$y_1[n] = x[n-n_0] - x[n-n_0-1]$$

If we delay the output $y[n]$ by n_0,

$$y[n-n_0] = x[n-n_0] - x[n-n_0-1] = y_1[n]$$

Hence, the system is time invariant.

Also, the output of the system depends only on the present and past values of input. Hence, the system is causal.

1.63 (d)

$$x(t) = e^{-2t}\cos t$$

The even-component of the signal is given by

$$x_e(t) = \frac{x(t)+x(-t)}{2}$$

$$x_e(t) = \frac{e^{-2t}\cos t + e^{2t}\cos(-t)}{2}$$

$$x_e(t) = \cos t\left[\frac{e^{-2t}+e^{2t}}{2}\right]$$

$$x_e(t) = \cos t\cdot\cosh(2t)$$

The odd-component of the signal is given by

$$x_0(t) = \frac{x(t)-x(-t)}{2}$$

$$x_0(t) = \frac{e^{-2t}\cos t - e^{2t}\cos(-t)}{2}$$

$$x_0(t) = \cos t\left[\frac{e^{-2t}-e^{2t}}{2}\right]$$

$$x_0(t) = -\cos t\left[\frac{e^{2t}-e^{-2t}}{2}\right] = -\cos t\cdot\sinh 2t$$

1.64 (b)

The power of unit-step sequence u(n) is calculated as

$$P = \lim_{N\to\infty}\frac{1}{2N+1}\sum_{n=-N}^{N}|u(n)|^2$$

$$P = \lim_{N\to\infty}\frac{N+1}{2N+1}$$

$$P = \frac{1}{2}\text{W}$$

The energy of unit-step sequence,

$$E = \sum_{n=-\infty}^{\infty}|u(n)|^2 = \text{infinite}$$

1.65 (c)

The linear system $\dot{x} = Ax$ has one equilibrium state if A is non-singular i.e. det(A) $\neq 0$, and there exists infinite equilibrium states if A is singular. Hence, linear system may have multiple equilibrium states. For asymptotic stability, we check if the system response goes to zero for zero input. Whereas BIBO stability refers to the property that a bounded input leads to bounded output. Hence, if a system is BIBO stable, it may not necessarily by asymptotically stable.

1.66 (b)

$$y(n) = x^2(n) + \frac{1}{x^2(n-1)}$$

Check for linearity :

For input $x_1(n)$,

$$y_1(n) = x_1^2(n) + \frac{1}{x_1^2(n-1)}$$

For input $x_2(n)$,

$$y_2(n) = x_2^2(n) + \frac{1}{x_2^2(n-1)}$$

For $x(n) = x_1(n) + x_2(n)$,

$$y(n) = [x_1(n)+x_2(n)]^2 + \frac{1}{[x_1(n-1)+x_2(n-1)]^2}$$

Hence, $y(n) \neq y_1(n) + y_2(n)$

Therefore, the given system is non-linear.

Check for causality:

$$y(n) = \underbrace{x^2(n)}_{\text{Present input}} + \underbrace{\frac{1}{x^2(n-1)}}_{\text{Delayed or past input}}$$

Since, the present output depends on present and past values of input. Therefore, the given system is causal.

Check for shift invariance:

For the delayed input $x(n-n_o)$,

$$y_1(n) = x^2(n-n_0) + \frac{1}{x^2(n-n_o-1)}$$

The delayed output,

$$y(n-n_o) = x^2(n-n_0) + \frac{1}{x^2(n-n_o-1)}$$

Since, $y_1(n) = y(n-n_o)$. Therefore, the given system is shift invariant.

1.67 (c)

1. The unit impulse is an even function of time i.e., $\delta(-t) = \delta(t)$.
2. We know that,

$$x(t)\cdot\delta(t) = x(0)\cdot\delta(t)$$

$$\therefore \int_{-\infty}^{\infty} x(t)\cdot\delta(t)\cdot dt = x(0)\cdot\int_{-\infty}^{\infty}\delta(t)\cdot dt = x(0)$$

No information is given about $x(0)$. Hence, statement 2 is incorrect.

3. We have,

$$\delta(t) = \begin{cases} +\infty, & t = 0 \\ 0, & \text{otherwise} \end{cases}$$

$$\Rightarrow |a|\delta(at) = \begin{cases} +\infty, & at = 0 \\ 0, & \text{otherwise} \end{cases}$$

$$\Rightarrow |a|\delta(at) = \delta(t)$$

$$\Rightarrow \delta(at) = \frac{1}{|a|}\cdot\delta(t)$$

1.68 (c)

Since the signal $x(n)$ takes infinite values, hence energy in one period is infinite. Therefore, both the power and the total energy of the signal are infinite. Hence, it is neither an energy signal nor power signal.

Note:

- The energy signal has finite energy and zero average power.
- The power signal has infinite energy and finite average power.

■■■

2 Continuous and Discrete Time LTI Systems

2.1 If the step response of a causal, linear time-invariant system is $a(t)$, then the response of the system to a general input $x(t)$ would be

(a) $\int_{0^+}^{t} \frac{da(\tau)}{d\tau} x(t-\tau)d\tau$

(b) $a(0)\,x(t) + \int_{0^+}^{t} \frac{da(\tau)}{d\tau} x(t-\tau)d\tau$

(c) $x(0)\,a(t) + \int_{0^+}^{t} x(\tau)\,a(t-\tau)d\tau$

(d) $x(0)\,a(t) + \int_{0^+}^{t} \frac{da(\tau)}{d\tau} x(t-\tau)d\tau$

[ESE-2000]

2.2 **Assertion (A):** The 'zero-state' response of a linear constant-parameter continuous-time system can have components having the 'natural frequencies' of the system.

Reason (R): The 'forced frequency' components in the response of the state for any given input may not add up to the given zero initial value of the state. The 'natural frequency' components may be needed to bridge the gap.

(a) Both A and R are true and R is the correct explanation of A
(b) Both A and R are true but R is NOT the correct explanation of A
(c) A is true but R is false
(d) A is false but R is true [ESE-2000]

2.3 The impulse response of a system consists of two delta functions as shown in the given figure

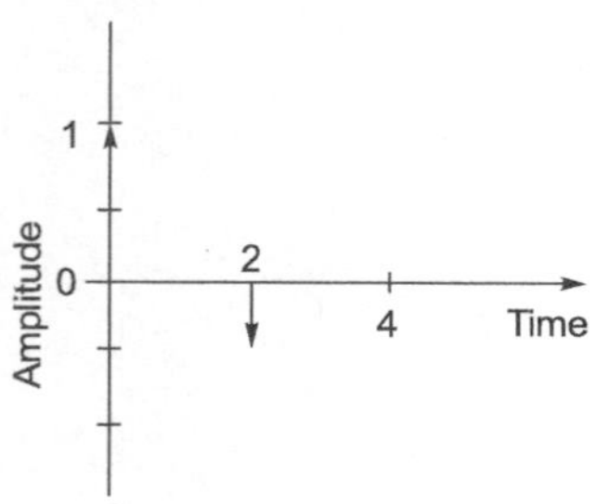

The input to the system is a unit amplitude square pulse of one unit time duration. Which one of the following diagrams depicts the correct output?

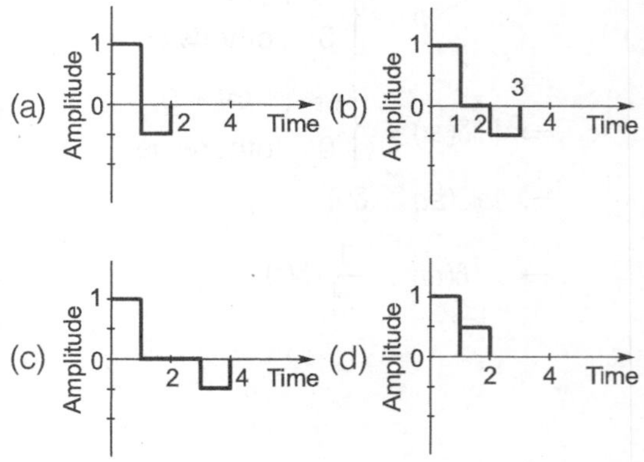

[ESE-2001]

2.4 The impulse response of a system is $h(t) = \delta(t-0.5)$

If two such systems are cascaded, the impulse response of the overall system will be

(a) $0.5\delta(t-0.25)$ (b) $\delta(t-0.25)$
(c) $\delta(t-1)$ (d) $0.5\delta(t-1)$

[ESE-2001]

2.5 The response of a linear, time-invariant, discrete-time system to a unit step input $u(n)$ is the unit impulse $\delta(n)$. The system response to a ramp input n $u(n)$ would be

(a) $u(n)$ (b) $u(n-1)$

(c) $n\delta(n)$ (d) $\sum_{k=0}^{\infty} k\delta(n-k)$

[ESE-2002]

2.6 The range of values of a and b for which the linear time invariant system with impulse response $h(n) = a^n,\ n \geq 0;\ b^n,\ n < 0$ will be stable is

(a) $|a|<1, |b|>1$ (b) $|a|<1, |b|<1$
(c) $|a|>1, |b|>1$ (d) $|a|>1, |b|<1$

[ESE-2002]

2.7 Two rectangular waveforms of duration T_1 and T_2 seconds are convolved. What is the shape of the resulting waveform?
(a) Triangular (b) Rectangular
(c) Trapezoidal (d) Semi-circular
[ESE-2003]

2.8 A discrete LTI system is non-causal if its impulse response is
(a) $a^n u(n-2)$ (b) $a^{n-2} u(n)$
(c) $a^{n+2} u(n)$ (d) $a^n u(n+2)$
[ESE-2003]

2.9 A discrete time system has impulse response $h(n) = a^n u(n+2)$, $|a| < 1$. Which one of the following statements is correct? The system is
(a) stable, causal and memoryless
(b) stable, non-causal and memoryless
(c) stable, non-causal and has memory
(d) unstable, non-causal and memoryless
[ESE-2004]

2.10 $y[n] = \sum_{k=-\infty}^{n} x[k]$

Which one of the following systems is inverse of the system given above?
(a) $x[n] = y[n] - y[n-1]$
(b) $x[n] = y[n]$
(c) $x[n] = y[n+4]$
(d) $x[n] = ny[n]$
[ESE-2004]

2.11 To which one of the following difference equations, the impulse response $h(n) = \delta(n+2) - \delta(n-2)$ corresponds?
(a) $y(n+2) = x(n) - x(n-2)$
(b) $y(n-2) = x(n) - x(n-4)$
(c) $y(n) = x(n+2) + x(n-2)$
(d) $y(n) = -x(n+2) + x(n-2)$ [ESE-2004]

2.12 The impulse response of a linear time-invariant system is a rectangular pulse of duration T. It is excited by an input of a pulse of duration T. What is the filter output waveform?
(a) Rectangular pulse of duration T
(b) Rectangular pulse of duration $2T$
(c) Triangular pulse of duration T
(d) Triangular pulse of duration $2T$
[ESE-2004]

2.13 Which one of the following gives the cross-correlation $[y_{xy}(k)]$ of two finite length sequences $x(n) = \{1, 3, 1, 3\}$ and $y(n) = \{1, 2, 1, 3\}$?
(a) {3, 10, 8, 14, 7, 5, 2}
(b) {2, 10, 7, 14, 6, 6, 3}
(c) {3, 9, 8, 14, 7, 5, 2}
(d) {2, 10, 6, 14, 7, 5, 3} [ESE-2004]

2.14 The lengths of two discrete time sequence $x_1(n)$ and $x_2(n)$ are 5 and 7, respectively. What is the maximum length of a sequence $x_1(n) * x_2(n)$?
(a) 5 (b) 6
(c) 7 (d) 11
[ESE-2005]

2.15 Consider the following statements about linear time invariant (LTI) continuous time systems:
1. The output signal in an LTI system with known input and known impulse response can always be determined.
2. A causal LTI system is always stable.
3. A stable LTI system has an impulse response, $h(t)$ which has a finite value when integrated over whole of the time axis, i.e.

$\int_{-\infty}^{+\infty} h(\lambda)d\lambda$ is finite.

Which of the statements given above are correct?
(a) 1 and 3 (b) 1 and 2
(c) 2 and 3 (d) 1, 2 and 3
[ESE-2005]

2.16 The unit sample response of a discrete system is 1 1/2 1/4 0 0 0 ..... For an input sequence 1 0 1 0 0 0 ...., what is the output sequence?
(a) 1 1/2 1/4 1/2 1/4 0 0 ...
(b) 1 0 1/4 0 0 ...
(c) 2 1/2 5/4 0 0 ...
(d) 1 1/2 5/4 1/2 1/4 0 0 ...
[ESE-2005]

2.17 Let $x[n] = a^n u[n]$, $h[n] = b^n u[n]$
What is the expression for y[n], for a discrete-time system?

(a) $\sum_{k=-\infty}^{\infty} a^k u[k] b^{n-k} u[n-k]$

(b) $\sum_{k=-\infty}^{\infty} a^n u[k] b^{n-k} u[n-k]$

(c) $\sum_{k=-\infty}^{\infty} a^n u[n-k] b^n u[k]$

(d) $\sum_{k=-\infty}^{\infty} a^{n-k} u[k] b^{n-k} u[n-k]$ [ESE-2005]

2.18 The graph shown below represents a waveform obtained by convolving two rectangular waveforms of duration

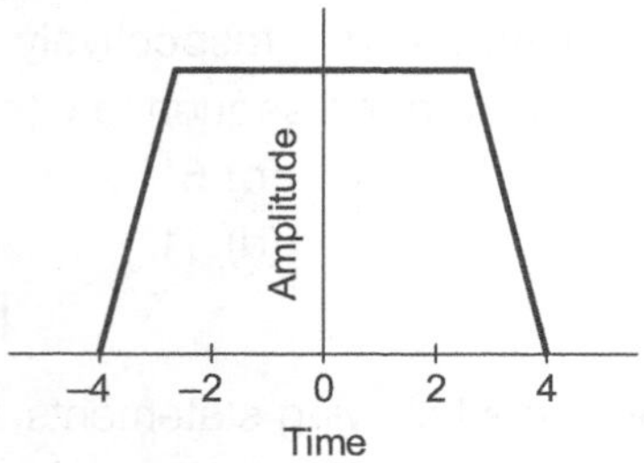

(a) Four units each
(b) Four and two units respectively
(c) Six and three units respectively
(d) Six and two units respectively [ESE-2005]

2.19 The impulse response of a system $h(n) = a^n u(n)$. What is the condition for the system to be BIBO stable?
(a) a is real and positive
(b) a is real and negative
(c) $|a| > 1$
(d) $|a| < 1$

[ESE-2006]

2.20 For continuous-time systems, state variable representation is

$$\frac{d}{dt}q(t) = Aq(t) + bx(t)$$

What is the corresponding representation for discrete – time system?

(a) $\frac{d}{dt}q[n] = Aq[n] + bx[n]$
(b) $q[n+1] = Aq[n] + bx[n]$
(c) $q[n] = Aq[n-1] + bx[n]$
(d) $\frac{d}{dn}[q(n+1)] = Aq[n-1] + bx[n-1]$

[ESE-2006]

2.21 The response of a linear, time-invariant system to a unit step is $s(t) = (1 - e^{-t/RC})\, u(t)$, where $u(t)$ is the unit step. What is the impulse response of this system?

(a) $e^{-t/RC}$ (b) $e^{-t/RC} u(t)$
(c) $1/RC\{e^{-t/RC} u(t)\}$ (d) $\delta(t)$

[ESE-2007]

2.22 The output of a linear system for any input can be computed in which one of the following ways?
(a) Only by summation of impulse responses by convolution integral
(b) Only by summation of step responses by superposition integral
(c) Neither (a) nor (b)
(d) Using either (a) or (b)

[ESE-2007]

2.23 The convolution of f(t) with itself is given to be $\int_0^t f(\tau)\, d\tau$. Then what is $f(t)$?
(a) The unit ramp function
(b) Equal to 1
(c) The unit step function
(d) The unit impulse function

[ESE-2008]

2.24 Which one of the following is the impulse response of the system whose step response is given as $c(t) = 0.5(1 - e^{-2t})\, u(t)$?
(a) $e^{-2t} u(t)$
(b) $0.5\,\delta(t) + e^{-2t} u(t)$
(c) $0.5\,\delta(t) - 0.5e^{-2t} u(t)$
(d) $0.5\, e^{-2t} u(t)$

[ESE-2008]

2.25 If the response of LTI continuous time system to unit step input is $\left(\frac{1}{2} - \frac{1}{2}e^{-2t}\right)$, then impulse response of the system is

(a) $\left(\frac{1}{2} - \frac{1}{2}e^{-2t}\right)$ (b) e^{-2t}
(c) $(1 - e^{-2t})$ (d) Constant

[ESE-2010]

2.26 The impulse response $h[n]$ of an LTI system is
$$h[n] = u[n+3] + u[n-2] - 2u[n-7].$$
Then the system is
1. Stable 2. Casual
3. Unstable 4. Not casual
Which of these are correct?
(a) 1 and 2 only (b) 2 and 3 only
(c) 3 and 4 only (d) 1 and 4 only

[ESE-2010]

2.27 With the following equations, the time-invariant systems are

1. $\dfrac{d^2y(t)}{dt^2} + 2t\dfrac{d}{dt}y(t) + 5y(t) = x(t)$
2. $y(t) = e^{-2x(t)}$
3. $y(t) = \left[\dfrac{d}{dt}x(t)\right]^2$
4. $y(t) = \dfrac{d}{dt}\left[e^{-2t}x(t)\right]$

(a) 1 and 2 (b) 1 and 4
(c) 2 and 3 (d) 3 and 4

[ESE-2012]

2.28 The convolution $x(n) * \delta(n - n_0)$ is equal to
(a) $x(n - n_0)$ (b) $x(n + n_0)$
(c) $x(n_0)$ (d) $x(n)$

[ESE-2013]

2.29 **Statement (I):** Zero state response is usually referred as particular solution of the filter.
Statement (II): Particular solution depends on only the input but not on the initial conditions.
(a) Both Statement (I) and Statement (II) are individually true and Statement (II) is the correct explanation of Statement (I).
(b) Both Statement (I) and Statement (II) are individually true but Statement (II) is not the correct explanation of Statement (I).
(c) Statement (I) is true but Statement (II) is false.
(d) Statement (I) is false but Statement (II) is true.

[ESE-2013]

2.30 An LTI system is causal if and only if
(a) $h(t) = 0$ for $t < 0$
(b) $h(t)$ is finite for $0 < t < \infty$
(c) $h(t)$ is finite for $t < 0$
(d) $h(t)$ is non-zero for all t

[ESE-2014]

2.31 Consider an LTI system representing a passive electrical network. If the input is a sinusoidal signal, then the steady state output of the network is
(a) sinusoidal with the same amplitude, frequency and phase
(b) sinusoidal with the same frequency, but possibly different amplitude and phase
(c) non-sinusoidal
(d) sinusoidal with a different frequency

[ESE-2014]

2.32 **Statement (I):** The standard definition of stability precludes $\sin\omega_0 t$ term in impulse response.
Statement (II): $\sin\omega_0 t$ is a periodic function.
(a) Both Statement (I) and Statement (II) are individually true and Statement (II) is the correct explanation of Statement (I).
(b) Both Statement (I) and Statement (II) are individually true but Statement (II) is not the correct explanation of Statement (I).
(c) Statement (I) is true but Statement (II) is false.
(d) Statement (I) is false but Statement (II) is true.

[ESE-2014]

2.33 The discrete LTI system is represented by impulse response

$h(n) = \left(\dfrac{1}{2}\right)^n u(n)$, then the system is

(a) Causal and stable
(b) Causal and unstable
(c) Non causal and stable
(d) Non causal and unstable

[ESE-2015]

2.34 Consider a continuous time system A, modelled by the equation $y(t) = tx(t) + 4$ and a discrete time system B modelled by the equation $y[n] = x^2[n]$. These systems are
(a) A-time invariant and B-time invariant
(b) A-time varying and B-time invariant
(c) A-time invariant and B-time varying
(d) A-time varying and B-time varying

[ESE-2015]

2.35 Consider a discrete time accumulator system $y[n] = \sum_{k=-\infty}^{n} x[k]$ and the backward difference system $y[n] = x[n] - x[n-1]$ where $x[\cdot]$ represents the input and $y[\cdot]$ represents the output of the individual systems.

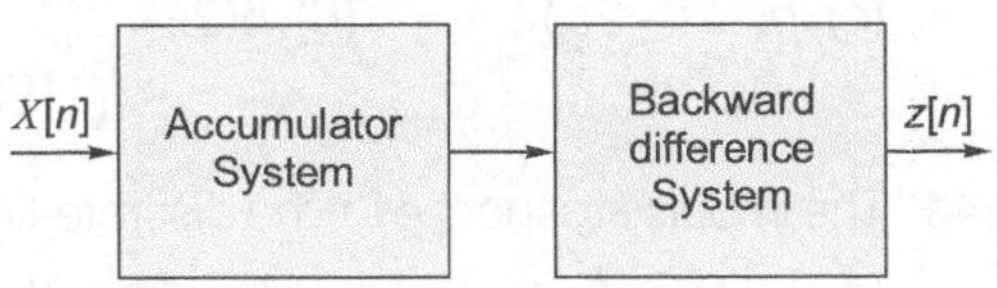

When these two systems are cascaded as in figure, the impulse response of combined system with output $z[n]$ is
(a) Unit impulse sequence
(b) Unit step sequence
(c) Unit ramp sequence
(d) None of the above

[ESE-2015]

2.36 The response of a system to a complex input $x(t) = e^{j2t}$ is specified as $y(t) = t \cdot e^{j2t} + e^{-j2t}$. The system
(a) is definitely LTI
(b) is definitely non-LTI
(c) may be LTI
(d) information is insufficient
[ESE-2015]

2.37 If a system produces frequencies in the output that are not present in the input, then the system cannot be
(a) Minimum phase system
(b) Linear shift invariant
(c) Stable and causal
(d) Stable and linear [ESE-2015]

2.38 Consider a system, which computes the 'MEDIAN' of signal values in a window of size '*N*'. Such a discrete time system is
(a) Linear
(b) Non-linear
(c) Sometimes linear
(d) Sometimes non-linear
[ESE-2016]

2.39 Consider a discrete time system which satisfies the additivity property, i.e., if the output for $u_1[n]$ is $y_1[n]$ and that for $u_2[n]$ is $y_2[n]$, then output for $u_1[n] + u_2[n]$ is $y_1[n] + y_2[n]$. Such a system is
(a) Linear (b) Sometimes linear
(c) Non-linear (d) Sometimes non-linear
[ESE-2016]

2.40 The result of $h(2t) * \delta(t - t_o)$ ("$*$" denotes convolution and "$\delta(\cdot)$" denotes the Dirac delta function) is
(a) $h(2t - 2t_0)$ (b) $h(2t_0 - 2t)$
(c) $h(-2t - 2t_0)$ (d) $h(2t + 2t_0)$
[ESE-2016]

2.41 The impulse response for the discrete-time system $y[n] = 0.24\,(x[n] + x[n-1] + x[n-2] + x[n-3])$ is given by
(a) 0 for $0 \le n \le 3$ and 0.24 otherwise
(b) 0.24 for $0 \le n \le 3$ and 0 otherwise
(c) 0.24 for $n = 0$ to $n = \infty$
(d) 0 for $n = 0$ to $n = \infty$
[ESE-2016]

2.42 The response of a linear, time-invariant, discrete-time system to a unit step input $u[n]$ is $\delta[n]$. The system response to a ramp input $n\,u[n]$ would be
(a) $\delta[n-1]$ (b) $u[n-1]$
(c) $n\delta[n-1]$ (d) $n\,u[n-1]$
[ESE-2016]

2.43 A signal x_n is given by $x_0 = 3, x_1 = 2, x_2 = 5, x_3 = 1, x_4 = 0, x_5 = 1, x_6 = 2, x_7 = 2, x_8 = 4$, where the subscript '*n*' denotes time. The peak value of the auto correlation of x_{2n-11}, is
(a) 0 (b) 10
(c) 54 (d) 64 [ESE-2016]

2.44 A system has impulse response $h[n] = \cos(n)u[n]$. The system is
(a) Causal and stable
(b) Non causal and stable
(c) Non causal and not stable
(d) Causal and not stable [ESE-2016]

2.45 Consider the following statements:
The output of a linear circuit, driven with a sine wave at a frequency *f*, is itself a sine wave
1. at, the same frequency
2. with chance of changed amplitude
3. with chances of changed amplitude and phase

Which of the above statements is/are correct?
(a) 1 and 2 (b) 1 only
(c) 1 and 3 (d) 2 only [ESE-2017]

2.46 An inductor is described by input-output relation as

$$y(t) = \frac{1}{L}\int_{-\infty}^{t} x(\tau)d\tau$$

The operation representing the inverse system $x(t)$ will be
(a) $L\frac{d}{dt}y(t)$ (b) L
(c) $\frac{d}{dt}y(t)$ (d) $Ly(t)$
[ESE-2020]

2.47 Which of the following statements are correct?
1. A continuous time system is said to be time invariant if the parameters of the system do not change with time.
2. The characteristics of time-invariant system are fixed over a time.

3. If the input to the time invariant system is delayed by t_0 seconds, the characteristics of the output response is also delayed by t_0 seconds.

(a) 1 and 2 only (b) 1 and 3 only
(c) 2 and 3 only (d) 1, 2 and 3

[ESE-2020]

2.48 What is the simplified value of $y(n)$, if

$$y(n) = \sum_{n=-5}^{5} \sin(2n)\,\delta(n+7)?$$

(a) sin10 (b) –sin10
(c) 1 (d) 0 [ESE-2021]

2.49 Consider the following statements for a system given by

$$y(n) = x(n)\sum_{k=-\infty}^{\infty} \delta(n-3k)$$

1. The system is linear.
2. The system is non-linear.
3. The system is causal.
4. The system is non-causal.

Which of the above statements is/are correct?

(a) 1 only (b) 2 only
(c) 2 and 4 only (d) 1 and 3 only

[ESE-2021]

2.50 Which one of the following is the zero-input response of the system

$$y[n] - 3y[n-1] - 4y[n-2] = 0$$

described by the homogeneous second-order difference equation if $y[-2] = 0$ and $y[-1] = 5$?

(a) $y_{zi}(n) = (-1)^{n+1} + (-4)^{n+2}, n \geq 0$

(b) $y_{zi}(n) = (1)^{n+1} + (4)^{n+2}, n \geq 0$

(c) $y_{zi}(n) = (-1)^{n+1} + (4)^{n+2}, n \geq 0$

(d) $y_{zi}(n) = (1)^{n+1} + (-4)^{n+2}, n \geq 0$ [ESE-2021]

2.51 What is the convolution integral $c(t)$ for a system with input $x(t)$ and impulse response $h(t)$, where $x(t) = u(t-1) - u(t-3)$ and $h(t) = u(t) - u(t-2)$?

(a) $c(t) = \begin{cases} 0, & t < 1 \\ t-1, & 1 \leq t < 3 \\ 5-t, & 3 \leq t < 5 \\ 0, & t \geq 5 \end{cases}$

(b) $c(t) = \begin{cases} 0, & t < 1 \\ t-\frac{1}{2}, & 1 \leq t < 2 \\ \frac{3}{2}-t, & 2 \leq t < 5 \\ 0, & t \geq 5 \end{cases}$

(c) $c(t) = \begin{cases} 0, & t < 1 \\ 5-t, & 1 \leq t < 4 \\ 0, & t \geq 4 \end{cases}$

(d) $c(t) = \begin{cases} 2, & 1 \leq t \leq 2 \\ 1, & 3 \leq t \leq 5 \\ 0, & \text{otherwise} \end{cases}$

[ESE-2021]

2.52 Consider the following statements:

1. Impulse response $h_1(n) = 3^n u(n)$ indicates the system is non-causal system.
2. Impulse response $h_2(n) = \cos\frac{n\pi}{2}$ indicates the system is causal system.
3. Impulse response $h_3(n) = e^{-5|n|}$ indicates the system is causal system.

Which of the above statements is/are not correct?

(a) 3 only (b) 1 and 2
(c) 1 and 3 (d) 2 and 3

[ESE-2024]

2.53 What is the input signal $x(n)$ that will generate the output $y(n) = \{1, 5, 10, 11, 8\}$ for a system with impulse response $h(n) = \{1, 2, 1\}$?

(a) {1, 2, 1} (b) {3, 3, 1}
(c) {1, 3, 3} (d) {1, 3, 1}

[ESE-2024]

2.54 The convolution sum is used to give the response of

(a) linear systems only
(b) time invariant systems only
(c) linear time invariant systems
(d) causal systems only

[ESE-2024]

■■■

Answers Continuous and Discrete Time LTI Systems

2.1 (a)	2.2 (b)	2.3 (b)	2.4 (c)	2.5 (b)	2.6 (a)	2.7 (c)	2.8 (d)	2.9 (c)
2.10 (a)	2.11 (b)	2.12 (d)	2.13 (a)	2.14 (d)	2.15 (a)	2.16 (d)	2.17 (a)	2.18 (d)
2.19 (d)	2.20 (b)	2.21 (c)	2.22 (a)	2.23 (c)	2.24 (a)	2.25 (b)	2.26 (d)	2.27 (c)
2.28 (a)	2.29 (a)	2.30 (a)	2.31 (b)	2.32 (a)	2.33 (a)	2.34 (b)	2.35 (a)	2.36 (b)
2.37 (b)	2.38 (b)	2.39 (b, d)	2.40 (a)	2.41 (b)	2.42 (b)	2.43 (b)	2.44 (d)	2.45 (c)
2.46 (a)	2.47 (d)	2.48 (d)	2.49 (d)	2.50 (c)	2.51 (a)	2.52 (*)	2.53 (*)	2.54 (c)

Explanations Continuous and Discrete Time LTI Systems

2.1 (a)

$\because$ step response of LTI system is a(t)

$\therefore$ Impulse response of LTI system is $\dfrac{da(t)}{dt}$

$$y(t) = \int_{0+}^{t} \frac{da(\tau)}{d\tau}\, x(t-\tau)d\tau$$

2.2 (b)

A solution to the differential equation with the input set to zero is often referred to as the natural response of the system.

Note:

(i) Auxiliary equation
 ⇒ Complementary Function
 ⇒ Transient Response
 ⇒ Natural Response

(ii) Particular Integral
 ⇒ Forced Response
 ⇒ Steady State Response

2.3 (b)

$$y(t) = x(t) * h(t)$$

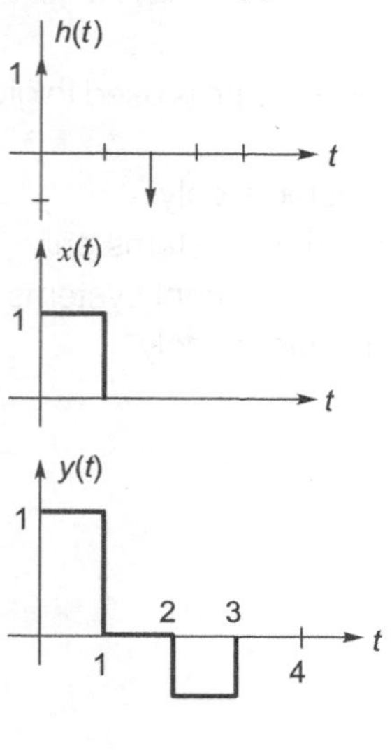

2.4 (c)

$\because \delta(t-t_1) * \delta(t-t_2) = \delta(t-t_1-t_2)$

$\therefore \delta(t-0.5) * \delta(t-0.5) = \delta(t-1)$

2.6 (a)

For the LTI system to be stable

$$\sum_{n=-\infty}^{\infty} |h[k]| < \infty$$

So, for the sum to be finite,

$$|a| < 1 \text{ and } |b| > 1$$

2.7 (c)

Convolution of two different duration rectangular waveform is trapezoidal waveform and if duration is same then resulting waveform will be triangular.

2.8 (d)

A causal discrete-time LTI system satisfies the condition $h[n] = 0$ for $n < 0$.
But $a^n u(n+2) \neq 0$ for $n = -1, -2$.

2.9 (c)

$$h(n) = a^n u(n+2), |a| < 1$$

(i) A system is causal if $h(n) = 0$ for $n < 0$.
The given $h(n)$ does not satisfy this condition, so it is non-causal.

(ii) Since $|a| < 1$, $h(n)$ is absolutely summable. Hence, it is stable.

(iii) A system is said to be memoryless if its output for each value of the independent variable at a given time is dependent only on the input at the same time,
Since given h(n) does not satisfy the condition, it is having memory.

2.10 (a)

$$y[n] = \sum_{k=-\infty}^{n} x[k] \quad \text{... (i)}$$

$$y[n-1] = \sum_{k=-\infty}^{n-1} x[k] \quad \text{... (ii)}$$

Subtracting equation (ii) from (i),

$$x[n] = y[n] - y[n-1]$$

2.11 (b)

$$h(n) = \delta(n+2) - \delta(n-2)$$
$$y(n) = h(n) * x(n)$$
$$= \{\delta(n+2) - \delta(n-2)\} * x(n)$$
$$\Rightarrow \quad y(n) = x(n+2) - x(n-2)$$
$$\Rightarrow \quad y(n-2) = x(n) - x(n-4)$$

2.12 (d)

$$y(t) = x(t) * h(t)$$
$$y(t) = \int_{-\infty}^{\infty} x(\tau)x(\tau)\, h(t-\tau)d\tau$$

for $t < 0$, $y(t) = 0$

for $0 < t < T$,

$$y(t) = \int_0^t A_1 A_2 d\tau$$
$$= A_1 A_2 t$$

for $T < t < 2T$,

$$y(t) = \int_{-T}^{T} A_1 A_2 d\tau$$
$$= A_1 A_2 (2T - t)$$

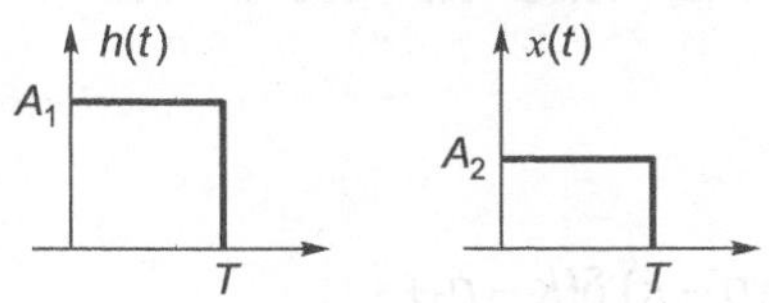

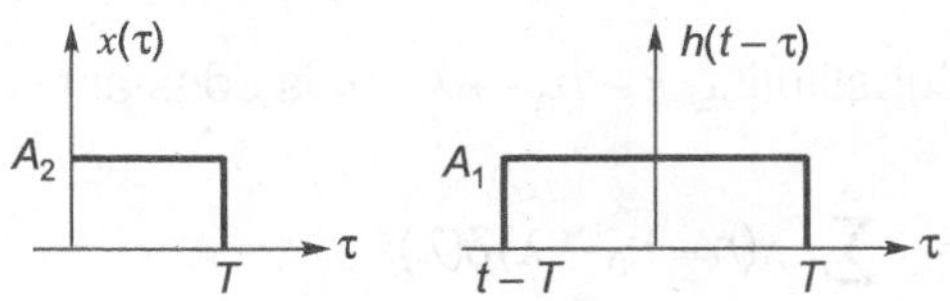

For $t > 2T$, $y(t) = 0$

$y(t)$ is plotted as below

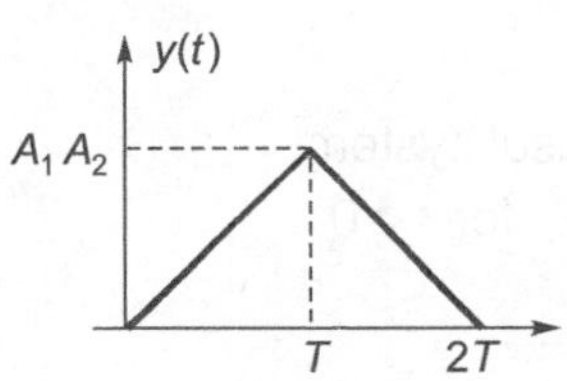

2.13 (a)

$$R_{xy}(k) = x(k) * y(-k)$$

y(−k) \ x(k)	1	3	1	2
3	3	9	3	6
1	1	3	1	2
2	2	6	2	4
1	1	3	1	2

Therefore, $R_{xy}(k) = \{3, 10, 8, 14, 7, 5, 2\}$

2.14 (d)

Maximum length of $x_1(n) * x_2(n)$ is

$L = L_1 + L_2 - 1$

where L_1 is length of $x_1(n)$

and L_2 is length of $x_2(n)$.

2.15 (a)

A causal LTI system is not always stable.

For example, let $h(n) = a^n u(n)$, $|a| > 1$.

The system is causal but not stable.

2.16 (d)

$$y(n) = x(n) * h(n)$$

By add and shift method

x(n) \ h(n)	1	1/2	1/4	0	0	0
1	1	1/2	1/4	0	0	0
0	0	0	0	0	0	0
1	1	1/2	1/4	0	0	0
0	0	0	0	0	0	0
0	0	0	0	0	0	0
0	0	0	0	0	0	0

Output sequence is

$$y(n) = \left\{ 1 \quad \frac{1}{2} \quad \left(1 + \frac{1}{4}\right) \quad \frac{1}{2} \quad \frac{1}{4} \quad 0 \quad 0 \cdots \right.$$

$$= \left\{ 1 \quad \frac{1}{2} \quad \frac{5}{4} \quad \frac{1}{2} \quad \frac{1}{4} \quad 0 \quad 0 \cdots \right.$$

2.17 (a)

$$y[n] = x[n] * h[n]$$

$$= \sum_{n=-\infty}^{\infty} a^k u[k] b^{n-k} u[n-k]$$

2.18 (d)

(i) When two rectangular waveforms of different lengths L_1 and L_2 are convolved, the resultant waveform will be trapezoidal of length $L_1 + L_2$.

(ii) When two rectangular waveforms of same length are convolved, the resultant waveform will be triangular of length $2L$.

2.19 (d)

$h(n) = a^n u(n)$

The condition for the system to be BIBO stable is

$$\sum_{n=-\infty}^{\infty} |h(n)| < \infty$$

$$\sum_{n=-\infty}^{\infty} a^n u(n) < \infty$$

$$\Rightarrow \quad |a| < 1$$

2.20 (b)

First derivative in continuous time systems is equivalent to first difference in discrete time systems.

So, $\frac{d}{dt}q(t) = Aq(t) + bx(t)$ in continuous time systems is equivalent to $q[n+1] = Aq[n] + bx[n]$ in discrete time systems.

2.21 (c)

$$\text{Impulse response} = \frac{d}{dt}[\text{Step response}]$$

$$= \frac{d}{dt}\left[1 - e^{-t/RC}\right]$$

$$= 1/RC\,\{e^{-t/RC}\,u(t)\}$$

2.23 (c)

Convolution of $f(t)$ is

$$\int_{-\infty}^{\infty} f(\tau)\, f(t-\tau)\, d\tau$$

Taking $f(t) = u(t)$

$$\int_{-\infty}^{\infty} f(\tau)\, f(t-\tau)\, d\tau = \int_0^t f(\tau)\, d\tau$$

2.24 (a)

Impulse response function

$$= \frac{d}{dt}\{\text{step response function}\}$$

$$= \frac{d}{dt}\{0.5\,(1 - e^{-2t})\,u(t)\}$$

$$= 0.5[0 - (-2)e^{-2t}]u(t) = e^{-2t}\,u(t)$$

2.25 (b)

$$\text{Impulse response} = \frac{d}{dt}[\text{Step response}]$$

$$= \frac{d}{dt}\left(\frac{1}{2} - \frac{1}{2}e^{-2t}\right) = e^{-2t}$$

2.26 (d)

$h(n) \neq 0$ for $n < 0$

so system is not causal.

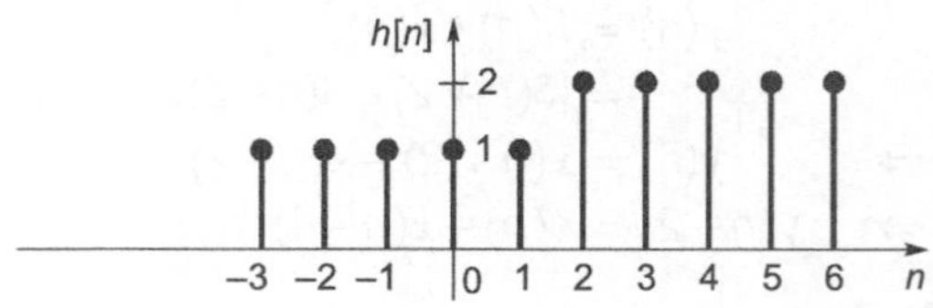

$\because \sum |h[n]| = 15 < \infty$

So the system is stable.

2.27 (c)

Time invariance property states that the delayed response should be equal to the response of the system to the delayed input. Thus by the observation we can see that

$$y(t, t_0) = e^{-2x(t-t_0)}$$

$$y(t - t_0) = e^{-2x(t-t_0)}$$

So, $y(t, t_0) = y(t - t_0)$ time invariant

In a similar fashion all the options can be checked.

2.28 (a)

$$= \sum_{k=-\infty}^{\infty} x(n-k)\,\delta(k-n_0)$$

Substituting $k - n_0 \rightarrow \lambda$; n_0 is constant

$$= \sum_{\lambda=-\infty}^{\infty} x(n - n_0 - \lambda)\,\delta(\lambda)$$

$$= x(n - n_0)$$

2.30 (a)

For a causal system,

$h(t) = 0$; for $t < 0$

2.31 (b)

If the input applied to an LTI system is a sinusoidal signal then the output of the system is also sinusoidal with different amplitude and frequency.

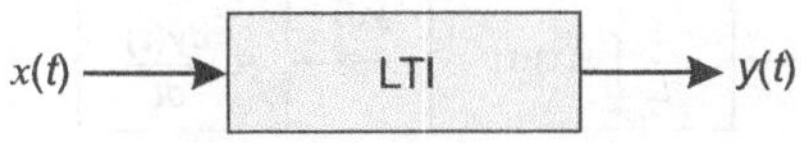

Here, $x(t) = \sin \omega t$

$y(t) = |H(t)|\, x(t) = A \sin(\omega t \pm \theta)$

2.32 (a)

The $\sin(\omega_0 t)$ term in the impulse response indicates poles on $(j\omega)$ axis which indicates the system to be marginally stable. Thus the definition omits if when we talk of stability of the system.

Due to the presence of sinusoidal term, it will not converge.

2.33 (a)

$$h(n) = \left(\frac{1}{2}\right)^n u(n)$$

$h(n) = 0 \quad n < 0 \quad$ causal

$$\sum_{n=-\infty}^{\infty} |h(n)| = \frac{1}{1-\frac{1}{2}} = 2 < \infty \quad \text{stable}$$

2.34 (b)

$y(t) = tx(t) + 4$

for $x(t) = x(t - t_0)$

$y'(t) = tx(t - t_0) + 4 \quad$ ...(i)

for $y(t) = y(t - t_0)$

$y(t - t_0) = (t - t_0)\, x(t - t_0) + 4 \quad$ ...(ii)

equation (i) $\neq$ equation (ii)

Hence A is time varying

$y(n) = x^2[n]$

$y(n) \to x(n - n_0)$

$y'(n) = x^2(n - n_0) \quad$ ...(iii)

for $y(n) = (y(n - n_0))$

$y(n - n_0) = x^2(n - n_0) \quad$ ...(iv)

equation (iii) = equation (iv)

Hence B is time invariant

2.35 (a)

For Accumulator system

$$y(n) = \sum_{k=-\infty}^{n} x[k]$$

for $x(n) = \delta(n)$

$$y(n) = \sum_{k=-\infty}^{n} \delta(k) = h_1(n)$$

$h_1(n) = u(n) \quad$ ...(1)

for backward difference

$y(n) = X(n) - X(n-1)$

$x(n) = \delta(n)$

$y(n) = h_2(n)$

$h_2(n) = \delta(n) - \delta(n-1) \quad$ ...(2)

Cascaded response becomes

$h(n) = h_1(n) * h_2(n)$

$= u(n) * (\delta(n) - \delta(n-1))$

$= u(n) - u(n-1)$

$h(n) = \delta(n)$

2.36 (b)

$x(t) = e^{j2t}$

$y(t) = te^{j2t} + e^{-j2t}$

$y(t)$ can be written as

$y(t) = t\,x(t) + x(-t)$

$x(t) \to x(t - t_0)$

$y'(t) = t\,x(t - t_0) + x(-t - t_0) \quad$ ...(i)

$y(t) \to y(t - t_0)$

$y(t - t_0) = (t - t_0)\, x(t - t_0) + x(-t + t_0) \quad$ ...(ii)

equation (i) $\neq$ equation (ii) $\to$ time varying

Hence it is not LTI.

OR

Response of LTI system to eigen function is an eigen function itself.

2.37 (b)

For linear shift invariant system, the frequencies produced in the output are same as that of input and if some other frequencies are present, then system can't be linear shift invariant.

$x(n) \to y(n) \quad$ ...(1)

$x(n - k) \to y(n - k) \quad$ ...(2)

Above illustrates the definition of linear shift invariant system.

2.38 (b)

If F is function, then it is said to be linear if

$F(A + B) = F(A) + F(B)$

So now say there are two sets of numbers.

$A = [2\ 5\ 9\ 6\ 3]$

$B = [1\ 0\ 4\ 7\ 4]$

mean$(A) = 5$

mean$(B) = 3.2$

mean$(A + B) = 8.2 = $ mean(A) + mean(B)

So you can see that mean is a linear operation and median is not. Now think in terms of signals or images and the same rule applies. In real life you will care about this in image processing only in some particularly.

2.39 (b, d)

- Linearity is combination of homogeneity principle and additivity principle.
- Even though the system is satisfying the additivity principle, it may not satisfy the homogeneity principle.
- Hence, system can be sometimes linear [when homogeneity principle is also satisfied] and sometimes non linear [when homogeneity principle is not satisfied].

2.40 (a)

According to the convolution property

$x(t) * \delta(t-t_o) = x(t-t_o)$

Hence, $h(2t) * \delta(t-t_o) = h(2(t-t_o)) = h(2t-2t_o)$

2.41 (b)

When input $x[n] = \delta[n]$, response

$$y[n] = h[n] \rightarrow \text{unit impulse response}$$

$$h[n] = 0.24[\delta[n]+\delta[n-1]+\delta[n-2]+\delta[n-3]]$$

i.e. $h[n] = 0.24 \quad 0 \le n \le 3$

$= 0 \quad$ otherwise

2.42 (b)

From LTI system, response, for $u[n] \rightarrow \delta[n]$, for ramp input $nu[n] = r[n]$

i.e.,

$u[n-1] + u[n-2] + \cdots = \delta[n-1] + \delta[n-2] + \cdots$

Hence, $u[n-1]$.

2.44 (d)

$$h[n] = \cos(n)\, u[n]$$

- Multiplication by $u[n]$ ensures,

  $h[n] = 0, n < 0$ Hence, causal
- $h[n]$ must be absolutely summable $\left(\text{if } \sum_{n=-\infty}^{\infty} |h[n]| < \infty\right)$ which is not verified by above $h[n]$.

So it is not stable.

2.45 (c)

Linear circuit provides output at same frequency with chances of change in amplitude and phase.

2.46 (a)

If an invertible system produces the output $y(t)$ for the input $x(t)$, then its inverse produces the output $x(t)$ for the input $y(t)$.

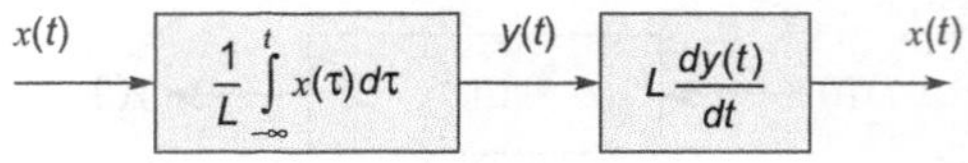

2.47 (d)

A system is said to be time-invariant if its behaviour or characteristics do not change with time. The parameters of such a system do not change with time.

The response of a time invariant system is independent of the time at which the input is applied. A time-invariant system responds identically no matter when the input is applied. In order words a system is time-invariant if a time shift of the input signal produces the same time shift in the output signal but without. Any change in the characteristics of the output.

2.48 (d)

$$y(n) = \sum_{n=-5}^{5} \sin(2n)\delta(n+7)$$

Using the property,

$x(n) \cdot \delta(n+n_0) = x(-n_0)\delta(n+n_0)$

We get, $$y(n) = \sum_{n=-5}^{5} \sin(-14)\delta(n+7)$$

$$y(n) = -\sin(14) \cdot \sum_{n=-5}^{5} \delta(n+7)$$

Since, impulse exists at $n = -7$ which is outside $-5 \le n \le 5$. Hence,

$$y(n) = 0$$

2.49 (d)

Check for linearity:

If $\quad x(n) = x_1(n)$, then

$$y_1(n) = x_1(n) \sum_{k=-\infty}^{\infty} \delta(n-3k)$$

If $x(n) = x_2(n)$, then

$$y_2(n) = x_2(n) \sum_{k=-\infty}^{\infty} \delta(n-3k)$$

For $x(n) = \alpha_1 x_1(n) + \alpha_2 x_2(n)$, we have

$$y(n) = (\alpha_1 x_1(n) + \alpha_2 x_2(n)) \cdot \sum_{k=-\infty}^{\infty} \delta(n-3k)$$

$$y(n) = \alpha_1 x_1(n) \sum_{k=-\infty}^{\infty} \delta(n-3k) + \alpha_2 x_2(n) \sum_{k=-\infty}^{\infty} \delta(n-3k)$$

$$y(n) = \alpha_1 y_1(n) + \alpha_2 y_2(n)$$

Hence, the system is linear.

Check for causal system:

For causal system, the output is independent of future values of input.

Here, $y(n) = x(n) \cdot \sum_{k=-\infty}^{\infty} \delta(n-3k)$

which is independent of future value of x(n). Hence, the system is causal.

2.50 (c)

Given difference equation

$y(n) - 3y(n-1) - 4y(n-2) = 0$

Applying unilateral z-transform, we get,

$Y(z) - 3[z^{-1}Y(z) + y(-1)] - 4[z^{-2}Y(z) + z^{-1}y(-1) + y(-2)] = 0$

$\Rightarrow \quad Y(z)[1 - 3z^{-1} - 4z^{-2}] = (3 + 4z^{-1})y(-1)$

Given $y(-1) = 5$, we get,

$$Y(z) = \frac{15 + 20z^{-1}}{1 - 3z^{-1} - 4z^{-2}}$$

$$Y(z) = \frac{15 + 20z^{-1}}{(1 + z^{-1})(1 - 4z^{-1})}$$

Taking partial fractions,

$$Y(z) = \frac{-1}{1 + z^{-1}} + \frac{16}{1 - 4z^{-1}}$$

Taking inverse z-transform, we get,

$$y(n) = -1(-1)^n u(n) + 16(4)^n u(n)$$
$$y(n) = [(-1)^{n+1} + (4)^{n+2}]u(n)$$
$$y(n) = (-1)^{n+1} + (4)^{n+2},\ n \geq 0$$

2.51 (a)

$$x(t) = u(t-1) - u(t-3)$$
$$h(t) = u(t) - u(t-2)$$

We have

$$c(t) = x(t) * h(t)$$
$$c(t) = [u(t-1) - u(t-3)] * [u(t) - u(t-2)]$$

Using the property $u(t-t_0) * u(t-t_1) = r(t - t_0 - t_1)$, we get

$$c(t) = r(t-1) - r(t-3) - r(t-3) + r(t-5)$$
$$c(t) = r(t-1) - 2r(t-3) + r(t-5)$$

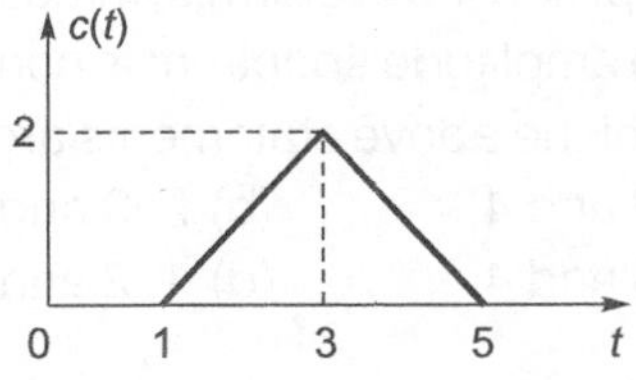

Therefore,

$$c(t) = \begin{cases} 0, & t < 1 \\ (t-1), & 1 \leq t < 3 \\ 5 - t & 3 \leq t < 5 \\ 0 & t \geq 5 \end{cases}$$

2.52 (*)

For a causal system, $h(n) = 0$ for $n < 0$. We have

(i) $h_1(n) = 3^n u(n) = \begin{cases} 3^n ; & n \geq 0 \\ 0 ; & n < 0 \end{cases}$ i.e., it is a causal system.

(ii) $h_2(n) = \cos\left(\frac{n\pi}{2}\right) \neq 0$ for $n < 0$ i.e., it represents a non-causal system.

(iii) $h_3(n) = e^{-5|n|} \neq 0$ for $n < 0$ i.e., it represents a non-causal system.

Hence, all the statements are not correct.

2.53 (*)

The convolution of discrete-time sequences $x(n)$ of length 'm' and $h(n)$ of length 'n' is $y(n)$ of length '$m + n - 1$'. We have,

$$5 = 3 + m - 1 \quad \Rightarrow \quad m = 3$$

Hence, $x(n)$ is a sequence of length '3'. Assume $x(n) = \{a, b, c\}$ and using tabular method of convolution,

h(n) \ x(n)	a	b	c
1	a	b	c
2	2a	2b	2c
1	a	b	c

We get, $y(n) = \{a, 2a + b, a + 2b + c, b + 2c, c\}$

Comparing with given expression of

$$y(n) = \{1, 5, 10, 11, 8\}$$

we can see that for none of the values of a, b and c, the given $y(n)$ can be obtained.

2.54 (c)

For a LTI system with input $x(n)$ and impulse response $h(n)$, the output is given by

$$y(n) = x(n) * h(n),$$

where $*$ denotes convolution.

■■■

3 Fourier Series

3.1 The Fourier series representations of a periodic current

$$[2+6\sqrt{2}\cos\omega t+\sqrt{48}\sin 2\omega t]A$$

The effective value of the current is

(a) $[2+6+\sqrt{24}]A$ (b) 8 A
(c) 6 A (d) 2 A

[ESE-2000]

3.2 Match **List-I** (Properties) with **List-II** (Characteristics of the trigonometric form) in regard to Fourier series of periodic f(t) and select the correct answer using the codes given below the lists:

	List-I		List-II
A.	$f(t)+f(-t)=0$	**1.**	Even harmonics can exist
B.	$f(t)-f(-t)=0$	**2.**	Odd harmonics can exist
C.	$f(t)+f(t-T/2)=0$	**3.**	The dc and cosine terms can exist
D.	$f(t)-f(t-T/2)=0$	**4.**	sine terms can exist
		5.	cosine terms of even harmonics can exist

Codes:

	A	B	C	D
(a)	4	5	3	1
(b)	3	4	1	2
(c)	5	4	2	3
(d)	4	3	2	1

[ESE-2000]

3.3 Consider the following statements regarding the fundamental component $f_1(t)$ of an arbitrary periodic signal $f(t)$:

It is possible for

1. the amplitude of $f_1(t)$ to exceed the peak value of $f(t)$.
2. $f_1(t)$ to be identically zero for a non-zero $f(t)$.
3. the effective value of $f_1(t)$ to exceed the effective value of $f(t)$.

Which of these statements is/are correct?

(a) 1 alone (b) 1 and 2
(c) 2 and 3 (d) 1 and 3 **[ESE-2000]**

3.4 Match **List-I** with **List-II** and select the correct answer using the codes given below the lists:

	List-I		List-II
A.	$f(t)=-f(-t)$	**1.**	Exponential form of Fourier series
B.	$\sum_{n=-\infty}^{\infty} C_n e^{jn\omega_0 t}$	**2.**	Fourier transform
C.	$\int_{-\infty}^{\infty} f(t)e^{-j\omega t}dt$	**3.**	Convolution integral
D.	$\int_0^t f_1(\tau)f_2(t-\tau)d\tau$	**4.**	z-transform
		5.	Odd function wave symmetry

Codes:

	A	B	C	D
(a)	5	1	2	3
(b)	2	1	5	3
(c)	5	4	2	1
(d)	4	5	1	2

[ESE-2000]

3.5 Consider the following statements related to Fourier series of a periodic waveform:

1. It expresses the given periodic waveform as a combination of DC component, sine and cosine waveforms of different harmonic frequencies.
2. The amplitude spectrum is discrete.
3. The evaluation of Fourier coefficients gets simplified if waveform symmetries are used.
4. The amplitude spectrum is continuous.

Which of the above statements are correct?

(a) 1, 2 and 4 (b) 2, 3 and 4
(c) 1, 3 and 4 (d) 1, 2 and 3

[ESE-2002]

3.6 **Assertion (A):** A periodic function satisfying Dirichlet conditions can be expanded into Fourier series.
Reason (R): A periodic function can be reconstructed from $\frac{a_0}{2}+\sum_{n=1} a_n \cos n\,\omega_0 t + \sum_{n=1} b_n \sin n\omega_0 t$ for very large n, excluding infinity.
(a) Both A and R are true and R is the correct explanation of A
(b) Both A and R are true but R is NOT the correct explanation of A
(c) A is true but R is false
(d) A is false but R is true [ESE-2003]

3.7 **Assertion (A):** In the exponential Fourier representation of a real-valued periodic function $f(t)$ of frequency f_0 the coefficients of the terms $e^{j2\pi nf_0t}$ and $e^{-j2\pi nf_0t}$ are negatives of each other.
Reason (R): The discrete magnitude spectrum of f(t) is even and the phase spectrum is odd.
(a) Both A and R are true and R is the correct explanation of A
(b) Both A and R are true but R is NOT the correct explanation of A
(c) A is true but R is false
(d) A is false but R is true [ESE-2003]

3.8 Match **List-I** (Nature of Periodic Function) with **List-II** (Properties of Spectrum Function) and select the correct answer using the codes given below the lists:
List-I

A. Impulse train
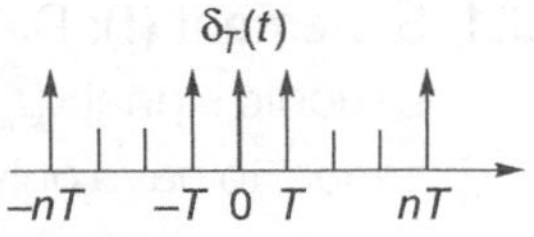

B. Full-wave rectified sine function
C. $2\sin(2\pi t/6)\cos(4\pi t/6)$
D.
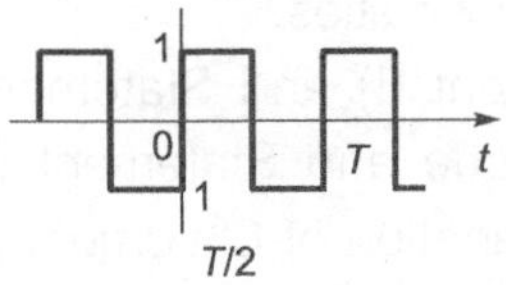

List-II
1. Only even harmonics are present
2. Impulse train with strength 1/T
3. $\alpha_3 = 1/2j;\ \alpha_{-3} = -1/2j;\ \alpha_1 = -1/2j;\ \alpha_{-1} = 1/2j$
4. Only odd harmonics are present
5. Both even and odd harmonics are present

Codes:

	A	B	C	D
(a)	5	2	3	4
(b)	2	1	4	3
(c)	5	2	4	3
(d)	2	1	3	4

[ESE-2003]

3.9 A square wave is defined by

$$x(t) = \begin{cases} A, & 0 < t < T_0/2 \\ -A, & T_0/2 < t < T_0 \end{cases}$$

It is periodically extended outside this interval. What is the general coefficient a_n in the Fourier expansion of this wave?
(a) 0
(b) $\frac{2A(1-\cos n\pi)}{n\pi}$
(c) $\frac{2A(1+\cos n\pi)}{n\pi}$
(d) $\frac{2A(1-\cos n\pi)}{[(n+1)\pi]}$
[ESE-2004]

3.10 **Assertion (A):** A periodic function satisfying Dirichlet's condition can be expanded into a Fourier series.
Reason (R): A Fourier series is a summation of weighted sine and cosine waves of the fundamental frequency and its harmonics.
(a) Both A and R are true and R is the correct explanation of A
(b) Both A and R are true but R is NOT the correct explanation of A
(c) A is true but R is false
(d) A is false but R is true [ESE-2004]

3.11 For half-wave (odd) symmetry, with T_0 = period of $x(t)$, which one of the following is correct?
(a) $x(t \pm T_0/2) = -x(t)$ (b) $x(t \pm T_0/2) = x(t)$
(c) $x(t \pm T_0) = -x(t)$ (d) $x(t \pm T_0) = x(t)$
[ESE-2004]

3.12 What is the average power of periodic non-sinusoidal voltage and currents?
(a) The average power of the fundamental component alone
(b) The sum of the average powers of the harmonics excluding the fundamental
(c) The sum of the average powers of the sinusoidal components including the fundamental
(d) The sum of the root mean square power of the sinusoidal components including the fundamental
[ESE-2008]

3.13 A signal $x_1(t)$ and $x_2(t)$ constitute the real and imaginary parts respectively of a complex valued signal $x(t)$. What form of waveform does $x(t)$ possess?
(a) Real symmetric
(b) Complex symmetric
(c) Asymmetric
(d) Conjugate symmetric [ESE-2009]

3.14 Consider the following statements:
Fourier series of any periodic function $x(t)$ can be obtained if

1. $\int_0^1 |x(t)|dt < \infty$
2. Finite number of discontinuous exist within finite time interval t.

Which of these statements is/are correct?
(a) 1 only (b) 2 only
(c) Both 1 and 2 (d) Neither 1 nor 2
[ESE-2010]

3.15 Which one of the following is a Dirichlet condition?
(a) $\int_{t_1}^{\infty} |x(t)| < \infty$
(b) Signal $x(t)$ must have a finite number of maxima and minima in the expansion interval
(c) $x(t)$ can have an infinite number of finite discontinuities in the expansion interval
(d) $x^2(t)$ must be absolutely summable
[ESE-2010]

3.16 A waveform is given by $v(t) = 10\sin 2\pi\, 100t$ V. What will be the magnitude of the second harmonic in its Fourier series representation?
(a) 0 V (b) 20 V
(c) 100 V (d) 200 V [ESE-2012]

3.17 A periodic function satisfies Dirichlet's conditions. This implies that the function
(a) is non-linear
(b) is not absolutely integrable
(c) guarantees that Fourier series representation of the function exists
(d) has infinite number of maxima and minima within a period
[ESE-2014]

3.18 Consider Fourier representation of continuous and discrete-time systems. The complex exponential (i.e., signals), which arise in such representation, have
(a) same properties always
(b) different properties always
(c) non-specific properties
(d) mostly same properties
[ESE-2014]

3.19 The Fourier series of a periodic function $x_T(t)$ with a period T is given by

$\sum_{k=-\infty}^{\infty} X_s(k)e^{jk\omega_0 t}$, where $\omega_0 = 2\pi/T$ and the Fourier coefficient $X_s(k)$ is defined as,

$$X_s(k) = \frac{1}{T}\int x_T(t)e^{-jk\omega_0 t}dt$$

If $x_T(t)$ is real and odd, the Fourier coefficients $X_s(k)$ are
(a) Real and odd (b) Complex
(c) Real (d) Imaginary
[ESE-2015]

3.20 Consider a continuous time periodic signal $x(t)$ with fundamental period T and Fourier series coefficient $X[k]$. What is the Fourier series coefficient of the signal

$$y(t) = x(t - t_0) + x(t + t_0)?$$

(a) $2\cos\left(\frac{2\pi}{T}kt_0\right)X[k]$

(b) $2\sin\left(\frac{2\pi}{T}Kt_0\right)X[k]$

(c) $e^{-t_0}X[k] + e^{t_0}X[-k]$

(d) $e^{-t_0}X[-k] + e^{t_0}X[k]$

[ESE-2015]

3.21 **Statement (I):** Dirichlet's conditions restrict the periodic signal $x(t)$, to be represented by Fourier series, to have only finite number of maxima and minima.
Statement (II): $x(t)$ should possess only a finite number of discontinuities.
(a) Both Statement (I) and Statement (II) are individually true and Statement (II) is the correct explanation of Statement (I)
(b) Both Statement (I) and Statement (II) are individually true but Statement (II) is **NOT** the correct explanation of Statement (I)
(c) Statement (I) is true but Statement (II) is false
(d) Statement (I) is false but Statement (II) is true
[ESE-2015]

3.22 The complex exponential power form of Fourier series of $x(t)$ is

$$x(t) = \sum_{K=-\infty}^{\infty} a_k \cdot e^{j\frac{2\pi}{T_0}\cdot kt}$$

if $x(t) = \sum_{b=-\infty}^{\infty} \delta(t-b)$, then the value of a_k is

(a) $1-(-1)^k$ (b) $1+(-1)^k$
(c) 1 (d) -1 [ESE-2016]

3.23 $x(t) = \frac{1}{T_o} + \sum_{k=1}^{N} \frac{2}{T_o}\cos k\omega_o t$, is the combined trigonometric form of Fourier series for

(a) Half rectified wave
(b) Saw-tooth wave
(c) Rectangular wave
(d) Impulse train

[ESE-2016]

■■■

Answers Fourier Series

3.1 (b)	3.2 (d)	3.3 (b)	3.4 (a)	3.5 (d)	3.6 (c)	3.7 (d)	3.8 (d)	3.9 (a)
3.10 (b)	3.11 (a)	3.12 (d)	3.13 (c)	3.14 (b)	3.15 (b)	3.16 (a)	3.17 (c)	3.18 (a)
3.19 (d)	3.20 (a)	3.21 (b)	3.22 (c)	3.23 (d)				

Explanations Fourier Series

3.1 (b)

$$I_{eff} = \sqrt{I_{dc}^2 + I_{1rms}^2 + I_{2rms}^2}$$

$$= \sqrt{(2)^2 + \left(\frac{6\sqrt{2}}{\sqrt{2}}\right)^2 + \left(\frac{\sqrt{48}}{\sqrt{2}}\right)^2}$$

$$= \sqrt{4+36+24} = \sqrt{64} = 8 \text{ A}$$

3.2 (d)

(i) $f(t) + f(-t) = 0$

$\Rightarrow f(t) = -f(-t)$

$\Rightarrow$ $f(t)$ is an odd function

$\Rightarrow$ sine terms exist in Fourier series of $f(t)$

(ii) $f(t) - f(t) = 0$

$\Rightarrow f(t) = f(-t)$

$\Rightarrow$ $f(t)$ is an even function

$\Rightarrow$ The dc and cosine terms exist in Fourier series of $f(t)$

(iii) $f(t) + f\left(t-\frac{T}{2}\right) = 0$

$\Rightarrow f\left(t-\frac{T}{2}\right) = -f(t)$

$\Rightarrow$ $f(t)$ is a half wave symmetric function

$\Rightarrow$ Odd harmonics exist

(iv) $f(t) = f(t-T/2)$

$\Rightarrow$ Even harmonics exist.

3.3 (b)

Let $f(t) = Af_1(t) + Bf_2(t) + Cf_3(t) + ...$

Let amplitude of $f(t)$ be P.

(i) Then, it is possible that $A > P$ as

$$P = A + B + C + ...$$

and other coefficients may be negative

(ii) $\because P^2 = A^2 + B^2 + C^2 + ...$

$\therefore$ Effective value of $f_1(t)$ can never exceed to that of $f(t)$.

(iii) $f_1(t)$ can not be identically zero for a non-zero $f(t)$.

3.6 (c)

$$f(t) = \frac{a_0}{2} + \sum_{n=1}^{\infty} a_n \cos n\omega_0 t \sum_{n=1}^{\infty} b_n \sin n\omega_0 t$$

3.8 (d)

Let $f(t)$ be impulse train.

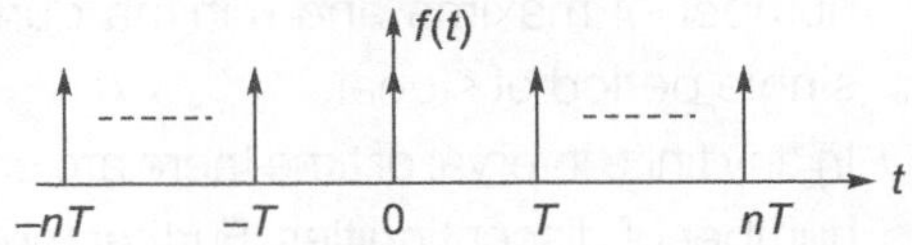

$f(t) = \delta(t);\ \frac{-T}{2} < t < \frac{T}{2}$

Fourier coefficient

$$F_n = \frac{1}{T}\int_{-T/2}^{T/2} f(t)e^{-jn\omega_0 t}dt$$

$$= \frac{1}{T}\int_{-T/2}^{T/2} \delta(t)\, e^{-jn\omega_0 t} dt = \frac{A}{T}$$

So, $f(t) = \sum_{n=-\infty}^{\infty} F_n . e^{jn\omega_0 t}$

The spectrum function will be the impulse train with strength $1/T$.

(ii) Let $f(t)$ be full wave rectified sine function.

$f(t) = \sin \omega_0 t + \sin(\omega_0 t - \pi)$

$F(\omega) = \frac{1}{2j}[\delta(\omega - \omega_0) - \delta(\omega + \omega_0)]$

3.9 (a)

Since the wave has odd symmetry, hence $a_n = 0$.

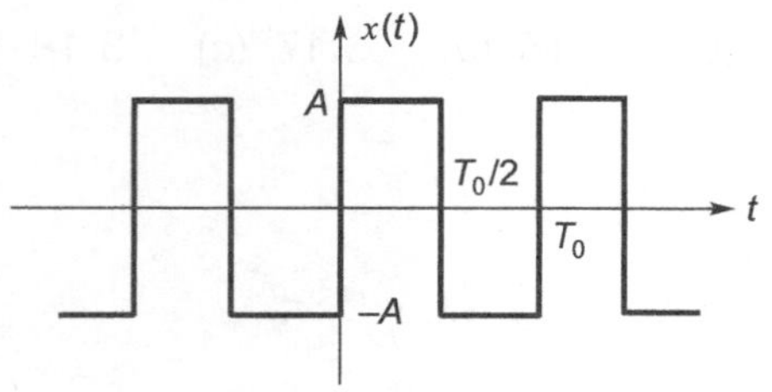

3.11 (a)

If $x\left(t \pm \frac{T_o}{2}\right) = -x(t)$ then $x(t)$ have half-wave odd symmetry.

3.13 (c)

Complex signal don't (need to) have any symmetry property in general.

3.14 (b)

Dirichlet's condition states that:

1. Over any period signal $x(t)$ must be absolutely integrable i.e.

$$\int_T |x(t)|dt < \infty$$

2. In any finite interval of time $x(t)$ is of bounded variations i.e. there are no more than a finite number of maxima and minima during any single period of signal.
3. In any finite interval of time there are only finite number of discontinuities. Further more each of these discontinuities is finite.

3.15 (b)

Dirichlet's condition states that:

1. Over any period signal $x(t)$ must be absolutely integrable i.e.

$$\int_T |x(t)|dt < \infty$$

2. In any finite interval of time $x(t)$ is of bounded variations i.e. there are no more than a finite number of maxima and minima during any single period of signal.
3. In any finite interval of time there are only finite number of discontinuities. Further more each of these discontinuities is finite.

3.16 (a)

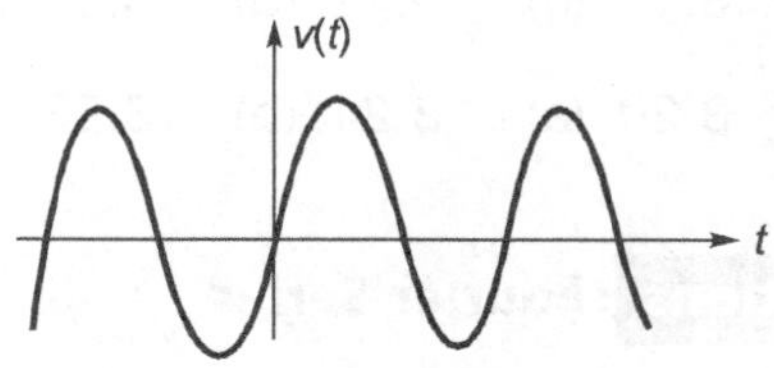

$$v(t) = 10 \sin 2\pi\, 100\, t$$
$$= \frac{10}{2j}(e^{j2\pi 100t} - e^{-j2\pi 100t})$$

$C_1 = -5j$ $\quad C_{-1} = 5j$

$C_n = 0$ for all n except $n = 1, -1$

3.17 (c)

For the existence of Fourier series of a given function the following condition must be satisfy:

(a) The function should have finite number of maxima or minima.

(b) The function should have finite number of discontinuities

(c) The function should be absolutely integrable.

3.19 (d)

For real function

$x(t) \xleftrightarrow{F.S.} a_k$

$x^*(t) \xleftrightarrow{F.S} a^*_{-k}$

For real

$\therefore \quad a^*_{-k} = a_k$

$a^*_k = a_{-k}$ ...(i)

For odd signal

$x(-t) = -x(t)$

$a_{-k} = -a_k$ ...(ii)

Put (ii) in (i) (For real and odd)

$a^*_k = -a_k \rightarrow$ Imaginary

3.20 (a)

$$y(t) = x(t - t_0) + x(t + t_0) \quad \text{...(i)}$$

By time shifting property,

$$x(t \pm t_0) = e^{\pm jk\omega_0 t_0} X(k)$$

Taking Fourier series of both the sides of (i)

$$Y(k) = e^{-jk\omega_0 t_0} X(k) + e^{jk\omega_0 t_0} X(k)$$

$$Y(k) = (e^{jk\omega_0 t_0} + e^{-jk\omega_0 t_0})X(k)$$

$$Y(k) = 2\cos(k\omega_0 t_0)\, X(k)$$

$$Y(k) = 2\cos\left(k\frac{2\pi}{T}t_0\right)X(k)$$

3.21 (b)

Dirichlet's conditions for convergence of Fourier series of $x(t)$.

(i) over any period, $x(t)$ must be integrable

$$\int_T x(t)\,dt < \infty$$

(ii) There are no more than a finite number of minima and maxima during any single period of signal.

(iii) In any finite interval of time, there are only a finite number of discontinuities.

3.22 (c)

$$x(t) = \sum_{b=-\infty}^{\infty} \delta(t - b)$$

$$\cdots + \delta(t+2) + \delta(t+1) + \delta(t) + \delta(t-1) + \delta(t-2) + \cdots$$

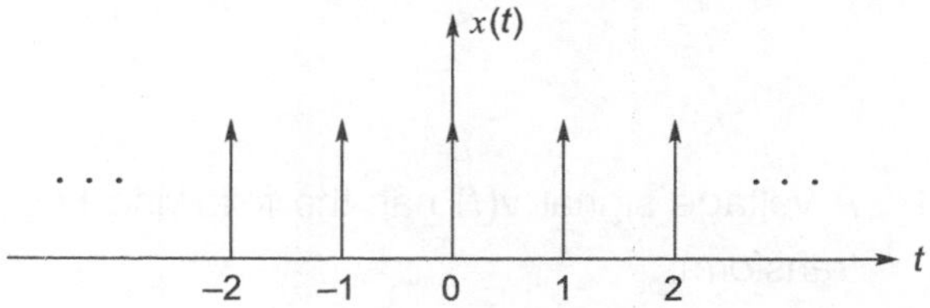

$$a_k = \frac{1}{T_o}\int_{-T_o/2}^{T_o/2} x(t)\, e^{-\frac{j2\pi}{T_o}kt}\, dt \quad [\because T_o = 1]$$

$$= \frac{1}{1}\int_{-1/2}^{1/2} \delta(t)\, e^{-\frac{j2\pi}{1}kt}\, dt$$

$$= 1$$

3.23 (d)

Given that,

$$x(t) = \frac{1}{T_o} + \sum_{K=1}^{N} \frac{2}{T_o}\cos K\omega_o t$$

$$a_0 = \frac{1}{T_o},\ a_k = \frac{2}{T_o}$$

As the Fourier series coefficient a_n is independent of 'K' signal cannot be sawtooth, half rectified (or) rectangular. Hence, impulse train.

(or)

The otherway is evaluating Fourier series coefficients are verifying.

■■■

4 Continuous Time Fourier Transform

4.1 A voltage signal $v(t)$ has the following Fourier transform:

$$V(j\omega) = \begin{cases} e^{-j\omega d} & \text{for } |\omega| < 1 \\ 0 & \text{for } |\omega| > 1 \end{cases}$$

The energy that would be dissipated in a 1 Ω resistor fed from $v(t)$ is

(a) $2/\pi$ Joules (b) $2e^{-2d}/\pi$ Joules
(c) $1/\pi$ Joules (d) $1/2\pi$ Joules

[ESE-2000]

4.2 The Fourier transform of a double-sided exponential signal $x(t) = e^{-b|t|}$

(a) is $2b/(b^2 + \omega^2)$

(b) is $\dfrac{e^{-j\tan^{-1}(\omega/b)}}{(b^2 + \omega^2)}$

(c) does not exist

(d) exists only when it is single sided

[ESE-2001]

4.3 The Fourier transform of $u(t)$ is

(a) $1/j\omega$ (b) $j\omega$
(c) $1/(1 + j\omega)$ (d) $\pi\delta(\omega) + 1/j\omega$

[ESE-2001]

4.4 Match **List-I** (Functions) with **List-II** (Fourier transforms) and select the correct answer using the codes given below the lists:

List-I	List-II
A. $\exp(-\alpha t)\, u(t),\ \alpha > 0$	1. $\dfrac{1}{(\alpha + j2\pi f)^2}$
B. $\exp(-\alpha\|t\|),\ \alpha > 0$	2. $\dfrac{1}{\alpha + j2\pi f}$
C. $t\exp(-\alpha t)\, u(t),\ \alpha > 0$	3. $\delta\left(f - \dfrac{\alpha}{t_0}\right)$
D. $\exp(j2\pi\alpha t/t_0)$	4. $\dfrac{2\alpha}{\alpha^2 + (2\pi f)^2}$

Codes:

	A	B	C	D
(a)	3	1	4	2
(b)	2	4	1	3
(c)	3	4	1	2
(d)	2	1	4	3

[ESE-2002]

4.5 Match **List-I** (Type of signal) with **List-II** (Property of Fourier transform) and select the correct answer using the codes given below the lists:

List-I	List-II
A. Real and even symmetric	1. Imaginary and even symmetric
B. Real and odd symmetric	2. Real and even symmetric
C. Imaginary and even symmetric	3. Real and odd symmetric
D. Imaginary and odd symmetric	4. Imaginary and odd symmetric

Codes:

	A	B	C	D
(a)	1	4	2	3
(b)	2	4	1	3
(c)	1	3	2	4
(d)	2	3	1	4

[ESE-2002]

4.6 The Fourier transform of $e^{-\pi t^2}$ is $e^{-\pi f^2}$; then the Fourier transform of $e^{-\alpha t^2}$ is

(a) $\left(\dfrac{1}{\alpha}\right)e^{-\alpha f^2}$ (b) $\sqrt{\dfrac{\pi}{\alpha}}\, e^{-\frac{\pi^2 f^2}{\alpha}}$

(c) $\dfrac{1}{\sqrt{\pi\alpha}}e^{-\pi^2 f^2}$ (d) $\sqrt{\pi\alpha}\, e^{-\frac{f^2}{\pi^2\alpha}}$

[ESE-2002]

4.7 Match **List-I** (Functions in the time domain) with **List-II** (Fourier transform of the function) and select the correct answer using the codes given below the lists:

	List-I		List-II
A.	Delta function	1.	Delta function
B.	Gate function	2.	Gaussian function
C.	Normalized Gaussian function	3.	Constant function
D.	Sinusoidal function	4.	Sampling function

Codes:

	A	B	C	D
(a)	1	2	4	3
(b)	3	4	2	1
(c)	1	4	2	3
(d)	3	2	4	1

[ESE-2002]

4.8 The Fourier transform $X(f)$ of the periodic delta functions, $x(t) = \sum_{k=-\infty}^{\infty} \delta(t - kT)$ is

(a) $T \sum_{k=-\infty}^{\infty} \delta(f - kT)$ (b) $T \sum_{k=-\infty}^{\infty} \delta\left(f - \frac{k}{T}\right)$

(c) $\frac{1}{T} \sum_{k=-\infty}^{\infty} \delta\left(f - \frac{k}{T}\right)$ (d) $\frac{1}{T} \sum_{k=-\infty}^{\infty} \delta(f - kT)$

[ESE-2002]

4.9 Match **List-I** with **List-II** and select the correct answer using the codes given below the lists:

List-I

A. Free and Forced response
B. z-transforms
C. Probability theory
D. Fourier series

List-II

1. Discrete time systems
2. Dirichlet conditions
3. Non-homogenous differential equation
4. Random processes

Codes:

	A	B	C	D
(a)	1	3	2	4
(b)	3	1	2	4
(c)	1	3	4	2
(d)	3	1	4	2

[ESE-2003]

4.10 The inverse Fourier transform of $\delta(f)$ is

(a) $u(t)$ (b) 1
(c) $\delta(t)$ (d) $e^{j2\pi t}$

[ESE-2003]

4.11 What one of the following represents the phase response of the function:

$$H(s) = \frac{s^2 + \omega_0^2}{s^2 + (\omega_0 / Q)s + \omega_0^2}?$$

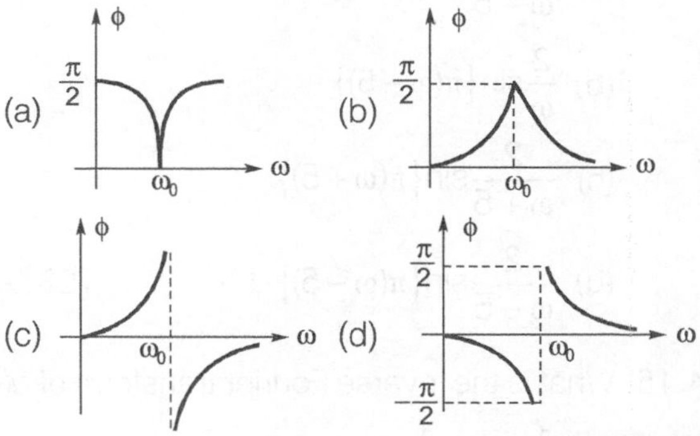

[ESE-2005]

4.12 Match **List-I** (Time Function) with **List-II** (Fourier Spectrum/Fourier Transform) and select the correct answer using the codes given below the lists:

	List-I		List-II
A.	Periodic Function	1.	Continuous spectrum at all frequencies
B.	Aperiodic Function	2.	$\delta(\omega)$
C.	Unit Impulse $\delta(t)$	3.	Line discrete spectrum
D.	$\sin \omega(t)$	4.	1

Codes:

	A	B	C	D
(a)	4	2	3	1
(b)	3	1	4	2
(c)	4	1	3	2
(d)	3	2	4	1

[ESE-2005]

4.13 Match **List-I** (CT Function) with **List-II** (CT Fourier Transform) and select the correct answer using the code given below the lists:

	List-I		List-II
A.	$e^{-t}u(t)$	1.	$2\pi e^{-\lvert\omega\rvert}$
B.	$x(t) = \begin{cases} 1, \lvert t\rvert \le 1 \\ 0, \lvert t\rvert > 1 \end{cases}$	2.	$j\omega X(j\omega)$
C.	$\frac{dx(t)}{dt}$	3.	$\frac{1}{1 + j\omega}$
D.	$\frac{2}{1 + t^2}$	4.	$\frac{2\sin\omega}{\omega}$

Codes:

	A	B	C	D
(a)	1	4	2	3
(b)	3	2	4	1
(c)	1	2	4	3
(d)	3	4	2	1

[ESE-2006]

4.14 If the Fourier transform of $x(t)$ is $\frac{2}{\omega}\sin(\pi\omega)$, then what is the Fourier transform of $e^{j5t}x(t)$?

(a) $\frac{2}{\omega-5}\sin(\pi\omega)$

(b) $\frac{2}{\omega}\sin\{\pi(\omega-5)\}$

(c) $\frac{2}{\omega+5}\sin\{\pi(\omega+5)\}$

(d) $\frac{2}{\omega-5}\sin\{\pi(\omega-5)\}$ [ESE-2006]

4.15 What is the inverse Fourier transform of $u(\omega)$?

(a) $\frac{1}{2}\delta(t)+\frac{1}{\pi t}$ (b) $\frac{1}{2}\delta(t)$

(c) $2\delta(t)+\frac{1}{\pi t}$ (d) $2\delta(t)+\text{sgn}(t)$

[ESE-2006]

4.16 A real signal $x(t)$ has Fourier transform $X(f)$. Which one of the following is correct?

(a) Magnitude of $X(f)$ has even symmetry while phase of $X(f)$ has odd symmetry

(b) Magnitude of $X(f)$ has odd symmetry while phase of $X(f)$ has even symmetry

(c) Both magnitude and phase of $X(f)$ have even symmetry

(d) Both magnitude and phase of $X(f)$ have odd symmetry

[ESE-2007]

4.17 Which one of the following is the correct relation?

(a) $F(at) \leftrightarrow aF(\omega/a)$

(b) $F(at) \leftrightarrow aF(a\omega)$

(c) $F(t/a) \leftrightarrow aF(\omega/a)$

(d) $F(at) \leftrightarrow (1/a)F(\omega/a)$ [ESE-2008]

4.18 The Fourier transform of a function is equal to its two-sided Laplace transform evaluated

(a) on the real axis of the s-plane

(b) on a line parallel to the real axis of the *s*-plane

(c) on the imaginary axis of the s-plane

(d) on a line parallel to the imaginary axis of the *s*-plane

[ESE-2008]

4.19 If $f(t)$ is an even function, then what is its Fourier transform $F(j\omega)$?

(a) $\int_0^{\infty} f(t)\cos(2\omega t)dt$

(b) $2\int_0^{\infty} f(t)\cos(\omega t)dt$

(c) $2\int_0^{\infty} f(t)\sin(\omega t)dt$

(d) $2\int_0^{\infty} f(t)\sin(2\omega t)dt$ [ESE-2008]

4.20 If the Fourier transform of $f(t)$ is $F(j\omega)$, then what is the Fourier transform of $f(-t)$?

(a) $F(j\omega)$

(b) $F(-j\omega)$

(c) $-F(j\omega)$

(d) Complex conjugate of $F(j\omega)$ [ESE-2008]

4.21 When $y(t)\xleftrightarrow{FT}Y(j\omega)$; $x(t)\xleftrightarrow{FT}X(j\omega)$; $h(t)\xleftrightarrow{FT}H(j\omega)$. What is $Y(j\omega)$?

(a) $\frac{X(j\omega)}{H(j\omega)}$ (b) $X(j\omega)\,H(j\omega)$

(c) $X(j\omega)+H(j\omega)$ (d) $X(j\omega)-H(j\omega)$

[ESE-2009]

4.22 For good quality signal transmission all frequency components should have the same transmission delay, t_d and same phase shift-ϕ_s. What can be said about the statement ?

(a) Correct

(b) True for t_d but not for ϕ_s

(c) Not true for t_d but true for ϕ_s

(d) Both t_d and ϕ_s are not involved in quality

[ESE-2009]

4.23 If $X(\omega)=\delta(\omega-\omega_0)$ then $x(t)$ is

(a) $e^{-j\omega_0 t}$ (b) $\delta(t)$

(c) $\frac{1}{2\pi}e^{j\omega_0 t}$ (d) 1 [ESE-2010]

4.24 The Fourier transform of a rectangular pulse is

(a) another rectangular pulse

(b) triangular pulse

(c) sinc function

(d) impulse function [ESE-2012]

4.25 The property of Fourier transforms which states that the compression in time domain is equivalent to expansion in the frequency domain is

(a) duality

(b) scaling

(c) time scaling

(d) frequency shifting

[ESE-2012]

4.26 If $f(t)$ is a real and odd function, then its Fourier transform $F(\omega)$ will be
(a) real and even function of ω
(b) real and odd function of ω
(c) imaginary and odd function of ω
(d) imaginary function of ω [ESE-2013]

4.27 A unit impulse function $\delta(t)$ is defined by
1. $\delta(t) = 0$ for all t except $t = 0$
2. $\int_{-\infty}^{\infty} \delta(t)dt = 1$

The Fourier transform $F(\omega)$ of $\delta(t)$ is
(a) 1 (b) $\frac{1}{\omega}$
(c) 0 (d) $\frac{1}{j\omega}$ [ESE-2013]

4.28 Which of the following Dirichlet's conditions are correct for convergence of Fourier transform of the function $x(t)$?
1. $x(t)$ is square integrable.
2. $x(t)$ must be periodic
3. $x(t)$ should have finite number of maxima and minima within any finite interval.
4. $x(t)$ should have finite number of discontinuities within any finite interval.

(a) 1, 2, 3 and 4 (b) 1, 2 and 4 only
(c) 1, 3 and 4 only (d) 2, 3 and 4 only
[ESE-2013]

4.29 The Fourier transform of a rectangular pulse of a period

$$t = -\frac{T}{2} \quad \text{to} \quad t = \frac{T}{2}$$

is
(a) a sinc function
(b) a sine function
(c) a cosine function
(d) a sine-squared function [ESE-2014]

4.30 Consider the following transfer functions:
1. $\frac{1}{j\omega + 1}$ 2. $\frac{1}{(j\omega + 1)^2}$ 3. $\frac{1}{(j\omega + 1)(j\omega + 2)}$

The transfer functions which have a non linear phase are:
(a) 1 and 2 only (b) 1 and 3 only
(c) 2 and 3 only (d) 1, 2 and 3
[ESE-2015]

4.31 Consider the following transforms:
1. Fourier transform 2. Laplace transform

Which of the above transforms is/are used in signal processing?
(a) 1 only (b) 2 only
(c) Both 1 and 2 (d) Neither 1 nor 2
[ESE-2016]

4.32 **Statement (I):** Non-stationary signals such as an image require time-frequency analysis.
Statement (II): The short time Fourier transform (STFT) can map a one dimensional function $f(t)$ into the two-dimensional function, STFT (f).
(a) Both Statement (I) and Statement (II) are individually true and Statement (II) is the correct explanation of Statement (I)
(b) Both Statement (I) and Statement (II) are individually true but Statement (II) is **NOT** the correct explanation of Statement (I)
(c) Statement (I) is true but Statement (II) is false
(d) Statement (I) is false but Statement (II) is true
[ESE-2020]

4.33 The cross-correlation of two finite length sequences $x(n) = \{1, 2, 1, 1\}$ and $y(n) = \{1, 1, 2, 1\}$ is
(a) {1, 2, 5, 6, 6, 4, 1}
(b) {1, 1, 4, 6, 6, 5, 2}
(c) {5, 2, 1, 1, 4, 6, 6}
(d) {1, 4, 6, 6, 5, 2, 1}

[ESE-2024]

■■■

Answers Continuous Time Fourier Transform

4.1 (c)	4.2 (a)	4.3 (d)	4.4 (b)	4.5 (b)	4.6 (b)	4.7 (b)	4.8 (c)	4.9 (d)
4.10 (b)	4.11 (d)	4.12 (b)	4.13 (d)	4.14 (d)	4.15 (∗)	4.16 (a)	4.17 (d)	4.18 (c)
4.19 (b)	4.20 (b)	4.21 (b)	4.22 (b)	4.23 (c)	4.24 (c)	4.25 (c)	4.26 (c)	4.27 (a)
4.28 (c)	4.29 (a)	4.30 (d)	4.31 (c)	4.32 (a)	4.33 (d)			

Explanations Continuous Time Fourier Transform

4.1 (c)

$$E = \frac{1}{2\pi}\int_{-\infty}^{\infty} |V(j\omega)|^2\, d\omega$$

$$E = \frac{1}{2\pi}\int_{-1}^{1} 1\, d\omega$$

$\because$ magnitude of $e^{-j\omega t} = 1$.

$$E = \frac{1}{2\pi}\times 2 = \frac{1}{\pi}$$

4.2 (a)

$$x(t) = e^{-b|t|}$$

$$X(\omega) = \int_{-\infty}^{\infty} x(t)\, e^{-j\omega t} dt$$

$$= \int_{-\infty}^{\infty} e^{-b|t|}\, e^{-j\omega t} dt$$

$$= \int_{-\infty}^{0} e^{bt}\, e^{-j\omega t} dt + \int_{0}^{\infty} e^{-bt}\, e^{-j\omega t} dt$$

$$= \frac{1}{(b-j\omega)} e^{(b-j\omega)t}\Big|_{-\infty}^{0} + \frac{1}{-(b+j\omega)} e^{-(b+j\omega)t}\Big|_{0}^{\infty}$$

$$= \frac{1}{b-j\omega} + \frac{1}{b+j\omega} = \frac{b+j\omega+b-j\omega}{b^2+\omega^2}$$

$$\Rightarrow X(\omega) = \frac{2b}{b^2+\omega^2}$$

4.3 (d)

$$u(t) \xleftrightarrow{\text{F.T.}} \frac{1}{j\omega} + \pi\delta(\omega)$$

4.5 (b)

Note: Trick

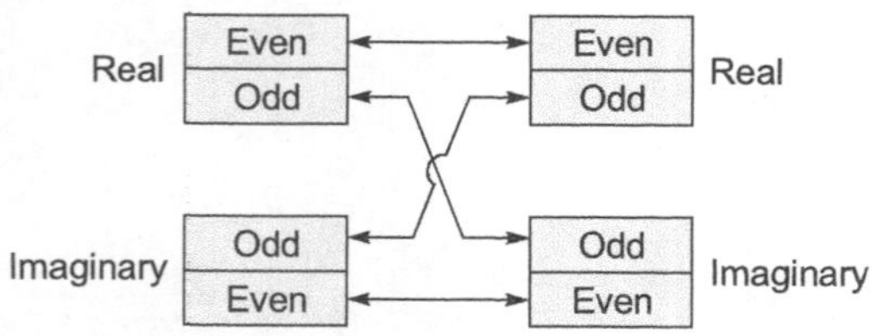

From above diagram:

Type of Signal	Properties DFFT
• Real & even symmetric	Even & real symmetric
• Real & odd symmetric	Odd & Img symmetric
• Img. & odd symmetric	Odd & real symmetric
• Img. & even symmetric	Even & Img. symmetric

4.6 (b)

$$e^{-\pi t^2} \xleftrightarrow{\text{FT}} e^{-\pi f^2}$$

$$e^{-\alpha t^2} = e^{-\pi\left(\sqrt{\frac{\alpha}{\pi}}\, t\right)^2}$$

By the property of Fourier transform pair

If $\quad x(t) \xrightarrow{\text{FT}} X(j\omega)$

then $\quad x(at) \xrightarrow{\text{FT}} \frac{1}{|a|} X\left(\frac{j\omega}{a}\right)$

So, $e^{-\pi\left(\sqrt{\frac{\alpha}{\pi}}t\right)^2} \xrightarrow{\text{FT}} \sqrt{\frac{\pi}{\alpha}}\, e^{-\pi f^2 \frac{\pi}{\alpha}}$

$$= \sqrt{\frac{\pi}{\alpha}}\, e^{\frac{-\pi^2 f^2}{\alpha}}$$

4.7 (b)

$$\delta(t) \xleftrightarrow{\text{FT}} 1$$

$$\sin\omega_0 t \xleftrightarrow{\text{FT}} j\pi[\delta(\omega+\omega_0) - \delta(\omega-\omega_0)]$$

$$\text{rect}\left(\frac{t}{\tau}\right) \xleftrightarrow{\text{FT}} \tau\,\text{sinc}\left(\frac{\omega\tau}{2}\right)$$

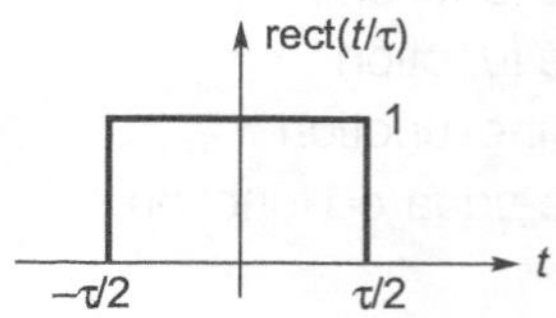

4.8 (c)

$$x(t) = \sum_{k=-\infty}^{\infty} \delta(t-kT)$$

F.T. of $x(t)$ is $\omega_0 \sum_{k=-\infty}^{\infty} \delta(\omega - k\omega_0)$

where, $\quad \omega_0 = \frac{2\pi}{T}$ and $\delta(\omega) = \frac{1}{2\pi}\delta(f)$

$$\therefore \quad \frac{2\pi}{T}\sum_{k=-\infty}^{\infty}\delta\left(2\pi\left(f-\frac{k}{T}\right)\right) = \frac{1}{T}\sum_{k=-\infty}^{\infty}\delta\left(f-\frac{k}{T}\right)$$

4.11 (d)

Phase response, $H(s) = \dfrac{s^2+\omega_0^2}{s^2+(\omega_0/Q)s+\omega_0^2}$

$$H(j\omega) = \frac{-\omega^2+\omega_0^2}{-\omega^2+j\dfrac{\omega\,\omega_0}{Q}+\omega_0^2}$$

$$\phi = \angle H(j\omega) = -\tan^{-1}\frac{\omega\omega_0/Q}{\omega_0^2-\omega^2}$$

At $\omega = \omega_0$, $\angle H(j\omega) = -\tan^{-1}\infty = -90°$

The plot of ϕ is shown figure.

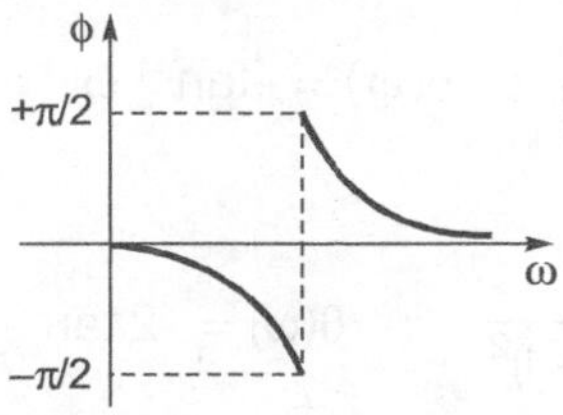

4.13 (d)

$$e^{-t}u(t) \xrightarrow{FT} \frac{1}{1+j\omega}$$

$$x(t) \to X(j\omega)$$

$$\frac{dx(t)}{dt} \to X(j\omega)\times j\omega$$

B = 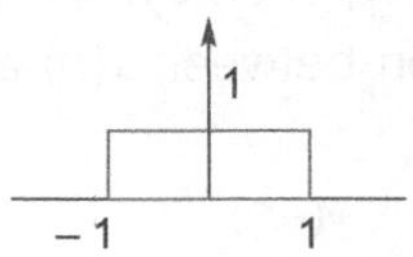 $\xrightarrow{FT} \dfrac{2\sin\omega}{\omega}$

$$e^{-|t|} \xrightarrow{FT} \frac{2}{\omega^2+1}$$

By duality $\dfrac{2}{1+t^2} \longrightarrow 2\pi e^{-|\omega|}$

4.14 (d)

If $x(t) \xleftrightarrow{FT} X(j\omega)$

$e^{j\omega_0 t}x(t) \xleftrightarrow{FT} X(j(\omega-\omega_0))$

Therefore, if $x(t) \xleftrightarrow{FT} \dfrac{2}{\omega}\sin(\pi\omega)$

then $e^{j5t}x(t) \xleftrightarrow{FT} \dfrac{2}{\omega-5}\sin\{\pi(\omega-5)\}$

4.15 (*)

By duality property,

if $x(t) \xleftrightarrow{FT} X(\omega)$

then $x(t) \xleftrightarrow{FT} 2\pi X(-\omega)$

By time reversal property,

if $x(t) \xleftrightarrow{FT} X(\omega)$

then $x(-t) \xleftrightarrow{FT} X(-\omega)$

$$u(t) \xleftrightarrow{FT} \pi\delta(\omega)+\frac{1}{j\omega}$$

$$\Rightarrow \pi\delta(t)+\frac{1}{jt} \xleftrightarrow{FT} 2\pi\,u(-\omega)$$

$$\Rightarrow \frac{\pi}{2\pi}\delta(-t)+\frac{-1}{j2\pi t} \xleftrightarrow{FT} u(\omega)$$

But $\delta(-t) = \delta(t)$

So, $\dfrac{1}{2}\delta(t)+\dfrac{j}{2\pi t} \xleftrightarrow{FT} u(\omega)$

4.16 (a)

Magnitude of $X(f)$ has even symmetry while phase of $X(f)$ has odd symmetry.

4.18 (c)

For real part = 0 i.e. on the imaginary axis of the s-plane, the Laplace transform is equal to the Fourier transform.

4.19 (b)

$$F(j\omega) = \int_{-\infty}^{\infty} f(t)\,e^{-j\omega t}dt$$

$$= \int_{-\infty}^{\infty} f(t)\{\cos\omega t - j\sin\omega t\}\,dt$$

$\because$ Even function × even function = even function and even function × odd function = odd function

$$F(j\omega) = \int_{-\infty}^{\infty} f(t)\cos\omega t\,dt$$

$$= 2\int_{0}^{\infty} f(t)\cos\omega t\,dt$$

4.20 (b)

This is the time reversal property of Fourier transform.

4.21 (b)

In frequency domain, $Y(j\omega) = X(j\omega)\,H(j\omega)$

In time domain, $y(t) = x(t) * h(t)$

4.22 (b)

For good quality signal transmission, all frequency components should have the same transmission delay but even same phase shift will distort the signal.

4.23 (c)

As $e^{j\omega_0 t}x(t) \xleftrightarrow{F.T.} X(\omega-\omega_0)$

As we know

$$1 \xleftrightarrow{\text{F.T.}} 2\pi\,\delta(\omega)$$

$$\therefore \quad e^{j\omega_0 t}1 \xleftrightarrow{\text{F.T.}} 2\pi\,\delta(\omega-\omega_0)$$

$$\text{or } \frac{1}{2\pi}e^{j\omega_0 t} \xleftrightarrow{F.T.} \delta(\omega-\omega_0)$$

4.24 (c)

Fourier transform of the rectangular function is a sinc function

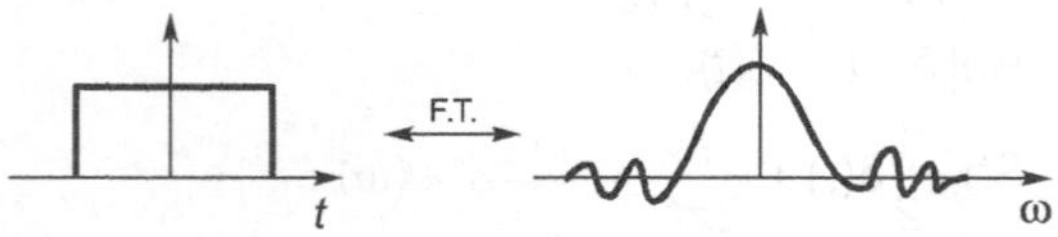

4.25 (c)

Time scaling is the property which states that compression in the time domain is equivalent to expansion in frequency domain.

4.26 (c)

$f(t)$ is real and odd function then $F(\omega)$ will be imaginary and odd function of ω.

4.27 (a)

Fourier, Laplace and z-transform of $\delta(t)$ is equal to 1.

4.28 (c)

Periodicity is not the necessary condition for the convergence of Fourier transform. So second condition is false.

4.29 (a)

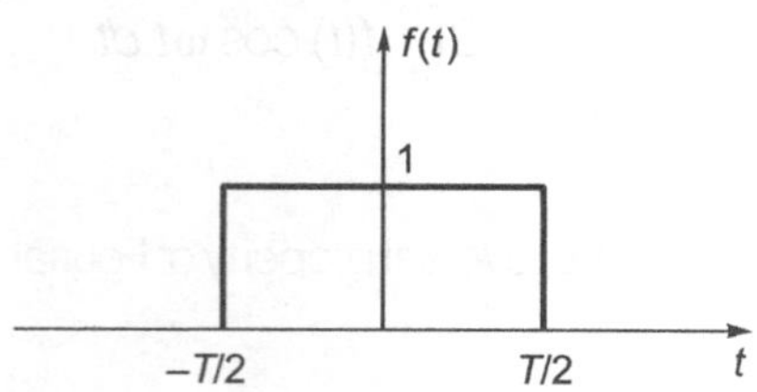

$\dfrac{df(t)}{dt} =$

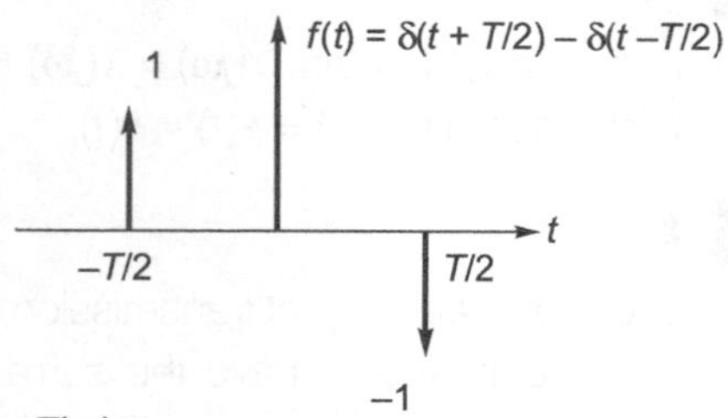

$\because f(t) \leftrightarrow F(\omega)$

By differentiation property in time domain, we know

$$\frac{df(t)}{dt} \leftrightarrow j\omega\,F(\omega)$$

$$\leftrightarrow 1/j\omega\,[e^{j\omega T/2} - e^{-j\omega T/2}]$$

$$\leftrightarrow \frac{1}{j\omega}\left[\cos\omega\frac{T}{2} + j\sin\omega\frac{T}{2} - \cos\omega\frac{T}{2} + j\sin\omega\frac{T}{2}\right]$$

$$f(t) \leftrightarrow \frac{1}{j\omega}\left[2j\sin\frac{\omega T}{2}\right]$$

$$= \frac{2}{\omega}\left[\sin\frac{\omega T}{2}\right] = 2\times\frac{\sin\frac{\omega T}{2}}{\frac{\omega T}{2}}\times\frac{T}{2} = \frac{2T}{2}Sa\left(\frac{\omega T}{2}\right)$$

$$f(t) \leftrightarrow T\,Sa\left(\frac{\omega T}{2}\right)$$

4.30 (d)

Linear phase refers to the condition where the phase response is a linear function of frequency.

1. $\dfrac{1}{j\omega+1}$; $\theta(\omega) = -\tan^{-1}\omega$

Similarly,

2. $\dfrac{1}{(j\omega+1)^2}$; $\theta(\omega) = -2\tan^{-1}\omega$

3. $\dfrac{1}{(j\omega+1)(j\omega+2)}$; $\theta(\omega) = -\tan^{-1}\dfrac{3\omega}{2-\omega^2}$

All are non linear functions of ω and $\dfrac{\theta(\omega)}{\omega} \neq$ constant.

Hence, all have non-linear phase.

4.33 (d)

Given: $x(n) = \{1, 2, 1, 1\}$ and $y(n) = \{1, 1, 2, 1\}$

The cross-correlation between $x(n)$ and $y(n)$ is given by

$$R_{XY}(n) = x(n) * y(-n)$$

$$R_{XY}(n) = \{\underset{\uparrow}{1}, 2, 1, 1\} * \{1, 2, 1, \underset{\uparrow}{1}\}$$

Using tabular method:

y(–n) \ x(n)	1	2	1	1
1	1	2	1	1
2	2	4	2	2
1	1	2	1	1
1	1	2	1	1

$$R_{XY}(n) = \{1, 4, 6, \underset{\uparrow}{6}, 5, 2, 1\}$$

■■■

5 Laplace Transform

5.1 The output of a linear system to a unit step input u(t) is t^2e^{-2t}. The system function $H(s)$ is

(a) $\frac{2}{s^2(s+2)}$ (b) $\frac{2}{(s+2)^2}$

(c) $\frac{2}{(s+2)^3}$ (d) $\frac{2s}{(s+2)^3}$

[ESE-2000]

5.2 Match **List-I** (System function) with **List-II** (Impulse response) and select the correct answer using the codes given below the lists:

List-I

A. $\frac{e^{-s}}{s+1}$

B. $\frac{1}{s^2+s+1}$

C. $\frac{1}{(s+1)^2}$

D. $\frac{1}{s^2+s}$

List-II

1. 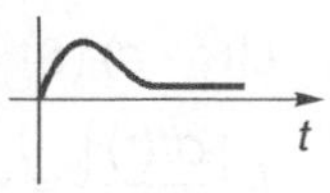

2. 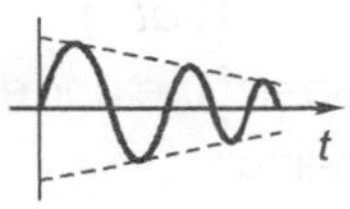

3. 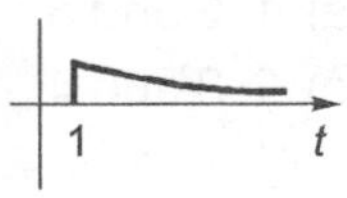

4. 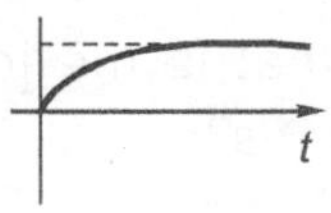

5. 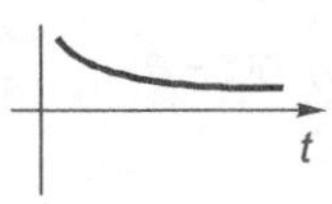

Codes:

	A	B	C	D
(a)	3	4	1	2
(b)	5	2	3	4
(c)	3	2	1	4
(d)	5	4	3	2

[ESE-2000]

5.3 **Assertion (A):** When a linear time-invariant system having a system function $H(s)$ is driven by an input $x(t) = e^{s_0 t}$ the forced part of the output is $H(s_0)\, e^{s_0 t}$.

Reason (R): In the partial fraction expansion of the Laplace transform of the output, the term corresponding to the pole at s_0 is $\frac{H(s_0)}{s-s_0}$. The inverse transform of $\frac{H(s_0)}{s-s_0}$ is $H(s_0)e^{s_0 t}$.

(a) Both A and R are true and R is the correct explanation of A
(b) Both A and R are true but R is NOT the correct explanation of A
(c) A is true but R is false
(d) A is false but R is true

[ESE-2000]

5.4 If $F(s) = \frac{2s+5}{s^2+5s+6}$, then $f(t)$ is given by

(a) $e^{-2.5t}\,[\cosh 0.5\,t + \sinh 0.5t]$
(b) $e^{2.5t}\,[\cosh 0.5\,t + \sinh 0.5t]$
(c) $\sqrt{2}\, e^{-2.5t}\sin(0.5t + 45°)$
(d) $e^{-2.5t}\,[\cos 0.5t + \sin 0.5t]$

[ESE-2001]

5.5 Which one of the following transfer functions does correspond to a non-minimum phase system?

(a) $\frac{s}{s^2+2s+1}$ (b) $\frac{s+1}{s^2+2s+1}$

(c) $\frac{s+1}{s^2+2s-1}$ (d) $\frac{s-1}{s^2+2s+1}$

[ESE-2001]

5.6 Consider the following statements:

1. Fourier transform is special case of Laplace transform
2. Region of convergence need not be specified for Fourier transform
3. Laplace transform is not unique unless the region of convergence is specified
4. Laplace transform is a special case of Fourier transform

Which of these statements are correct?

(a) 2 and 4 (b) 4 and 1
(c) 4, 3 and 2 (d) 1, 2 and 3

[ESE-2002]

5.7 The relationship between the input $x(t)$ and the output $y(t)$ of a system is

$$\frac{d^2y}{dt^2} = x(t-2)\,u(t-2) + \frac{d^2x}{dt^2}$$

The transfer function of the system is

(a) $1+\frac{s^2}{e^{2s}}$ (b) $1+\frac{e^{-2s}}{s^2}$

(c) $1+\frac{e^{2s}}{s^2}$ (d) $1+\frac{s^2}{e^{-2s}}$

[ESE-2003]

5.8 If $\left(\frac{27s+97}{s^2+33s}\right)$ is the Laplace transform of $f(t)$, then $f(0^+)$ is

(a) Zero (b) 97/33
(c) 27 (d) Infinity [ESE-2003]

5.9 The Laplace transform of the waveform shown in the below figure is

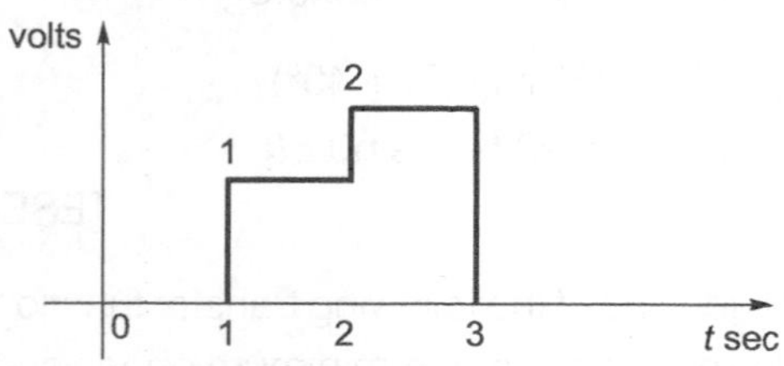

(a) $1/s[e^{s} + e^{2s} + 2e^{3s}]$
(b) $1/s[e^{s} + e^{2s} + 2e^{-3s}]$
(c) $1/s[e^{-s} + e^{-2s} + 2e^{-3s}]$
(d) $1/s[e^{-s} + e^{-2s} - 2e^{-3s}]$ [ESE-2003]

5.10 Match **List-I** [$F(s)$] with **List-II** [$f(t)$] and select the correct answer using the codes given below the lists:

List-I	List-II
A. $\frac{10}{s(s+10)}$	1. $10\,\delta(t)$
B. $\frac{10}{(s^2+100)}$	2. $(e^{-10t}\cos 10t)\cdot u(t)$
C. $\frac{(s+10)}{(s+10)^2+100}$	3. $(\sin 10t)\cdot u(t)$
D. 10	4. $(1-e^{-10t})\cdot u(t)$

Codes:

	A	B	C	D
(a)	3	4	1	2
(b)	4	3	1	2
(c)	3	4	2	1
(d)	4	3	2	1

[ESE-2003]

5.11 The Laplace transform of $\sin \omega t$ is

(a) $\frac{s}{s^2+\omega^2}$ (b) $\frac{\omega^2}{s^2+\omega^2}$

(c) $\frac{s^2}{s^2+\omega^2}$ (d) $\frac{\omega}{s^2+\omega^2}$

[ESE-2003]

5.12 Given $L[f(t)] = F(s) = \int_0^\infty f(t)e^{-st}dt$, which of the following expressions are correct?

1. $L\{f(t-a)u(t-a)\} = F(s)e^{-sa}$
2. $L\{tf(t)\} = \frac{-dF(s)}{ds}$
3. $L\{(t-a)f(t)\} = asF(s)\}$
4. $L\left\{\frac{df(t)}{dt}\right\} = sF(s) - f(0^+)$

Select the correct answer using the codes given below:

(a) 1, 2 and 3 (b) 1, 2 and 4
(c) 2, 3 and 4 (d) 1, 3 and 4

[ESE-2003]

5.13 What is the inverse Laplace transform of e^{-as}/s?

(a) e^{-at} (b) $u(t-a)$
(c) $\delta(t-a)$ (d) $(t-a)u(t-a)$

[ESE-2004]

5.14 Given that $x_1(t) = e^{k_1 t}u(t)$ and $x_2(t) = e^{-k_2 t}u(t)$. Which one of the following gives their convolution?

(a) $\frac{\left[e^{k_1 t} - e^{-k_2 t}\right]}{[k_1+k_2]}$ (b) $\frac{\left[e^{k_1 t} - e^{-k_2 t}\right]}{[k_2-k_1]}$

(c) $\frac{\left[e^{k_1 t} + e^{-k_2 t}\right]}{[k_2+k_1]}$ (d) $\frac{\left[e^{k_1 t} + e^{-k_2 t}\right]}{[k_2-k_1]}$

[ESE-2004]

5.15 Laplace transforms of $f(t)$ and $g(t)$ are $F(s)$ and $G(s)$, respectively. Which one of the following expressions gives the inverse Laplace transform of $F(s)\,G(s)$?

(a) $f(t)\,g(t)$ (b) $f(t)\,/\,g(t)$
(c) $f(t) + g(t)$ (d) $f(t) * g(t)$

[ESE-2004]

5.16 What does the transfer function of a system describe for the system?

(a) Only zero-input response
(b) Only zero-state response
(c) Both zero-input and zero-state responses
(d) Neither zero-input response nor zero-state response

[ESE-2005]

5.17 What is the Laplace transform of the waveform shown below?

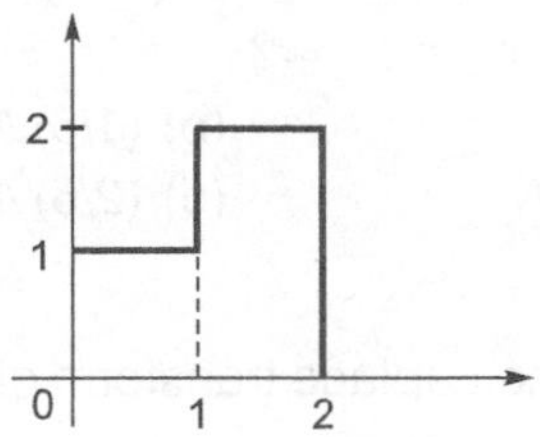

(a) $F(s) = \frac{1}{s} + \frac{1}{s}e^{-s} - \frac{2}{s}e^{-2s}$

(b) $F(s) = \frac{1}{s} - \frac{1}{s}e^{-s} + \frac{2}{s}e^{-2s}$

(c) $F(s) = \frac{1}{s} + \frac{1}{s}e^{s} + \frac{2}{s}e^{2s}$

(d) $F(s) = \frac{1}{s} - \frac{1}{s}e^{-2s} - \frac{2}{s}e^{-s}$

[ESE-2005]

5.18 For the signal shown below

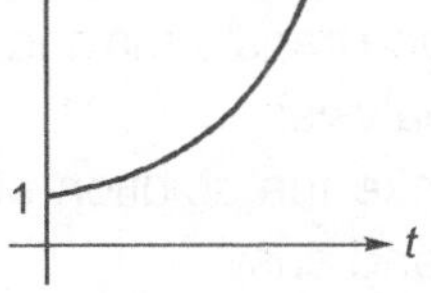

(a) Only Fourier transform exists
(b) Only Laplace transform exist
(c) Both Laplace and Fourier transforms exist
(d) Neither Laplace nor Fourier transform exists

[ESE-2005]

5.19 What is the Laplace transform of $x(t) = -e^{2t}u(t) * tu(t)$?

(a) $\frac{-1}{s^2(s+2)}$ (b) $\frac{-1}{s^2(s-2)}$

(c) $\frac{1}{s^2(s-2)}$ (d) $\frac{-1}{s(s-2)}$

[ESE-2006]

5.20

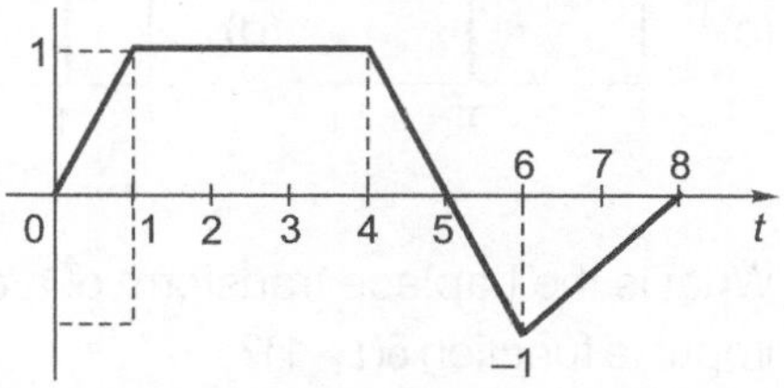

The Laplace transform of the waveform shown in the figure is

$\frac{1}{s^2}(1 + Ae^{-s} + Be^{-4s} + Ce^{-6s} + De^{-8s})$.

What is the value of D?

(a) –0.5 (b) –1.5
(c) 0.5 (d) 2.0

[ESE-2006]

5.21 What is $F(s) = \frac{8s+10}{(s+1)(s+2)^3}$ equal to?

(a) $\frac{2}{s+1} + \frac{4}{(s+2)^3} - \frac{4}{(s+2)^2} - \frac{2}{s+2}$

(b) $\frac{2}{s+1} + \frac{6}{(s+2)^3} - \frac{2}{(s+2)^2} - \frac{2}{s+2}$

(c) $\frac{2}{s+1} + \frac{6}{(s+2)^3} - \frac{2}{(s+2)^2} + \frac{2}{s+2}$

(d) $\frac{4}{s+1} + \frac{6}{(s+2)^3} - \frac{2}{(s+2)^2} - \frac{4}{s+2}$

[ESE-2006]

5.22 For the function $x(t)$, $X(s)$ is given by:

$$X(s) = e^{-s}\left[\frac{-2}{s(s+2)}\right]$$

Then, what are the initial and final values of $x(t)$, respectively?

(a) 0 and 1 (b) 0 and –1
(c) 1 and 1 (d) –1 and 0

[ESE-2006]

5.23 The Laplace transform $X(s)$ of a function $x(t)$ is

$$X(s) = \frac{1 - e^{-sT}}{s}$$

What is the wave shape of $x(t)$?

(a) 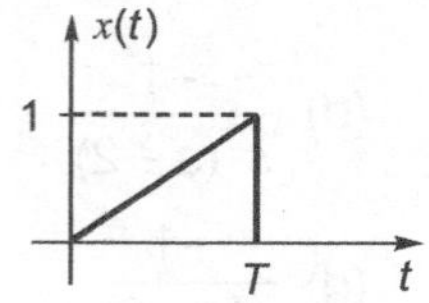

(b) 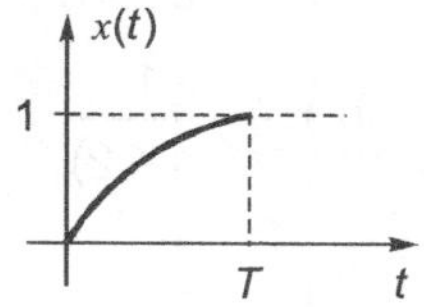

(c) 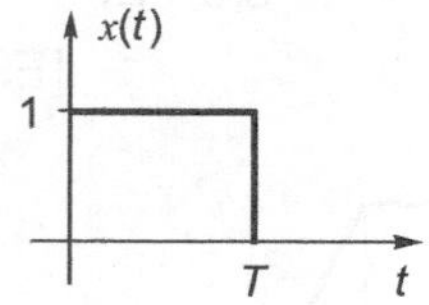

(d) 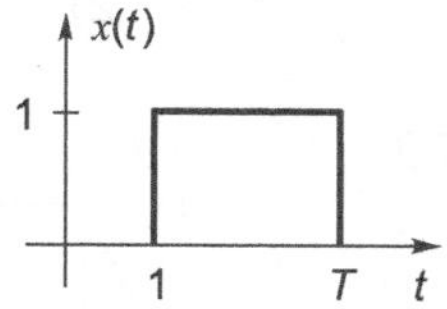

[ESE-2006]

5.24 What is the Laplace transform of a delayed unit impulse function $\delta(t-1)$?

(a) 1 (b) zero
(c) exp(–s) (d) s

[ESE-2007]

5.25 What is the output $t \to \infty$ for a system that has a transfer function $G(s) = \dfrac{2}{s^2 - s - 2}$; when subjected to a step input?

(a) –1 (b) 1
(c) 2 (d) Unbounded

[ESE-2007]

5.26 Which one of the following is correct? A system can be completely described by a transfer function if it is

(a) non-linear and continuous
(b) linear and time-varying
(c) non-linear and time-invariant
(d) linear and time invariant [ESE-2007]

5.27 Match **List-I** [Function in time domain $f(t)$] with **List -II** (Property) and select the correct answer using the code given below the lists:

List-I

A. $\sin \omega_0 t u(t-t_0)$
B. $\sin \omega_0 (t-t_0)\, u(t-t_0)$
C. $\sin \omega_0 (t-t_0)\, u(t)$
D. $\sin \omega_0 t\, u(t)$

List-II

1. $\dfrac{\omega_0}{s^2+\omega_0^2}$
2. $\left\{\dfrac{\omega_0}{s^2+\omega_0^2}\right\} e^{-t_0 s}$
3. $\dfrac{e^{-t_0 s}}{\sqrt{s^2+\omega_0^2}} \sin\left(\omega_0 t_0 + \tan^{-1}\dfrac{\omega_0}{s}\right)$
4. $-\dfrac{1}{\sqrt{s^2+\omega_0^2}} \sin\left(\omega_0 t_0 - \tan^{-1}\dfrac{\omega_0}{s}\right)$

Codes:

	A	B	C	D
(a)	3	1	4	2
(b)	4	2	3	1
(c)	3	2	4	1
(d)	4	1	3	2

[ESE-2008]

5.28 A voltage having the Laplace transform $\dfrac{4s^2+3s+2}{7s^2+6s+5}$ is applied across a 2H inductor having zero initial current. What is the current in the inductor at $t = \infty$?

(a) Zero (b) (1/5) A
(c) (2/7) A (d) (2/5) A

[ESE-2008]

5.29 What is the Laplace transform of $\cos\omega_0 t$?

(a) $\dfrac{\omega_0}{s^2+\omega_0^2}$ (b) $\dfrac{s}{s^2+\omega_0^2}$

(c) $\dfrac{\omega_0}{(s+\omega_0)^2}$ (d) $\dfrac{s}{(s+\omega_0)^2}$

[ESE-2008]

5.30 Laplace transform of $\sin(\omega t + \alpha)$ is

(a) $\dfrac{s\cos\alpha + \omega \sin\alpha}{s^2+\omega^2}$ (b) $\dfrac{\omega}{s^2+\omega^2}\cos\alpha$

(c) $\dfrac{s}{s^2+\omega^2}\sin\alpha$ (d) $\dfrac{s\sin\alpha + \omega\cos\alpha}{s^2+\omega^2}$

[ESE-2008]

5.31 Consider the following statements regarding the use of Laplace transforms and Fourier transforms in circuit analysis:

1. Both make the solution of circuit problems simple and easy.
2. Both are applicable for the study of circuit behaviour for $t - \alpha$ to α.
3. Both convert differential equations to algebraic equations.
4. Both can be used for transient and steady state analysis.

Which of the above statements are correct?
(a) 1, 2, 3 and 4 (b) 2, 3 and 4 only
(c) 1, 2 and 4 only (d) 1, 3 and 4 only
[ESE-2010]

5.32 The output of a linear system for step input is, $t^2 e^{-t}$. Then the transfer function is

(a) $\frac{s}{(s+1)^2}$ (b) $\frac{2s}{(s+1)^3}$

(c) $\frac{s}{s^2(s+1)}$ (d) $\frac{1}{(s+1)^3}$

[ESE-2010]

5.33 Consider the following statements:
1. The Laplace transform of the unit impulse function is s × Laplace transform of the unit ramp function.
2. The impulse function is a time derivative of the ramp function.
3. The Laplace transform of the unit impulse function is s × Laplace transform of the unit step function.
4. The impulse function is a time derivative of the unit step function.

Which of these statements is/are correct?
(a) 1 and 2 only (b) 3 and 4 only
(c) 2 and 3 only (d) 1, 2, 3 and 4
[ESE-2010]

5.34 If $x(t)$ is of finite duration and is absolutely integrable, then the "region of convergence" is
(a) Entire s plane
(b) From $\sigma = -1$ to $\sigma = +\infty$
(c) From $\sigma = -1$ to $\sigma = +\infty$
(d) Entire right half plane [ESE-2010]

5.35 Laplace transform of the function $f(t)$ shown in the figure is

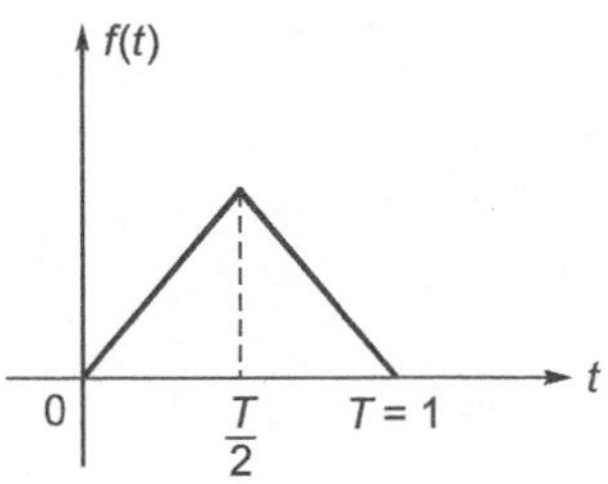

(a) $\frac{2}{s^2}\left[1-e^{-0.5s}\right]^2$

(b) $\frac{2}{s^2}\left[1+e^{-0.5s}\right]^2$

(c) $\frac{2}{s^2}\left[1-e^{0.5s}\right]^2$

(d) $\frac{2}{s^2}\left[1+e^{0.5s}\right]^2$

[ESE-2011]

5.36 A linear time-invariant system has an impulse response of e^{2t}, $t > 0$. If the initial conditions are zero and the input is e^{3t}, the output for $t > 0$ is
(a) $e^{3t} - e^{2t}$ (b) e^{5t}
(c) $e^{3t} + e^{2t}$ (d) e^{t} [ESE-2012]

5.37 A system described by the following differential equation is initially at rest and then excited by the input $x(t) = 3u(t)$:

$$\frac{d^2y}{dt^2} + 4\frac{dy}{dt} + 3y = x(t)$$

The output $y(t)$ is
(a) $1 - 1.5\,e^{-t} + 0.5e^{-3t}$
(b) $1 - 0.5\,e^{-t} + 1.5\,e^{-3t}$
(c) $1 + 1.5\,e^{-t} - 0.5\,e^{-3t}$
(d) $1 + 0.5\,e^{-t} - 0.5\,e^{-3t}$
[ESE-2012]

5.38 If $F(s)$ and $G(s)$ are the Laplace transform of $f(t)$ and $g(t)$, then their product $F(s)\,G(s) = H(s)$, where $H(s)$ is the Laplace transform of $h(t)$, is defined as
(a) $(f \cdot g)(t)$
(b) $\int_0^t f(\tau)g(t-\tau)d\tau$
(c) Both (a) and (b) are correct
(d) $f(t) \cdot g(t)$ [ESE-2012]

5.39 Consider a system with transfer function

$$H(s) = \frac{3s^2 - 2}{s^2 + 3s + 2}$$

The step response of the system is given by
(a) $c(t) = 5e^{-2t} - e^{-t} - 1$
(b) $c(t) = 3\delta(t) - 10e^{-2t} + e^{-t}$
(c) $c(t) = 4e^{-t} - e^{-2t} - 1$
(d) $c(t) = 2(1 - e^{-2t})$
[ESE-2012]

5.40 The function which has its Fourier transform, Laplace transform and z-transform unity is
(a) Gaussian (b) impulse
(c) sinc (d) pulse
[ESE-2012]

5.41 The unit impulse response of system is $-4e^{-t} + 6e^{-2t}$. The step response of the same system for $t \geq 0$ is $Ae^{-t} + Be^{-2t} + C$, where A, B and C are respectively
(a) –4, –3 and +1 (b) +4, –3 and –1
(c) –4, –3 and –1 (d) +4, –3 and +1
[ESE-2014]

5.42 The network function, $H(s)$ is equal to

(a) $\frac{y(s)}{x(s)}$ (b) $\frac{x(s)}{y(s)}$

(c) $x(s)\, y(s)$ (d) $\frac{1}{x(s)\, y(s)}$

[ESE-2014]

5.43 Laplace transform of the function $v(t)$ shown in the figure is

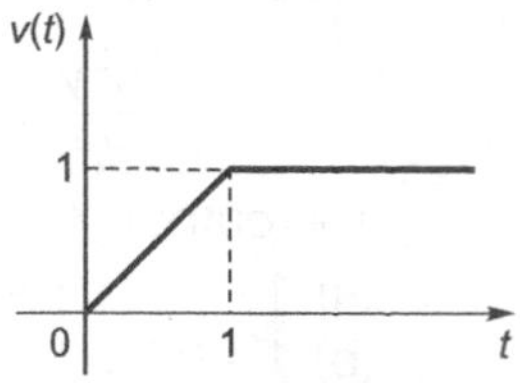

(a) $s^2[1 - e^s]$
(b) $s^2[1 - e^{-s}]$
(c) $\frac{1}{s^2}[1 - e^s]$
(d) $\frac{1}{s^2}[1 - e^{-s}]$

[ESE-2016]

5.44 A function $c(t)$ satisfies the differential equation $\dot{c}(t) + c(t) = \delta(t)$. For zero initial condition $c(t)$ can be represented by

(a) $\in^{-t}$ (b) $\in^{t}$
(c) $\in^{t} u(t)$ (d) $\in^{-t} u(t)$

where $u(t)$ is a unit step function

[ESE-2016]

5.45 If $x(t)$ is as shown in the figure, its Laplace transform is

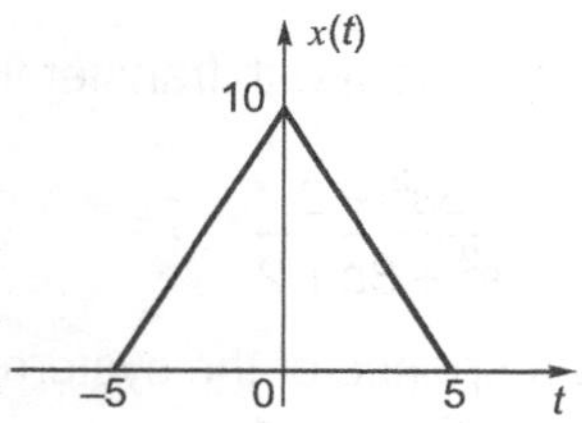

(a) $\frac{2e^{+5s} + 2e^{-5s}}{s^2}$ (b) $\frac{2e^{+5s} - 4 + 2e^{-5s}}{s^2}$

(c) $\frac{2e^{+5s} - 2 + 2e^{-5s}}{s^2}$ (d) $\frac{2e^{+5s} + 4 - 2e^{-5s}}{s^2}$

[ESE-2018]

5.46 The Laplace transform of a function $f(t)$ is

$$F(s) = \frac{s+2}{(s+2)^2 + 10^2}$$

The value of $f(0)$ will be

(a) –1 (b) 0
(c) 1 (d) 2

[ESE-2020]

5.47 A function, in Laplace domain is given by

$$F(s) = \frac{2}{s} - \frac{1}{s+3}$$

Its value by final value theorem in t domain will be

(a) $\lim_{t\to\infty} f(t) = 3$ (b) $\lim_{t\to\infty} f(t) = 2$

(c) $\lim_{t\to\infty} f(t) = 1$ (d) $\lim_{t\to\infty} f(t) = 4$

[ESE-2020]

5.48 The Laplace transform of $f(t) = 1 - e^{-2t}$ is

(a) $\frac{2}{s(s+2)}$ (b) $\frac{1}{s(s+2)}$

(c) $\frac{2}{s(s-2)}$ (d) $\frac{1}{s(s-2)}$

[ESE-2020]

5.49 What is the Laplace transform of the periodic waveform shown in the figure, where a = 1, 2a = 2, 3a = 3 and 4a = 4?

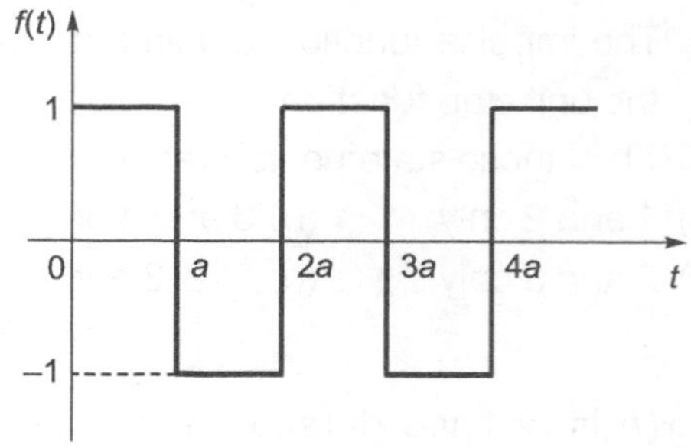

(a) $\frac{1}{s}\tanh\left(\frac{s}{2}\right)$ (b) $\frac{1}{2s}\tanh\left(\frac{s}{2}\right)$

(c) $\frac{1}{s}\tanh\left(\frac{1}{2}\right)$ (d) $\frac{1}{s}\tanh\left(\frac{3}{2}\right)$

[ESE-2023]

■■■

Answers Laplace Transform

5.1 (d)	5.2 (c)	5.3 (a)	5.4 (*)	5.5 (d)	5.6 (d)	5.7 (b)	5.8 (c)	5.9 (d)
5.10 (d)	5.11 (d)	5.12 (b)	5.13 (b)	5.14 (a)	5.15 (d)	5.16 (b)	5.17 (a)	5.18 (b)
5.19 (b)	5.20 (a)	5.21 (b)	5.22 (b)	5.23 (c)	5.24 (c)	5.25 (d)	5.26 (d)	5.27 (c)
5.28 (b)	5.29 (b)	5.30 (d)	5.31 (*)	5.32 (b)	5.33 (b)	5.34 (a)	5.35 (a)	5.36 (a)
5.37 (a)	5.38 (b)	5.39 (a)	5.40 (b)	5.41 (b)	5.42 (a)	5.43 (d)	5.44 (d)	5.45 (b)
5.46 (c)	5.47 (b)	5.48 (a)	5.49 (a)					

Explanations Laplace Transform

5.1 (d)

$$x(t) = u(t) \text{ and } y(t) = t^2 \cdot e^{-2t}$$

$$X(s) = \frac{1}{s}$$

$$Y(s) = \frac{2}{(s+2)^3}$$

$$H(s) = \frac{Y(s)}{X(s)} = \frac{2s}{(s+2)^3}$$

5.2 (c)

Inverse Laplace transforms of the functions are as follows:

(i) $\dfrac{e^{-s}}{s+1} \xleftrightarrow{(LT)^{-1}} e^{-(t-1)}\, u(t-1)$

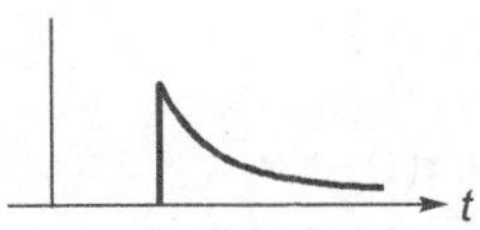

(ii) $\dfrac{1}{s^2+s+1} \xleftrightarrow{(LT)^{-1}} \dfrac{2}{\sqrt{3}} e^{-t/2} \sin\left(\dfrac{\sqrt{3}}{2} t\right)$

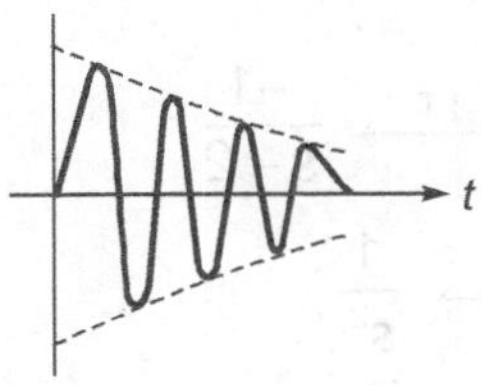

(iii) $\dfrac{1}{(s+1)^2} \xleftrightarrow{(LT)^{-1}} te^{-t}\, u(t)$

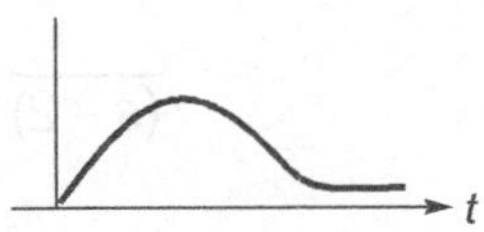

(iv) $\dfrac{1}{s^2+s} \xleftrightarrow{(LT)^{-1}} (1-e^{-t})\, u(t)$

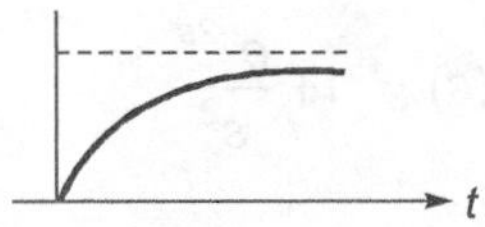

5.3 (a)

Input, $x(t) = e^{s_0 t}$

$\Rightarrow \quad X(s) = \dfrac{1}{s-s_0}$

Output, $Y(s) = X(s)\, H(s) = \dfrac{H(s)}{s-s_0}$

On partial fraction expansion,

$$Y(s) = \frac{H(s_0)}{s-s_0}$$

So, $y(t) = H(s_0)\, e^{s_0 t}$

5.4 (*)

$$F(s) = \frac{2s+5}{s^2+5s+6} = \frac{1}{s+2} + \frac{1}{s+3}$$

$$f(t) = e^{-2t} + e^{-3t}$$

$$= e^{-2.5t}[e^{0.5t} + e^{-0.5t}]$$

$$= 2\, e^{-2.5\,t} \cos h\, 0.5\, t$$

$$\left[\because \cosh t = \frac{e^t + e^{-t}}{2}\right]$$

Hence, no option is correct.

5.5 (d)

When transfer function has at least one pole or zero in the RHS of s-plane, it is called non-minimum phase transfer function.

Transfer function $\dfrac{s-1}{s^2+2s+1}$ has one zero (i.e., $s = 1$) at RHS of s-plane, hence it is a non-minimum phase system.

5.6 (d)

Fourier transform is special case of Laplace

$FT\{x(t)\} = X(s)|_{s=j\omega}$

Laplace transform is not a special case of Fourier transform.

5.7 (b)

$$\frac{d^2y}{dt^2} = x(t-2)\,u(t-2) + \frac{d^2x}{dt^2}$$

Taking Laplace transform of both sides,

$$s^2Y(s) = e^{-2s}X(s) + s^2X(s)$$

$$\Rightarrow \quad \frac{Y(s)}{X(s)} = \frac{e^{-2s}+s^2}{s^2}$$

$$\Rightarrow \quad H(s) = 1+\frac{e^{-2s}}{s^2}$$

5.8 (c)

$$f(0^+) = \lim_{s\to\infty} sF(s)$$

$$= \lim_{s\to\infty} s\left(\frac{27s+97}{s^2+33s}\right) = \lim_{s\to\infty}\frac{27+\frac{97}{s}}{1+\frac{33}{s}} = 27$$

5.9 (d)

$$f(t) = u(t-1) + u(t-2) - 2u(t-3)$$

$$F(s) = \frac{1}{s}[e^{-s} + e^{-2s} - 2e^{-3s}]$$

5.10 (d)

$$\delta(t) \xleftrightarrow{LT} 1$$

$$\sin\omega_0 t\, u(t) \xleftrightarrow{LT} \frac{\omega_0}{s^2+\omega_0^2}$$

$$(e^{-at}\cos bt)\,u(t) \xleftrightarrow{LT} \frac{s+a}{(s+a)^2+b^2}$$

$$u(t) \xleftrightarrow{LT} \frac{1}{s}$$

$$e^{at}\,u(t) \xleftrightarrow{LT} \frac{1}{s-a}$$

5.12 (b)

These are the properties of Laplace transform.

5.13 (b)

$$u(t-a) \xleftrightarrow{LT} \frac{e^{-as}}{s}$$

5.14 (a)

$$x_1(t) * x_2(t) = X_1(s)\cdot X_2(s)$$

$$x_1(t) = e^{k_1t}u(t)$$

$$\Rightarrow \quad X_1(s) = \frac{1}{s-k_1}$$

$$x_2(t) = e^{-k_2t}u(t)$$

$$\Rightarrow \quad X_2(s) = \frac{1}{s+k_2}$$

$$X_1(s)X_2(s) = \frac{1}{(s-k_1)}\cdot\frac{1}{(s+k_2)}$$

$$= \frac{1}{(k_1+k_2)}\left[\frac{1}{s-k_1}-\frac{1}{s+k_2}\right]$$

$$\Rightarrow x_1(t) * x_2(t) = \frac{e^{k_1t}-e^{-k_2t}}{k_1+k_2}$$

5.15 (d)

$$f(t) * g(t) \xleftrightarrow{LT} F(s)\,G(s)$$

5.16 (b)

The transfer function of a system describes only zero-state response for the system.

5.17 (a)

$$f(t) = u(t) + u(t-1) - 2u(t-2)$$

$$F(s) = \frac{1}{s}+\frac{e^{-s}}{s}-\frac{2e^{-2s}}{s}$$

5.18 (b)

For $t\to\infty$, $x(t)\to\infty$

Therefore F.T. cannot exist.

Whereas L.T. exist for some ROC.

5.19 (b)

$$-e^{2t}\,u(t) \xleftrightarrow{LT} \frac{-1}{s-2}$$

$$t\,u(t) \xleftrightarrow{LT} \frac{1}{s^2}$$

$$\because x_1(t) * x_2(t) \xleftrightarrow{LT} X_1(s)\cdot X_2(s)$$

$$\therefore x(t) = -e^{2t}\,u(t) * t\,u(t) \xleftrightarrow{LT}$$

$$\frac{-1}{(s-2)}\cdot\frac{1}{s^2} = \frac{-1}{s^2(s-2)}$$

5.20 (a)

$$x(t) = r(t) - r(t-1) - r(t-4) + 1.5r(t-6) - 0.5r(t-8)$$

$$X(s) = \frac{1}{s^2}(1 - e^{-s} - e^{-4s} + 1.5e^{-6s} - 0.5e^{-8s})$$

So, $D = -0.5$

5.21 (b)

$$F(s) = \frac{8s+10}{(s+1)(s+2)^3}$$

$$= \frac{A}{s+1} + \frac{B}{(s+2)^3} + \frac{C}{(s+2)^2} + \frac{D}{s+2}$$

$$= A(s+2)^3 + B(s+1) + C(s+1)(s+2) + D(s+1)(s+2)^2$$

At $s = -1$, $A = 10 - 8 = 2$

At $s = -2$, $B(-1) = -16 + 10 \Rightarrow B = 6$

At $s = 0$, $8A + B + 2C + 4D = 10$

$\Rightarrow 16 + 6 + 2C + 4D = 10$

$\Rightarrow \quad 2C + 4D = -12$

$\Rightarrow \quad C + 2D = -6$ ... (i)

At $s = 1$, $27A + 2B + 6C + 18D = 18$

$\Rightarrow 54 + 12 + 6C + 18D = 18$

$\Rightarrow 6C + 18D = -48$

$\Rightarrow \quad C + 3D = -8$ ... (ii)

Solving (i) & (ii),

$C = -2$

$D = -2$

Putting values

$$F(s) = \frac{2}{s+1} + \frac{6}{(s+2)^3} - \frac{2}{(s+2)^2} - \frac{2}{s+2}$$

5.22 (b)

$$x(0^+) = \lim_{s\to\infty} sX(s)$$

$$= \lim_{s\to\infty} se^{-s}\left(\frac{-2}{s(s+2)}\right)$$

$$= 0$$

$$x(\infty) = \lim_{s\to 0} sX(s)$$

$$= \lim_{s\to 0} se^{-s}\left(\frac{-2}{s(s+2)}\right)$$

$$= e^0\left[\frac{-2}{2}\right]$$

$$= -1$$

5.23 (c)

$$X(s) = \frac{1}{s} - \frac{e^{-sT}}{s}$$

$$\Rightarrow \quad x(t) = u(t) - u(t-T)$$

The shape of $x(t)$ is

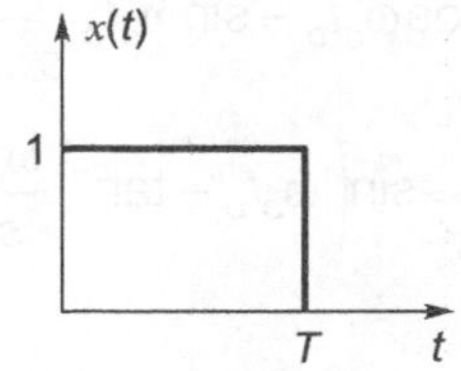

5.24 (c)

$$x(t) = \delta(t-1)$$

$$X(s) = e^{-s}\cdot 1$$

$$X(s) = \exp(-s)$$

5.25 (d)

$$G(s) = \frac{Y(s)}{X(s)} = \frac{2}{s^2 - s - 2}$$

Roots of $Y(s)$ $\quad s = 0, -1, 2$

Since, one root lies in RHP therefore system is unstable and thus final value theorem cannot be applied to it.

5.27 (c)

$$\sin\omega_o t\, u(t) \to \frac{\omega_o}{s^2 + \omega_o^2}$$

$$\sin\omega_o(t - t_o)u(t - t_o) \to \left(\frac{\omega_o}{s^2+\omega_o^2}\right)e^{-t_o s}$$

$$\sin\omega_o t\, u(t - t_o) = \sin\omega_o(t - t_o + t_o)u(t - t_o)$$

$$= \cos\omega_o t_o e^{-t_o s}\frac{\omega_o}{s^2+\omega_o^2} + \sin\omega_o e^{-t_o s}\frac{s}{s^2+\omega^2}$$

$$= \cos\omega_o t_o e^{-t_o s}\frac{\omega_o}{s^2+\omega_o^2} + \sin\omega_o t_o e^{-t_o s}\frac{s}{s^2+\omega^2}$$

$$= \frac{1}{s^2\omega_o^2}e^{-t_o s}\{s\cdot\sin\omega_o t_o + \omega_o\cos\omega_o t_o\}$$

Let $\quad s = \cos\theta$

$\omega_o = \sin\theta$

$$\theta = \tan^{-1}\left(\frac{\omega_o}{s}\right)$$

$$\sin\omega_o t\, u(t - t_o) \to \frac{e^{-t_o s}}{\sqrt{s^2+\omega_o^2}}\sin\left[\omega_o t_o + \tan^{-1}\left(\frac{\omega_o}{s}\right)\right]$$

Similarly we can find $\sin\omega_o(t-t_o)u(t)$

$= \sin\omega_o t\cos\omega_o t_o u(t) - \cos\omega_o t\sin\omega_o t_o u(t)$

$$= \frac{\omega_o}{\omega_o^2+s^2}\cos\omega_o t_o - \sin\omega_o t_o\frac{s}{s^2+\omega_o^2}$$

$$= -\frac{1}{\sqrt{s^2+\omega_o^2}}\sin\left(\omega_o t_o - \tan^{-1}\frac{\omega_o}{s}\right)$$

5.28 (b)

$$i(\infty) = \lim_{s\to 0} sI(s)$$

$\because$ $V(s) = LsI(s)$

$$\therefore \quad i(\infty) = \lim_{s\to 0}\frac{V(s)}{L}$$

$$= \lim_{s\to 0}\frac{1}{2}\left[\frac{4s^2+3s+2}{7s^2+6s+5}\right]$$

$= (1/5)$ A

5.30 (d)

$$\sin(\omega t+\alpha) = \sin\omega t\cos\alpha + \cos\omega t\sin\alpha$$

$$L\{\sin(\omega t+\alpha)\} = \frac{\omega\cos\alpha}{s^2+\omega^2} + \frac{s\sin\alpha}{s^2+\omega^2}$$

$$= \frac{s\sin\alpha + \omega\cos\alpha}{s^2+\omega^2}$$

5.31 (*)

Laplace transform and Fourier transform are applicable for the study of circuit behaviour for $t = -\infty$ to $t = \infty$.

5.32 (b)

Given, $r(t) = u(t)$

$c(t) = t^2 e^{-t}$

Taking Laplace transform of both

$$R(s) = \frac{1}{s}$$

$$C(s) = \frac{2}{(s+1)^3}$$

$$\left[\text{Using } \frac{t^{n-1}}{(n-1)!}e^{-at}u(t) \xleftrightarrow{LT} \frac{1}{(s+a)^n}\right]$$

$\therefore$ Transfer function of the system is

$$H(s) = \frac{Y(s)}{X(s)} = \frac{C(s)}{R(s)} = \frac{\frac{2}{(s+1)^3}}{\frac{1}{s}}$$

$$\Rightarrow \quad H(s) = \frac{2s}{(s+1)^3}$$

5.33 (b)

$$\delta(t) = \frac{d}{dt}u(t)$$

$$\therefore \quad LT(\delta(t)) = sLT(u(t))$$

5.35 (a)

As convolution of a rectangular function with itself is a triangular function i.e.

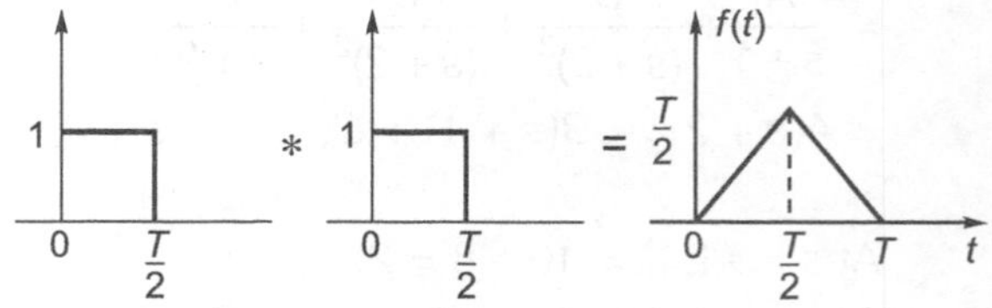

So

$$f(t) = \frac{2}{T}\left\{\left[u(t)-u\left(t-\frac{T}{2}\right)\right] * \left[u(t)-u\left(t-\frac{T}{2}\right)\right]\right\}$$

Now,

$$u(t)-u\left(t-\frac{T}{2}\right) \xleftrightarrow{LT} \frac{1}{s} - \frac{e^{-sT/2}}{s}$$

$$= \frac{1-e^{-sT/2}}{s} = \frac{1-e^{-0.5s}}{s}$$

As we know that convolution in time domain results in multiplication in s-domain; so

$$F(s) = 2\left[\frac{1-e^{-0.5s}}{s}\right]^2$$

$$\Rightarrow \quad F(s) = \frac{2}{s^2}\left[1-e^{-0.5s}\right]^2$$

5.36 (a)

Impulse response, $h(t) = e^{2t}$

so, $$H(s) = \frac{1}{(s-2)}$$

Input, $x(t) = e^{3t}$

$$\text{LT}[(x(t)] = X(s) = \frac{1}{(s-3)}$$

Output $y(t) = x(t) * h(t)$

Taking Laplace transform

$$Y(s) = X(s)H(s)$$

$$= \frac{1}{(s-2)}\cdot\frac{1}{(s-3)}$$

$$= \frac{1}{(s-3)} - \frac{1}{(s-2)}$$

Taking inverse Laplace transform.

$$y(t) = e^{3t} - e^{2t}$$

Note: e^{3t} is a real exponential.

5.37 (a)

$$\frac{d^2y(t)}{dt^2} + 4\frac{dy}{dt} + 3y = x(t)$$

Taking Laplace transform of both sides having initially relaxed conditions,

$\therefore (s^2 + 4s + 3)\, Y(s) = X(s)$

$$Y(s) = \frac{X(s)}{(s^2+4s+3)} \qquad \left[\text{here } X(s) = \frac{3}{s}\right]$$

$$= \frac{3}{s(s+1)(s+3)}$$

$$= \frac{1}{s} - \frac{3/2}{(s+1)} + \frac{+1/2}{(s+3)}$$

$$\Rightarrow \quad y(t) = 1 - 1.5e^{-t} + 0.5e^{-3t}$$

5.38 (b)

This is a convolution property in the time domain

$$y(t) = \int_0^t f(\tau)g(t-\tau)d\tau$$

Note: As per RTI filed and reply of **UPSC** answer is option (c).

5.39 (a)

$$H(s) = \frac{3s^2 - 2}{s^2 + 3s + 2} = \frac{C(s)}{R(s)}$$

Since the input signal is step so $R(s) = \frac{1}{s}$

so, $$C(s) = \frac{(3s^2-2)}{s(s+1)(s+2)}$$

$$= \frac{-1}{s} - \frac{1}{(s+1)} + \frac{5}{(s+2)}$$

Taking its inverse Laplace transform

$$c(t) = -1 - e^{-t} + 5e^{-2t}$$

5.40 (b)

$$\delta(t) \xleftrightarrow{\text{F.T.}} 1$$

$$\delta(t) \xleftrightarrow{\text{L.T.}} 1$$

$$\delta(t) \xleftrightarrow{\text{Z.T.}} 1$$

5.41 (b)

Impulse response,

$$y(t) = -4e^{-t} + 6\, e^{-2t} \qquad \text{...(i)}$$

using Laplace transform in equation (i)

$$y(s) = \frac{-4}{(s+1)} + \frac{6}{(s+2)} \qquad \text{...(ii)}$$

$\therefore$ Step response for the above system is

$$Y(s) = \frac{-4(s+2)+6(s+1)}{(s+1)(s+2)} \times U(s)$$

$$\Rightarrow \quad Y(s) = \frac{-4s-8+6s+6}{(s+1)(s+2)} \times \frac{1}{s} \qquad \text{...(iii)}$$

using partial fraction,

$$Y(s) = \frac{4-8-6+6}{(s+1)(-1)} + \frac{8-8-12+6}{(s+2)(-1)(-2)} + \frac{-2}{s \times 2}$$

$$= \frac{4}{(s+1)} + \frac{(-3)}{(s+2)} + \frac{(-1)}{s} \qquad \text{...(iv)}$$

using inverse Laplace transform

$$y(t) = (4e^{-t} - 3e^{-2t} - 1)\, u(t)$$

$\therefore A = 4, B = -3$ and $C = -1$

Alternate method:

$$\text{Unit step response} = \int \text{Impulse response} \qquad \text{...(i)}$$

$$\therefore \text{Unit step response} = \int \left(-4e^{-t} + 6e^{-2t}\right) dt$$

or $$y(t) = 4e^{-t} - 3e^{-2t} + C \qquad \text{...(ii)}$$

for $$t = 0, \quad 0 = 4 - 3 + C \qquad \text{...(iii)}$$

from equation (ii) and (iii),

$$C = -1, \quad A = 4, \quad B = -3$$

5.43 (d)

From the figure,

$$v(t) = r(t) - r(t-1)$$

Apply LT, $$V(s) = \frac{1}{s^2} - \frac{1}{s^2}e^{-s} \qquad \left[\because r(t) \longleftrightarrow \frac{1}{s^2}\right]$$

$$= \frac{1}{s^2}[1 - e^{-s}]$$

$\because$ Time shifting property

$$\delta(t - t_o) \longleftrightarrow X(s)\, e^{-st_o}$$

5.44 (d)

$$\frac{dc}{dt} + c(t) = \delta(t)$$

$$sC(s) + C(s) = 1$$

$$C(s) = \frac{1}{s+1}$$

$$c(t) = \in^{-t} u(t)$$

5.45 (b)

$$x(t) = 2r(t+5) - 4r(t) + 2r(t-5)$$

$$X(s) = \frac{2e^{+5s} - 4 + 2e^{-5s}}{s^2}$$

5.46 (c)

$$F(s) = \frac{s+2}{(s+2)^2+10^2} = \frac{s+2}{s^2+4s+104}$$

$$\lim_{S\to\infty} SF(s) = \lim_{S\to\infty} \frac{s(s+2)}{s^2+4s+104}$$

$$= \lim_{S\to\infty} \frac{s^2\left(1+\frac{2}{s}\right)}{s^2\left(1+\frac{1}{s}+\frac{104}{s^2}\right)} = 1$$

5.47 (b)

$$F(s) = \frac{2s+6-s}{s(s+3)} = \frac{s+6}{s(s+3)}$$

$$\lim_{s\to 0} sF(s) = \frac{s(s+6)}{s(s+3)} = \frac{6}{3} = 2$$

5.48 (a)

$$F(s) = \frac{1}{s} - \frac{1}{s+2} = \frac{s+2-s}{s(s+2)} = \frac{2}{s(s+2)}$$

5.49 (a)

The given waveform can be transformed into,

$$F(s) = \frac{X(s)}{1-e^{-sT}} \quad \text{where, } T = 2 \quad \text{...(i)}$$

and $x(t)$

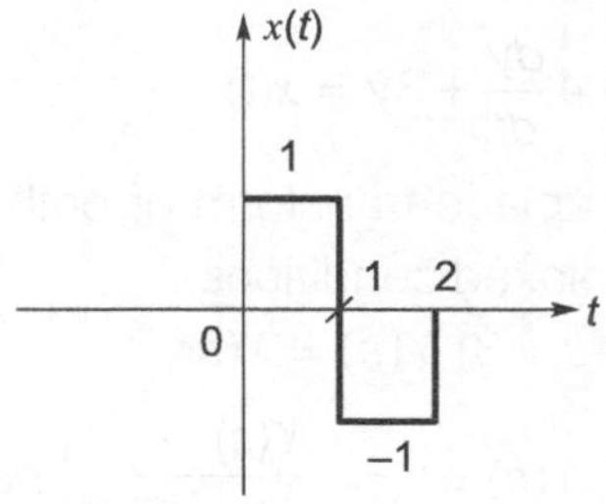

From the figure,

$$x(t) = u(t) - 2u(t-1) + u(t-2)$$

$$\Big\Downarrow \text{L.T.}$$

$$X(s) = \frac{1}{s}(1-2e^{-s}+e^{-2s}) = \frac{(1-e^{-s})^2}{s}$$

From equation (i),

$$F(s) = \frac{(1-e^{-s})^2}{s(1-e^{-2s})} = \frac{1-e^{-s}}{s(1+e^{-s})}$$

$$= \frac{e^{s/2}-e^{-s/2}}{s(e^{s/2}+e^{-s/2})} = \frac{1}{s}\frac{\sinh\left(\frac{s}{2}\right)}{\cosh\left(\frac{s}{2}\right)}$$

$$= \frac{1}{s}\tanh\left(\frac{s}{2}\right)$$

■■■

6 Sampling

6.1 What is the Nyquist rate for the signal
$x(t) = \cos 2000\pi t + 3 \sin 6000\pi t$?
(a) 2 kHz (b) 4 kHz
(c) 12 kHz (d) 6 kHz

[ESE-2004]

6.2 A signal represented by $x(t) = 5 \cos 400\ \pi t$ is sampled at a rate 300 samples/s. The resulting samples are passed through an ideal low pass filter of cut-off frequency 150 Hz. Which of the following will be contained in the output of the LPF?
(a) 100 Hz
(b) 100 Hz, 150 Hz
(c) 50 Hz, 100 Hz
(d) 50 Hz, 100 Hz, 150 Hz

[ESE-2005]

6.3 Match **List-I** (Application of Signals) with **List-II** (Definition) and select the correct answer using the code given below the lists:

List-I

A. Reconstruction
B. Over sampling
C. Interpolation
D. Decimation

List-II

1. Sampling rate is chosen significantly greater than the Nyquist rate
2. A mixture of continuous and discrete time signals
3. To convert the discrete time sequence back to a continuous time signal and then resample
4. Assign values between samples and signals

Codes:

	A	B	C	D
(a)	3	4	1	2
(b)	2	1	4	3
(c)	3	1	4	2
(d)	2	4	1	3

[ESE-2006]

6.4 **Statement (I):** Aliasing occurs when the sampling frequency is less than twice the maximum frequency in the signal.
Statement (II): Aliasing is a reversible process.
(a) Both Statement (I) and Statement (II) are individually true and Statement (II) is the correct explanation of Statement (I).
(b) Both Statement (I) and Statement (II) are individually true but Statement (II) is not the correct explanation of Statement (I).
(c) Statement (I) is true but Statement (II) is false.
(d) Statement (I) is false but Statement (II) is true.

[ESE-2013]

6.5 **Statement (I):** Sampling in one domain makes the signal to be periodic in the other domain.
Statement (II): Multiplication in one domain is the convolution in the other domain.
(a) Both Statement (I) and Statement (II) are individually true and Statement (II) is the correct explanation of Statement (I).
(b) Both Statement (I) and Statement (II) are individually true but Statement (II) is not the correct explanation of Statement (I).
(c) Statement (I) is true but Statement (II) is false.
(d) Statement (I) is false but Statement (II) is true.

[ESE-2013]

6.6 A half duty cycle rectangular clock output is sampled at 5 times its rate. Spectrum of the sampled clock will be having mirror image in negative frequency domain of a
(a) Sampled sinc pulse
(b) Periodic sinc pulse
(c) Periodic sampled sinc pulse
(d) None of the above [ESE-2015]

6.7 Consider the following statements:
1. The general formula for the minimum possible sampling rate for a bandpass signal can be recovered from the samples.

$$f_s = \frac{2f_h}{\text{Largest integer not exceeding } \frac{f_h}{f_h - f_l}}$$

2. If we sample at a rate higher than twice the highest frequency, the aliases do not overlap and the original signal can be recovered.
3. The absolute minimum sampling rate in most favourable situation is twice the highest frequency, not the bandwidth of the signal.

Which of the above statements are correct?

(a) 1 and 2 only (b) 1 and 3 only
(c) 2 and 3 only (d) 1, 2 and 3

[ESE-2024]

■■■

Answers Sampling

6.1 (d) 6.2 (a) 6.3 (c) 6.4 (c) 6.5 (b) 6.6 (c) 6.7 (d)

Explanations Sampling

6.1 (d)

$x(t) = \cos 2000\,\pi t + 3 \sin 6000\,\pi t$

Maximum frequency component

$$f_m = \frac{6000\pi}{2\pi} = 3000 \text{ Hz}$$

$$f_{\text{Nyquist}} = 2f_m = 2 \times 3000 = 6 \text{ kHz}$$

6.2 (a)

$$x(t) = 5\cos 400\pi t$$

$$f_m = \frac{400\pi}{2\pi} = 200 \text{ Hz}$$

The input to the LPF will contain frequency components $nf_s \pm f_m$, $n = 0, 1, 2, \ldots$

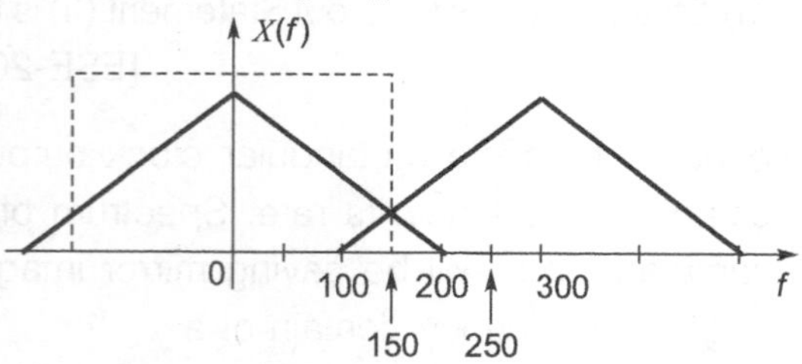

The output of the LPF will contain frequencies which are less than f_c i.e. 150 Hz.

So $f_s - f_m = 300 - 200 = 100$ Hz is the only component present in the output of LPF.

6.4 (c)

Aliasing is an **irreversible** process. Once aliasing has occurred then signal can not be recovered back.

6.5 (b)

⇒ Relationship between two domains:

Continuous ⟷ Aperiodic

Discrete ⟷ Periodic

⇒ Multiplication in one domain leads to convolution in other domain.

6.6 (c)

The sampled signal will be a discrete time periodic signal thus the Fourier transform will be a discrete frequency periodic signal and the rectangular function has sinc function as its transform hence option (c) is correct.

6.7 (d)

- For a base-band signal with highest frequency component f_m, the sampling frequency, $f_s \geq 2f_m$, to avoid aliasing effect and recover the signal.
- A bandpass signal $m(t)$ can be recovered from it's sampled signal, if,

$$f_s \geq \frac{2f_h}{k}; \text{ where } k = \left\lfloor \frac{f_h}{f_h - f_l} \right\rfloor$$

Here, f_h and f_l are the highest and the lowest frequency components in the bandpass signal respectively.

Hence, all the given statements are correct.

■■■

7 DTFS and DTFT

7.1 Match **List-I** (Fourier series and Fourier transforms) with **List-II** (Their properties) and select the correct answer using the codes given below the lists:

List-I	List-II
A. Fourier series	**1.** Discrete, periodic
B. Fourier transform	**2.** Continuous, periodic
C. Discrete time Fourier series	**3.** Discrete, aperiodic
D. Discrete Time Fourier transform	**4.** Continuous, aperiodic

Codes:

	A	B	C	D
(a)	3	4	1	2
(b)	1	2	4	3
(c)	3	2	4	1
(d)	1	4	2	3

[ESE-2002]

7.2 Match **List-I** (Time Domain Property) with **List-II** (Frequency Domain Property pertaining to Fourier Representation Periodicity Properties) and select the correct answer using the codes given below the lists:

List-I	List-II
A. Continuous	**1.** Periodic
B. Discrete	**2.** Continuous
C. Periodic	**3.** Non-periodic
D. Non-periodic	**4.** Discrete

Codes:

	A	B	C	D
(a)	3	4	1	2
(b)	2	4	1	3
(c)	2	1	4	3
(d)	3	1	4	2

[ESE-2004]

7.3 What is the output of the system with $h[n] = (1/2)^n u[n]$ in response to the input

$$x[n] = 3 + \cos\left(\pi n + \frac{\pi}{3}\right)$$

(a) $y[n] = 3 + \frac{1}{3}\left[\cos\left(\pi n + \frac{\pi}{3}\right)\right]$

(b) $y[n] = 3 + \frac{2}{3}\left[\cos\left(\pi n + \frac{\pi}{3}\right)\right]$

(c) $y[n] = 1 + \frac{2}{3}\left[\sin\left(\pi n + \frac{\pi}{3}\right)\right]$

(d) $y[n] = 6 + \frac{2}{3}\left[\cos\left(\pi n + \frac{\pi}{3}\right)\right]$ [ESE-2004]

7.4 The discrete time Fourier transform (DTFT) of $x[n] = 2(3)^n u[-n]$ is equal to

(a) $\frac{2}{1 - 1/3e^{j\Omega}}$ (b) $\frac{2}{1 + 1/3e^{j\Omega}}$

(c) $2\left(\frac{1 + 1/3e^{j\Omega}}{1 - 1/3e^{j\Omega}}\right)$ (d) $2\left(\frac{1 - 1/3e^{j\Omega}}{1 + 1/3e^{j\Omega}}\right)$

[ESE-2005]

7.5 A discrete-time signal $x[n]$ has Fourier transform $X(e^{j\omega})$. Match **List-I** (Signal) with **List-II** (Fourier Transform) and select the correct answer using the code given below the lists:

List-I	List-II
A. $x[-n]$	**1.** $X^*(e^{-j\omega})$
B. $nx[n]$	**2.** $X(e^{-j\omega})$
C. $x^*[n]$	**3.** $e^{-j\omega}X(e^{j\omega})$
D. $x[n-1]$	**4.** $j\frac{d}{d\omega}X(e^{j\omega})$

Codes:

	A	B	C	D
(a)	1	3	2	4
(b)	2	4	1	3
(c)	1	4	2	3
(d)	2	3	1	4

[ESE-2007]

7.6 **Assertion (A) :** There are no convergence issues with the discrete-time Fourier series in general.
Reason (R) : A discrete-time signal is always obtained by sampling a continuous-time signal.

(a) Both A and R are true and R is the correct explanation of A
(b) Both A and R are true but R is NOT the correct explanation of A
(c) A is true but R is false
(d) A is false but R is true [ESE-2009]

7.7 Decimation is the process of
(a) Retaining sequence values of $X_p[n]$ other than zeroes
(b) Retaining all sequence values of $X_p[n]$
(c) Dividing the sequence value by 10
(d) Multiplying the sequence value by 10
[ESE-2010]

7.8 Frequency scaling [relationship between discrete time frequency (Ω) and continuous time frequency (ω)] is defined as
(a) $\omega = 2\,\Omega$ (b) $\omega = 2\,T_s/\Omega$
(c) $\Omega = 2\,\omega/T_s$ (d) $\Omega = \omega\,T_s$
[ESE-2010]

7.9 The Fourier transform of unit step sequence is
(a) $\pi\delta(\Omega)$
(b) $\dfrac{1}{1-e^{-j\Omega}}$
(c) $\pi\delta(\Omega) + \dfrac{1}{1-e^{-j\Omega}}$
(d) $1 - e^{-j\Omega}$ [ESE-2010]

7.10 The frequency response $H(\Omega)$ of a system for impulse sequence response

$$h[n] = \delta[n] + \delta[n-1]$$

(a) $H(\Omega) = 2\cos\left(\dfrac{\Omega}{2}\right)\angle -\dfrac{\Omega}{2}$
(b) $H(\Omega) = 2\cos\Omega\,\angle -\Omega$
(c) $H(\Omega) = 2\cos\Omega\,\angle -\dfrac{\Omega}{2}$
(d) $H(\Omega) = 2\,\angle -\dfrac{\Omega}{2}$ [ESE-2010]

7.11 The impulse response of a discrete time system is given by

$$h(n) = \frac{1}{2}(\delta[n] + \delta[n-2])$$

The magnitude of the response can be expressed as
(a) $|\cos\Omega|$ (b) $\cos\Omega$
(c) $|\sin\Omega|$ (d) $\sin\Omega$ [ESE-2012]

7.12 $H(e^{j\omega})$ is the frequency response of a discrete time LTI system and $H_1(e^{j\omega})$ is the frequency response of its inverse function. Then
(a) $H(e^{j\omega})\,H_1(e^{j\omega}) = 1$
(b) $H(e^{j\omega})\,H_1(e^{j\omega}) = \delta(\omega)$
(c) $H(e^{j\omega}) * H_1(e^{j\omega}) = 1$
(d) $H(e^{j\omega}) * H_1(e^{j\omega}) = \delta(\omega)$ [ESE-2012]

7.13 For certain sequences which are neither absolutely summable nor square summable, it is possible to have a Fourier Transform (FT) representation if we
(a) take short time FT
(b) evaluate FT only the real part of the sequence
(c) allow DTFT to contain impulses
(d) evaluate FT over a limited time span
[ESE-2013]

7.14 Ideal low pass filter as a discrete time system is
(a) Causal, realizable
(b) Non causal, physically/computationally unrealizable
(c) Non causal, physically realizable
(d) None of the above [ESE-2015]

7.15 In a discrete-time Low Pass Filter, the frequency response is
(a) A periodic
(b) A periodic with response restricted to $(-\omega_c, +\omega_c)$
(c) Periodic with period 2π
(d) Quasi periodic with response extending to infinity
[ESE-2015]

7.16 Let $x_a(t)$ be an analog signal with bandwidth $B = 6$ kHz. We wish to use an $N = 2^m$ point DFT to compute the spectrum of the signal with resolution less than or equal to 200 Hz. What is the minimum length of the analog signal recorded?
(a) 60 seconds (b) 0.05 seconds
(c) 0.005 second (d) 6000 seconds
[ESE-2023]

Answers Discrete Time Fourier Transform

7.1 (a) 7.2 (d) 7.3 (d) 7.4 (a) 7.5 (b) 7.6 (d) 7.7 (a) 7.8 (d) 7.9 (c)
7.10 (a) 7.11 (a) 7.12 (a) 7.13 (c) 7.14 (b) 7.15 (c) 7.16 (c)

Explanations Discrete Time Fourier Transform

7.1 (a)

Summary of Fourier series and Fourier transforms:

	CONTINUOUS TIME		DISCRETE TIME	
	Time domain	Frequency domain	Time domain	Frequency domain
Fourier series	Continuous time periodic in time	Discrete frequency aperiodic in frequency	Discrete time periodic in time	Discrete frequency periodic in frequency
		duality	duality	duality
Fourier Transform	Continuous time aperiodic in time	Continuous frequency aperiodic in frequency	Discrete time aperiodic in time	Continuous frequency periodic in frequency

7.2 (d)

- Discrete in one domain, periodic in other domain.
- Continuous in one domain, aperiodic in other domain.

7.3 (d)

$$x(n) = 3+\cos\left(\pi n+\frac{\pi}{3}\right) \longrightarrow \boxed{H(e^{j\omega}) = \frac{e^{j\omega}}{e^{j\omega}-1/2}} \longrightarrow y[n]$$

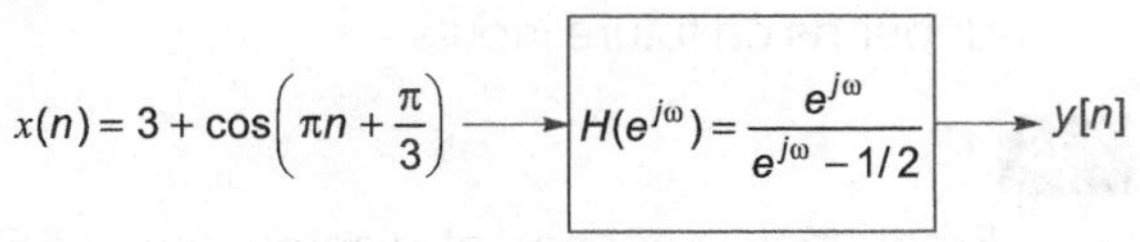

$$h[n] = \left(\frac{1}{2}\right)^n u(n)$$

$$x[n] = 3+\cos\left(\pi n+\frac{\pi}{3}\right) = x_1[n]+x_2[n]$$

From the principle of superposition,

$$y[n] = y_1[n]+y_2[n]$$

DTFT of h(n)

$$H[e^{j\omega}] = \frac{e^{j\omega}}{e^{j\omega}-\frac{1}{2}} \quad \text{...(i)}$$

Now $$y_1[n] = \left.\frac{e^{j\omega}}{e^{j\omega}-1/2}\right|_{\omega=0} \times 3$$

$$= \frac{1}{1/2}\times 3 = 6 \quad \text{...(ii)}$$

and $$y_2[n] = \left.\frac{e^{j\omega}}{e^{j\omega}-1/2}\right|_{\omega=\pi} \cdot \cos\left(\pi n+\frac{\pi}{3}\right)$$

$$= \frac{2}{3}\cos\left(\pi n+\frac{\pi}{3}\right) \quad \text{...(iii)}$$

From equation (ii) and (iii)

Now $$y(n) = y_1[n]+y_2[n]$$

$$= 6+\frac{2}{3}\cos\left(\pi n+\frac{\pi}{3}\right)$$

7.4 (a)

If $a^n u[n] \xrightarrow{\text{DTFT}} \dfrac{1}{1-ae^{-j\Omega}}$

then $a^{-n} u[-n] \xrightarrow{\text{DTFT}} \dfrac{1}{1-ae^{j\Omega}}$

$$x[n] = 2(3)^n u[-n]$$

or $$x[n] = 2\left(\frac{1}{3}\right)^{-n} u[-n]$$

So, $2\left(\dfrac{1}{3}\right)^{-n} u[-n] \xrightarrow{\text{DTFT}} \dfrac{2}{1-\frac{1}{3}e^{j\Omega}}$

7.5 (b)

Properties of Fourier transform:

$$x[-n] \xleftrightarrow{\text{F.T.}} X(e^{-j\omega})$$

$$nx[n] \xleftrightarrow{\text{F.T.}} j\frac{d}{d\omega}X(e^{j\omega})$$

$$x^*[n] \xleftrightarrow{\text{F.T.}} X^*(e^{-j\omega})$$

$$x[n-1] \xleftrightarrow{\text{F.T.}} e^{-j\omega}X(e^{j\omega})$$

7.6 (d)

Convergence is an important issue with the discrete-time Fourier series in general as discrete-time Fourier series exists only for convergent signals.

7.7 (a)

Decimation is the process of time scaling in discrete time sequence when scaling factor is greater than '1'. i.e. $\alpha > 1$.

7.8 (d)

Consider a continuous time signal $x_c(t)$

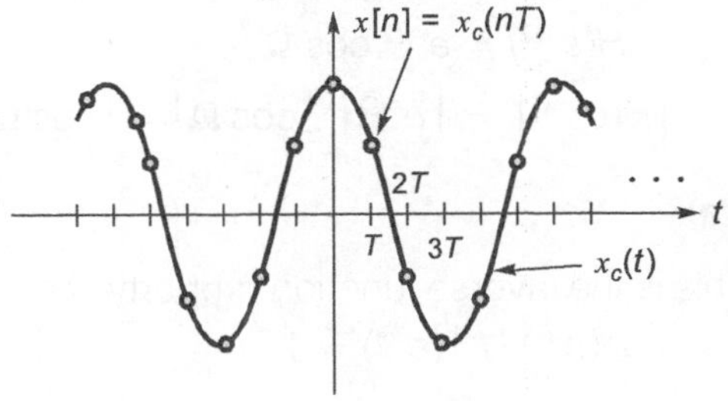

Mathematically, the relationship between the discrete-time signal and the continuous-time signal is given by

$$x[n] = x_c(nT)$$

The sampling frequency is

$$f_s = 1/T \text{ (in Hz)}$$

or $\Omega_s = 2\pi/T$ (in radians per second)

The discrete-time Fourier transform of $x[n]$ is related to the continuous-time Fourier transform of $x_c(t)$ as follows:

$$X[\omega] = \frac{1}{T}\sum_{k=-\infty}^{\infty} X_c\left(\frac{\omega}{T} + \frac{2\pi k}{T}\right)$$

Scaling relationship between ω and Ω: $\omega = \Omega T$.

Radian/sample = (Radian/sec) × (sec/sample)

7.9 (c)

$$u(n) \xleftrightarrow{F.T.} \frac{1}{1-e^{-j\Omega}} + \pi\delta(\Omega)$$

7.10 (a)

Given, $h[n] = \delta[n] + \delta[n-1]$

Taking Fourier transform

$$H(\Omega) = 1 + e^{-j\Omega}$$

$$\Rightarrow \quad H(\Omega) = 2e^{-j\Omega/2}\left(\frac{e^{j\Omega/2} + e^{-j\Omega/2}}{2}\right)$$

$$\Rightarrow \quad H(\Omega) = 2\cos\left(\frac{\Omega}{2}\right)e^{-\frac{j\Omega}{2}}$$

$$\Rightarrow \quad H(\Omega) = 2\cos\left(\frac{\Omega}{2}\right)\angle -\Omega/2$$

7.11 (a)

$$h(n) = \frac{1}{2}[\delta(n) + \delta(n-2)]$$

Taking z-transform

$$H(z) = \frac{1}{2}[1 + z^{-2}]$$

Put $z = e^{j\Omega}$

$$H(e^{j\Omega}) = \frac{1}{2}\left[1 + e^{-j2\Omega}\right] = \frac{1}{2}e^{-j\Omega}\left[e^{j\Omega} + e^{-j\Omega}\right]$$

$$H(e^{j\Omega}) = e^{-j\Omega}\cos\Omega$$

$$|H(e^{j\Omega})| = |e^{-j\Omega}|\,|\cos\Omega| = |\cos\Omega|$$

7.12 (a)

This is the inverse function property

$$H(e^{j\omega})\,H^{-1}(e^{j\omega}) = 1$$

Here, $H_1(e^{j\omega}) = \dfrac{1}{H(e^{j\omega})}$

Thus, $H(e^{j\omega})\,H_1(e^{j\omega}) = 1$

7.13 (c)

If any sequence is neither absolutely summable nor square summable then it will not converge and direct Fourier transform formula will not be applicable in this case. For this type of sequence Fourier transform representation will be possible if we allow DTFT to contain impulses.

7.14 (b)

Impulse response of LPF is given by

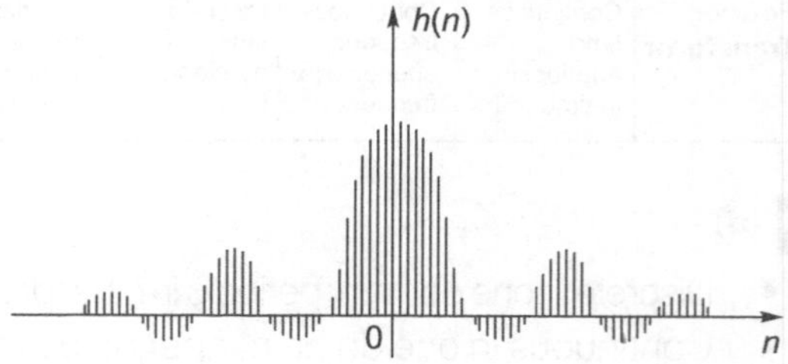

$h(n) \neq 0$ for $n < 0$

Therefore, discrete time LPF is non causal and it can't be realized because here output depends on future inputs.

7.15 (c)

Frequency response of discrete time LPF is as shown below figure

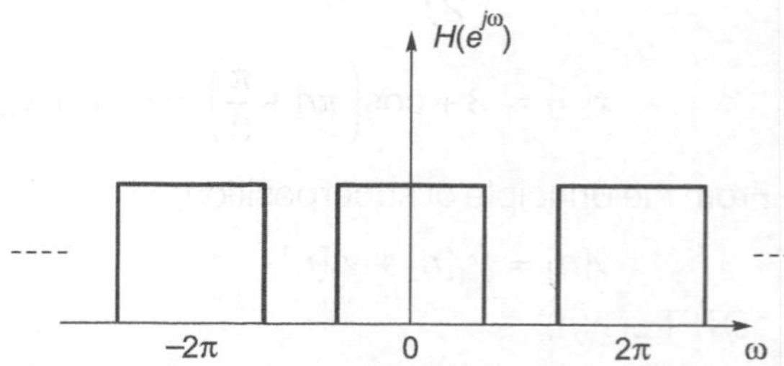

Also sampling in one domain leads to periodicity in other domain.

So it will be periodic with period 2π.

7.16 (c)

We know that, for N-point DFT,

$$\frac{1}{NT_s} \le 200$$

$$\Rightarrow \quad NT_s \ge \frac{1}{200}$$

$$\Rightarrow \quad NT_s \ge 0.005$$

8 Z-Transform

8.1 The impulse response of a discrete system with a simple pole shown in the below figure.
The pole of the system must be located on the

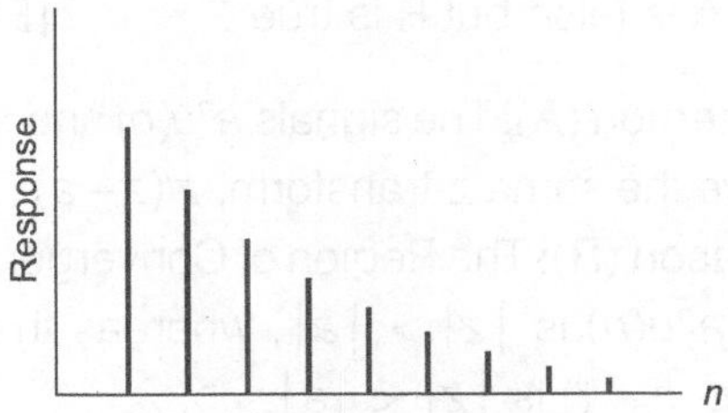

(a) real axis at $z = -1$
(b) real axis between $z = 0$ and $z = 1$
(c) imaginary axis at $z = j$
(d) imaginary axis between $z = 0$ and $z = j$

[ESE-2000]

8.2 Consider the compound system shown in the below figure. Its output is equal to input with a delay of two units.

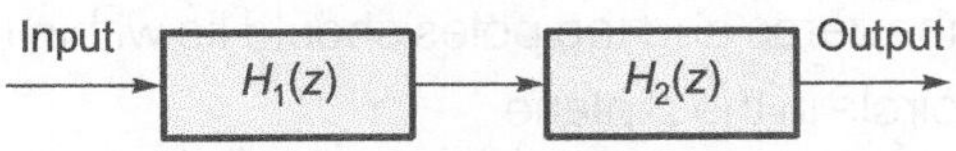

If the transfer function of the first system is given by

$$H_1(z) = \frac{z - 0.5}{z - 0.8},$$

then the transfer function of the second system would be

(a) $H_2(z) = \dfrac{z^{-2} - 0.2z^{-3}}{1 - 0.4z^{-1}}$

(b) $H_2(z) = \dfrac{z^{-2} - 0.8z^{-3}}{1 - 0.5z^{-1}}$

(c) $H_2(z) = \dfrac{z^{-1} - 0.2z^{-3}}{1 - 0.4z^{-1}}$

(d) $H_2(z) = \dfrac{z^{-2} + 0.8z^{-3}}{1 + 0.5z^{-1}}$ **[ESE-2000]**

8.3 Match **List-I** with **List-II** and select the correct answer using the codes given below the lists:

List-I
A. $\alpha^n u(n)$
B. $-\alpha^n u(-n-1)$
C. $-n\alpha^n u(-n-1)$
D. $n\alpha^n u(n)$

List-II

1. $\dfrac{\alpha z^{-1}}{(1-\alpha z^{-1})^2}$ ROC: $|z| > |\alpha|$

2. $\dfrac{1}{(1-\alpha z^{-1})}$ ROC: $|z| > |\alpha|$

3. $\dfrac{1}{(1-\alpha z^{-1})}$ ROC: $|z| < |\alpha|$

4. $\dfrac{\alpha z^{-1}}{(1-\alpha z^{-1})^2}$ ROC: $|z| < |\alpha|$

Codes:

	A	B	C	D
(a)	2	4	3	1
(b)	1	3	4	2
(c)	1	4	3	2
(d)	2	3	4	1

[ESE-2000]

8.4 Two linear time-invariant discrete time systems S_1 and S_2 are cascaded as shown in the below figure. Each system is modelled by a second order difference equation. The difference equation of the overall cascaded system can be of the order of

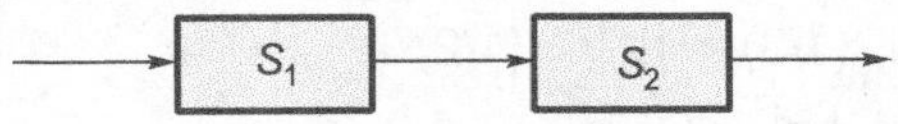

(a) 0, 1, 2, 3 or 4
(b) either 2 or 4
(c) 2
(d) 4 **[ESE-2000]**

8.5 The minimum number of delay elements required realizing a digital filter with the transfer function

$$H(z) = \frac{1 + az^{-1} + bz^{-2}}{1 + cz^{-1} + dz^{-2} + ez^{-3}}$$

(a) 2
(b) 3
(c) 4
(d) 5

[ESE-2001]

8.6 The impulse response of a discrete system with a simple pole is shown in the given figure

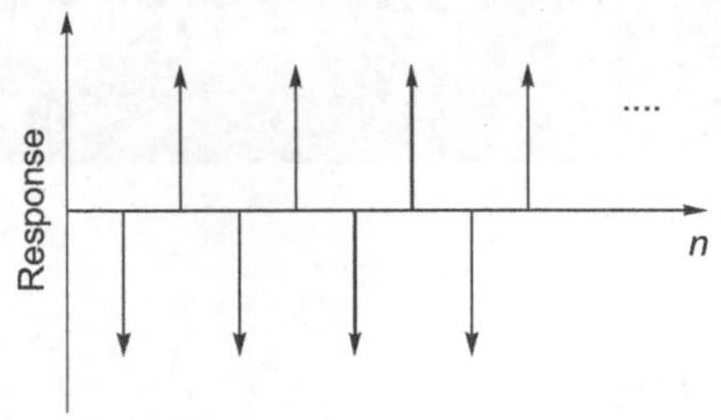

The pole must be located
(a) on the real axis at $z = 1$
(b) on the real axis at $z = -1$
(c) at the origin of the z-plane
(d) at $z = \infty$ [ESE-2001]

8.7 If $X(z) = \dfrac{z + z^{-3}}{z + z^{-1}}$, then $x(n)$ series has
(a) alternate 0s (b) alternate 1s
(c) alternate 2s (d) alternate −1s
[ESE-2002]

8.8 For a z-transform

$$X(z) = \frac{z\left(2z - \frac{5}{6}\right)}{\left(z - \frac{1}{2}\right)\left(z - \frac{1}{3}\right)}$$

Match **List-I** (The sequences) with **List-II** (The region of convergence) and select the correct answer using the codes given below the lists:

List-I
A. $[(1/2)^n + (1/3)^n]\, u(n)$
B. $(1/2)^n u(n) - (1/3)^n u(-n-1)$
C. $-(1/2)^n u(-n-1) + (1/3)^n u(n)$
D. $-[(1/2)^n + (1/3)^n]\, u(-n-1)$

List-II
1. $(1/3) < |z| < (1/2)$
2. $|z| < (1/3)$
3. $|z| < 1/3$ & $|z| > 1/2$
4. $|z| > 1/2$

Codes:

	A	B	C	D
(a)	4	2	1	3
(b)	1	3	4	2
(c)	4	3	1	2
(d)	1	2	4	3

[ESE-2002]

8.9 **Assertion (A):** The stability of the system is assured if the Region of Convergence (ROC) includes the unit circle in the z-plane.
Reason (R): For a causal stable system all the poles should be outside the unit circle in the z-plane.
(a) Both A and R are true and R is the correct explanation of A
(b) Both A and R are true but R is NOT the correct explanation of A
(c) A is true but R is false
(d) A is false but R is true [ESE-2002]

8.10 **Assertion (A):** The signals $a^n u(n)$ and $a^n u(-n-1)$ have the same z-transform, $z/(z-a)$.
Reason (R): The Region of Convergence (ROC) for $a^n u(n)$ is $|z| > |a|$, whereas the ROC for $a^n u(-n-1)$ is $|z| < |a|$.
(a) Both A and R are true and R is the correct explanation of A
(b) Both A and R are true but R is NOT the correct explanation of A
(c) A is true but R is false
(d) A is false but R is true [ESE-2002]

8.11 **Assertion (A):** For a rational transfer function $H(z)$ to be causal, stable and causally invertible, both the zeros and the poles should lie within the unit circle in the z-plane.
Reason (R): For a rational system, ROC is bounded by poles.
(a) Both A and R are true and R is the correct explanation of A
(b) Both A and R are true but R is NOT the correct explanation of A
(c) A is true but R is false
(d) A is false but R is true [ESE-2002]

8.12 The poles of an analog system are related to the corresponding poles of the digital system by the relation $z = e^{sT}$. Consider the following statements.
1. Analog system poles in the left half plane map onto digital system poles inside the circle $|z| = 1$.
2. Analog system zeros in the left half of s-plane map onto digital system zeros inside the circle $|z| = 1$.

3. Analog system poles on the imaginary axis of s-plane map onto digital system poles on the unit circle $|z| = 1$.
4. Analog system zeros on the imaginary axis of *s*-plane map onto digital system zeros on the unit circle $|z| = 1$.

Which of these statements are correct?

(a) 1 and 2 (b) 1 and 3
(c) 3 and 4 (d) 2 and 4

[ESE-2002]

8.13 What is the number of roots of the polynomial $F(z) = 4z^3 - 8z^2 - z + 2$, lying outside the unit circle?

(a) 0 (b) 1
(c) 2 (d) 3 [ESE-2004]

8.14 The output $y[n]$ of a discrete time LTI system is related to the input $x[n]$ as given below:

$$y[n] = \sum_{k=0}^{\infty} x[n-k]$$

Which one of the following correctly relates the *z*-transforms of the input and output, denoted by $X(z)$ and $Y(z)$, respectively?

(a) $Y(z) = (1 - z^{-1})\,X(z)$
(b) $Y(z) = z^{-1}\,X(z)$
(c) $Y(z) = \dfrac{X(z)}{1 - z^{-1}}$
(d) $Y(z) = \dfrac{dX(z)}{dz}$ [ESE-2005]

8.15 Match **List-I** (Discrete Time Signal) with **List-II** (Transform) and select the correct answer using the codes given below the lists:

List-I	List-II
A. Unit step function	1. 1
B. Unit impulse function	2. $\dfrac{z - \cos\omega T}{z^2 - 2z\cos\omega T + 1}$
C. $\sin\omega t$, $t = 0, T, 2T$	3. $\dfrac{z}{z-1}$
D. $\cos\omega t$, $t = 0, T, 2T, \ldots$	4. $\dfrac{z - \sin\omega T}{z^2 - 2z\cos\omega T + 1}$

Codes:

	A	B	C	D
(a)	2	4	1	3
(b)	3	1	4	2
(c)	2	1	4	3
(d)	3	4	1	2

[ESE-2005]

8.16 Which one of the following is the inverse *z*-transform of

$$X(z) = \frac{z}{(z-2)(z-3)},\ |z| < 2?$$

(a) $[2^n - 3^n]u(-n-1)$ (b) $[3^n - 2^n]u(-n-1)$
(c) $[2^n - 3^n]u(n+1)$ (d) $[2^n - 3^n]u(n)$

[ESE-2005]

8.17 Assertion (A): The discrete time system described by $y[n] = 2x[n] + 4x[n-1]$ is unstable, (here $y[n]$ is the output and $x[n]$ the input)

Reason (A): It has an impulse response with a finite number of non-zero samples.

(a) Both A and R are true and R is the correct explanation of A
(b) Both A and R are true but R is NOT the correct explanation of A
(c) A is true but R is false
(d) A is false but R is true

[ESE-2005]

8.18 What is the inverse *z*-transform of $X(z)$?

(a) $\dfrac{1}{2\pi j}\oint X(z)z^{n-1}dz$
(b) $2\pi j\oint X(z)z^{n+1}dz$
(c) $\dfrac{1}{2\pi j}\oint X(z)z^{1-n}dz$
(d) $2\pi j\oint X(z)z^{-(n+1)}dz$

[ESE-2006]

8.19 Which one of the following is the correct statement?

The region of convergence of *z*-transform of $x[n]$ consists of the values of z for which $x[n]r^{-n}$ is

(a) absolutely integrable
(b) absolutely summable
(c) unity
(d) < 1 [ESE-2006]

8.20

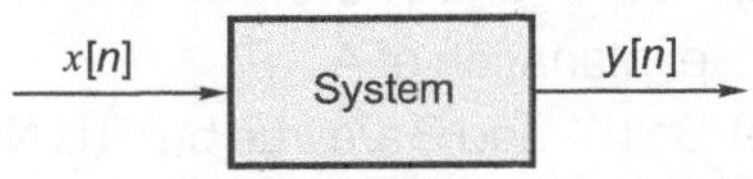

For the system shown, $x[n] = k\delta[n]$, and $y[n]$ is related to $x[n]$ as

$$y[n] - \frac{1}{2}y[n-1] = x[n]$$

What is $y[n]$ equal to?

(a) k (b) $(1/2)^n k$

(c) nk (d) 2^n

[ESE-2006]

8.21 What is the z-transform of the signal $x[n] = \alpha^n u(n)$?

(a) $X(z) = \dfrac{1}{z-1}$ (b) $X(z) = \dfrac{1}{1-z}$

(c) $X(z) = \dfrac{z}{z-\alpha}$ (d) $X(z) = \dfrac{1}{z-\alpha}$

[ESE-2007]

8.22 Algebraic expression for z-transform of $x[n]$ is $X[z]$. What is the algebraic expression for z-transform of $e^{j\omega_o n} x[n]$?

(a) $X(z - z_0)$ (b) $X(e^{-j\omega_0} z)$

(c) $X(e^{j\omega_o} z)$ (d) $X(z)e^{j\omega_o} z$

[ESE-2007]

8.23 Which one of the following rules determines the mapping of s-plane to z-plane?

(a) Right half of the s-plane maps into outside of the unit circle in z-plane

(b) Left half of the s-plane maps into inside of the unit circle

(c) Imaginary axis in s-plane maps into the circumference of the unit circle

(d) All of the above

[ESE-2007]

8.24 Assertion (A): The z-transform of the output of an ideal sampler is given by

$$Z[f(t)] = K_0 + \frac{K_1}{z} + \frac{K_2}{z^2} + \cdots + \frac{K_n}{z^n}$$

Reason (R): The relationship is the result of application of $z = e^{-sT}$, where T stands for the time gap between the samples.

(a) Both A and R are true and R is the correct explanation of A

(b) Both A and R are true but R is NOT the correct explanation of A

(c) A is true but R is false

(d) A is false but R is true

[ESE-2008]

8.25 If $X(z)$ is $\dfrac{1}{1-z^{-1}}$ with $|z| > 1$, then what is the corresponding $x(n)$?

(a) e^{-n} (b) e^n

(c) $u(n)$ (d) $\delta(n)$

[ESE-2008]

8.26 z and Laplace transform are related by

(a) $s = \ln z$ (b) $s = \dfrac{\ln z}{T}$

(c) $s = z$ (d) $s = \dfrac{T}{\ln z}$

[ESE-2010]

8.27 Convolution of two sequences $X_1[n]$ and $X_2[n]$ is represented as

(a) $X_1(z) * X_2(z)$ (b) $X_1(z) X_2(z)$

(c) $X_1(z) + X_2(z)$ (d) $\dfrac{X_1(z)}{X_2(z)}$

[ESE-2010]

8.28 Unit step response of the system described by the equation $y(n) + y(n-1) = x(n)$ is

(a) $\dfrac{z^2}{(z+1)(z-1)}$ (b) $\dfrac{z}{(z+1)(z-1)}$

(c) $\dfrac{z+1}{z-1}$ (d) $\dfrac{z(z-1)}{(z+1)}$

[ESE-2010]

8.29 Given $X(z) = \dfrac{1}{1-az^{-1}}$; $|z| > |a|$ then the region of convergence is (Shaded region)

(a) 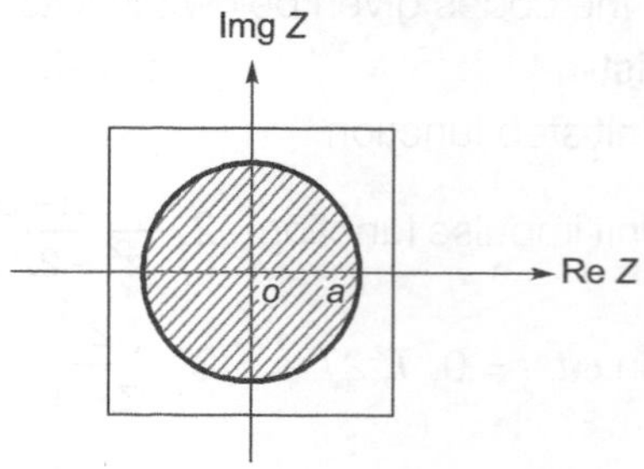

(b) 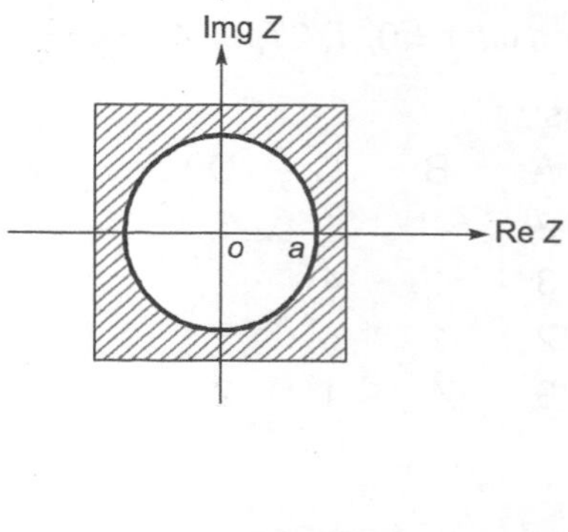

(c) 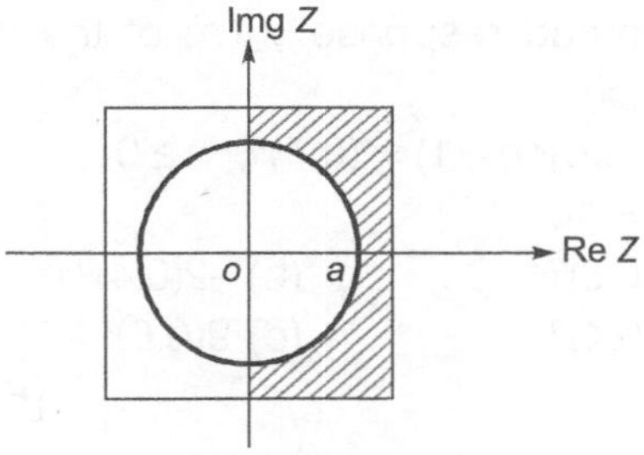

(d) 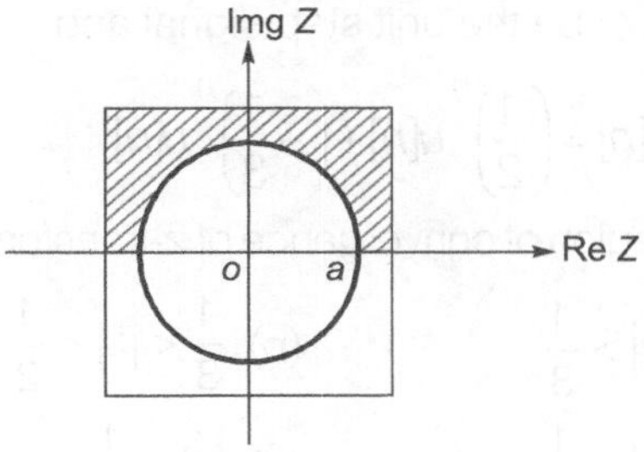

[ESE-2010]

8.30 The z-transform of $-u(-n-1)$ is

(a) $\frac{z}{z-1}$ with $|z|>1$

(b) $\frac{z}{z-1}$ with $|z|<1$

(c) $\frac{z}{z-1}$ with $|z|=1$

(d) $\frac{z}{z-1}$ with $|z|=0$ [ESE-2010]

8.31 Unit step response of the system described by difference equation $y(n) + y(n-1) = x(n)$ is

(a) $\frac{z^2}{(z+1)(z-1)}$ (b) $\frac{z}{(z+1)(z-1)}$

(c) $\frac{(z+1)}{(z-1)}$ (d) $\frac{z(z-1)}{(z+1)}$

[ESE-2011]

8.32 Assertion (A): The system function

$$H(z) = \frac{z^3 - 2z^2 + z}{z^2 + \frac{1}{4}z + \frac{1}{8}} \text{ is not causal.}$$

Reason (R): If the numerator of $H(z)$ is of lower order than the denominator, the system may be causal.

(a) Both A and R are true and R is the correct explanation of A

(b) Both A and R are true but R is not a correct explanation of A

(c) A is true but R is false

(d) A is false but R is true [ESE-2011]

8.33 Statement (I): z-transform approach is used to analyze the discrete time systems and is also called as pulse transfer function approach.

Statement (II): The sampled signal is assumed to be a train of impulses whose strengths, or areas, are equal to the continuous time signal at the sampling instants.

(a) Both Statement (I) and Statement (II) are individually true and Statement (II) is the correct explanation of Statement (I).

(b) Both Statement (I) and Statement (II) are individually true but Statement (II) is not the correct explanation of Statement (I).

(c) Statement (I) is true but Statement (II) is false.

(d) Statement (I) is false but Statement (II) is true.

[ESE-2012]

8.34 The natural response of an LTI system described by the difference equation

$y(n) - 1.5\, y(n-1) + 0.5 y(n-2) = x(n)$ is

(a) $y(n) = 0.5\, u(n) - 2(0.5)^n u(n)$

(b) $y(n) = 0.5\, u(n) - (0.5)^n u(n)$

(c) $y(n) = 2\, u(n) - 0.5\, (0.5)^n u(n)$

(d) $y(n) = 2\, u(n) - (0.5)^n u(n)$

[ESE-2012]

8.35 The step response of a discrete time system with transfer function

$$H(z) = \frac{10}{(z-1)(z+2)} \text{ is given by}$$

(a) $-\frac{10}{9} + \frac{10}{3}n + \frac{10}{9}(-2)^n$

(b) $-5 + \frac{n}{2} + (-2)^n$

(c) $-\frac{7}{9} + \frac{5}{3}n + (-3)^n$

(d) $-2 + 5(1 - 2^n)$ [ESE-2012]

8.36 The z-transform corresponding to the Laplace transform function

$$G(s) = \frac{10}{s(s+5)} \text{ is}$$

(a) $\frac{2ze^{-5z}}{(z-1)(z-e^{-T})}$ (b) $\frac{2(1-e^{-5T})z}{(z-1)(z-e^{-5T})}$

(c) $\frac{e^{-5T}}{(z-1)^2}$ (d) $\frac{e^{-T}}{z(z-e^{-3T})}$

[ESE-2012]

8.37 The difference equation for a system is given by
$y(n+2)+y(n+1)+0.16y(n)=x(n+1)+0.32\,x(n)$
The transfer function of the system is

(a) $\frac{z+0.32}{z^2+z+0.16}$ (b) $\frac{1}{z^2+z+0.16}$

(c) $\frac{z+0.32}{z^2+0.16}$ (d) $\frac{z+0.32}{(z-1)(z^2+z+0.16)}$

[ESE-2012]

8.38 Homogeneous solution of

$y(n)-\frac{9}{16}y(n-2)=x(n-1)$ is

(a) $C_1\left(\frac{3}{4}\right)^n+C_2\left(-\frac{3}{4}\right)^n$

(b) $C_1\left(-\frac{3}{4}\right)^{n-1}+C_2\left(\frac{3}{4}\right)^{n-1}$

(c) $C_1\left(\frac{3}{4}\right)^n$

(d) $C_1\left(-\frac{3}{4}\right)^n$ [ESE-2013]

8.39 If the z-transform of $x(n)$ is

$X(z)=\frac{z(8z-7)}{4z^2-7z+3}$, then the $\lim_{n\to\infty} x(n)$ is

(a) 1 (b) 2
(c) ∞ (d) 0

[ESE-2013]

8.40 The final value theorem is

(a) $\lim_{k\to\infty} x(k)=\lim_{z\to 1}(z-1)X^+(z)$

(b) $\lim_{k\to\infty} x(k)=\lim_{z\to 1}X^+(z)$

(c) $\lim_{k\to\infty} x(k)=\lim_{z\to 0}(z^{-1})\,X^+(z)$

(d) $\lim_{k\to\infty} x(k)=\lim_{z\to 0}(z-1)^{-1}\,X^+(z^{-1})$

[ESE-2013]

8.41 For the discrete signal $x[n]=a^n u[n]$ the z-transform is

(a) $\frac{z}{z+a}$ (b) $\frac{z-a}{z}$

(c) $\frac{z}{a}$ (d) $\frac{z}{z-a}$

[ESE-2013]

8.42 The forced response $y_F(n)$ of the differential equation
$y(n)-0.6y(n-1)=(0.4)^n,\ n\geq 0$
$y(-1)=10$
(a) $9(0.6)^n$ (b) $-2(0.4)^n$
(c) $9(0.4)^n$ (d) $9(0.6)^n-2(0.4)^n$

[ESE-2013]

8.43 Let $u[n]$ be the unit step signal and

$$x[n]=\left(\frac{1}{2}\right)^n u[n]+\left(-\frac{1}{3}\right)^n u[n]$$

The region of convergence of z-transform of $x[n]$ is

(a) $|z|>\frac{1}{3}$ (b) $\frac{1}{3}<|z|<\frac{1}{2}$

(c) $|z|>\frac{1}{2}$ (d) $|z|<\frac{1}{2}$

[ESE-2014]

8.44 If the z-transform of a sequence $x[n]=\{1, 1, -1, -1\}$ is $X(z)$, then the value of $X\left(\frac{1}{2}\right)$ is

(a) 9 (b) 1.875
(c) −1.125 (d) 15 [ESE-2014]

8.45 If the z-transform of a system is given by

$$H(z)=\frac{\alpha+z^{-1}}{1+\alpha z^{-1}}$$

where a is real valued, $|\alpha|<1$, ROC: $|z|>|\alpha|$, then the system is
(a) a low-pass filter (b) a band-pass filter
(c) an all-pass filter (d) a high-pass filter

[ESE-2014]

8.46 Consider the following system function of a discrete-time LTI system:

$$H(z)=\frac{z^{-1}-a^*}{1-az^{-1}}$$

where, a^* is the complex conjugate of a. The frequency response of such a system is
(a) aperiodic; depends on frequency ω
(b) aperiodic; does not depends on frequency ω
(c) periodic; depends on frequency ω
(d) periodic; does not depends on frequency ω

[ESE-2014]

8.47 Consider two infinite duration input sequences $\{x_1[n],\ x_2[n]\}$. When will the Region of Convergence [ROC] of z-transform of their superposition i.e. $\{x_1[n]+x_2[n]\}$ be entire z-plane except possibly at $z=0$ or $z=\infty$?

(a) When their linear combination is of finite duration
(b) When they are left sided sequences
(c) When they are right sided sequences
(d) When their linear combination is causal
[ESE-2015]

8.48 If the lower limit of Region of Convergence (ROC) is greater than the upper limit of ROC, the series

$$X(z) = \sum_{n=-\infty}^{\infty} x(n)z^{-n}$$

(a) Converges
(b) Zero
(c) Does not converge
(d) None of the above [ESE-2015]

8.49 Consider a two-sided discrete-time signal (neither left sided, nor right sided). The region of convergence (ROC) of the z-transform of the sequence is
1. All region of z-plane outside a unit circle (in z-plane)
2. All region of z-plane inside a unit circle (in z-plane)
3. Ring in z-plane

Which of the above is/are correct?
(a) 1 only (b) 2 only
(c) 3 only (d) 1 and 3 [ESE-2016]

8.50 Z-transform deals with discrete time systems for their
1. Transient behaviour
2. Steady-state behavior

Which of the above behaviours is/are correct?
(a) 1 only (b) 2 only
(c) Both 1 and 2 (d) Neither 1 nor 2
[ESE-2016]

8.51 What is the range of values of a and b for which the linear time-invariant system with impulse response

$$h(n) = \begin{cases} a^n, & n \geq 0 \\ b^n, & n < 0 \end{cases}$$

is stable?
(a) Both $|a| < 1$ and $|b| > 1$ are satisfied
(b) Both $|a| > 1$ and $|b| < 1$ are satisfied
(c) Both $|a| > 1$ and $|b| > 1$ are satisfied
(d) Both $|a| < 1$ and $|b| < 1$ are satisfied
[ESE-2019]

8.52 A discrete-time LTI system with rational system function $H(z)$ is causal if and only if
(a) the ROC is the exterior of a circle outside the outermost pole.
(b) the ROC is the interior of a circle outside the outermost pole.
(c) the ROC is the exterior of a circle outside the innermost pole.
(d) the ROC is the interior of a circle outside the innermost pole.
[ESE-2021]

8.53 If the impulse response $h(n)$ and the output $y(n)$ of a system are given as
$h(n) = \{2, 1, 0, -1, 3\}$ and
$y(n) = \{2, -5, 1, 1, 6, -11, 6\}$
then the input $x(n)$ of the system is
(a) $\{1, -3, 2\}$ (b) $\{2, -3, 1\}$
(c) $\{1, 3, 2, 1\}$ (d) $\{2, -3, 1, 1\}$
[ESE-2022]

8.54 The z-transform of the impulse response of a causal LTI system is $H(z) = \dfrac{1}{2}\dfrac{z^{-1}}{z^{-2} - 4.5z^{-1} + 5}$.
What is an input $x(n)$ that would produce the output $y(n) = u(-n) + (0.5)^n u(n)$?
(a) $x(n) = [0.5^n - 0.4^n]u(n)$
(b) $x(n) = 10u(n+1) - 14u(n) + 2u(n-1) + 3u(-n)$
(c) $x(n) = [0.5^n + 0.4^n]u(n)$
(d) $x(n) = 10u(n+1) - 14u(-n) + 2u(n+1) + 3u(n)$
[ESE-2023]

8.55 Sample $x(t)$ at the rate of 10 samples per second. What is the ROC of its z-transform directly from $x(n)$, if the input signal $x(t) = [-2e^{-3t} + 3e^{-4t}]u(t)$?
(a) $Z = 0.7408$ (b) $Z = 0.6703$
(c) $|Z| > 0.6703$ (d) $|Z| > 0.7408$
[ESE-2024]

8.56 What is the particular solution of the difference equation

$$y(n) = \frac{5}{6}y(n-1) - \frac{1}{6}y(n-2) + x(n)$$

when the forcing function $x(n) = 2^n$, $n \geq 0$ and zero elsewhere?
(a) $y_p(n) = (-8)2^n, n \geq 0$
(b) $y_p(n) = (192/5)2^n, n \geq 0$
(c) $y_p(n) = (8/5)2^n, n \geq 0$
(d) $y_p(n) = (-8/5)2^n, n \geq 0$

[ESE-2024]

Answers Z-Transform

8.1 (b)	8.2 (b)	8.3 (d)	8.4 (a)	8.5 (b)	8.6 (b)	8.7 (a)	8.8 (c)	8.9 (c)
8.10 (d)	8.11 (d)	8.12 (b)	8.13 (b)	8.14 (c)	8.15 (b)	8.16 (a)	8.17 (d)	8.18 (a)
8.19 (b)	8.20 (b)	8.21 (c)	8.22 (b)	8.23 (d)	8.24 (c)	8.25 (c)	8.26 (b)	8.27 (b)
8.28 (a)	8.29 (b)	8.30 (b)	8.31 (a)	8.32 (a)	8.33 (a)	8.34 (*)	8.35 (*)	8.36 (b)
8.37 (a)	8.38 (a)	8.39 (a)	8.40 (a)	8.41 (d)	8.42 (*)	8.43 (c)	8.44 (*)	8.45 (c)
8.46 (d)	8.47 (a)	8.48 (c)	8.49 (c)	8.50 (c)	8.51 (a)	8.52 (a)	8.53 (a)	8.54 (*)
8.55 (d)	8.56 (c)							

Explanations Z-Transform

8.1 (b)

The given impulse response has the form

$$H(z) = \frac{z}{z-a}, \quad |z| > 0$$

$$\Rightarrow \quad h(n) = a^n u(n), \; 0 < a < 1$$

So, the pole of the system must be located on the real axis between $z = 0$ and $z = 1$.

8.2 (b)

$$y(n) = x(n-2)$$

$$Y(z) = z^{-2} X(z)$$

$$\Rightarrow \quad H(z) = H_1(z)\, H_2(z) = \frac{Y(z)}{X(z)} = z^{-2}$$

$$\Rightarrow \left(\frac{z-0.5}{z-0.8}\right) H_2(z) = z^{-2}$$

$$\Rightarrow \quad H_2(z) = \frac{z^{-1} - 0.8z^{-2}}{z-0.5}$$

$$\Rightarrow \quad H_2(z) = \frac{z^{-2} - 0.8z^{-3}}{1-0.5z^{-1}}$$

8.4 (a)

$$H(s) = H_1(s) \cdot H_2(s)$$
$$= (as^2 + \ldots)\,(ms^2 + \ldots)$$
$$= (ps^4 + \ldots)$$

where a, m and p are constants

Thus the overall cascaded system will have the difference equation of the order of 4.

8.5 (b)

$$H(z) = \frac{1 + az^{-1} + bz^{-2}}{1 + cz^{-1} + dz^{-2} + ez^{-3}}$$

Minimum number of delay elements

= (Maximum power of z – minimum power of z)

= {0 – (–3)} = 3

8.6 (b)

From diagram we have impulse response,

$$h(n) = (-1)^n\, u(n)$$

$$= \frac{1}{\left[1-(-1)z^{-1}\right]} = \frac{1}{1+z^{-1}}$$

$$z = -1$$

8.7 (a)

$$X(z) = \frac{z + z^{-3}}{z + z^{-1}} = \frac{1 + z^{-4}}{1 + z^{-2}}$$
$$= (1 + z^{-4})\,(1 + z^{-2})^{-1}$$
$$= (1 + z^{-4})\,(1 - z^{-2} + z^{-4} - z^{-6} + \ldots)$$
$$= 1 - z^{-2} + 2z^{-4} - 2z^{-6} + \ldots$$

So, $x(n) = \delta(n) - \delta(n-2) + 2\delta(n-4) - 2\delta(n-6) + \ldots$

Hence, $x(n)$ series has alternate zeros, i.e. at $n = 1, 3, 5, \ldots$

8.8 (c)

$$X(z) = \frac{z\left(2z - \frac{5}{6}\right)}{\left(z - \frac{1}{2}\right)\left(z - \frac{1}{3}\right)}$$

$$= \frac{z}{z - \frac{1}{2}} + \frac{z}{z - \frac{1}{3}}$$

We know that

$$X(z) = \frac{z}{z-a}$$

$\Rightarrow$ $x(n) = a^n u(n),\ |z| > |a|$

or $x(n) = -a^n u(-n-1),\ |z| < |a|$

If $|z| > \frac{1}{2}$,

then $x(n) = \left\{\left(\frac{1}{2}\right)^n + \left(\frac{1}{3}\right)^n\right\} u(n)$

If $\frac{1}{3} < |z| < \frac{1}{2}$,

then $x(n) = -\left(\frac{1}{2}\right)^n u(-n-1) + \left(\frac{1}{3}\right)^n u(n)$

If $|z| < \frac{1}{3}$,

then $x(n) = -\left(\frac{1}{2}\right)^n u(-n-1) - \left(\frac{1}{3}\right)^n u(-n-1)$

If $|z| < \frac{1}{3}$ & $|z| > \frac{1}{2}$,

then $x(n) = \left(\frac{1}{2}\right)^n u(n) - \left(\frac{1}{3}\right)^n u(-n-1)$

8.9 (c)

A causal LTI system is stable if and only if all of poles of $H(z)$ lie inside the unit circle.

8.10 (d)

$$a^n u(n) \xleftrightarrow{\text{Z-transform}} \frac{z}{z-a};\ |z| > |a|$$

$$a^n u(-n-1) \xleftrightarrow{\text{Z-transform}} \frac{-z}{z-a};\ |z| < |a|$$

So assertion is wrong but reason is true.

8.13 (b)

$$F(z) = 4z^3 - 8z^2 - z + 2$$
$$= 4z^2(z-2) - (z-2)$$
$$= (4z^2 - 1)(z-2)$$
$$= (2z+1)(2z-1)(z-2)$$

Roots of the equation are

$$z = \frac{-1}{2}, \frac{1}{2}, 2$$

Therefore, only one root is lying outside the unit circle.

8.14 (c)

$$y[n] = \sum_{k=0}^{\infty} x[k] = \sum_{k=-\infty}^{\infty} x[k]u[k]$$

$$\frac{Y(z)}{X(z)} = \sum_{k=-\infty}^{\infty} u[n]z^{-n} = \sum_{k=0}^{\infty} (z^{-1})^n$$

$$\Rightarrow \quad \frac{Y(z)}{X(z)} = \frac{1}{1-z^{-1}}$$

$$\Rightarrow \quad Y(z) = \frac{X(z)}{1-z^{-1}}$$

8.16 (a)

$$X(z) = \frac{z}{(z-2)(z-3)},\ |z| < 2$$

$$= \frac{z}{z-3} - \frac{z}{z-2}$$

$\Rightarrow$ $x(n) = -3^n u(-n-1) + 2^n u(-n-1)$

$\Rightarrow$ $x(n) = (2^n - 3^n)\, u(-n-1)$

8.17 (d)

$$y[n] = 2x[n] + 4x[n-1]$$

$$H(z) = \frac{Y(z)}{X(z)} = 2 + 4z^{-1}$$

$\because$ $h[n]$ is absolutely summable.

$\therefore$ System is stable.

8.18 (a)

By definition of z-transform as Fourier transform of exponential sequence

$$X(z) = X(re^{j\omega})\Big|_{z=re^{j\omega}} = FT\left[x[n]r^{-n}\right]$$

$$x[n]r^{-n} = IFT\left[X(re^{j\omega})\right]$$

$$= \frac{1}{2\pi}\int_{2\pi} X(re^{j\omega})e^{j\omega n}\, d\omega$$

$$x[n] = \frac{1}{2\pi}\int_{2\pi} X(re^{j\omega})\left(\frac{e^{j\omega n}}{r^{-n}}\right) d\omega$$

$$= \frac{1}{2\pi}\int_{2\pi} X(re^{j\omega})(re^{j\omega})^n\, d\omega \quad \text{...(i)}$$

$\because$ $z = re^{j\omega}$

$$dz = jre^{j\omega}\, d\omega$$

$$dz = jz\, d\omega$$

$$d\omega = \frac{dz}{jz} \quad \text{...(ii)}$$

Putting $d\omega$ in equation (i)

$$= \frac{1}{2\pi}\int_{2\pi} X(re^{j\omega})(re^{j\omega})^n \frac{dz}{jz}$$

$$= \frac{1}{2\pi j}\oint X(z)\frac{z^n}{z}dz$$

$$x[n] = \frac{1}{2\pi j}\oint X(z)z^{n-1}dz$$

8.20 (b)

$$y[n]-\frac{1}{2}y[n-1] = x[n] = k\delta[n]$$

Taking z-transform

$$Y(z)-\frac{1}{2}z^{-1}Y(z) = k$$

$$Y(z) = \frac{k}{1-\frac{1}{2}z^{-1}}$$

$$\Rightarrow \quad y[n] = \left(\frac{1}{2}\right)^n k$$

$$\because a^n u[n] \xrightarrow{Z} \frac{1}{1-az^{-1}},\ |z| > |a|$$

8.22 (b)

Property of z-transform

$$x(n) \longleftrightarrow X(z)$$

$$(e^{j\omega_o})^n x(n) \longleftrightarrow X\left(\frac{z}{e^{j\omega_o}}\right)$$

8.24 (c)

Ideal sampling function is

$$f(t) = \sum_{n=0}^{\infty} K_n\delta[t-nT]$$

where T = sampling period

since, $\delta[n+n_0] \xleftrightarrow{\text{Z.T.}} z^{n_0}$

So, $z[f(t)] = K_0 + \frac{K_1}{z} + \frac{K_2}{z^2} + \cdots \frac{K_n}{z^n}$

8.25 (c)

$$a^n u[n] \xleftrightarrow{Z} \frac{1}{1-az^{-1}}, |z|>|a|$$

Putting $a = 1$

$$u[n] \xleftrightarrow{Z} \frac{1}{1-z^{-1}}, |z|>1$$

8.26 (b)

$$z = e^{sT}$$

$$\therefore \quad \ln z = \ln e^{sT}$$

$$\Rightarrow \quad \ln z = sT\ln e$$

$$\Rightarrow \quad \ln z = sT$$

$$\therefore \quad \boxed{s = \frac{\ln z}{T}}$$

8.27 (b)

Convolution in time domain results multiplication in frequency domain.

8.28 (a)

Given, $y(n) + y(n-1) = x(n)$

Taking z-transform

$$Y(z) + z^{-1}Y(z) = X(z)$$

$$\therefore \quad \frac{Y(z)}{X(z)} = \frac{1}{(1+z^{-1})}$$

and $x(n) = u(n)$

$$\therefore \quad X(z) = \frac{1}{1-z^{-1}}$$

Hence $Y(z) = \frac{1}{(1+z^{-1})}X(z)$

$$\Rightarrow Y(z) = \frac{1}{(1+z^{-1})}\times\frac{1}{(1-z^{-1})} = \frac{z^2}{(z-1)(z+1)}$$

$$\Rightarrow \quad \boxed{Y(z) = \frac{z^2}{(z+1)(z-1)}}$$

8.31 (a)

$$y(n) + y(n-1) = x(n)$$

Taking z-transform of above equation

$$Y(z) + z^{-1}Y(z) = X(z)$$

$$\Rightarrow \quad \frac{Y(z)}{X(z)} = \frac{1}{1+z^{-1}} = \frac{z}{z+1}$$

or $Y(z) = \left(\frac{z}{z+1}\right)X(z)$

But given $x(n) = u(n)$

or $X(z) = \frac{1}{1-z^{-1}} = \frac{z}{z-1}$

So $Y(z) = \left(\frac{z}{z+1}\right)\left(\frac{z}{z-1}\right)$

$$= \frac{z^2}{(z+1)(z-1)}$$

8.32 (a)

For causality of $H(z)$, degree of numerator must be less than the degree of denominator along with other conditions.

8.34 (*)

Natural response cann't be evaluated because initial conditions are not given.

Note: As per RTI filed and reply of **UPSC** answer is option (c).

8.35 (*)

$$H(z) = \frac{10}{(z-1)(z+2)}$$

Given,

$$X(z) = \frac{1}{1-z^{-1}} = \frac{z}{(z-1)}$$

$$\therefore \quad Y(z) = H(z)X(z)$$

$$= \frac{10}{(z-1)(z+2)} \cdot \frac{z}{(z-1)}$$

$$\Rightarrow \quad Y(z) = \frac{10z}{(z-1)^2(z+2)}$$

$$= \frac{A}{(z-1)} + \frac{B}{(z-1)^2} + \frac{C}{(z+2)}$$

$$= \frac{20/9}{(z-1)} + \frac{10/9}{(z-1)^2} - \frac{20/9}{(z+2)}$$

$$= \frac{20/9z^{-1}}{1-z^{-1}} + \frac{10/9z^{-2}}{(1-z^{-1})^2} - \frac{20/9z^{-1}}{\left[1-(-2)z^{-1}\right]}$$

$$= \frac{20}{9}u(n) + \frac{10}{9}nu(n) - \frac{20}{9}(-2)^n u(n)$$

$$= \left[\frac{20}{9} + \frac{10}{9}n - \frac{20}{9}(-2)^n\right]u(n)$$

Note: As per RTI filed and reply of **UPSC** answer is option (a).

8.36 (b)

$$G(s) = \frac{10}{s(s+5)}$$

$$= \frac{2}{s} - \frac{2}{s+5}$$

$$g(t) = 2u(t) - 2e^{-5t}u(t)$$

Converting the continuous time into discrete time at $t = nT$; T = constant

We have,

$$g(nT) = 2u(nT) - 2e^{-5nT}u(nT)$$

$$u(nT) = u(n)$$

$$g(n) = 2u(n) - 2(e^{-5T})^n u(n)$$

$$= \frac{2}{1-z^{-1}} - \frac{2}{1-e^{-5T}z^{-1}} = 2z\left[\frac{1}{z-1} - \frac{1}{z-e^{-5T}}\right]$$

$$= 2z\left[\frac{z-e^{-5T}-z+1}{(z-1)(z-e^{-5T})}\right] = \frac{2(1-e^{-5T})z}{(z-1)(z-e^{-5T})}$$

8.37 (a)

$y(n+2) + y(n+1) + 0.16y(n) = x(n+1) + 0.32x(n)$

Taking z-transform of both sides

$(z^2 + z + 0.16)\,Y(z) = (z + 0.32)\,X(z)$

$$\text{Transfer function, } H(z) = \frac{(z+0.32)}{(z^2+z+0.16)}$$

8.38 (a)

Taking z-transform of given difference equation

$$Y(z) - \frac{9}{16}z^{-2}Y(z) = z^{-1}X(z)$$

$$\Rightarrow \quad Y(z) = \frac{z^{-1}}{1-\frac{9}{16}z^{-2}} = \frac{z}{\left(z^2 - \frac{9}{16}\right)}$$

So for homogeneous solution or complementary solution.

$$z^2 - \frac{9}{16} = 0$$

$$\Rightarrow \quad z = \pm\frac{3}{4}$$

So homogeneous solution is given as

$$C_1\left(\frac{3}{4}\right)^n + C_2\left(-\frac{3}{4}\right)^n$$

8.39 (a)

$$X(z) = \frac{z(8z-7)}{(z-1)(4z-3)}$$

Using final value theorem

$$\lim_{n\to\infty} x(n) = \lim_{z\to1}(z-1)\frac{z(8z-7)}{(z-1)(4z-3)}$$

$$= \frac{1(8-7)}{(4-3)} = 1$$

8.40 (a)

Final value theorem for z-transform is

$$\lim_{k\to\infty} x(k) = \lim_{z\to1}(z-1)X^+(z)$$

8.41 (d)

$$X(z) = \sum_{n=-\infty}^{\infty} x[n]z^{-n}$$

$$= \sum_{n=0}^{\infty} a^n z^{-n} = 1 + (az^{-1}) + (az^{-1})^2 + \cdots$$

$$= \frac{1}{1-az^{-1}} = \frac{z}{z-a}$$

8.42 (*)

Force response is calculated when input is applied and initial condition is zero i.e. $y(-1) = 10$ will be zero.

$y(n) - 0.6y(n-1) = (0.4)^n$

Taking z-transform

$$Y(z)(1-0.6z^{-1}) = \frac{1}{1-0.4z^{-1}}$$

$$Y(z) = \frac{1}{(1-0.4z^{-1})(1-0.6z^{-1})}$$

$$Y(z) = \frac{-2}{(1-0.4z^{-1})} + \frac{3}{(1-0.6z^{-1})}$$

Taking inverse z-transform

$$y(n) = -2(0.4)^n u(n) + 3(0.6)^n u(n)$$

Note: given system is not an LTI system.

8.43 (c)

$$x(n) = \left(\frac{1}{2}\right)^n u(n) + \left(-\frac{1}{3}\right)^n u(n) \quad \text{...(i)}$$

Taking z transform of equation (i) we get

$$x(z) = \frac{z}{z-\frac{1}{2}} + \frac{z}{z+\frac{1}{3}}$$

So, the region of convergence is

$$z - \frac{1}{2} > 0 \text{ and } z + \frac{1}{3} > 0$$

$$\text{or } |z| > \frac{1}{2} \text{ and } |z| > -\frac{1}{3}$$

Thus, the common ROC will be

$$|z| > \frac{1}{2}$$

8.44 (*)

$$x(n) = \{1, 1, -1, -1\}$$

$$X(z) = z^0 + z^{-1} - z^{-2} - z^{-3}$$

$$X\left(\frac{1}{2}\right) = \left(\frac{1}{2}\right)^0 + \left(\frac{1}{2}\right)^{-1} - \left(\frac{1}{2}\right)^{-2} \left(\frac{1}{2}\right)^{-3}$$

$$= 1 + 2 - 4 - 8 = -9$$

8.45 (c)

$$H(z) = \frac{\alpha + z^{-1}}{1+\alpha z^{-1}} \quad \text{...(i)}$$

$$= \frac{z^{-1}(\alpha z + 1)}{z^{-1}(z+\alpha)}$$

$$= \frac{(1+\alpha z)}{(z+\alpha)} \quad \text{...(ii)}$$

Also, $|\alpha| < 1$ and ROC $= |z| < |\alpha|$

(Left side signal)

From equation (ii), it is clear that the system $H(z)$ has a zero at $-1/\alpha$ and pole at $-\alpha$.

$\because$ $|\alpha| < 1$ thus zero > pole

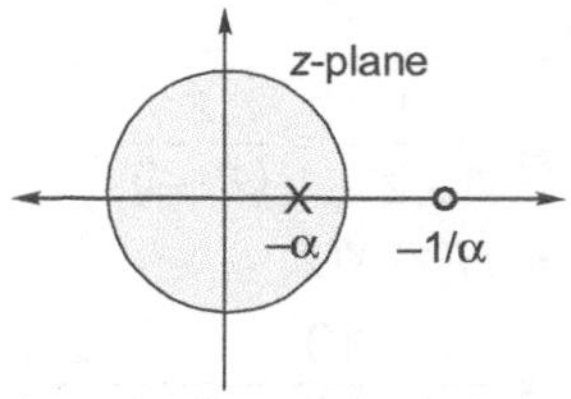

The above system represents an all pass filter.

8.46 (d)

Given, $H(z) = \dfrac{z^{-1} - a^*}{1 - az^{-1}}$

Put $z = re^{j\Omega}$

The discrete time frequency response will be aperiodic and does not depends upon the frequency.

8.47 (a)

$x_1(n) + x_2(n)$

For finite duration, the ROC is entire z-plane except $z = 0$ and $z = \infty$

example:

$\delta(n-1) + \delta(n+1) \leftrightarrow z^{-1} + z$

$$= \frac{1}{z} + z$$

Discontinuity occurs at $z = 0$ and $z = \infty$

Hence ROC = entire z-plane except

$z = 0$ and $z = \infty$

8.48 (c)

$$X(z) = \sum_{n=-\infty}^{\infty} x(n)z^{-n}$$

If ROC's lower limit is greater than upper limit of ROC, then there is no common region of convergence. Hence signal doesn't converge.

8.49 (c)

From the properties of ROC of z-transform, for a two sided sequences the ROC of its z-transform is in the form of circular strip or annular strip i.e., in the form of ring.

8.50 (c)

Using the unilateral transforms [both Laplace and Z-transform] the analysis of continuous time and discrete time systems can be analysed both for transient [using time differentiation property (Laplace transform), using time shifting property (Z-transform)] and steady state responses [using final value theorem].

8.51 (a)

$$h[n] = a^n u[n] + b^n u[-n-1]$$

$$H(z) = \frac{1}{1-az^{-1}} - \frac{1}{1-bz^{-1}} \qquad |a|<|z|<|b|$$

For system to be stable, $|a|<1$ and $|b|>1$.

8.52 (a)

For the system to be causal, ROC should be right-sided and since, ROC is bounded by the poles and cannot include any poles, therefore, the ROC should be the exterior of a circle outside the outermost pole.

8.53 (a)

The output $y(n)$ of a system is given as the convolution of input $x(n)$ and impulse response $h(n)$.

$$y(n) = x(n) \otimes h(n)$$

Taking z-transform,

$$Y(z) = X(z) \cdot H(z)$$

$$\therefore \quad X(z) = \frac{Y(z)}{H(z)} \qquad \text{...(i)}$$

Given : $h(n) = \{2, 1, 0, -1, 3\}$

Taking z-transform of $h(n)$,

$$H(z) = 2 + z^{-1} + (-1)z^{-3} + 3z^{-4} \qquad \text{...(ii)}$$

$$y(n) = \{2, -5, 1, 1, 6, -11, 6\}$$

Taking z-transform of $y(n)$,

$$Y(z) = 2 - 5z^{-1} + z^{-2} + z^{-3} + 6z^{-4} - 11z^{-5} + 6z^{-6} \qquad \text{...(iii)}$$

From equations (i), (ii) and (iii)

$$X(z) = \frac{Y(z)}{H(z)}$$

$$= \frac{2 - 5z^{-1} + z^{-2} + z^{-3} + 6z^{-4} - 11z^{-5} + 6z^{-6}}{2 + z^{-1} - z^{-3} + 3z^{-4}}$$

$$\begin{array}{r} 1 - 3z^{-1} + 2z^{-2} \\ 2 + z^{-1} - z^{-3} + 3z^{-4} \,\big)\, 2 - 5z^{-1} + z^{-2} + z^{-3} + 6z^{-4} - 11z^{-5} + 6z^{-6} \\ 2 + z^{-1} + 0 - z^{-3} + 3z^{-4} \\ \hline -6z^{-1} + z^{-2} + 2z^{-3} + 3z^{-4} - 11z^{-5} + 6z^{-6} \\ -6z^{-1} - 3z^{-2} + 0 + 3z^{-4} - 9z^{-5} \\ \hline 4z^{-2} + 2z^{-3} - 2z^{-5} + 6z^{-6} \\ 4z^{-2} + 2z^{-3} - 2z^{-5} + 6z^{-6} \\ \hline \times \end{array}$$

$$\therefore \quad X(z) = 1 - 3z^{-1} + 2z^{-2}$$

Taking inverse z-transform,

$$x(n) = \{1, -3, 2\}$$

8.54 (*)

Given output of an LTI system

$$\Rightarrow \quad y(n) = u(-n) + (0.5)^n u(n)$$

$$\Big\| \text{Z.T}$$

$$Y(z) = \frac{1}{1-z} + \frac{z}{z-0.5} = \frac{(2z - 0.5 - z^2)}{(1-z)(z-0.5)}$$

Now,

$$X(z) = \frac{Y(z)}{H(z)} = \left[\frac{2z-0.5-z^2}{(1-z)(z-0.5)}\right] 2\left[\frac{z^{-2} - 4.5z^{-1} + 5}{z^{-1}}\right]$$

$$= \frac{2z-0.5-z^2}{(1-z)(z-0.5)}\left(\frac{2z^{-2} - 9z^{-1} + 10}{z^{-1}}\right)$$

$$= \frac{2z-0.5-z^2}{(1-z)(z-0.5)} \cdot \left(\frac{2 - 9z + 10z^2}{z}\right)$$

$$= \frac{(2z-0.5-z^2)}{(1-z)(z-0.5)} \cdot \frac{2(z-0.5)(5z-2)}{z}$$

$$= \frac{(2z-0.5-z^2)(10z-4)}{z(1-z)}$$

$$X(z) = \frac{24z - 13 + 2z^{-1} - 10z^2}{(1-z)} \quad ; \quad \text{ROC}: |z| < 1$$

Because $x(n)$ should be left-sided signal so that $y(n)$ would become both-sided.

So, all options are wrong.

8.55 (d)

Given:

Sampling rate $f_s = 10$ Hz

$$x(t) = \left[-2e^{-3t} + 3e^{-4t}\right] u(t)$$

So, sampled signal,

$$x(n) = x(t)\big|_{t=nT_s} = x(t)\big|_{t=\frac{n}{f_s}}$$

$$\therefore \quad x(n) = \left[-2e^{-\frac{3n}{10}} + 3e^{-\frac{4n}{10}}\right]u(n)$$

$$= -2e^{-\frac{3}{10}n}u(n) + 3e^{-\frac{2}{5}n}u(n)$$

For a signal $y(n) = a^n u(n)$, ROC: $|z| > |a|$.

Hence,

$$\text{ROC of } x(n): \left(|z| > e^{-\frac{3}{10}}\right) \cap \left(|z| > e^{-\frac{4}{10}}\right)$$

$$: \left(|z| > 0.7408\right) \cap \left(|z| > 0.6703\right)$$

$$: |z| > 0.7408$$

8.56 (c)

By applying ZT on given difference equation

$$Y(z) = \frac{5}{6}z^{-1}Y(z) - \frac{1}{6}z^{-2}Y(z) + X(z)$$

$$\Rightarrow \quad \frac{Y(z)}{X(z)} = \frac{1}{1 - \frac{5}{6}z^{-1} + \frac{1}{6}z^{-2}}$$

$$\Rightarrow \quad Y(z) = \frac{1}{\left(1 - \frac{5}{6}z^{-1} + \frac{1}{6}z^{-2}\right)} \cdot \frac{1}{\left(1 - 2z^{-1}\right)}$$

$$\Rightarrow \quad Y(z) = \frac{1}{\left(1 - \frac{1}{3}z^{-1}\right)\left(1 - \frac{1}{2}z^{-1}\right)(1 - 2z^{-1})}$$

$$= \frac{A}{1 - 2z^{-1}} + \frac{B}{1 - \frac{1}{3}z^{-1}} + \frac{C}{1 - \frac{1}{2}z^{-1}}$$

$$\left[\because A = \frac{1}{\left(1 - \frac{1}{3} \times \frac{1}{2}\right)\left(1 - \frac{1}{2} \times \frac{1}{2}\right)} = \frac{1}{\frac{5}{6} \times \frac{3}{4}} = \frac{8}{5}\right]$$

$$\Rightarrow \quad Y(z) = \frac{8/5}{1 - 2z^{-1}} + \frac{B}{1 - \frac{1}{3}z^{-1}} + \frac{C}{1 - \frac{1}{2}z^{-1}}$$

$$\Rightarrow y(n) = \frac{8}{5}2^n \cdot u(n) + B\left(\frac{1}{3}\right)^n u(n) + C\left(\frac{1}{2}\right)^n u(n)$$

$$\Rightarrow \quad y_p(n) = \text{Particular solution} = \frac{8}{5}(2)^n u(n)$$

■■■

9 Digital Filters: Design and Applications

9.1 Which one of the following is the correct state-space realization of a discrete system given by the difference equation
$c(k + 2) + \alpha c(k + 1) + \beta c(k) = u(k)$?

(a)

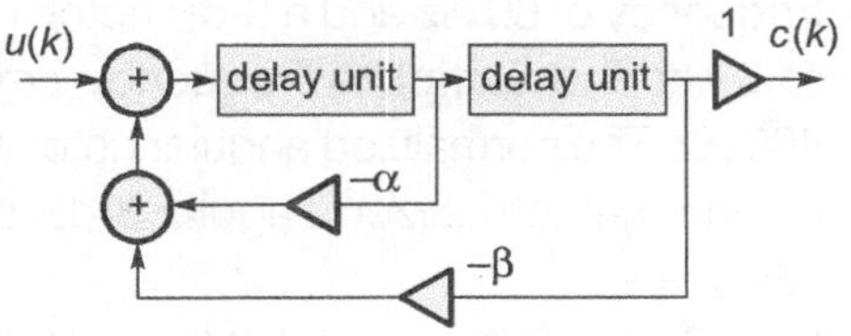

(b)

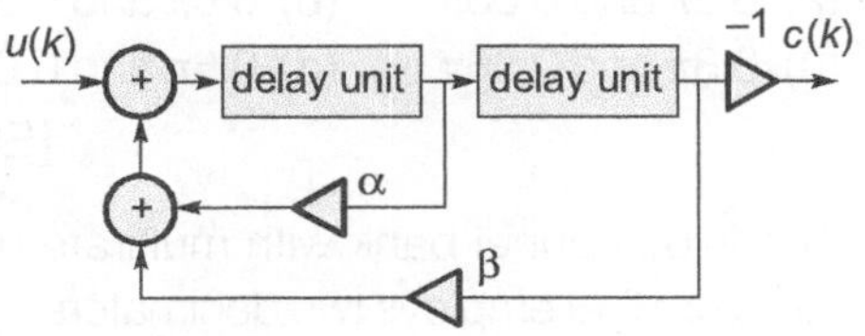

(c)

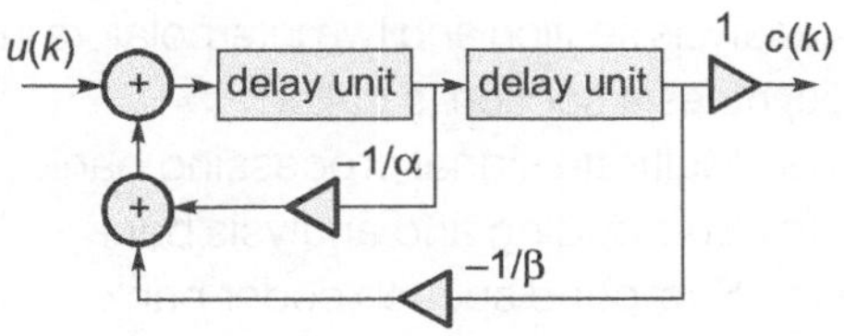

(d)

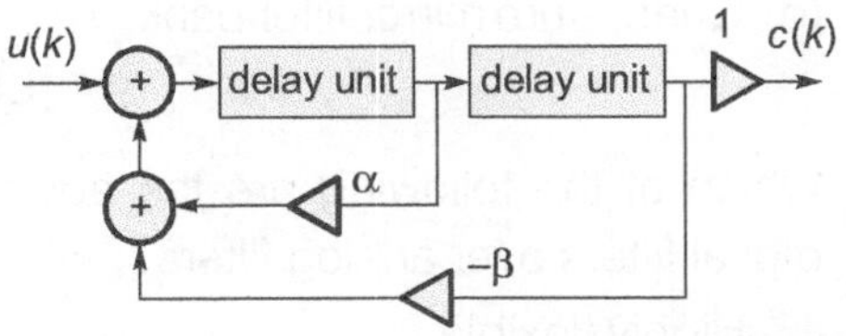

[ESE-2000]

9.2 Which one of the following digital filters does have a linear phase response?
(a) $y(n) + y(n - 1) = x(n) - x(n - 1)$
(b) $y(n) = 1/6[3x(n) + 2x(n - 1) + x(n - 2)]$
(c) $y(n) = 1/6[x(n) + 2x(n - 1) + 3x(n - 2)]$
(d) $y(n) = 1/4[x(n) + 2x(n - 1) + x(n - 2)]$

[ESE-2001]

9.3 The poles of a digital filter with linear phase response can lie
(a) only at $z = 0$
(b) only on the unit circle
(c) only inside the unit circle but not at $z = 0$
(d) on the left side of Real $(z) = 0$ line

[ESE-2001]

9.4 Consider the following systems:
1. $y[k] = x[k] + a_1x[k - 1] - b_1y[k - 1] - b_2y[k - 2]$
2. $y[k] = x[k] + a_1x[k - 1] - a_2x[k - 2]$
3. $y[k] = x[k + 1] + a_1x[k] + a_2x[k - 1]$
4. $y[k] = a_1x[k] + a_2x[k + 1] - b_1y[k - 2]$

Which of the systems given above represent recursive discrete system?
(a) 1 and 4 (b) 1 and 2
(c) 1, 2 and 3 (d) 2, 3 and 4

[ESE-2004]

9.5 An ideal low-pass filter has a cut-off frequency of 100 Hz. If the input to the filter in volts is $v(t) = 30\sqrt{2}\sin 1256t$, the magnitude of the output of the filter will be
(a) 0 V (b) 20 V
(c) 100 V (d) 200 V

[ESE-2012]

9.6 Consider an ideal low pass filter. Such a discrete-time system is
(a) always realizable physically
(b) never realizable physically
(c) a non linear system
(d) a linear, causal system

[ESE-2016]

9.7 Consider the following statements with respect to bilinear transformation method of digital filter design:
1. It preserves the number of poles and thereby the order of the filter.
2. It maintains the phase response of the analog filter.
3. The impulse response of the analog filter is not preserved.

Which of the above statements are correct?
(a) 1, 2 and 3 (b) 1 and 2 only
(c) 1 and 3 only (d) 2 and 3 only

[ESE-2017]

Direction (9.8 and 9.9): The following items consists of two statements, one labelled as **Statement (I)** and the other labelled as **Statement (II)**. You have to examine these two statements carefully and select your answers to these items using the codes given below:

Codes:

(a) Both Statement (I) and Statement (II) are individually true and Statement (II) is the correct explanation of Statement (I).
(b) Both Statement (I) and Statement (II) are individually true but Statement (II) is not the correct explanation of Statement (I).
(c) Statement (I) is true but Statement (II) is false.
(d) Statement (I) is false but Statement (II) is true.

9.8 **Statement (I):** FIR filters are always stable.
Statement (II): IIR filters require less memory and are less complex.

[ESE-2017]

9.9 **Statement (I):** Speech enhancement techniques are used to make a processed speech signal sound superior to the unprocessed one.
Statement (II): A 'perfect signal' is required as reference for speech enhancement.

[ESE-2017]

9.10 Consider the following statements pertaining to FIR filters:

1. These are non-recursive and hence stable.
2. These have high coefficient sensitivity.
3. These have linear phase characteristics.
4. These are realized using feedback structures.

Which of the above statements are correct?

(a) 1 and 4 (b) 2 and 3
(c) 1 and 3 (d) 2 and 4 [ESE-2018]

9.11 Which one of the following refers to the frequency ω_k in the frequency response of an FIR filter?

(a) $\frac{16\pi}{M}(k+\alpha)$ (b) $\frac{8\pi}{M}(k+\alpha)$
(c) $\frac{4\pi}{M}(k+\alpha)$ (d) $\frac{2\pi}{M}(k+\alpha)$

[ESE-2018]

9.12 The two advantages of FIR filters over IIR- filters are

(a) they are guaranteed to be stable and non-linear
(b) they are marginally stable and linear.
(c) they are. guaranteed to be stable and may be constrained to have linear phase
(d) they are marginally stable and non-linear

[ESE-2019]

9.13 The frequency response and the main lobe width for rectangular window are

(a) $\dfrac{\sin\frac{\omega N}{2}}{\sin\frac{\omega}{2}}$ and $\frac{4\pi}{N}$ (b) $\dfrac{\sin\frac{\omega N}{2}}{\frac{\omega}{2}}$ and $\frac{\pi}{N}$
(c) $\dfrac{\sin\frac{\omega}{2}}{\sin\frac{\omega N}{2}}$ and $\frac{2\pi}{N}$ (d) $\dfrac{\sin\frac{\omega N}{4}}{\sin\frac{\omega}{2}}$ and $\frac{8\pi}{N}$

[ESE-2019]

9.14 In a second order digital notch filter having notch frequency of 60 Hz and a 3-dB notch bandwidth of 6 Hz and sampling frequency employed is 400 Hz. The normalized angular notch frequency ω_o and the normalized angular 3-dB bandwidth $\Delta\omega_{3\text{ dB}}$ are

(a) 0.3π and 0.03π (b) 0.6π and 0.03π
(c) 0.3π and 0.06π (d) 0.6π and 0.06π

[ESE-2020]

9.15 The two channel bank with multirate digital filter structure that employs two decimators in the signal analysis section and two interpolators in the signal synthesis section is called

(a) Multirate signal processing bank
(b) Sub-coding and analysis bank
(c) Sub-band speech coder bank
(d) Quadrature mirror filter bank

[ESE-2020]

9.16 Which of the following are the advantages of digital filters over analog filters?

1. Highly flexible
2. Portable
3. Negligible noise interference
4. Lumped RLC components

(a) 1, 2 and 3 only (b) 1, 2 and 4 only
(c) 1, 3 and 4 only (d) 2, 3 and 4 only

[ESE-2020]

9.17 The realization of a length *M FIR* filter for a linear phase structure, the number of multipliers required is

(a) $\left[\frac{M+1}{2}\right]$ (b) $2M$
(c) M (d) $M-1$

[ESE-2020]

9.18 FIR filter having anti-symmetrical impulse response with even filter order can be used to design
(a) low-pass, high-pass, band-pass and band-stop.
(b) low-pass and band-pass only.
(c) high-pass and band-stop only.
(d) differentiator and Hilbert transformer.
[ESE-2021]

9.19 An IIR filter having numerator order M and denominator order N is to be realized using direct form II structure. How much total number of multiplications, additions and memory locations are required respectively
(a) $M + N$, $M + N$ and $\{M + N\}$
(b) $M + N$, $M + N$ and maximum of $[M, N]$
(c) $M + N + 1$, $M + N + 1$ and $M + N$
(d) $M + N + 1$, $M + N$ and maximum of $\{M, N\}$
[ESE-2021]

9.20 The approximate transition width of main lobe in Blackman window is
(a) $4\pi/N$ (b) $8\pi/N$
(c) $12\pi/N$ (d) $16\pi/N$
[ESE-2022]

9.21 Consider the following statements regarding filters:
1. Cauer filter has equiripple passband and stopband.
2. For a given filter order, passband and stopband deviations, Cauer filters have more transition bandwidth as compared to Chebyshev filters.
3. Linear phase characteristics cannot be achieved in IIR filters.

Which of the above statements are correct?
(a) 1 and 2 only (b) 1 and 3 only
(c) 2 and 3 only (d) 1, 2 and 3
[ESE-2022]

9.22 In the design of Digital IIR filters by means of Bilinear Transform, the design specifications are given. Match the following lists:

List-I
P. N and Δf fixed
Q. Δf and δ fixed
R. N and δ fixed

List-II
1. The design procedure has to start with the evaluation of the order of the filter necessary to meet the specifications in terms of the desired attenuation, transition bandwidth and pass-band deviation.
2. The filter is completely specified and the transition bandwidth is directly obtainable during the design procedure.
3. The design is completely determined for the Butterworth filter case by obtaining the value of the attenuation at f_a directly.

Select the correct answer using the code given below:

(a) P Q R: 2 3 1
(b) P Q R: 3 2 1
(c) P Q R: 1 2 3
(d) P Q R: 3 1 2

[ESE-2023]

9.23 Matched-z transformation method is used to design
(a) Inverse Chebyshev filter
(b) Non-recursive filters
(c) All pole filters
(d) Recursive filters
[ESE-2024]

9.24 The recursive system described by a linear constant-coefficient difference equation is
(a) linear and time invariant
(b) linear and time variant
(c) non-linear and time invariant
(d) non-linear and time variant
[ESE-2024]

■■■

Answers Digital Filters: Design and Applications

9.1 (a)	9.2 (d)	9.3 (a)	9.4 (a)	9.5 (a)	9.6 (b)	9.7 (c)	9.8 (b)	9.9 (c)
9.10 (c)	9.11 (d)	9.12 (c)	9.13 (a)	9.14 (a)	9.15 (d)	9.16 (a)	9.17 (a)	9.18 (d)
9.19 (d)	9.20 (c)	9.21 (b)	9.22 (d)	9.23 (d)	9.24 (a)			

Explanations Digital Filters: Design and Applications

9.1 (a)

$c(k+2) = -\alpha c(k+1) - \beta c(k) + u(k)$

The correct state-space realization is shown below.

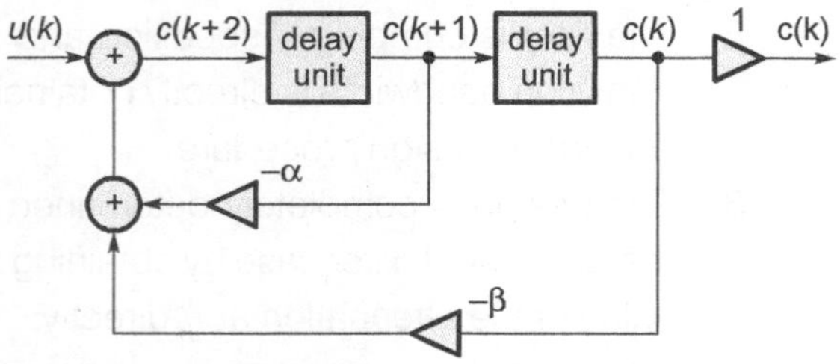

9.3 (a)

For linear phase response,

$$y[n] = Kx[n-n_0]$$

$$\Rightarrow \quad Y(z) = Kz^{-n_0}X(z)$$

⇒Transfer function,

$$H(z) = Kz^{-n_0}$$

which has the poles only at $z = 0$.

9.4 (a)

System 2 and 3 are not recursive because it is not in the form $y(n-k) = \Sigma x(n-k)$.

9.5 (a)

An ideal low pass filter has a cut-off frequency of 100 Hz, so it pass the signals which have their frequency components below or equal to 100 Hz. When the signal of 200 Hz (as given in the problem) will try to pass through LPF, it will not pass and hence the output will be zero.

9.6 (b)

Ideal LPF magnitude response

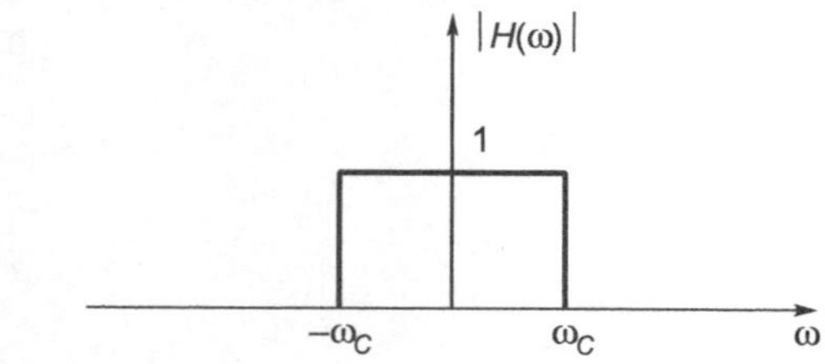

$$H(\omega) = |H(\omega)|\, e^{j\angle H(\omega)}$$

Inverse Fourier transform of above frequency response i.e. $h(t)$ will be a sinc function which is non zero for $t < 0$. Therefore system cannot be realized physically.

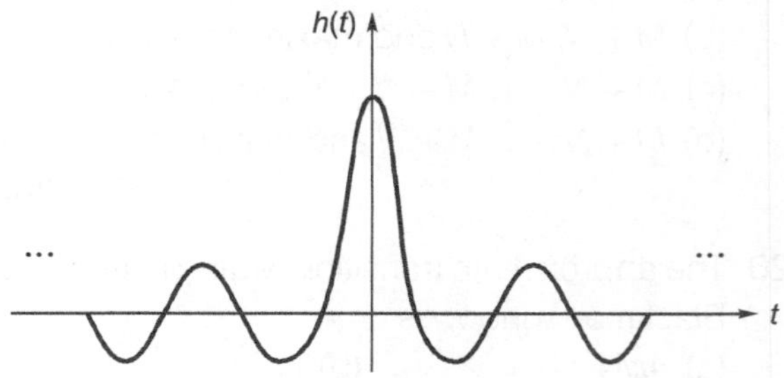

9.7 (c)

Bilinear transformation does not maintain phase response of analog filter.

9.8 (b)

FIR are always stable. IIR requires less memory and are less complex.

9.12 (c)

FIR filters are always stable and can have linear phase.

9.13 (a)

For rectangular window, main lobe width is $\frac{4\pi}{N}$

and frequency response is $\dfrac{\sin\frac{\omega N}{2}}{\sin\frac{\omega}{2}}$.

9.14 (a)

$$\omega = \Omega \cdot T_s = \frac{\Omega}{f_s} = \frac{2\pi 60}{400} = 0.3\pi$$

$$\text{Bandwidth} = \frac{0.3\pi}{10} = 0.03\pi$$

9.18 (d)

The given FIR filter having anti-symmetrical impulse response with even filter order is a Type-III FIR filter.

Both Type-III and Type-IV FIR filters (i.e. filters with anti-symmetrical impulse response) introduce a 90° phase shift and are used to design differentiators and Hilbert transformers.

The frequency response of a Type-II filter is always zero at $f = 0.5$ (half the sampling frequency as all the frequencies are normalized to sampling frequency) and thus, is unsuitable as a high pass filter.

The frequency response of a Type-III filter is always 0 at $f = 0$ and 0.5, while that of Type-IV filter is zero at $f = 0$. Thus, Type-III filter cannot be used as either a low-pass or high pass filter whereas Type-IV cannot be used as a low-pass filter.

9.19 (d)

For, $$H(z) = \frac{b_0 + b_1 z^{-1} + ... + b_M z^{-M}}{1 + a_1 z^{-1} + ... + a_N z^{-N}},$$

the direct form-II realization is given as:

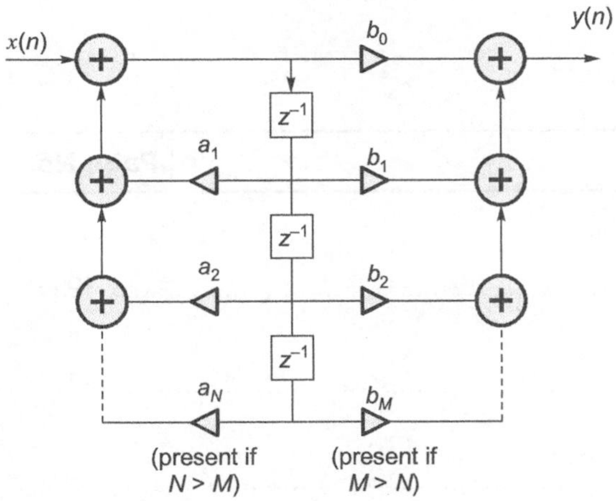

Hence, it requires,

$(M + N + 1)$ multiplications

$(M + N)$ additions

max(M, N) delay elements/memory locations.

9.20 (c)

The approximate transition width of the main lobe in the case of Blackman window is equal to 12 π/M where M is the length of the window.

9.21 (b)

- The elliptic filter, also known as cauer filter has equiripple passband and stopband.
- For a given filter order, passband and stopband deviations, cauer filters have less transition bandwidth as compared to Chebyshev filters. Hence, statement 2 is not correct.
- In order to get linear phase, an IIR filter would need conjugate reciprocal poles outside of the unit circle making it BIBO unstable. Hence, linear phase characteristics cannot be achieved in IIR filters.

9.23 (d)

The matched z-transform method is used to design recursive (IIR) filters by mapping all poles and zeros of the s-plane to z-plane locations $z = e^{sT}$, for a sample interval $T = 1/f_s$.

9.24 (a)

The linear constant coefficient difference equation can be represented as below:

$$y(n) = -\sum_{k=1}^{N} a_k y(n-k) + \sum_{k=0}^{M} b_k x(n-k)$$

$$\Rightarrow \quad \sum_{k=0}^{N} a_k y(n-k) = \sum_{k=0}^{M} b_k x(n-k), \text{ where, } a_0 = 1$$

It represents a linear and time-invariant system.

UNIT

III

Electromagnetics

Syllabus

Elements of vector calculus, Maxwell's equations-basic concepts; Gauss, Stokes theorems; Wave propagation through different media; Transmission Lines-different types, basics, Smith's chart, impedance matching/transformation, S-parameters, pulse excitation, uses; Waveguides-basics, rectangular types, modes, cut-off frequency, dispersion, dielectric types; Antennas-radiation pattern, monopoles/dipoles, gain, arrays-active/passive, theory, uses.

Contents

Sl.	Topic	Page No.
1.	Basic of Electromagnetics	195
2.	Uniform Plane Waves	231
3.	Transmission Lines	250
4.	Waveguides	269
5.	Antennas and Radar	284

OOOO

1 Basic of Electromagnetics

1.1 Match **List-I** (Laws) with **List-II** (Applications) and select the correct answer:

List-I

A. Ampere's law
B. Biot's law
C. Coulomb's law
D. Gauss's law

List-II

To find the
1. force on a charge
2. field due to a current element
3. electric flux density at a point
4. magnetic flux density at a point

	A	B	C	D
(a)	3	2	1	4
(b)	4	2	1	3
(c)	4	1	2	3
(d)	3	1	2	4

[ESE-2000]

1.2 A solid cylindrical conductor of radius 'R' has a uniform current density. The magnetic field 'H' inside the conductor at a distance 'r' from the axis of the conductor is
(a) $I/2\pi r$
(b) $I/4\pi r$
(c) $Ir/2\pi R^2$
(d) $Ir/4\pi R^2$ [ESE-2000]

1.3 The equation $\nabla \cdot \vec{J} = 0$ is known as
(a) Poisson's equation
(b) Laplace equation
(c) Continuity equation
(d) Maxwell equation [ESE-2000]

1.4 In a hundred-turn coil, if the flux through each turn is $(t^3 - 2t)$mWb, the magnitude of the induced emf in the coil at a time of 4s is
(a) 46 mV
(b) 56 mV
(c) 4.6 V
(d) 5.6 V [ESE-2000]

1.5 The dimensions of magnetic flux density are
(a) $MT^{-1}Q^{-1}$
(b) $MT^{-2}Q^{-2}$
(c) $MT^{1}Q^{1}$
(d) $MT^{-1}Q^{-2}$

[ESE-2000]

1.6 **Assertion (A):** The solution of Poisson's equation is the same as the solution of Laplace's equation.
Reason (R): Laplace's equation is a special case of Poisson's equation for source-free region.
(a) Both A and R are true and R is the correct explanation of A
(b) Both A and R are true but R is NOT the correct explanation of A
(c) A is true but R is false
(d) A is false but R is true [ESE-2000]

1.7 A boundary separates two magnetic materials of permeabilities μ_1 and μ_2. The magnetic field vector in μ_1 is H_1 with a normal component H_{n_1} and tangential component H_{t_1} while that in μ_2 is H_2 with a normal component H_{n_2} and a tangential component H_{t_2}. Then the derived conditions would be
(a) $H_1 = H_2$ and $H_{t_1} = H_{t_2}$
(b) $H_{t_1} = H_{t_2}$ and $\mu_1 H_{n_1} = \mu_2 H_{n_2}$
(c) $H_1 = H_2$ and $\mu_1 H_{n_1} = \mu_2 H_{n_2}$
(d) $H_1 = H_2$, $H_{t_1} = H_{t_2}$ and $\mu_1 H_{n_1} = \mu_2 H_{n_2}$

[ESE-2001]

1.8 The energy stored per unit volume in an electric field (with usual notations) is given by
(a) $\frac{1}{2}\varepsilon H^2$
(b) $\frac{1}{2}\varepsilon E$
(c) $\frac{1}{2}\varepsilon E^2$
(d) εE^2 [ESE-2001]

1.9 A positive charge of Q Coulombs is located at point $A(0, 0, 3)$ and a negative charge of magnitude Q Coulombs is located at point $B(0, 0, -3)$. The electric field intensity at point $C(4, 0, 0)$ is in the
(a) negative x-direction
(b) negative z-direction
(c) positive x-direction
(d) positive z-direction [ESE-2001]

1.10 The force between two point charges of 1 nC each with a 1 mm separation in air is
(a) 9×10^{-3}N (b) 9×10^{-6}N
(c) 9×10^{-9}N (d) 9×10^{-12}N
[ESE-2001]

1.11 A 3 μF capacitor is charged by a constant current of 2 μA for 6 seconds. The voltage across the capacitor to the end of charging will be
(a) 3 V (b) 4 V
(c) 6 V (d) 9 V [ESE-2001]

1.12 Consider the following statements:
A parallel plate capacitor is filled with a dielectric of relativity ϵ_{r_1} and connected to a d.c. voltage of V volts. If the dielectric is changed to another with relative permittivity $\epsilon_{r_2} = 2\epsilon_{r_1}$, keeping the voltage constant, then
1. the electric field intensity E within the capacitor doubles.
2. the displacement flux density D doubles.
3. the charge Q on the plates is reduced to half.
4. the energy stored in the capacitor is doubled.

Select the correct answer using the codes given below:
(a) 1 and 2 (b) 2 and 3
(c) 2 and 4 (d) 3 and 4
[ESE-2001]

1.13 Laplace equation in cylindrical coordinates is given by

(a) $\nabla^2 V = \frac{1}{\rho}\frac{\partial}{\partial \rho}\left(\frac{\rho \partial V}{\partial \rho}\right) + \frac{1}{\rho^2}\left(\frac{\partial^2 V}{\partial \phi^2}\right) + \frac{\partial^2 V}{\partial z^2} = 0$

(b) $\nabla^2 V = \frac{\partial^2 V}{\partial x^2} + \frac{\partial^2 V}{\partial y^2} + \frac{\partial^2 V}{\partial z^2}$

(c) $\nabla^2 V = -\frac{\rho}{\varepsilon}$

(d) $\nabla^2 V = \frac{1}{r}\frac{\partial}{\partial r}\left(\frac{r^2 \partial V}{\partial r}\right) + \frac{1}{r^2 \sin\theta}$
$\frac{\partial}{\partial \theta}\left(\sin\theta \frac{\partial V}{\partial \theta}\right) + \frac{1}{r^2 \sin^2\theta}\frac{\partial^2 V}{\partial \phi^2} = 0$

[ESE-2001]

1.14 Gauss's law relates the electric field intensity $\bar{E}$ with the volume charge density ρ at a point as
(a) $\nabla \times \bar{E} = \varepsilon_0 \rho$ (b) $\nabla \cdot \bar{E} = \varepsilon_0 \rho$
(c) $\nabla \times \bar{E} = \rho/\varepsilon_0$ (d) $\nabla \cdot \bar{E} = \rho/\varepsilon_0$
[ESE-2001]

1.15 A coil of resistance of 5 Ω and inductance 0.4 H is connected to a 50 V d.c. supply. The energy stored in the field is
(a) 10 Joules (b) 20 Joules
(c) 40 Joules (d) 80 Joules
[ESE-2001]

1.16 A field $\vec{A} = 3x^2yz\hat{a}_x + x^3z\hat{a}_y + (x^3y - 2z)\hat{a}_z$ can be termed as
(a) Irrotational (b) Divergence less
(c) Solenoidal (d) Rotational
[ESE-2002]

1.17 The circulation of $\vec{H}$ around the closed contour C, shown in the figure is

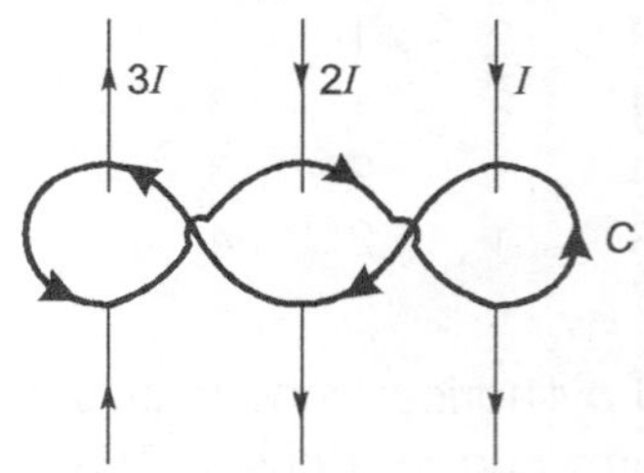

(a) 0 (b) $2I$
(c) $4I$ (d) $6I$
[ESE-2002]

1.18 The total length of the curve $\mu = \cos^2\theta$ (Cylindrical co-ordinates) from $\theta = 0$ to $\theta = \pi$ is
(a) 0.5π (b) 1.0
(c) 2.0 (d) π [ESE-2002]

1.19 An electric charge Q is placed in a dielectric medium. Which of the following quantities are independent of the dielectric constant $\in$ of the medium?
(a) Electric potential V and Electric field intensity E
(b) Displacement density D and Displacement ψ
(c) Electric field intensity E and Displacement density D
(d) Electric potential V and Displacement ψ
[ESE-2003]

1.20 Two coaxial cylindrical sheets of charge are present in free space, $\rho_s = 5$ C/m^2 at $r = 2$ m and $\rho_s = -2$ C/m^2 at $r = 4$ m. The displacement flux density $\bar{D}$ at r = 3 m is

(a) $\bar{D} = 5\hat{a}_r$ C/m^2 (b) $\bar{D} = \frac{2}{3}\hat{a}_r$ C/m^2

(c) $\bar{D} = \frac{10}{3}\hat{a}_r$ C/m^2 (d) $\bar{D} = \frac{18}{3}\hat{a}_r$ C/m^2

[ESE-2003]

1.21 Two thin parallel wires are carrying current along the same direction. The force experienced by one due to the other is
(a) Parallel to the lines
(b) Perpendicular to the lines and attractive
(c) Perpendicular to the lines and repulsive
(d) Zero

[ESE-2003]

1.22 An electric potential field is produced in air by point charges 1 μC and 4 μC located at (– 2, 1, 5) and (1, 3, – 1) respectively. The energy stored in the field is
(a) 2.57 mJ (b) 5.14 mJ
(c) 10.28 mJ (d) 12.50 mJ

[ESE-2003]

1.23 Which one of the following potentials does NOT satisfy Laplace's Equation?
(a) $V = 10\,xy$ (b) $V = r\cos\phi$
(c) $V = 10/r$ (d) $V = \rho\cos\phi + 10$

[ESE-2003]

1.24 Laplacian of a scalar function V is
(a) Gradient of V
(b) Divergence of V
(c) Gradient of the gradient of V
(d) Divergence of the gradient of V

[ESE-2003]

1.25 An air condenser of capacitance of 0.005 μF is connected to a d.c. supply of 500 Volts, disconnected and then immersed in oil with a dielectric constant of 2.5. Energy stored in the capacitor before and after immersion, respectively is
(a) 5.00 × 10^{-4} Joules and 2.50 × 10^{-4} Joules
(b) 2.50 × 10^{-4} Joules and 5.00 × 10^{-4} Joules
(c) 6.25 × 10^{-4} Joules and 2.50 × 10^{-4} Joules
(d) 2.50 × 10^{-4} Joules and 6.25 × 10^{-4} Joules

[ESE-2003]

1.26 A point charge of +10 μC placed at a distance of 5 cm from the centre of a conducting grounded sphere of radius 2 cm is shown in the diagram given below:

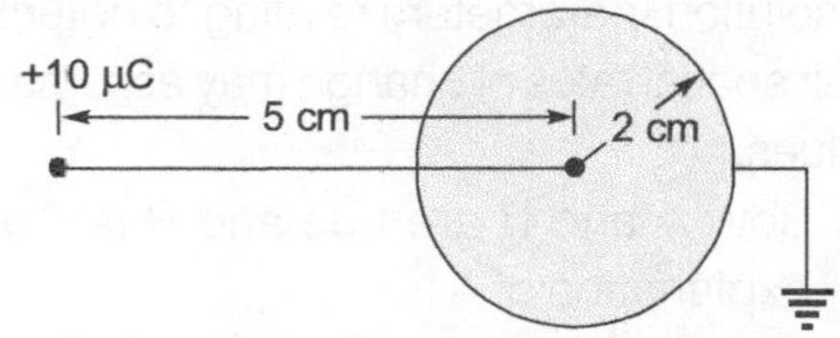

What is the total induced charge on the conducting sphere?
(a) 10 μC (b) 4 μC
(c) 5 μC (d) 12.5 μC

[ESE-2004]

1.27 An infinitely long uniform charge of density 30 nC/m is located at $y = 3$, $z = 5$. The field intensity at (0, 6, 1) is $\vec{E} = 64.7\hat{a}_y - 86.3\hat{a}_z$ V/m.

What is the field intensity at (5, 6, 1)?
(a) E

(b) $\left(\frac{6^2+1^2}{5^2+6^2+1^2}\right)E$

(c) $\left(\frac{6^2+1^2}{5^2+6^2+1^2}\right)^{1/2}E$

(d) $\left(\frac{5^2+6^2+1^2}{6^2+1^2}\right)^{1/2}E$

[ESE-2004]

1.28 What is the magnetic dipole moment in A.m^2 for a square current loop having the vertices at the points A(10, 0, 0), B(0, 10, 0), C(–10, 0, 0) and D(0, –10, 0) and with current 0.01 A flowing in the sense $ABCDA$?

(a) $2\hat{a}_z$ (b) $-2\hat{a}_z$

(c) $4\hat{a}_z$ (d) $4(\hat{a}_x + \hat{a}_y)$

[ESE-2004]

1.29 An infinitely long line charge of uniform charge density ρ_0 C/m is situated parallel to and at a distance from the grounded infinite plane conductor. This field problem can be solved by which one of the following?
(a) By conformal transformation
(b) By method of images
(c) By Laplace's equation
(d) By Poisson's equation [ESE-2004]

1.30 **Assertion (A):** Even if two different approaches are used to solve Laplace equation, we will end up in the same solution.

Reason (R): It is mainly because in the process of solution, parameters relating to potentials and their spatial rates of change may assume different values.

(a) Both A and R are true and R is the correct explanation of A
(b) Both A and R are true but R is NOT the correct explanation of A
(c) A is true but R is false
(d) A is false but R is true [ESE-2005]

1.31 The capacitance of a parallel plate capacitor is given by $\varepsilon_0 \varepsilon_r (A/d)$ where A is the area of each plate and d is the distance between the plates. Considering fringing field, under which one of the following conditions is the above expression valid?

(a) $\frac{A}{d}$ is tending towards zero
(b) $\frac{A}{d}$ is tending towards infinity
(c) $\frac{A}{d}$ is 1
(d) $\frac{A}{d}$ is $\frac{1}{\epsilon_r \epsilon_0}$ [ESE-2005]

1.32 Current density $(\vec{J})$, in cylindrical co-ordinate system is given as

$$\vec{J}(r, \theta, z) = 0 \text{ for } 0 < r < a$$
$$= J_0(r/a^2)\vec{I}_z \text{ for } a < r < b$$

Where $\vec{I}_z$ is the unit vector along z-axis. In the region, $a < r < b$, what is the expression for the magnitude of magnetic field intensity vector $(\vec{H})$?

(a) $\frac{J_0}{r^2}(r^3 - a^3)$ (b) $\frac{J_0}{r^2}(r^3 + a^3)$
(c) $\frac{J_0(r^3 - a^3)}{3a^2 r}$ (d) $\frac{J_0}{2\pi r}(r^3 - a^3)$

[ESE-2005]

1.33 The potential (scalar) distribution is given as $V = 10y^4 + 20x^3$. If ϵ_0 is the permittivity of free space, what is the charge density ρ at the point (2, 0)?

(a) $-200\,\epsilon_0$ (b) $-200/\epsilon_0$
(c) $200\,\epsilon_0$ (d) $-240\,\epsilon_0$

[ESE-2005]

1.34 What is the expression for capacitance of an infinitely conducting solid sphere of radius 'R' in free space?

(a) $2\pi\,\epsilon_0 R$ (b) $4\pi\,\epsilon_0 R$
(c) $8\pi\,\epsilon_0 R$ (d) $0.5\pi\,\epsilon_0 R$

[ESE-2005]

1.35 **Assertion (A):** Potential everywhere on a conducting surface of infinite extent is zero.

Reason (R): Displacement density on a conducting surface is normal to the surface.

(a) Both A and R are true and R is the correct explanation of A
(b) Both A and R are true but R is NOT the correct explanation of A
(c) A is true but R is false
(d) A is false but R is true [ESE-2006]

1.36 A parallel plate capacitor has a homogeneous dielectric slab of area 4 cm^2 and thickness 1 cm with $\varepsilon_r = 4$. If a cubic air void of 1 mm^3 is developed in the volume of the dielectric slab, then

(a) the average voltage stress in the dielectric material will increase and discharge free rating of the capacitor will increase
(b) the average voltage stress in the dielectric material will increase and discharge free rating of the capacitor will decrease
(c) the average voltage stress in the dielectric material will decrease and discharge free rating of the capacitor will decrease
(d) the average voltage stress in the dielectric material will decrease and discharge free rating of the capacitor will increase

[ESE-2006]

1.37 A parallel plate capacitor of 5 pF capacitance has a charge of 0.1 μC on its plates. What is the energy stored in the capacitor?

(a) 1 mJ (b) 1 μJ
(c) 1 nJ (d) 1 pJ [ESE-2006]

1.38 Equipotential surfaces about a point charge are in which one of the following forms?

(a) Spheres (b) Planes
(c) Cylinders (d) Cubes [ESE-2006]

1.39 Which one of the following is the correct expression for torque on a loop in magnetic field $\vec{B}$? (Here $\vec{M}$ is the loop moment)

(a) $\vec{T} = \nabla \cdot \vec{B}$ (b) $\vec{T} = \vec{M} \cdot \vec{B}$

(c) $\vec{T} = \vec{M} \times \vec{B}$ (d) $\vec{T} = \vec{B} \times \vec{M}$

[ESE-2006]

1.40 What does the expression $\frac{1}{2}\vec{J} \cdot \vec{A}$ represent?

(a) Electric energy density
(b) Magnetic energy density
(c) Power density
(d) Radiation resistance [ESE-2006]

1.41 Consider the following statements regarding an electrostatic field:

1. It is irrotational
2. It is solenoidal
3. It is static only from a macroscopic view point.
4. Work done in moving a charge in the field from one point to another is independent of the path of movement.

Which of these statements are correct?

(a) 1, 2 and 3 (b) 1, 2 and 4
(c) Only 2 and 4 (d) 1, 3 and 4

[ESE-2006]

1.42 Which one of the following statements is correct?

(a) Both Laplace's and Poisson's equations are non-linear equations.
(b) Laplace's equation is non-linear but Poisson's equation is linear.
(c) Laplace's equation is linear but Poisson's equation is non-linear
(d) Both Laplace's and Poisson's equations are linear [ESE-2006]

1.43 A charge of 1 Coulomb is placed near a grounded conducting plate at a distance of 1 m. What is the force between them?

(a) $\frac{1}{4\pi\varepsilon_0}N$ (b) $\frac{1}{8\pi\varepsilon_0}N$

(c) $\frac{1}{16\pi\varepsilon_0}N$ (d) $4\pi e_0 N$ [ESE-2006]

1.44 Match **List-I** (Electromagnetic Law) with **List-II** (Differential Form) and select the correct answer using the codes given below the lists:

List-I	List-II
A. Ampere's law	1. $\nabla \cdot \vec{D} = \rho_v$
B. Faraday's law	2. $\nabla \cdot \vec{J} = -\frac{\partial \vec{H}}{\partial t}$
C. Gauss law	3. $\nabla \times \vec{H} = \vec{J} + \frac{\partial \vec{D}}{\partial t}$
D. Current continuity equation	4. $\nabla \times \vec{E} = -\frac{\partial \vec{B}}{\partial t}$

Codes:

	A	B	C	D
(a)	1	2	3	4
(b)	3	4	1	2
(c)	1	4	3	2
(d)	3	2	1	4

[ESE-2006]

1.45 Which one of the following statements is correct? On a conducting surface boundary, electric field lines are

(a) always tangential
(b) always normal
(c) neither tangential nor normal
(d) at an angle depending on the field intensity

[ESE-2007]

1.46 Magnetic current is composed of which of the following?

(a) Only conduction component
(b) Only displacement component
(c) Both conduction and displacement components
(d) Neither conduction component nor displacement component

[ESE-2007]

1.47 Application of the method of images to a boundary value problem in electrostatics involves which one of the following?

(a) Introduction of an additional distribution of charges and removal of a set of conducting surfaces
(b) Introduction of an additional distribution of charges and an additional set of conducting surfaces
(c) Removal of a charge distribution and introduction of an additional set of conducting surfaces
(d) Removal of a charge distribution as well as a set of conducting surfaces

[ESE-2007]

1.48 What is the value of the magnetic vector potential due to an infinitesimally small current element, evaluated at infinite distance from it?
(a) Infinity
(b) Unity
(c) Zero
(d) Any number between zero and infinity depending on the strength of the current element

[ESE-2007]

1.49 The potential distribution V between two infinite flat metal sheets meeting at an angle θ is to be found. The left edges of the sheets are separated by an infinitesimal gap. The lower sheet is at zero potential and the upper sheet is at potential V_1. If α be an angle from $V = 0$ plate, what is the potential distribution?

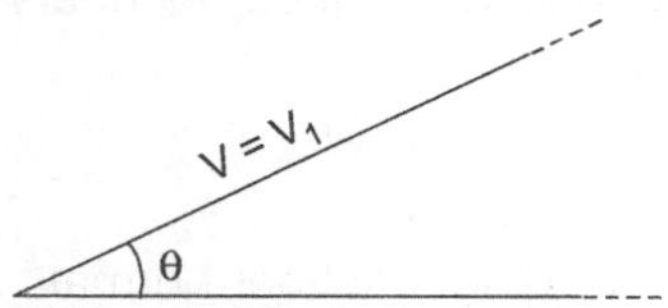

(a) $V = V_1 \sin\left(\frac{\pi\alpha}{2\theta}\right)$

(b) $V = V_1 \sin\left(\frac{\alpha}{2\theta}\right)$

(c) $V = V_1 \sin\left(\frac{3\pi\alpha}{2\theta}\right)$

(d) $V = V_1 \sin\left(\frac{\alpha}{\theta}\right)$

[ESE-2007]

1.50 W_1 is the electrostatic energy stored in a system of three equal point charges arranged in a line with 0.5 m separation between them. If W_2 is the energy stored with 1 m separation between them, then which one of the following is correct?
(a) $W_1 = 0.5\ W_2$ (b) $W_1 = W_2$
(c) $W_1 = 2\ W_2$ (d) $W_1 = 4\ W_2$

[ESE-2007]

1.51 What is the magnetic field intensity vector $\vec{H}$ between two parallel sheets with separation 'd' along z-axis both sheets carrying surface current $\vec{k} = k_y\, \hat{a}_y$?
(a) $-k_y\, \hat{a}_y$ (b) $+k_y\, \hat{a}_y$
(c) $-k_y\, \hat{a}_x$ (d) Zero [ESE-2007]

1.52 The electric field intensity $\vec{E}$ at a point P is given by $10\hat{i} + 10\hat{j} + 10\hat{k}$ where $\hat{i}, \hat{j}$ and $\hat{k}$ are unit vectors in x, y and z directions respectively. If α, β, γ are respectively the angles the $\vec{E}$ vector makes with x, y and z axis respectively, they are given by which of the following?
(a) $\alpha = \beta = \gamma = 30°$
(b) $\alpha = \beta = \gamma = 60°$
(c) $\alpha = \beta = \gamma = \cos^{-1}\frac{1}{\sqrt{3}}$
(d) $\alpha = \beta = \gamma = \cos^{-1}\frac{1}{3}$ [ESE-2007]

1.53 The line integral of the vector potential $\vec{A}$ around the boundary of a surface S represents which one of the following?
(a) Flux through the surface S
(b) Flux density in the surface S
(c) Magnetic field intensity
(d) Current density [ESE-2008]

1.54 What is the electric flux density (in $\mu C/m^2$) at a point (6, 4, –5) caused by a uniform surface charge density of 60 $\mu C/m^2$ at a plane $x = 8$?
(a) $-30\ \hat{a}_x$ (b) $-60\ \hat{a}_x$
(c) $30\ \hat{a}_x$ (d) $60\ \hat{a}_x$

[ESE-2008]

1.55 An infinitely long straight conductor located along z-axis carries a current I in the positive z-direction. The magnetic field at any point P in the x-y plane is in which direction?
(a) In the positive z-direction
(b) In the negative z-direction
(c) In the direction perpendicular to the radial line OP (in x-y plane) joining the origin O to the point P
(d) Along the radial line OP [ESE-2008]

1.56 A 5 A current enters a right circular cylinder of 5 cm radius. What is the linear surface current density at the end surface?
(a) $(50/\pi)$ A/m (b) $(100/\pi)$ A/m
(c) $(1000/\pi)$ A/m (d) $(2000/\pi)$ A/m

[ESE-2008]

1.57 Of two concentric long conducting cylinders, the inner one is kept at a constant positive potential $+V_0$ and the outer one is grounded. What is the electric field in the space between the cylinders?

(a) Uniform and directed radially outwards
(b) Uniform and directed radially inwards
(c) Non-uniform and directed radially outwards
(d) Non-uniform and directed parallel to the axis of the cylinders

[ESE-2008]

1.58 In a charge-free space, the Poisson's equation results in which one of the following?
(a) Continuity equation
(b) Maxwell's equation
(c) Laplace equation
(d) None of the above [ESE-2008]

1.59 What is the unit of measurement of surface/ sheet resistivity?
(a) Ohm/metre (b) Ohm metre
(c) Ohm/sq. meter (d) Ohm [ESE-2008]

1.60 In the field of a charge Q at the origin, the potentials at $A(2, 0, 0)$ and $B(1/2, 0, 0)$ are $V_A = 15$ volt and $V_B = 30$ volt respectively. What will be the potential at $C(1, 0, 0)$?
(a) 25 volt (b) 22·5 volt
(c) 20 volt (d) 17·5 volt

[ESE-2009]

1.61 What will be the equipotential surfaces for a pair of equal and opposite line charges?
(a) Spheres
(b) Concentric cylinders
(c) Non-concentric cylinders
(d) None of the above

[ESE-2009]

1.62 If the potential functions V_1 and V_2 satisfy Laplace's equation within a closed region and assume the same values on its surface, then which of the following is correct?
(a) V_1 and V_2 are identical
(b) V_1 is inversely proportional to V_2
(c) V_1 has the same direction as V_2
(d) V_1 has the same magnitude as V_2 but has different direction

[ESE-2009]

1.63 If $V = \sinh x \cdot \cos ky \cdot e^{pz}$ is a solution of Laplace's equation, what will be the value of k?
(a) $\dfrac{1}{\sqrt{1+p^2}}$ (b) $\sqrt{1+p^2}$
(c) $\dfrac{1}{\sqrt{1-p^2}}$ (d) $\sqrt{1-p^2}$ [ESE-2009]

1.64 By what name is the equation $\nabla \cdot \vec{J} = 0$ frequently known?
(a) Poisson's equation
(b) Laplace's equation
(c) Continuity equation for steady currents
(d) Displacement equation [ESE-2009]

1.65 Method of images is applicable to which fields?
(a) Electrostatic fields only
(b) Electrodynamic fields only
(c) Neither electrostatic fields nor electrodynamic fields
(d) Both electrostatic fields and electrodynamic fields

[ESE-2009]

1.66 A single turn loop is situated in air, with a uniform magnetic field normal to its plane. The area of the loop is 5 m^2 and the rate of change of flux density is 2 Wb/m^2/s. What is the emf appearing at the terminals of the loop?
(a) –5 V (b) –2 V
(c) –0·4 V (d) –10 V [ESE-2009]

1.67 Assertion (A) : Capacitance of a solid conducting spherical body of radius 'a' is given by $4\pi\varepsilon_0 a$ in free space.
Reason (R) : $\nabla \times \vec{H} = j\omega\varepsilon\vec{E} + \vec{J}$
(a) Both A and R are true and R is the correct explanation of A
(b) Both A and R are true but R is NOT the correct explanation of A
(c) A is true but R is false
(d) A is false but R is true [ESE-2009]

1.68 Assertion (A) : The expression $\vec{E} = -\nabla V$, where $\vec{E}$ is the electric field and V, is the potential is not valid for time varying fields.
Reason (R) : The curl of a gradient is identically zero.
(a) Both A and R are true and R is the correct explanation of A
(b) Both A and R are true but R is NOT the correct explanation of A
(c) A is true but R is false
(d) A is false but R is true

[ESE-2009]

1.69 There are two conducting plates of sizes 1 m × 1 m and 2 m × 2 m. Ratio of the capacitance of the second one with respect to that of the first one is

(a) 4 (b) 2
(c) 1/2 (d) 1/4

[ESE-2010]

1.70 The flux and potential functions due to a line charge and due to two concentric circular conductors are of the following form:

(a) Concentric circular equipotential lines and straight radial flux lines
(b) Concentric circular flux lines and straight equipotential lines
(c) Equipotential due to line charge are concentric cylinders and equipotential due to two conductors are straight lines
(d) Equipotentials due to line charge are straight flat surfaces and those due to two conductors are concentric cylinders

[ESE-2010]

1.71 Consider points *A*, *B*, *C* and *D* on a circle of radius 2 units as in the below figure. The items in List-II are the values of $\hat{a}_\phi$ at different points on the circle. Match **List-I** with **List-II** and select the correct answer using the codes given below the lists:

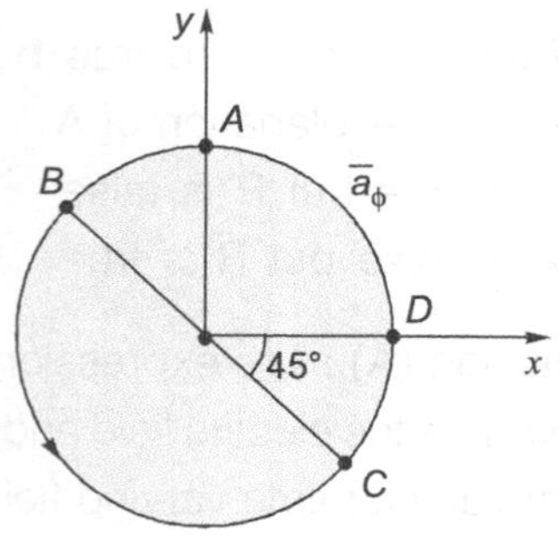

List-I	List-II
A.	1. $\hat{a}_x$
B.	2. $\hat{a}_y$
C.	3. $-\hat{a}_x$
D.	4. $(\hat{a}_x + \hat{a}_y)/\sqrt{2}$
	5. $-(\hat{a}_x + \hat{a}_y)/\sqrt{2}$
	6. $(\hat{a}_x - \hat{a}_y)/\sqrt{2}$

Codes:

	A	B	C	D
(a)	3	4	5	2
(b)	1	6	5	2
(c)	1	6	2	4
(d)	3	5	4	2

[ESE-2010]

1.72 An electric charge of *Q* coulombs is located at the origin. Consider electric potential *V* and electric field intensity *E* at any point (*x*, *y*, *z*). Then

(a) *E* and *V* are both scalars
(b) *E* and *V* are both vectors
(c) *E* is a scalar and *V* is a vector
(d) *E* is a vector and *V* is scalar [ESE-2010]

1.73 Consider the following:
In a parallel plate capacitor, let the charge be held constant while the dielectric material is replaced by a different dielectric. Consider

1. Stored energy.
2. Electric field intensity.
3. Capacitance.

Which of these changes?

(a) 1 only (b) 1 and 2 only
(c) 2 and 3 only (d) 1, 2 and 3

[ESE-2010]

1.74 A long straight non-magnetic conductor of radius 8 mm is carrying a uniform current density of 100 kA/m² in the a_z direction. For this case, which one of the following is not correct?

(a) $\nabla \times \vec{H} = 10^5 \hat{a}_z$ A/m² for $0 < \rho < 8$ mm
(b) $\nabla \cdot \vec{B} = 0$
(c) The magnetic field intensity for $\rho > 8$ mm is $10^5/2\ \rho \hat{a}_\phi$ A/m
(d) The total current carried by the conductor is 6.4π A [ESE-2010]

1.75 Consider the following statements relating the electrostatic and magnetostatic fields:

1. The relative distribution of charges on an isolated conducting body is dependent on the total charge of the body.
2. The magnetic flux through any closed surface is zero.

Which of these statement(s) is/are correct?

(a) Neither 1 nor 2 (b) 1 only
(c) 2 only (d) Both 1 and 2

[ESE-2010]

1.76 The magnetic flux density $\vec{B}$ and the vector magnetic potential A are related as

(a) $\vec{B} = \nabla \times \vec{A}$ (b) $\vec{A} = \nabla \times \vec{B}$

(c) $\vec{B} = \nabla \cdot \vec{A}$ (d) $\vec{A} = \nabla \cdot \vec{B}$

[ESE-2010]

1.77 Assertion (A): Capacitance between two parallel plates of area 'A' each and distance of separation 'd' is $\varepsilon A/d$ for large A/d ratio.

Reason (R): Fringing electric field can be neglected for large A/d ratio.

(a) Both A and R are true and R is the correct explanation of A
(b) Both A and R are true but R is NOT the correct explanation of A
(c) A is true but R is false
(d) A is false but R is true

[ESE-2010]

1.78 Assertion (A): In solving boundary value problems, the method of images is used.

Reason (R): By this technique, conducting surface can be removed from the solution domain.

(a) Both A and R are true and R is the correct explanation of A
(b) Both A and R are true but R is NOT the correct explanation of A
(c) A is true but R is false
(d) A is false but R is true [ESE-2010]

1.79 A varying magnetic flux linking a coil is given by $\phi = \frac{1}{3}\lambda t^3$. If at time $t = 3$ s, the emf induced is 9 V, then the value of λ is

(a) zero (b) 1 Wb/s^3

(c) –1 Wb/s^3 (d) 9 Wb/s^3

[ESE-2011]

1.80 If the potential, $V = 4x + 2$ volts, the electric field is

(a) 6 V/m (b) 2 V/m

(c) 4 V/m (d) $-4\bar{a}_x$ V/m

[ESE-2011]

1.81 If the current element represented by $4 \times 10^{-4}\hat{a}_y$ Amp-m is placed in a magnetic field of $\vec{H} = \frac{5\hat{a}_x}{\mu}$ A/m, the force on the current element is

(a) $-2.0\,\hat{a}_z$ mN (b) $2.0\,\hat{a}_z$ mN

(c) $-2.0\,\hat{a}_z$ N (d) $2.0\,\hat{a}_z$ N

[ESE-2011]

1.82 Match **List-I** with **List-II** and select the correct answer using codes given below the lists:

List-I	List-II
A. MMF	1. Conductivity
B. Magnetic flux	2. Electric current
C. Reluctance	3. EMF
D. Permeability	4. Resistance

Codes:

	A	B	C	D
(a)	3	4	2	1
(b)	1	2	4	3
(c)	3	2	4	1
(d)	1	4	2	3

[ESE-2011]

1.83 Given that the electric flux density $\vec{D} = z\rho\,(\cos^2\phi)\hat{a}_z$ C/m^2.
The charge density (in C/m^3) at point $(1, \pi/4, 3)$ is

(a) 3 (b) 1

(c) 0.5 (d) $0.5\,\bar{a}_z$ [ESE-2011]

1.84 Assertion (A): Under static conditions, the surface of conductor is an equipotential surface.

Reason (R): The tangential component of electric field on conductor surface is zero.

(a) Both A and R are true and R is the correct explanation of A
(b) Both A and R are true but R is NOT the correct explanation of A
(c) A is true but R is false
(d) A is false but R is true [ESE-2011]

1.85 Two dielectric media with permittivities 3 and $\sqrt{3}$ are separated by a charge-free boundary as shown in figure. The electric field intensity in medium 1 at point P_1 has magnitude E_1 and makes an angle $\alpha_1 = 60°$ with the normal. The direction of the electric field intensity at point P_2, α_2 is

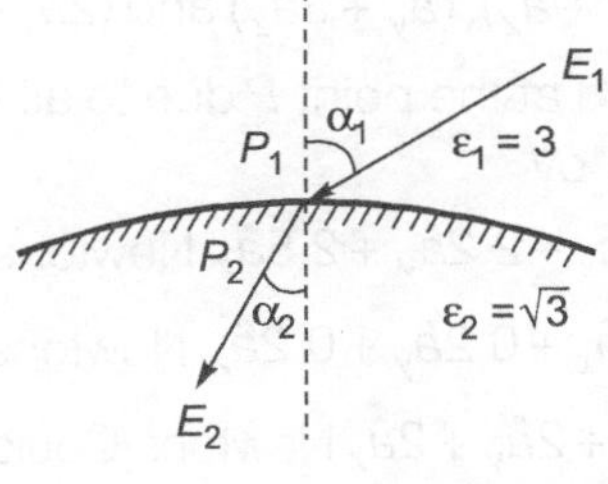

(a) $\sin^{-1}\left(\frac{\sqrt{3}E_1}{2}\right)$ (b) 45°

(c) $\cos^{-1}\left(\frac{\sqrt{3}E_1}{2}\right)$ (d) 30° [ESE-2011]

1.86 **Assertion (A):** The ratio of the tangential components of current density at two sides of an interface are equal to the conductivities.

Reason (R): The normal component of current density is continuous.

(a) Both A and R are true and R is the correct explanation of A
(b) Both A and R are true but R is NOT the correct explanation of A
(c) A is true but R is false
(d) A is false but R is true [ESE-2011]

1.87 The space between the plates of a parallel-plate capacitor of capacitance C is filled with three dielectric slabs of identical size as shown in the below figure. If dielectric constants are ε_1, ε_2 and ε_3, the new capacitance is

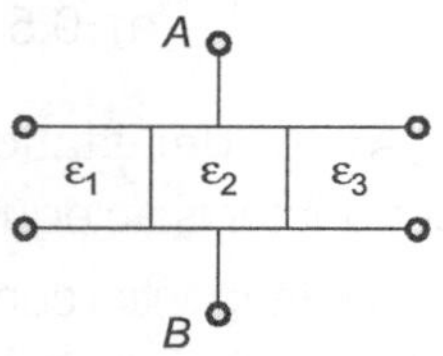

(a) $\frac{C}{3}$ (b) $\frac{(\varepsilon_1+\varepsilon_2+\varepsilon_3)C}{3}$

(c) $(\varepsilon_1+\varepsilon_2+\varepsilon_3)C$ (d) $\frac{9(\varepsilon_1+\varepsilon_2+\varepsilon_3)}{\varepsilon_1\varepsilon_2\varepsilon_3}$

[ESE-2012]

1.88 If the potential difference between points $A(1, 0, 0)$ and $B(2, 0, 0)$ is 10 V, determined for point $C(d, 0, 0)$, when V_{BC} is 6 V in a uniform field.

(a) 1 m (b) 2 m
(c) 6 m (d) 5 m [ESE-2012]

1.89 There are three charges, which are given by $Q_1 = 1\ \mu C$, $Q_2 = 2\ \mu C$ and $Q_3 = 3\ \mu C$. The field due to each charge at a point P in free space is $(\hat{a}_x + \hat{a}_y + \hat{a}_z)$, $(\hat{a}_y + 3\hat{a}_z)$ and $(2\hat{a}_x - \hat{a}_y)$ N/C. The total field at the point P due to all three charges is given by

(a) $1.6\hat{a}_x + 2.2\hat{a}_y + 2.5\hat{a}_z$ Newtons/Coulomb
(b) $0.3\hat{a}_x + 0.2\hat{a}_y + 0.2\hat{a}_z$ Newtons/Coulomb
(c) $3\hat{a}_x + 2\hat{a}_y + 2\hat{a}_z$ Newtons/Coulomb
(d) $0.6\hat{a}_x + 0.2\hat{a}_y + 0.5\hat{a}_z$ Newtons/Coulomb

[ESE-2012]

1.90 The ratio of the charges stored by two metallic spheres raised to the same potential is 6. The ratio of the surface area of the sphere is

(a) 6 (b) $\frac{1}{6}$

(c) 36 (d) $\frac{1}{\sqrt{6}}$ [ESE-2013]

1.91 A conducting plane at $z = 0$ has a voltage of 100 V on it. V_1 and V_2 are two solutions, with $V_1 = 5z + 100$ and $V_2 = 100$; which satisfy Laplace equation as well as the boundary condition $V = 100$ at $z = 0$. Which of the following is the most correct option?

(a) Both V_1 and V_2 are correct and solution is not unique
(b) Unique solution cannot be found
(c) Only V_1 is the correct solution
(d) The given plane does not serve as a proper boundary as the reference is not given

[ESE-2013]

1.92 Electric displacement current density D at any point on a spherical surface of radius r centred at the isolated charge q is

(a) $\frac{q^2}{r^2}$ (b) $\frac{q}{r^2}$

(c) $\frac{q}{4\pi r^2}$ (d) $\frac{q}{4\pi^2 r^2}$

[ESE-2013]

1.93 Consider a uniform sphere of charge density ρ_0 and radius b centred at the origin. The electric field at a radial distance $r(r < b)$, according to Gauss's law, is

(a) $\frac{r\rho_0}{3\epsilon}$ (b) $\frac{\rho_0}{4\pi\epsilon r^2}$

(c) $\frac{b^3\rho_0}{3\epsilon r^2}$ (d) $\frac{\rho_0}{r}$

where ϵ is the permittivity. [ESE-2013]

1.94 At a point (x, y, z) potential is given by $A(x^2 + y^2 + z^2)$. The potential difference between points $P(1, 0, 2)$ and $Q(1, 1, 2)$ is

(a) 8 V (b) 8 AV
(c) 9 AV (d) 9 V [ESE-2013]

1.95 Two charges are placed at a small distance apart. If a glass slab is placed between them, the force between the charges will

(a) not change (b) increase
(c) decrease (d) reduce to zero

[ESE-2013]

1.96 Which of the following are electromagnetic in nature?

1. Alpha rays 2. X-rays
3. Gamma rays 4. Cathode rays

(a) 1 and 2 (b) 2 and 3
(c) 3 and 4 (d) 1 and 4 [ESE-2013]

1.97 For a sheet with resistivity ρ, width w, length l and thickness y, the resistance per square (sheet resistance) R_s is

(a) $\frac{\rho}{y}$ (b) ρy

(c) $\frac{\rho l}{y}$ (d) $\frac{\rho l}{ly}$

[ESE-2013]

1.98 Consider a long line charge of λ Coulomb/meter perpendicular to the plane of a paper. The electrical field lines and equipotential surfaces are respectively

(a) radial, cylindrical concentric with line charge
(b) cylindrical concentric with line charge, radial
(c) radial, radial but opposite in direction
(d) concentric with line charge, parallel to line charge

[ESE-2014]

1.99 Which of the following statements about electric field lines associated with electric charges is false?

(a) Electric field lines can be either straight or curved.
(b) Electric field lines form closed loops.
(c) Electric field lines begin on positive charges and end on negative charges.
(d) Electric field lines do not intersect.

[ESE-2014]

1.100 Which of the following represents Maxwell's divergence equation for static electric field?

(a) $\nabla \cdot \vec{B} = 0$ (b) $\nabla \cdot \vec{H} = 0$

(c) $\nabla \cdot \vec{B} = \mu$ (d) $\nabla \cdot \vec{H} = \mu$

[ESE-2014]

1.101 A current of 5 A passes along the axis of a cylinder of 5 cm radius. The flux density at the surface of the cylinder is

(a) 2 μT (b) 20 μT
(c) 200 μT (d) 2000 μT

[ESE-2014]

1.102 Maxwell's major contribution to EM theory was to assert

(a) that an electric field varying with time in free space gives rise to a current
(b) that a magnetic field varying with time gives rise to an electric field
(c) that a magnetic field varying with space gives rise to an electric field
(d) that energy density due to an electric field is $1/2\, \varepsilon E^2$

[ESE-2014]

1.103 Consider the following statements regarding Maxwell's equation in differential form

1. For free space
$\nabla \times \vec{H} = (\sigma + j\omega\varepsilon)\vec{E}$
2. For free space, $\nabla \cdot \vec{D} = \rho$
3. For steady current, $\nabla \times \vec{H} = \vec{J}$
4. For static electric field, $\nabla \cdot \vec{D} = \rho$

Which of the above statements are correct?

(a) 1 and 2 (b) 2 and 3
(c) 3 and 4 (d) 4 and 1 [ESE-2014]

1.104 The equation which states the non-existence of isolated magnetic pole is

(a) $\nabla \cdot \vec{D} = \rho$ (b) $\nabla \cdot \vec{B} = 0$

(c) $\nabla \cdot \vec{J} = -\frac{\partial \rho}{\partial t}$ (d) $\nabla \times \vec{H} = \vec{J}$

[ESE-2014]

1.105 A voltage of 2000 V exists across 1 cm insulating space between two parallel conducting plates. An electron of charge 1.6×10^{-19} Coulomb is introduced into the space. The force on the electron is

(a) 18.2×10^{-26} N (b) 3.2×10^{-14} N
(c) 1.6×10^{-19} N (d) 4.5×10^{26} N

[ESE-2014]

1.106 The solutions to many problems involving electric fields are simplified by making use of equipotential surfaces. An equipotential surface is a surface.

1. On which the potential is same everywhere
2. The movement of charge over such a surface would require no work
3. The tangential electric field is zero
4. The normal electric field is zero

Which of the above statements are correct?

(a) 1, 2 and 3 only
(b) 1, 2 and 4 only
(c) 2, 3 and 4 only
(d) 1, 2, 3 and 4 [ESE-2015]

1.107 A charge 'Q' is divided between two point charges. What should be the values of this charge on the objects so that the force between them is maximum?

(a) $\frac{Q}{3}$ (b) $\frac{Q}{2}$

(c) $(Q-2)$ (d) $2Q$

[ESE-2015]

1.108 What is the force developed per meter length between two current-carrying conductors 10 cm apart and carrying 1000 A and 1500 A currents, respectively?

(a) 3 N (b) $\frac{1}{3}$N

(c) 2 N (d) $\frac{1}{2}$N [ESE-2015]

1.109 A point charge is located at origin. At point (a, a) electric field is $\vec{E}_1$. At point $(-a, a)$ the electric field is $\vec{E}_2$ and at a point $(-a, -a)$ the electric field is $\vec{E}_3$.

Which one of the following is correct?

(a) $\vec{E}_1 \cdot \vec{E}_2 = 0$

(b) $\left|\vec{E}_1 \times \vec{E}_3\right| = 0$

(c) Both $\vec{E}_1 \cdot \vec{E}_2 = 0$ and $\left|\vec{E}_1 \times \vec{E}_3\right| = 0$

(d) Neither $\vec{E}_1 \times \vec{E}_2$ nor $\left|\vec{E}_1 \times \vec{E}_3\right| = 0$

[ESE-2015]

1.110 Two point charges $q_1 = 2\mu C$ and $q_2 = 1\ \mu C$ are placed at distances $b = 1$ cm and $a = 2$ cm from the origin on the Y and X axes as shown in figure. The electric field vector at point $P(a, b)$ subtend an angle θ with x-axis, Then,

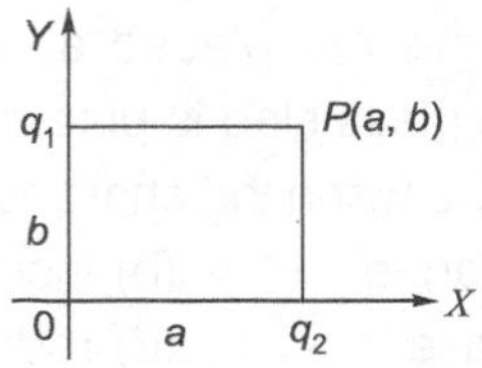

(a) $\tan\theta = 1$ (b) $\tan\theta = 2$

(c) $\tan\theta = 3$ (d) $\tan\theta = 4$

[ESE-2015]

1.111 **Statement (I):** A cylindrical conductor of radius R carries a current I. The magnetic field intensity within the conductor increases linearly with the radial distance $r < R$.

Statement (II): The current enclosed increases as the square of the radial distance while the circumference increases linearly with the radius distance.

(a) Both Statement (I) and Statement (II) are individually true and Statement (II) is the correct explanation of Statement (I)
(b) Both Statement (I) and Statement (II) are individually true but Statement (II) is **NOT** the correct explanation of Statement (I)
(c) Statement (I) is true but Statement (II) is false
(d) Statement (I) is false but Statement (II) is true

[ESE-2015]

1.112 In a coaxial transmission line ($\varepsilon_r = 1$), the electric field intensity is given by:

$$E = \frac{100}{\rho}\cos(10^9 t - 6z)u_p \text{ V/m}$$

The displacement current density is

(a) $-\frac{100}{\rho}\sin(10^9 t - 6z)u_p$ A/m^2

(b) $\frac{116}{\rho}\sin(10^9 t - 6z)u_p$ A/m^2

(c) $-\frac{0.9}{\rho}\sin(10^9 t - 6z)u_p$ A/m^2

(d) $-\frac{216}{\rho}\cos(10^9 t - 6z)u_p$ A/m^2

[ESE-2015]

1.113 Poisson's equation is derived with the following assumption about the medium. The medium is

(a) Non-homogeneous and isotropic
(b) Non-homogeneous and non-isotropic
(c) Homogeneous and non-isotropic
(d) Homogeneous and isotropic [ESE-2016]

1.114 Consider the following statements:
1. The magnetic field at the centre of a circular coil of a wire carrying current is inversely proportional to the radius of the coil.
2. Lifting power of a magnet is proportional to square of magnetic flux density.
3. A static electric field is conservative (irrotational).
4. If the divergence of a vector 'A' is zero, then vector 'A' can be expressed as Curl of a vector F.

Which of the above statements are correct?
(a) 1, 2 and 3 only (b) 3 and 4 only
(c) 1,2 and 4 only (d) 1, 2, 3 and 4
[ESE-2016]

1.115 Consider the following:
1. Electric current flowing in a conducting wire
2. A moving charged belt
3. An electron beam in a cathode ray tube
4. Electron movement in a vacuum tube

Which of the above are examples of convection current?
(a) 2, 3 and 4 only (b) 1,2 and 4 only
(c) 1 and 3 only (d) 1, 2, 3 and 4
[ESE-2016]

1.116 Consider the following sources :
1. A permanent magnet
2. A charged disc rotating at uniform speed
3. An accelerated charge
4. An electric field which changes linearly with time

Which of the above are the sources of steady magnetic field?
(a) 1, 2 and 3 only (b) 3 and 4 only
(c) 1, 2 and 4 only (d) 1, 2, 3 and 4
[ESE-2016]

1.117 A charge Q is enclosed by a Gaussian spherical surface of radius R. If R is doubled then the outward flux is
(a) Doubled
(b) Increased four times
(c) Reduced to a quarter
(d) Remains unaltered [ESE-2016]

1.118 Divergence of a vector div D in the cylindrical coordinate system is
(a) $\frac{1}{\rho}\frac{\partial}{\partial\rho}(D_\rho)+\frac{1}{\rho}\frac{\partial D_\phi}{\partial\phi}+\frac{\partial D_z}{\partial z}$
(b) $\frac{1}{\rho}\frac{\partial}{\partial\rho}(\rho D_\rho)+\frac{1}{\rho}\frac{\partial(\phi D_\phi)}{\partial\phi}+\frac{1}{z}\frac{\partial(ZD_z)}{\partial z}$
(c) $\frac{1}{\rho}\frac{\partial}{\partial\rho}(\rho D_\rho)+\frac{1}{\rho}\frac{\partial D_\phi}{\partial\phi}+\frac{\partial D_z}{\partial z}$
(d) $\frac{\partial D_\rho}{\partial\rho}+\frac{\partial D_\phi}{\partial\phi}+\frac{\partial D_z}{\partial z}$
[ESE-2016]

1.119 What is the value of work required to move a +8 nC charge from infinity to a point P which is at 2 m distance from a point charge $Q = +5\ \mu C$?
(a) 180 μJ (b) 180 nJ
(c) 18 μJ (d) 18 nJ
[ESE-2016]

1.120 An electrostatic force between two point charges increases when they are
(a) More apart and dielectric constant of the medium between them decreases
(b) Less apart and dielectric constant of the medium between them decreases
(c) More apart and dielectric constant of the medium between them increases
(d) Less apart and dielectric constant of the medium between them increases
[ESE-2016]

1.121 A plane $Y = 2$ carries infinite sheet of charge 6 nC/m^2. If medium is free space then force on a point charge of 10 mC located at the origin is
(a) $-1080\,\pi\hat{a}_y$ N (b) $-108\,\pi\hat{a}_y$ N
(c) $-10.8\,\pi\hat{a}_y$ N (d) $-1.08\,\pi\hat{a}_y$ N
[ESE-2016]

1.122 The potential at the centroid of an equilateral triangle of side $r\sqrt{3}$ due to three equal positive point charges each of value q and placed at the vertices of the triangle would be
(a) $\frac{q}{2\pi\epsilon_0 r}$ (b) $\frac{\sqrt{3}q}{8\pi\epsilon_0 r}$
(c) $\frac{3q}{4\pi\epsilon_0 r}$ (d) zero [ESE-2016]

1.123 The point form of the relation connecting vector magnetic potential $\vec{A}$ and current density $\vec{J}$ is
(a) $\nabla\times\vec{A}=\vec{J}+\frac{\partial\vec{D}}{\partial t}$ (b) $\vec{A}=\int\frac{\mu_0\vec{J}}{4\pi\in R}dv$
(c) $\nabla^2\vec{A}=-\mu_0\vec{J}$ (d) $\frac{\partial\vec{A}}{\partial t}=-\frac{\vec{J}}{\sigma}$
[ESE-2016]

1.124 In the region $Z < 0$, $\in_{r1} = 2$, $\bar{E}_1 = -3\bar{a}_x + 4\bar{a}_y - 2\bar{a}_z$ V/m. For region $Z > 0$, where $\varepsilon_{r2} = 6.5$, $\bar{E}_2$ is

(a) $-3\bar{a}_x + 4\bar{a}_y + \frac{6.5}{4}\bar{a}_z$ V/m

(b) $-3\bar{a}_x + 4\bar{a}_y + \frac{4}{6.5}\bar{a}_z$ V/m

(c) $-3\bar{a}_x + 4\bar{a}_y - \frac{6.5}{4}\bar{a}_z$ V/m

(d) $-3\bar{a}_x + 4\bar{a}_y - \frac{4}{6.5}\bar{a}_z$ V/m [ESE-2016]

1.125 Consider the following statements regarding a conductor and free space boundary
1. No charge and no electric field can exist at any point within the interior of a conductor
2. Charge may appear on the surface of a conductor

Which of the above statement(s) are correct?

(a) 1 only (b) 2 only
(c) Both 1 and 2 (d) Neither 1 nor 2 [ESE-2016]

1.126 A sphere of homogeneous linear dielectric material of dielectric constant ≥ 1 is placed in a uniform electric field E_0, then the electric field E that exists inside the sphere is

(a) Uniform and $E \le E_0$
(b) Uniform and $E \ge E_0$
(c) Varies but $E < E_0$ always
(d) Varies but $E > E_0$ always [ESE-2016]

1.127 Which of the following Maxwell's equations represents Ampere's law with correction made by Maxwell?

(a) $\nabla \cdot \vec{E} = \frac{\rho}{\epsilon_o}$

(b) $\nabla \cdot \vec{B} = 0$

(c) $\nabla \times \vec{E} = \frac{\partial \vec{B}}{\partial t}$

(d) $\nabla \times \vec{B} = \mu_0 \vec{J} + \mu_0 \epsilon_0 \frac{\partial \vec{E}}{\partial t}$ [ESE-2016]

1.128 If phasors $P_1 = 3 + j4$ and $P_2 = 6 - j8$, then $|P_1 - P_2|$ is

(a) 5 (b) $\sqrt{53}$
(c) $\sqrt{73}$ (d) $\sqrt{153}$ [ESE-2016]

1.129 Consider the following statements:
Stokes' theorem is valid irrespective of
1. shape of closed curve C
2. type of vector A
3. type of coordinate system
4. whether the surface is closed or open

Which of the above statements are correct?

(a) 1, 2 and 4 (b) 1, 3 and 4
(c) 2, 3 and 4 (d) 1, 2 and 3 [ESE-2017]

1.130 A plane $y = 2$ carries an infinite sheet of charge 4 nC/m^2. If the medium is free space, what is the force on a point charge of 5 mC located at the origin?

(a) $0.54\,\pi\hat{a}_y$ N (b) $0.18\,\pi\hat{a}_y$ N
(c) $-0.36\,\pi\hat{a}_y$ N (d) $-0.18\,\pi\hat{a}_y$ N [ESE-2017]

1.131 The angle θ_{AB} between the vectors $A = 3a_x + 4a_y + a_z$ and $B = 2a_y - 5a_z$ is nearly

(a) 83.7° (b) 73.7°
(c) 63.7° (d) 53.7° [ESE-2018]

1.132 The value of $\nabla \cdot A$, where $A = 3xy\vec{a}_x + x\vec{a}_y + xyz\vec{a}_z$ at a point (2, –2, 2) is

(a) –10 (b) –6
(c) 2 (d) 4 [ESE-2018]

1.133 If $V = \frac{10}{r^2}\sin\theta \cdot \cos\phi$ the electric flux density at $\left(2, \frac{\pi}{2}, 0\right)$ is

(a) $32.1\vec{a}_r$ pC/m^2 (b) $22.1\vec{a}_r$ pC/m^2
(c) $10.2\vec{a}_r$ pC/m^2 (d) $5.8\vec{a}_r$ pC/m^2 [ESE-2018]

1.134 The vector R_{AB} extends from $A(1, 2, 3)$ to B. If the length of R_{AB} is 10 units and its direction is given by

$$a = 0.6\hat{a}_x + 0.64\hat{a}_y + 0.48\hat{a}_z$$

the coordinates of B will be

(a) $7\hat{a}_x + 4.8\hat{a}_y + 4.8\hat{a}_z$
(b) $6\hat{a}_x + 6.4\hat{a}_y + 4.8\hat{a}_z$
(c) $7\hat{a}_x + 8.4\hat{a}_y + 7.8\hat{a}_z$
(d) $6\hat{a}_x + 8.4\hat{a}_y + 7.8\hat{a}_z$ [ESE-2019]

1.135 What is the value for the total charge enclosed in an incremental volume of 10^{-9} m^3 located at the origin if

$D = e^{-x} \sin y \hat{a}_x - e^{-x} \cos y \hat{a}_y + 2z\hat{a}_z$ C/m^2?

(a) 8 nC (b) 4 nC
(c) 2 nC (d) 1 nC

[ESE-2019]

1.136 The unit vector extending from origin toward the point $G(2, -2, -1)$ is

(a) $\frac{2}{3}\hat{a}_x + \frac{2}{3}\hat{a}_y + \frac{1}{3}\hat{a}_z$ (b) $-\frac{2}{3}\hat{a}_x + \frac{2}{3}\hat{a}_y + \frac{1}{3}\hat{a}_z$

(c) $\frac{2}{3}\hat{a}_x - \frac{2}{3}\hat{a}_y - \frac{1}{3}\hat{a}_z$ (d) $-\frac{2}{3}\hat{a}_x - \frac{2}{3}\hat{a}_y - \frac{1}{3}\hat{a}_z$

[ESE-2019]

1.137 In a pair of straight parallel bus bars of circular cross-section spaced 23 cm between centres, each carry a current of 70,000 A. The force required to withstand will be nearly

(a) 4,800 N/m (b) 4,620 N/m
(c) 4,400 N/m (d) 4,260 N/m

[ESE-2020]

1.138 Consider the following two points

$M(2, 5, -3)$ and $N(-3, 1, 4)$

The distance from the origin to the mid-point of the line MN will be nearly

(a) 3.1 units (b) 2.3 units
(c) 1.5 units (d) 0.7 units

[ESE-2020]

1.139 Consider $\vec{D} = 10x\hat{a}_x - 4y\hat{a}_y + kz\hat{a}_z$ μC/m^2 and $\vec{B} = 2\hat{a}_y$ mT, to satisfy the Maxwell's equation for region $\sigma = 0$ and $\rho_v = 0$, the value of k will be

(a) −8 μC/m^3 (b) −6 μC/m^3
(c) −4 μC/m^3 (d) −2 μC/m^3

[ESE-2020]

1.140 Consider a cube defined by

$x, y, z \in [1, 3]$

If vector, $\vec{A} = 2x^2y\hat{a}_x + 3x^2y^2\hat{a}_y,$

$\nabla \cdot \vec{A}$ at the centre of the cube will be

(a) 72 (b) 64
(c) 60 (d) 48 [ESE-2020]

1.141 Two vectors V_1 and V_2 are orthogonal if their dot product is

(a) 1 (b) 0
(c) infinity (d) 0.5 [ESE-2021]

1.142 The component of $\vec{P} = 2\hat{a}_x - \hat{a}_z$ along $\vec{Q} = 2\hat{a}_x - \hat{a}_y + 2\hat{a}_z$ is

(a) $0.745\hat{a}_x + 0.298\hat{a}_y - 0.596\hat{a}_z$

(b) $4\hat{a}_x - 2\hat{a}_y + 4\hat{a}_z$

(c) $0.2222\hat{a}_x - 0.1111\hat{a}_y + 0.2222\hat{a}_z$

(d) $0.4444\hat{a}_x - 0.2222\hat{a}_y + 0.4444\hat{a}_z$

[ESE-2021]

1.143 Consider the following equations with time factor $e^{j\omega t}$.

1. $\oint D_s \cdot dS = \int \rho_{vs} dv$
2. $\oint E_s \cdot dl = -j\omega B_s$
3. $\oint B_s \cdot dS = 0$
4. $\oint H_s \cdot dS = \int (J_s + j\omega D_s) \cdot dl$

Which of the above Time-Harmonic Maxwell's equations are correct?

(a) 1 and 2 only (b) 1 and 3 only
(c) 1, 2 and 3 only (d) 3 and 4 only

[ESE-2021]

1.144 If $\vec{A} = \rho\cos\phi\hat{a}_p + \sin\phi\hat{a}_\phi$, then the surface integration of curl of $\vec{A}$ (for $30° \le \phi \le 60°$ and $2 \le \rho \le 5$), is

(a) 6.750 (b) 4.941
(c) 0.732 (d) 1.765

[ESE-2021]

1.145 Which one of the following is ***not*** the basic rule for boundary conditions at the surface between two different materials?

(a) The tangential components of electric field intensity are continuous across the boundary.
(b) The normal components of electric flux density are discontinuous at the boundary by an amount equal to the surface-charge density on the boundary.
(c) The tangential components of magnetic field intensity are discontinuous at the boundary by an amount equal to the surface-current density on the boundary.
(d) The normal components of electric field intensity are continuous across the boundary.

[ESE-2021]

1.146 **Statement (I):** The total flux out of a closed surface is equal to the net charge enclosed within the surface.

Statement (II): An electric field is completely specified by its intensity vector.

(a) Both Statement (I) and Statement (II) are individually true and Statement (II) is the correct explanation of Statement (I).
(b) Both Statement (I) and Statement (II) are individually true but Statement (II) is **NOT** the correct explanation of Statement (I).
(c) Statement (I) is true, but Statement (II) is false.
(d) Statement (I) is false, but Statement (II) is true.

[ESE-2021]

1.147 Which one of the following laws states that the line integral of the tangential component of H around a closed path is the same as the net current I_{enc} enclosed by the path?

(a) Biot-Savart's law (b) Lenz's law
(c) Gauss's Law (d) Ampere's circuit law

[ESE-2021]

1.148 If V is a scalar, then the given equation (in spherical coordinate system)

$$\frac{1}{r^2}\frac{\partial}{\partial r}\left(r^2\frac{\partial V}{\partial r}\right)+\frac{1}{r^2\sin\theta}\frac{\partial}{\partial\theta}\left(\sin\theta\frac{\partial V}{\partial\theta}\right)+\frac{1}{r^2\sin^2\theta}\frac{\partial^2 v}{\partial\phi^2}=-\frac{\rho}{\varepsilon}$$

is referred to as

(a) Laplace's equation (b) Poisson's equation
(c) gradient of V (d) divergence of V

[ESE-2022]

1.149 The average value of the electrostatic field over the volume V of a sphere due to a point charge q somewhere within the sphere, where r_0 is the position of the charge, is

(a) $E_{av}=-\dfrac{qr_0}{3\varepsilon_0 V}$ (b) $E_{av}=-\dfrac{qr_0}{4\pi\varepsilon_0 V}$

(c) $E_{av}=\dfrac{qr_0}{3\varepsilon_0 V}$ (d) $E_{av}=\dfrac{qr_0}{4\pi\varepsilon_0 V}$

[ESE-2022]

1.150 In the relation between the field and the potential, if $r < a$, then the potential at a point r inside a uniformly charged sphere of radius a, is

(a) $\dfrac{a^3\rho}{3\varepsilon_0 r}$ (b) $\dfrac{\rho}{6\varepsilon_0}(3a^2-r^2)$

(c) $\dfrac{4\pi a^3}{3r}$ (d) $\dfrac{2\pi}{3}(3a^2-r^2)$

[ESE-2022]

1.151 Vector potential is a vector

(a) whose curl is equal to the magnetic flux density.
(b) whose curl is equal to the electric field intensity.
(c) whose divergence is equal to the electric potential.
(d) which is equal to the vector product $E\times H$.

[ESE-2023]

1.152 If the magnetic vector potential $A=-\dfrac{\rho^2}{4}a_z$ Wb/m, what is the total magnetic flux crossing the surface $\phi=\dfrac{\pi}{2}$, $1\le\rho\le 2$ m, $0\le z\le 5$ m?

(a) 3.25 Wb (b) 3.50 Wb
(c) 3.75 Wb (d) 4.00 Wb

[ESE-2023]

1.153 A vector $\vec{P}$ is given by
$\vec{P}=x^3\overrightarrow{a_x}-x^2y^2\overrightarrow{a_y}-x^2yz\overrightarrow{a_z}$. Which one of the following statements is correct?

(a) $\vec{P}$ is solenoidal, but not irrotational.
(b) $\vec{P}$ is irrotational, but not solenoidal.
(c) $\vec{P}$ is neither solenoidal nor irrotational.
(d) $\vec{P}$ is both solenoidal and irrotational.

[ESE-2023]

1.154 The electric field on the surface of a perfect conductor is 2 V/m. The conductor is immersed in water with $\varepsilon=80\varepsilon_0$. The surface charge density on the conductor is

(a) 0 C/m^2 (b) 2 C/m^2
(c) 1.8×10^{-11} C/m^2 (d) 1.41×10^{-9} C/m^2

[ESE-2023]

1.155 If the electric field intensity is given by $\bar{E}=(xu_x+yu_y+zu_z)$ V/m, the potential difference between $X(2, 0, 0)$ and $Y(1, 2, 3)$ is

(a) +1 V (b) –1 V
(c) +5 V (d) +6 V [ESE-2023]

1.156 Consider a parallel-plate capacitor, each of the plates ha an area S and they are separated by a distance d. Assume that plates 1 and 2 carry charges $+Q$ and $-Q$ uniformly distributed on them. The energy stored in the capacitor is

(a) $-\frac{Q}{\varepsilon S}a_x$ (b) $\frac{Qd}{\varepsilon S}$

(c) $\frac{1}{2C}Q^2$ (d) $\frac{1}{2}Q \cdot C$

[ESE-2023]

1.157 Two dipoles with dipole moments $-5a_z$ nC-m and $9a_z$ nC-m are located at points (0, 0, –2) and (0, 0, 3) respectively. What is the potential at the origin?

(a) –24.25 V (b) –22.25 V
(c) –20.25 V (d) –18.25 V

[ESE-2023]

1.158 If $\nabla \cdot D = \varepsilon \nabla \cdot E$ and $\nabla \cdot J = \sigma \nabla \cdot E$ in a given material, the material is said to be

(a) linear and isotropic
(b) linear and homogeneous
(c) isotropic and homogeneous
(d) homogeneous and dielectric

[ESE-2023]

1.159 A parallel-plate air-filled capacitor has plate area of 10^{-4} m^2 and plate separation of 10^{-3} m. It is connected to a 0.5 V, 4.5 GHz source. The magnitude of the displacement current is (take $\varepsilon_0 = \frac{1}{36\pi \times 10^9}$ F/m)

(a) 10 mA (b) 10 A
(c) 12.5 mA (d) 50 A

[ESE-2023]

1.160 In the international system of units (SI), charges Q_1, Q_2 are expressed in Coulombs, distance R in meters, and the force in Newtons. The constant of proportionality k is

(a) 6×10^{-7} m/F (b) 9×10^{9} m/F
(c) 5×10^{8} m/F (d) 7×10^{-9} m/F

[ESE-2024]

1.161 Which one of the following does **not** indicate the space pervasiveness of electromagnetic effects?

(a) The presence of current in two circuits in proximity to each other is accompanied by electrical forces on each conductor, forces which change if either current is changed.
(b) A changing of the current in either of two such circuits is accompanied by an induced voltage in the other.
(c) Capacitors consisting of metallic spheres or other conducting bodies suspended in vacuum or in an insulating medium may be charged and later discharged. During these processes wire-borne current flows onto one sphere and off the other.
(d) The presence of electric charges on two bodies is accompanied by a mechanical force on each, forces which change if either charge is changed.

[ESE-2024]

1.162 A non-uniform electric field is given by $E = ya_x + xa_y + 2ya_z$ V/m. A charge of $2C$ is transferred from point $A(1, 0, 3)$ to point $B(2, 1, 3)$ along the straight line from point A to point B. What is the work done?

(a) –1 J (b) –2 J
(c) –3 J (d) –4 J

[ESE-2024]

1.163 The differential magnetic field at any point is proportional to the product of the current, differential length and the sine of the angle between the element and the line joining to that point. This is known as

(a) Coulomb's law
(b) Ampere's Circuital Law
(c) Gauss's law
(d) Biot-Savart's Law

[ESE-2024]

■■■

Answers Basic of Electromagnetics

1.1	(b)	1.2	(c)	1.3	(c)	1.4	(c)	1.5	(a)	1.6	(d)	1.7	(b)	1.8	(c)	1.9	(b)
1.10	(a)	1.11	(b)	1.12	(c)	1.13	(a)	1.14	(d)	1.15	(b)	1.16	(a)	1.17	(c)	1.18	(a)
1.19	(b)	1.20	(c)	1.21	(b)	1.22	(b)	1.23	(b)	1.24	(d)	1.25	(c)	1.26	(b)	1.27	(a)
1.28	(a)	1.29	(b)	1.30	(a)	1.31	(b)	1.32	(c)	1.33	(d)	1.34	(b)	1.35	(b)	1.36	(b)
1.37	(a)	1.38	(a)	1.39	(c)	1.40	(b)	1.41	(d)	1.42	(c)	1.43	(c)	1.44	(b)	1.45	(b)
1.46	(d)	1.47	(a)	1.48	(c)	1.49	(a)	1.50	(c)	1.51	(d)	1.52	(c)	1.53	(a)	1.54	(a)
1.55	(c)	1.56	(a)	1.57	(c)	1.58	(c)	1.59	(c)	1.60	(c)	1.61	(c)	1.62	(a)	1.63	(b)
1.64	(c)	1.65	(a)	1.66	(d)	1.67	(b)	1.68	(a)	1.69	(a)	1.70	(a)	1.71	(d)	1.72	(d)
1.73	(d)	1.74	(c)	1.75	(d)	1.76	(a)	1.77	(a)	1.78	(a)	1.79	(c)	1.80	(d)	1.81	(a)
1.82	(c)	1.83	(c)	1.84	(a)	1.85	(b)	1.86	(b)	1.87	(b)	1.88	(d)	1.89	(c)	1.90	(c)
1.91	(c)	1.92	(c)	1.93	(a)	1.94	(*)	1.95	(c)	1.96	(b)	1.97	(a)	1.98	(a)	1.99	(b)
1.100	(a)	1.101	(b)	1.102	(a)	1.103	(c)	1.104	(b)	1.105	(b)	1.106	(a)	1.107	(b)	1.108	(a)
1.109	(c)	1.110	(b)	1.111	(a)	1.112	(c)	1.113	(d)	1.114	(d)	1.115	(a)	1.116	(c)	1.117	(d)
1.118	(c)	1.119	(a)	1.120	(b)	1.121	(d)	1.122	(c)	1.123	(c)	1.124	(d)	1.125	(c)	1.126	(a)
1.127	(d)	1.128	(d)	1.129	(d)	1.130	(c)	1.131	(a)	1.132	(a)	1.133	(b)	1.134	(c)	1.135	(c)
1.136	(c)	1.137	(d)	1.138	(a)	1.139	(b)	1.140	(b)	1.141	(b)	1.142	(d)	1.143	(b)	1.144	(b)
1.145	(d)	1.146	(b)	1.147	(d)	1.148	(b)	1.149	(c)	1.150	(b)	1.151	(a)	1.152	(c)	1.153	(c)
1.154	(d)	1.155	(c)	1.156	(c)	1.157	(c)	1.158	(b)	1.159	(c)	1.160	(b)	1.161	(a)	1.162	(d)
1.163	(d)																

Explanations Basic of Electromagnetics

1.2 (c)

$$I_1 = \oint \vec{H}\cdot d\vec{l}$$

$$\frac{I}{\pi R^2}\cdot \pi r^2 = H \times 2\pi r$$

$$\Rightarrow \quad H = \frac{Ir}{2\pi R^2}$$

1.3 (c)

$$\nabla\cdot\vec{J} = -\frac{\partial\rho_v}{\partial t} \text{ general continuity equation}$$

The equation of continuity for steady currents is

$\nabla\cdot\vec{J} = 0$ for charge free region

1.4 (c)

$$V_{ind} = -N\frac{d\phi}{dt} = -100\frac{d(t^3-2t)}{dt}\times 10^{-3}$$

$$V_{ind}\big|_{t=4s} = -100\cdot(3t^2-2)\Big|_{t=4}\times 10^{-3}$$

$$= 100 \times (3\times 4^2 - 2)\times 10^{-3}$$

$$= 4.6 \text{ V}$$

1.5 (a)

Lorentz law, $F = Q(V\times B)$

$$B = \frac{\text{Newton}}{\text{Coulomb}\times\dfrac{\text{meters}}{\text{second}}} = \frac{\text{Mass}\times\text{Acceleration}}{\text{Coulomb}\times\dfrac{\text{meters}}{\text{second}}}$$

$$B = \frac{MLT^{-2}}{Q\dfrac{L}{T}} = MT^{-1}Q^{-1}$$

1.6 (d)

The solution of Poisson's equation is not same as that of Laplace's equation.

1.9 (b)

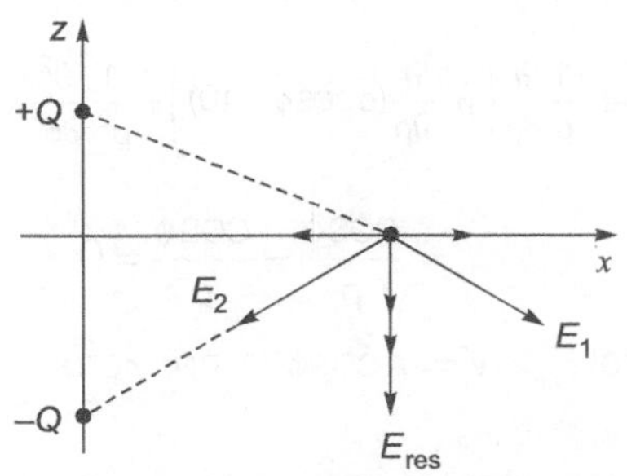

The directions of electric field E_1 due to $+Q$ and E_2 due to $-Q$ and the resultant electric field are shown in the figure.

1.10 (a)

$$F = \frac{q_1 q_2}{4\pi\epsilon_o r^2}$$

$$\Rightarrow \quad F = \frac{9\times 10^9\times(1\times 10^{-9})^2}{(1\times 10^{-3})^2}$$

$$\Rightarrow \quad F = 9\times 10^{-3} \text{ N}$$

1.11 (b)

$$i = C\frac{dV}{dt}$$

$$\Rightarrow \quad \text{Voltage} = \frac{\text{Current}\times\text{time}}{\text{Capacitance}}$$

$$\Rightarrow \quad = \frac{2\times 10^{-6}\times 6}{3\times 10^{-6}} = 4 \text{ V}$$

1.12 (c)

$$E = \frac{\rho_s}{\epsilon} = \frac{V}{d}$$

As V is same, E is same.

$$E' = \frac{\rho_s'}{\epsilon'} = \frac{\rho_s'}{2\epsilon} = E = \frac{\rho_s}{\epsilon}$$

$\rho' = 2\rho_s$ charge is doubled, D doubled.

Energy $W = \frac{1}{2}\epsilon E^2$, doubled due to ϵ.

1.13 (a)

Laplace equation in Cartesian coordinates

$$\nabla^2 V = \frac{\partial^2 V}{\partial x^2} + \frac{\partial^2 V}{\partial y^2} + \frac{\partial^2 V}{\partial z^2}$$

Laplace equation in cylindrical coordinates

$$\nabla^2 V = \frac{1}{\rho}\frac{\partial}{\partial\rho}\left(\frac{\rho\partial V}{\partial\rho}\right) + \frac{1}{\rho^2}\left(\frac{\partial^2 V}{\partial\phi^2}\right) + \frac{\partial^2 V}{\partial z^2}$$

Laplace equation in spherical coordinates

$$\nabla^2 V = \frac{1}{r}\frac{\partial}{\partial r}\left(\frac{r^2\partial V}{\partial r}\right) + \frac{1}{r^2\sin\theta}\frac{\partial}{\partial\theta}\left(\sin\theta\frac{\partial V}{\partial\theta}\right) + \frac{1}{r^2\sin^2\theta}\frac{\partial^2 V}{\partial\phi^2}$$

1.14 (d)

$$\nabla\cdot\bar{D} = \rho \text{ and } \bar{D} = \varepsilon_o\bar{E}$$

1.15 (b)

Current in the coil

$$I = \frac{V_{dc}}{R} = \frac{50}{5} = 10 \text{ A}$$

Energy stored in the field

$$E = \frac{1}{2}LI^2 = \frac{1}{2}\times 0.4\times(10)^2 = 20 \text{ Joules}$$

1.16 (a)

$$\vec{A} = 3x^2yz\hat{a}_x + x^3z\hat{a}_y + (x^3y - 2z)\hat{a}_z$$

$$\nabla\cdot\vec{A} = \frac{\partial}{\partial x}(3x^2yz) + \frac{\partial}{\partial y}(x^3z) + \frac{\partial}{\partial z}(x^3y - 2z)$$

$$= 6xyz - 2$$

$\nabla\cdot\vec{A} \neq 0 \Rightarrow \vec{A}$ is not a solenoidal.

$$\nabla\times\vec{A} = \begin{vmatrix} \hat{a}_x & \hat{a}_y & \hat{a}_z \\ \frac{\partial}{\partial x} & \frac{\partial}{\partial y} & \frac{\partial}{\partial z} \\ 3x^2yz & x^3z & x^3y - 2z \end{vmatrix}$$

$$= (x^3 - x^3)\hat{a}_x - (3x^2y - 3x^2y)\hat{a}_y + (3x^2z - 3x^2z)\hat{a}_z$$

$$= 0$$

$\Rightarrow \vec{A}$ is irrotational.

1.17 (c)

$$\oint \vec{H}\cdot d\vec{l} = I$$

Seeing the direction of loops, for H and current direction along thumb.

$$\oint \vec{H}\cdot d\vec{l} = 3I + 2I - I = 4I$$

1.18 (a)

$$\int_{\theta=0}^{\pi} \cos^2\theta d\theta = \int_{\theta=0}^{\pi}\left(\frac{1+\cos 2\theta}{2}\right)d\theta = \frac{\pi}{2}$$

1.19 (b)

$$\iint_s \vec{D}\cdot\overline{ds} = Q = \Psi$$

$\Rightarrow D$ and Ψ are independent of $\in$.

where, D = Displacement density

Q = Electric charge

Ψ = Displacement

Electric field intensity, $E = \frac{D}{\in}$

$\Rightarrow E$ is dependent of $\in$

$$V = \int \vec{E}\cdot\vec{dl}$$

$\Rightarrow V$ is also dependent of $\in$.

1.20 (c)

Using Gauss law

$$\int \bar{D}\cdot\overline{ds} = Q = \int \rho ds$$

$$D\cdot 2\pi(3)h = 5.2\pi(2)h$$

$$\Rightarrow \quad D = \frac{10}{3}\bar{a}_r \text{ C/m}^2$$

1.21 (b)

$F = \frac{\mu_o I_1 I_2}{2\pi d}$ magnitude of force between infinitely long parallel wires.

Force is perpendicular to the lines and attractive if the currents in both wires are flowing in the same direction. Force is repulsive if the currents are in the opposite direction.

1.22 (b)

Energy stored in the field,

$$E = \frac{9\times10^9\times q_1q_2}{r} = \frac{9\times10^9\times1\times10^{-6}\times4\times10^{-6}}{\sqrt{(1+2)^2+(3-1)^2+(-1-5)^2}}$$

E = 5.14 mJ

1.23 (b)

$\nabla^2 V = 0$ is Laplace equation

$$\nabla^2 V = \frac{\partial^2 V}{\partial x^2} + \frac{\partial^2}{\partial y^2} \text{ in Cartesian}$$

$V = 10\,xy$ given zero value

$\nabla^2 V$ for $\quad V = \frac{10}{r}$ given

$$\nabla^2 V = \frac{1}{r^2}\frac{\partial}{\partial r}\left(r^2\cdot\frac{\partial}{\partial r}\left(\frac{10}{r}\right)\right)$$

$$= \frac{1}{r^2}\frac{\partial}{\partial r}\left(r^2\cdot\left(\frac{-10}{r^2}\right)\right) = 0$$

$\nabla^2 V$ for $\quad V = \rho\cos\phi + 10$ given

$$\nabla^2 V = \frac{1}{\rho}\frac{\partial}{\partial\rho}\left(\rho\cdot\frac{\partial}{\partial\rho}(\rho\cos\phi + 10)\right) + \frac{1}{\rho^2}\frac{\partial^2}{\partial\phi^2}(\rho\cos\phi + 10)$$

$$= \frac{\cos\phi}{\rho} - \frac{\cos\phi}{\rho} = 0$$

$\nabla^2 V$ for $\quad V = r\cos\phi$ is not zero

1.24 (d)

$$\nabla^2 V = \nabla.\nabla V$$

This means the Laplacian of a scalar function V is the divergence of the gradient of V.

1.25 (c)

Before immersion,

$$C = 0.005\ \mu F \text{ and } V = 500\ V$$

Energy stored,

$$E_1 = \frac{1}{2}CV^2$$

$$= \frac{1}{2} \times 0.005 \times 10^{-6} \times (500)^2$$

$$\Rightarrow \quad E_1 = 6.25 \times 10^{-4} \text{ Joules}$$

After immersion, $C_1 = \in_r C$, but charge on capacitor remains same as before.

$$q = CV = 0.005 \times 10^{-6} \times 500$$

$$\Rightarrow \quad q = 2.5 \times 10^{-6}$$

Energy stored,

$$E_2 = \frac{1}{2}\frac{q^2}{C_1} = \frac{1}{2} \cdot \frac{(2.5 \times 10^{-6})^2}{2.5 \times 0.005 \times 10^{-6}}$$

$$= 2.5 \times 10^{-4} \text{ Joules}$$

1.26 (b)

Let Q' be induced charge or image charge.

Consider A point,

$$V = \frac{Q'}{4\pi\in(R-b)} + \frac{Q}{4\pi\in(d-R)} = 0$$

Consider B point,

$$V = \frac{Q'}{4\pi\in(R+b)} + \frac{Q}{4\pi\in(d+R)} = 0$$

Solving, $Q' = \dfrac{QR}{d} = 4\ \mu C$

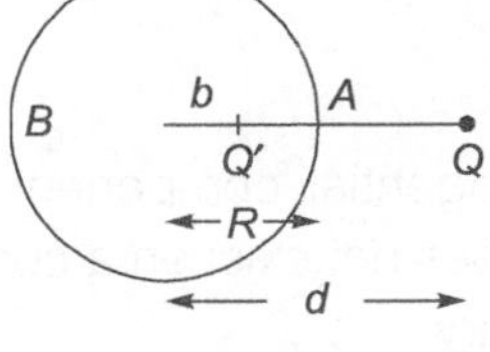

1.27 (a)

$$E = \frac{\rho_L}{2\pi\in\rho}\hat{a}_\rho$$

Distance ρ from line charge to points decides the E field.

For (0, 6, 1) to (x, 3, 5)

$$\rho = \sqrt{(6-3)^2 + (1-5)^2} = 5$$

It is the same for (5, 6, 1) to (x, 3, 5)

1.28 (a)

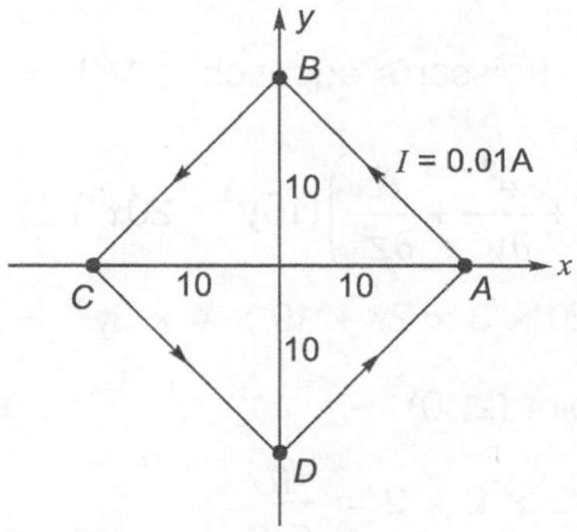

Area of the square

$$A = (10\sqrt{2})^2 = 200\ m^2$$

Magnetic dipole moment

$$M = IA = 0.01 \times 200 = 2$$

The direction of the magnetic dipole moment can be found out by right Hand rule and that is $\bar{a}_z$.

So, $\vec{M} = 2\hat{a}_z$

1.30 (a)

Potential is unique at a point at a time and Laplace or Poisson's equation have unique solutions due to uniqueness theorem.

Parameters of potentials and the spatial rates of change means voltage values and electric field values change but the net value is fixed.

Solution of Laplace's equation is always unique.

1.31 (b)

For large area of cross-section, we can ignore the fringing effect assuming uniform fields.

1.32 (c)

Current flows in the entire cylinder having non-uniform density.

$$\oint H \cdot dl = I = \iint J \cdot ds$$

$$H \cdot 2\pi r = \int_{r=a}^{r}\int_{\theta=0}^{2\pi} \frac{J_o r}{a^2} r\, d\theta\, dr$$

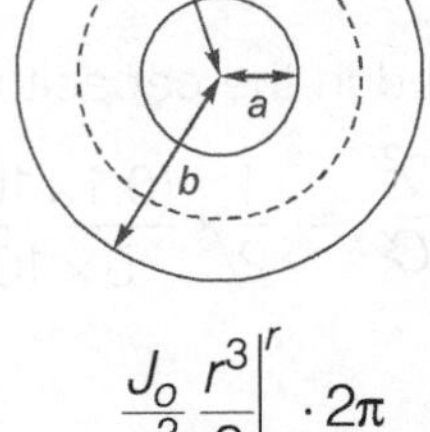

$$H = \frac{\dfrac{J_o}{a^2}\dfrac{r^3}{3}\Big|_a^r \cdot 2\pi}{2\pi r} = \frac{J_o(r^3 - a^3)}{3a^2 r}$$

1.33 (d)

Poisson's equation is $\nabla^2 V = \frac{-\rho}{\epsilon_o}$

$$\left(\frac{\partial^2}{\partial x^2}+\frac{\partial^2}{\partial y^2}+\frac{\partial^2}{\partial z^2}\right)(10y^4+20x^3) = -\frac{\rho}{\epsilon_o}$$

$$\Rightarrow \quad 20 \times 3 \times 2x + 10 \times 4 \times 3y^2 = \frac{-\rho}{\epsilon_o}$$

At point (2, 0)

$$20 \times 3 \times 2 \times 2 = \frac{-\rho}{\epsilon_o}$$

$$\Rightarrow \rho = -240\,\epsilon_o$$

1.34 (b)

Voltage on an infinitely conducting solid sphere of radius R due to charge Q on it is

$$V = \frac{Q}{4\pi\epsilon_o R}$$

$$Q = CV$$

$\Rightarrow$ Capacitance

$$C = \frac{Q}{V} = \frac{Q}{\frac{Q}{4\pi\epsilon_o R}} = 4\pi\epsilon_o R$$

1.35 (b)

Electric field is always normal to equipotential surfaces and conductor is an excellent equipotential surface.

Displacement density on a conducting surface is normal to the surface. Therefore, the conducting surface is an equipotential surface. Since potential at infinity is zero, therefore, potential everywhere on a conducting surface of infinite extent is zero.

1.36 (b)

Voltage stress means $\frac{V}{d}$ = Electric field intensity

Discharge free rating = change on the capacitor

= Displacement flux density

1.37 (a)

Energy stored in the capacitor

$$E = \frac{1}{2}\frac{Q^2}{C} = \frac{1}{2}\times\frac{(0.1\times10^{-6})^2}{5\times10^{-12}} = 1 \text{ mJ}$$

1.40 (b)

$$\frac{1}{2}J\cdot A = \frac{1}{2}(\nabla\times H)\cdot A = \frac{1}{2}H\cdot(\nabla\times A)$$

$$= \frac{1}{2}B\cdot H = \frac{1}{2}\mu H^2$$

1.41 (d)

Macroscopic view point means looking over a region of length or area.

Electric field has $\nabla\times\vec{E} = 0$, implies irrotational.

$\nabla\cdot\vec{E} \neq 0$ implies not solenoidal.

1.42 (c)

$$\nabla^2 V = -\frac{\rho_v}{\epsilon}$$

Second order derivative being zero is linear function Poisson's there force has non-linear solution.

$\nabla^2 V = 0$ is linear only in Cartesian and for single dimensional solution for a general view in cylindrical or spherical it may or may not be linear.

1.43 (c)

Use the method of image charge.

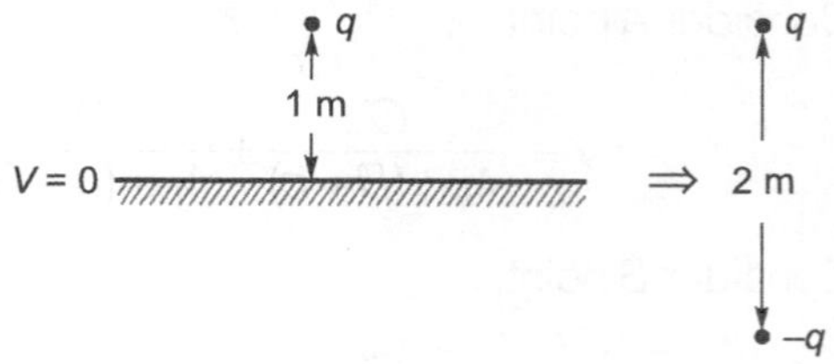

Force, $F = \frac{1}{4\pi\epsilon_o}\cdot\frac{q^2}{r^2}$

$$= \frac{1}{4\pi\epsilon_o}\cdot\frac{(1)^2}{(2)^2} = \frac{1}{16\pi\epsilon_o}N$$

1.45 (b)

(i) The tangential component of electric field lines does not exist on a conducting surface boundary.

(ii) Any conducting surface always supports the normal component of the electric field such that the normal component of electric flux density $\vec{D}$ is equal to the magnitude of the surface charge density ρ_s present on the conducting surface.

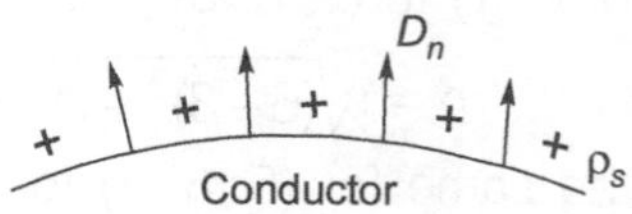

1.46 (d)

Magnetic current is the cause of magnetization effect in magnetic material. It is called as bound current which is due to atomic dipoles causing increasing magnetic field in ferrite materials.

1.47 (a)

According to the method of images, an infinite conducting surface can be replaced by an image charge as shown below without affecting electric field or voltage distribution

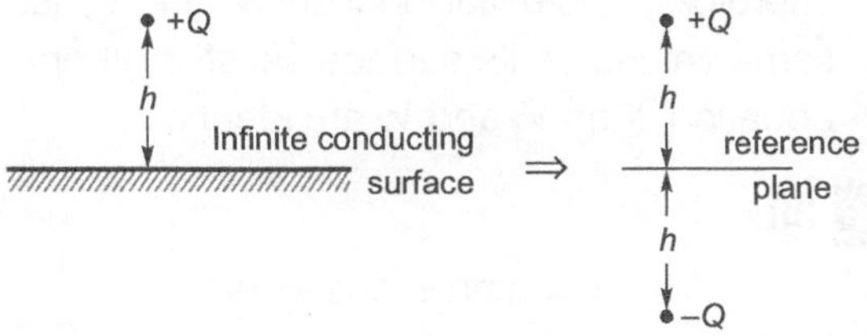

1.48 (c)

Magnetic vector potential due to a current element Idl is

$$A = \int \frac{\mu I dl}{4\pi R}$$

Since, in the question,

$R \to \infty$

Therefore, $A = 0$

1.49 (a)

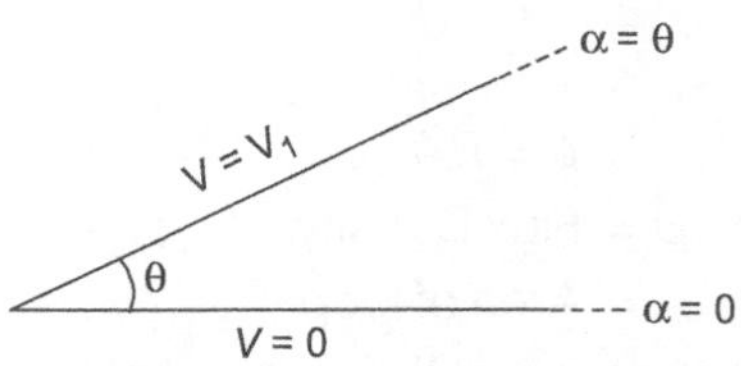

From option (a)

$$V = V_1 \sin\left(\frac{\pi a}{2\theta}\right)$$

when, $\alpha = 0$

$V = 0$

when, $\alpha = \theta$

$V = V_1$

So only option (a) matches.

1.50 (c)

Work depends on potential

$$\omega \propto V \propto \frac{1}{r}$$

When r is doubled, energy becomes half.

1.51 (d)

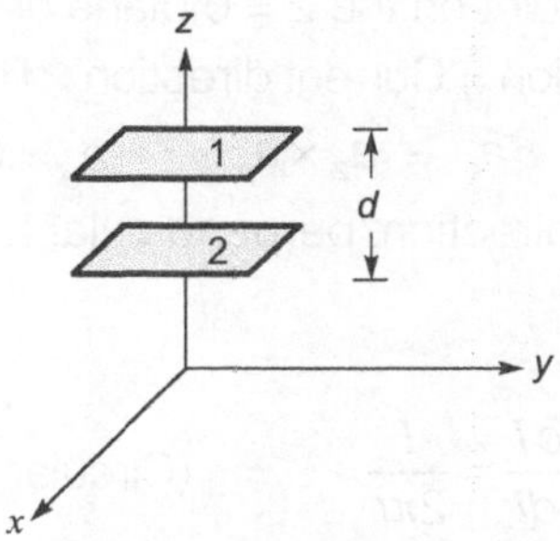

We know that magnetic field

$$\vec{H} = \frac{1}{2}\vec{k}_s \times \hat{a}_n$$

due to sheet (i)

$$\vec{H}_1 = \frac{1}{2}(k_y \hat{a}_y) \times (-\hat{a}_z) = -\frac{1}{2}k_y \hat{a}_x$$

due to sheet (ii)

$$\vec{H}_2 = \frac{1}{2}(k_y \hat{a}_y) \times \hat{a}_z = \frac{1}{2}k_y \hat{a}_x$$

Total magnetic field $\vec{H} = \vec{H}_1 + \vec{H}_2$

$$\vec{H} = 0$$

1.52 (c)

$$\alpha = \beta = \gamma$$

$$= \cos^{-1}\frac{10}{\sqrt{10^2 + 10^2 + 10^2}}$$

$$= \cos^{-1}\frac{10}{10\sqrt{3}} = \cos^{-1}\left(\frac{1}{\sqrt{3}}\right)$$

1.53 (a)

$$\oint \vec{A} \cdot \vec{dl} = \int (\nabla \times \vec{A}) \cdot \vec{ds} = \int \vec{B} \cdot \vec{ds} = \psi_m$$

Flux crossing the surface.

1.54 (a)

Electric field intensity due to a uniform surface charge density ρ_s is

$$\vec{E} = \frac{\rho_s}{2\epsilon}\hat{a}_n$$

Where $\hat{a}_n$ = unit vector normal to the surface charge density and directed outward or away from it

So, $$\vec{E} = \frac{-60}{2\epsilon}\hat{a}_x = \frac{-30}{\epsilon}\hat{a}_x$$

But electric flux density

$$\vec{D} = \epsilon \vec{E} = -30\,\hat{a}_x$$

1.55 (c)

At any point on the $z = 0$ plane or XY plane.

H-direction = Current direction × Radial direction

$$\hat{a}_\phi = \hat{a}_z \times \hat{a}_\rho \quad ; \quad a_p = \text{OP vector}$$

$\hat{a}_\phi$ = H-direction, perpendicular to OP vector

1.56 (a)

$$K = \frac{dI}{dl} = \frac{I}{2\pi r} \qquad \text{(Circular cross-section)}$$

1.57 (c)

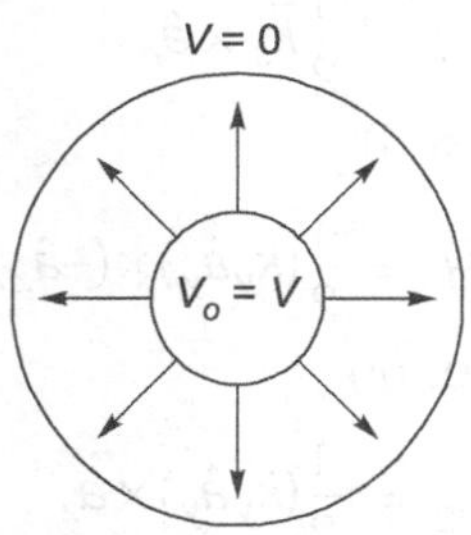

Non-uniform as the field depends on radial distance.

1.58 (c)

By Poisson's equation: $\nabla^2 V = -\dfrac{\rho_v}{\varepsilon_0}$

In charge free space $\rho_v = 0$

Hence $\nabla^2 V = 0$

which is Laplace's equation.

1.59 (c)

E volts/m drives a current of K Amps/m on a surface such that

$$E \propto K$$

$$E = \rho K$$

where, ρ = Surface resistivity = ohms

1.60 (c)

$$V = \frac{Q}{4\pi \in r} + C$$

where C is the initial reference voltage for not assuming $V = 0$ at infinity

$$V\left(\frac{1}{2}m\right) = 30 = \frac{Q}{4\pi \in \frac{1}{2}} + C$$

$$30 = \frac{2Q}{4\pi \in} + C \qquad \text{...(i)}$$

$$V(2m) = 15 = \frac{Q}{4\pi \in 2} + C \qquad \text{...(ii)}$$

The two equation now given

$$V(1\text{ m}) = 20\text{ V}$$

1.61 (c)

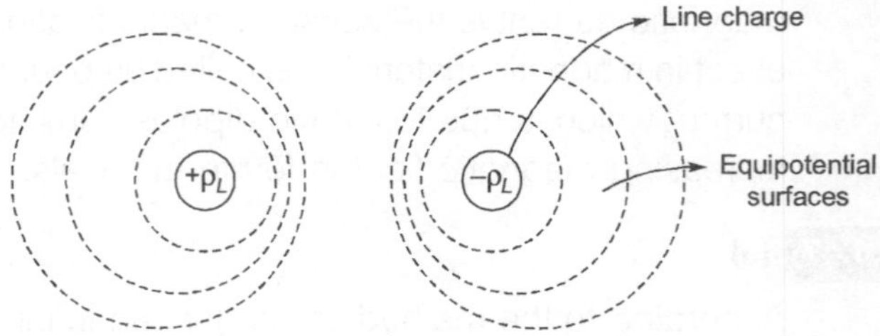

1.62 (a)

Laplace's equation gives a unique solution. Therefore, if potential functions V_1 and V_2 assume same values on its surface satisfying Laplace's equation, then V_1 and V_2 are identical.

1.63 (b)

$$V = \sinh x \cdot \cos ky \cdot e^{pz}$$

Laplace equation

$$\nabla^2 V = 0$$

$$\frac{\partial^2}{\partial x^2}(V) + \frac{\partial^2}{\partial y^2}(V) + \frac{\partial^2}{\partial z^2}(V) = 0$$

$$\sinh x \cdot \cos ky \cdot e^{pz} - k^2 \sinh x \cdot \cos ky \cdot e^{pz} + p^2 \sinh x \cdot \cos ky \cdot e^{pz} = 0$$

$$\sinh x \cdot \cos ky \cdot e^{pz}(1 - k^2 + p^2) = 0$$

$$k^2 = 1 + p^2$$

$$k = \sqrt{1 + p^2}$$

1.66 (d)

Flux, $\phi = BA$

where, B = Flux Density

A = Area of loop

EMF appearing at the terminal of loop,

$$E_{\text{ind}} = -\frac{d\phi}{dt} = \frac{-d(BA)}{dt}$$

$$= -A\frac{dB}{dt} = -5 \times 2 = -10\text{ V}$$

1.67 (b)

Potential on a solid conducting spherical body of radius 'a' due to charge Q,

$$V = \frac{Q}{4\pi \epsilon_0 a}$$

Capacitance of spherical body,

$$C = \frac{Q}{V} = 4\pi\epsilon_0 a$$

1.68 (a)

$$\vec{B} = \nabla\times\vec{A}$$

and $$\nabla\times\vec{E} = -\frac{\partial\vec{B}}{\partial t}$$

$$\nabla\times\vec{E} = -\frac{\partial}{\partial t}(\nabla\times\vec{A})$$

$$\nabla\times\vec{E}+\frac{\partial}{\partial t}(\nabla\times\vec{A}) = 0$$

$$\nabla\times\left(\vec{E}+\frac{\partial\vec{A}}{\partial t}\right) = 0 \quad \text{...(i)}$$

From vector identity
Curl of a gradient is identically zero

$$\nabla\times(-\nabla V) = 0 \quad \text{(ii)}$$

Comparing equation (i) and (ii), we have

$$E+\frac{\partial\vec{A}}{\partial t} = -\nabla V$$

$$E = -\nabla V-\frac{\partial\vec{A}}{\partial t}$$

1.70 (a)

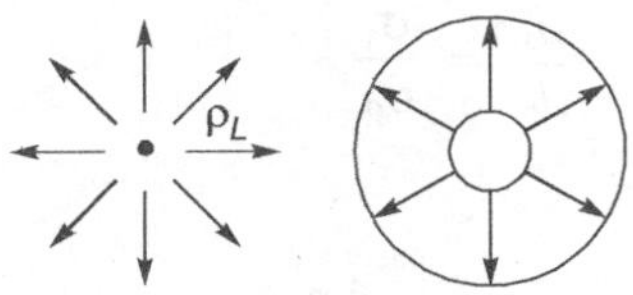

1.71 (d)

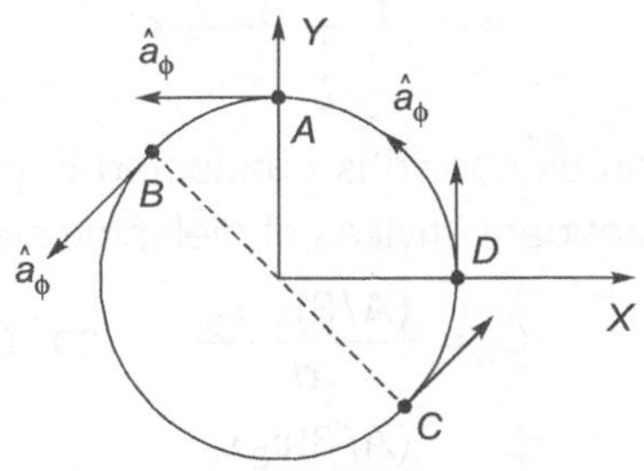

At A: $\hat{a}_\phi = -\hat{a}_x$

At B: $\hat{a}_\phi = -\dfrac{\hat{a}_x+\hat{a}_y}{\sqrt{2}}$

At C: $\hat{a}_\phi = \dfrac{\hat{a}_x+\hat{a}_y}{\sqrt{2}}$

At D: $\hat{a}_\phi = \hat{a}_y$

1.72 (d)

Electric field is a vector quantity while potential is a scalar quantity.

1.73 (d)

$$Q = \text{Constant}$$

$$W_E = \frac{1}{2}\frac{Q^2}{C} \quad \text{changes due to } C$$

$$C = \frac{\in A}{d} \quad \text{changes due to } \in$$

$$E = \frac{\rho_s}{\in} \quad \text{changes due to } \in$$

1.74 (c)

(a) $\nabla\times\vec{H} = \vec{J}_c = 100\ \text{kA/m}^2 = 10^5\ \text{A/m}^2$

(b) $\nabla\cdot\vec{B} = 0$

(d) $I = J_cA = 10^5\times\pi(8\times10^{-3})^2$ Amp.

$\Rightarrow$ $I = \pi\times10^5\times64\times10^{-6}$ Amp.

$\Rightarrow$ $I = 6.4\,\pi$ Amp.

1.75 (d)

Charge density depends on charge on the conductor. This is called as relative distribution of charge.

1.76 (a)

As $\nabla\cdot\vec{B} = 0$

and $\nabla\cdot(\nabla\times\vec{A}) = 0$ [Vector identity]

$\therefore$ $\boxed{\vec{B}=\nabla\times\vec{A}}$

1.77 (a)

For a large A/d ratio uniform field can be assumed and fringe effect can be neglected.

1.78 (a)

According to the method of images, an infinite conducting surface can be replaced by an image charge as shown below without affecting electric field or voltage distribution

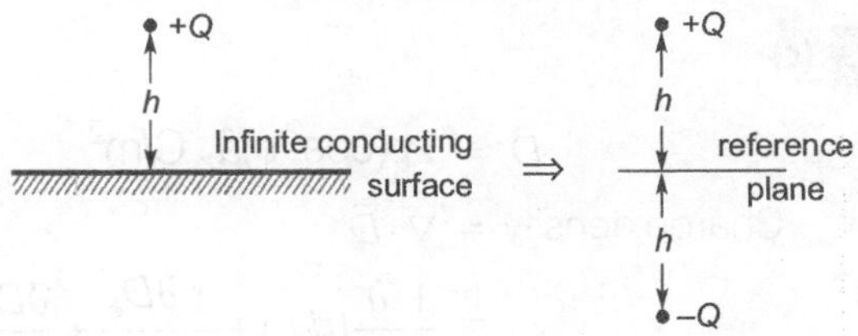

1.79 (c)

Given $\phi = \dfrac{1}{3}\lambda t^3$

emf at $t = 3s$ is 9 V

$$\text{Induced emf} = -\frac{d\phi}{dt} = -\frac{d}{dt}\left[\frac{1}{3}\lambda t^3\right]$$

$$\Rightarrow \quad \text{emf} = -\lambda t^2$$

$$9 = -\lambda(3)^2$$

$$\Rightarrow \quad \lambda = -1 \text{ Wb/s}^3$$

1.80 (d)

$$V = 4x + 2 \text{ volts}$$

$$\vec{E} = -\frac{dV}{dx}\hat{a}_x$$

[Since voltage is a function of x only. So electric field will be directed towards x-axis]

$$\vec{E} = -\frac{d}{dx}(4x+2)\,\hat{a}_x$$

$$\vec{E} = -4\hat{a}_x$$

1.81 (a)

$$I\,\vec{dl} = 4\times10^{-4}\,\hat{a}_y \text{ Amp-m}$$

$$\vec{H} = \frac{5\hat{a}_x}{\mu} \text{ A/m}$$

$$\therefore \quad \vec{B} = \mu\vec{H} = 5\,\hat{a}_x \,Wb.$$

$$\vec{F} = I\,\vec{dl}\times\vec{B}$$

$$= 4\times10^{-4}\hat{a}_y \times 5\,\hat{a}_x$$

$$= -2.0\,\hat{a}_z \text{ mN.}$$

1.82 (c)

$$I = \psi_m \times \text{Reluctance}$$

$$V = I \times \text{Resistance}$$

We can reach upto the answer by observing analogous quantities of electric and magnetic field.

Electric field	Magnetic field
Conductivity	Permeability
Electric current	Magnetic flux
EMF	MMF
Resistance	Reluctance

1.83 (c)

$$\vec{D} = z\rho(\cos^2\phi)\hat{a}_z \text{ C/m}^2$$

$$\text{Charge density} = \nabla\cdot\vec{D}$$

$$= \frac{1}{r}\frac{\partial}{\partial r}(rD_r) + \frac{1}{r}\frac{\partial D_\phi}{\partial\phi} + \frac{\partial D_z}{\partial z}$$

$$= \frac{\partial}{\partial z}(z\,\rho\cos^2\phi)$$

$$= \rho\cos^2\phi$$

So charge density at point $(1, \pi/4, 3)$ is

$$= 1\cdot\cos^2\frac{\pi}{4} = 0.5\ (\text{C/m}^3)$$

1.84 (a)

Conductor surface is an equipotential as tangential component of electric field on conductor surface is zero and only normal component of electric field exists.

1.85 (b)

Using boundary condition at dielectric-dielectric interface for oblique incidence, we have

$$E_{t_1} = E_1\sin\alpha_1 = E_2\sin\alpha_2 = E_{t_2}$$

$$D_{n_1} = \epsilon_1 E_1\cos\alpha_1 = \epsilon_2 E_2\cos\alpha_2 = D_{n_2}$$

$$\frac{\tan\alpha_1}{\tan\alpha_2} = \frac{\epsilon_1}{\epsilon_2}$$

$$\Rightarrow \quad \frac{\tan 60^\circ}{\tan\alpha_2} = \frac{3}{\sqrt{3}}$$

$$\Rightarrow \quad \tan\alpha_2 = \frac{\sqrt{3}}{3}\tan 60^\circ = \frac{\sqrt{3}\times\sqrt{3}}{3} = 1$$

$$\therefore \quad \alpha_2 = 45^\circ$$

1.86 (b)

$$J = \sigma E$$

$$E_{t_1} = E_{t_2}$$

$$\frac{J_1}{J_2} = \frac{\sigma_1}{\sigma_2}$$

1.87 (b)

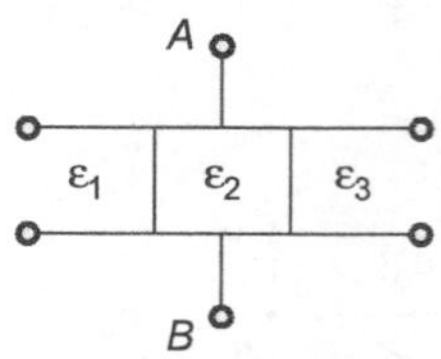

The circuit shown is connected in parallel with different permittivities of dielectric slabs.

So, $C_1 = \dfrac{(A/3)\varepsilon_1\varepsilon_0}{d}$ Also, $C = \dfrac{A\varepsilon_0}{d}$

$$C_2 = \frac{(A/3)\varepsilon_2\varepsilon_0}{d}$$

$$C_3 = \frac{(A/3)\varepsilon_3\varepsilon_0}{d}$$

Since all the three slabs are connected in parallel, so net capacitance is

$$= C_1 + C_2 + C_3$$

$$= \left(\frac{A\varepsilon_1}{3d} + \frac{A\varepsilon_2}{3d} + \frac{A\varepsilon_3}{3d}\right)\varepsilon_0$$

$$= \frac{A\varepsilon_0}{3d}(\varepsilon_1+\varepsilon_2+\varepsilon_3)$$

$$= \frac{C}{3}(\varepsilon_1+\varepsilon_2+\varepsilon_3)$$

1.88 (d)

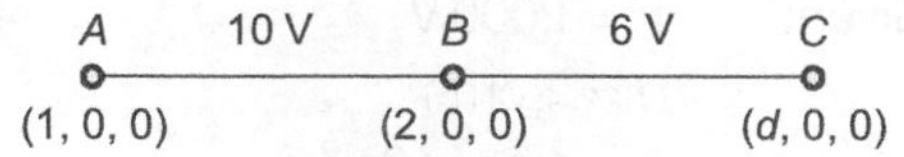

$$E = \frac{10 \text{ Volts}}{1\text{m}} = 10 \text{ V/m}$$

As seen from *A-B* distance

$$6 = V = E(d-2)$$
$$6 = 10\,(d-2)$$
$$0.6 = d-2$$
$$d = 2.6 \text{ m}$$

1.89 (c)

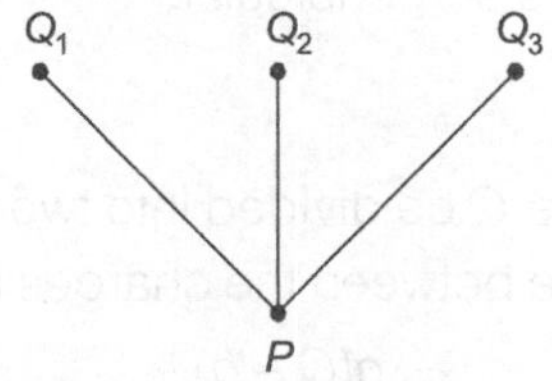

Electric field at P due to Q_1

$$= (\hat{a}_x + \hat{a}_y - \hat{a}_z)$$

Electric field at P due to $Q_2 = \hat{a}_y + 3\hat{a}_z$

Electric field at P due to $Q_3 = 2\hat{a}_x - \hat{a}_y$

So, electric field at P due to all Q_1, Q_2 and Q_3 is

$$= (\hat{a}_x + 2\hat{a}_y - \hat{a}_z) + (\hat{a}_y + 3\hat{a}_z) + (2\hat{a}_x - \hat{a}_y)$$
$$= 3\hat{a}_x + 2\hat{a}_y + 2\hat{a}_z$$

1.90 (c)

Charge stored on a metallic sphere is

$$Q = CV$$

where capacitance $C = 4\pi \in_o r$

Hence

$$Q_1 = 4\pi \in_o r_1 V_1$$
$$Q_2 = 4\pi \in_o r_2 V_2$$

So, $$\frac{Q_1}{Q_2} = \frac{r_1 V_1}{r_2 V_2} = \frac{r_1}{r_2} \quad \text{[given } V_1 = V_2]$$

$$\Rightarrow \quad \frac{r_1}{r_2} = 6$$

$$\Rightarrow \quad \frac{r_1^2}{r_2^2} = 36$$

1.91 (c)

Laplace equation

$$\nabla^2 V = 0$$

From the given problem it is clear that V is variable only in z-direction, so

$$\frac{d^2V}{dz^2} = 0$$

$$\Rightarrow \quad \frac{dV}{dz} = C$$

$$\Rightarrow \quad V = Cz + D \quad \text{...(i)}$$

at $\quad z = 0, \quad V = 100$

So, from equation (i), $\quad V = Cz + 100$

Which is in the form of V_1 and also note that Laplace equation gives unique solution.

1.92 (c)

Electric field intensity,

$$E = \frac{1}{4\pi \in} \frac{q}{r^2}$$

$$\therefore \quad D = \in E = \frac{q}{4\pi r^2}$$

1.93 (a)

$$\int_s \vec{D} \cdot \vec{ds} = \rho_o \times \frac{4}{3}\pi r^3$$

$$\Rightarrow \quad D \times 4\pi r^2 = \rho_o \times \frac{4\pi}{3} r^3$$

$$\Rightarrow \quad D = \frac{\rho_o r}{3}$$

$$\therefore \quad E = \frac{\rho_o r}{3\in}$$

1.94 (*)

Potential at point $P = V_p = 5$ A

Potential at point $Q = V_Q = 6$ A

So potential difference $= V_Q - V_p = 1$ A

1.95 (c)

Force between the two charges is given by Coulomb's law,

$$F = \frac{1}{4\pi \in_r \in_o} \frac{q_1 q_2}{r^2}$$

For air $\quad \in_r = 1$

For glass slab $\in_r > 1$, hence force will decrease between the two charges.

1.96 (b)

α and cathode rays are not electromagnetic in nature.

1.97 (a)

$E \rightarrow$ Volts/m field induces $\bar{K}$-Amp/m surface current due to surface resistivity

$$K \propto E$$

$$K = \rho E$$

where ρ = ohms = surface resistivity

1.98 (a)

The electric field lines for a long line charge is always radial also equipotential surfaces will be formed by concentric cylinders which cover both the line charges as shown in the diagram.

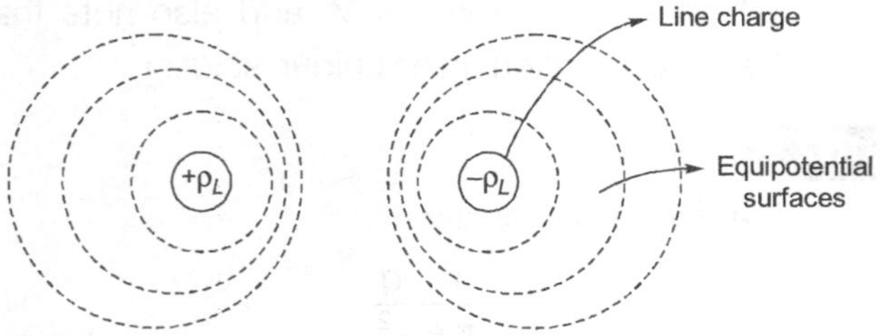

1.99 (b)

Electric field ($\vec{E}$) is irrotational vector. Hence, electric field lines does not form closed loops.

1.101 (b)

According to Ampere's law

$$H = \frac{I}{2\pi r}$$

Here, I = 5 A ; r = 5 cm

$$\therefore \quad H = \frac{5}{2\pi \times 5} = \frac{1}{2\pi} \text{A/m}$$

$$B = \mu H = \frac{4\pi \times 10^{-7}}{2\pi} = 20\ \mu\text{T}$$

1.102 (a)

$$\nabla \times \vec{H} = \vec{J}_c + \vec{J}_d$$

where, $\vec{J}_d = \in \dfrac{\partial \vec{E}}{\partial t}$

1.103 (c)

For free space,

$$\sigma = 0,\ \mu = \mu_0,\ \varepsilon = \varepsilon_0$$

$$\therefore \quad \nabla \times \vec{H} = j\omega\varepsilon_o \vec{E} = \vec{J}$$

and, $\nabla \cdot \vec{D} = \rho$ for static electric field.

1.104 (b)

$\nabla \cdot B = 0$ is due to

$$\oiint B \cdot ds = \iiint \nabla \cdot B dv = 0$$

$\oiint B \cdot ds = 0$ is due to

Magnetic monopoles not existing.

1.105 (b)

Given: $V = 2000$ V

$d = 1$ cm

$q = 1.6 \times 10^{-19}$ C

$\because \quad F = q \times E$

where, $E = \dfrac{V}{d} = \dfrac{2000}{10^{-2}} = 2 \times 10^5$ V/m

$\therefore \quad F = 1.6 \times 10^{-19} \times 200 \times 10^3$

$= 3.2 \times 10^{-14}$ N

1.106 (a)

For an equipotential surface, electric field lines are always perpendicular.

1.107 (b)

Let charge Q be divided into two parts as q and Q-q. Force between the charges is given by,

$$F = \frac{q(Q-q)}{4\pi\epsilon_o r^2}$$

For force to be maximum,

$$\frac{dF}{dq} = 0 \Rightarrow Q - 2q = 0$$

$$\Rightarrow \quad Q = 2q$$

$$\Rightarrow \quad q = \frac{Q}{2}$$

1.108 (a)

$I_1 = 1000$ A

$I_2 = 1500$ A

$R = 10$ cm

$$F = \frac{\mu_0 I_1 I_2 l}{2\pi R}$$

$$\Rightarrow \quad \frac{F}{l} = \frac{4\pi \times 10^{-7} \times 15 \times 10^5}{2\pi \times 0.1}$$

$F = 3\ N$

1.109 (c)

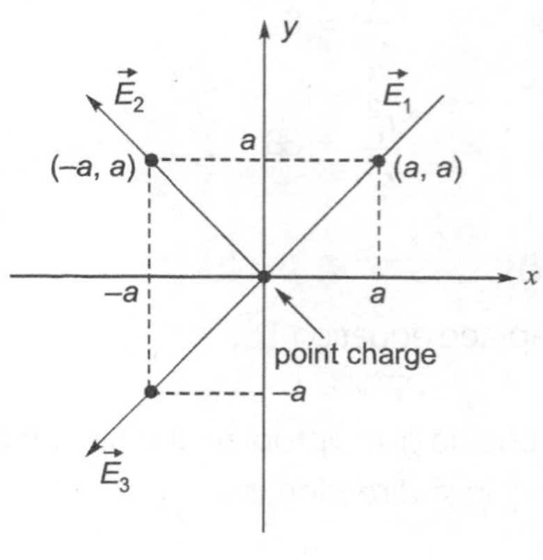

$$\vec{E}_1 = K(a\,\hat{a}_x + a\,\hat{a}_y) \quad (K = \text{constant})$$
$$\vec{E}_2 = K(-a\,\hat{a}_x + a\,\hat{a}_y)$$
$$\vec{E}_3 = K(-a\,\hat{a}_x + -a\,\hat{a}_y)$$
$$\vec{E}_1 . \vec{E}_2 = K^2(-a^2 + a^2) = 0$$
$$|\vec{E}_1 \times \vec{E}_3| = K^2(0 + 0) = 0$$

1.110 (b)

$$\vec{E}_1 \text{ due to } q_1 = \frac{q_1}{4\pi\,\varepsilon_0 R_1^2}\hat{a}_{R_1}$$

where $q_1 = 2\,\mu C$,
$R_1 = a = 2$ cm,
$\hat{a}_{R_1} = \hat{a}_x$

$$\Rightarrow \quad \vec{E}_1 = \frac{2\times 10^{-6}}{4\pi\,\varepsilon_0 . 2^2}\hat{a}_x$$

$$\text{Now, } \vec{E}_2 \text{ due to } q_2 = \frac{q_2}{4\pi\,\varepsilon_0 R_2^2}\hat{a}_{R2}$$

where, $q_2 = 1\,\mu C$,
$R_2 = b = 1$ cm,
$\hat{a}_{R2} = \hat{a}_y$

$$\Rightarrow \quad \vec{E}_2 = \frac{1\times 10^{-6}}{4\pi\,\varepsilon_0 1^2}\hat{a}_y$$

$$\text{Total } \vec{E} = \vec{E}_1 + \vec{E}_2 = \frac{10^{-6}}{4\pi\,\varepsilon_0}\left[\frac{1}{2}\hat{a}_x + \hat{a}_y\right]$$

$$\therefore \quad \tan\theta = \frac{1}{1/2} = 2$$

1.111 (a)

$$\oint \vec{H}\cdot\vec{dl} = I_{\text{enclosed}}$$
$$H\times 2\pi r = \frac{J}{\pi R^2}\times \pi r^2$$

⇒ Statement (II) is correct explanation of statement (I).

1.112 (c)

$$\vec{E} = \frac{100}{\rho}\cos(10^9 t - 6z)\hat{u}_p \text{ V/m}$$
$$\vec{J} = \frac{\partial\vec{D}}{\partial t} = \varepsilon_0\frac{\partial\vec{E}}{\partial t}$$
$$= \frac{10^{-9}}{36\pi}\times\frac{100}{\rho}\times 10^9(-)\sin(10^9 t - 6z)\hat{u}_p$$
$$= -\frac{100}{36\pi\,\rho}\sin(10^9 t - 6z)\hat{u}_p \text{ A/m}^2$$
$$\approx \frac{-0.9}{\rho}\sin(10^9 t - 6z)\hat{u}_p \text{ A/m}^2$$

1.119 (a)

Work done = ω

$$= -Q\int_{\text{initial}}^{\text{final}}\vec{E}\cdot\vec{dl} = Q\left[-\int_{\infty}^{\text{final}}\vec{E}\,\vec{dl}\right] = QV$$

Potential at 2 m distance from a point change Q at the origin is

$$V = \frac{Q}{4\pi t_o r} = \frac{5\times 10^{-6}}{2}9\times 10^9$$
$$W = QV$$
$$= 8\times 10^{-9}\,\frac{5\times 10^{-6}}{2}(9\times 10^9)$$
$$= 4\times 5\times 9\times 10^{-6} = 180\ \mu\text{J}$$

1.120 (b)

Force between two point charges Q_1 and Q_2 is

$$F = \frac{Q_1 Q_2}{4\pi\in d^2}$$

If d, ∈ both decreases than F increases.

1.121 (d)

Electric field at origin due to $P_s = 6\frac{nC}{m^2}$ infinite sheet charge on $y = 2$ surface is

$$\vec{E} = \frac{P_s}{2\in_o}\hat{a}_N$$
$$= \frac{6\times 10^{-9}}{2\frac{1}{36\pi}\times 10^{-9}}(-\hat{a}_y) = 3\times 36\pi(-\hat{a}_y)$$

Force on 10 mC charge = $\vec{F} = Q\vec{E}$

$$\vec{F} = 10\times 10^{-3}\left[3\times 36\pi(-\hat{a}_y)\right] = 1.08\pi(-\hat{a}_y)$$

1.122 (c)

If a is the side of the equilateral triangle than potential at the centre due to 3 point charges each having 'q' charge at corners is

$$V = \frac{3\sqrt{3}\,q}{4\pi\in_o a}$$

given side of equilateral triangle = $a = r\sqrt{3}$

$$V = \frac{3\sqrt{3}\,q}{4\pi\in_o r\sqrt{3}} = \frac{3q}{4\pi\in_o r}$$

1.124 (d)

For $z = 0$ boundary $\hat{a}_z$ component of the vector is normal.

$$\vec{E}_1 = -3\hat{a}_x + 4\hat{a}_y - 2\hat{a}_z$$

$$\vec{E}_{t_1} = -3\hat{a}_x + 4\hat{a}_y \; ; \; \vec{E}_{N1} = -2\hat{a}_z$$

First boundary condition $\Rightarrow \vec{E}_{t_1} = \vec{E}_{t_2}$

$$\vec{E}_{t_2} = -3\hat{a}_x + 4\hat{a}_y$$

Second boundary condition $\Rightarrow \vec{D}_{N1} - \vec{D}_{N2}$

$$\epsilon_1 \vec{E}_{N1} = \epsilon_2 \vec{E}_{N2}$$

$$\vec{E}_{N2} = \frac{\epsilon_1}{\epsilon_2}\vec{E}_{N1} = \frac{2\epsilon_o}{6.5\epsilon_o}(-2\overrightarrow{a_z})$$

$$\vec{E}_{N2} = -\frac{4}{6.5}\hat{a}_z$$

$$\vec{E}_2 = \vec{E}_{t2} + \vec{E}_{N2} = -3\bar{a}_x + 4\bar{a}_y - \frac{4}{6.5}\bar{a}_z \text{ V/m}$$

1.128 (d)

$$|P_1 - P_2| = \sqrt{3^2 + 12^2} = \sqrt{9 + 144} = \sqrt{153}$$

1.129 (d)

According to Stokes theorem

$$\oint \vec{A}\cdot\overline{dl} = \iint (\vec{\nabla}\times\vec{A})\cdot\overline{dS}$$

Stokes theorem is defined for any vector $\vec{A}$, for closed line [closed curve], open surface inside a closed line, for any coordinate system. It is not defined for closed surface.

1.130 (c)

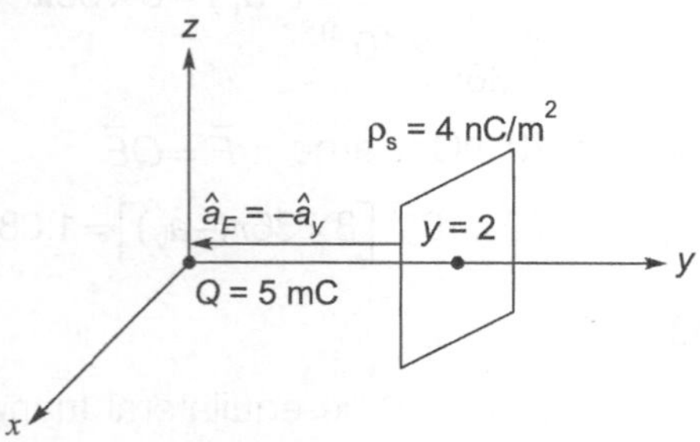

Electric field at origin due to infinite area charge surface is

$$\vec{E} = \frac{|\rho_S|}{2\varepsilon_0}\hat{a}_E = \frac{4\times10^{-9}}{2\left(\frac{1}{36\pi}\times10^{-9}\right)}(-\hat{a}_y)$$

$$= 72\pi(-\hat{a}_y) \text{ Volt/m}$$

Force on 'Q' point charge is $\vec{F} = Q\vec{E}$

$$\vec{F} = 5\times10^{-3}[72\pi(-\hat{a}_y)]$$

$$\vec{F} = 0.36\pi(-\hat{a}_y)$$

$$\vec{F} = -0.36\pi(\hat{a}_y) \text{ (Newton)}$$

1.131 (a)

$$\vec{A}\cdot\vec{B} = |\vec{A}||\vec{B}|\cos\theta_{AB}$$

$$|\vec{A}| = \sqrt{3^2 + 4^2 + 1^2} = \sqrt{26}$$

$$|\vec{B}| = \sqrt{2^2 + 5^2} = \sqrt{29}$$

$$\vec{A}\cdot\vec{B} = (4\times2) - (1\times5) = 3$$

$$\cos\theta_{AB} = \frac{3}{\sqrt{26\times29}}$$

$$\theta_{AB} = \cos^{-1}\left[\frac{3}{\sqrt{26\times29}}\right] = 83.7°$$

1.132 (a)

$$\nabla\cdot\vec{A} = \frac{\partial A_x}{\partial x} + \frac{\partial A_y}{\partial y} + \frac{\partial A_z}{\partial z} = 3y + xy$$

At (2, –2, 2),

$$\nabla\cdot\vec{A} = 3(-2) + (2)(-2) = -10$$

1.133 (b)

$$V = \frac{10}{r^2}\sin\theta\cdot\cos\phi$$

$$\vec{E} = -\vec{\nabla}V$$

$$\Rightarrow \vec{E} = -\left[\frac{\partial V}{\partial r}\hat{a}_r + \frac{1}{r}\frac{\partial V}{\partial\theta}\hat{a}_\theta + \frac{1}{r\sin\theta}\frac{dv}{d\phi}\hat{a}_\phi\right]$$

$$\vec{E} = -\left[10\sin\theta\cos\phi\frac{\partial}{\partial r}(r^{-2}) + \frac{1}{r}\frac{10}{r^2}\cos\phi\frac{\partial}{\partial\theta}(\sin\theta)\hat{a}_\theta + \frac{1}{r\sin\theta}\frac{10}{r^2}\sin\theta\frac{\partial}{\partial\phi}(\cos\phi)\hat{a}_\phi\right]$$

$$\vec{E} = -\left[10\sin\theta\cos\phi\left(\frac{-2}{r^3}\right)\hat{a}_r + \frac{10}{r^3}\cos\phi\cos\theta\hat{a}_\theta + \frac{10}{r^3}(-\sin\phi)\hat{a}_\phi\right]$$

$$\vec{D} = \epsilon_o\vec{E}$$

$$\underset{\substack{(2,\pi/2,0)\\(r,\theta,\phi)}}{\vec{D}}$$

$$= -8.85a\times10^{-12}\left[10\sin\left(\frac{\pi}{2}\right)\cos(\theta)\frac{-2}{2^3}\hat{a}_r + 0 + 0\right]$$

$$= 8.854\times10^{-12}\left(-\frac{5}{2}\hat{a}_r\right) = 22.135\times10^{-12}\hat{a}_r \text{ C/m}^2$$

$$\simeq 22.1\hat{a}_r \text{ pC/m}^2$$

1.134 (c)

As R_{AB} length is 10 units

$$\vec{R}_{AB} = \left|\vec{R}_{AB}\right| \vec{a}$$

$$\vec{R}_{AB} = 10\vec{a} = 6\hat{a}_x + 6.4\hat{a}_y + 4.8\hat{a}_z$$

$\vec{A}$ radial vector $= \hat{a}_x + 2\hat{a}_y + 3\hat{a}_z$

$$\vec{R}_{AB} = \vec{B} - \vec{A}$$

$$\vec{B} = \vec{R}_{AB} + \vec{A}$$

$$\therefore \quad \vec{B} = 10a + \vec{A} = 7\hat{a}_x + 8.4\hat{a}_y + 7.8\hat{a}_z$$

1.135 (c)

$$\vec{\nabla} \cdot \vec{D} = \rho_v$$

$$\Rightarrow \quad \rho_v = \frac{\partial D_x}{\partial x} + \frac{\partial D_y}{\partial y} + \frac{\partial D_z}{\partial z}$$

$$\vec{\nabla} \cdot \vec{D} = -e^{-x}\sin y + e^{-x}\sin y + 2$$

At origin,

$$\vec{\nabla} \cdot \vec{D} = 2$$

$$\therefore \quad Q = \rho_v\, dV = 2 \times 10^{-9} = 2 \text{ nC}$$

1.136 (c)

$$\overline{OG} = 2\hat{a}_x - 2\hat{a}_y - \hat{a}_z$$

$$\vec{a} = \frac{\overline{OG}}{\left|\overline{OG}\right|} = \frac{2\hat{a}_x - 2\hat{a}_y - \hat{a}_z}{\sqrt{2^2 + 2^2 + 1^2}} = \frac{2}{3}\hat{a}_x - \frac{2}{3}\hat{a}_y - \frac{1}{3}\hat{a}_z$$

1.137 (d)

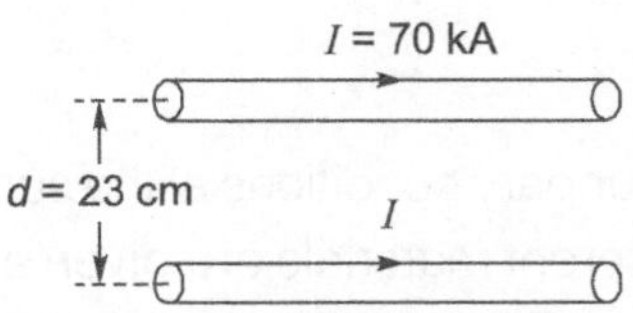

Force between two parallel conductor

$$F = \frac{\mu_o I_1 I_2}{2\pi d} h \text{ (Newton)}$$

$$\frac{f}{h} = \frac{4\pi \times 10^{-7}(70 \times 70)(10^6)}{2\pi(23 \times 10^{-2})}$$

$$= \frac{2(490)(100)}{23} = 4260.86 \text{ (N/m)}$$

1.138 (a)

$M(2, 5, -3)$ and $N(-3, 1, 4)$

Mid-point of MN

$$= \left(\frac{2-3}{2}, \frac{5+1}{2}, \frac{-3+4}{2}\right) = \left(\frac{-1}{2}, 3, \frac{1}{2}\right)$$

Distance from the origin to mid-point of MN is

$$= \sqrt{\left(0 - \frac{1}{2}\right)^2 + (0-3)^2 + \left(0 - \frac{1}{2}\right)^2}$$

$$= \sqrt{\frac{1}{4} + 9 + \frac{1}{4}} = \sqrt{9.5} = 3.08 \simeq 3.1$$

1.139 (b)

$$\vec{D} = 10x\hat{a}_x - 4y\hat{a}_y + kz\hat{a}_z \ \mu\text{C/m}^2$$

Given, $\rho_v = 0$ and satisfies Maxwell equation

$$\vec{\nabla} \cdot \vec{D} = 0$$

$$\frac{\partial D_x}{\partial x} + \frac{\partial D_y}{\partial y} + \frac{\partial D_z}{\partial z} = 0$$

$$\frac{\partial}{\partial x}(10x) + \frac{\partial}{\partial y}(-4y) + \frac{\partial}{\partial z}(kz) = 0$$

$$10 - 4 + k = 0$$

$$\Rightarrow \quad k = -6 \frac{\mu\text{C}}{\text{m}^3}$$

1.140 (b)

$$A = 2x^2y\bar{i} + 3x^2y^2\bar{j}$$

$$\nabla \cdot \vec{A} = \frac{\partial}{\partial x}(2x^2y) + \frac{\partial}{\partial y}(3x^2y^2)$$

$$= 4xy + 6x^2y$$

when $(x, y, z) \in [1, 3]$

$$1 \le x \le 3$$

$$1 \le y \le 3$$

$$1 \le z \le 3$$

$\therefore$ Mid-point of the cube is [2, 2, 2]

$$\therefore \nabla \cdot \vec{A}_{(2,2,2)} = 4(2)(2) + 6(2)^2(2)$$

$$= 16 + 48 = 64$$

1.141 (b)

Two vectors are said to be orthogonal if the angle between them is 90°.

Dot product of $\vec{V_1}$ and $\vec{V_2}$ is

$$\vec{V_1} \cdot \vec{V_2} = |V_1||V_2|\cos\theta$$

For $\theta = 90°$, $\vec{V_1} \cdot \vec{V_2} = 0$

1.142 (d)

$$\vec{P} = 2\hat{a}_x - \hat{a}_z$$

$$\vec{Q} = 2\hat{a}_x - \hat{a}_y + 2\hat{a}_z$$

Projection of $\vec{P}$ on $\vec{Q} = P\cos\theta$, where θ = angle between $\vec{P}$ and $\vec{Q}$.

$\because\ \vec{P}\cdot\vec{Q} = PQ\cos\theta$

$$P\cos\theta = \frac{\vec{P}\cdot\vec{Q}}{|\vec{Q}|}$$

The component of $\vec{P}$ along $\vec{Q}$ is, therefore

$$= \left(\frac{\vec{P}\cdot\vec{Q}}{|\vec{Q}|}\right)\hat{a}_Q$$

$$= \left(\frac{(2\hat{a}_x - \hat{a}_z)\cdot(2\hat{a}_x - \hat{a}_y + 2\hat{a}_z)}{\sqrt{2^2+1^2+2^2}}\right)\cdot\frac{(2\hat{a}_x - \hat{a}_y + 2\hat{a}_z)}{\sqrt{2^2+1^2+2^2}}$$

$$= \left(\frac{4-2}{3}\right)\cdot\left(\frac{2}{3}\hat{a}_x - \frac{1}{3}\hat{a}_y + \frac{2}{3}\hat{a}_z\right)$$

$$= 0.4444\hat{a}_x - 0.2222\hat{a}_y + 0.4444\hat{a}_z$$

1.143 (b)

(i) From Gauss's Law,

$$\oint \vec{D}\cdot d\vec{s} = Q = \int_{\text{Volume}} \rho_{Vs}\cdot dV$$

Hence, equation 1 is correct.

(ii) From Maxwell's equation,

$$\oint \vec{E_s}\,d\vec{l} = -\int \frac{\partial \vec{B_s}}{\partial t}\cdot d\vec{s}$$

$$\vec{B_s} = \vec{B_0}e^{j\omega t} \Rightarrow \frac{\partial B}{\partial t} = j\omega\vec{B_s}$$

$$\oint \vec{E_s}\cdot d\vec{l} = \int -j\omega\vec{B_s}\,d\vec{s}$$

(iii) From Maxwell's equation,

$$\oint \vec{B_s}\cdot d\vec{s} = 0$$

Hence, equation 3 is correct.

(iv) From Maxwell's equation

$$\oint \vec{H_s}\cdot d\vec{l} = \iint_S \left(\vec{J_s} + \frac{\partial \vec{D_s}}{\partial t}\right)d\vec{s}$$

$$\frac{\partial \vec{D_s}}{\partial t} = j\omega\vec{D_s}$$

$$\therefore\ \oint \vec{H_s}\cdot d\vec{l} = \iint_S \left(\vec{J_s} + j\omega\vec{D_s}\right)d\vec{s}$$

Hence, equation 4 is incorrect.

1.144 (b)

$$\vec{A} = \rho\cos\phi\,\hat{a}_\rho + \sin\phi\,\hat{a}_\phi$$

$$\nabla\times\vec{A} = \frac{1}{\rho}\begin{vmatrix} \hat{a}_\rho & \hat{a}_\phi & \hat{a}_z \\ \frac{\partial}{\partial\rho} & \frac{\partial}{\partial\phi} & \frac{\partial}{\partial z} \\ \rho\cos\phi & \rho\sin\phi & 0 \end{vmatrix}$$

$$\nabla\times\vec{A} = \left[\frac{1}{\rho}\sin\phi + \sin\phi\right]\hat{a}_z$$

The surface integration of curl of $\vec{A}$ is

Given, $30° < \phi < 60°$, $2 \le \rho \le 5$

$$\iint(\vec{\nabla}\times\vec{A})\cdot\vec{ds}$$

$$= \int_{\phi=30°}^{60°}\int_{\rho=2}^{5}\frac{1}{\rho}\sin\phi\,\rho\,d\rho\,d\phi + \int_{\phi=30°}^{60°}\int_{\rho=2}^{5}\sin\phi\,\rho\,d\rho\,d\phi$$

$$= \int_{\rho=2}^{5} \rho\,d\rho\cdot\int_{\phi=30°}^{60°}\sin\phi\,d\phi + \int_{\rho=2}^{5}\rho\,d\phi\cdot\int_{\phi=30°}^{60°}\sin\phi\,d\phi$$

$$= (\rho)_2^5(-\cos\phi)_{\phi=30°}^{60°} + \left(\frac{\rho^2}{2}\right)_{\rho=2}^{5}(-\cos\phi)_{\phi=30°}^{60°}$$

$$= (5-2)\left[-\left(\frac{1}{2}-\frac{\sqrt{3}}{2}\right)\right] + \left(\frac{5^2-2^2}{2}\right)\left[-\left(\frac{1}{2}-\frac{\sqrt{3}}{2}\right)\right]$$

$$= \left(\frac{\sqrt{3}}{2}-\frac{1}{2}\right)\left[3+\frac{(25-4)}{2}\right]$$

$$= \left(\frac{\sqrt{3}-1}{2}\right)(13.5) = 4.941$$

1.145 (d)

The boundary conditions at the surface between two different materials are given as:

(i) $E_{t1} = E_{t2}$ i.e. the tangential components of electric field intensity are continuous across the boundary.

(ii) $D_{n1} - D_{n2} = \rho_s$, i.e., the normal components of electric flux density are discontinuous at the boundary by an amount equal to the surface-charge density on the boundary.

(iii) $H_{t1} - H_{t2} = K$ i.e., the tangential components of magnetic field intensity are discontinuous at the boundary by an amount equal to the surface-current density on the boundary.

(iv) $B_{n1} = B_{n2}$ i.e., the normal components of magnetic flux density are continuous across the boundary.

Hence, option (d) is incorrect.

1.146 (b)

According to Gauss law,

$$\oint D \cdot ds = Q$$

i.e. the total electric flux out of a closed surface is equal to the charge enclosed within the surface.

Electric field is a vector quantity and it can be completely specified when its magnitude and direction both are known.

Hence, both statements (I) and (II) are true but statement (II) is not the correct explanation of statement (I).

1.147 (d)

(i) **Biot-Savart's law :**

$$\overrightarrow{dB} = \frac{\mu_o I \, \overrightarrow{dl} \times \hat{r}}{4\pi r^2}$$

(ii) **Lenz's law :** An induced current always flow in a direction such that it opposes the change which produced it.

(iii) Gauss's law :

$$\oint \vec{D} \cdot \overrightarrow{ds} = Q$$

(iv) **Ampere's circuit law :**

$$\oint \vec{H} \cdot \overrightarrow{dl} = I$$

Hence, (d) is the correct option.

1.148 (b)

In spherical coordinate system,

Laplace equation : $\nabla^2 V = 0$

$$\frac{1}{r^2}\frac{\partial}{\partial r}\left(r^2 \frac{\partial V}{\partial r}\right) + \frac{1}{r^2 \sin\theta}\frac{\partial}{\partial \theta}\left(\sin\theta \frac{\partial V}{\partial \theta}\right) + \frac{1}{r^2 \sin^2\theta}\cdot\frac{\partial^2 V}{\partial \phi^2} = 0$$

Poisson's equation :

$$\nabla^2 V = -\frac{\rho}{\epsilon}$$

where, ρ = Volume charge density

$$\frac{1}{r^2}\frac{\partial}{\partial r}\left(r^2 \frac{\partial V}{\partial r}\right) + \frac{1}{r^2 \sin\theta}\frac{\partial}{\partial \theta}\left(\sin\theta \cdot \frac{\partial V}{\partial \theta}\right) + \frac{1}{r^2 \sin^2\theta}\cdot\frac{\partial^2 V}{\partial \phi^2} = -\frac{\rho}{\epsilon}$$

Gradient of V :

$$\nabla V = \frac{\partial V}{\partial r}\hat{a}_r + \frac{1}{r}\cdot\frac{\partial V}{\partial \theta}\hat{a}_\theta + \frac{1}{r\sin\theta}\cdot\frac{\partial V}{\partial \phi}\hat{a}_\phi$$

Divergence of V :

$$\vec{\nabla}\cdot\vec{V} = \frac{1}{r^2}\frac{\partial}{\partial r}(r^2 V_r) + \frac{1}{r\sin\theta}\cdot\frac{\partial}{\partial \theta}(\sin\theta \cdot V_\theta) + \frac{1}{r\sin\theta}\cdot\frac{\partial V_\phi}{\partial \phi}$$

Hence, given equation is referred to as Poisson's equation.

1.149 (c)

Using Gauss's law,

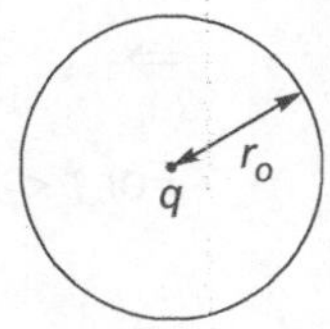

$$\oint \vec{D} \cdot \overrightarrow{ds} = q$$

$$\Rightarrow \vec{D} \cdot (4\pi r_0^2)\hat{a}_r = q$$

$$\Rightarrow \quad \vec{D} = \frac{q}{4\pi r_0^2}\hat{a}_r$$

$$\Rightarrow \quad \epsilon_o \vec{E} = \frac{q}{4\pi r_0^2}\hat{a}_r$$

$$\Rightarrow \quad \vec{E} = \frac{q}{4\pi\epsilon_o r_o^2}\hat{a}_r$$

The average value of electrostatic field over the volume V of sphere is given as

$$E_{avg} = \frac{1}{V}\int \vec{E}\cdot dV$$

$$E_{avg} = \frac{q}{V 4\pi \epsilon_0 r_0^2}\frac{4\pi r_0^3}{3}$$

$$E_{avg} = \frac{q}{V 4\epsilon_0 r_0^2}\times\frac{4 r_0^3}{3} = \frac{q r_0}{3\epsilon_0 V}$$

Hence, option (c) is the correct answer.

1.150 (b)

Consider a uniformly charged sphere of radius 'a' as shown below:

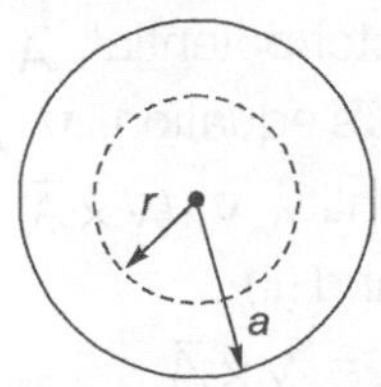

Let ρ be the volume charge density for $r < a$, using Gauss's law, we get,

$$\oint \vec{D} \cdot \overrightarrow{ds} = Q_{inc} = \int \rho dV$$

$$\Rightarrow \epsilon_0 E(4\pi r^2) = \rho \cdot \frac{4}{3}\pi r^3$$

$$\Rightarrow \quad E = \frac{\rho r}{3\epsilon_0}$$

$$\Rightarrow \quad \vec{E} = \frac{\rho r}{3\epsilon_0}\hat{a}_r$$

For $r > a$, using Gauss's law

$$\oint \vec{D}\cdot\vec{ds} = Q_{inc} = \frac{4}{3}\pi a^3$$

$$\Rightarrow \epsilon_0 E(4\pi r^2) = \frac{4}{3}\pi a^3$$

$$\Rightarrow \quad E = \frac{\rho a^3}{3\epsilon_0 r^2}$$

$$\Rightarrow \quad \vec{E} = \frac{\rho a^3}{3\epsilon_0 r^2}\hat{a}_r$$

For $r < a$, the potential at point r is given as

$$V = -\int_{r=\infty}^{r} \vec{E}\cdot\vec{dl}$$

$$\Rightarrow \quad V = -\left[\int_{r=\infty}^{a} \frac{\rho a^3}{3\epsilon_0 r^2}dr + \int_{r=a}^{r}\frac{\rho r}{3\epsilon_0}dr\right]$$

$$\Rightarrow \quad V = -\left[\frac{\rho a^3}{3\epsilon_0}\left(-\frac{1}{r}\right)_{\infty}^{a} + \frac{\rho}{3\epsilon_0}\left(\frac{r^2}{2}\right)_a^r\right]$$

$$\Rightarrow \quad V = -\left[-\frac{\rho a^3}{3\epsilon_0}\left(\frac{1}{a}-\frac{1}{\infty}\right) + \frac{\rho}{6\epsilon_0}\left[r^2 - a^2\right]\right]$$

$$\Rightarrow \quad V = -\left[-\frac{\rho a^2}{3\epsilon_0} + \frac{\rho}{6\epsilon_0}(r^2 - a^2)\right]$$

$$\Rightarrow \quad V = \frac{\rho a^2}{3\epsilon_0} - \frac{\rho}{6\epsilon_0}\left[r^2 - a^2\right]$$

$$\Rightarrow \quad V = \frac{\rho}{6\epsilon_0}\left[3a^2 - r^2\right]$$

1.151 (a)

We know that,

Magnetic vector potential, $\bar{A}$

From Maxwell's equation, $\nabla\cdot\bar{B} = 0$ ...(i)

From Null identity, $\nabla\cdot(\nabla\times\bar{A}) = 0$ ...(ii)

$\therefore$ From (i) and (ii)

$$\bar{B} = \nabla\times\bar{A}$$

1.152 (c)

Given, magnetic vector potential,

$$\bar{A} = \frac{-\rho^2}{4}\hat{a}_z$$

As we know that,

Magnetic flux, $\psi_m = \int \bar{B}\cdot\overline{dS}$

$$\Rightarrow \quad \psi_m = \int(\nabla\times\bar{A})\cdot\overline{dS} \qquad [\because \bar{B} = \nabla\times\bar{A}]$$

$$\text{Hence, } \nabla\times\bar{A} = \frac{1}{h_1h_2h_3}\begin{vmatrix} h_1\hat{a}_u & h_2\hat{a}_v & h_3\hat{a}_w \\ \frac{\partial}{\partial u} & \frac{\partial}{\partial v} & \frac{\partial}{\partial w} \\ h_1A_u & h_2A_v & h_3A_w \end{vmatrix}$$

For cylindrical co-ordinate system:

$$h_1\,h_2\,h_3 = |\rho|$$

$$\therefore \quad \nabla\times\bar{A} = \frac{1}{\rho}\begin{vmatrix} \hat{a}_\rho & \rho\hat{a}_\phi & \hat{a}_z \\ \frac{\partial}{\partial\rho} & \frac{\partial}{\partial\phi} & \frac{\partial}{\partial z} \\ 0 & 0 & -\frac{\rho^2}{4} \end{vmatrix}$$

$$= \frac{1}{\rho}\left[\frac{\partial}{\partial\rho}\left(-\frac{\rho^2}{4}\right)\right](-\rho\hat{a}_\phi) = \frac{2\rho}{4}\hat{a}_\phi = \frac{\rho}{2}\hat{a}_\phi$$

$$\therefore \quad \psi_m = \int\frac{\rho}{2}\hat{a}_\phi\cdot\overline{dS};\ \overline{dS} = d\rho\, dz\,\hat{a}_\phi$$

$$\text{Hence,} \quad \psi_m = \int_S\frac{\rho}{2}d\rho\, dz$$

$$= \frac{1}{2}\int_{\rho=1}^{2}\rho\, d\rho\int_{z=0}^{5}dz = \frac{1}{2}\cdot\frac{\rho^2}{2}\Bigg|_1^2\cdot z\Big|_0^5$$

$$= \frac{1}{2}\cdot\frac{3}{2}\cdot 5 = \frac{15}{4} = 3.75 \text{ Wb}$$

1.153 (c)

Given, vector, $\bar{P} = x^3\hat{a}_x - x^2y^2\hat{a}_y - x^2yz\,\hat{a}_z$

(i) For solenoid, $\nabla\cdot\bar{P} = 0$

$$\therefore \quad \nabla\cdot\bar{P} = 3x^2 - 2x^2y - x^2y = 3x^2 - 3x^2y \neq 0$$

(ii) For irrotational,

$$\nabla\times\bar{P} = 0$$

$$\therefore \nabla\times\bar{P} = \begin{vmatrix} \hat{a}_x & \hat{a}_y & \hat{a}_z \\ \frac{\partial}{\partial x} & \frac{\partial}{\partial y} & \frac{\partial}{\partial z} \\ x^3 & -x^2y^2 & -x^2yz \end{vmatrix}$$

$$= -x^2z\cdot\hat{a}_x - (-2xyz)\hat{a}_y + (-2xy^2)\hat{a}_z$$

$$= -x^2z\,\hat{a}_x + 2xyz\,\hat{a}_y - 2xy^2\,\hat{a}_z \neq 0$$

1.154 (d)

Given data electric field,

$$E\big|_{surface} = 2 \text{ V/m}$$

As we know that

Electric flux density,

$$D_n = \rho_s$$

Surface charge density,

$$\rho_S = \varepsilon E = \varepsilon_0 \varepsilon_r E$$

$$\Rightarrow \quad \rho_S = 8.85 \times 10^{-12} * 80 * 2$$

$$= 1.41 * 10^{-9}\ \text{c/m}^2$$

1.155 (c)

Given electric field intensity,

$$\bar{E} = x\hat{u}_x + y\hat{u}_y + z\hat{u}_z$$

As we know that

Potential difference,

$$V_X - V_Y = -\int_Y^X \bar{E}\cdot\overline{dl} = -\int_{1,2,3}^{2,0,0} \bar{E}\cdot\overline{dl}$$

Here, $\bar{E} = x\hat{u}_x + y\hat{u}_y + z\hat{u}_z$

$$\overline{dl} = dx\,\hat{u}_x + dy\,\hat{u}_y + dz\,\hat{u}_z$$

$$\therefore \quad \bar{E}\cdot\overline{dl} = xdx + ydy + zdz$$

So, $$V_X - V_Y = -\int_{1,2,3}^{2,0,0} xdx + ydy + zdz$$

$$= -\int_{x=1}^{2} xdx - \int_{y=2}^{0} ydy - \int_{z=3}^{0} zdz$$

$$= -\frac{x^2}{2}\Bigg|_1^2 - \frac{y^2}{2}\Bigg|_2^0 - \frac{z^2}{2}\Bigg|_3^0 = -\frac{3}{2}+\frac{4}{2}+\frac{9}{2} = 5\ \text{V}$$

1.156 (c)

The energy stored in the capacitor, $E = \dfrac{1}{2C}Q^2$.

1.157 (c)

We know that, potential due to dipole is given as

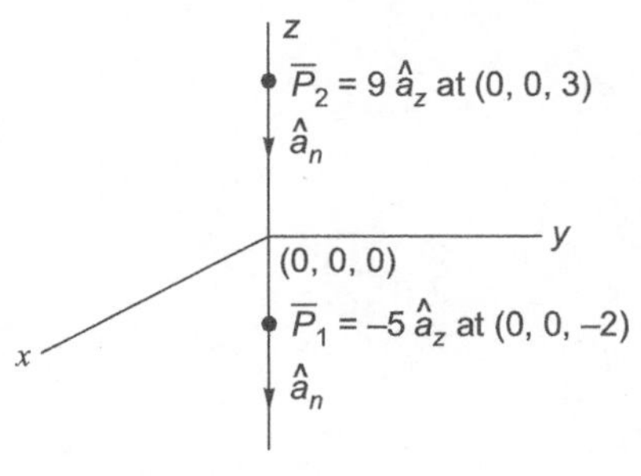

Potential, $V = \dfrac{K\bar{p}.\hat{a}_n}{r^2}$; $K = \dfrac{1}{4\pi\epsilon_0} = 9\times10^9$ F/m

For dipole - (1), $\bar{P}_1$:

$$V_1 = \frac{9\times10^9\times -5\times10^{-9}(\hat{a}_z.\hat{a}_z)}{(2)^2} = -11.25\ \text{V}$$

For dipole - (2) $\bar{P}_2$:

$$V_2 = \frac{9\times10^9\times9\times10^{-9}(\hat{a}_z(-\hat{a}_z))}{(3)^2} = -9\ \text{V}$$

$$\therefore \quad V = V_1 + V_2 = -11.25 - 9 = -20.25\ \text{V}$$

1.158 (b)

We know that,

$\nabla.\bar{D} = \nabla.\epsilon\,\bar{E} = \epsilon\nabla.\bar{E} \quad \Rightarrow \quad \bar{D}\propto\bar{E}$; Linear

$\nabla.\bar{J} = \nabla.\sigma\bar{E} = \sigma\nabla.\bar{E} \quad \Rightarrow \quad \bar{J}\propto\bar{E}$; Linear

As ϵ, σ does not depends on space i.e. uniform in space hence, it is homogeneous medium.

So, medium is linear and homogeneous.

1.159 (c)

As we know that,

Displacement current density,

$$\bar{J}_d = \frac{\partial\bar{D}}{\partial t} = \varepsilon\frac{\partial\bar{E}}{\partial t}$$

$$\Rightarrow \quad \bar{J}_d = j\omega\varepsilon\bar{E}$$

$$\Rightarrow \quad \bar{J}_d = \omega\varepsilon EA = \omega\varepsilon\left(\frac{V}{d}\right)A$$

$$= 2\pi\times4.5\times10^9\times\frac{10^{-9}}{36\pi}\times\left(\frac{0.5}{10^{-3}}\right)\times10^{-4}$$

$$= 12.5\ \text{mA}$$

1.160 (b)

From Coulomb's law, the force of attraction between two charged particles Q_1 and Q_2 separated by a distance R is given by

$$F = k\frac{Q_1Q_2}{R^2}$$

where $$k = \frac{1}{4\pi\,\epsilon_0}$$

$$= \frac{1}{4\pi\times8.854\times10^{-12}\text{F/m}}$$

$$= 9\times10^9\ \text{m/F}$$

1.161 (a)

- The presence of current in two circuits in proximity to each other is accompanied by **mechanical** forces on each conductor, forces which change if either current is changed.

- Option (b) states an application of Faraday's Law.
- Option (c) describes a situation in which the displacement current, postulated by Maxwell, complements a discontinuous conduction current to yield, as a resultant, a composite current which is continuous.
- Option (d) describes mechanical-force effects and help in assigning directions to the fields.

Hence, option (a) does not indicate the space pervasiveness of electromagnetic effects.

1.162 (d)

Work done,

$$W = -q\int_i^t \bar{E}\cdot\overline{dl}\,;\ \bar{E} = y\hat{a}_x + x\hat{a}_y + 2y\hat{a}_z$$

$$\overline{dl} = dx\hat{a}_x + dy\hat{a}_y + dz\hat{a}_z$$

$$\bar{E}\cdot\overline{dl} = ydx + xdy + 2ydz$$

Work done in moving a charge from point A to point B,

$$W = -q\int_A^B (ydx + xdy + 2ydz)$$

The point A and B lies on $z = 3$ plane

$\Rightarrow\quad dz = 0$

$$\therefore\ W = -q\int_A^B ydx + xdy = -2\int_{1,0}^{2,1} ydx + xdy$$

$\because$ The path is a straight line

$$\therefore\quad \text{slope; } m = \frac{y_2 - y_1}{x_2 - x_1} = \frac{1-0}{2-1} = 1$$

so, $\quad y - y_1 = m(x - x_1)$

$\Rightarrow\quad y - 0 = (x - 1)$

$\Rightarrow\quad y = (x - 1)$

$\Rightarrow\quad dy = dx$

$$\therefore\quad W = -2\int_{x=1}^{2} (x-1)dx + xdx$$

$$\Rightarrow\quad W = -2\left[\int_{x=1}^{2} 2x\,dx - \int_{x=1}^{2} dx\right]$$

$$\Rightarrow\quad W = -2\left[\frac{2x^2}{2}\Big|_1^2 - x\Big|_1^2\right]$$

$$= -2[3 - 1] = -4\text{J}$$

1.163 (d)

Biot-savart law,

$$d\bar{H} = \frac{i\overline{dl}\times\hat{a}_r}{4\pi r^2}\ \text{A/m}$$

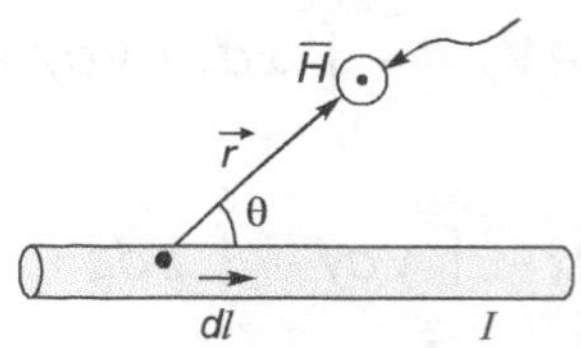

■■■

2 Uniform Plane Waves

2.1 Polystyrene has a relative permittivity of 2.7. If the wave is incident at an angle 'θ_1' of 30° from air onto polystyrene, the angle of transmission will be nearly

(a) 0.2° (b) 2°
(c) 18° (d) 48°

[ESE-2000]

2.2 Match **List-I** with **List-II** and select the correct answer using the codes given below: (Notations have their usual meaning):

List-I
A. Permeability
B. Force
C. Velocity of EM wave
D. Displacement density

List-II
1. QE
2. $\frac{1}{\sqrt{\mu_0 \varepsilon_0}}$
3. $\mu_0 \mu_r$
4. $\sqrt{\frac{\mu}{\epsilon}}$
5. εE

Codes:

	A	B	C	D
(a)	3	2	4	5
(b)	3	1	2	5
(c)	4	1	2	3
(d)	4	2	5	3

[ESE-2000]

2.3 The Poynting vector $\vec{P} = \vec{E} \times \vec{H}$ has the dimensions of

(a) Power/unit area (b) Volts
(c) Power (d) Volt/unit length

[ESE-2000]

2.4 Consider the following statements regarding a plane wave propagating through free space: The direction of field

1. 'E' is perpendicular to the direction of propagation
2. 'H' is perpendicular to the direction of propagation
3. 'E' is perpendicular to the direction of field 'H'

Which of these statements are correct?

(a) 1 and 2 (b) 2 and 3
(c) 1 and 3 (d) 1, 2 and 3

[ESE-2000]

2.5 If the velocity of electromagnetic wave in free space is 3×10^8 m/s, the velocity in a medium with ε_r of 4.5 and μ_r of 2 would be

(a) 1×10^8 m/s (b) 3×10^8 m/s
(c) 9×10^8 m/s (d) 27×10^8 m/s

[ESE-2000]

2.6 If $\vec{H} = 0.1\sin(10^8 \pi t + \beta y)\hat{a}_x$ A/m for a plane wave propagating in free space, then the time average Poynting vector is

(a) $(0.6\pi \sin^2 \beta y)\hat{a}_y$ W/m²
(b) $-0.6\pi\,\hat{a}_y$ W/m²
(c) $1.2\pi\,\hat{a}_x$ W/m²
(d) $-1.2\pi\,\hat{a}_x$ W/m²

[ESE-2000]

2.7 Which one of the following is NOT a correct Maxwell equation?

(a) $\nabla \times \bar{H} = \frac{\partial \bar{D}}{\partial t} + \bar{J}$ (b) $\nabla \times \bar{E} = \frac{\partial \bar{H}}{\partial t}$

(c) $\bar{\nabla} \cdot \bar{D} = \rho$ (d) $\bar{\nabla} \cdot \bar{B} = 0$

[ESE-2001]

2.8 Match **List-I** (Maxwell equation) with **List-II** (Description) and select the correct answer:

List-I

A. $\oint \vec{B} \cdot d\vec{a} = 0$

B. $\oint \vec{D} \cdot d\vec{a} = \int_V \rho dv$

C. $\oint \vec{E} \cdot d\vec{l} = -\iint \frac{\partial \vec{B}}{\partial t} \cdot \overrightarrow{da}$

D. $\oint \vec{H} \cdot \overrightarrow{dl} = \iint \left(\frac{\partial \vec{D}}{\partial t} + \vec{J} \right) \cdot \overrightarrow{da}$

List-II

1. The mmf around a closed path is equal to the conduction current plus the time derivative of the electric displacement current through any surface bounded by the path.
2. The emf around a closed path is equal to the time derivative of the magnetic displacement through any surface bounded by the path.
3. The total electric displacement through the surface enclosing a volume is equal to total charge within the volume
4. The net magnetic flux emerging through any closed surface is zero

Codes:

	A	B	C	D
(a)	1	3	2	4
(b)	4	3	2	1
(c)	4	2	3	1
(d)	1	2	3	4

[ESE-2001]

2.9 In a uniform plane wave, the value of $|E|/|H|$ is

(a) $\sqrt{\frac{\mu}{\varepsilon}}$ (b) $\sqrt{\frac{\varepsilon}{\mu}}$

(c) 1 (d) $\sqrt{\mu\varepsilon}$

[ESE-2001]

2.10 Which one of the following statements is NOT correct for a plane wave with

$$\vec{H} = 0.5e^{-0.1x}\cos(10^6 t - 2x)\hat{a}_z \text{ A/m?}$$

(a) The wave frequency is 10^6 r.p.s.
(b) The wavelength is 3.14 m
(c) The wave travels along + x-direction
(d) The wave is polarized in the z-direction

[ESE-2001]

2.11 **Assertion (A):** For an E.M. wave normally incident on a conductor surface the magnetic field $\bar{H}$ undergoes a 180° phase reversal and the phase of electric field $\bar{E}$ remains same.
Reason (R): The direction of propagation of incident wave will reverse after striking a conductor surface.

(a) Both A and R are true and R is the correct explanation of A
(b) Both A and R are true but R is NOT the correct explanation of A
(c) A is true but R is false
(d) A is false but R is true

[ESE-2002]

2.12 Match **List-I** with **List-II** and select the correct answer:

List-I	List-II
A. Propagation constant	1. $\sqrt{\frac{\omega\mu\sigma}{2}}$
B. Radiation intensity	2. $\|E\|^2 \times \frac{r^2}{2\eta}$
C. Wave impedance	3. E_t/H_t
	4. $\vec{E} \times \vec{H}$

Codes:

	A	B	C
(a)	1	2	3
(b)	4	3	2
(c)	1	3	2
(d)	4	2	3

[ESE-2002]

2.13 Consider the following statements with reference to Brewster angle:

1. For oblique incidence at a boundary, there is no reflected wave if incident wave is vertically polarized.
2. Brewster angle can occur only at the boundary of a perfect conductor.
3. For a horizontally polarized wave, there is no Brewster angle.
4. When the incident wave is not fully vertically polarized, there will be a reflected component which is horizontally polarized.

Which of these statements are correct?

(a) 1, 2 and 3 (b) 2, 3 and 4
(c) 1, 3 and 4 (d) 1, 2 and 4

[ESE-2002]

2.14 If the $\vec{E}$ field of a plane polarized e.m. wave travelling in the z-direction is:

$\vec{E} = \hat{a}_x E_x + \hat{a}_y E_y$, then its $\vec{H}$ field is

(a) $\hat{a}_y \dfrac{E_x}{Z_o} - \hat{a}_x \dfrac{E_y}{Z_o}$ (b) $\hat{a}_y \dfrac{E_x}{Z_o} + \hat{a}_x \dfrac{E_y}{Z_o}$

(c) $\hat{a}_x \dfrac{E_y}{Z_o} - \hat{a}_y \dfrac{E_x}{Z_o}$ (d) $-\hat{a}_x \dfrac{E_y}{Z_o} - \hat{a}_y \dfrac{E_x}{Z_o}$

[ESE-2002]

2.15 Consider the following statements:

For electromagnetic waves propagating in free space:

1. electric field is perpendicular to direction of propagation
2. electric field is along the direction of propagation
3. magnetic field is perpendicular to direction of propagation
4. magnetic field is along the direction of propagation

Which of these statements are correct?

(a) 1 and 3 (b) 1 and 4

(c) 2 and 3 (d) 2 and 4 **[ESE-2002]**

2.16 A TEM wave impinges obliquely on a dielectric-dielectric boundary ($\epsilon_{r1} = 2$, $\epsilon_{r2} = 1$). The angle of incidence for total reflection is

(a) 30° (b) 45°

(c) 60° (d) 75° **[ESE-2003]**

2.17 A plane electromagnetic wave travelling in a perfect dielectric medium of characteristic impedance η_1 is incident normally on its boundary with another perfect dielectric medium of characteristic impedance η_2. The electric and magnetic field strengths of the incident wave are denoted by E_i and H_i respectively whereas E_r and H_r denote these quantities for the reflected wave, and E_t and H_t for the transmitted wave. Which of the following relations are correct?

1. $E_i = \eta_1 H_i$
2. $E_r = -\eta_1 H_r$
3. $E_t = -\eta_2 H_t$

Select the correct answer using the codes given below:

(a) 1, 2 and 3 (b) 1 and 2

(c) 1 and 3 (d) 2 and 3

[ESE-2003]

2.18 A plane electromagnetic wave travelling in a perfect dielectric medium of dielectric constant ϵ_1 is incident on its boundary with another perfect dielectric medium of dielectric constant ϵ_2. The incident ray makes an angle of θ_1 with the normal to the boundary surface. The ray transmitted into the other medium makes an angle of θ_2 with the normal.

If $\epsilon_1 = 2\epsilon_2$ and $\theta_1 = 60°$, which one of the following is correct?

(a) $\theta_2 = 45°$

(b) $\theta_2 = \sin^{-1} 0.433$

(c) $\theta_2 = \sin^{-1} 0.612$

(d) There will be no transmitted wave

[ESE-2003]

2.19 Which one of the following Maxwell's equations gives the basic idea of radiation?

(a) $\left.\begin{array}{l} \nabla \times \vec{H} = \partial \vec{D}/\partial t \\ \nabla \times \vec{E} = -\partial \vec{B}/\partial t \end{array}\right\}$

(b) $\left.\begin{array}{l} \nabla \times \vec{E} = -\partial \vec{D}/\partial t \\ \nabla \cdot \vec{D} = 0 \end{array}\right\}$

(c) $\left.\begin{array}{l} \nabla \cdot \vec{D} = \rho \\ \nabla \cdot \vec{B} = 0 \end{array}\right\}$

(d) $\left.\begin{array}{l} \nabla \cdot \vec{B} = 0 \\ \nabla \times \vec{H} = \left(\partial \vec{D}/\partial t\right) \end{array}\right\}$

[ESE-2003]

2.20 Match **List-I** (Nature of Polarization) with **List-II** (Relationship Between X and Y components) for a propagating wave having cross-section in the XY plane and propagating along z-direction and select the correct answer:

List-I

A. Linear

B. Left circular

C. Right circular

D. Elliptical

List-II

1. X and Y components are in same phase

2. X and Y components have arbitrary phase difference

3. X component leads Y by 90°

4. X component lags behind Y by 90°

Codes:

	A	B	C	D
(a)	1	4	2	3
(b)	4	1	2	3
(c)	1	4	3	2
(d)	4	1	3	2

[ESE-2003]

2.21 Which one of the following statements is correct?

A right circularly polarised wave is incident from air onto a polystyrene (ε_r = 2.7). The reflected wave is

(a) right circularly polarised
(b) left circularly polarised
(c) right elliptically polarised
(d) left elliptically polarised [ESE-2004]

2.22 For sea water with σ = 5 mho/m and ε_r = 80, what is the distance for which radio signal can be transmitted with 90% attenuation at 25 kHz?

(a) 0.322 m (b) 3.22 m
(c) 32.2 m (d) 322 m

[ESE-2004]

2.23 The electric field of a wave propagating through a lossless medium (μ_0, $81\varepsilon_0$) is

$$\bar{E} = 10\cos(6\pi \times 10^8 t - \beta x)\bar{a}_y$$

What is the phase constant β of the wave?

(a) 2π rad/m (b) 9π rad/m
(c) 18π rad/m (d) 81π rad/m

[ESE-2004]

2.24 Which one of the following gives the values of the attenuation factor α and phase shift factor β for a wave propagated in a good dielectric having $\sigma/(\omega\varepsilon) >> 1$?

(a) $\alpha = \dfrac{\left[\sigma\sqrt{\mu/\varepsilon}\right]}{2}; \beta = \omega\sqrt{\mu\varepsilon}$

(b) $\alpha = \sqrt{\mu/\varepsilon}\,; \beta = \sqrt{\omega\mu/\varepsilon}$

(c) $\alpha = \dfrac{\left[\sigma\sqrt{\mu/\varepsilon}\right]}{2}; \beta = \omega\sqrt{\omega\mu\varepsilon}$

(d) α = 0, $\beta = \omega^2\,\mu\varepsilon$ [ESE-2004]

2.25 If the phase velocity of a plane wave in a perfect dielectric is 0.4 times its value in free space, then what is the relative permittivity of the dielectric?

(a) 6.25 (b) 4.25
(c) 2.5 (d) 1.25 [ESE-2004]

2.26 In free space $\bar{E}(x,t) = 60\cos(\omega t - 2x)\bar{a}_y$ V/m. What is the average power crossing a circular area of radius 4 m in the plane x = constant?

(a) 480 W (b) 240 W
(c) 120 W (d) 60 W [ESE-2004]

2.27 For an electric field $E = E_0\sin\omega t$, what is the phase difference between the conduction current and the displacement current?

(a) 0° (b) 45°
(c) 90° (d) 180° [ESE-2004]

2.28 Consider the following statements:

Electromagnetic waves are refracted when they
1. pass through a medium of different permittivity
2. pass through a small slot in a dielectric slab
3. are incident on a perfectly conducting surface

Which of these statement(s) is/are correct?

(a) 1 only (b) 2 only
(c) 3 only (d) 1 and 3

[ESE-2004]

2.29 Two metal rings 1 and 2 are placed in a uniform magnetic field which is decreasing with time with their planes perpendicular to the field. If the rings are identical except that ring 2 has a thin air gap in it, which one of the following statements is correct?

(a) No e.m.f is induced in ring 1
(b) An e.m.f. is induced in both the rings
(c) Equal Joule heating occurs in both the rings
(d) Joule heating does not occur in either ring

[ESE-2004]

2.30 Maxwell's equations are obeyed by an EM wave while it is propagating

(a) Only in the free space
(b) Only in water and free space
(c) Only in ionosphere, water and free space
(d) In all the solids, liquids, gases and all the media mentioned above at (a), (b) and (c)

[ESE-2005]

2.31 The x-directed electric field E_x having sinusoidal time variation $e^{j\omega t}$ and space variation in z-direction satisfies the equation $\nabla^2 E_x + k^2 E_x = 0$ under source free condition in a lossless medium. What is the solution representing propagation in positive z-direction?

(a) $E_x = E_0 e^{-kz}$ (b) $E_x = E_0 e^{+jkz}$
(c) $E_x = E_0 e^{-jkz}$ (d) $E_x = E_0 e^{+kz}$

[ESE-2005]

2.32 For an electromagnetic wave incident on a conducting medium, the depth of penetration
(a) is directly proportional to the attenuation constant
(b) is inversely proportional to the attenuation constant
(c) has a logarithmic relationship with the attenuation constant
(d) is independent of the attenuation constant
[ESE-2005]

2.33 A plane wave having x-directed electric field propagating in free space along the z-direction is incident on an infinite electrically conducting (perfect conductor) sheet at $z = 0$ plane. Which one of the following is correct?
(a) The sheet will absorb the wave
(b) There will be x-directed surface electric current on the sheet
(c) There will be y-directed surface electric current on the sheet
(d) There will be magnetic current in the sheet
[ESE-2005]

2.34 **Assertion (A):** Time varying electric field produces magnetic fields.
Reason (R): Time varying magnetic field produces electric fields.
(a) Both A and R are true and R is the correct explanation of A
(b) Both A and R are true but R is NOT the correct explanation of A
(c) A is true but R is false
(d) A is false but R is true
[ESE-2006]

2.35 Match **List-I** with **List-II** and select the correct answer using the codes given below the lists:

List-I	List-II
A. Line charge	**1.** Maxwell's equation
B. Magnetic flux density	**2.** Poynting vector
C. Displacement current	**3.** Biot-Savart's law
D. Power flow	**4.** Gauss's law

Codes:

	A	B	C	D
(a)	1	2	4	3
(b)	4	3	1	2
(c)	1	3	4	2
(d)	4	2	1	3

[ESE-2006]

2.36 The electric field component of a wave in free space is given by

$$\vec{E} = 50\sin(10^7 t + kz)\hat{j} \text{ V/m}$$

Which one of the following is the correct inference that can be drawn from this expression?
(a) The wave propagates along y-axis
(b) The wavelength is 188.5 m
(c) The wave number k = 0.33 rad/m
(d) The wave attenuates as it travels
[ESE-2006]

2.37 A uniform plane electromagnetic wave is normally incident upon a thick magnetic material such that its complex permittivity is equal to its complex permeability i.e. $\mu^* = \varepsilon^*$.
Which one of the following is the correct statement?
(a) There will not be any reflection from the material
(b) A part of the wave will be reflected such that reflection coefficient would be 0.5
(c) Reflection coefficient would be close to 1
(d) Transmission coefficient would be 0.9
[ESE-2006]

2.38 If $\vec{E} = (\hat{x} + \hat{j}y)e^{-j\beta z}$, then the wave is said to be which one of the following?
(a) Right circularly polarized
(b) Right elliptically polarized
(c) Left circularly polarized
(d) Left elliptically polarized **[ESE-2007]**

2.39 Which one of the following is correct? As frequency increases, the surface resistance of a metal
(a) decreases
(b) increases
(c) remains unchanged
(d) varies in an unpredictable manner
[ESE-2007]

2.40 What must be the angle θ of a corner reflector, such that an incident wave is reflected in the same direction?

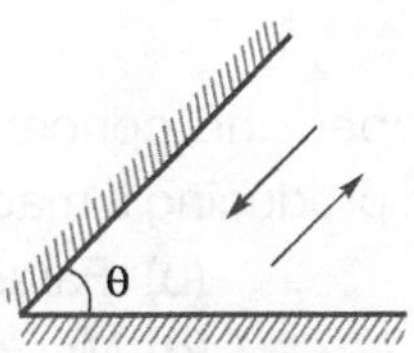

(a) 30° (b) 45°
(c) 60° (d) 90°
[ESE-2007]

2.41 Poynting vector is a measure of which one of the following?
(a) Maximum power flow through a surface surrounding the source
(b) Average power flow through the surface
(c) Instantaneous power flow through the surface
(d) Power dissipated by the surface
[ESE-2007]

2.42 Two conducting thin coils X and Y (identical except for a thin cut in coil Y) are placed in a uniform magnetic field which is decreasing at a constant rate. If the plane of the coils is perpendicular to the field lines, which of the following statement is correct? As a result, emf is induced in
(a) both the coils
(b) coil Y only
(c) coil X only
(d) none of the two coils [ESE-2007]

2.43 Consider the following statements associated with boundary conditions between two media:
1. Normal component of B is continuous at the surface of discontinuity.
2. Normal component of D may or may not be continuous.

Which of the statement(s) given above is/are correct?
(a) 1 only (b) 2 only
(c) Both 1 and 2 (d) Neither 1 nor 2
[ESE-2008]

2.44 According to Poynting theorem, the vector product $\vec{E} \times \vec{H}$ is a measure of which one of the following?
(a) Stored energy density of the electric field
(b) Stored energy density of the magnetic field
(c) Power dissipated per unit volume
(d) Rate of energy flow per unit area
[ESE-2008]

2.45 Who developed the concept of time varying electric field producing a magnetic field?
(a) Gauss (b) Faraday
(c) Hertz (d) Maxwell [ESE-2009]

2.46 Which of the following equations results from the circuital form of Ampere's law?
(a) $\nabla \times \vec{E} = -\frac{\partial \vec{B}}{\partial t}$ (b) $\nabla \cdot \vec{B} = 0$
(c) $\nabla \cdot \vec{D} = \rho$ (d) $\nabla \times \vec{H} = \vec{J} + \frac{\partial \vec{D}}{\partial t}$
[ESE-2009]

2.47 In which direction is the plane wave $\bar{E} = 50 \sin(10^8 t + 2z)\hat{a}_y$ V/m, (where $\hat{a}_y$ is the unit vector in y-direction), travelling?
(a) along y direction (b) along $-y$ direction
(c) along z direction (d) along $-z$ direction
[ESE-2009]

2.48 Consider the following statements:
1. (Electric or magnetic) field must have two orthogonal linear components.
2. The two components must have the same magnitude.
3. The two components must have a time-phase difference of odd multiples of 90°.

Which of these are the necessary and sufficient conditions for a time-harmonic wave to be circularly polarized at a given point in space?
(a) 1 and 2 only (b) 2 and 3 only
(c) 1, 2 and 3 (d) 1 and 3 only
[ESE-2010]

2.49 Consider the following statements regarding depth of penetration or skin depth in a conductor:
1. It increase as frequency increases.
2. It is inversely proportional to square root of μ and σ.
3. It is inversely proportional to square root of f.
4. It is directly proportional to square root of μ and σ.

Which of the above statements are correct?
(a) 1 and 2 only (b) 3 and 4 only
(c) 2 and 3 only (d) 1, 2, 3 and 4
[ESE-2010]

2.50 The intrinsic impedance of copper at 3 GHz (with parameters : $\mu = 4\pi \times 10^{-7}$ H/m; $\varepsilon = 1/36\pi \times 10^{-9}$ and $\sigma = 5.8 \times 10^7$ mho/m) will be
(a) $0.02\, e^{j\pi/4}$ ohm (b) $0.02\, e^{j\pi/2}$ ohm
(c) $0.2\, e^{j\pi/2}$ ohm (d) $0.2\, e^{j\pi/4}$ ohm
[ESE-2010]

2.51 An elliptically polarized wave travelling in the positive z-direction in air has x and y components
$E_x = 3 \sin(\omega t - \beta z)$ V/m

$E_y = 6\sin(\omega t - \beta z + 75°)$ V/m

The magnitude of Poynting vector is

(a) 8 W/m^2 (b) 4 W/m^2

(c) 62.5 mW/m^2 (d) 125 mW/m^2

[ESE-2010]

2.52 **Assertion (A):** The velocity of light in any medium is slower than that of vacuum.

Reason (R): The dielectric constant of the vacuum is unity and is lesser than that of any other medium.

(a) Both A and R are true and R is the correct explanation of A

(b) Both A and R are true but R is NOT the correct explanation of A

(c) A is true but R is false

(d) A is false but R is true [ESE-2010]

2.53 For no reflection condition a vertically polarized wave should be incident at the interface between two dielectrics having $\varepsilon_1 = 4$ and $\varepsilon_2 = 9$, with an incident angle of

(a) $\tan^{-1}\left(\frac{9}{4}\right)$ (b) $\tan^{-1}\left(\frac{3}{2}\right)$

(c) $\tan^{-1}\left(\frac{2}{3}\right)$ (d) $\tan^{-1}\left(\frac{4}{9}\right)$

[ESE-2011]

2.54 Assuming that each loop is stationary and time varying magnetic field $\bar{B}$, induces current I, which of the configurations in the figure are correct?

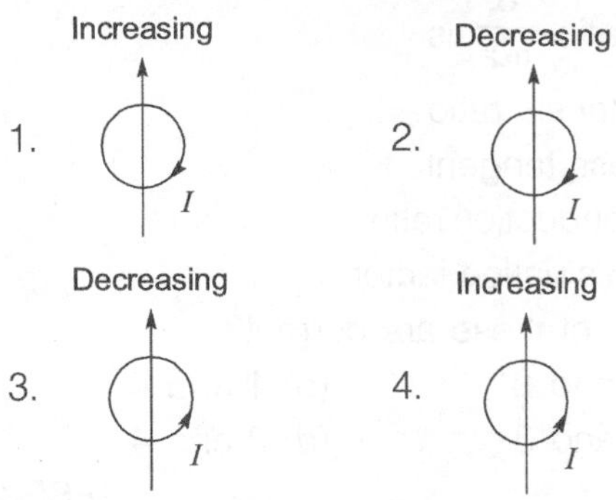

(a) 1, 2, 3 and 4 (b) 1 and 3 only

(c) 2 and 4 only (d) 3 and 4 only

[ESE-2011]

2.55 The electric field component of a wave in free space is given by

$$\bar{E} = 10\cos(10^7 t + kz)\bar{a}_y \text{ V/m.}$$

Following is a list of possible inferences:

1. Wave propagates along $\bar{a}_y$
2. Wavelength $\lambda = 188.5$ m
3. Wave amplitude is 10 V/m
4. Wave number = 0.33 rad/m
5. Wave attenuates as it travels

Which of these inferences can be drawn from $\bar{E}$?

(a) 1, 2, 3, 4 and 5 (b) 2 and 3 only

(c) 3 and 4 only (d) 4 and 5 only

[ESE-2011]

2.56 An electromagnetic wave of frequency 3 MHz passes from vacuum into a dielectric medium with permittivity $\varepsilon = 4.0$. Then

(a) Wavelength is doubled and the frequency remains unchanged

(b) Wavelength is doubled and the frequency becomes half

(c) Wavelength is halved and the frequency remains unchanged

(d) Both the wavelength and frequency remain unchanged

[ESE-2011]

2.57 A plane wave is generated under water ($\varepsilon = 81\varepsilon_0$ and $\mu = \mu_0$). The wave is parallel polarized. At the interface between water and air, the angle α for which there is no reflection is

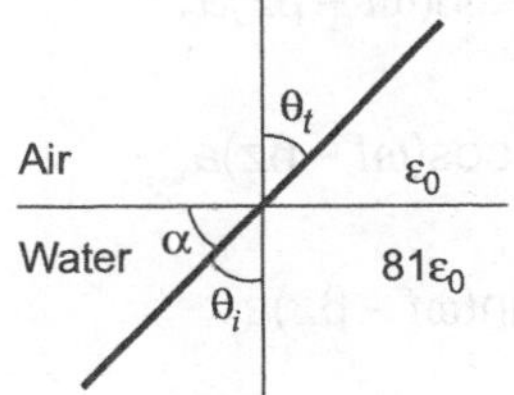

(a) 83.88° (b) 83.66°

(c) 84.86° (d) 84.08° [ESE-2011]

2.58 **Assertion (A):** For time varying field the relation $\bar{E} = -\nabla V$ is inadequate.

Reason (R): Faraday's law, states that for time varying field $\vec{\nabla} \times \vec{E} = 0$.

(a) Both A and R are true and R is the correct explanation of A

(b) Both A and R are true but R is NOT the correct explanation of A

(c) A is true but R is false

(d) A is false but R is true

[ESE-2011]

2.59 The credit of defining the following current is due to Maxwell
(a) Conduction current
(b) Drift current
(c) Displacement current
(d) Diffusion current [ESE-2012]

2.60 A plane wave travelling in a medium of $\varepsilon_r = 1$, $\mu_r = 1$ (free space) has an electric field intensity of $100\sqrt{\pi}$ V/m. Determine the total energy density of this magnetic field.
(a) 13.9 nJ/m^3 (b) 27.8 nJ/m^3
(c) 139 nJ/m^3 (d) 278 nJ/m^3
[ESE-2012]

2.61 For a plane wave propagating in an unbounded medium (say, free space), the minimum angle between electric field and magnetic field vectors is
(a) 0° (b) 60°
(c) 90° (d) 180° [ESE-2012]

2.62 The field strength of a plane wave is 2 V/m. The strength of the magnetic field (H) in free space is
(a) 5.2 mA/m (b) 2.25 mA/m
(c) 250 mA/m (d) 520 mA/m
[ESE-2013]

2.63 If $E = E_m \sin(\omega t - \beta z)\hat{a}_y$ in free space, then B is given by
(a) $\frac{-E_m\beta}{\omega}\sin(\omega t - \beta z)\hat{a}_z$
(b) $\frac{-E_m\beta}{\omega}\cos(\omega t - \beta z)\hat{a}_x$
(c) $\frac{E_m\beta}{\omega}\sin(\omega t - \beta z)\hat{a}_y$
(d) $\frac{E_m\beta}{\omega}\cos(\omega t - \beta z)\hat{a}_y$ [ESE-2013]

2.64 A uniform plane wave with an intensity of electric field 1 V/m is travelling in free space. The magnitude of associated magnetic field is
(a) 2.65 mA/m (b) 2.65 A/m
(c) 2.65 μA/m (d) 26.5 A/m
[ESE-2013]

2.65 The intrinsic impedance η of a conducting medium for which σ = 58 MS/m, $\mu_r = 1$ at a frequency of 100 MHz is
(a) $2.14 \times 10^5 \angle 45°\ \Omega$
(b) $1.84 \times 10^{-3} \angle 45°\ \Omega$
(c) $3.69 \times 10^{-3} \angle 45°\ \Omega$
(d) $3.69 \times 10^{-3} \angle -45°\ \Omega$ [ESE-2013]

2.66 A plane wave is travelling in the positive X-direction in a lossless unbounded medium having permeability the same as the free space and a permittivity 9 times that of the free space, the phase velocity of the wave will be
(a) 3×10^8 m/s (b) 10^8 m/s
(c) $\frac{1}{3} \times 10^8$ m/s (d) $\sqrt{3} \times 10^8$ m/s
[ESE-2013]

2.67 The electric field in an electromagnetic wave (in vacuum) is described by

$$E = E_{max}\sin(kx - \omega t)$$

where, $E_{max} = 100$ N/C and $k = 1 \times 10^7$ m^{-1}
Speed of light is 3×10^8 m/s. What is the amplitude of the corresponding magnetic wave?
(a) 4.57×10^{-7} T (b) 2.99×10^{-7} T
(c) 3.33×10^{-7} T (d) 2.99×10^7 T
[ESE-2014]

2.68 The numerical value of the ratio of electric field intensity E and magnetic field intensity H is
(a) 350 Ω (b) 377 Ω
(c) 37.7 Ω (d) 35 Ω [ESE-2014]

2.69 The ratio σ/ωε is called
1. Intrinsic ratio 2. Loss tangent
3. Conduction ratio 4. Dissipation ratio
Which of the above statements is/are correct?
(a) 1 only (b) 2 only
(c) 2 and 4 (d) 2 and 3
[ESE-2014]

2.70 The ratio $\frac{\sigma}{\omega\varepsilon}$ is
1. Intrinsic ratio
2. Loss tangent
3. Conduction ratio
4. Dissipation factor
Which of these are correct?
(a) 1 and 2 (b) 1 and 4
(c) 2 and 3 (d) 2 and 4
[ESE-2015]

2.71 Consider the following properties of electromagnetic waves:
1. These waves do not require any material medium to propagate
2. Both electric and magnetic field vectors are parallel to each other and perpendicular to the direction of propagation of waves

3. The energy in electromagnetic wave is divided equally between electric and magnetic vectors
4. Both electric and magnetic field vectors attain the maxima and minima at the same place and same time

Which of the above properties of electromagnetic waves are correct?

(a) 1, 2 and 3 only (b) 1, 3 and 4 only
(c) 2, 3 and 4 only (d) 1, 2, 3 and 4

[ESE-2015]

2.72 Electromagnetic waves are transverse in nature due to

(a) Reflection (b) Diffraction
(c) Interference (d) Polarization

[ESE-2015]

2.73 A plane wave in free space has a magnetic field intensity of 0.2 A/m in the Y-direction. The wave is propagating in the Z-direction with a frequency of 3 GHz. The wavelength and amplitude of the electric field intensity are, respectively :

(a) 0.05 m and 75 V/m
(b) 0.10 m and 75 V/m
(c) 0.05 m and 150 V/m
(d) 0.10 m and 150 V/m [ESE-2016]

2.74 The characteristic impedance η_0 of a free space is

(a) $\frac{\mu_0}{\varepsilon_0}$ (b) $\sqrt{\frac{\mu_0}{\varepsilon_0}}$

(c) $\sqrt{\mu_0 \varepsilon_0}$ (d) $\mu_0 \varepsilon_0$ [ESE-2016]

2.75 When a plane wave travelling in free-space is incident normally on a medium having $\varepsilon_r = 9.0$ and $\mu_r = 1.0$, the fraction of power transmitted into the medium is

(a) 4/3 (b) 3/4
(c) 1/2 (d) 1/3 [ESE-2016]

2.76 An electromagnetic wave is transmitted into a conducting medium of conductivity σ. The depth of penetration is

(a) directly proportional to frequency
(b) directly proportional to square root of frequency
(c) inversely proportional to frequency
(d) inversely proportional to square root of frequency

[ESE-2017]

2.77 Which of the following are the properties of TEM mode in a lossless medium?

1. Its cut-off frequency is zero.
2. Its transmission line is a hollow waveguide.
3. Its wave impedance is the impedance in a bounded dielectric.
4. Its phase velocity is the velocity of light in an unbounded dielectric.

Select the correct answer using the code given below.

(a) 1, 2 and 3 (b) 1, 3 and 4
(c) 1, 2 and 4 (d) 2, 3 and 4

[ESE-2017]

2.78 Consider the following statements:

1. For an isotropic medium, ε is a scalar constant.
2. For a homogeneous medium, ε, μ and σ are constant throughout the region.
3. In an anisotropic medium, D and E have the same direction.
4. For certain crystalline medium, ε varies with the direction of E.

Which of the above statements are correct?

(a) 1, 2 and 3 (b) 1, 3 and 4
(c) 1, 2 and 4 (d) 2, 3 and 4

[ESE-2018]

2.79 The depth of penetration δ of a plane electromagnetic wave incident normally on a good conductor is

(a) $\frac{1}{\sqrt{2\pi f \mu \sigma}}$ (b) $\frac{1}{\sqrt{\pi f \mu \sigma}}$

(c) $\frac{2}{\sqrt{3\pi f \mu \sigma}}$ (d) $\frac{2}{\sqrt{\pi f \mu \sigma}}$

[ESE-2020]

where: f = Frequency in Hz
σ = Conductivity in Siemens per meter

2.80 A uniform plane wave propagating in a medium has

$$\vec{E} = 2e^{-\alpha z} \sin\left(10^8 t - \beta z\right) \hat{a}_y \text{ V/m},$$

If the medium is characterized by $\varepsilon_r = 1$, $\mu_r = 20$ and $\sigma = 3$ S/m. The value of α and β are respectively.

(a) 30.70 Np/m, 30.70 rad/m
(b) 61.40 Np/m, 61.40 rad/m
(c) 122.80 Np/m, 122.80 rad/m
(d) 15.35 Np/m, 15.35 rad/m

[ESE-2021]

2.81 The intrinsic impedance of free space is
(a) 107 Ω (b) 214 Ω
(c) 377 Ω (d) 754 Ω

[ESE-2022]

2.82 Which one of the following is an angle at which there is no reflected wave when the incident wave is parallel (or vertically) polarized?
(a) Critical angle (b) Reference angle
(c) Relative angle (d) Brewster angle

[ESE-2022]

2.83 The electric field of a uniform plane electromagnetic wave in free space, along the positive x-direction, is given by $\overline{E} = 10(a_y + ja_z)e^{-j25x}$. The frequency and polarization of the wave respectively are
(a) 1.2 GHz and right circular
(b) 1.2 GHz and left circular
(c) 4 GHz and right circular
(d) 4 GHz and left circular

[ESE-2023]

2.84 In electromagnetic field, which one of the following does **not** satisfy the wave equation?
(a) $25e^{j(\omega t - 3z)}$ (b) $\sin(\omega(27z + 15t))$
(c) $\sin(x)\cos(t)$ (d) $\cos(y^2 + 5t)$

[ESE-2023]

2.85 The depth of penetration of a wave in a lossy dielectric increases with increasing.
(a) conductivity (b) permeability
(c) wavelength (d) permittivity

[ESE-2023]

2.86 Which one of the following can wave propagate in a conducting medium before its amplitude becomes insignificant?
(a) Characteristic impedance
(b) Skip distance
(c) Line of sight
(d) Skin depth

[ESE-2023]

2.87 Copper behaves as a
(a) conductor always.
(b) conductor or dielectric depending on the applied electric field strength.
(c) conductor or dielectric depending on the frequency.
(d) conductor or dielectric depending on the dielectric current density.

[ESE-2023]

2.88 Which one of the following points is **not** related to Poynting's Theorem?
(a) The variations in the time of a term corresponding to the sum of the energies associated to the electric and magnetic fields, and stored along the line.
(b) The power dissipated along the line due to the Joule effect in the conductors and the losses in the dielectric.
(c) The flux of the Poynting vector is coming out from the ends of the line, $p(d, t) - p(0, t)$.
(d) This theorem is not applicable to the more general case of transmission lines with frequency-dependent parameters.

[ESE-2024]

2.89 Which one of the following properties is not correct for skin depth?
(a) If the frequency is higher, then the skin depth is smaller.
(b) If the permeability is larger, then the skin depth is smaller.
(c) If the conductivity is larger, then the skin depth is smaller.
(d) If the permeability is larger, then the skin depth is smaller.

[ESE-2024]

■■■

Answers Uniform Plane Waves

2.1 (c)	2.2 (b)	2.3 (a)	2.4 (d)	2.5 (a)	2.6 (b)	2.7 (b)	2.8 (b)	2.9 (a)
2.10 (d)	2.11 (d)	2.12 (a)	2.13 (c)	2.14 (a)	2.15 (a)	2.16 (b)	2.17 (b)	2.18 (d)
2.19 (a)	2.20 (c)	2.21 (b)	2.22 (b)	2.23 (c)	2.24 (a)	2.25 (a)	2.26 (b)	2.27 (c)
2.28 (a)	2.29 (b, d)	2.30 (d)	2.31 (c)	2.32 (b)	2.33 (b)	2.34 (b)	2.35 (b)	2.36 (b)
2.37 (c)	2.38 (c)	2.39 (b)	2.40 (d)	2.41 (b)	2.42 (b)	2.43 (c)	2.44 (d)	2.45 (b)
2.46 (d)	2.47 (d)	2.48 (c)	2.49 (c)	2.50 (a)	2.51 (c)	2.52 (a)	2.53 (b)	2.54 (b)
2.55 (b)	2.56 (c)	2.57 (b)	2.58 (c)	2.59 (c)	2.60 (c)	2.61 (c)	2.62 (a)	2.63 (b)
2.64 (a)	2.65 (c)	2.66 (b)	2.67 (c)	2.68 (b)	2.69 (c)	2.70 (d)	2.71 (b)	2.72 (d)
2.73 (b)	2.74 (b)	2.75 (b)	2.76 (d)	2.77 (*)	2.78 (c)	2.79 (b)	2.80 (b)	2.81 (c)
2.82 (d)	2.83 (b)	2.84 (d)	2.85 (c)	2.86 (d)	2.87 (a)	2.88 (d)	2.89 (*)	

Explanations Uniform Plane Waves

2.1 (c)

By Snell's law

$$n_1 \sin \theta_i = n_2 \sin \theta_r$$

But $n_2 = \sqrt{\epsilon_r}$

So, $\sin 30 = \sqrt{2.7} \cdot \sin \theta_r$

$\Rightarrow \sin \theta_r = 0.3$

$\Rightarrow \theta_r = 17.7$

$\Rightarrow \theta_r \approx 18°$

2.3 (a)

Dimension of $\vec{E}$ = Volt/m

Dimension of $\vec{H}$ = Ampere/m

Dimension of $\vec{P} = \vec{E} \times \vec{H}$ is

$$\frac{\text{Volt}}{\text{m}} \cdot \frac{\text{Ampere}}{\text{m}} = \frac{\text{Watt}}{\text{m}^2}$$

or dimension of $\vec{P}$ = Power/Unit area

2.5 (a)

Velocity, $v = \dfrac{c}{\sqrt{\mu_r \epsilon_r}}$

$$= \frac{3 \times 10^8}{\sqrt{4.5 \times 2}} = 1 \times 10^8 \text{ m/s}$$

2.6 (b)

Magnitude of time average Poynting vector

$$|\vec{P}| = \frac{1}{2}\eta H^2 = \frac{1}{2} \times 120\pi \times (0.1)^2$$

$$|\vec{P}| = 0.6\pi$$

The direction of propagation is $-y$ direction.

So, $\vec{P} = -0.6\pi \hat{a}_y$ W/m^2

2.7 (b)

Maxwell's equation is

$$\vec{\nabla} \times \vec{E} = -\frac{\partial \vec{B}}{\partial t} = -\mu \frac{\partial \vec{H}}{\partial t}$$

2.8 (b)

1. mmf $= \int \vec{H} \cdot \vec{dl}$

   in closed path $\oint \vec{H} \cdot \vec{dl} = I$

2. Emf $= \int \vec{E} \cdot \vec{dl}$

3. Electric displacement = flux crossing

4. $\oint \vec{B} \cdot \vec{ds} = 0$

2.10 (d)

$$\vec{H} = 0.5 e^{-0.1x} \cos(10^6 t - 2x) \hat{a}_z \text{ A/m}$$

Comparing with

$$\vec{H} = A \cos(\omega t - \beta x)$$

Wave frequency

$$\omega = 10^6 \text{ radians/s}$$

Phase constant

$$\beta = 2$$

$$\beta = \frac{2\pi}{\lambda} = 2$$

$\Rightarrow \quad \lambda = \pi = 3.14$ m

The wave is travelling along $+x$ direction.

Polarization is E field direction which is Y direction so wave is polarized in Y direction.

2.11 (d)

$$\Gamma = \frac{n_2 - n_1}{n_2 + n_1} \text{ for } E \text{ fields}$$

$$\text{with } n_2 = \sqrt{\frac{j\omega\mu}{\sigma}} \text{ and } \sigma \to \infty\,;\, n_2 \to 0$$

$$\Gamma \cong -1 = \frac{E_r}{E_i}$$

In the case of a plane wave in air incident normally upon the surface of a perfect conductor, the wave is entirely reflected.

The amplitude of the reflected electric field strength remains equal to that of the incident electric field strength, but its phase reverses on reflection.

The phase of the reflected magnetic field strength is same as that of the incident magnetic field strength at the surface of reflection.

2.13 (c)

$$\tan\theta_B = \sqrt{\frac{\epsilon_2}{\epsilon_1}}$$

Brewster's angle of incidence for p polarized waves gives zero reflections and complete transmissions.

Vertical polarization: p-polarization.

Horizontal polarization: s-polarization.

2.14 (a)

$$\vec{H} = \frac{1}{Z_o}(\hat{a}_z \times \vec{E})$$

$$= \frac{1}{Z_o}\begin{vmatrix} \hat{a}_x & \hat{a}_y & \hat{a}_z \\ 0 & 0 & 1 \\ E_x & E_y & E_z \end{vmatrix} = \hat{a}_y \frac{E_x}{Z_o} - \frac{\hat{a}_x E_y}{Z_o}$$

2.16 (b)

For total reflection, $n_1 \sin\theta_i = n_2 \sin 90°$

But $\quad n = \sqrt{\epsilon_r}$

So, $\quad \sqrt{\epsilon_{r_1}} \sin\theta_i = \sqrt{\epsilon_{r_2}}$

$$\Rightarrow \quad \sin\theta_i = \sqrt{\frac{\epsilon_{r_2}}{\epsilon_{r_1}}}$$

$$\Rightarrow \quad \sin\theta_i = \frac{1}{\sqrt{2}}$$

$$\Rightarrow \quad \theta_i = 45°$$

2.17 (b)

$$\Gamma_E = \frac{1-\sqrt{\epsilon_r}}{1+\sqrt{\epsilon_r}} < 0$$

When a plane electromagnetic wave is incident normally on the surface of a perfect dielectric, the following relations hold:

$$E_i = \eta_1 H_i$$

$$E_r = -\eta_1 H_r$$

$$E_t = \eta_2 H_t$$

2.18 (d)

$$\frac{\sin\theta_1}{\sin\theta_2} = \sqrt{\frac{\epsilon_2}{\epsilon_1}} = \frac{1}{\sqrt{2}}$$

$$\Rightarrow \quad \sin\theta_2 = \sqrt{2}.\sin 60$$

$$= \sqrt{2}\cdot\frac{\sqrt{3}}{2} = 1.22$$

$\Rightarrow \quad \sin\theta_2 > 1$ which is not possible.

$\Rightarrow$ So, there will be no transmitted wave.

2.19 (a)

Radiation means EM wave generation. The cross product equations give the basic links for transverse EM waves.

Maxwell's equations are

$$\nabla\times\vec{H} = \vec{J} + \frac{\partial\vec{D}}{\partial t}$$

$$\nabla\times\vec{E} = \frac{-\partial\vec{B}}{\partial t}$$

$$\nabla\cdot\vec{D} = \rho$$

$$\nabla\cdot\vec{B} = 0$$

The first two equations give the basic idea of radiation.

2.21 (b)

The wave is normally incident. Equal amplitudes of both the circular components reflect equal in magnitude and the wave is circular polarization. Right becomes left due to changing propagation direction. The question cannot be form oblique incidence when the angles are not mentioned.

2.22 (b)

$$\frac{\sigma}{\omega\epsilon} = \frac{5}{2\pi\times25\times10^3\times80\times8.854\times10^{-12}}$$
$$= 44938.7$$

since $\frac{\sigma}{\omega\epsilon} >> 1$, hence sea water is acting as a good conductor.

since attenuation is 90%, transmission is 10%.

So, $e^{-\alpha x} = 0.1$

where α = attenuation constant

$$\alpha = \sqrt{\frac{\omega\mu\sigma}{2}}$$
$$= \sqrt{\frac{2\pi\times25\times10^3\times4\pi\times10^{-7}\times5}{2}}$$
$$\Rightarrow \alpha = 0.7025$$
$$-\alpha x = ln(0.1)$$
$$-0.7025\,x = -2.3$$
$$\Rightarrow x = 3.27 \text{ m}$$

2.23 (c)

$$v = \frac{1}{\sqrt{\mu\epsilon}} = \frac{1}{\sqrt{\mu_o\times81\epsilon_o}}$$
$$v = \frac{1}{9}\frac{1}{\sqrt{\mu_o\epsilon_o}} = \frac{3\times10^8}{9}$$
$$\vec{E} = 10\cos(6\pi\times10^8 t - \beta x)\bar{a}_y$$

Comparing with

$$E = A\cos(\omega t - \beta x)$$
$$\omega = 6\pi\times10^8$$

Phase constant,

$$\beta = \frac{\omega}{v} = \frac{6\pi\times10^8\times9}{3\times10^8}$$
$$\Rightarrow \beta = 18\pi \text{ rad/m}$$

2.24 (a)

Propagation constant $(\gamma) = \sqrt{j\omega\mu(\sigma + j\omega\varepsilon)}$

$$\gamma = \sqrt{j\omega\mu\cdot j\omega\varepsilon\left(1+\frac{\sigma}{j\omega\varepsilon}\right)}$$
$$= j\omega\sqrt{\mu\varepsilon}\left[1+\frac{\sigma}{2j\omega\varepsilon}\right]$$
$$= \frac{\sigma\sqrt{\mu/\varepsilon}}{2} + j\omega\sqrt{\mu\varepsilon}$$

On comparing with $\gamma = \alpha + j\beta$

$$\alpha = \frac{\sigma\sqrt{\mu/\varepsilon}}{2} \text{ and } \beta = \omega\sqrt{\mu\varepsilon}$$

2.25 (a)

$$v_p = \frac{c}{\sqrt{\epsilon_r}}$$
$$\frac{v_p}{c} = \frac{1}{\sqrt{\epsilon_r}}$$
$$0.4 = \frac{1}{\sqrt{\epsilon_r}}$$
$$\Rightarrow \sqrt{\epsilon_r} = 2.5$$
$$\Rightarrow \epsilon_r = 6.25$$

2.26 (b)

$$\vec{E}(x,t) = 60\cos(\omega t - 2x)\bar{a}_y \text{ V/m}$$

Comparing with

$$\vec{E} = E_o\cos(\omega t - \beta x)$$
$$E_0 = 60$$

Average Power,

$$P = \frac{1}{2}\frac{E_0^2}{\eta}\times\text{Area}$$
$$= \frac{1}{2}\times\frac{(60)^2}{120\pi}\times\pi(4)^2 = 240 \text{ W}$$

2.27 (c)

Maxwell's equation is

$$\nabla\times H = J + \frac{\partial D}{\partial t} = \sigma E + \frac{\epsilon\,\partial E}{\partial t}$$

where, J = conduction current density

$\frac{\partial D}{\partial t}$ = Displacement current density.

But $E = E_0 \sin\omega t$

So, $\nabla\times H = \sigma E_0\sin\omega t + \epsilon E_0\cos\omega t$

$$= \underbrace{\sigma E_0\sin\omega t}_{\text{Conduction current density}} + \underbrace{\epsilon\,\omega E_0\sin(\omega t + 90)}_{\text{Displacement current density}}$$

Therefore, the phase difference between the conduction current and displacement current is 90°.

2.28 (a)

When em waves pass through a medium of different permittivity, the refraction phenomena occurs.

Note:

- When EM waves pass through a small slot, the diffraction phenomenon occurs.
- When EM waves are incident on a perfectly conducting surface, reflection phenomenon occurs.

2.29 (b, d)

Given: $B(t) \rightarrow$ decreases with time.

$B \rightarrow$ Uniform with distance

$$v = -\frac{d}{dt}(B \cdot A)$$

where A = Area of loop

Voltage induced $v(t)$ = AC voltage

No load resistance being given for consideration, no joules heating in both rings.

EMF is induced in both rings.

Both option (b) and (d) are correct.

2.30 (d)

Maxwell's equation are obeyed by an EM wave everywhere.

2.31 (c)

$\nabla^2 E_x + k^2 E_x = 0$ is the standard differential equation form of $D^2 + m^2 = 0$ whose solutions are e^{-jkz} or e^{jkz}.

For $+z$ solution e^{-jkz} is correct solution.

2.32 (b)

Depth of penetration

$$\delta = \frac{1}{\alpha}$$

where, α = attenuation constant.

2.33 (b)

$E \rightarrow$ Volts/m drives

$K \rightarrow$ Amps/m in the conductor due to induction

Electric field moves electrons when it is tangential to the conductor surface this is surface current.

2.34 (b)

$$\nabla \times E = -\mu \frac{\partial H}{\partial t} \quad ; \quad \nabla \times H = J + \epsilon \frac{\partial E}{\partial t}$$

2.35 (b)

(i) According to Biot-Savart's law the magnetic flux density

$$B = \frac{\mu I}{4\pi} \oint_c \frac{dl \times \hat{a}_r}{r^2}$$

(ii) Poynting vector,

$$\vec{P} = \vec{E} \times \vec{H}$$

Poynting theorem states that the vector product $\vec{P} = \vec{E} \times \vec{H}$ at any point is a measure of the power flow per unit area at that point. The direction of flow is perpendicular to $\vec{E}$ and $\vec{H}$ in the direction of the vector $\vec{E} \times \vec{H}$.

2.36 (b)

$$E = 50\sin(10^7 t + kz)\hat{i} \text{ V/m}$$

comparing with

$$E = E_o \sin(\omega t - \beta z)$$

following conclusions can be drawn:

(i) The wave propagates along $-z$ axis

(ii) $\omega = 10^7$ rad/s

$$\Rightarrow \quad 2\pi f = 10^7$$

$$\Rightarrow \quad \frac{2\pi c}{\lambda} = 10^7$$

$$\Rightarrow \quad \lambda = \frac{2\pi \times 3 \times 10^8}{10^7} = 188.5 \text{ m}$$

(iii) Wave number

$$k = \frac{2\pi}{\lambda} = \frac{2\pi}{188.5} = 0.033$$

2.37 (c)

$$\mu^* = j\omega\mu$$

$$\epsilon^* = \sigma + j\omega\epsilon$$

$$\eta = \sqrt{\frac{\mu^*}{\epsilon^*}} = 1 \quad ; \quad \Gamma = \frac{1-377}{1+377} \cong -1$$

2.38 (c)

Two equal E field components out of phase by 90° gives circular polarization.

Let, $E_x = \sin\omega t$

$E_y = \cos\omega t$

Tracing with time for sense of rotation as left or right.

$t = 0 \quad E_x = 0 \quad E_y = 1$

$t = \frac{T}{4} \quad E_x = 1 \quad E_y = 0$

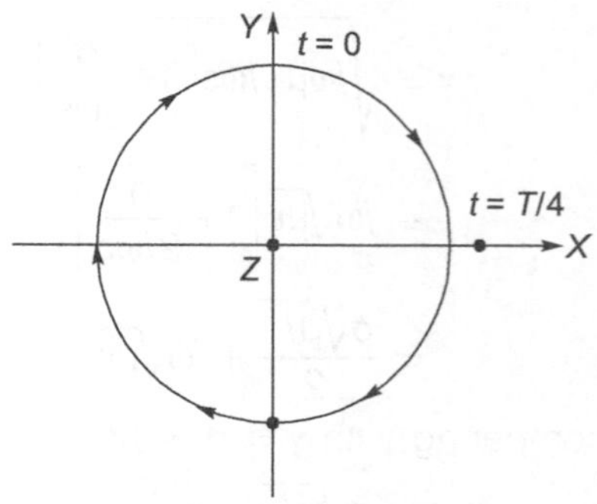

Left circular for $+z$ propagation.

2.39 (b)

Surface resistance of a metal

$$R_s \approx \sqrt{\frac{\omega\mu}{2\sigma}}$$

Therefore, as f increases, surface resistance increases.

2.40 (d)

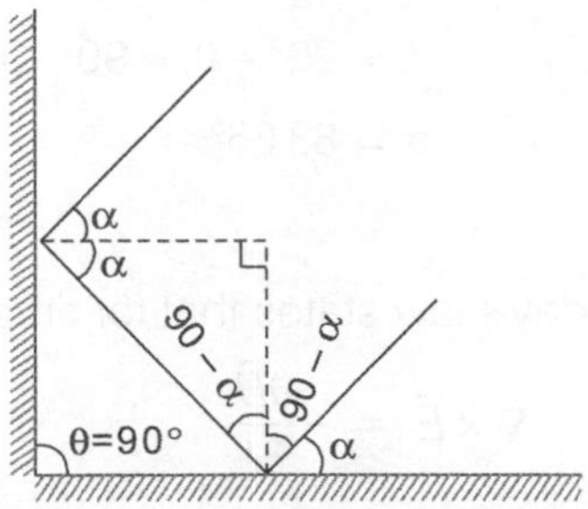

It is clear from the above figure that the angle θ must be 90° for an incident wave to reflect in the same direction.

2.42 (b)

B-linearly decreases with time

$$B(t) = -Kt$$

$$V = \frac{-d}{dt}(B(t)\cdot A) = \text{DC voltage}$$

Only in split coil or open circuit there is emf induced.

2.48 (c)

Consider,

$$E_x = E_1 \sin(\omega t - \beta z)$$
$$E_y = E_2 \sin(\omega t - \beta z + \delta)$$

ϕ ... time phase difference between E_x and E_y.

at $z = 0$

$$E_x = E_1 \sin \omega t$$
$$E_y = E_2 \sin(\omega t + \phi)$$

For circular polarization

$$\boxed{\begin{matrix} E_1 = E_2 = E_0 \\ \phi = \pm 90^\circ \end{matrix}}$$

Hence all the three statements are correct.

2.49 (c)

Skin depth, $\delta = \dfrac{1}{\sqrt{\pi f \sigma \mu}}$

$$\boxed{\delta \propto \frac{1}{\sqrt{f\sigma\mu}}}$$

2.50 (a)

Given, $\mu = 4\pi \times 10^{-7}$ H/m

$$\in = \frac{1}{36\pi}\times 10^{-9}$$
$$\sigma = 5.8\times 10^7 \text{ mho/m}$$
$$f = 3\text{ GHz}$$
$$= 3\times 10^9 \text{ Hz}$$

Intrinsic impedance of copper

$$\eta = \sqrt{\frac{j\omega\mu}{\sigma + j\omega\in}} \quad ...(1)$$

$$\omega\in = 2\pi\times 3\times 10^9 \times \frac{1}{36\pi}\times 10^{-9}$$

$\Rightarrow$ $\omega\in = 0.166$

$\because$ $\omega\in << \sigma \cdots$ for a good conductor

$$\therefore \quad \eta = \sqrt{\frac{j\omega\mu}{\sigma}} = \sqrt{\frac{\omega\mu}{\sigma}}\sqrt{j}$$

$$\eta = \sqrt{\frac{2\pi f\mu}{\sigma}}\, e^{j\pi/4}$$

$$\eta = \sqrt{\frac{2\pi\times 3\times 10^9\times 4\pi\times 10^{-7}}{5.8\times 10^7}}\, e^{j\pi/4}$$

$$\eta = 0.02\, e^{j\pi/4}\ \Omega$$

2.51 (c)

E_x, E_y orthogonal components have 3 and 6 amplitudes.

$$|E| = \sqrt{3^2+6^2} = 3\sqrt{5}$$

$$\text{Power} = \frac{1}{2}(E\times H^*) = \frac{1}{2}\frac{E_0^2}{n}$$

$$= 59.5 \text{ mW/m}^2$$

2.52 (a)

Velocity of light in any medium is

$$v = \frac{c}{\sqrt{\in_r}}$$

where, c = velocity of light in vacuum.

$\in_r$ = dielectric constant of medium.

as $\in_r > 1$

$$\therefore \quad \boxed{v < c}$$

2.53 (b)

For no reflection incident wave should be completely transmitted to another medium. For this condition incident wave should strike at the boundary with Brewster angle which is given as

$$\tan\theta_B = \sqrt{\frac{\epsilon_2}{\epsilon_2}} = \sqrt{\frac{9}{4}}$$

$$\therefore \quad \theta_B = \tan^{-1}\frac{3}{2}$$

2.54 (b)

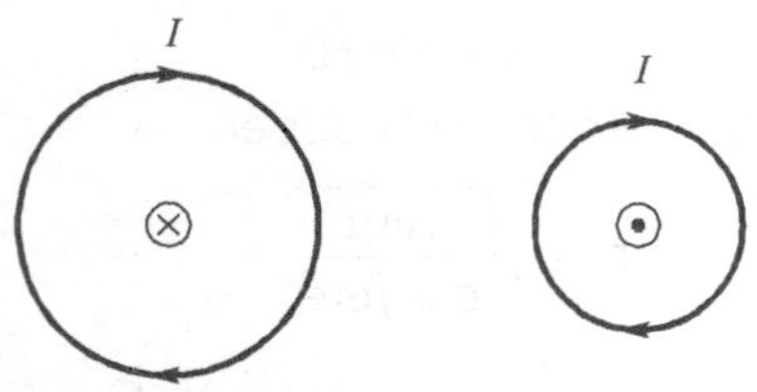

Current-clockwise
B into paper
(South pole)
Figure (a)

Anti-clockwise current
B into paper
(North pole)
Figure (b)

Figure (a) B increasing.
Induced current has to decrease B i.e. oppose the existing B clockwise current of the induction results in opposing B Correct diagram.
Similarly check for Lenz's law in all other cases.

2.55 (b)

Given, $\vec{E} = 10\cos(10^7 t + kz)\hat{a}_y$ V/m ...(i)

Standard equation

$$\vec{E} = Ae^{-\alpha t}\cos(\omega t + kz)\hat{a}_y \text{ V/m} \quad ...(ii)$$

Here,

- Wave propagates along negative z direction
- Comparing equations (i) and (ii)

$$\alpha = 0;$$

Hence wave propagates without any attenuation.

$$\omega = 10^7$$

$$\Rightarrow \quad f = \frac{10^7}{2\pi}\text{ Hz}$$

Since wave is propagating in free space. So,

$$\lambda = \frac{c}{f} = \frac{3\times10^8}{10^7/2\pi} = 188.5 \text{ m}$$

2.56 (c)

Velocity of wave in dielectric medium

$$v = \frac{c}{\sqrt{\epsilon_r}} = \frac{c}{\sqrt{4}} = \frac{c}{2}$$

$$v = f\lambda = \frac{c}{2}$$

When a wave travels from one medium to another medium then frequency remains same; so in the given case wavelength will become half.

2.57 (b)

For no reflection incident angle should be equal to Brewster angle. So,

$$\tan\theta_B = \sqrt{\frac{\epsilon_2}{\epsilon_1}} = \sqrt{\frac{\epsilon_0}{81\epsilon_0}} = \frac{1}{9}$$

$$\therefore \quad \theta_B = \tan^{-1}\left(\frac{1}{9}\right) = 6.34°$$

$$\Rightarrow \quad \theta_i = \theta_B = 6.34°$$

So, $\alpha = 90° - \theta_i = 90° - 6.34°$

$$\Rightarrow \quad \alpha = 83.66°$$

2.58 (c)

Faraday's law states that for time varying field

$$\nabla\times\vec{E} = -\frac{\partial\vec{B}}{\partial t}$$

2.60 (c)

$$\varepsilon_r = 1, \mu_r = 1 \quad \text{(free space)}$$

Total magnetic energy density

$$= \frac{1}{2}\mu H^2$$

Electric field intensity,

$$E = 100\sqrt{\pi} \text{ V/m}$$

η (intrinsic impedance)

$$= \frac{E}{H} = 377\ \Omega \text{ (for free space)}$$

$$\therefore \quad H = 0.470 \text{ A/m}$$

So, energy density $= \frac{1}{2}\mu_o H^2$

$$= \frac{1}{2}\times4\pi\times10^{-7}\times(0.470)^2 = 139 \text{ nJ/m}^3$$

2.61 (c)

For a plane wave propagating in an unbounded medium, the angle between the electric field and magnetic field is always 90°.

2.62 (a)

$$\frac{E}{H} = \eta = 120\pi \text{ for free space}$$

$$\therefore \quad H = \frac{E}{120\pi} = \frac{2}{120\pi} = 5.3 \text{ mA/m}$$

2.63 (b)

$$E = E_m\sin(\omega t - \beta z)\hat{a}_y$$

$$\bar{P} = \bar{E}\times\bar{H}$$

$$\hat{a}_z = \hat{a}_y\times(-\hat{a}_x)$$

So $\bar{H}$ will be in $-\bar{a}_x$ direction

$$\bar{H} = \frac{\bar{E}}{\eta} = -\frac{(E_m \sin\omega t - \beta z)}{\eta}\hat{a}_x$$

$$\bar{B} = \mu\bar{H}$$

$$= -\frac{\mu}{\eta}\cdot[E_m \sin(\omega t - \beta z)]\hat{a}_x$$

$$= \mu / \frac{1}{\sqrt{\mu/\in}}[E_m \sin(\omega t - \beta z)]\hat{a}_x$$

$$= -\sqrt{\mu\in}\cdot E_m[\sin(\omega t - \beta z)]\hat{a}_x$$

But $\quad v_p = \frac{1}{\sqrt{\mu\in}} = \frac{\omega}{\beta}$

So $\quad \bar{B} = -\frac{\beta}{\omega}\cdot E_m \sin(\omega t - \beta z)\hat{a}_x$

Hence $\quad \boxed{\bar{B} = -\frac{E_m\beta}{\omega}\sin(\omega t - \beta z)\hat{a}_x}$

2.64 (a)

$$\frac{E}{H} = 120\pi \cdots \text{ for free space}$$

$$\therefore \quad H = \frac{E}{120\pi} = \frac{1}{120\pi} = 2.65 \text{ mA/m}$$

2.65 (c)

Intrinsic impedance for a conducting medium

$$\eta = \sqrt{\frac{\omega\mu}{\sigma}}\angle 45°$$

$$= \sqrt{\frac{2\pi\times100\times10^6\times4\pi\times10^{-7}}{58\times10^6}}\angle 45°$$

$$= 3.69\times10^{-3}\angle 45°\ \Omega$$

2.66 (b)

$$\text{Phase velocity} = \frac{c}{\sqrt{\in_r}} = \frac{3\times10^8}{\sqrt{9}} = 10^8 \text{ m/s}$$

2.67 (c)

$$E = E_{max}\sin(kx - \omega t) \qquad \text{...(i)}$$

For free space

$$\frac{|E|}{|H|} = \eta = 120\,\pi \qquad \text{...(ii)}$$

$$\therefore \quad |H| = \frac{100}{120\pi} = \frac{5}{6\pi} \qquad \text{...(iii)}$$

Also, $\quad B = \mu H = 4\times\pi\times10^{-7}\times\frac{5}{6\pi}$

$$= 3.33\times10^{-7}\ T$$

2.68 (b)

$$\frac{E}{H} = \eta = \sqrt{\frac{\mu}{\in}}$$

in free space

$$\frac{E}{H} = \sqrt{\frac{\mu_0}{\in_0}} = 120\ \Omega = 377\ \Omega$$

2.71 (b)

$\vec{E}$ and $\vec{H}$ fields are always perpendicular in EM wave propagation i.e $\vec{E}.\vec{H} = 0$

So, statement 2 is wrong.

2.72 (d)

Polarization of EM waves is due to the transverse nature of EM waves i.e both E and H waves are perpendicular to the propagation of wave.

2.73 (b)

$$\vec{H} = 0.2\ a_y : f = 3 \text{ GHz}$$

$$\lambda = \frac{c}{f} = \frac{3\times10^8}{3\times10^9} = 0.1 \text{m}$$

$$\frac{E}{H} = 120\pi \text{ for free space}$$

$$E = 120\pi(H)$$

$$\Rightarrow \quad E = 120\pi(0.2) = 24\pi = 75 \text{ V/m}$$

2.74 (b)

$$\eta_0 = \sqrt{\frac{\mu_0}{\in_o}} = 120\pi$$

2.75 (b)

$\in_{r1} = 1; \mu_{r1} = 1$ (air) | $\in_{r2} = 9; \mu_{r2} = 1$ (dielectric)

$$\Gamma_E = \frac{1-\sqrt{9}}{1+\sqrt{9}} = -\frac{1}{2}$$

Hence, reflection coefficient for the power

$$\Gamma_P = -\frac{1}{4}$$

$$\text{Transmitted power fraction} = \frac{3}{4}$$

2.76 (d)

$$\text{Depth of penetration} = \delta = \frac{1}{\sqrt{\pi f\mu\sigma}} \quad ; \quad \delta \propto \frac{1}{\sqrt{f}}$$

2.77 (*)

- For TEM mode $f_c = 0$ [f_c is cut of frequency].
- TEM does not exist in hollow waveguide.
- TEM wave impedance is same as unbounded dielectric impedance.
- TEM mode phase velocity is the velocity of light in an unbounded dielectric.

The above are properties of TEM mode in a lossless medium ($\sigma = 0$).

2, 3 statements are wrong. 1, 4 statements are correct.

2.79 (b)

For a good conductor:

$$\text{Depth of penetration} = \delta = \frac{1}{\alpha} = \frac{1}{\sqrt{\pi f \mu \sigma}}$$

2.80 (b)

$$\vec{E} = 2e^{-\alpha z}\sin(10^8 t - \beta z)\,\hat{a}_y \text{ V/m}$$

Given: $\epsilon_r = 1$, $\mu_r = 20$, $\sigma = 3$ S/m

Here, $$\frac{\sigma}{\omega \in} = \frac{3}{(10^8)\left(\frac{1}{36\pi}\times 10^{-9}\right)}$$

$$= 3 \times 36\pi \times 10$$
$$= 3391.2$$

Since, $\frac{\sigma}{\omega \in} >> 1$. Hence, the given medium is a good conductor.

In a good conductor,

$$\alpha = \beta = \sqrt{\frac{\omega\mu\sigma}{2}}$$

$$\alpha = \beta = \sqrt{\frac{(10^8)(4\pi\times 10^{-7})(20)(3)}{2}}$$

$$\alpha = \beta = \sqrt{2\pi\times 20\times 3\times 10}$$

$$\alpha = \beta = \sqrt{3768} = 61.384 \simeq 61.40$$

$\therefore$ $\alpha = 61.40$ Np/m and

$\beta = 61.40$ rad/m

2.81 (c)

The intrinsic impedance of free space is given as

$$\eta = \sqrt{\frac{\mu_o}{\epsilon_o}}$$

where, μ_o = Permeability = $4\pi \times 10^{-7}$ H/m

ϵ_o = Permittivity = $\frac{1}{36\pi\times 10^9}$ F/m

$$\therefore \quad \eta = \sqrt{\frac{4\pi\times 10^{-7}}{\frac{1}{36\pi\times 10^9}}}$$

$$\eta = 120\pi = 377\ \Omega$$

2.82 (d)

Brewster angle is the angle of incidence where the ray of light having a *p*-polarization (vertically polarized) is transmitted through a dielectric without any reflection.

2.83 (b)

Given, $\bar{E} = (10\hat{a}_y + j10\hat{a}_z)e^{-j25x}$

$$\bar{E}(x,t) = (10\hat{a}_y + j10\hat{a}_z)e^{j(\omega t - 25x)}$$

at $x = 0$, $\bar{E}(t) = (10\hat{a}_y + j10\hat{a}_z)e^{j\omega t}$

Taking real terms, we get

$$\bar{E}(t) = 10\cos\omega t\,\hat{a}_y + 10\cos(\omega t + \pi/2)\hat{a}_z$$

$$\Rightarrow \quad \bar{E}(t) = 10\cos\omega t\,\hat{a}_y - 10\sin\omega t\,\hat{a}_z$$

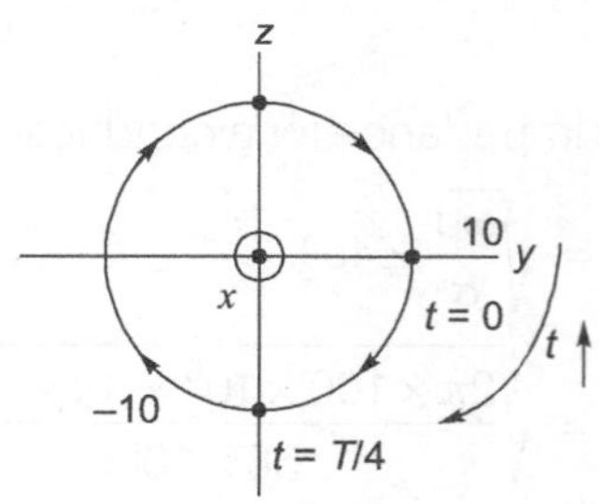

at $t = 0$, $\omega t = 0$, $\bar{E}(t) = 10\hat{a}_y$

at $t = \pi/4$, $\omega t = \pi/2$, $\bar{E}(t) = -10\hat{a}_z$

$\therefore$ it is LCP.

Also $\beta = 25$

$\therefore$ $\beta = \frac{\omega}{c}$; free space given

$\Rightarrow$ $\omega = \beta c = 25\times 3\times 10^8 = 75 \times 10^8$ rad/s

$\Rightarrow$ $f = \frac{\omega}{2\pi} = 1.2$ GHz

2.85 (c)

For a lossy medium,
Depth of penetration

$$\delta \propto \frac{1}{\sqrt{f}}$$

As $v = f\lambda \Rightarrow f \propto \frac{1}{\lambda}$

$\therefore \quad \delta \propto \sqrt{\lambda}$

So, if $\lambda\uparrow$, $\delta\uparrow$

2.86 (d)

Skin depth is defined as wave propagate in a conducting medium before its amplitude becomes insignificant.

2.87 (a)

Due to very large loss tangent of copper because of conductivity, Cu behaves as a conductor in all conditions.

2.88 (d)

For the two-conductor transmission line including suitable distributed sources along the line,

$$\frac{-\partial v}{\partial x} = L\frac{\partial i}{\partial t} + Ri - e_s \quad \text{...(i)}$$

$$\frac{-\partial i}{\partial x} = C\frac{\partial v}{\partial t} + Gv - j_s \quad \text{...(ii)}$$

where $e_s(x, t)$ and $j_s(x, t)$ are distributed sources. Here, the line parameters L, R, C, G are considered frequency-independent.

Multiplying equation (i) by i and equation (ii) by v and adding them, we get,

$$\frac{\partial p}{\partial x} + \frac{\partial W_{em}}{\partial t} + S_d = S_s \quad \text{...(iii)}$$

where,

$$W_{em} = \frac{Li^2}{2} + \frac{Cv^2}{2}$$

represents the energy per unit length associated to the electromagnetic field along the guiding structure described by the transmission line.

$$s_d = Ri^2 + Gv^2$$

represents the electrical power dissipated per unit length.

$$s_s = e_s i + v j_s$$

where s_s represents the per-unit length electrical power generated by the distributed sources.

The quantity $p(x, t)$ represents the flux of the Poynting vector through the transversal section of the line at absissa x and at time t in forward direction.

Integrating equation (iii), we get

$$\frac{d}{dt}\int_0^d W_{em}(x,t)dx + \int_0^d s_d(x,t)dx + [p(d,t) - p(0,t)]$$

$$= \int_0^d s_s(x,t)dx$$

It represents the Poynting's theorem. The power generated by the distributed sources along the line is equal to the sum of three terms.

(i) the variations in the time of a term corresponding to the sum of energies associated to electric and magnetic fields, and stored along the line.

(ii) the power dissipated along the line due to the Joule effect in the conductors and the losses in the dielectric, and

(iii) the flux of the Poynting vector coming out from ends of the line, $p(d, t) - p(0, t)$.

Hence, option (d) is not related to Poynting's theorem.

2.89 (*)

The skin-depth is given by

$$\delta = \frac{1}{\sqrt{\pi f \mu \sigma}}$$

Hence, if the frequency (f), permeability (μ) or conductivity (σ) is larger, then the skin-depth is smaller. Hence, none of the given options are correct.

■■■

3 Transmission Lines

3.1 A transmission line has primary constants R, L, G and C, and secondary constants Z_0 and $\gamma(= \alpha + j\beta)$. If the line is lossless, then
(a) $R = 0$, $G \neq 0$ and $\alpha = 0$
(b) $R = 0$, $G \neq 0$ and $\beta = |\gamma|$
(c) $G = 0$ and $\alpha = \beta$
(d) $R = 0$, $G = 0$, $\alpha = 0$ and $\beta = |\gamma|$
[ESE-2000]

3.2 A transmission line having 50 Ω impedance is terminated in a load of $(40 + j30)$ Ω. The VSWR is
(a) $j0.033$ (b) $0.8 + j0.6$
(c) 1 (d) 2 [ESE-2000]

3.3 A$(75 - j40)\Omega$ load is connected to a coaxial line of $Z_0 = 75\ \Omega$ at 6 MHz. The load matching on the line can be accomplished by connecting.
(a) a short-circuited stub at the load
(b) an inductance at the load
(c) a short-circuited stub at a specific distance from the load
(d) a capacitance at a specific distance from the load
[ESE-2000]

3.4 A quarter-wave transformer matching a 75 Ω source with a 300 Ω load should have a characteristic impedance of
(a) 50 Ω (b) 100 Ω
(c) 150 Ω (d) 200 Ω [ESE-2000]

3.5 If R, L, C and G are the resistance, inductance, capacitance and conductance of a transmission line respectively, then the condition for distortionless transmission line is
(a) $R/C = G/L$ (b) $RC = LG$
(c) $R^2C = G^2L$ (d) $RC^2 = GL^2$
[ESE-2001]

3.6 The input impedance of $\lambda/8$ long short-circuited section of a lossless transmission line is
(a) zero (b) inductive
(c) capacitive (d) infinite [ESE-2001]

3.7 Match **List-I** (Parameters) with **List-II** (Values) for a transmission line with a series impedance $Z = R + j\omega L$ Ω/m and a shunt admittance $Y = G + j\omega C$ mho/m, and select the correct answer:

List-I
A. Characteristic impedance Z_0
B. Propagation constant γ
C. The sending-end input impedance Z_s when the line is terminated in its characteristic impedance Z_0

List-II
1. $\sqrt{ZY}$
2. $\sqrt{Z/Y}$
3. $\sqrt{Y/Z}$

Codes:

	A	B	C
(a)	3	1	1
(b)	2	3	3
(c)	2	1	2
(d)	1	2	2

[ESE-2001]

3.8 Which of the following conditions will not guarantee a distortionless transmission line?
(a) $R = G = 0$
(b) $RC = GL$
(c) Very low frequency range ($R >> \omega L$, $G >> \omega C$)
(d) Very high frequency range ($R << \omega L$, $G << \omega C$)
[ESE-2001]

3.9 In an air line, adjacent maxima are found at 12.5 cm and 37.5 cm. The operating frequency is
(a) 1.5 GHz (b) 600 MHz
(c) 300 MHz (d) 1.2 GHz [ESE-2001]

3.10 Figure I shows an open circuited transmission line. The switch is closed at time $t = 0$ and after a time t, the voltage distribution on the line reaches that shown in figure II. If c is the velocity in the line, then

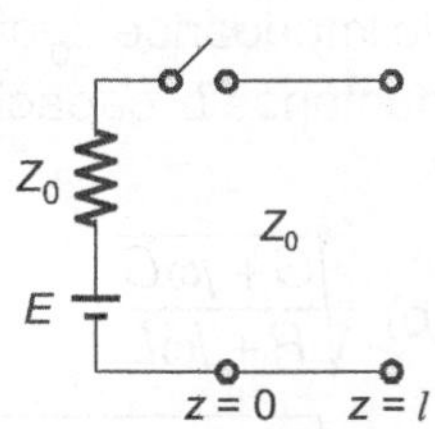

Figure I

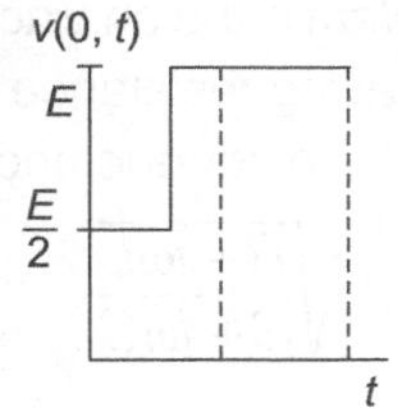

Figure II

(a) $t < l/c$ (b) $t = l/c$
(c) $t > 2l/c$ (d) $2l/c > t > l/c$

[ESE-2001]

3.11 In a VSWR measurement, a square law detector is used to detect the signal level. The voltmeter reads maximum and minimum as 64 mV and 16 mV respectively. The VSWR of the system will be
(a) 4 (b) 2
(c) 8 (d) 16

[ESE-2002]

3.12 **Assertion (A):** A $\lambda/4$ short circuited line behaves as a series resonant circuit.
Reason (R): At a particular frequency the input impedance becomes infinite.
(a) Both A and R are true and R is the correct explanation of A
(b) Both A and R are true but R is NOT the correct explanation of A
(c) A is true but R is false
(d) A is false but R is true [ESE-2002]

3.13 Consider the following:
For a lossless transmission line we can write:
1. $Z_{in} = -jZ_0$ for a shorted line with $l = \lambda/8$
2. $Z_{in} = \pm j\infty$ for a shorted line with $l = \lambda/4$
3. $Z_{in} = -jZ_0$ for an open line with $l = \lambda/2$
4. $Z_{in} = Z_0$ for a matched line of any length

Select the correct answer using the codes given below:
(a) 1 and 2 (b) 2 and 3
(c) 1 and 3 (d) 2 and 4 [ESE-2002]

3.14 The input impedance of a short circuited quarter wave long transmission line is
(a) purely reactive
(b) purely resistive
(c) dependent on the characteristic impedance of the line
(d) none of the above

[ESE-2002]

3.15 A transmission line of input impedance 400 Ω is to be matched to a load of 25 Ω through a quarter wavelength line. The quarter wave line characteristic impedance must be
(a) 40 Ω (b) 100 Ω
(c) 400 Ω (d) 425 Ω

[ESE-2002]

3.16 A lossless transmission line is connected to two identical voltage sources as shown in the figure.

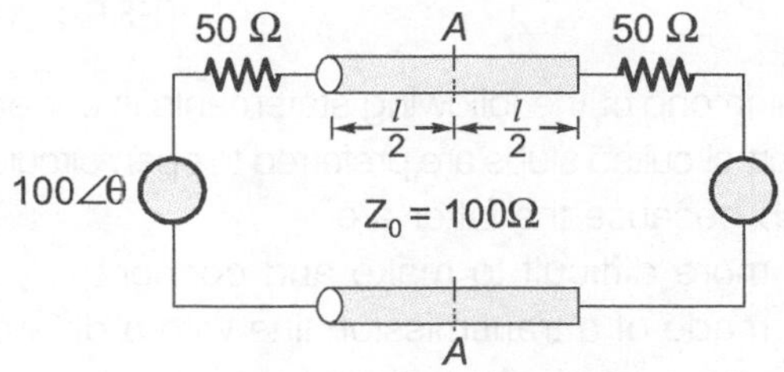

The impedance measured at the mid point *AA* of the line is
(a) ∞ (b) 100
(c) 50 (d) 25

[ESE-2002]

3.17 In a line the VSWR of a load is 6 dB. The reflection coefficient will be
(a) 0.033 (b) 0.33
(c) 0.66 (d) 3.3

[ESE-2003]

3.18 $Z_L = 200\ \Omega$ and it is desired that $Z_i = 50\ \Omega$. The quarterwave transformer should have a characteristic impedance of
(a) 100 Ω (b) 40 Ω
(c) 10,000 Ω (d) 4 Ω [ESE-2003]

3.19 Consider the following statements regarding Smith charts:
1. A normalized Smith chart applies to a line of any characteristic resistance and serves as well for normalized admittance
2. A polar coordinate Smith chart contains circles of constant $|Z|$ and circles of constant $\angle Z$
3. In Smith chart, the distance towards the load is always measured in clockwise direction.

Which of the statements given above are correct?
(a) 1, 2 and 3 (b) 2 and 3
(c) 1 and 3 (d) 1 and 2

[ESE-2004]

3.20 A $(100 - j75)\Omega$ load is connected to a co-axial cable of characteristic impedance 75 ohms at 12 GHz. In order to obtain the best matching, which one of the following will have to be connected?
(a) A short-circuited stub at load
(b) Inductance at load
(c) A capacitance at a specific distance at load
(d) A short-circuited stub at some specific distance from load

[ESE-2004]

3.21 Which one of the following statements is correct? Short-circuited stubs are preferred to open-circuited stub because the latter are
(a) more difficult to make and connect
(b) made of a transmission line with a different characteristic impedance
(c) liable to radiate energy
(d) incapable of giving a full range of frequencies

[ESE-2004]

3.22 In a transmission line the reflection coefficient at the load end is given by $0.3e^{-j30°}$. What is the reflection coefficient at a distance of 0.1 wavelength towards source?
(a) $0.3e^{+j30°}$ (b) $0.3e^{+j102°}$
(c) $0.3e^{+j258°}$ (d) $0.3e^{+j66°}$ [ESE-2005]

3.23 To couple a coaxial line to a parallel wire, it is best to use a
(a) Balun
(b) Slotted line
(c) Directional coupler
(d) Quarter wave transformer [ESE-2005]

3.24 A uniform transmission line of characteristic impedance 100 Ω and feeding a purely resistive load of 500 Ω uses single stub matching. The stub is placed at a distance d from the load. The VSWR on the length d and on the stub itself will be
(a) 0 and 0, respectively
(b) 1 and 1, respectively
(c) 5 and 5, respectively
(d) 5 and ∞, respectively [ESE-2005]

3.25 A quarter wave transformer can be used for matching accurately
(a) Inductive loads only
(b) Capacitive loads only
(c) Loads with any nature of reactance
(d) Purely resistive loads only [ESE-2005]

3.26 What is the characteristic impedance Z_0 of a line having resistance R, inductance L, capacitance C and conductance G?
(a) $\sqrt{\dfrac{R+j\omega L}{G+j\omega C}}$ (b) $\sqrt{\dfrac{G+j\omega C}{R+j\omega L}}$
(c) $R+j\omega L-\dfrac{G}{\omega C}$ (d) $\sqrt{R^2+\left(\omega L-\dfrac{1}{\omega C}\right)^2}$

[ESE-2006]

3.27 In general, attenuation per unit length in a coaxial cable
(a) increases with frequency
(b) decreases with frequency
(c) remains constant with frequency
(d) depends upon the type of coaxial cable. It can either increase or decrease

[ESE-2006]

3.28 What are the characteristic impedance and the phase velocity on radio frequency lines?
(a) $Z=\sqrt{1/LC}\,\Omega$ and $v=\sqrt{L/C}$ m/s
(b) $Z=R+(1/\sqrt{LC})\,\Omega$ and $v=1/(\sqrt{LC})$m/s
(c) $Z=(\sqrt{L/C})\,\Omega$ and $v=(1/\sqrt{LC})$m/s
(d) $Z=R+(\sqrt{L/C})\,\Omega$ and $v=(1/\sqrt{LC})$m/s

[ESE-2006]

3.29 Which one of the following is the characteristic impedance of lossless transmission line?
(a) $\sqrt{R/G}$ (b) $\sqrt{L/G}$
(c) $\sqrt{L/C}$ (d) $\sqrt{R/C}$

[ESE-2006]

3.30 Match **List-I** (Quantity) with **List-II** (Range of Values) and select the correct answer using the codes given below the lists:

List-I	List-II
A. Input Impedance	1. –1 to +1
B. Reflection coefficient	2. 1 to ∞
C. VSWR	3. 0 to ∞

Codes:

	A	B	C
(a)	2	3	1
(b)	3	2	1
(c)	3	1	2
(d)	2	1	3

[ESE-2006]

3.31 A quarter wave impedance transformer is terminated by a short circuit. What would its input impedance be equal to?

(a) The line characteristic impedance
(b) Zero
(c) Infinity
(d) Square root of the line characteristic impedance

[ESE-2006]

3.32 Which one of the following statements for a short circuited loss free line is not correct?

(a) The line appears as a pure reactance when viewed from the sending end
(b) It can be either inductive or capacitive
(c) There are no reflections in the line
(d) Standing waves of voltage and current are set up along length of the lines

[ESE-2007]

3.33 Match **List-I** (Load Impedance) with **List-II** (Value of Reflection Coefficient) and select the correct answer using the codes given below the lists:

List-I	List-II
A. Short Circuit	1. 0
B. Open Circuit	2. –1
C. Line characteristic impedance	3. +1
D. 2 × line characteristic impedance	4. +1/3

Codes:

	A	B	C	D
(a)	2	1	3	4
(b)	4	3	1	2
(c)	2	3	1	4
(d)	4	1	3	2

[ESE-2007]

3.34 When the reflection coefficient equals $1\angle 0°$, what is the VSWR?

(a) Zero (b) 1
(c) 3 (d) Infinite

[ESE-2007]

3.35 Match **List-I** (Transmission Line) with **List-II** (Characteristic Impedance) and select the correct answer using the codes given below the lists:

List-I

A. Coaxial line

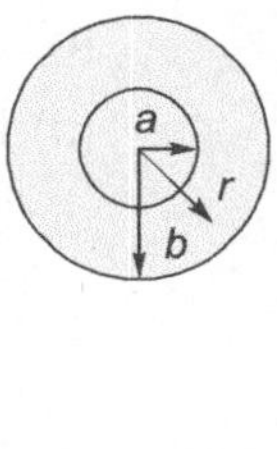

B. Two wire line

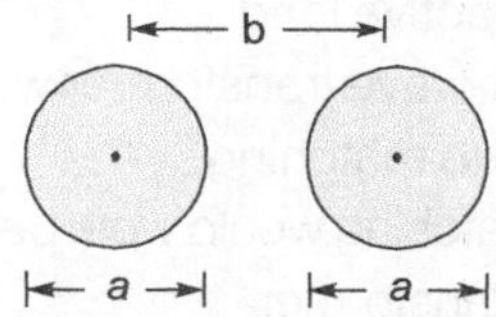

C. Parallel plate line

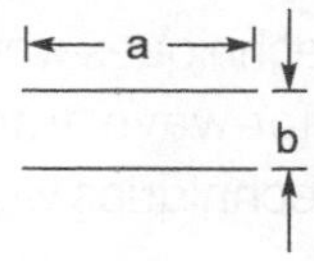

D. Collinear plate

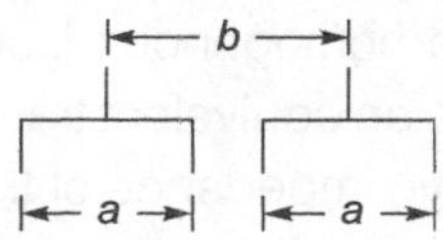

List-II

1. $R_0 = \dfrac{\eta b}{a}, a >> b$

2. $R_0 = \dfrac{\eta}{\pi} ln\left(\dfrac{4b}{a}\right), b >> a$

3. $R_0 = \dfrac{\eta}{2\pi} ln\,(b/a)$

4. $R_0 = \dfrac{\eta}{\pi} ln\left(\dfrac{2b}{a}\right), b >> a$

Codes:

	A	B	C	D
(a)	3	4	1	2
(b)	2	1	4	3
(c)	3	1	4	2
(d)	2	4	1	3

[ESE-2007]

3.36 If the reflection coefficient is 1/5, what is the corresponding VSWR?

(a) 3/2 (b) 2/3
(c) 5/2 (d) 2/5

[ESE-2007]

3.37 Why is delay distortion in coaxial cables caused?

(a) The dielectric constant (ε_r) of the dielectric filling the coaxial cable varies with frequency
(b) The amplitude of the signal on the cable varies with frequency
(c) The velocity of electromagnetic waves propagating on the cable varies with frequency
(d) Because of both factors given at (a) and (c) above

[ESE-2007]

3.38 For matching of a radio frequency coaxial line to a purely reactive load
(a) quarter-wave transformer would work better than stub matching
(b) stub matching would work better than quarter-wave transformer
(c) both quarter-wave transformer and stub matching techniques would work equally well
(d) both quarter-wave transformer and stub matching techniques would fail
[ESE-2008]

3.39 If an infinite homogenous isotropic medium is modelled as an equivalent transmission line, the characteristic impedance of the corresponding transmission line is known as its
(a) characteristic impedance
(b) iterative impedance
(c) wave impedance
(d) intrinsic impedance [ESE-2008]

3.40 A lossless transmission line of characteristic impedance Z_0 and length $l < \lambda / 4$ is terminated at the load end by an open circuit. What is its input impedance Z_s?
(a) $Z_s = j Z_0 \tan \beta l$ (b) $Z_s = j Z_0 \cot \beta l$
(c) $Z_s = -j Z_0 \tan \beta l$ (d) $Z_s = -j Z_0 \cot \beta l$
[ESE-2008]

3.41 The reflection coefficient on a 500 m long transmission line has a phase angle of –150°. If the operating wavelength is 150 m, what will be the number of voltage maxima on the line?
(a) 8 (b) 3
(c) 6 (d) 7
[ESE-2009]

3.42 Consider the following statements :
For a 10 m long common power line connecting a switch to a light bulb
1. It is a distributed circuit.
2. Time delay for propagation through it is negligible.
3. It is in the form of a shielded coaxial cable of circular cross-section.
4. As the intensity of the lamp varies, input impedance of this line also changes.
Which of the above statement(s) is/are correct?
(a) 1 only (b) 1 and 2
(c) 2 and 3 (d) 2 and 4
[ESE-2009]

3.43 With regard to a transmission line, which of the following statements is correct?
(a) Any impedance repeats itself every λ/4 on the Smith chart.
(b) The S.W.R. = 2 circle and the magnitude of reflection coefficient = 0.5 circle coincide on the Smith chart.
(c) At any point on a transmission line, the current reflection coefficient is the reciprocal of the voltage reflection coefficient.
(d) Matching eliminates the reflected wave between the source and the matching device location.
[ESE-2009]

3.44 It is required to match a 200 Ω load to a 450 Ω transmission line. To reduce the SWR along the line to 1, what must be the characteristic impedance of the quarter-wave transformer used for this purpose, if it is connected directly to the load?
(a) 90 kΩ (b) 300 Ω
(c) $\frac{9}{4}\Omega$ (d) $\frac{3}{2}\Omega$
[ESE-2009]

3.45 The load end of a quarter wave transformer gets disconnected thereby causing an open-circuited load. What will be the input impedance of the transformer?
(a) Zero
(b) Infinite
(c) Finite and Positive
(d) Finite and Negative [ESE-2009]

3.46 What is the minimum value of VSWR that may exist on a transmission line?
(a) Less than zero (b) Zero
(c) One (d) 10 [ESE-2009]

3.47 Consider the transmission line of length 37.5 cm, which is terminated into zero resistance. This line is being excited by a source of 1 GHz which has an internal impedance of 50 Ω. What is the input impedance of the line as seen by the source?
(a) 50 Ω (b) Zero Ω
(c) 100 Ω (d) Infinite Ω
[ESE-2009]

3.48 Given a range of frequencies, which of the following systems is best for transmission line load matching?
(a) Single stub
(b) Double stub
(c) Single stub with adjustable position
(d) Quarter wave transformer [ESE-2010]

3.49 A line of characteristic impedance 50 ohms is terminated at one end by $+j50$ ohms. The VSWR on the line is
(a) 1 (b) ∞
(c) 0 (d) j [ESE-2010]

3.50 At UHF short-circuited lossless transmission lines can be used to provide appropriate values of impedance. Match **List-I** with **List-II** and select the correct answer using the codes given below the lists:

	List-I		List-II
A.	$l < \lambda/4$	1.	Capacitive
B.	$l = \lambda/4 < l < \lambda/2$	2.	Inductive
C.	$l = \lambda/4$	3.	0
D.	$l = \lambda/2$	4.	∞

Codes:

	A	B	C	D
(a)	2	1	4	3
(b)	3	1	4	2
(c)	2	4	1	3
(d)	3	4	1	2

[ESE-2010]

3.51 Consider the following statements regarding a transmission line:
1. Its attenuation is constant and independent of frequency.
2. Its attenuation varies linearly with frequency.
3. Its phase shift varies linearly with frequency.
4. Its phase shift is constant and is independent of frequency.

Which of the above statements are correct for distortionless line?
(a) 1, 2, 3 and 4 (b) 2 and 3 only
(c) 1 and 3 only (d) 3 and 4 only
[ESE-2010]

3.52 When waves travel along a transmission line from a generator to a load, through which region is power transmission taking place?
(a) Only through the conducting region
(b) Only through the non-conducting
(c) Both through conducting and non-conducting regions
(d) Through the conducting regions for half a cycle and through the non-conducting regions for the next half cycle
[ESE-2010]

3.53 **Assertion (A):** The impedance of a matched load is equal to characteristic impedance of line.
Reason (R): A matched termination absorbs the entire power incident on it.
(a) Both A and R are true and R is the correct explanation of A
(b) Both A and R are true but R is NOT the correct explanation of A
(c) A is true but R is false
(d) A is false but R is true [ESE-2010]

3.54 The characteristic impedance of TV receiving antenna cable is 300 Ω. If the conductors are made of copper separated by air and are 1 mm thick, what is the phase velocity and phase constant when receiving VHF channel 3(63 MHz) and VHF 69 (803 MHz)?
(a) 1.32 rad/m and 17.82 rad/m
(b) 1.52 rad/m and 16.82 rad/m
(c) 1.52 rad/m and 17.82 rad/m
(d) 1.32 rad/m and 16.82 rad/m
[ESE-2011]

3.55 If maximum and minimum voltages on a transmission line are 4 V and 2 V respectively, VSWR is
(a) 0.5 (b) 2
(c) 1 (d) 8 [ESE-2011]

3.56 An ideal lossless transmission line of $Z_0 = 60\ \Omega$ is connected to unknown Z_L. If SWR = 4, find Z_L.
(a) 240 Ω (b) 480 Ω
(c) 120 Ω (d) 100 Ω [ESE-2011]

3.57 Loading of a cable is done to
1. Increase its inductance
2. Increase its leakage resistance
3. Decrease its capacitance
4. Achieve distortionless condition

(a) 1, 2, 3 and 4 (b) 1 and 2 only
(c) 2 and 3 only (d) 1 and 4 only
[ESE-2011]

3.58 Consider the following statements about the Smith chart:

1. A complete revolution (360°) around the Smith chart represents a distance of half wavelength on the line.
2. Clockwise movement on the chart is regarded as moving toward the generator.
3. Although three scales around the periphery of Smith chart are used, but only one scale can be sufficient.
4. Smith chart cannot be used as admittance chart.

Which of these statements are correct?

(a) 1, 2, 3 and 4 (b) 1, 2 and 4 only
(c) 2, 3 and 4 only (d) 1, 2 and 3 only

[ESE-2011]

3.59 A $\lambda/4$ line, shorted at one end, presents impedance at the other end equal to

(a) Z_o (b) $\sqrt{2}Z_o$
(c) ∞ (d) 0

Where Z_o is characteristic impedance of the line.

[ESE-2012]

3.60 A 75 Ω transmission line is first short-terminated and the minima locations are noted. When the short is replaced by a resistive load R_L, the minima locations are not altered and the VSWR is measured to be 3. The value of R_L is

(a) 25 Ω (b) 50 Ω
(c) 225 Ω (d) 250 Ω [ESE-2012]

3.61 A transmission line, has a characteristic impedance (Z_0) of 600 Ω. Its length is 500 m. If the line is cut into half what will be the Z_0 for each half?

(a) $\frac{Z_0}{4}$ (b) $\frac{Z_0}{2}$
(c) Z_0 (d) $2Z_0$ [ESE-2013]

3.62 A loss-less transmission line of length l is open-circuited and has characteristic impedance Z_0. The input impedance is

(a) $+jZ_0 \tan\beta l$ (b) $-jZ_0 \tan\beta l$
(c) $-jZ_0 \cot\beta l$ (d) $+jZ_0 \cot\beta l$

[ESE-2014]

3.63 Conditions for a transmission line to be of low loss are

(a) $R >> \omega L$, $G >> \omega C$
(b) $R << \omega L$, $G >> \omega C$
(c) $R << \omega L$, $G << \omega C$
(d) $R >> \omega L$, $G << \omega C$ [ESE-2014]

3.64 A transmission line has characteristic impedance of 500 Ω. It has been terminated in a 200 Ω load. If the load is dissipating a continuous power of 100 Ω, its reflection coefficient is

(a) $\frac{6}{7}$ (b) $\frac{4}{7}$
(c) $\frac{3}{7}$ (d) $\frac{2}{7}$ [ESE-2014]

3.65 A transmission line having a characteristic impedance of 50 Ω is terminated at one end by j 50 Ω. The voltage standing wave ratio produced will be

(a) ∞ (b) $+j$
(c) 1 (d) 0 [ESE-2015]

3.66 For energy propagation in a lossless transmission line, the characteristic impedance of the line is expressed in ohm as below (where notations have usual meanings).

(a) $\sqrt{LC}\,\Omega$ (b) $\sqrt{\frac{L}{C}}\,\Omega$
(c) $\sqrt{\frac{C}{L}}\,\Omega$ (d) $\sqrt{\frac{R+j\omega L}{G-j\omega L}}\,\Omega$

[ESE-2016]

3.67 A quarter wavelength transformer is used to match a load of 200 Ω to a line with input impedance of 50 Ω. The characteristic impedance of the transformer would be

(a) 40 Ω (b) 100 Ω
(c) 400 Ω (d) 1000 Ω

[ESE-2016]

3.68 For a lossless transmission line L = 0.35 μH/m, C = 90 pF/m and frequency = 500 MHz. Then the magnitude of propagation constant is

(a) 14.48 (b) 17.63
(c) 19.59 (d) 21.20 [ESE-2016]

3.69 Consider the following statements:

As compared to short-circuited stubs, open-circuited stubs are not preferred because the latter

1. Are of different characteristic impedance
2. Have a tendency to radiate

Which of the above statements is/are correct?
(a) 1 only (b) 2 only
(c) Both 1 and 2 (d) Neither 1 nor 2
[ESE-2016]

3.70 A quarter-wave transformer of characteristic impedance 60 Ω has been used to match a transmission line of characteristic impedance Z_0 with a load of 72 Ω. What is the characteristic impedance of the transformer, when the load of 72 Ω is replaced by 98 Ω?
(a) 98 Ω (b) 80 Ω
(c) 70 Ω (d) 60 Ω [ESE-2017]

3.71 What is the reflection coefficient for the line $Z_0 = 300\angle 0°$ Ω and $Z_L = 150\angle 0°$ Ω?
(a) 0.5 (b) 0.333
(c) –0.333 (d) –0.5 [ESE-2017]

3.72 If a 75 Ω line is terminated by a load of $(120 + j80)$Ω, the maximum and minimum impedances over the line are nearly
(a) 135 Ω and 28 Ω (b) 190.5 Ω and 28 Ω
(c) 135 Ω and 16 Ω (d) 190.5 Ω and 16 Ω
[ESE-2018]

3.73 A measure of the mismatch between the maximum and minimum voltage and current variations along the transmission line is called SWR, i.e., standing wave ratio. SWR indicates how much power is delivered to the load, and how much is lost in the line. When SWR is 1, the percent reflected power is zero. When SWR is 1.5, the percent reflected power will be
(a) 4 (b) 8
(c) 25 (d) 40 [ESE-2018]

3.74 When a transmission line section is first short-circuited, and then open-circuited, it shows input impedances of 25 Ω and 100 Ω, respectively. The characteristic impedance of the transmission line is
(a) 25 Ω (b) 50 Ω
(c) 75 Ω (d) 100 Ω [ESE-2018]

3.75 For a lossless line terminated in a short circuit, the stationary voltage minima and maxima are separated by
(a) $\frac{\lambda}{8}$ (b) $\frac{\lambda}{2}$
(c) $\frac{\lambda}{3}$ (d) $\frac{\lambda}{4}$ [ESE-2019]

3.76 The characteristic impedance of an 80 cm long lossless transmission line having $L = 0.25$ μH/m and $C = 100$ pF/m will be
(a) 25 Ω (b) 40 Ω
(c) 50 Ω (d) 80 Ω [ESE-2019]

3.77 It is required to match a 200 Ω load to a 300 Ω transmission line to reduce the SWR along the line to 1. If it is connected directly to the load, the characteristic impedance of the quarterwave transformer used for this purpose will be
(a) 275 Ω (b) 260 Ω
(c) 245 Ω (d) 230 Ω
[ESE-2019]

3.78 A 4-pole, wave wound armature having 45 slots with 18 conductors/slot is driven at 1200 rpm. If the flux per pole is 0.016 Wb, the generated emf will be
(a) 534.4 V (b) 526.8 V
(c) 518.4 V (d) 502.8 V
[ESE-2020]

3.79 For a terminated uniform transmission line, the impedance Z_x at a distance x from the load will be

(a) $Z_0 \dfrac{Z_L + Z_0 \tanh \gamma x}{Z_0 + Z_L \tanh \gamma x}$ Ω

(b) $Z_L \dfrac{Z_L + Z_0 \tanh \gamma x}{Z_0 + Z_L \tanh \gamma x}$ Ω

(c) $Z_0 \dfrac{Z_L + jZ_0 \tanh \gamma x}{Z_0 + jZ_L \tanh \gamma x}$ Ω

(d) $Z_L \dfrac{Z_L + jZ_0 \tanh \gamma x}{Z_0 + jZ_L \tanh \gamma x}$ Ω

[ESE-2020]

where: Z_0 = Characteristics impedance of line, Ω
Z_L = Load impedance, Ω
γ = Propagation constant = $\alpha + j\beta$, m^{-1}
α = Attenuation constant, Np m^{-1}
β = Phase constant, rad m^{-1}

3.80 A distortionless transmission line has the following parameters:
Characteristic impedance = 60 Ω, wave velocity = 0.6c, where *c* is the speed of light in a vacuum, α = 20 mNp/m. The values of transmission line parameters R, L, G and C at 100 MHz are respectively,

(a) 1.2 Ω/m, 333 nH/m, 333 μS/m, 92.59 pF/m
(b) 1.2 Ω/m, 111 H/m, 333 μS/m, 92.59 F/m
(c) 2.4 Ω/m, 333 nH/m, 333 μS/m, 92.59 F/m
(d) 2.4 Ω/m, 111 H/m, 333 S/m, 92.59 pF/m

[ESE-2021]

3.81 Consider the following statements regarding the Smith's chart:

1. Smith's chart is a graphical indication of the impedance of a transmission line and of the corresponding reflection coefficient as one moves along the line.
2. λ distance on the line corresponds to a 720° movements on the Smith's chart.
3. The admittance chart can be obtained by shifting each and every point on the impedance chart by 90°.
4. Counter-clockwise movement on the chart corresponds to moving towards the generator.

Which of the above statements are correct?

(a) 1 and 2 only (b) 1 and 3 only
(c) 1, 2 and 4 only (d) 2, 3 and 4 only

[ESE-2021]

3.82 Consider the following statements regarding the load matching and impedance measurements :

1. A mismatched load can be properly matched to a line by inserting prior to the load a transmission line $\lambda/4$ long.
2. For matching of 120 Ω load to a 75 Ω line, the quarter-wave transformer must have a characteristic impedance of 190 Ω.
3. The main disadvantage of single-stub matching is that it is a narrow-band or frequency-sensitive device.

Which of the above statements are ***not*** correct?

(a) 1 and 2 only (b) 2 and 3 only
(c) 1 and 3 only (d) 1, 2 and 3

[ESE-2021]

3.83 The condition for the low-loss transmission line is

(a) $R >> \omega C$ and $G >> \omega L$
(b) $R << \omega L$ and $G << \omega C$
(c) $\frac{R}{G} = \frac{L}{C}$
(d) $RG >> LC$

[ESE-2022]

3.84 Consider the following statements regarding impedance matching :

1. When the line is terminated in an impedance other than its characteristic impedance, reflection will occur and there will be standing waves of voltage and current along the line which may be very large if there is considerable mismatch.
2. The single-stub device has the advantage that it will match any load.
3. The double stub has the disadvantage that the line length needs to be adjustable.

Which of the above statements are correct?

(a) 2 and 3 only (b) 1 and 2 only
(c) 1 and 3 only (d) 1, 2 and 3

[ESE-2022]

3.85 The VSWR can have any value between

(a) 0 and 1 (b) –1 and 1
(c) 1 and ∞ (d) 0 and ∞

[ESE-2023]

3.86 Consider the following statements regarding impedance matching:

1. The single-stub tuner (matching) consists of an open or shorted section of transmission line of length *d* connected in parallel with the main line at some distance from the load.
2. An open-circuited stub radiates some energy at high frequencies.
3. Double-stub matching allows for the adjustment of the load impedance.
4. At very high frequencies, lumped inductances and capacitances can be used as circuit elements.

Which of the above statements are correct?

(a) 1 and 2 only (b) 1, 2 and 3
(c) 2, 3 and 4 (d) 1, 2 and 4

[ESE-2023]

3.87 A coaxial cable with an inner diameter of 1 mm and outer diameter of 2.4 mm is filled with a dielectric of relative permittivity 10.89. Given $\mu_0 = 4\pi \times 10^{-7}$ H/m, $\varepsilon_0 = \frac{1}{36\pi \times 10^9}$ F/m. The characteristic impedance of the cable is

(a) 33 Ω (b) 43.4 Ω
(c) 143.3 Ω (d) 16 Ω

[ESE-2023]

3.88 The intrinsic impedance of copper at high frequency is
(a) purely resistive
(b) purely inductive
(c) complex with an inductive component
(d) complex with a capacitive component
[ESE-2023]

3.89 A transmission line has a characteristic impedance of 50 Ω and a resistance of 0.1 Ω/m. If the line is distortionless, the attenuation constant is
(a) 500 (b) 5
(c) 0.01 (d) 0.002 [ESE-2023]

Answers Transmission Lines

3.1 (d)	3.2 (d)	3.3 (c)	3.4 (c)	3.5 (b)	3.6 (b)	3.7 (c)	3.8 (c)	3.9 (b)
3.10 (c)	3.11 (a)	3.12 (d)	3.13 (d)	3.14 (d)	3.15 (b)	3.16 (a)	3.17 (b)	3.18 (a)
3.19 (a)	3.20 (d)	3.21 (c)	3.22 (c)	3.23 (a)	3.24 (d)	3.25 (d)	3.26 (a)	3.27 (a)
3.28 (c)	3.29 (c)	3.30 (c)	3.31 (c)	3.32 (c)	3.33 (c)	3.34 (d)	3.35 (a)	3.36 (a)
3.37 (d)	3.38 (b)	3.39 (d)	3.40 (d)	3.41 (d)	3.42 (d)	3.43 (d)	3.44 (b)	3.45 (a)
3.46 (c)	3.47 (d)	3.48 (b)	3.49 (b)	3.50 (a)	3.51 (c)	3.52 (c)	3.53 (a)	3.54 (d)
3.55 (b)	3.56 (a)	3.57 (d)	3.58 (d)	3.59 (c)	3.60 (a)	3.61 (c)	3.62 (c)	3.63 (c)
3.64 (c)	3.65 (a)	3.66 (b)	3.67 (b)	3.68 (b)	3.69 (b)	3.70 (c)	3.71 (c)	3.72 (b)
3.73 (a)	3.74 (b)	3.75 (d)	3.76 (c)	3.77 (c)	3.78 (c)	3.79 (a)	3.80 (a)	3.81 (a)
3.82 (b)	3.83 (b)	3.84 (d)	3.85 (c)	3.86 (b)	3.87 (d)	3.88 (c)	3.89 (d)	

Explanations Transmission Lines

3.1 (d)

Propagation constant,

$$\gamma = \sqrt{(R + j\omega L)(G + j\omega C)}$$

In a lossless line,

$$R = 0$$

and $G = 0$

For a lossless line,

$$\gamma = \sqrt{(j\omega L)(j\omega C)}$$

or $\gamma = j\omega\sqrt{LC}$

$$\gamma = \alpha + j\beta$$

So, $\alpha = 0$

$$\beta = \omega\sqrt{LC}$$

So, $\beta = |\gamma|$

3.2 (d)

Reflection coefficient

$$\Gamma = \frac{Z_L - Z_0}{Z_L + Z_0}$$

$$\Gamma = \frac{40 + j30 - 50}{40 + j30 + 50}$$

$$\Gamma = \frac{-10 + j30}{90 + j30}$$

$$|\Gamma| = \frac{\sqrt{(10)^2 + (30)^2}}{\sqrt{(90)^2 + (30)^2}}$$

$$\rho = \frac{1}{3}$$

$$\text{VSWR} = \frac{1+\rho}{1-\rho} = \frac{1+\frac{1}{3}}{1-\frac{1}{3}}$$

$\Rightarrow$ VSWR = 2

3.3 (c)

The best way for the impedance matching is to use a short-circuited stub at a specific distance from the load.

3.4 (c)

Characteristic impedance of quarter-wave transformer is

$$Z_o = \sqrt{Z_S Z_L} = \sqrt{75 \times 300} = 150\ \Omega$$

3.6 (b)

$$Z_L = 0$$

$$\beta l = \frac{2\pi}{\lambda} \cdot \frac{\lambda}{8} = \frac{\pi}{4}$$

$$\Rightarrow \quad \tan\beta l = \tan\frac{\pi}{4} = 1$$

Input impedance,

$$Z_{in} = Z_o\left(\frac{Z_L + jZ_o \tan\beta l}{Z_o + jZ_L \tan\beta l}\right)$$

$$Z_{in} = Z_o\left(\frac{0 + jZ_o}{Z_o + 0}\right)$$

$$Z_{in} = jZ_o$$

Since the transmission line is lossless, Z_o is real. Hence Z_{in} is inductive.

3.8 (c)

For distortionless transmission line, phase response should be linear, i.e. β should be linear function of frequency ω.

(i)
$$\gamma = \alpha + j\beta = \sqrt{(R + j\omega L)(G + j\omega C)}$$

when, $R = G = 0$,

$$\gamma = \sqrt{(j\omega L)(j\omega C)}$$

$$\Rightarrow \quad \alpha + j\beta = j\omega\sqrt{LC}$$

$$\Rightarrow \quad \beta = \omega\sqrt{LC}$$

(ii)
$$\gamma = \sqrt{(R + j\omega L)(G + j\omega C)}$$

But $LG = RC$

$$\Rightarrow \quad R = \frac{LG}{C}$$

$$\gamma = \sqrt{\left(\frac{LG}{C} + j\omega L\right)(G + j\omega C)} = \frac{\sqrt{L}}{\sqrt{C}}\sqrt{(G + j\omega C)(G + j\omega C)}$$

$$\Rightarrow \quad \alpha + j\beta = \sqrt{\frac{L}{C}} \cdot (G + j\omega C)$$

$$\Rightarrow \quad \beta = \omega\sqrt{LC}$$

(iii) When $R >> \omega L$ and $G >> \omega C$

$$\gamma = \sqrt{(R + j\omega L)(G + j\omega C)}$$

$$\gamma = \sqrt{RG} = \alpha + j\beta$$

$$\Rightarrow \quad \beta = 0$$

Since β is not a linear function of ω at very low frequency range, hence this condition does not guarantee a distortionless transmission line.

(iv) $R << \omega L$ and $G << \omega C$

$$\gamma = \sqrt{(R + j\omega L)(G + j\omega C)} = j\omega\sqrt{LC}$$

$$\Rightarrow \quad \beta = \omega\sqrt{LC}$$

3.9 (b)

Distance between adjacent maxima

$$d = \frac{\lambda}{2}$$

$$\frac{\lambda}{2} = (37.5 - 12.5)$$

$$\Rightarrow \quad \lambda = 50 \text{ cm}$$

Operating frequency

$$f = \frac{c}{\lambda} = \frac{3 \times 10^{10}}{50} = 600 \text{ MHz}$$

3.10 (c)

At $t = 0$, $\quad V_{IN} = \frac{Z_0}{2Z_0} \times E = \frac{E}{2}$ at $Z = 0$

At $t = \frac{l}{C} = T_0 \quad V_L = \frac{E}{2}$

At $t = \frac{2l}{C} = 2T_0$

Wave reflects, $V_{IN} = \frac{E}{2} + \frac{E}{2} = E$

3.11 (a)

$$\text{VSWR} = \frac{|V_{max}|}{|V_{min}|} = \frac{64}{16} = 4$$

3.12 (d)

$Z_{SC} = Z_{in} = jZ_0 \tan\beta l$ with $l = \lambda/4$

$Z_{in} = \infty$ - parallel resonant circuit

As $Z_{in} = jZ_0 \tan\frac{2\pi}{\lambda} l$ with changing λ, Z_{in} changes.

A λ/4 short circuited line behaves as a parallel resonant circuit.

$$Z_{sc} = jz_o \tan\beta l\big|_{l = \lambda/4}$$

$$= jz_o \tan\frac{\pi}{2}$$

$$= \infty \cdots \text{parallel resonant.}$$

3.13 (d)

$$Z_{in} = Z_o\left[\frac{Z_L + jZ_o\tan\beta l}{Z_o + jZ_L\tan\beta l}\right]$$

(i) For a shorted line with $l = \frac{\lambda}{8}$,

$$\beta l = \frac{2\pi}{\lambda}\cdot\frac{\lambda}{8} = \frac{\pi}{4},$$

$$Z_{in} = Z_o\left[\frac{0 + jZ_o}{Z_o + 0}\right]$$

$$Z_{in} = jZ_o$$

(ii) For a shorted line with $l = \frac{\lambda}{4}$, $\beta l = \frac{\pi}{2}$,

$$Z_{in} = \frac{Z_o^2}{Z_L} = \frac{Z_o^2}{0}$$

$$Z_{in} = j\infty$$

(iii) For an open line with $l = \frac{\lambda}{2}$, $\beta l = \pi$,

$$Z_{in} = Z_o\left[\frac{Z_L + jZ_o\tan\pi}{Z_o + jZ_L\tan\pi}\right]$$

$$Z_{in} = Z_o\frac{Z_L}{Z_L}\left[\frac{1 + j\frac{Z_o}{Z_L}\tan\pi}{\frac{Z_o}{Z_L} + \tan\pi}\right]$$

$$Z_{in} = -j\infty$$

(iv) For a matched line of any length,

$$Z_L = Z_o$$

$$Z_{in} = Z_o\left[\frac{Z_o + jZ_o\tan\beta l}{Z_o + jZ_o\tan\beta l}\right]$$

3.14 (d)

For a short circuited quarter wave long transmission line, i.e.

$$Z_L = 0 \ \& \ l = \frac{\lambda}{4},$$

$$Z_{in} = Z_o\left[\frac{Z_L + jZ_o\tan\beta l}{Z_o + jZ_L\tan\beta l}\right]$$

$$= Z_o\left[\frac{\frac{Z_L}{\tan\beta l} + jZ_o}{\frac{Z_o}{\tan\beta l} + jZ_L}\right]$$

$$= Z_o\cdot\frac{Z_o}{Z_L} = \frac{Z_o^2}{Z_L}$$

Since, $Z_L = 0$ so $Z_{in} = \infty$

3.15 (b)

The quarter wave line characteristic impedance,

$$Z_o = \sqrt{Z_{in}Z_L} = \sqrt{400\times 25}$$

$$\Rightarrow \quad Z_o = 20 \times 5 = 100\ \Omega$$

3.16 (a)

V_1 from left side of the line and V_2 from right side travel in opposite directions and add up by being in phase.

V_{max} at centre results in I_{min} at centre in the standing in wave.

Z_{max} results at this centre with infinite values.

3.17 (b)

$$20\log(\text{VSWR}) = 6$$

$$\text{VSWR} = (10)^{6/20} = 2$$

Reflection coefficient,

$$\rho = |\Gamma| = \frac{\text{VSWR}-1}{\text{VSWR}+1}$$

$$= \frac{2-1}{2+1} = \frac{1}{3} = 0.33$$

3.18 (a)

Characteristic impedance of quarter wave transformer is

$$Z_o = \sqrt{Z_i Z_L} = \sqrt{50\times 200} = 100\ \Omega$$

3.19 (a)

$Z(x)$ and $Y(x)$ have almost identical formulae and the same chart seems for both.

$$Z = R + jX = |Z|\angle\theta$$

Rectangular graph can always be replaced with polar graph.

$$\Gamma(x) = \Gamma_L e^{-j2\beta x}$$

This equation is clear that with increasing x a clockwise movement or negative angular increase can be seen towards source.

3.20 (d)

In order to obtain the best matching, a short-circuited stub will have to be connected at some specific distance from load.

3.21 (c)

Theoretically, short circuit and open circuit lines are similar in all aspects. However, open circuit has radiation possibility due to the openers of the line.

3.22 (c)

$$\Gamma(x) = \Gamma_L e^{-j2\beta x}$$

$$\Gamma(0.1\lambda) = 0.3e^{-j30°}e^{-j2\frac{2\pi}{\lambda}\frac{\lambda}{10}}$$

$$\Gamma(0.1\lambda) = 0.3e^{-j30°}e^{-j72°}$$

$$= 0.3e^{-j102°} = 0.3e^{j258°}$$

3.23 (a)

Balun: Stands for balanced unbalanced line. Equal conductor sizes of the line means balanced line (parallel wire) the equal conductor sizes means unbalanced line (coaxial cable).

3.24 (d)

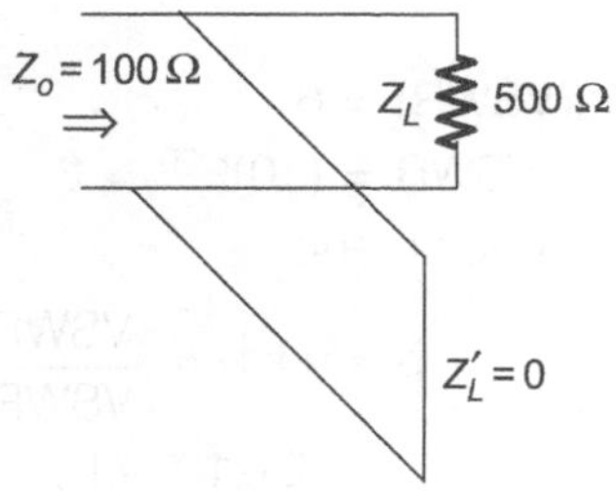

$$\text{VSWR} = \frac{Z_L}{Z_o}, Z_L > Z_o = \frac{Z_o}{Z_L}, Z_o < Z_L$$

On the length d,

$$Z_L = 500\ \Omega\ ;\quad Z_o = 100\ \Omega$$

$$\text{VSWR} = \frac{Z_L}{Z_o} = \frac{500}{100} = 5$$

On the stub itself,

$$Z_L' = 0,\ Z_o = 100\ \Omega$$

$$\text{VSWR} = \frac{Z_o}{Z_L'} = \frac{100}{0} = \infty$$

3.25 (d)

$$Z_0' = \sqrt{Z_0 \cdot Z_L}$$

Characteristic impedance of matching device. If Z_0 or Z_L are complex then Z_0' is complex, then matching device is lossy. So, Z_0 and Z_L have to real i.e. Z_L should be resistive.

3.26 (a)

Characteristic impedance

$$Z_o = \sqrt{\frac{R + j\omega L}{G + j\omega C}}$$

3.27 (a)

Higher the frequency $\omega L >> R\ ;\ \omega C >> G$ Lower the losses.

3.28 (c)

$$Z = \sqrt{\frac{R + j\omega L}{G + j\omega C}}$$

At Radio frequencies, $j\omega L >> R$ & $j\omega C >> G$

So, $$Z = \sqrt{\frac{j\omega L}{j\omega C}} = \sqrt{\frac{L}{C}}\ \Omega$$

Phase velocity $v = \frac{\omega}{\beta}$

Propagation constant,

$$\gamma = \sqrt{(R + j\omega L)(G + j\omega C)}$$

At, radio frequencies,

$$\gamma \cong \sqrt{j\omega L \cdot j\omega C}$$

$$\gamma = j\omega\sqrt{LC}$$

$$\Rightarrow \quad j\beta = j\omega\sqrt{LC}$$

where β is phase constant.

So, $$\beta = \omega\sqrt{LC}$$

$$v = \frac{\omega}{\beta} = \frac{1}{\sqrt{LC}}\ \text{m/s}$$

3.29 (c)

Characteristic impedance, $Z_o = \sqrt{\frac{L}{C}}$

3.30 (c)

Range of reflection coefficient is

$$-1 \le |\Gamma| \le 1$$

$$\text{VSWR} = \frac{1 + |\Gamma|}{1 - |\Gamma|}$$

So, range of VSWR is

$$1 \le \text{VSWR} < \infty$$

3.31 (c)

In quarter wave impedance transformer, input impedance

$$Z_{in} = \frac{Z_o^2}{Z_L}$$

where, $Z_o \rightarrow$ characteristic impedance
$Z_L \rightarrow$ Load impedance

Given that $Z_L = 0$

So, $$Z_{in} = \frac{Z_o^2}{0} = \infty$$

3.32 (c)

$$Z_{SC} = Z_{in} = jZ_0 \tan\beta l$$

Input impedance for lossless line.
It can be inductive or capacitive depending on length.

$$\Gamma = \frac{Z_L - Z_0}{Z_L + Z_0} = -1$$

Complete reflection occur on the line with standing wave.

3.33 (c)

Reflection coefficient

$$\rho = |\Gamma|$$

$$\Gamma = \frac{Z_L - Z_o}{Z_L + Z_o}$$

(i) For short circuit,

$$Z_L = 0$$

$$\Gamma = \frac{0 - Z_o}{0 + Z_o} = -1$$

(ii) For open circuit,

$$Z_L = \infty$$

$$\Gamma = \frac{1 - Z_o/Z_L}{1 + Z_o/Z_L} = 1$$

(iii) If $Z_L = Z_o$, then

$$\Gamma = \frac{Z_o - Z_o}{Z_o + Z_o} = 0$$

(iv) If $Z_L = 2Z_o$, then

$$\Gamma = \frac{2Z_o - Z_o}{2Z_o + Z_o} = \frac{1}{3}$$

3.34 (d)

$$\text{VSWR} = \frac{1+|\Gamma|}{1-|\Gamma|}$$

Given that $\Gamma = 1\angle 0°$

So, $$\text{VSWR} = \frac{1+1}{1-1} = \frac{2}{0} = \infty$$

3.35 (a)

Co-axial cable $R_o = Z_o = 60\ \ln(b/a)$

$$Z_o = \frac{120\pi}{2\pi}\ln(b/a)$$

Parallel wire $R_o = Z_o = 120\ln(D/r)$

where $r = \frac{a}{2}$ radius of the cable

$$R_o = \frac{120\pi}{\pi}\ln\left(\frac{b}{a/2}\right)$$

with $D = b$ = distance between the cables

parallel plate $\frac{c}{l} = \frac{\epsilon a}{b}$

$$\frac{L}{l} = \frac{\mu b}{a}$$

$$R_o = Z_o = \sqrt{\frac{L}{C}} = \sqrt{\frac{\mu}{\epsilon}}\cdot\frac{b}{a} = n\frac{b}{a}$$

3.36 (a)

Reflection coefficient, $|\Gamma| = 1/5$

$$\text{VSWR} = \frac{1+|\Gamma|}{1-|\Gamma|} = \frac{1+1/5}{1-1/5} = \frac{6}{4} = \frac{3}{2}$$

3.37 (d)

Delay distortion is a condition where v_p-phase velocity is frequency dependent.

$$v_p = \frac{\omega}{\beta} = \frac{1}{\sqrt{LC}} = \frac{1}{\sqrt{\mu\epsilon}}$$

3.38 (b)

Reactive loads cannot be used with quarterwave transformer.
Stub matching is a suitable option.

3.40 (d)

Input impedance,

$$Z_s = Z_0\left(\frac{Z_L + jZ_0\tan\beta l}{Z_0 + jZ_L\tan\beta l}\right)$$

But given that $Z_L = \infty$

$$Z_s = Z_0\frac{\left(1 + \frac{jZ_0}{Z_L}\tan\beta l\right)}{\left(\frac{Z_0}{Z_L} + j\tan\beta l\right)} = -jZ_0\cot\beta l$$

3.41 (d)

Number of maximas depend on first maxima from the load.

$$\Gamma\text{'s phase} = -150° = 210° = \theta$$

$$2\beta x_{max} = 2n\pi + \theta$$

For $n = 0$,

$$2\cdot\frac{2\pi}{150}\cdot x_{max} = \frac{7\pi}{6} = 210°$$

$$x_{max} = 43.75\ \text{m} \rightarrow 1^{st}\ \text{maxima}$$

Spacing between maximas being $\lambda/2 = 75$ m
2nd maxima occurs at 43.75 + 75 = 118.75 m
3rd maxima occurs at 193.75 m
And so on.
7th maxima at 493.75 m.

3.42 (d)

Power line has $\lambda = \frac{3\times10^8}{50} = 6000\ \text{km}$

$l = 10$ m $<< \lambda$ the line is a short circuit with R, L, G and C almost ignorable.

$$\text{Time delay} = \frac{10}{3\times10^8} \cong 0 \text{ Negligible.}$$

It is most often a parallel wire or twisted pair.
Intensity means $I(x)$, as this varies $Z(x)$ varies.
Such a small power line is considered as a lumped network.
Time delay is considered as negligible. Input impedance of this line remains constant.
Intensity of lamp varies according to applied voltage or current.

3.43 (d)

$Z(x)$ repeats for every $\lambda/2$ length

$$\text{SWR} = 2 = \frac{1+|\Gamma|}{1-|\Gamma|} \qquad |\Gamma| = \frac{1}{3}$$

$$\Gamma_{voltage} = -\Gamma_{currents}$$

Matching techniques most often protect sources from reflected waves.

3.44 (b)

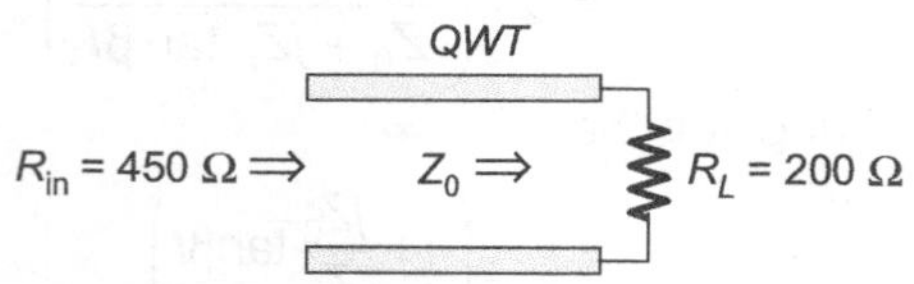

Input impedance to quarter-wave transformer is 450 Ω.
Characteristic impedance of QWT,

$$Z = \sqrt{R_{in}\,R_L} = \sqrt{450\times200} = 300\ \Omega$$

3.45 (a)

Open-circuited load means $Z_L = \infty$
Input impedance of QWT,

$$Z_{in} = \frac{Z_0^2}{Z_L} = \frac{Z_0^2}{\infty} = 0$$

3.46 (c)

The range of VSWR is $1 \le \text{VSWR} < \infty$

3.47 (d)

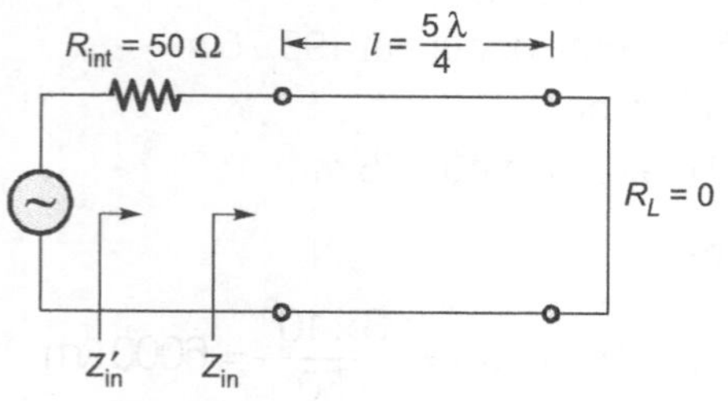

$$\lambda = \frac{3\times10^8}{1\times10^9} = 0.3 \text{ m} = 30 \text{ cm}$$

$$\text{Length of transmission line} = 37.5 \text{ cm} = \frac{5\lambda}{4}$$

$$Z_{in} = Z_0\left(\frac{Z_L + jZ_0\tan\beta l}{Z_0 + jZ_L\tan\beta l}\right)$$

$$= Z_0\left(\frac{0 + jZ_0\tan\frac{2\pi}{\lambda}\cdot\frac{5\lambda}{4}}{Z_0 + 0}\right)$$

$$= jZ_0\tan\frac{5\pi}{2} = j\infty$$

$$Z_{in}' = R_{int} + Z_{in} = 50 + j\infty = j\infty$$

3.49 (b)

Given, $Z_o = 50\ \Omega$; $Z_L = j50\ \Omega$
Reflection coefficient,

$$\Gamma = \frac{Z_L - Z_0}{Z_L + Z_0} = \frac{j50-50}{j50+50}$$

Now, $\rho = |\Gamma| = 1$

Hence, $\text{VSWR} = \dfrac{1+\rho}{1-\rho} = \dfrac{1+1}{1-1} = \infty$

3.50 (a)

$$Z_{in} = Z_0\frac{Z_L + jZ_0\tan\beta l}{Z_0 + jZ_L\tan\beta l}$$

As line is short circuited.

$\therefore$ $Z_L = 0$

$\Rightarrow$ $Z_{in} = jZ_o\tan\beta l$ ...(1)

$$\beta l = \frac{2\pi}{\lambda}\cdot l$$

1. $l < \frac{\lambda}{4} \Rightarrow \tan\beta l = +\text{ve}$

   $\therefore$ $Z_{in} \Rightarrow$ Inductive

2. $l > \frac{\lambda}{4}$ and $l < \frac{\lambda}{2} \Rightarrow \tan\beta l = -\text{ve}$

   $\therefore$ $Z_{in} \Rightarrow$ capacitive

3. $l = \frac{\lambda}{4} \Rightarrow \beta l = \frac{2\pi}{\lambda}\times\frac{\lambda}{4} = \frac{\pi}{2}$

   $\therefore$ $Z_{in} = jZ_0\tan\frac{\pi}{2} = \infty$

4. $l = \frac{\lambda}{2} \Rightarrow \beta l = \pi \Rightarrow \tan\pi = 0$

   $\therefore$ $Z_{in} = 0$

3.51 (c)

1. To avoid the frequency distortion the attenuation constant must be independent of frequency i.e.

$$\alpha \neq f(\omega)$$

2. To avoid phase distortion the phase constant must vary linearly with frequency i.e.

$$\beta \propto \omega$$
$$\Rightarrow \beta = k\omega$$
$$\text{or } \beta = \frac{\omega}{v_p}$$

3.52 (c)

Energy exists in the conductor as voltage and current (*V*/*I* wave) energy equal exists outside the conductor as electromagnetic fields (*E*/*H* wave)
At low frequencies *VI* energy is strong and higher frequencies *EH* energy is strong.
At transmission line operational frequencies both are almost equally strong.

3.54 (d)

Phase constant,

$$\beta = \frac{2\pi}{\lambda}$$

For VHF channel-3 (63 MHz)

$$\lambda = \frac{3\times10^8}{63\times10^6} = 4.761\,\text{m}$$

So $$\beta = \frac{2\times3.14}{4.761} = 1.319 \text{ rad/m}$$

or $$\beta \simeq 1.32 \text{ rad/m}$$

For VHF channel 69 (803 MHz)

$$\lambda = \frac{3\times10^8}{803\times10^6} = 0.3735\,\text{m}$$

So $$\beta = \frac{2\times3.14}{0.3735} = 16.82 \text{ rad/m}$$

Note: Although in the problem phase velocity is also asked to calculate but in the given options only value of phase constant is provided. If students are interested to calculate phase velocity then they can use following relation.

$$v_p = \frac{\omega}{\beta}$$

3.55 (b)

$$\text{VSWR} = \left|\frac{V_{max}}{V_{min}}\right|$$

$$\Rightarrow \quad \text{VSWR} = \frac{4}{2} = 2$$

3.56 (a)

For resistive load and characteristic impedance

$$\text{SWR} = \frac{Z_L}{Z_0} \text{ If } Z_L > Z_0 = \frac{Z_0}{Z_L} \text{ if } Z_o > Z_L$$

Here $$4 = \frac{Z_L}{Z_0}$$

$$\Rightarrow \quad Z_L = 4Z_o = 4 \times 60 = 240\ \Omega$$

3.57 (d)

Loading is increasing inductance to achieve $\omega L >> R$ or $\omega C >> G$ and make the line distortionless.

3.58 (d)

Impedance is periodic with $\lambda/2$ and one complete revolution is a periodicity.

$$\Gamma(x) = \Gamma_L e^{-j2\beta x}$$

Negative angle means clockwise movement with increasing distance x from load to generator.
Three scales around the boundary or periphery are reactance.
Γ's phase angle θ, distance towards chart can sufficiently have reactance scale only.
$Z(x)$ and $Y(x)$ have similar formulae and so Smith chart can be used as admittance chart also.

3.59 (c)

For $\lambda/4$ line,

$$\text{Input impedance, } Z_{in} = \frac{Z_o^2}{Z_L}$$

Where Z_o is characteristic impedance.
If the line is shorted, i.e. $Z_L = 0$

So, $$Z_{in} = \infty$$

3.60 (a)

$Z_o = 75\ \Omega$ and VSWR = 3
When the line was initially shorted, minima occurs at the load end.
Now, the load is connected but the minima still holds at same position. This interprets that R_L is less than Z_o.

So, $$\text{VSWR} = \frac{Z_o}{R_L};\ R_L < Z_o$$

$$\therefore \quad \text{VSWR} = \frac{Z_o}{R_L} = 3$$

$$\frac{75}{R_L} = 3$$

$$\therefore \quad R_L = 25\ \Omega$$

3.61 (c)

Characteristic impedance is independent of length of transmission line so it will remain Z_o for the half section.

3.62 (c)

$$Z_{in} = Z_0\left(\frac{Z_L + jZ_0 \tan\beta l}{Z_0 + jZ_L \tan\beta l}\right)$$

for open circuited load $Z_L = \infty$

$$Z_{in} = Z_0\left(\frac{Z_L}{Z_L}\left[\frac{1 + j\dfrac{Z_0}{Z_L}\tan\beta l}{\dfrac{Z_0}{Z_L} + j\dfrac{Z_0}{Z_L}\tan\beta l}\right]\right) = Z_0 \times \frac{1}{j\tan\beta l}$$

$$= -j Z_0 \cot\beta l$$

3.64 (c)

$$Z_0 = 500\ \Omega$$

$$Z_L = 200\ \Omega$$

$$\Gamma = \frac{Z_L - Z_0}{Z_L + Z_0}$$

$$|\Gamma| = \frac{200 - 500}{700} = \frac{3}{7}$$

3.65 (a)

$$Z_L = j50\ \Omega$$

$$Z_0 = 50\ \Omega$$

$$\Gamma = \frac{Z_L - Z_0}{Z_L + Z_0} = \frac{j50 - 50}{j50 + 50}$$

$$\Rightarrow \quad |\Gamma| = 1$$

$$\text{VSWR} = \frac{1 + |\Gamma|}{1 - |\Gamma|} = \frac{2}{0} = \infty$$

3.67 (b)

$$Z_q = \sqrt{(50)(200)} = 100\ \Omega$$

3.68 (b)

For lossless line,

$$\gamma = j\omega\sqrt{LC}$$

$$|\gamma| = \omega\sqrt{LC} = 2\pi f\sqrt{LC}$$

$$= 2\pi(500 \times 10^6)\sqrt{0.35 \times 10^{-6} \times 90 \times 10^{-12}}$$

$$= 2\pi(500 \times 10^6)\,5.61\,(10^{-9})$$

$$= 176.32 \times 10^{-1} = 17.63$$

3.69 (b)

Theoretically, short circuit and open circuit lines are similar in all aspects. However, open circuit has radiation possibility due to the openers of the line.

3.70 (c)

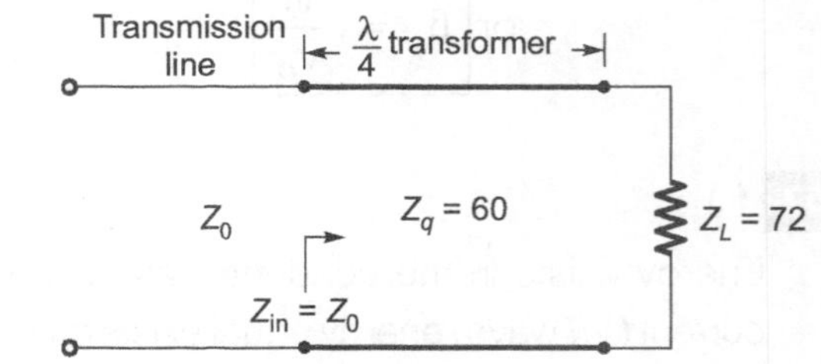

$$Z_q^2 = Z_{in} Z_L$$

$$(60)^2 = Z_0\, 72$$

$$Z_0 = \frac{(60)^2}{72} = 50\ \Omega$$

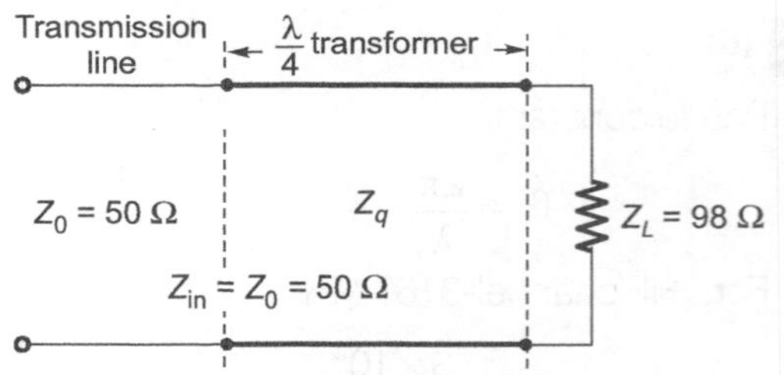

$$Z_q^2 = Z_{in} Z_L = (50)(98)$$

$$Z_q = \sqrt{(50)(98)} = 70\ \Omega$$

3.71 (c)

Given $Z_0 = 300\angle 0°$; $Z_L = 150\angle 0°$

Reflection coefficient,

$$\Gamma_L = \frac{Z_L - Z_0}{Z_L + Z_0} = \frac{150 - 300}{150 + 300} = \frac{-150}{450} = -0.3333$$

3.72 (b)

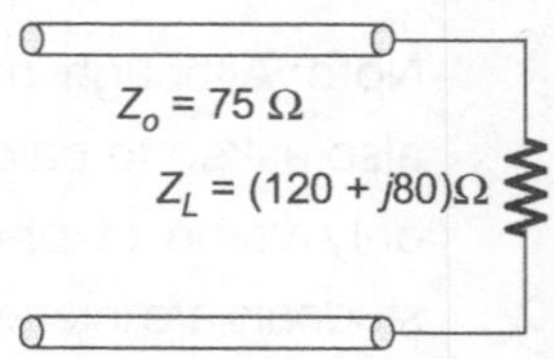

$$\Gamma_l = \frac{Z_L - Z_o}{Z_L + Z_o}$$

$$= \frac{120 + j80 - 75}{120 + j80 + 75}$$

$$= \frac{45 + j80}{195 + j80}$$

$$|\Gamma_l| = \frac{\sqrt{(45)^2+(80)^2}}{\sqrt{(195)^2+(80)^2}} = \sqrt{\frac{8425}{44425}} = \sqrt{0.1896}$$

$$= 0.43548$$

$$\text{SWR} = \frac{1+|\Gamma_l|}{1-|\Gamma_l|} = \frac{1+0.435}{1-0.435} = \frac{1.435}{0.565} = 2.54$$

$$\frac{Z_{max}}{Z_o} = \text{SWR}$$

$$Z_{max} = (\text{SWR})Z_o = 2.54\ (75) = 190.5\ \Omega$$

$$\frac{Z_{min}}{Z_o} = \frac{1}{SWR}$$

$$Z_{min} = \frac{75}{2.54} = 29.527\ \Omega$$

3.73 (a)

$$SWR = 1.5$$

$$|\Gamma_l| = \frac{SWR-1}{SWR+1} = \frac{1.5-1}{1.5+1} = \frac{0.5}{2.5} = 0.2$$

$$\frac{P_r}{P_{inc}} \times 100 |\Gamma_l|^2 \times 100 = (0.2)^2 \times 100\% = 4\%$$

3.74 (b)

$$Z_o = \sqrt{Z_{OC}\ Z_{SC}} = \sqrt{25 \times 100} = 50\ \Omega$$

3.75 (d)

Periodicity of standing wave is $\lambda/2$, thus the distrace between maxima to maxima is $\lambda/2$ and the distance between maxima and minima is $\lambda/4$.

3.76 (c)

$$Z_o = \sqrt{\frac{L}{C}} = \sqrt{\frac{0.25\times10^{-6}}{100\times10^{-12}}} = \sqrt{0.25}\times 100 = 50$$

3.77 (c)

$$Z_0' = \sqrt{200\times300} = 245\ \Omega$$

3.78 (c)

Total conductors = 45 × 18

N = 1200 rpm

Flux per pole = ϕ = 0.016 Wb, A = 2

P = 4

$$E_g = \frac{\phi PN}{60}\times\frac{Z}{A}$$

$$= \frac{0.016\times4\times1200}{60}\times\frac{45\times18}{2}\ \text{V} = 518.4\ \text{V}$$

3.79 (a)

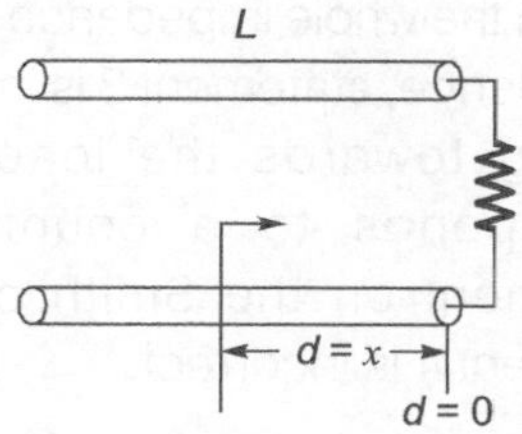

$$Z(d=x) = Z_0\frac{Z_L + Z_0\tanh\gamma x}{Z_0 + Z_L\tanh\gamma x}\ \Omega$$

3.80 (a)

Given: A distortionless transmission line with

$Z_o = 60\ \Omega$; $v_p = 0.6c = 0.6 \times 3 \times 10^8$ m/s

$\alpha = 20 \times 10^{-3}$ (Np/m)

f = 100 MHz

For a distortionless transmission line, $LG = RC$ and

$$Z_o = \sqrt{\frac{L}{C}},\quad \alpha = \sqrt{RG}$$

$$v_p = \frac{1}{\sqrt{LC}},\quad \beta = \omega\sqrt{LC}$$

(i) $$Z_o v_p = \sqrt{\frac{L}{C}}\frac{1}{\sqrt{LC}}$$

$$\Rightarrow\quad C = \frac{1}{Z_o v_p} = \frac{1}{60\,(0.6\times3\times10^8)}$$

$$\Rightarrow\quad C = \frac{1}{10.8}10^{-9} = 92.59\times10^{-12}\ \text{F/m}$$

$$\Rightarrow\quad C = 92.59\ \text{pF/m}$$

(ii) $$L = Z_0^2C = (60)^2 \times 92.59 \times 10^{-12}$$

$$L = 333.324 \times 10^{-9} = 333\ \text{nH/m}$$

(iii) $Z_o = \sqrt{\frac{L}{C}} = \sqrt{\frac{R}{G}}$. Therefore,

$$\alpha Z_o = \sqrt{RG}\sqrt{\frac{R}{G}}$$

$$\Rightarrow\quad R = 20 \times 10^{-3} \times 60 = 1.2\ \Omega/\text{m}$$

(iv) $$G = \frac{\alpha}{Z_o} = \frac{20\times10^{-3}}{60}$$

$$\Rightarrow\quad G = 333.3 \times 10^{-6}\ \text{S/m}$$

$$\Rightarrow\quad G \approx 333\ \mu\text{S/m}$$

3.81 (a)

- One complete rotation of Smith chart corresponds to $\lambda/2$. Hence, λ distance on the line corresponds to two rotations i.e., 720° movements on the Smith chart.

- Admittance Smith Chart can be obtained by rotating the whole impedance Smith Chart by 180°. Hence, statement 3 is incorrect.
- Moving towards the load impedance corresponds to a counter-clockwise movement on the Smith chart. Hence, statement 4 is incorrect.

3.82 (b)

A mismatched load Z_L can be properly matched to a line with characteristic impedance Z_0 by connecting $\lambda/4$ line with characteristic impedance Z_0' such that

$$Z_0' = \sqrt{Z_0 \cdot Z_L}$$

For matching 120 Ω load to a 75 Ω line, the quarter-wave transformer must have a characteristic impedance,

$$Z_0' = \sqrt{120 \times 75} = 94.8\,\Omega$$

Hence, statement 2 is incorrect.

The main drawback of quarter-wave transformer is that it is a narrow-band or frequency sensitive device. Hence, statement 3 is incorrect.

3.83 (b)

The propagation constant for the transmission line is given by

$$\gamma = \alpha + j\beta = \sqrt{(R + j\omega L)(G + j\omega C)}$$

For low loss line, α should be negligible.

If $R << \omega L$ and $G << \omega C$,

$$\gamma \approx j\omega\sqrt{LC} \qquad \text{(Lossless line)}$$

3.84 (d)

- If the line is terminated in an impedance other than characteristic impedance, reflection will occur along the line with reflection coefficient given by

$$\Gamma = \frac{Z_L - Z_o}{Z_L + Z_o}$$

If $Z_L = Z_0$, there is no reflection ($\Gamma = 0$). For $Z_L \neq Z_0$, the incident and reflected waves superimpose to form a standing wave along the line.

- Stub tuning is the process of adding a length of transmission line to the existing length in shunt, either in open circuit or short circuit configuration to match the line to the load. Single stub tuners have the ability to match any load impedance by placing them at a specific distance from the load to maintain matching.
- Once the stub locations fixed, a certain range of load admittance can not be matched.

Hence, all statements are correct.

3.85 (c)

$$S = \frac{1+|\Gamma|}{1-|\Gamma|}$$

At $|\Gamma| = 0$, $S = 1$

At $|\Gamma| = 1$, $S = \infty$

$\therefore$ Range of S : $1 \leq S \leq \infty$

3.87 (d)

Given that

Inner diameter, $a = 0.5 * 10^{-3}$ m

Outer diameter, $b = 1.2 * 10^{-3}$ m

As we know that

characteristic impedance of a co-axial cable is given as

$$Z_0 = \frac{60}{\sqrt{\epsilon_r}} \ln\left(\frac{b}{a}\right); \text{ a, b = Radius of}$$

inner and outer cable

$$\Rightarrow Z_0 = \frac{60}{\sqrt{10.89}} \ln\left(\frac{1.2}{0.5}\right) = 15.91\,\Omega \simeq 16\,\Omega$$

3.88 (c)

Copper is a good conductor,

$$\therefore \eta|_{cond} = \sqrt{\frac{\omega\mu}{\sigma}} \angle 45°$$

$$\Rightarrow \qquad \eta = \sqrt{\frac{\omega\mu}{2\sigma}} + j\sqrt{\frac{\omega\mu}{2\sigma}} \equiv R_S + j\omega_s$$

3.89 (d)

Given that:

Characteristics impedance, $Z_0 = 50\,\Omega$

Resistance, $R = 0.1$ Ω/m

Given, line is distortionless.

As we know that,

Attenuation constant, $\alpha = \sqrt{RG} = R\sqrt{\frac{C}{L}}$

$$Z_0 = \sqrt{\frac{C}{L}}$$

$$\therefore \qquad \alpha Z_0 = R$$

$$\therefore \qquad \alpha = \frac{R}{Z_0} = \frac{0.1}{50} = 0.002$$

■■■

4 Waveguides

4.1 Consider the following statements:
If the narrow dimension of a standard rectangular waveguide carrying the dominant mode is reduced, then the
1. wave impedance will increase
2. attenuation will increase
3. guide wavelength will decrease
4. power handling capability will decrease

Which of these statements are correct?
(a) 1 and 2 (b) 2 and 4
(c) 3 and 4 (d) 1 and 3 **[ESE-2000]**

4.2 The cut-off wavelength λ_c for TE_{20} mode for a standard rectangular waveguide is
(a) $2/a$ (b) $2a$
(c) a (d) $2a^2$ **[ESE-2000]**

4.3 Phase velocity 'v_p' and the group velocity 'v_g' in a waveguide ('v' is velocity of light) are related as
(a) $v_p\, v_g = c^2$
(b) $v_p + v_g = c$
(c) $v_p/v_g = a$ constant
(d) $v_p + v_g = a$ constant
[ESE-2000]

4.4 For a hollow waveguide, the axial current must necessarily be
(a) a combination of conduction and displacement currents
(b) conduction current only
(c) time-varying conduction current and displacement current
(d) displacement current only
[ESE-2001]

4.5 For a wave propagating in an air filled rectangular waveguide.
(a) guided wavelength is never less than free space wavelength
(b) wave impedance is never less than the free space impedance
(c) TEM mode is possible if the dimensions of the waveguide are properly chosen
(d) Propagation constant is always a real quantity
[ESE-2002]

4.6 When a particular mode is excited in a waveguide, there appears an extra electric component in the direction of propagation. The resulting mode is
(a) transverse-electric
(b) transverse-magnetic
(c) longitudinal
(d) transverse-electromagnetic **[ESE-2002]**

4.7 **Assertion (A):** TE_{10} mode is the dominant mode in a rectangular wave guide.
Reason (R): TE_{10} mode has the lowest cut-off frequency for which the wavelength is double the broadwall dimension.
(a) Both A and R are true and R is the correct explanation of A
(b) Both A and R are true but R is NOT the correct explanation of A
(c) A is true but R is false
(d) A is false but R is true **[ESE-2002]**

4.8 A rectangular wave guide measure 3 × 4.5 cm internally and has a 9 GHz signal propagated in it. The cut off wavelength for TE_{10} mode is
(a) 5 cm (b) 10 cm
(c) 15 cm (d) 9 cm **[ESE-2002]**

4.9 If the height of the waveguide is halved, its cut-off wavelength will
(a) Be halved
(b) Be doubled
(c) Remain unchanged
(d) Be one-fourth of the previous value
[ESE-2003]

4.10 In a rectangular waveguide with broader dimension a and narrow dimension b, the dominant mode of microwave propagation would be
(a) TE_{10} (b) TM_{10}
(c) TE_{01} (d) TM_{01}
[ESE-2003]

4.11 Match **List-I** (Dominant Mode of Propagation) with **List-II** (Type of transmission Structure) and select the correct answer:

List-I	List-II
A. Coaxial line	**1.** TE
B. Rectangular waveguide	**2.** Quasi TEM
C. Microstrip line	**3.** Hybrid
D. Coplanar waveguide	**4.** TEM

Codes:

	A	B	C	D
(a)	1	4	2	3
(b)	4	1	3	2
(c)	1	4	3	2
(d)	4	1	2	3

[ESE-2003]

4.12 For TE or TM modes of propagation in bounded media, the phase velocity
(a) is independent of frequency
(b) is a linear function of frequency
(c) is a non-linear function of frequency
(d) Can be frequency-dependent or frequency-independent depending on the source

[ESE-2003]

4.13 A waveguide operated below cut-off frequency can be used as
(a) A phase shifter
(b) An attenuator
(c) An isolator
(d) None of the above

[ESE-2003]

4.14 **Assertion (A):** The quality factor Q of a waveguide is closely related to its attenuation factor α.
Reason (R): Normally attenuation factors obtainable in waveguides are much higher than those obtainable in transmission lines.
(a) Both A and R are true and R is the correct explanation of A
(b) Both A and R are true but R is NOT the correct explanation of A
(c) A is true but R is false
(d) A is false but R is true

[ESE-2003]

4.15 The cut-off frequency of the dominant mode of a rectangular waveguide having aspect ratio more than 2 is 10 GHz. The inner broad wall dimension is given by
(a) 3 cm (b) 2 cm
(c) 1.5 cm (d) 2.5 cm

[ESE-2005]

4.16 In a waveguide, the evanescent modes are said to occur if
(a) The propagation constant is real
(b) The propagation constant is imaginary
(c) Only the TEM waves propagate
(d) The signal has a constant frequency

[ESE-2005]

4.17 Scattering parameters are more suited than impedance parameters to describe a waveguide junction because
(a) the scattering parameters are frequency invariant whereas the impedance parameters are not so
(b) scattering matrix is always unitary
(c) impedance parameters vary over unacceptably wide ranges
(d) scattering parameters are directly measurable but impedance parameters are not so

[ESE-2006]

4.18 In an air-filled waveguide of dimensions a cm × b cm, at a given frequency, the longitudinal component of electric field of TM_{32} mode is of the form
$E_Z = 20 \sin(60\pi x)\sin(100\pi y)$.
Which form would E_Z have for the lowest order TM mode?
(a) $E_Z = 20\sin(20\pi x)$
(b) $E_Z = 20\sin(20\pi y)$
(c) $E_Z = 20\sin(20\pi x)\sin(50\pi y)$
(d) $E_Z = 20\sin(20\pi x)\sin(100\pi y)$

[ESE-2005]

4.19 For a rectangular waveguide of dimensions: $a\sqrt{3}$ cm × a cm, the cut off frequency for the TE_{10} mode is 2 GHz.
What is the cut off frequency for TM_{11} mode in the waveguide?
(a) 1 GHz (b) 3.46 GHz
(c) 4 GHz (d) 6 GHz

[ESE-2005]

4.20 Which one of the following modes has the highest cut-off wavelength in a rectangular waveguide?
(a) TE_{10} `(b) TE_{01}
(c) TM_{01} (d) TM_{11}

[ESE-2006]

4.21

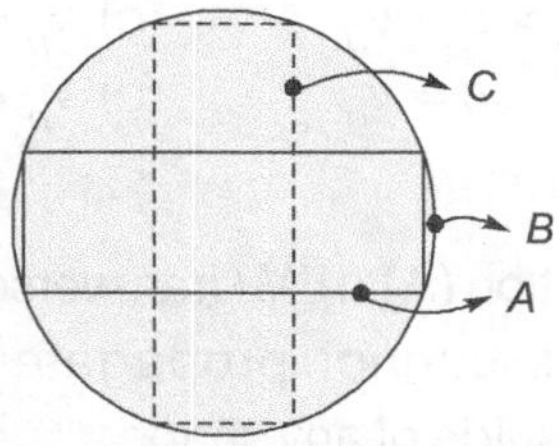

A rectangular waveguide (A) is gradually deformed first into a circular waveguide (B) and back again into a rectangular waveguide (C) which is oriented through 90° with respect to (A). If the input mode is TE_{10}, which mode is excited in the output waveguide (C)?

(a) TE_{10} (b) TE_{01}
(c) TE_{11} (d) TM_{11} **[ESE-2006]**

4.22 Which one of the following is the correct statement? A rectangular coaxial line can support

(a) only TEM mode of propagation
(b) both TEM and TE modes of propagation
(c) either TE or TM mode of propagation
(d) TEM, TE or TM mode of propagation

[ESE-2006]

4.23 The waveguide (a = 1.5 cm, b = 1 cm) is loaded with dielectric (ε_r = 4). Which one of the following is correct?

The 8 GHz signal will

(a) pass through the waveguide
(b) not pass through the waveguide
(c) be absorbed in the guide
(d) none of the above

[ESE-2007]

4.24 Inside a waveguide with perfectly conducting walls, any current present is in the form of

(a) displacement current only
(b) conduction current only
(c) partially displacement current and partially conduction current
(d) sometimes displacement current and sometimes conduction current

[ESE-2008]

4.25 Match **List-I** (Medium Type) and **List-II** (Frequency Range) and select the correct answer using the code given below the lists:

List-I	List-II
A. Open wire	**1.** 100 kHz
B. Twisted pair	**2.** 100 MHz
C. Coaxial-cable	**3.** 100 GHz
D. Waveguide	**4.** 5.0 kHz

Codes:

	A	B	C	D
(a)	4	3	2	1
(b)	2	1	4	3
(c)	2	3	4	1
(d)	4	1	2	3

[ESE-2008]

4.26 TEM wave cannot propagate in which one of the following?

(a) Two wire transmission line
(b) Coaxial cable
(c) Rectangular waveguide
(d) Microstrip line

[ESE-2008]

4.27 A standard waveguide WR90 has inside wall dimensions of a = 2·286 cm and b = 1.016 cm. What is the cut-off wavelength for TE_{01} mode?

(a) 4·572 cm (b) 2·286 cm
(c) 2·032 cm (d) 1·857 cm

[ESE-2009]

4.28 When a particular mode is excited in a waveguide, there appears an extra electric component, in the direction of propagation. In what mode is the wave propagating?

(a) Transverse electric
(b) Transverse magnetic
(c) Transverse electromagnetic
(d) Longitudinal **[ESE-2009]**

4.29 Which is the dominant mode in rectangular waveguides?

(a) TE_{10} (b) TE_{11}
(c) TM_{01} (d) TM_{11} **[ESE-2009]**

4.30 Consider the following statements :

For a square waveguide of cross-section 3 cm × 3 cm it has been found

1. At 6 GHz dominant mode will propagate.
2. At 4 GHz all the modes are evanescent.
3. At 11 GHz only dominant modes and no higher order mode will propagate.
4. At 7 GHz degenerate modes will propagate.

Which of the above statements are correct?

(a) 1 and 2 only (b) 1, 2 and 4
(c) 2 and 3 only (d) 2, 3 and 4

[ESE-2009]

4.31 Match **List-I** (Modes) with **List-II** (Characteristic) and select the correct answer using the code given below the lists:

List-I

A. Evanescent mode
B. Dominant mode
C. TM_{10} and TM_{01}

List-II

1. Rectangular waveguide does not support.
2. No wave propagation
3. Lowest cut-off frequency

Codes:

	A	B	C
(a)	1	2	3
(b)	2	3	1
(c)	1	3	2
(d)	2	1	3

[ESE-2009]

4.32 Which of the following modes can exist in a rectangular wave guide?

(a) TM_{10} (b) TE_{10}
(c) TM_{00} (d) TM_{01}

[ESE-2009]

4.33 For parallel plane waveguides, which is the mode with lowest cut-off frequency?

(a) TE_{10} (b) TM_{10}
(c) TEM (d) TE_{11}

[ESE-2009]

4.34 **Assertion (A) :** A *z*-directed rectangular waveguide with cross-sectional dimensions 3 cm × 1 cm will support propagation at 4 GHz.

Reason (R) : $k_z^2 + \left(\frac{m\pi}{3}\right)^2 + \left(\frac{n\pi}{1}\right)^2 = \left(\frac{2\pi}{\lambda}\right)^2$

where λ is the wavelength

(a) Both A and R are true and R is the correct explanation of A
(b) Both A and R are true but R is NOT the correct explanation of A
(c) A is true but R is false
(d) A is false but R is true

[ESE-2009]

4.35 For plane wave propagating in free space or two conductor transmission line, what must be the relationship between the phase velocity v_p, the group velocity v_g and speed of light c?

(a) $v_p > c > v_g$ (b) $v_p < c < v_g$
(c) $v_p = c = v_g$ (d) $v_p < v_g < c$

[ESE-2009]

4.36 **Assertion (A):** TEM (Transverse Electro Magnetic) waves cannot propagate within a hollow waveguide of any shape.

Reason (R): For a TEM wave to exist within the waveguide, lines of *H* field must be closed loops which requires an axial components of *E* which is not present in a TEM wave.

(a) Both A and R are true and R is the correct explanation of A
(b) Both A and R are true but R is NOT the correct explanation of A
(c) A is true but R is false
(d) A is false but R is true

[ESE-2010]

4.37 Consider the following statements:

For a rectangular waveguide with dimensions $a \times b$ where b is the narrow dimension, small value of b

1. Gives a larger separation between cut-off frequencies of TE_{01} and TE_{10} modes.
2. Gives increased attenuation.
3. Limits power handling capabilities because of breakdown field limits.

Which of these statements is/are correct?

(a) 1 and 2 only (b) 1, 2 and 3
(c) 2 only (d) 3 only

[ESE-2010]

4.38 The mode with lowest cut-off frequency for an electromagnetic wave propagating between two perfectly conducting parallel plates of infinite extent is

(a) TE_{10} (b) TM_{10}
(c) TM_{01} (d) TEM

[ESE-2010]

4.39 A standard air filled waveguide WR-187 has inside wall dimensions of a = 4.755 cm and b = 2.215 cm. At 12 GHz, it will support

(a) TE_{10} mode only
(b) TE_{10} and TE_{20} modes only
(c) TE_{10}, TE_{20} and TE_{01} modes only
(d) TE_{10}, TE_{20}, TE_{01} and TE_{11} modes

[ESE-2010]

4.40 An air-filled rectangular waveguide has dimensions of $a = 6$ cm and $b = 4$ cm. The signal frequency is 3 GHz. Match **List-I** with **List-II** and select the correct answer using the code given below the lists:

List-I	List-II
A. TE_{10}	1. 2.5 GHz
B. TE_{01}	2. 3.75 GHz
C. TE_{11}	3. 4.506 GHz
D. TM_{11}	4. 4.506 GHz

Codes:

	A	B	C	D
(a)	1	2	3	4
(b)	4	2	3	1
(c)	1	3	2	4
(d)	4	3	2	1

[ESE-2010]

4.41 A dominant mode of a waveguide is characterized by lowest
1. Cut-off frequency
2. Cut-off wavelength
3. Attenuation

(a) 1 only (b) 1 and 3 only
(c) 2 and 3 only (d) 1, 2 and 3

[ESE-2011]

4.42 The ratio of the transverse electric field to the transverse magnetic field is called as
(a) waveguide impedance
(b) waveguide wavelength
(c) phase velocity
(d) Poynting vector **[ESE-2012]**

4.43 Consider a rectangular waveguide of internal dimensions 8 cm × 4 cm. Assuming an TE_{10} mode of propagation, the critical wavelength would be
(a) 8 cm (b) 16 cm
(c) 4 cm (d) 32 cm

[ESE-2012]

4.44 $\gamma = \sqrt{\left(\frac{m\pi}{a}\right)^2 + \left(\frac{n\pi}{b}\right)^2 - \omega^2 \mu\varepsilon}$ represents the propagation constant in a rectangular waveguide for
(a) TE waves only (b) TM waves only
(c) TEM waves (d) TE and TM waves

[ESE-2012]

4.45 With the symbols having their standard meanings, cut-off frequency (frequency below which wave propagation will not occur) for a rectangular waveguide is
(a) $\frac{1}{\sqrt{\mu\varepsilon}}\sqrt{\left(\frac{m\pi}{a}\right)+\left(\frac{n\pi}{b}\right)}$
(b) $\frac{1}{2\pi\sqrt{\mu\varepsilon}}\sqrt{\left(\frac{m\pi}{a}\right)^2+\left(\frac{n\pi}{b}\right)^2}$
(c) $\frac{1}{2\pi\sqrt{\mu\varepsilon}}\left(\frac{m\pi}{a}\right)+\left(\frac{n\pi}{b}\right)$
(d) $\frac{1}{\sqrt{\mu\varepsilon}}\sqrt{\left(\frac{m\pi}{a}\right)^2+\left(\frac{n\pi}{b}\right)^2}$ **[ESE-2012]**

4.46 When electromagnetic waves are propagated in a waveguide
(a) They travel along the walls of the waveguide
(b) They travel through the dielectric without touching the walls
(c) The are reflected from the walls but do not travel along the walls
(d) None of these

[ESE-2012]

4.47 The minimum cut-off frequencies for a rectangular waveguide are obtained when
(a) $a > b$, $m = 0$ and $n = 1$
(b) $a > b$, $m = 1$ and $n = 0$
(c) $a < b$, $m = 0$ and $n = 1$
(d) $a < b$, $m = 1$ and $n = 0$ **[ESE-2013]**

4.48 A waveguide section in microwave circuits will act as a
(a) low-pass filter (b) band-pass filter
(c) band-stop filter (d) high-pass filter

[ESE-2013]

4.49 The waveguide modes are usually excited from a signal source through
(a) an antenna
(b) an aperture
(c) a coaxial cable
(d) free-space coupling **[ESE-2013]**

4.50 In a waveguide, attenuation near the cut-off frequency is
(a) low (b) high
(c) very high (d) zero **[ESE-2014]**

4.51 The phase velocity of waves propagating in a hollow metal waveguide is
(a) equal to the group velocity
(b) equal to the velocity of light in free space
(c) less than the velocity of light in free space
(d) greater than the velocity of light in free space
[ESE-2014]

4.52 For transverse electric waves between parallel plates, the lowest value of *m*, without making all the field components zero is equal to
(a) 3 (b) 2
(c) 1 (d) 0 [ESE-2014]

4.53 A rectangular waveguide has the dimensions of 5.1 cm × 2.4 cm. For the dominant mode TE_{10} the cut-off frequency is
(a) 2.94 GHz (b) 5.88 GHz
(c) 6.25 GHz (d) 68.99 GHz
[ESE-2014]

4.54 Phase velocity v_p and group velocity v_g in a waveguide are related to the velocity of light *c* as
(a) $v_p v_g = c^2$ (b) $v_p + v_g = c$
(c) $\frac{v_p}{v_g} = \text{constant}$ (d) $v_p + v_g = \text{constant}$
[ESE-2014]

4.55 **Statement (I):** In TE_{mn} modes *z*-component of magnetic field is non-zero.
Statement (II): In TE_{mn} modes *z*-component of electric field is non-zero.
(a) Both Statement (I) and Statement (II) are individually true and Statement (II) is the correct explanation of Statement (I).
(b) Both Statement (I) and Statement (II) are individually true but Statement (II) is not the correct explanation of Statement (I).
(c) Statement (I) is true but Statement (II) is false.
(d) Statement (I) is false but Statement (II) is true.
[ESE-2014]

4.56 The cut-off frequency of a rectangular waveguide in dominant mode is 10 GHz. The width of the wave guide is
(a) 2 cm (b) 1.5 cm
(c) 1 cm (d) 2.5 cm [ESE-2015]

4.57 Consider the following statements for TEM (Transverse Electro Magnetic) waves:
1. Only for TEM wave, could the conductor separation be small compared with a wavelength.
2. TEM mode needs to be considered at low frequencies.
3. TEM waves are usually not obeying Maxwell's equations.
4. TEM waves are usually treated like 'ordinary transmission line signals'.

Which of the above statements are correct?
(a) 1, 2 and 3 only (b) 1 and 2 only
(c) 2, 3 and 4 only (d) 1, 2, 3 and 4
[ESE-2015]

4.58 In a waveguide with perfectly conducting flat wall, the angle of reflection is equal to the angle of
(a) Diffraction (b) Incidence
(c) Refraction (d) Penetration
[ESE-2016]

4.59 In microwave system, waveguides have the advantages of
(a) High power-handling capability and low loss
(b) Thin dielectric substrate
(c) Low power-handling and adequate stability
(d) Positive phase shift [ESE-2016]

4.60 What is the voltage attenuation provided by a 25 cm length of waveguide having *a* = 1 cm and *b* = 0.5 cm in which a 1 GHz signal is propagated in the dominant mode?
(a) 721 dB (b) 681 dB
(c) 521 dB (d) 481 dB [ESE-2016]

4.61 Consider the following statements:
Plane wave propagation through a circular waveguide results in
1. TE modes 2. TM modes

Which of the above statements is/are correct?
(a) 1 only (b) 2 only
(c) Either 1 or 2 (d) Both 1 and 2
[ESE-2017]

4.62 For a WR 90 waveguide, the cut-off frequency for TE_{20} mode is 16 GHz. Then the cut-off frequency for TE_{11} mode will be
(a) $4\sqrt{3}$ GHz (b) $6\sqrt{3}$ GHz
(c) $8\sqrt{5}$ GHz (d) $16\sqrt{5}$ GHz
[ESE-2018]

4.63 During wave propagation in an air-filled rectangular waveguide
1. wave impedance is never less than the free-space impedance.
2. propagation constant is an imaginary number.
3. TEM mode is possible if the dimensions of the waveguide are properly chosen.

Which of the above are correct?
(a) 1 and 2 only (b) 1 and 3 only
(c) 2 and 3 only (d) 1, 2 and 3
[ESE-2018]

4.64 For a standard rectangular waveguide having an aspect ratio of 2 : 1, the cutoff wavelength for $TM_{1,1}$ mode will be nearly
(a) 0.9a (b) 0.7a
(c) 0.5a (d) 0.3a [ESE-2019]

4.65 The irises in the rectangular metallic waveguide may be
1. inductive 2. resistive 3. capacitive

Select the correct answer using the code given below.
(a) 1, 2 and 3 (b) 1 and 2 only
(c) 1 and 3 only (d) 2 and 3 only
[ESE-2019]

4.66 A 10 GHz signal is propagated in a waveguide whose wall separation is 6 cm. The greatest number of half-waves of electric intensity will be possible to establish between the two walls. The guide wavelength for this mode of propagation will be
(a) 6.48 cm (b) 4.54 cm
(c) 2.48 cm (d) 1.54 cm
[ESE-2019]

4.67 In $TE_{m,n}$ mode m and n are integers denoting the number of
(a) $\frac{1}{2}$ the wavelengths of intensity between each pair of walls
(b) $\frac{1}{3}$ the wavelengths of intensity between each pair of walls
(c) $\frac{1}{4}$ the wavelengths of intensity between each pair of walls
(d) $\frac{1}{8}$ the wavelengths of intensity between each pair of walls
[ESE-2019]

4.68 A rectangular waveguide is 5.1 cm by 2.4 cm (inside measurements). The cutoff frequency of the dominant mode will be nearly
(a) 5.38 GHz (b) 4.54 GHz
(c) 3.78 GHz (d) 2.94 GHz
[ESE-2020]

4.69 Consider the following statements regarding the waveguides :
1. Cut-off frequency is the operating frequency below which attenuation occurs and above which propagation takes place.
2. The dominant mode is the mode with the lowest cut-off wavelength.
3. The dominant mode in the rectangular waveguide (for $a > b$) will be TE_{01}.

Which of the above statements is/are ***not*** correct?
(a) 1 and 2 only (b) 1 and 3 only
(c) 2 and 3 only (d) 3 only
[ESE-2021]

4.70 The wave impedance in z direction for TE wave is
(a) $Z_z(TE) = \dfrac{\eta}{\sqrt{\left(1+\dfrac{\omega_c^2}{\omega^2}\right)}}$
(b) $Z_z(TE) = \dfrac{2\eta}{\sqrt{\left(1-\dfrac{\omega_c^2}{2\omega^2}\right)}}$
(c) $Z_z(TE) = \dfrac{\eta}{\sqrt{\left(1-\dfrac{\omega_c^2}{\omega^2}\right)}}$
(d) $Z_z(TE) = \dfrac{2\eta}{\sqrt{\left(1+\dfrac{\omega_c^2}{2\omega^2}\right)}}$
[ESE-2022]

4.71 Which one of the following is the ratio of power lost per unit length to the twice of power transmitted?
(a) Attenuation factor
(b) Power efficiency
(c) Quality factor
(d) Transmission efficiency [ESE-2022]

4.72 A section of X-band waveguide with dimensions a = 2.286 cm and b = 1.016 cm has perfectly conducting walls and is filled with a lossy dielectric $\left(\sigma_d = 367.5\frac{\mu S}{m}, \varepsilon_r = 2.1, \mu_r = 1\right)$. The cutoff frequency of TE_{10} for the dominant mode of propagation at a frequency of 9 GHz is

(a) 2.27 GHz (b) 4.53 GHz
(c) 6.80 GHz (d) 9.07 GHz

[ESE-2022]

4.73 Which of the following modes has the solution of $H_z = 0$, but $E_z \neq 0$?

(a) TEM only (b) TE only
(c) TM only (d) Both TE and TM

[ESE-2023]

■■■

Answers Waveguides

4.1 (b)	4.2 (c)	4.3 (a)	4.4 (d)	4.5 (a)	4.6 (b)	4.7 (a)	4.8 (d)	4.9 (c)
4.10 (a)	4.11 (d)	4.12 (c)	4.13 (b)	4.14 (b)	4.15 (c)	4.16 (a)	4.17 (d)	4.18 (c)
4.19 (c)	4.20 (a)	4.21 (a)	4.22 (d)	4.23 (a)	4.24 (a)	4.25 (d)	4.26 (c)	4.27 (c)
4.28 (b)	4.29 (a)	4.30 (b)	4.31 (b)	4.32 (b)	4.33 (a)	4.34 (d)	4.35 (c)	4.36 (b)
4.37 (b)	4.38 (d)	4.39 (d)	4.40 (a)	4.41 (b)	4.42 (a)	4.43 (b)	4.44 (d)	4.45 (b)
4.46 (c)	4.47 (b)	4.48 (d)	4.49 (c)	4.50 (c)	4.51 (d)	4.52 (c)	4.53 (a)	4.54 (a)
4.55 (c)	4.56 (b)	4.57 (b)	4.58 (b)	4.59 (a)	4.60 (b)	4.61 (d)	4.62 (c)	4.63 (c)
4.64 (a)	4.65 (c)	4.66 (b)	4.67 (a)	4.68 (d)	4.69 (c)	4.70 (c)	4.71 (a)	4.72 (b)
4.73 (c)								

Explanations Waveguides

4.1 (b)

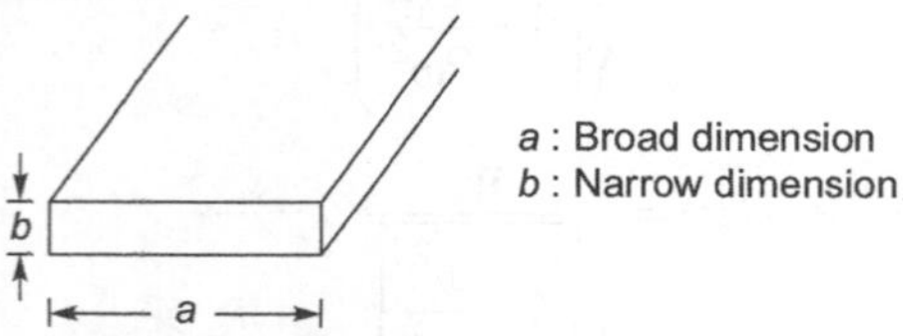

Cut-off frequency of rectangular waveguide, for dominant mode (TE_{10})

$$f_c = \frac{c}{2a}$$

Wave impedance for dominant mode in rectangular waveguide,

$$\eta_{TE} = \frac{\eta}{\sqrt{1-\left(\frac{f_c}{f}\right)^2}}$$

Guided wavelength in rectangular waveguide for dominant mode,

$$\lambda_g = \frac{\lambda}{\sqrt{1-\left(\frac{f_c}{f}\right)^2}}$$

When narrow side dimension decrease neither η_{TE} nor λ_g changes.

Therefore statement (1) and statement (3) is wrong

Narrow side dimension determine attenuation with inverse dependency as $\alpha = k/b$. Attenuation increases with narrow side dimension.

Power handling = Density × ab (area)

This decreases with narrow side dimension.

4.2 (c)

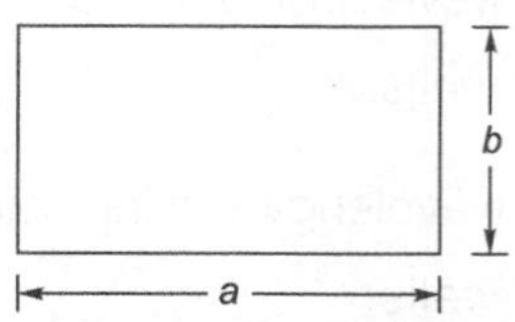

Cutoff wavelength in a rectangular waveguide for TE_{mn} mode is

$$\lambda_c = \frac{2}{\sqrt{\left(\frac{m}{a}\right)^2 + \left(\frac{n}{b}\right)^2}}$$

For TE_{20}, $$\lambda_c = \frac{2}{\sqrt{\left(\frac{2}{a}\right)^2 + 0}} = \frac{2a}{2} = a$$

4.3 (a)

$$v_p = \frac{c}{\cos\theta} \quad v_g = c \cdot \cos\theta$$

where θ = path angle inside the guide.

$$v_p \cdot v_g = c^2$$

4.4 (d)

A waveguide carries electromagnetic wave power which is displacement current.

4.5 (a)

(i) Guided wavelength, $$\lambda_g = \frac{\lambda_0}{\sqrt{1-\left(\frac{\lambda_0}{\lambda_c}\right)^2}}$$

where, λ_0 = free space wavelength

λ_c = cut-off wavelength

(ii) Wave impedance for TM mode,

$$\eta_{TM} = \eta\sqrt{1-\left(\frac{\lambda_o}{\lambda_c}\right)^2}$$

where, η = free space impedance

(iii) TEM mode is never possible in rectangular waveguide or single conductor guides.

(iv) Above cut-off frequency, γ becomes imaginary and wave propagates.

4.6 (b)

In a waveguide, the propagating wave is either transverse electric or transverse magnetic. Since an extra electric component is in the direction of propagation, it can't be TE mode. Therefore, it is TM mode.

4.8 (d)

$$a = 4.5 \text{ cm}$$

Cut off wavelength for TE_{10} mode is

$$\lambda_c = 2a = 2 \times 4.5 = 9 \text{ cm}$$

4.9 (c)

The cut-off wavelength of a dominant mode in a rectangular waveguide depends only on its width, not on height.

4.10 (a)

The dominant mode of a rectangular waveguide is TE_{10}.

4.11 (d)

Micro-strip line is a asymmetric structure and has both EM waves and VI waves.

It has heavy radiation losses, hence the mode is called as Quasi TEM mode.

4.12 (c)

Phase velocity, $$v_p = \frac{c}{\sqrt{1-\left(\frac{f_c}{f}\right)^2}}$$

where , c = velocity of wave in free space (i.e. 3×10^8 m/s).

f = operating frequency

f_c = cut-off frequency.

4.13 (b)

A waveguide operated below cut-off frequency can be used as an attenuator where the cut-off attenuation constant α is given by

$$\alpha = \frac{54.6}{\lambda_c}\sqrt{1-\left(\frac{f}{f_c}\right)^2} \text{ dB/length}$$

4.15 (c)

The cut-off frequency of the dominant mode of a rectangular waveguide is

$$f_c = \frac{c}{2a}$$

where a is the dimension of inner broad wall.

$$\Rightarrow \quad a = \frac{3\times10^{10}}{2\times10\times10^9} = 1.5 \text{ cm}$$

4.16 (a)

In waveguide, the evanescent or non-propagating modes are those modes for which waves have $f/f_c < 1$.

Propagation constant is real for evanescent modes.

4.17 (d)

Scattering parameters have waveforms at the input and output as dependent and independent variables.

This suit for high frequency guide and junction.

4.18 (c)

$$\frac{m\pi}{a} = 60\pi \quad \text{with } m = 3$$

$$\frac{n\pi}{b} = 100\pi \quad \text{with } n = 3$$

The lowest TM mode is TM_{11}.

4.19 (c)

Cutoff frequency for TE_{10} mode is

$$f_c = \frac{3\times10^{10}}{2}\sqrt{\left(\frac{1}{a\sqrt{3}}\right)^2 + 0}$$

$$= 2 \text{ GHz}$$

$$\Rightarrow \quad \frac{3\times10^{10}}{2\sqrt{3}\,a} = 2\times10^9$$

$$\Rightarrow \quad a = \frac{10\sqrt{3}}{4} = \frac{5\sqrt{3}}{2}\text{ cm}$$

Cutoff frequency for TM_{11} mode is

$$f_c = \frac{3\times10^{10}}{2}\sqrt{\left(\frac{1}{a\sqrt{3}}\right)^2 + \left(\frac{1}{a}\right)^2}$$

$$= \frac{3\times10^{10}}{2}\cdot\frac{2}{a\sqrt{3}}$$

$$= \frac{3\times10^{10}\times2}{5\sqrt{3}\,.\,\sqrt{3}} = 4 \text{ GHz}$$

4.21 (a)

The dominant mode in a rectangular waveguide is TE_{10}.

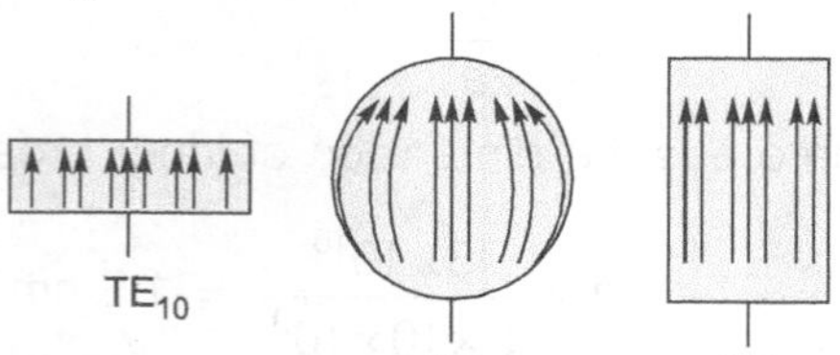

TE_{10} TE_{11} TE_{10}

A B C

Due to changing orientation of the guide, the field orientation does not change, but *a* becomes smaller dimension than *b*.

4.23 (a)

Cut-off frequency

$$f_c = \frac{c}{2\sqrt{\epsilon_r}\,a} = \frac{3\times10^8}{2\sqrt{4}\times1.5\times10^{-2}} = 5 \text{ GHz}$$

Therefore, the 8 GHz signal will pass through the waveguide.

4.24 (a)

A waveguide carries electromagnetic wave power which is displacement current.

4.25 (d)

Open wire due to heavy radiation is preferred only for kHz and audio frequencies.

Waveguides are used for micrwave frequency.

Co-axial cable is used for radio frequency.

4.26 (c)

In a rectangular waveguide, when the wave travels longitudinally down the guide, the plane waves are reflected from the wall to wall. This process results in a component of either direction of propagation of the resultant wave. Therefore, the wave cannot be TEM wave.

Comparison of modes in transmission lines and waveguide

Characteristic	Coaxial cable	Waveguide	Stripline	Microstripline
Modes : Preferred	TEM	TE_{10}	TEM	Quasi-TEM
Other	TE, TM	TM, TE	TM, TE	Hybrid TM, TE

4.27 (c)

Cut-off wavelength for TE_{01} mode,

$$\lambda_c = 2b = 2\times1.016 = 2.032 \text{ cm}$$

4.28 (b)

In a waveguide, the propagating wave is either transverse electric or transverse magnetic. Since an extra electric component is in the direction of propagation, it can't be TE mode. Therefore, it is TM mode.

4.29 (a)

Dominant mode is that mode for which the cut-off wavelength (λ_c) assumes a maximum value.

4.30 (b)

For dominant mode, i.e. TE_{10} mode cut-off

$$\text{frequency, } f_c = \frac{c}{3\times10^{-2}\times2}$$

$$= \frac{3\times10^8}{2\times3\times10^{-2}} = 5 \text{ GHz}$$

At 11 GHz even TE_{20} can also propagate.

At 7 GHz there are degenerate mode TE_{11} and TM_{11}.

4.31 (b)

(i) In evanescent mode, waveguide is excited under cut-off frequency. Therefore there is no wave propagation.

(ii) The dominant mode in a particular waveguide is the mode having the lower cut-off frequency.

(iii) In TM_{00}, TM_{01} and TM_{10} mode, all field components (E_x, E_y, H_x, H_y) vanish. Hence, these modes do not exist.

4.32 (b)

In TM_{00}, TM_{01} and TM_{10} modes, all field components (E_x, E_y, H_x, H_y) vanish. Hence, these modes do not exist.

4.34 (d)

Cut-off frequency of waveguide,

$$f_c = \frac{c}{\lambda} = \frac{c}{2a}$$

$$f_c = \frac{3\times10^8}{2\times3\times10^{-2}} = 5 \text{ GHz}$$

Therefore, the given waveguide can't support propagation at 4 GHz.

4.35 (c)

For plane wave propagating in free space or two conductor transmission line, $v_p = c = v_g$

Note: For plane wave propagating in waveguide, $v_p > c > v_g$.

4.36 (b)

H-field always closed loops even in TE/TM forms which needs *E* field in any direction.

4.37 (b)

1. For TE_{10} mode

$$f_{c_1} = \frac{v_0}{2a}$$

For TE_{01} mode

$$f_{c_2} = \frac{v_0}{2b}$$

If $b << a$;

then $f_{c_2} >> f_{c_1}$

Hence, small value of '*b*' gives a larger separation between cut-off frequencies of TE_{10} and TE_{01} modes.

2. Attenuation constant

$$\alpha = \frac{R_s}{b\eta_0} \text{ Np/length}$$

where, R_s = sheet resistivity in Ω/m^2

η_0 = intrinsic impedance of free space (377 Ω)

$$\therefore \quad \boxed{\alpha \propto \frac{1}{b}}$$

Hence a smaller value of '*b*' will result in increased attenuation.

3. $E = V/b$

From above relation it is clear that a small value of '*b*' limits power handling capabilities because of breakdown field limits.

4.38 (d)

Transverse Electromagnetic (TEM) wave will propagate between two perfectly conducting parallel plates of infinite extent.

4.39 (d)

Given, a = 4.755 cm

b = 2.215 cm

$$f_c = \frac{v_0}{2\pi}\sqrt{\left(\frac{m\pi}{a}\right)^2 + \left(\frac{n\pi}{b}\right)^2}$$

for TE_{10} mode

$$f_{c_1} = \frac{v_0}{2\pi}\times\frac{\pi}{a} = \frac{v_0}{2a} = \frac{3\times10^{10}}{2\times4.755}$$

$$\boxed{f_{c_1} = 3.15 \text{ GHz}}$$

for TE_{20} mode

$$f_{c_2} = \frac{v_0}{2\pi}\sqrt{\left(\frac{2\pi}{a}\right)^2}$$

$$= \frac{v_0}{a} = \frac{3\times10^{10}}{4.755}$$

$$\boxed{f_{c_2} = 6.3 \text{ GHz}}$$

for TE_{01} mode

$$f_{c_3} = \frac{v_0}{2\pi}\sqrt{\left(\frac{n\pi}{b}\right)^2}$$

$$= \frac{v_0}{2\pi}\cdot\frac{\pi}{b} = \frac{v_0}{2b}$$

$$f_{c_3} = \frac{3\times10^{10}}{2\times2.215}$$

$$\boxed{f_{c_3} = 6.7 \text{ GHz}}$$

for TE_{11} mode

$$f_{c_4} = \frac{v_0}{2\pi}\sqrt{\left(\frac{\pi}{a}\right)^2 + \left(\frac{\pi}{b}\right)^2}$$

$$f_{c_4} = \frac{3\times10^{10}}{2}\cdot\sqrt{\left(\frac{1}{4.755}\right)^2 + \left(\frac{1}{2.2155}\right)^2}$$

$$\boxed{f_{c_4} = 7.47 \text{ GHz}}$$

Hence at 12 GHz waveguide WR-187 will support all above modes.

4.40 (a)

Given, $a = 6$ cm; $b = 4$ cm

as $$f_c = \frac{v_0}{2\pi}\sqrt{\left(\frac{m\pi}{a}\right)^2 + \left(\frac{n\pi}{b}\right)^2}$$

(A) TE_{10}

$\therefore \quad m = 1; n = 0$

$$f_c = \frac{v_0}{2\pi}\sqrt{\left(\frac{\pi}{a}\right)^2 + 0}$$

$$\Rightarrow \quad f_c = \frac{3\times10^{10}}{2\times6} = 2.5 \text{ GHz}$$

(B) TE_{01}

$\therefore \quad m = 0; n = 1$

$$f_c = \frac{v_0}{2\pi}\sqrt{0 + \left(\frac{\pi}{b}\right)^2}$$

$$\Rightarrow \quad f_c = \frac{3\times10^{10}}{2\times4} = 3.75 \text{ GHz}$$

(C) TE_{11}

$\therefore \quad m = 1; n = 1$

$$f_c = \frac{v_0}{2\pi}\sqrt{\left(\frac{\pi}{a}\right)^2 + \left(\frac{\pi}{b}\right)^2}$$

$$\Rightarrow \quad f_c = \frac{v_0}{2}\sqrt{\frac{1}{a^2} + \frac{1}{b^2}}$$

$$f_c = \frac{3\times10^{10}}{2}\sqrt{\frac{1}{36} + \frac{1}{16}} = 4.506 \text{ GHz}$$

(D) TM_{11}

$\therefore \quad m = 1; n = 1$

$$f_c = \frac{3\times10^{10}}{2}\sqrt{\frac{1}{36} + \frac{1}{16}} = 4.506 \text{ GHz}$$

4.42 (a)

$$\eta = \frac{E}{H} = \text{waveguide impedance}$$

4.43 (b)

$$a = 8 \text{ cm}, \quad b = 4 \text{ cm}$$

$$\lambda_c = \frac{2\pi}{\sqrt{\left(\frac{m\pi}{a}\right)^2 + \left(\frac{n\pi}{b}\right)^2}}$$

For TE_{10} mode

$$\lambda_c = 2\times 8$$

$$\therefore \quad \lambda_c = 16 \text{ cm}$$

4.45 (b)

Cut-off frequency,

$$f_c = \frac{v_0}{2\pi}\sqrt{\left(\frac{m\pi}{a}\right)^2 + \left(\frac{n\pi}{b}\right)^2}$$

→ For free space $v_o = c$

If the medium is not a free space, then

$$v_o = \frac{1}{\sqrt{\mu\epsilon}}$$

Therefore $$f_c = \frac{1}{2\pi\sqrt{\mu\epsilon}}\sqrt{\left(\frac{m\pi}{a}\right)^2 + \left(\frac{n\pi}{b}\right)^2}$$

4.46 (c)

Electromagnetic waves propagated in waveguide are reflected from the walls but do not travel along them because the waveguides are made up of conductors.

4.47 (b)

For standard waveguide $a > b$ and in this case minimum cut-off frequency is given by TE_{10} mode which is given as below

$$f_c = \frac{1}{2a\sqrt{\mu\epsilon}}$$

for $a > b$ and $m = 1, \; n = 0$

4.48 (d)

A waveguide acts as high-pass filter because it allows the transmission of electromagnetic wave above a certain frequency called cut-off frequency given as

$$f_c = \frac{v_0}{2\pi}\sqrt{\left(\frac{m\pi}{a}\right)^2 + \left(\frac{n\pi}{b}\right)^2}$$

4.49 (c)

The waveguide are usually excited from a signal source cable. The outer conductor of the cable makes a 360° connection with the body of the waveguide and the central conductor is projected inside the guide either as a small probe or forming a small loop as shown in figure. All TE_{mo} modes with an odd m have a maximum field at centre of the cross-section in transverse direction. These modes can be excited by a single probe

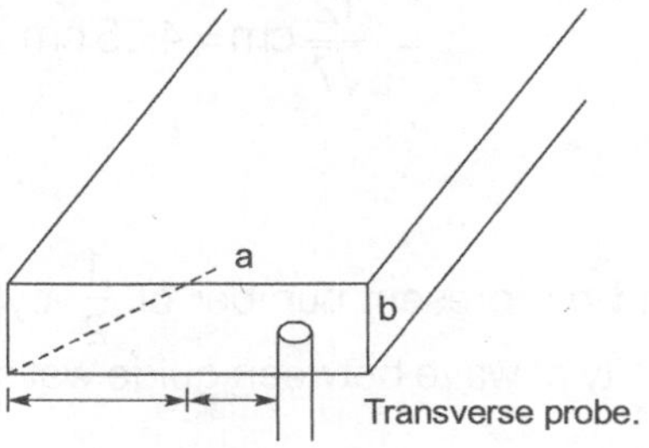

4.50 (c)

The attenuation near the cutoff is very high and it is nearly equal to ∞.

4.51 (d)

For wave propagation in waveguide,

$$v_p > c > v_g$$

4.52 (c)

For transverse electric waves the dominant mode for rectangular waveguide is TE_{10}

∴ $m = 1$ and $\lambda = 2a$

4.53 (a)

The cutoff frequency for dominant mode TE_{10} is

$$f_c = \frac{c}{2a}$$

where, $a = 5.1$ cm (given)

and $c = 3 \times 10^8$ m/s $= 3 \times 10^{10}$ cm/s

$$\therefore \quad f_c = \frac{3\times10^{10}}{2\times5.1} = 2.94\,\text{GHz}$$

4.55 (c)

In TE_{MN} (Transverse Electric Mode), the z-component of electric field is zero.

4.56 (b)

f_c for TE_{10} is 10 GHz

$$f_c = \frac{c}{2a}$$

$$10 \times 10^9 = \frac{3\times10^{10}}{2a}$$

$$\Rightarrow \quad a = \frac{3\times10^{10}}{10^{10}\times2} = \frac{3}{2} = 1.5 \text{ cm}$$

4.57 (b)

1. Guide dimensions and operational wave length have no relationship in TEM waves.
2. TEM waves have no cut-frequency.
3. TEM waves obey Maxwell's equations.
4. TEM waves are electromagnetic in nature unlike transmission line signals, which are *VI* waves.

4.60 (b)

For dominant mode,

$$f_{c10} = \frac{c}{2a} = \frac{3\times10^8}{2\times(1\times10^{-2})} = 15\,\text{GHz}$$

Given operating frequency, $f = 1$ GHz

When $f < f_c$,

$$\gamma = \alpha + j0$$

where, $$\alpha = \omega\sqrt{\mu_o\epsilon_o}\sqrt{\left(\frac{f_{c10}}{f}\right)^2 - 1}$$

$$= \frac{2\pi(10^9)}{3\times10^8}\sqrt{\left(\frac{15}{1}\right)^2 - 1} = 313.46$$

Voltage attenuation in

$$\text{dB} = 20\log_{10} e^{(313.46)(25\times10^{-2})} = 681 \text{ dB}$$

4.61 (d)

In Circular Waveguide TE (transverse electric) and TM (transverse magnetic) modes exist.

4.62 (c)

$$f_{c20} = 16 \times 10^9 \text{ Hz}$$

Find f_{c11} for WR 90 waveguide $\frac{a}{b} = 2$

$$f_{c20} = \frac{c}{2}\sqrt{\left(\frac{2}{a}\right)^2}$$

$$\Rightarrow \quad f_{c20} = \frac{c}{2}\left(\frac{2}{a}\right)$$

$$16\times10^9 = \frac{3\times10^8}{a}$$

$$\Rightarrow \quad a = \frac{3\times10^8}{16\times10^9} = 1.875 \text{ cm}$$

$$\Rightarrow \quad f_{c_{11}} = \frac{c}{2}\sqrt{\frac{1}{a^2}+\frac{1}{b^2}}$$

$$f_{c_{11}} = \frac{c}{2a}\sqrt{1+\left(\frac{a}{b}\right)^2}$$

$$f_{c_{11}} = \frac{3\times10^8}{2\times(1.875\times10^{-2})}\sqrt{1+(2)^2}$$

$$= 8\times\sqrt{5}\text{ GHz}$$

4.63 (c)

For TE mode $\eta_g = \dfrac{\eta_0}{\sqrt{1-\left(\frac{f_c}{f}\right)^2}}$ $\eta_g > \eta_0$

For TM mode $\eta_g = \eta_0\sqrt{1-\left(\frac{f_c}{f}\right)^2}$; $\eta_g < \eta_0$

So, statement 1 is wrong.

4.64 (a)

$$f_{c_{mn}} = \frac{c}{2}\left(\sqrt{\left(\frac{m}{a}\right)^2+\left(\frac{n}{b}\right)^2}\right) \quad \text{for TM}_{11}$$

$$\frac{a}{b} = \frac{2}{1} \Rightarrow b = \frac{a}{2}$$

$$f_{c_{11}}\lambda_{c_{11}} = c$$

$$\Rightarrow \quad \lambda_{c_{11}} = \frac{c}{f_{c_{11}}}$$

$$\lambda_c = \frac{2}{\sqrt{\left(\frac{m}{a}\right)^2+\left(\frac{n}{b}\right)^2}} = \frac{2}{\sqrt{\left(\frac{1}{a}\right)^2+\left(\frac{2}{a}\right)^2}} = \frac{2a}{\sqrt{5}}$$

$$\therefore \quad \lambda_c = 0.9\,a$$

4.65 (c)

Iris is effectively an obstruction within a waveguide that provides a capacitance or inductive element.

4.66 (b)

Given $a = 6$ cm

$$f_{c_1} = \frac{c}{2a} = \frac{3\times10^8}{2\times6\times10^{-2}}$$

$$= 2.5\times10^9 = 2.5\text{ GHz}$$

$$f_{c_2} = 2f_{c_1} = 5\text{ GHz}$$

$$f_{c_3} = 3f_{c_1} = 7.5\text{ GHz}$$

$$f_{c_4} = 10\text{ GHz does not exist}$$

Highest mode $m = 3$

$$\lambda_3 = \frac{\lambda}{\sqrt{1-\left(\frac{7.5}{10}\right)^2}} = \frac{3}{\sqrt{\frac{16-9}{16}}}\text{ cm}$$

$$= \frac{12}{\sqrt{7}}\text{ cm} = 4.55\text{ cm}$$

4.67 (a)

m and n represent number of $\frac{1}{2}$ cycles in the intensity of wave between guide walls.

4.68 (d)

In rectangular waveguide $a = 5.1$ cm : $b = 2.4$ cm

Find $f_{c_{10}}$: $f_{c_{10}} = \dfrac{c}{2a} = \dfrac{3\times10^8}{2(5.1\times10^{-2})} = \dfrac{3}{2(5.1)}\times10^{10}$

$= 2.94$ GHz

4.69 (c)

- The dominant mode is the mode with lowest cut-off frequency i.e., the highest cut-off wavelength.
- For $a > b$, the dominant mode in the rectangular waveguide is TE_{10} with cut-off frequency,

$$f_c = \frac{c}{2a}$$

which is less than the cut-off frequency of TE_{01} mode with $f_c = \dfrac{c}{2b}$.

Hence, statements 2 and 3 are incorrect.

4.70 (c)

For a TE wave, the wave impedance is given as

$$Z_z(\text{TE}) = \eta_{TE} = \frac{\eta_{TEM}}{\sqrt{1-\left(\frac{f_c}{f}\right)^2}}$$

$$\Rightarrow \quad \eta_{TE} = \frac{\eta_{TEM}}{\sqrt{1-\left(\frac{\omega_C}{\omega}\right)^2}}$$

4.71 (a)

The ratio of the power loss per unit length to twice the power transmitted through the guide is called as attenuation constant.

Mathematically, $\alpha = \dfrac{P_L}{2P_{tr}}$

where, P_L = Power loss per unit length
P_{tr} = Power transmitted

4.72 (b)

Given data: $a = 2.286$ cm; $b = 1.016$ cm;
$\sigma_d = 367.5\ \mu$S/m; $\epsilon_r = 2.1$; $\mu_r = 1$

The cut-off frequency for the TE_{mn} mode in rectangular waveguide is given as

$$f_c = \frac{c}{2\sqrt{\mu_r \epsilon_r}}\sqrt{\left(\frac{m}{a}\right)^2 + \left(\frac{n}{b}\right)^2}$$

For TE_{10} mode,

$$f_c = \frac{c}{2a\sqrt{\mu_r \epsilon_r}}$$

Substituting the values, we get,

$$f_c = \frac{3\times 10^{10}}{2\times\sqrt{2.1}\times 2.286}$$

$$f_c = 4.53 \times 10^9 \text{ Hz} = 4.53 \text{ GHz}$$

4.73 (c)

For a rectangular waveguide $H_z = 0$, $E_z \neq 0$, hence it is TM - mode.

■■■

5 Antennas and Radar

5.1 An altimeter is basically
(a) a CW radar
(b) a FM radar
(c) a Doppler radar
(d) a device to indicate the direction at height
[ESE-2000]

5.2 An antenna has 40 Ω antenna resistance and 60 Ω radiation resistance. The efficiency of the antenna is
(a) 30% (b) 40%
(c) 50% (d) 60%
[ESE-2000]

5.3 **Assertion (A):** A knowledge of the amplitude and the phase of the field at each point at the mouth of a horn antenna may be used to predict the expected radiation pattern.
Reason (R): The electromagnetic field at any point in space may be obtained by considering each point on a wavefront as a source of spherical waves and superimposing the effect of all those sources at that point.
(a) Both A and R are true and R is the correct explanation of A
(b) Both A and R are true but R is NOT the correct explanation of A
(c) A is true but R is false
(d) A is false but R is true
[ESE-2000]

5.4 Match **List-I** (Antennas) with **List-II** (Radiation patterns) and select the correct answer:

List-I	List-II
A. Simple dipole	1. 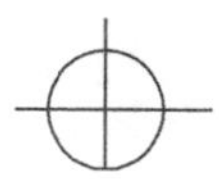
B. Omni-directional antenna	2. 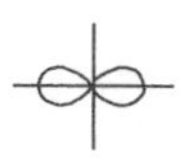
C. Loop antenna	3. 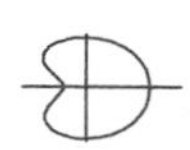

Codes:

	A	B	C
(a)	1	2	3
(b)	2	1	1
(c)	3	2	1
(d)	1	1	2

[ESE-2000]

5.5 A vertical wire of 1m length carries a current of 1 A at 10 MHz. The total radiated power is nearly
(a) 0.13 W (b) 0.88 W
(c) 7.3 W (d) 73 W [ESE-2000]

5.6 A Yagi antenna has a driven antenna
(a) only with a director
(b) only with a reflector
(c) with one or more directors
(d) with a reflector and one or more directors
[ESE-2000]

5.7 For frequencies up to 1650 kHz, the transmitting antenna used is a
(a) parabolic dish (b) vertical antenna
(c) Yagi antenna (d) turnstile antenna
[ESE-2001]

5.8 The radiation field of an antenna at a distance r varies as
(a) $1/r$ (b) $1/r^2$
(c) $1/r^3$ (d) $1/r^4$ [ESE-2001]

5.9 The wave radiated by a helical antenna is
(a) linearly polarized
(b) right circularly polarized
(c) left circularly polarized
(d) elliptically polarized [ESE-2001]

5.10 **Assertion (A):** For an end-fire array, the current in successive antennas must lag in phase.
Reason (R): Radiation of successive antennas will cancel along the axis.
(a) Both A and R are true and R is the correct explanation of A
(b) Both A and R are true but R is NOT the correct explanation of A
(c) A is true but R is false
(d) A is false but R is true [ESE-2001]

5.11 For a parabolic reflector antenna with diameter of 3 m, the far field pattern measurement at 10 GHz should be carried out at a distance of at least
(a) 30 m (b) 200 m
(c) 400 m (d) 600 m [ESE-2001]

5.12 For producing circularly polarized beams in microwave communication field, the type of antenna ideally suited is
(a) helical antenna
(b) parabolic disc with circular aperture
(c) pyramidal horn with symmetrical beam shapes in *E* and *H* plane
(d) circular loop antenna [ESE-2001]

5.13 Given figure shows a 90° sector of spherical reflector with vertex '*V*', centre of curvature '*C*' and equal distances *VD*, *DF*, *FE* and *EC*. When a beam of electromagnetic wave is incident parallel to the axis, all the energy will pass through the region

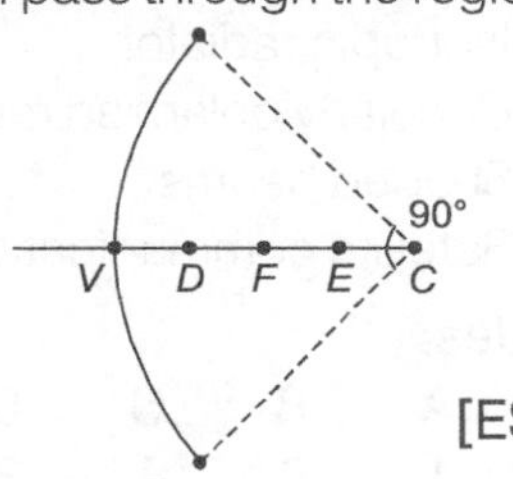

(a) *VD*
(b) *DF*
(c) *FE*
(d) *EC* [ESE-2001]

5.14 A radio station works at 800 kHz and uses AM. If this is a public broadcast system, it should transmit using
(a) parabolic reflector to transmit all round
(b) turnstile antenna for the required band
(c) half-wave long horizontal wire
(d) a vertical antenna less than quarter wavelength for practical reasons
[ESE-2001]

5.15 The false alarm probability (p_{fa}) in a radar is found to be 0.05 under conditions like large prf and pulse integration. Then the detection probability
(a) is independent of p_{fa}
(b) is equal to $(1-\sqrt{0.05})$
(c) is equal to $(1 - 0.05)$
(d) depends on the power received in echo
[ESE-2001]

5.16 Which of the following is/are the hyperbolic system(s) of the navigation?
1. VOR 2. LORAN
3. DECCA 4. TACAN
Select the correct answer using the codes given below:
(a) 1 and 2 (b) 2 alone
(c) 2 and 3 (d) 1 and 4
[ESE-2001]

5.17 A received signal in a Radar system occupies an IF bandwidth of 12 MHz. If the transmitted signal was a pulsed carrier of 10 GHz with duty ratio 1:0.5 at 400 Hz, the received signal power will contain the
(a) frequency component at 400 Hz
(b) frequency spread over the whole IF bandwidth
(c) dc or average value of the IF
(d) RF signals at 10 GHz and 12 MHz
[ESE-2001]

5.18 If the average power of radar transmitter is 2 kW and the peak power of the transmitter is 1000 kW, what will be the duty cycle?
(a) 0.002 (b) 0.02
(c) 0.001 (d) 0.01 [ESE-2002]

5.19 A radar receives an echo from a target 20 micro second after sending the signal. The approximate range of the target is
(a) 300 m (b) 3000 m
(c) 600 m (d) 6000 m [ESE-2002]

5.20 Doppler effect is employed in which of the following?
1. Moving target plotting on PPI
2. The MTI system
3. FM radar
4. CW radar
Select the correct answer using the codes given below:
(a) 2, 3 and 4 (b) 1, 2 and 3
(c) 1, 2 and 4 (d) 1, 3 and 4
[ESE-2002]

5.21 Match **List-I** with **List-II** and select the correct answer using the codes given below the lists:

List-I	List-II
A. Pulsed radar system	**1.** Phased array radar
B. Moving target radar	**2.** Doppler radar
C. CW radar	**2.** Magnetrons are in use
D. Phase shifter	**4.** Search radar

Codes:

	A	B	C	D
(a)	4	2	3	1
(b)	3	1	4	2
(c)	4	1	3	2
(d)	3	2	4	1

[ESE-2002]

5.22 Which of the following can be used to improve range resolution in a Radar?
(a) Short duration pulse
(b) Long duration pulse
(c) High frequency of operating signal
(d) Increasing pulsewidth [ESE-2002]

5.23 The type of antenna to be used for producing circularly polarized beams is
(a) pyramidal horn (b) log-periodic array
(c) paraboloid (d) helical antenna
[ESE-2002]

5.24 The current distribution along a travelling wave antenna can be written in the form
(a) $I(z) = I_0 e^{-j\beta z}$ (b) $I(z) = I_0 \sin\beta z$
(c) $I(z) = I_0$ (d) $I(z) = I_0 \cos(\omega t - \beta z)$
[ESE-2002]

5.25 A dipole produces an electric field intensity of 1 mV/m at a distance of 2 km. The field intensity at a distance of 4 km will be
(a) 1 mV/m (b) 0.75 mV/m
(c) 0.50 mV/m (d) 0.25 mV/m
[ESE-2002]

5.26 Following antenna is frequently used for local area transmission at UHF/VHF
(a) Ground monopole
(b) Turnstile antenna
(c) Slot antenna
(d) Loop antenna [ESE-2002]

5.27 In a four element Yagi-Uda antenna
(a) There is one driven element, one director and two reflectors
(b) There is one driven element, two directors and one reflector
(c) There are two driven elements, one director and two reflectors
(d) All the four elements are driven elements
[ESE-2003]

5.28 **Assertion (A):** For extremely high frequency ranges or above, compared to linear antennas, aperture antennas are more useful.
Reason (R): The larger the effective area of an antenna, the sharper is the radiated beam.
(a) Both A and R are true and R is the correct explanation of A
(b) Both A and R are true but R is NOT the correct explanation of A
(c) A is true but R is false
(d) A is false but R is true [ESE-2003]

5.29 Typical *f*/*D* ratio in practice for a paraboloid reflector (where f is the focal length and *D* is the aperture diameter) is in the range of
(a) 0.2 to 0.3 (b) 2 to 3
(c) 0.02 to 0.03 (d) 0.5 to 1 [ESE-2003]

5.30 A parabolic dish antenna has a diameter of 1 m. The maximum possible (ideal) gain of the antenna at a wavelength of 3.14 cm is
(a) 20 dB (b) 30 dB
(c) 40 dB (d) 50 dB [ESE-2003]

5.31 Match **List-I** with **List-II** and select the correct answer using the codes given below the lists:

List-I
A. Helical antennas
B. Dipoles
C. Parabolic dish
D. Array antennas

List-II
1. Isotropic radiator
2. Circularly polarized radiation
3. Shaped beams
4. Satellite communication systems

Codes:

	A	B	C	D
(a)	1	2	4	3
(b)	1	2	3	4
(c)	2	1	4	3
(d)	2	1	3	4

[ESE-2004]

5.32 Which is the most common antenna to obtain a predetermined radiation pattern?
(a) Array antenna (b) Corner reflector
(c) Sectoral horn (d) Helical antenna
[ESE-2004]

5.33 What is the beamwidth of a pyramidal horn antenna with aperture dimension of 9 cm × 8 cm and operating at a frequency of 5 GHz?
(a) 7.2° (b) 10°
(c) 15° (d) 16.4° [ESE-2004]

5.34 Match **List-I** (Characteristic to be Obtained) with **List-II** (Arrangement of Radiators) and select the correct answer using the codes given below:

List-I
A. Concentration of radiation in azimuth
B. Concentration of radiation in elevation
C. Bi-directional pattern

List-II
1. Two radiators spaced 3λ/4 apart and fed with currents having 180° phase difference

2. Two radiators spaced $\lambda/4$ apart and fed with currents having 90° phase difference
3. Radiators arranged to form broadside array with horizontal array line
4. Radiators arranged to form broadside array with vertical array line

Codes:

	A	B	C
(a)	2	4	3
(b)	3	4	2
(c)	3	1	2
(d)	4	3	1

[ESE-2004]

5.35 **Assertion (A):** In free-space transmission between a transmitting and a receiving antenna, a high frequency wave decreases more rapidly in magnitude than a low frequency wave.

Reason (R): For a given gain, the aperture of a high frequency antenna is smaller than that of a low frequency antenna.

(a) Both A and R are true and R is the correct explanation of A
(b) Both A and R are true but R is NOT the correct explanation of A
(c) A is true but R is false
(d) A is false but R is true

[ESE-2004]

5.36 Which one of the following antenna structures is best for generating a circularly polarized radiation?

(a) Helical antenna
(b) Log-periodic antenna
(c) Rhombic antenna
(d) Dipole antenna [ESE-2005]

5.37 On which of the following parameters does the maximum range of detection (by radar) for an isolated objection space depend?

1. Wavelength (as $\sqrt{\lambda}$)
2. Minimum detectable echo pulse power
3. Antenna gain (as reciprocal of gain)
4. Transmitted power
5. Radar cross section of the target

Select the correct answer using the code given below:

(a) 1, 2 and 3 (b) 2, 4 and 5
(c) 3, 4 and 5 (d) 1, 3 and 5

[ESE-2005]

5.38 To avoid grating lobe at any arbitrary angle of scan using a uniformly spaced array to identical radiators with inter-element spacing '*d*' and operating wavelength 'λ', which one of the following conditions should be satisfied?

(a) $d < 0.5\lambda$ (b) $d > \lambda$
(c) $0.7\lambda \le d < 2\lambda$ (d) $0.5\lambda < d < 2\lambda$

[ESE-2005]

5.39 A uniformly spaced linear array of identical radiators having uniform amplitude of excitation and linear phase variation with non zero gradient will produce

(a) Pencil beam at broadside
(b) Fan beam
(c) Scanned cosecant beam
(d) Scanned pencil beam [ESE-2005]

5.40 For electronic scanning, the beam width with scan angle θ_0 away from broadside is proportional to

(a) $\sec\theta_0$ (b) $\cos\theta_0$
(c) $\tan\theta_0$ (d) $\text{cosec}\theta_0$ [ESE-2005]

5.41 Which one of the following concepts is used to find the expression of radiated $\vec{E}$ and $\vec{H}$ field due to a magnetic current element?

(a) Concept of vector magnetic potential
(b) Concept of scalar electric potential
(c) Concept of scalar magnetic potential
(d) Concept of vector electric potential

[ESE-2005]

5.42 Match **List-I** (Antenna) with **List-II** (Property) and select the correct answer using the codes given below the lists:

List-I
A. Log periodic antenna
B. Helical antenna
C. Microstrip antenna
D. Dipole antenna

List-II
1. Circular polarization
2. Wide bandwidth
3. Omni directional pattern
4. Compact and light weight

Codes:

	A	B	C	D
(a)	3	1	4	2
(b)	2	4	1	3
(c)	3	4	1	2
(d)	2	1	4	3

[ESE-2005]

5.43 Match **List-I** with **List-II** and select the correct answer using the codes given below the lists:

List-I

A. Helical antenna
B. Pyramidal horn
C. Microstrip patch
D. Sectoral horn

List-II

1. Fan-shaped beam, high power handling
2. Low bandwidth, low antenna power handling
3. Narrow beam, high antenna power handling
4. Circularly polarised antenna beam, moderate power handling

Codes:

	A	B	C	D
(a)	4	3	2	1
(b)	2	1	4	3
(c)	4	1	2	3
(d)	2	3	4	1

[ESE-2005]

5.44 For a parabolic reflector of 5 metre diameter, the far field pattern measurement at 6 GHz should be carried out a distance of at least

(a) 0.5 km (b) 1.0 km
(c) 1.5 km (d) 2.0 km

[ESE-2005]

5.45 A dispersive channel is one in which

(a) Phase velocity of signal depends upon frequency
(b) Phase velocity of signal depends upon its electric field intensity in the medium
(c) Phase velocity of signal depends upon its magnetic field intensity in the medium
(d) Phase velocity of signal depends upon its amplitude in the medium

[ESE-2005]

5.46 A pyramidal horn antenna is best suited to be excited from a

(a) Microstrip line (b) Microwave waveguide
(c) Coaxial line feed (d) Parallel twin line

[ESE-2006]

5.47 A paraboloid of revolution is to have a power gain of 1000 at λ = 10 cm. What is the diameter of the antenna?

(a) 129.1 m (b) 12.91 m
(c) 1.291 m (d) 0.1291 m

[ESE-2006]

5.48 A half-wave dipole radiating at 100 MHz is situated 50 m above the ground. A similar receiving antenna is situated 20 km away at a height of 20 m above the ground. If the transmitted power is 100 W, what is the received power on direct path? (Assume maximum power gain of antenna 2.1 dB)

(a) –74.3 dBW (b) –90 dBW
(c) –47 dBW (d) –50 dBW

[ESE-2006]

5.49 What is the value of directivity of an isotropic antenna?

(a) 1 dB (b) 0
(c) 4π dB (d) $\pi/2$ dB [ESE-2006]

5.50 Match **List-I** (Type of Antenna) with **List-II** (Example) and select the correct answer using the code given below the lists:

List-I

A. Aperture antenna
B. Circularly polarized
C. Frequency independent
D. Isotropic antenna

List-II

1. Helical antenna
2. Point source antenna
3. Log periodic antenna
4. Microstrip antenna

Codes:

	A	B	C	D
(a)	3	2	4	1
(b)	4	1	3	2
(c)	3	1	4	2
(d)	4	2	3	1

[ESE-2006]

5.51 What is the approximate HPBW (Half Power Beam Width) of a paraboloid of 1 metre diameter radiating at 0.1 meter wavelength?

(a) 70° (b) 7°
(c) 35° (d) 78°

[ESE-2006]

5.52 A broadside array of four isotropic point sources of same amplitude and phase, placed along x-axis with a spacing of $\lambda/2$ between adjacent elements has a polar pattern consisting of which of the following?

(a) One major lobe in the +ve y direction and one major lobe in the negative y direction and no side lobes

(b) One major lobe in the +ve y-direction, one major lobe in the negative y-direction, one minor lobe in the +ve x-direction and one minor lobe in the –ve x-direction
(c) One major lobe in the +ve y direction, one major lobe in the negative y direction, one minor lobe in each of the four quadrants
(d) Four major lobes one each in $+y$; $-y$; $+x$ and $-x$ directions and no other side lobes

[ESE-2007]

5.53 Which one of the following is correct? Normal mode helical antenna has
(a) low radiation efficiency and high directive gain
(b) high radiation efficiency and low directive gain
(c) low radiation efficiency and low directive gain
(d) high radiation efficiency and high directive gain

[ESE-2007]

5.54 For taking antenna far field pattern, what must be the distance R, between transmitting and receiving antennas?
(a) $R > \frac{2D^2}{\lambda}$ (b) $R > \frac{4D^2\lambda^2}{3}$
(c) $R > \frac{D^2}{2\lambda^2}$ (d) $R > \frac{2D^2}{\lambda^2}$

[ESE-2007]

5.55 A transmitting antenna has a gain of 10. It is fed with a signal power of 1 W. Assuming free-space propagation, what power would be captured by a receiving antenna of effective area 1 m^2 in the bore sight direction at a distance of 1 m?
(a) 10 W (b) 1 W
(c) 2 W (d) 0.8 W [ESE-2007]

5.56 The Fraunhofer region where the pattern measurement of transmitting antenna has to be taken is from a distance of $\frac{2D^2}{\lambda}$, where D is the maximum aperture dimension and λ is the free-space wavelength. What is this region generally known as?
(a) The near field (b) The far field
(c) Quiet zone (d) Induction field

[ESE-2007]

5.57 If the total input power to an antenna is W_t, the radiated power is W_r, and the radiation intensity is ϕ, then match **List-I** with **List-II** and select the correct answer using the code given below the lists:

List-I	List-II
A. Power gain	1. W_r / W_t
B. Directive gain	2. $W_r / 4\pi$
C. Average power radiated	3. $4\pi\phi / W_t$
D. Efficiency of the antenna	4. $4\pi\phi / W_r$

Code:

	A	B	C	D
(a)	3	4	2	1
(b)	4	3	2	1
(c)	3	4	1	2
(d)	4	3	1	2

[ESE-2008]

5.58 Where does the maximum radiation for an end-fire array occur?
(a) Perpendicular to the line of the array only
(b) Along the line of the array
(c) At 45° to the line of the array
(d) Both perpendicular to and along the line of the array

[ESE-2008]

5.59 As the aperture area of an antenna increases, its gain
(a) increases
(b) decreases
(c) remains steady
(d) behaves unpredictably [ESE-2008]

5.60 For obtaining a beam of electromagnetic waves with circular polarization, what is the ideal antenna?
(a) A double conical reflector
(b) A helical antenna
(c) A pyramidal horn antenna
(d) A hog-horn antenna [ESE-2008]

5.61 In measuring pulsed microwave power, the pulse duration is 1.5 μs with PRF of 1 kHz. If the CW power sensor reads 15 mW, what is the average signal power during the pulse?
(a) 5 W (b) 10 W
(c) 15 W (d) 20 W [ESE-2008]

5.62 Which of the following is a circularly polarized antenna?
(a) Horn (b) Dipole
(c) Helical (d) Pyramidal

[ESE-2009]

5.63 Which of the following antennas uses a number of varying length parallel elements ?
(a) Helical antenna
(b) Pyramidal Horn

(c) Corner reflection antenna
(d) Yagi-Uda antenna [ESE-2009]

5.64 Multiple member of antennas are arranged in arrays in order to enhance what property?
(a) Both directivity and bandwidth
(b) Only directivity
(c) Only bandwidth
(d) Neither directivity nor bandwidth
[ESE-2009]

5.65 Consider the following statements regarding an antenna:
1. It is a transducer.
2. Its performance is essentially frequency sensitive.
3. It is a reciprocal device.
4. With increase in its effective aperture area, width of the radiated beam increases.

Which of the above statements are correct?
(a) 1, 2 and 3 (b) 1, 2 and 4
(c) 2 and 3 only (d) 1 and 4 only
[ESE-2009]

5.66 **Assertion (A) :** As the length of the resonant antenna is increased, the number of lobes increases and the direction of the major lobes is closer and closer to the direction of the dipole.
Reason (R) : As the length increases, the current distribution along the wire becomes more and more uniform.
(a) Both A and R are true and R is the correct explanation of A
(b) Both A and R are true but R is NOT the correct explanation of A
(c) A is true but R is false
(d) A is false but R is true [ESE-2009]

5.67 Which of the following does not apply to Yagi-Uda antenna?
(a) High gain
(b) Reasonably good bandwidth
(c) Folded dipole
(d) Parasitic elements [ESE-2009]

5.68 Which of the following antenna is obtained by modifying a wave guide ?
(a) Microstrip Antenna (b) Helical Antenna
(c) Horn Antenna (d) Dipole Antenna
[ESE-2009]

5.69 In a three elements Yagi antenna.
(a) All the three elements are of equal length
(b) The driven element and the director are of equal length but the reflector is longer than both of them
(c) The reflector is longer than the driven element which in turn in longer than the director
(d) The reflector is longer than the driven element which in turn in longer than the reflector
[ESE-2010]

5.70 Match **List-I** with **List-II** and select the correct answer using the codes given below the lists:

List-I

A. $\frac{d}{\lambda} = 0.25, \alpha = 90°$

B. $\frac{d}{\lambda} = 0.5, \alpha = 0°$

C. $\frac{d}{\lambda} = 0.5, \alpha = 180°$

D. $\frac{d}{\lambda} = 1.0, \alpha = 180°$

List-II

1. 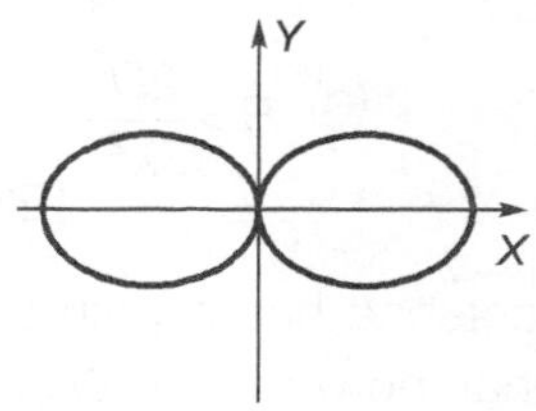

2. 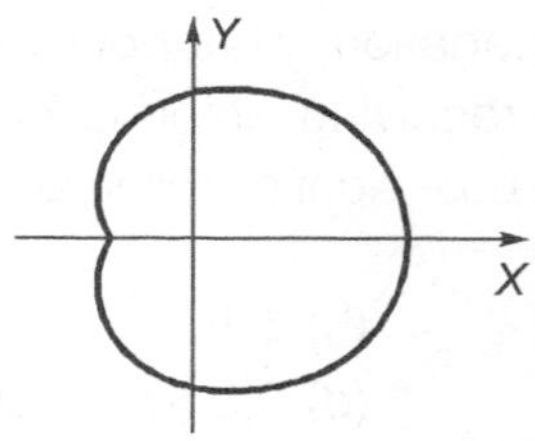

3. 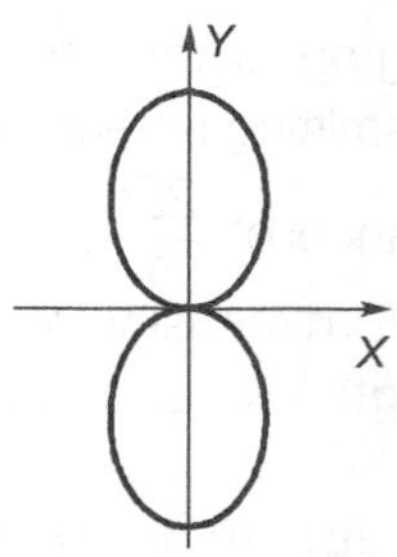

4. 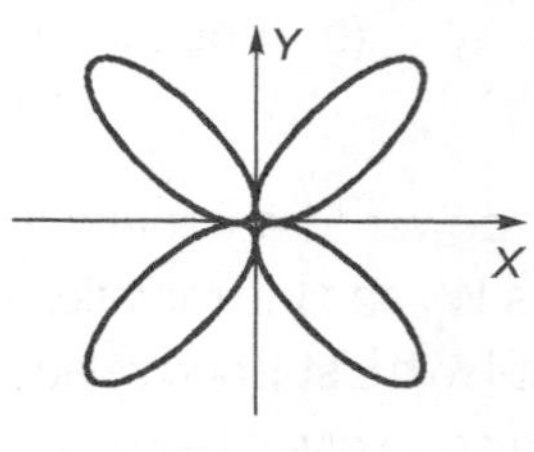

Codes:

	A	B	C	D
(a)	4	3	1	2
(b)	2	1	3	4
(c)	4	1	3	2
(d)	2	3	1	4

[ESE-2010]

5.71 A short current element has length $l = 0.03\,\lambda$, where λ is the wavelength. The radiation resistance for uniform current distribution is
(a) $0.072\,\pi^2\,\Omega$ (b) $80\,\pi^2\,\Omega$
(c) $72\,\Omega$ (d) $80\,\Omega$
[ESE-2010]

5.72 A half-wave dipole working at 100 MHz in free space radiates a power of 1000 Watts. The field strength at a distance of 10 kms in the direction of maximum radiation is
(a) 1.73 mV/m (b) 2.12 mV/m
(c) 2.22 mV/m (d) 22.2 mV/m
[ESE-2010]

5.73 Consider the following statements:
The inter element spacing of
1. More than λ in array antenna will produce grating lobes in unscanned condition.
2. More than $\lambda/2$ but less than λ in array antenna will produce grating lobes in unscanned condition.
3. Less than $\lambda/2$ in array antenna will not produce any grating lobes.
4. More than $1.5\,\lambda$ in array antenna will not produce any grating lobe.

Which of these statements are correct?
(a) 1 and 2 only (b) 3 and 4 only
(c) 1, 2 and 3 only (d) 1, 2, 3 and 4
[ESE-2010]

5.74 **Assertion (A):** In microwave point-to-point communication systems, parabolic reflector antennas are generally used.
Reason (R): A parabolic reflector antenna receiver has the property of focusing all axial rays to its focus and when used as transmitter with a feed at the focus, it will generate parallel beams along the axis with a pencil beam radiation pattern.
(a) Both A and R are true and R is the correct explanation of A
(b) Both A and R are true but R is NOT the correct explanation of A
(c) A is true but R is false
(d) A is false but R is true [ESE-2010]

5.75 **Assertion (A):** A Cassegrain antenna uses a main paraboloidal reflector and a relatively small hyperboloidal subreflector with a small horn-feed at the vertex of the main paraboloidal reflector.
Reason (R): The optical technique developed by William Cassegrain was used in telescope design to obtain large magnification with a physically short telescope. This configuration is found to be effective in the design of microwave antenna also.
(a) Both A and R are true and R is the correct explanation of A
(b) Both A and R are true but R is NOT the correct explanation of A
(c) A is true but R is false
(d) A is false but R is true
[ESE-2010]

5.76 In a communication system both transmitting and receiving antennas are vertically polarized. On a clear sunny day the power received at the Receiver is 1 mW. On a rainy day due to rain-induced depolarization the plane of polarization of the received wave gets rotated by 60° when it reaches to the receiving antenna. The received power at the receiver shall be
(a) 0.5 mW (b) 0.866 mW
(c) 1 mW (d) 0.25 mW
[ESE-2011]

5.77 Which of the following antennas exhibit circular polarization?
(a) Small circular loop
(b) Folded dipole
(c) Helical
(d) Parabolic dish
[ESE-2011]

5.78 Match **List-I** with **List-II** and select the correct answer using the code given below the lists:

List-I	List-II
A. Helical antenna	**1.** Fan shaped beams
B. Sectoral horn	**2.** Shaped beams
C. Phased arrays	**3.** Circular polarization
D. Parabolic reflector	**4.** Pencil beams

Codes:

	A	B	C	D
(a)	3	1	2	4
(b)	4	1	2	3
(c)	3	2	1	4
(d)	4	2	1	3

[ESE-2011]

5.79 In microwave communication systems, sometimes, the same frequency is used by separation of signals through vertical and horizontal polarizations. This technique is generally called

(a) steady frequency multiplexing
(b) variable frequency modulation technique
(c) frequency reconditioning technique
(d) frequency re-uses technique [ESE-2011]

5.80 **Assertion (A):** The power handling capacity of a receiver antenna could be very low compared to identical transmitting antenna.

Reason (R): A transmitter antenna has to radiate a large power. The receiver antenna will have to deal only with a very small fraction of the radiated power from the transmitter.

(a) Both A and R are true and R is the correct explanation of A
(b) Both A and R are true but R is NOT the correct explanation of A
(c) A is true but R is false
(d) A is false but R is true [ESE-2011]

5.81 Consider the following statements:

1. The antennas radiate energy.
2. An antenna is a transition device, or transducer between a guided wave and a free space wave or vice versa.
3. The resonators and transmission lines store energy.
4. An antenna converts electromagnetic signal to currents or vice versa.

Which of these statements are correct?

(a) 1, 2 and 4 only (b) 1, 2 and 3 only
(c) 2, 3 and 4 only (d) 1, 2, 3, and 4

[ESE-2011]

5.82 An antenna can be modeled as an electric dipole of length 5 m at 3 MHz. Find the radiation resistance of the antenna assuming uniform current over the length.

(a) 2 Ω (b) 1 Ω
(c) 4 Ω (d) 0.5 Ω [ESE-2011]

5.83 Match **List-I** with **List-II** and select the correct answer using the code given below the lists:

List-I **List-II**

A. Two-isotropes half wavelength apart fed in phase

1. 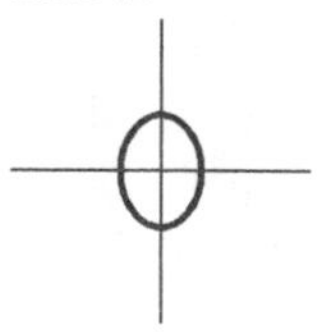

B. Two isotropes quarter wavelength apart 90° phase shift

2. 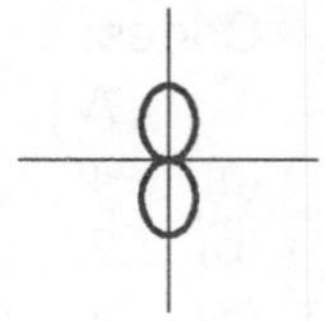

C. Two isotropes quarter wavelength apart fed in phase

3. 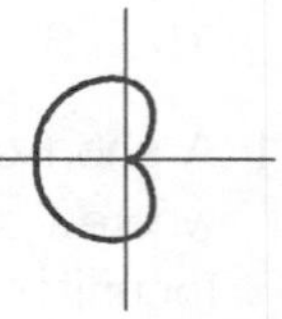

Codes:

	A	B	C
(a)	1	3	2
(b)	2	3	1
(c)	1	2	3
(d)	2	1	3

[ESE-2011]

5.84 An antenna located on the surface of a flat earth transmits an average power of 200 kW. Assuming that all the power is radiated uniformly over the surface of a hemisphere with the antenna at the center, the time average Poynting vector at 50 km is

(a) Zero (b) $\frac{2}{\pi}\hat{a}_r$ W/m^2

(c) $\frac{40}{\pi}$ μW/m^2 (d) $\frac{40}{\pi}\hat{a}_r$ μW/m^2

[ESE-2011]

5.85 A dipole with a length of 1.5 m operates at 100 MHz while the other has a length of 15m and operates at 10 MHz. The dipole are fed with same current. The power radiated by the two antennas will be

(a) the longer antenna will radiate 10 times more power than the shorter one
(b) both antennas radiate same power
(c) shorter antenna will radiate 10 times more power than the longer antenna
(d) longer antenna will radiate $\sqrt{10}$ times more power than the shorter antenna

[ESE-2011]

5.86 Which of the following antenna is used as a standard reference for calculating directive gain?

(a) Half wave dipole
(b) Infinitesimal dipole
(c) Elementary doublet
(d) Isotropic antenna

[ESE-2011]

5.87 Assertion (A): For producing radiation patterns with pre-determined characteristics like beam-width, side-lobe levels etc. phased array antennas are widely used in antenna technology.

Reason (R): In phased array antenna system, the resultant radiation pattern is formed by the superposition of electromagnetic waves radiated from various antenna elements which maintain specific, pre-determined phase conditions.

(a) Both A and R are true and R is the correct explanation of A
(b) Both A and R are true but R is NOT the correct explanation of A
(c) A is true but R is false
(d) A is false but R is true [ESE-2011]

5.88 An elliptically (arbitrarily) polarized wave can be broken up into
(a) Two circularly polarized components rotating in same direction
(b) Two circularly polarized components rotating in opposite directions
(c) Two stationary circularly polarized components
(d) None of these

[ESE-2012]

5.89 The directivity of a $\lambda/2$ long wire antenna is
(a) 1.5 (b) 1.66
(c) 2 (d) $\sqrt{2}$ [ESE-2012]

5.90 A communication link is to be set up between two stations 100 km apart using $\frac{\lambda}{2}$ antennae to transmit 1 kW power. The operating frequency is 100 MHz and the directivity of the two antennae is 1.64. The maximum received power would be
(a) 3.06×10^{-8} W (b) 1.53×10^{-8} W
(c) 6.12×10^{-9} W (d) 1×10^{-9} W

[ESE-2012]

5.91 For an antenna, Radiation Intensity is defined as
(a) the time-averaged radiated power per unit solid angle
(b) the peak radiated power per unit solid angle
(c) the peak radiated power per unit area
(d) the time-averaged radiated power per unit area

[ESE-2013]

5.92 Statement (I): Slot antenna is essentially a wave-guide with a slot length of one half lengths with a radiation pattern similar to that of a dipole with plane reflection.

Statement (II): Slot antenna has a gain much higher than that of a horn antenna.
(a) Both Statement (I) and Statement (II) are individually true and Statement (II) is the correct explanation of Statement (I).
(b) Both Statement (I) and Statement (II) are individually true but Statement (II) is not the correct explanation of Statement (I).
(c) Statement (I) is true but Statement (II) is false.
(d) Statement (I) is false but Statement (II) is true.

[ESE-2013]

5.93 Parabolic reflectors use Cassegrain feed to
(a) increase the gain of the system
(b) increase the beam width of the system
(c) reduce the size of the main reflector
(d) allow the feed to be placed at a convenient point

[ESE-2013]

5.94 Which one of the following statements is correct? An antenna array consists of 5 omni-directional elements carrying equal in-phase currents. The elements are equally spaced along x-axis with the spacing between adjacent elements being equal to one wavelength. The directivity pattern will have a maximum
(a) only in x-direction
(b) only in y-direction
(c) both x-and y-direction
(d) in the direction making an angle of 45° with x-axis

[ESE-2014]

5.95 If the receiving antenna is polarized at 90° with respect to transmitting antenna, it will receive
(a) no signal (b) minimum signal
(c) maximum signal (d) same signal

[ESE-2014]

5.96 Which of the following antenna gives circular polarization?
1. Yagi-Uda 2. Parabolic
3. Helical 4. Dipole
(a) 1, 2, 3 and 4 (b) 1, 2 and 3 only
(c) 3 only (d) 4 only

[ESE-2014]

5.97 If the diameter of a $\lambda/2$ dipole antenna is increased from $\lambda/100$ to $\lambda/50$, then its
(a) Bandwidth increases
(b) Bandwidth decreases
(c) Gain increases
(d) Gain decreases [ESE-2014]

5.98 The efficiency of an antenna having a resistance of 30 Ω and radiation resistance of 60 Ω is
(a) 33.3% (b) 50%
(c) 66.6% (d) 75% [ESE-2014]

5.99 An antenna behaves as a resonant circuit if
(a) its length is integral multiple of $n\lambda/2$
(b) its height is integral multiple of $n\lambda/2$
(c) its length is even multiple of $n\lambda/2$
(d) its length is odd multiple of $n\lambda/2$
[ESE-2014]

5.100 The discone antenna is
(a) a useful direction finding antenna
(b) used as a radar receiving antenna
(c) circularly polarized like other circular antennas
(d) useful as VHF receiving antenna
[ESE-2014]

5.101 If a dipole antenna has a radiation resistance of 73 Ω, the loss resistance of 7 Ω and the power gain is 16, then the directivity is
(a) 17.53 (b) 24.7
(c) 40 (d) 14.6 [ESE-2014]

5.102 Compensation theorem applicable to antennas is also called as
(a) Millman's theorem
(b) Superposition theorem
(c) Substitution theorem
(d) Power transfer theorem
[ESE-2014]

5.103 An isotropic radiator is one which radiates energy
(a) in a well-defined directions
(b) uniformly in all directions
(c) inside a hollow space
(d) uniformly in horizontal plane
[ESE-2014]

5.104 The effective length of an antenna is a measure of
(a) length of the antenna neglecting fringe effects
(b) effectiveness of the antenna as a radiator/collector of electromagnetic energy
(c) power consumed by the antenna
(d) range of the antenna [ESE-2014]

5.105 For a dipole antenna
(a) the radiation intensity is maximum along the normal to the dipole axis
(b) the current distribution along its length is uniform irrespective of the length
(c) the effective length equals its physical length
(d) the input impedance is independent of the location of the feed point
[ESE-2014]

5.106 **Statement (I):** Helical antenna has the largest bandwidth, high directivity and circular polarization.
Statement (II): Log-periodic antenna has a broad bandwidth.
(a) Both Statement (I) and Statement (II) are individually true and Statement (II) is the correct explanation of Statement (I).
(b) Both Statement (I) and Statement (II) are individually true but Statement (II) is not the correct explanation of Statement (I).
(c) Statement (I) is true but Statement (II) is false.
(d) Statement (I) is false but Statement (II) is true.
[ESE-2014]

5.107 Range resolution in RADAR is determined by
(a) The radiated power
(b) The bandwidth of transmitted pulse
(c) The antenna size
(d) The center frequency of RADAR
[ESE-2014]

5.108 For a microwave antenna
(a) Shape only depends on the frequency range used
(b) Size only depends on the frequency range used
(c) Neither shape nor size depend on the frequency range used
(d) Both shape and size depend on the frequency range used [ESE-2015]

5.109 The antenna efficiency of a $\lambda/8$ long dipole antenna is 89.159%. The equivalent loss resistance of the antenna is
(a) 1.5 Ω (b) 15 Ω
(c) 12.33 Ω (d) 125 Ω
[ESE-2015]

5.110 A small elemental wire antenna is excited with a sinusoidal current of frequency 1 MHz. The induction field and radiation field are at equal distance *d* from the antenna. The value of *d* will be nearly

(a) 300 m (b) 50 m
(c) 150 m (d) 20 m [ESE-2015]

5.111 A transmitting antenna with a 300 MHz carrier frequency produces 2 kW of power. If both transmitting and receiving antennas have unity power gain, the power received by the receiving antenna at a distance of 1 km is

(a) 11.8 mW (b) 18.4 mW
(c) 18.4 μW (d) 12.7 μW

[ESE-2015]

5.112 If the diameter of a $\frac{\lambda}{2}$ dipole antenna is increased from $\frac{\lambda}{100}$ to $\frac{\lambda}{50}$ then its

(a) Bandwidth increases
(b) Bandwidth decreases
(c) Gain increases
(d) Gain decreases [ESE-2015]

5.113 The ideal gain of a parabolic antenna of diameter 10 m for a wavelength of 316 cm is

(a) 20 dB (b) 30 dB
(c) 40 dB (d) 60 dB [ESE-2015]

5.114 An antenna consists of 4 identical Hertzian dipoles uniformly located along the *z*-axis and polarized in *z*-direction. The spacing between the dipoles is $\frac{\lambda}{4}$. The group pattern function is

(a) $4\cos\left(\frac{\pi}{4}\cos\theta\right)\cos\left(\frac{\pi}{2}\cos\theta\right)$

(b) $4\cos\left(\frac{\pi}{4}\cos\theta\right)\cos\left(\frac{\pi}{8}\cos\theta\right)$

(c) $4\cos\left(\frac{\pi}{4}\cos\theta\right)\sin\left(\frac{\pi}{2}\cos\theta\right)$

(d) $4\cos\left(\frac{\pi}{4}\cos\theta\right)\sin\left(\frac{\pi}{8}\cos\theta\right)$ [ESE-2015]

5.115 Current required to radiate 100 W of power at 100 MHz from 0.01 m dipole will be

(a) 131 A (b) 141 A
(c) 151 A (d) 161 A

[ESE-2015]

5.116 The length of half-wave dipole at 30 MHz will be

(a) 5 m (b) 10 m
(c) 15 m (d) 5.5 m

[ESE-2015]

5.117 A parabolic dish antenna of 1 m diameter and efficiency of 70%, operating at 20 GHz, will have a gain of nearly

(a) 20 dB (b) 25 dB
(c) 30 dB (d) 40 dB [ESE-2015]

5.118 If an antenna has a main beam with both half-power beam widths equal to 20°, its directivity (*D*) is nearly:

(a) 90.6 (b) 102.5
(c) 205 (d) 226 [ESE-2016]

5.119 A straight dipole radiator fed in the centre will produce maximum radiation at

1. The plane parallel to its axis
2. The plane normal to its axis
3. Extreme ends

Which of the above statements is/are correct?

(a) 1 only (b) 2 only
(c) 1 and 3 only (d) 2 and 3 only

[ESE-2016]

5.120 A microwave antenna with the absorbing cross-section area (A) and the power flux density(S) in the incident wave is employed as an absorber. The absorbed power (*P*) of the antenna is

(a) $\frac{A}{S}$ (b) $\frac{S}{A}$

(c) SA (d) $\frac{S+1}{A-1}$

[ESE-2016]

5.121 **Statement (I) :** An antenna of length $\frac{\lambda}{2}$ will have radiation pattern of two lobes.

Statement (II): An antenna of length $\frac{3\lambda}{2}$ will have radiation pattern of two lobes and two minor lobes.

(a) Both Statement (I) and Statement (II) are individually true and Statement (II) is the correct explanation of Statement (I).
(b) Both Statement (I) and Statement (II) are individually true but Statement (II) is not the correct explanation of Statement (I).
(c) Statement (I) is true but Statement (II) is false.
(d) Statement (I) is false but Statement (II) is true.

[ESE-2016]

5.122 The maximum radiation, for an endfire array occurs at

(a) $\phi_0 = 0$ (b) $\phi_0 = \frac{\pi}{2}$

(c) $\phi_0 = -\frac{\pi}{2}$ (d) $\phi_0 = \frac{3\pi}{2}$

[ESE-2017]

5.123 The field strength at the receiving antenna location at a distance of 28 km from a half-wave dipole transmitter radiating 0.1 kW is

(a) 1.5 mV/m (b) 2.5 mV/m
(c) 3.5 mV/m (d) 4.5 mV/m

[ESE-2017]

5.124 Consider the following statements:

If the maximum range of radar has to be doubled.

1. the peak transmitted power may be increased 16 folds.
2. the antenna diameter may be doubled.
3. the sensitivity of receiver may be doubled.
4. the transmitted pulse width may be doubled.

(a) 1 and 2 (b) 2 and 3
(c) 3 and 4 (d) 1 and 4

[ESE-2017]

5.125 In a certain material medium, a propagating electromagnetic wave attains 60% of the velocity of light. The distance at which the electromagnetic wave (f = 10 MHz) will have the same magnitude for the induction as well as the radiation fields is nearly

(a) 57.4 m (b) 29.0 m
(c) 5.8 m (d) 2.9 m

[ESE-2018]

5.126 A communication link operating at 3 GHz has a 22.5 W transmitter connected to an antenna of 2.5 m^2 effective aperture. The receiving antenna has effective aperture of 0.5 m^2 and is located at 15 km line-of-sight distance from the transmitting antenna (assume lossless, matched antennas). The power delivered to the receiver is

(a) 12.5 μW (b) 125 μW
(c) 12.5 mW (d) 125 mW

[ESE-2018]

5.127 A fixed radar station transmits 100 kW at 3 GHz. The smallest ocean-going ship has a radar cross-section of 200 m^2. The radar antenna gain is 16 dB. If detection requires a minimum of 1 nW/m^2 at the radar antenna, the effective range of the radar is nearly

(a) 7.1 km (b) 8.4 km
(c) 70.7 km (d) 84.0 km [ESE-2018]

5.128 For a 4-element broadside antenna array, Ψ equals

(a) $\sin\phi$ (b) $\pi\cos\phi$
(c) $\frac{\cot\phi}{\sqrt{2\pi}}$ (d) $\pi\sin\phi$ [ESE-2018]

5.129 For a system with a wavelength of 23.5 cm and a depression angle of $\sin^{-1} 0.94$, the surface relief below which the surface will appear smooth is determined as limited nearly to

(a) 2.8 cm (b) 3.1 cm
(c) 3.5 cm (d) 4.0 cm [ESE-2018]

5.130 Ground waves progress along the surface of the earth and must be polarized

(a) horizontally (b) circularly
(c) elliptically (d) vertically

[ESE-2019]

5.131 The light intensity 3 m from a lamp that emits 25 W of light energy will be

(a) 243 mW/m^2 (b) 232 mW/m^2
(c) 221 mW/m^2 (d) 210 mW/m^2

[ESE-2019]

5.132 Consider the following statements with reference to dipole arrays :

1. In broadside array, all the dipoles are fed in the same phase from the same source.
2. In end-fire array, the magnitude of the current in each element is same and there is no phase difference between these currents.

Which of the above statements is/are correct?

(a) 1 only (b) 2 only
(c) Both 1 and 2 (d) Neither 1 nor 2

[ESE-2019]

5.133 If aperture efficiency is 70%, the directivity D of a parabolic dish antenna as a function of its radius is

(a) $20\left(\frac{r}{\lambda}\right)^2$ (b) $28\left(\frac{r}{\lambda}\right)^2$

(c) $36\left(\frac{r}{\lambda}\right)^2$ (d) $44\left(\frac{r}{\lambda}\right)^2$

[ESE-2020]

5.134 An antenna radiates isotropically over a half-space above a perfectly conducting flat ground plane. If E = 50 mV m^{-1} rms at a distance of 1 km and the antenna terminal current I = 3.5 A, the radiation resistance will be

(a) 3.4 Ω (b) 4.3 Ω
(c) 5.2 Ω (d) 6.1 Ω [ESE-2020]

5.135 Which one of the following is the correct relationship between an antenna gain G and an effective area A_e?

(a) $G = \frac{4\pi f^2 A_e}{c^2}$ (b) $G = \frac{3\pi f^2 A_e}{c^2}$

(c) $G = \frac{2\pi f^2 A_e}{c^2}$ (d) $G = \frac{0.5\pi f^2 A_e}{c^2}$

[ESE-2020]

where: f = Carrier frequency
c = Speed of light

5.136 The signal-to-noise ratio $\frac{S}{N}$ for isotropic antenna is

(a) $\frac{\lambda^2}{16\pi^2 r^2 k T_{sys} B}$ (b) $\frac{\lambda^2}{14\pi^2 r^3 k T_{sys} B}$

(c) $\frac{\lambda^2}{12\pi^2 r^4 k T_{sys} B}$ (d) $\frac{\lambda^2}{10\pi^2 r^4 k T_{sys} B}$

[ESE-2020]

where: λ = Wavelength, m
r = Distance from transmitter to receiver, m
T_{sys} = System temperature, K
B = Bandwidth, Hz
k = Boltzmann's constant

5.137 At the higher frequencies, which one of the following becomes more confined to the region between the micro-strip and ground plane?
(a) Electric field (b) Magnetic field
(c) Dispersion (d) Skin effect

[ESE-2021]

5.138 A magnetic field strength of 5 μA/m is required at a point on $\theta = \pi/2$, which is 2 km from a half-wave dipole antenna in air. If the radiation resistance of the half-wave dipole antenna is 73 Ω, then the power radiated by this antenna (neglecting the ohmic loss) is
(a) 72 mW (b) 144 mW
(c) 158 mW (d) 316 mW

[ESE-2021]

5.139 Which of the following antennas are used directly as radiators aboard satellites to illuminate comparatively large areas of the Earth and they are widely used as primary feeds for reflector-type antennas both in transmitting and receiving modes?
(a) Dipole antennas
(b) End fire-array antennas
(c) Microstrip antennas
(d) Horn antennas

[ESE-2021]

5.140 The antenna or array designed to yield directive gain appreciably greater than that obtainable from uniform distribution has become known as
(a) superheterodyne array
(b) supergain array
(c) super antenna
(d) universal antenna

[ESE-2022]

5.141 **Statement (I) :** Antenna is a transformation device converting electromagnetic photons into circuit currents.
Statement (II) : An antenna converts photons to currents or vice versa.
(a) Both Statement (I) and Statement (II) are individually true and Statement (II) is the correct explanation of Statement (I).
(b) Both Statement (I) and Statement (II) are individually true, but Statement (II) is not the correct explanation of Statement (I).
(c) Statement (I) is true, but Statement (II) is false.
(d) Statement (I) is false, but Statement (II) is true.

[ESE-2022]

5.142 The radiation pattern of an antenna in spherical coordinates is given by $F(\theta) = \cos^4(\theta)$, $0 \le \theta \le \frac{\pi}{2}$. The directivity of the antenna is
(a) 16.42 (b) 18.02
(c) 20.42 (d) 22.02

[ESE-2023]

5.143 The directive gain $G_d(\theta, \phi)$ depends on antenna pattern. For the Hertzian dipole, P_{avg} is maximum at $\theta = \frac{\pi}{2}$ and minimum at $\theta = 0$ or π. For an isotropic antenna, $G_d(\theta, \phi) = 1$. The directive gain $G_d(\theta, \phi)$ can be defined as

(a) the measure of the concentration of the radiated power in a particular direction.
(b) the total radiated power divided by 4π.
(c) the ratio of the maximum radiation intensity to the average radiation intensity.
(d) the ratio of total power divided by array factor.

[ESE-2023]

5.144 What is the radiation resistance of an infinitesimal dipole whose overall length $l = \lambda/50$?
(a) 0.316 Ω (b) 1.5 Ω
(c) 1.25 Ω (d) 2.25 Ω

[ESE-2024]

5.145 When two antennas are near each other, some of the energy that is primarily intended for one ends up at the other. The amount does **not** depend on
(a) radiation characteristics of each
(b) relative separation between them
(c) relative orientation of each
(d) mutual coupling between them

[ESE-2024]

5.146 Match the following lists:

List-I
P. Cavity-backed printed antennas
Q. Triplate
R. Waveguide
S. Microstrip antennas

List-II
1. Difficult to make conformal
2. Low efficiency
3. Narrow bandwidth
4. Coupling occurs inside structure

Select the correct answer using the code given below:

	P	Q	R	S
(a)	2	3	4	1
(b)	3	4	1	2
(c)	2	1	3	4
(d)	4	1	3	2

[ESE-2024]

5.147 A slot antenna of dual monopole feed has
(a) 3-4 bands, wide bandwidth, low height, high fabrication complexity
(b) 1-4 bands, narrow bandwidth, large height, low fabrication complexity
(c) 2-3 bands, wide bandwidth, low height, low fabrication complexity
(d) 1-2 bands, narrow bandwidth, medium height, low fabrication complexity

[ESE-2024]

5.148 Consider the following statements regarding the antenna arrays :
1. A broadside array has its maximum radiation directed along the axis of the array.
2. An end-fire array has its maximum radiation directed along the axis of the array.
3. The resultant pattern for an array, which is known as pattern multiplication, is the product of the unit pattern and the group pattern.

Which of the above statements is/are correct?
(a) 1 and 2 (b) 1 and 3
(c) 2 and 3 (d) 2 only

[ESE-2024]

■■■

Answers Antennas and Radar

5.1 (a)	5.2 (d)	5.3 (b)	5.4 (b)	5.5 (b)	5.6 (d)	5.7 (b)	5.8 (a)	5.9 (d)
5.10 (c)	5.11 (d)	5.12 (a)	5.13 (d)	5.14 (d)	5.15 (d)	5.16 (c)	5.17 (b)	5.18 (a)
5.19 (b)	5.20 (a)	5.21 (d)	5.22 (a)	5.23 (d)	5.24 (c)	5.25 (c)	5.26 (a)	5.27 (b)
5.28 (a)	5.29 (a)	5.30 (c)	5.31 (c)	5.32 (a)	5.33 (a)	5.34 (d)	5.35 (d)	5.36 (a)
5.37 (b)	5.38 (d)	5.39 (d)	5.40 (d)	5.41 (a)	5.42 (d)	5.43 (c)	5.44 (b)	5.45 (a)
5.46 (b)	5.47 (c)	5.48 (a)	5.49 (b)	5.50 (b)	5.51 (b)	5.52 (c)	5.53 (a)	5.54 (a)
5.55 (d)	5.56 (b)	5.57 (a)	5.58 (b)	5.59 (a)	5.60 (b)	5.61 (b)	5.62 (c)	5.63 (d)
5.64 (b)	5.65 (a)	5.66 (b)	5.67 (a)	5.68 (c)	5.69 (c)	5.70 (d)	5.71 (a)	5.72 (d)
5.73 (c)	5.74 (a)	5.75 (b)	5.76 (d)	5.77 (c)	5.78 (a)	5.79 (d)	5.80 (d)	5.81 (a)
5.82 (a)	5.83 (b)	5.84 (d)	5.85 (b)	5.86 (d)	5.87 (a)	5.88 (b)	5.89 (b)	5.90 (b)
5.91 (a)	5.92 (c)	5.93 (d)	5.94 (d)	5.95 (a)	5.96 (c)	5.97 (c)	5.98 (c)	5.99 (b)
5.100 (d)	5.101 (a)	5.102 (c)	5.103 (b)	5.104 (b)	5.105 (a)	5.106 (b)	5.107 (b)	5.108 (d)
5.109 (a)	5.110 (b)	5.111 (d)	5.112 (c)	5.113 (a)	5.114 (a)	5.115 (c)	5.116 (a)	5.117 (d)
5.118 (b)	5.119 (b)	5.120 (c)	5.121 (b)	5.122 (a)	5.123 (a)	5.124 (a)	5.125 (d)	5.126 (a)
5.127 (b)	5.128 (b)	5.129 (b)	5.130 (d)	5.131 (c)	5.132 (a)	5.133 (b)	5.134 (a)	5.135 (a)
5.136 (a)	5.137 (a)	5.138 (b)	5.139 (d)	5.140 (b)	5.141 (b)	5.142 (b)	5.143 (c)	5.144 (a)
5.145 (d)	5.146 (b)	5.147 (c)	5.148 (c)					

Explanations Antennas and Radar

5.1 (a)

The FM-CW radar principle is used in the aircraft radio altimeter to measure height above the surface of the earth. Since the relative motion between the aircraft and ground is small, the effect of the Doppler frequency shift may usually be neglected.

5.2 (d)

Radiation resistance,

$$R_{rad} = 60\ \Omega$$

Antenna resistance,

$$R_{ant} = 40\ \Omega$$

Efficiency of the antenna,

$$\eta = \frac{R_{rad}}{R_{rad} + R_{ant}} \times 100$$

$$\Rightarrow \quad \eta = \frac{60}{60 + 40} \times 100 = 60\%$$

5.3 (b)

Radiation pattern is a function of field or power strength.

Every ray is a source of secondary emissions-Huygen's wave principle.

5.4 (b)

Simple dipole is vertically sinθ dependent.

Omin-directional antenna has spherical pattern.

Loop antenna is for direction finding on horizontal earth.

Therefore strictly uni-directional pattern.

5.5 (b)

$$\lambda = \frac{c}{f} = \frac{3 \times 10^8}{10 \times 10^6} = 30\ \text{m}$$

Radiation resistance,

$$R_{rad} = 80\pi^2 \left(\frac{dl}{\lambda}\right)^2$$

$$= 80\pi^2\left(\frac{1}{30}\right)^2 = 0.88\ \Omega$$

Total radiated power,

$$P_r = I^2 R_{rad} = (1)^2 \times 0.88 = 0.88\ \text{W}$$

5.6 (d)

Single reflector for uni-directional beam and multiple directors for increasing gain and decreasing beam angle.

5.7 (b)

Low frequency antennas are usually vertical antennas because of vertical polarization. Frequencies upto 1650 kHz are used for AM broadcasting.

5.8 (a)

To measure the antennas performance in any aspect the near field or $\frac{1}{r^2}$ and $\frac{1}{r^3}$ have to be neglected.

The radiation field $\frac{1}{r}$ term should be strong enough ignoring $\frac{1}{r^2}$ and $\frac{1}{r^3}$.

5.10 (c)

Maxima along the array axis

$$\psi = \alpha + \beta d\cos\theta_{max} = 0$$

For $\theta_{max} = 0°$ or $180°$

End fire array $\alpha = \pm\beta d$

When maximas are along the axis radiation has to add along the axis.

5.11 (d)

To measure the antennas performance in any aspect the near field or $\frac{1}{r^2}$ and $\frac{1}{r^3}$ have to be neglected. The radiation field $\frac{1}{r}$ term should be strong enough ignoring $\frac{1}{r^2}$ and $\frac{1}{r^3}$.

This distance is called as Fraunhofer zone where measurements are carried.

$$r > \frac{2D^2}{\lambda} = \frac{2\times(3)^2}{3\times10^8}\times10^{10} = 600\ \text{m}$$

Where D is the biggest dimension of the antenna.

5.12 (a)

Helical antenna produces circularly polarized beams in microwave communication fields.

5.13 (d)

When an electromagnetic wave is incident parallel to the axis, it passes through the focus point of spherical reflector.

5.14 (d)

At low frequencies for broadcasting a vertical monopole is preferred as it has horizontal isotropic radiation.

5.18 (a)

$$\text{Duty cycle} = \frac{\text{Average Power}}{\text{Peak Power}} = \frac{2}{1000} = 0.002$$

5.19 (b)

$$\text{Range,}\quad R = \frac{cT}{2} = \frac{3\times10^8\times20\times10^{-6}}{2}$$

$$= 3000\ \text{m}$$

5.20 (a)

The Doppler effect is not used in moving-target plotting on PPI.

5.21 (d)

Pulsed radar system → Magnetrons are in use
Moving target radar → Doppler radar
CW radar → Search radar
Phase shifter → Phased array radar

5.22 (a)

To improve range resolution in a radar, pulse repetition frequency should be increased or in other words, pulse should be used of short duration.

5.23 (d)

Helical antenna is circularly polarized.

5.24 (c)

Travelling waves mean matched conditions at load. Hence no maximas or minimas and uniform currents $I(z) = I_o$.

5.25 (c)

For a dipole, $E \propto \frac{1}{r}$

$$\frac{E_2}{E_1} = \frac{r_1}{r_2}$$

$$\Rightarrow \quad E_2 = 1\times\frac{2}{4} = 0.50 \text{ mV/m}$$

5.26 (a)

At low frequencies for broadcasting a vertical monopole is preferred as it has horizontal isotropic radiation.

5.27 (b)

Single active element
Single reflector
Multiple directors.
Yagi-Uda antenna has one reflector, one and more directors.

5.28 (a)

High frequency, radiation is stronger and so smaller apertures can radiate strongly.

$$\text{Gain} = \frac{4\pi}{\lambda^2}A_e = \frac{4\pi}{\Omega_A}$$

5.29 (a)

$$\frac{f}{D} = \frac{1}{4\tan(\theta/2)}$$

where θ = Beam angle of feed.

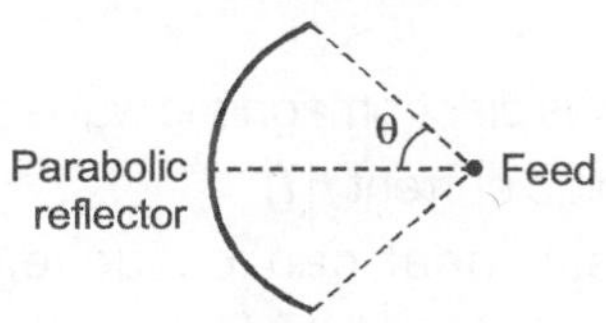

With θ approximately 90°.

$$\frac{f}{D} = \frac{1}{4} = 0.25$$

Typical f/D ratio in practice for a paraboloid reflector is between 0.2 to 0.3.

5.30 (c)

Maximum possible gain

$$= \pi^2\left(\frac{D}{\lambda}\right)^2 = (3.14)^2\left(\frac{1}{3.14\times10^{-2}}\right)^2 = 10^4$$

Maximum possible gain (dB)

$$= 10\log 10^4 = 40 \text{ dB}$$

5.32 (a)

Array antenna have their radiation pattern depending on currents phase difference between the elements. This makes it easy to pre-determine the radiation pattern.

5.33 (a)

$$\text{Gain} = \frac{4\pi}{\lambda^2}\cdot A_e = \frac{4\pi}{\Omega_A} = \frac{4\pi\cdot hW}{\lambda^2}$$

$$\Omega_A = \theta_{HPBW}\times\phi_{HPBW} = \frac{hW}{\lambda\cdot\lambda}$$

$$\text{HPBW} \cong \frac{\lambda}{h}$$

$$\lambda = \frac{3\times10^8}{5\times10^9} = 6 \text{ cm}$$

$$\text{HPBW} = \frac{6}{9}\times57° = 38°$$

5.34 (d)

$$d = \frac{3\lambda}{4},\ \alpha = \pi$$

$$\psi = \pi + \frac{2\pi}{\lambda}\frac{3\lambda}{4}\cos\theta$$

$$\psi = \pi + \frac{3\pi}{2}\cos\theta$$

For maximas, $\psi \to 0$ or 2π.

$$\pi + \frac{3\pi}{2}\cos\theta = 0$$

$$\cos\theta = \frac{-2}{3}$$

$$\pi + \frac{3\pi}{2}\cos\theta = 2\pi$$

$$\cos\theta = \frac{2}{3}$$

Bi-directional pattern.

$\lambda/4$, $\alpha = \frac{\pi}{2}$ is uni-directional strictly with maxima is θ = 180°.

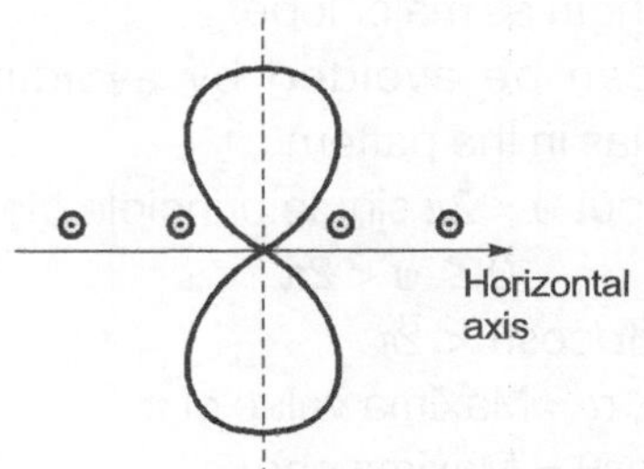

Vertical pattern forms

Similarly opposite for case 4.

5.35 (d)

Received power, $W_r = \dfrac{W_t \cdot A_{et} \cdot A_{er}}{(\lambda d)^2}$

$W_r \propto f^2$

where, A_e = Capture areas of the antennas

High frequency has higher magnitude than low frequency.

$$\text{Gain} = \frac{4\pi}{\lambda^2} A_e$$

Gain being constant

$$A_e \propto \lambda^2 \propto \frac{1}{f^2}$$

5.37 (b)

If P_{min} = minimum detectable power

P_t = transmitted power

σ = RADAR cross section

G = gain of the antenna

R_{max} = maximum range

Using RADAR equation

$$R_{max} = \left[\frac{P_t \cdot G^2 \cdot \sigma\lambda^2}{(4\pi)^3 P_{min}}\right]^{1/4}$$

We have,

- $R_{max} \propto \lambda^{1/2}$
- $R_{max} \propto \dfrac{1}{P_{min}^{1/4}}$
- $R_{max} \propto G^{1/2}$
- $R_{max} \propto P_t^{1/4}$
- $R_{max} \propto \sigma^{1/4}$

Since statements 1, 2, 4 and 5 are true and statement 3 is wrong, so the best possible answer is option (b).

5.38 (d)

Grating lobe is a maximum lobe or back lobe equal in strength as major lobe.

This can be avoided by avoiding multiple maximas in the pattern

$\psi = 0$ but $\psi \neq 2\pi$ single principle maxima

$$0 \leq \psi < 2\pi$$

$$\alpha + \beta d\cos\theta < 2\pi$$

where, α = Maxima value of π

and $\cos\theta$ = Maxima one

$$\Rightarrow \quad d < \frac{\lambda}{2}$$

5.39 (d)

Linear phase variations means α is constant with non-zero value ($\alpha \neq 0$)

$\psi = \alpha + \beta d\cos\theta$ decides the pattern for two elements $\dfrac{\sin(N\psi/2)}{\sin(\psi/2)}$ decides for N elements.

In general $\psi \to 0$ is the condition for maxima and d is important is deciding pattern.

5.40 (d)

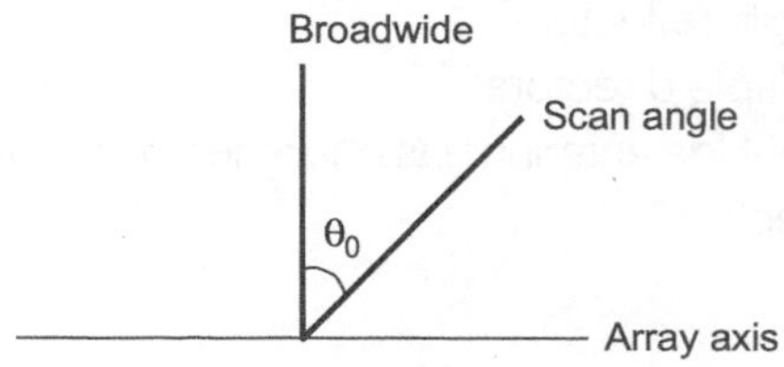

$$\psi = \alpha + \beta d\cos\theta = 0$$

where θ is angle from array axis.

$\psi = \alpha + \beta d\sin\theta_0 = 0$ or 2π for maxima

$\psi = \alpha + \beta d\sin\theta_{0H}$

$= \dfrac{\pi}{2}$ for half power direction

$\therefore$ Beamdwidth $\propto \cos\theta_0$.

Scanning array has $\psi \to 0$ for θ_0.

5.41 (a)

Radiation is electromagnetic wave generation due to harmonic current $I(t)$.

The basic ideal can be developed as $I(t)$ produces $A(t)$ or $H(t)$ which produces $E(t)$.

This $\bar{A}$ is called vector potential.

5.42 (d)

Wide band antenna are made out of multiple dipoles of multiple lengths as in log periodic array.

Dipole antennas are horizontally isotropic.

Microstrip or patch antennas are microwave antennas which are small and compact. They are used in mobile phone.

5.43 (c)

Sectorial horn is an opening in only E or H side with low bandwidth and low power handling due to reduced dimensions pyramidal horn has focussed and narrow beam.

5.44 (b)

Required distance

$$= \frac{2D^2}{\lambda} = \frac{2D^2 f}{c} = \frac{2\times 5^2\times 6\times 10^9}{3\times 10^8} = 1 \text{ km}$$

5.45 (a)

A dispersive channel is one in which the phase velocity of signal depends upon the frequency.

5.46 (b)

A pyramidal horn antenna is best suited to excite from a microwave waveguide.

5.47 (c)

$$G = 6\left(\frac{D}{\lambda}\right)^2$$

$$\Rightarrow \quad 1000 = 6\left(\frac{D}{0.1}\right)^2$$

$$\Rightarrow \quad D = 1.291 \text{ m}$$

5.48 (a)

$$[P_t]_{dB} = 10\log_{10}(100) = 20 \text{ dB}$$

$$[\text{Path loss}]_{dB} = 32.5 + 20\log_{10}(f_{MHz}\, d_{km}) \text{ dB}$$

$$= 32.5 + 20\log_{10}(2000) = 98.5 \text{ dB}$$

For direct path calculation,

$$[P_r]_{dB} = [P_t]_{dB} + [G_t]_{dB} + [G_r]_{dB} - [\text{Path loss}]_{dB}$$

$$[G_t]_{dB} = [G_r]_{dB} = 2.1 \text{ dB}$$

So, $\quad [P_r]_{dB} = 20 + 2(2.1) - 98.5 = -74.3 \text{ dB}$

5.49 (b)

Directivity of isotropic antenna, $D = 1$

$D(\text{dB}) = 10 \log 1 = 0 \text{ dB}$

5.50 (b)

(i) Helical antenna is broadband VHF and UHF antenna to provide circular polarization.

(ii) Log periodic antennas are frequency independent antennas.

5.51 (b)

$$\text{HPBW} = \frac{70\lambda}{D} = \frac{70\times 0.1}{1} = 7^\circ$$

5.52 (c)

4-element array,

$$E = E_o \frac{\sin(4\psi/2)}{\sin(\psi/2)}$$

$$= 4E_o \cos(\psi/2)\cos(\psi)$$

with $\quad \psi = \alpha + \beta d\cos\theta$

$$= \frac{2\pi}{\lambda}\times\frac{\lambda}{2}\cos\theta = \pi\cos\theta$$

$\cos\frac{\psi}{2}$ pattern $\Rightarrow$

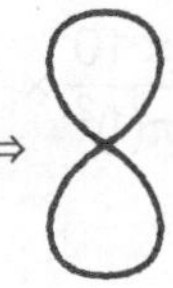

$\cos\psi$ pattern $\Rightarrow$

$\left(\cos\frac{\psi}{2}\times\cos\psi\right)$ pattern $\Rightarrow$

5.53 (a)

Normal mode is the most uncommon mode of operation of helical antenna when $Nl << \lambda$.

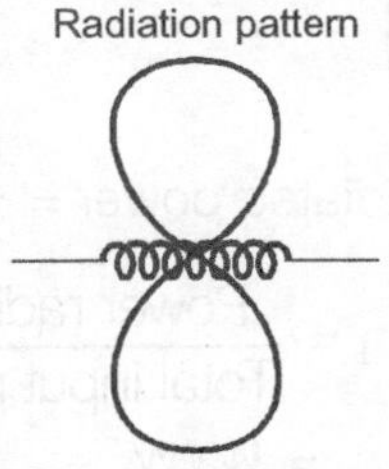

This mode has low radiation efficiency and low gain behaving like a wire antenna.

5.54 (a)

To measure the antennas performance in any aspect the near field or $\frac{1}{r^2}$ and $\frac{1}{r^3}$ have to be neglected. The radiation field $\frac{1}{r}$ term should be strong enough ignoring $\frac{1}{r^2}$ and $\frac{1}{r^3}$.

This distance is called as Fraunhofer zone where measurements are carried.

$$r > \frac{2D^2}{\lambda}$$

Where D is the biggest dimension of the antenna.

5.55 (d)

Received power, $P_r = \frac{P_t G_t}{4\pi r^2}\cdot A$

where, P_t = transmitted power

G_t = gain of transmitting antenna
r = distance
A = effective area

$$P_r = \frac{1\times 10}{4\pi(1)^2}\times 1 = 0.8 \text{ W}$$

5.57 (a)

$$W_t = W_r + W_l$$

Here, ϕ = radiation intensity
W_t = total input power
W_r = radiated power
W_l = power loss

(i) Power gain, $G_p = \dfrac{\phi}{\left(\dfrac{W_t}{4\pi}\right)} = 4\pi\phi/W_t$

(ii) Directive gain,

$$G_d = \frac{\text{Radiation intensity}}{\text{Average radiation intensity}}$$

$$= \frac{\phi}{\left(\dfrac{W_r}{4\pi}\right)} = 4\pi\phi/W_r$$

(iii) Average radiated power $= \dfrac{W_r}{4\pi} = W_r/4\pi$

(iv) Efficiency, $\eta = \dfrac{\text{Power radiated}}{\text{Total input power}}$
$= W_r/W_t$

5.58 (b)

Front view of an end-fire array is shown below:

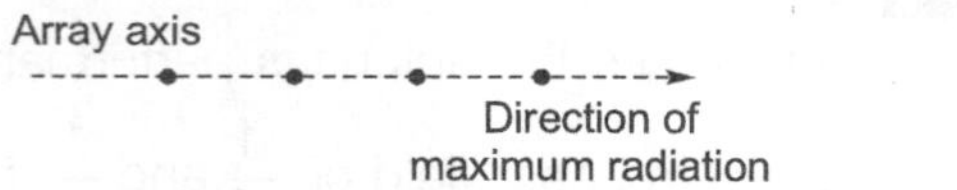

5.59 (a)

Gain of antenna, $G = \dfrac{4\pi A}{\lambda^2}$

where, A = Aperture area of antenna

5.60 (b)

Helical antenna is a broadband VHF or UHF antenna to provide circular polarization.

5.61 (b)

Pulse duration $\tau = 1.5\ \mu$s
PRF $f_p = 1$ kHz
Average signal power

$$P_{av} = \frac{15\times 10^{-3}}{\tau f_p} = \frac{15\times 10^{-3}}{1.5\times 10^{-6}\times 10^{3}} = 10 \text{ W}$$

5.62 (c)

The wide bandwidth, simplicity, highest directivity and circular polarization of the helical beam antenna have made it indispensable for space communication applications.

5.64 (b)

An antenna array is a system of similar antennas oriented similarly to get greater directivity in a desired direction.

5.65 (a)

Antenna is a transducer. It converts electrical energy into electromagnetic energy and vice-versa. Hence, it is a reciprocal device.

5.66 (b)

Increasing length increases number of maximas due to possibilities of multiple solutions of in phase points of $\psi \to 0$ or 2π or 4π and so on.
Depending on odd and even multiple of $\lambda/2$ the maximas may be along broad-side or end fire.
As the length increases to be quite larger than λ then current is uniform.

5.68 (c)

A horn antenna may be regarded as a flared out or opened out waveguide.

5.70 (d)

As $\Psi = \beta d \cos\theta + \alpha$
for maxima; $\Psi = 0$

(B) $\dfrac{d}{\lambda} = 0.5;\ \alpha = 0$

$$\Psi = \beta d\cos\theta + \alpha = 0$$

$$\Rightarrow \frac{2\pi}{\lambda} d\cos\theta + 0 = 0$$

$$\Rightarrow 2\pi(0.5)\cos\theta = 0$$

$$\Rightarrow \cos\theta = 0 \Rightarrow \boxed{\theta = 90^\circ}$$

(C) $\dfrac{d}{\lambda} = 0.5;\ \alpha = 180 = \pi$

$$\Psi = \frac{2\pi}{\lambda}\cdot d\cos\theta + \pi = 0$$

$$\Rightarrow \pi\cos\theta + \pi = 0$$

$$\therefore \cos\theta = -1$$

$$\Rightarrow \boxed{\theta = 180^\circ}$$

(D) $\frac{d}{\lambda} = 1.0;\ \alpha = 180°$

$$\Psi = \frac{2\pi}{\lambda}\cdot d\cos\theta + \pi = 0$$

$$\Rightarrow 2\pi\cos\theta + \pi = 0$$

$$\Rightarrow \cos\theta = -\frac{1}{2} \Rightarrow \boxed{\theta = 120°}$$

From above analysis we can reach upto the correct option which is option (d).

5.71 (a)

$$R_{rad} = 80\pi^2\left(\frac{l}{\lambda}\right)^2$$

$$\Rightarrow R_{rad} = 80\,\pi^2\,(0.03)^2$$

$$\Rightarrow R_{rad} = 0.072\,\pi^2$$

5.72 (d)

$$E_{rms} = \frac{\sqrt{30\,P_t\cdot G_t}}{d}$$

$G_t = 1.64$ for halve wave dipole

$$= \frac{\sqrt{30\times1000\times1.64}}{10^4} = 22.2\ \text{mV/m}$$

5.73 (c)

Grating lobe is a maximum lobe or back lobe equal in strength as major lobe.

This can be avoided by avoiding multiple maximas in the pattern

$\psi = 0$ but $\psi \neq 2\pi$ single principle maxima

$$0 \le \psi < 2\pi$$

$$\alpha + \beta d\cos\theta < 2\pi$$

where, α = Maxima value of π

and $\cos\theta$ = Maxima one

$$\Rightarrow d < \frac{\lambda}{2}$$

5.75 (b)

Cassegrain feed is as shown below:

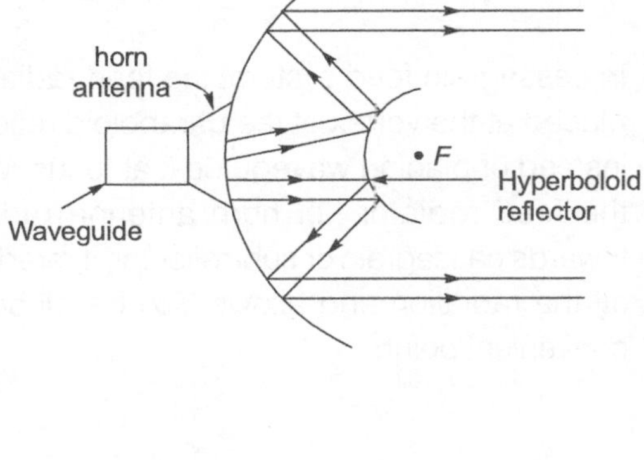

5.76 (d)

Clear sunny day received power

$$P_r = 1\ \text{mW}$$

angle of rotation ; $\theta = 60°$

So now the received power

$$P_r' = P_r\cos^2\theta = 1\ \text{mW}\cdot\cos^2 60°$$

$$P_r' = 0.25\ \text{mW}$$

5.77 (c)

Helical antenna exhibit circular polarization.

5.78 (a)

Helical antenna — Circular polarization

Sectoral horn — Fan Shaped beams

5.79 (d)

By definition the technique used is frequency reuse-technique.

5.80 (d)

Antenna is a reciprocal device and hence transmitting in receiving devices are identical in power handling (given identical antennas).

5.81 (a)

Transmission lines do not store energy.

5.82 (a)

Given: $l = 5$ m and $f = 3$ MHz

$$\therefore \quad \lambda = \frac{3\times10^8}{3\times10^6} = 100\ \text{m}$$

So radiation resistance of dipole antenna

$$R_{rad} = 80\pi^2\left(\frac{l}{\lambda}\right)^2 = 80\pi^2\left(\frac{5}{100}\right)^2$$

$$\Rightarrow R_{rad} = 1.97 \simeq 2\ \Omega$$

5.84 (d)

It is a hemisphere, so area = $2\pi r^2$

Now transmitted power

$$P_T = P_{avg}\times\text{area}$$

$$P_{avg} = \frac{P_T}{area} = \frac{200\ \text{kW}}{2\pi(50\times10^3)\ \text{m}^2}$$

$$= \frac{200\times10^3}{25\times50\times50\times10^6}\ \text{W/m}^2$$

$$= \frac{2\times10^5}{2\pi\times25\times10^8}\ \text{W/m}^2 = \frac{40}{\pi}\ \mu\text{W/m}^2$$

Poynting vector will be directed into a_r direction; So

$$P_{avg} = \frac{40}{\pi}\bar{a}_r \ \mu W/m^2$$

5.85 (b)

Power radiated, $P_{rad} = I^2 R_{rad.}$ ...(i)

$$R_{rad} = 80\pi^2\left(\frac{l}{\lambda}\right)^2 \quad ...(ii)$$

Now first dipole antenna

l = 1.5 m and

$$\lambda = \frac{3\times10^8}{100\times10^6} = 3\text{ m}$$

So $\frac{l}{\lambda} = \frac{1.5}{3} = \frac{1}{2}$

For second dipole antenna

l = 15 m and

$$\lambda = \frac{3\times10^8}{10\times10^6} = 30\text{ m}$$

So $\frac{l}{\lambda} = \frac{15}{30} = \frac{1}{2}$

As l/λ for both the antennas is same, so from equation (i) and (ii) we conclude that both dipoles will radiate same power.

5.86 (d)

Isotropic antenna is used as a reference for calculating directive gain.

5.87 (a)

Radiation pattern depends on phase difference between consecutive elements $\psi = \alpha + \beta d\cos\theta$.

5.88 (b)

$E_1 \sin\omega t\, \hat{a}_x + E_1 \cos\omega\, \hat{a}_y$ left circulating polarized.

$E_2 \sin\omega t\, \hat{a}_x + E_2 \cos\omega\, \hat{a}_y$ right circulating polarized.

$(E_1 + E_2)\sin\omega t\, \hat{a}_x \quad ; \quad (E_1 - E_2)\sin\omega t\, \hat{a}_y$

This is elliptical polarization.

5.89 (b)

$$\text{Gain} = \frac{4\pi}{\Omega_A}$$

$$\Omega_A = \theta_{HPBW}\times\phi_{HPBW}$$

$$\theta_{HPBW} \cong 84°, \quad \phi_{HPBW} = 360°$$

$$\text{Gain} \cong 1.63$$

5.90 (b)

Power transmitted = 1000 W

frequency = 100 MHz

Wavelength = 3 m

Directivity of both the antennae is 1.64 and as we know that maximum directivity is said to be the gain of the antenna.

$$P_R = \frac{P_T G_T G_R}{(4\pi R/\lambda)^2} = \frac{P_T G_T G_R}{(4\pi R)^2}\cdot\lambda^2$$

$$= \frac{10^3\times1.64\times1.64\times(3)^2}{(4\times\pi\times100\times10^3)^2}$$

$$= 1.53\times10^{-8}\text{ W}$$

5.91 (a)

Radiation intensity of an antenna is defined as the total radiated power per unit solid angle.

5.92 (c)

Slot antenna is a discontinuity in a conductor and it is a RF antenna with gain smaller than microwave antenna.

Horn antenna is a MW antenna.

The slot antenna is a very efficient radiator. A $\lambda/2$ slot is cut in a flat metal sheet in such a way that the width of the slot is small i.e. $w << \lambda$. Therefore the currents are not confined to the edges of the slot but spread out over the sheet.

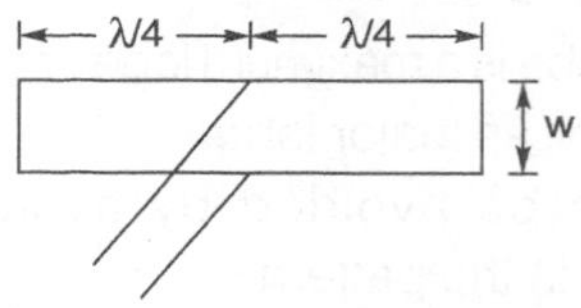

Gain of slot antenna is in range of 10-15 dBi and of Horn is in range of 25 dBi.

5.93 (d)

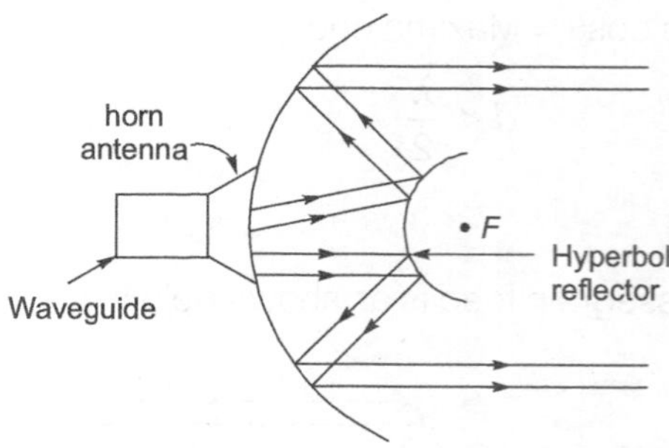

In cassegrain feed system, the feed radiator is placed at the vertex of the paraboloid reflector, instead of placing waveguide it at focus. When the feed radiator i.e. horn antenna radiates towards cassegrain or sub-reflector, it reradiates all the radiation and allows it to be placed at convenient point.

5.94 (d)

$\psi \to 0$ is condition for maxima with $\alpha = 0$ θ_{max} is 90°.
Y-direction maxima occurs.

Array factor $\dfrac{\sin(N\psi/2)}{\sin(\psi/2)}$ should be maximum.

$$\sin\left(\frac{\psi}{2}\right) \neq 0$$

$$\frac{N\psi}{2} = \frac{\pi}{2} \text{ or } (2n+1)\frac{\pi}{2} \text{ and } \psi \neq 2n\pi$$

$$\psi = \frac{(2n+1)\pi}{5} \text{ and } \psi \neq 2n\pi$$

$$\alpha + \beta d\cos\theta = \frac{\pi}{5} \text{ or } \frac{2\pi}{5} \text{ or } \frac{3\pi}{5}$$

5.95 (a)

If there is mismatch in polarization of transmitter and receiver by 90° then there will be no signal at the output.

5.96 (c)

Helical antenna has circular polarization.

5.97 (c)

Diameter of the conductor of the dipole effects only losses and loss resistance

$$R_l \propto \frac{1}{A}$$

Losses reduce when area increases and gain increases. Bandwidth is unaffected as the gain increase is due to loss aspects but not beam angle aspects.
Unlike an amplifier we cannot relate gain × bandwidth relationship in antennas.

5.98 (c)

$$\eta\% = \frac{R_{rad}}{R_{rad} + R_{Loss}} = \frac{60}{60+30} \times 100$$

$$= \frac{60}{90} \times 100\% = 66.66\%$$

5.99 (b)

To behave as a resonator circuit the height of an antenna is an important factor and it should be integral multiple of $\lambda/2$.

5.100 (d)

Discone antenna is a monopole and disc + cone conductors.

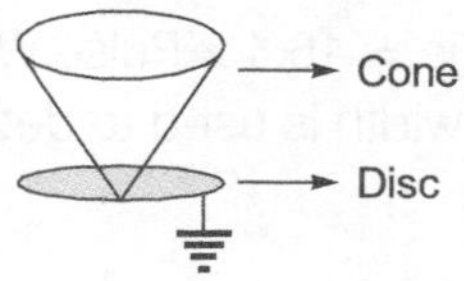

5.101 (a)

$$\eta = \frac{R_{rad}}{R_{rad} + R_{loss}} = \frac{73\Omega}{73\Omega + 7\Omega}$$

$$= 91.25\%$$

Also, $D = \dfrac{G_p}{\eta}$

Here, $G_p = 16$

$\therefore$ $D = \dfrac{16}{0.9125} = 17.53$

5.102 (c)

Compensation theorem is also known as substitution theorem.

5.103 (b)

An isotropic radiator radiates equally in all directions.

5.104 (b)

The effective length is used in radiation resistance and power calculations, hence it is measure of effectiveness of an antenna as a radiator/collector.

5.105 (a)

For a dipole antenna

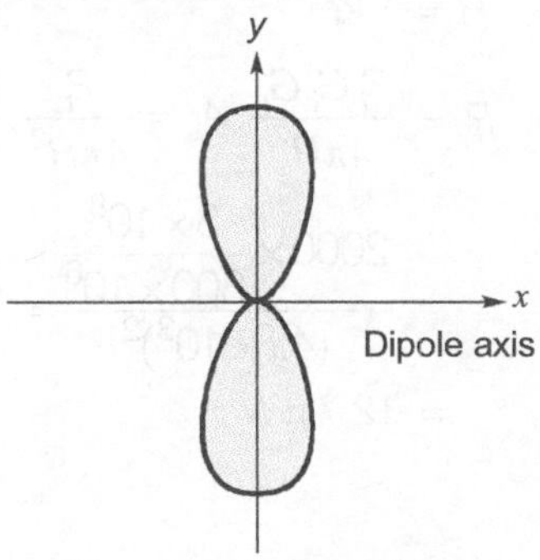

Radiation intensity is maximum along normal to the dipole axis.

5.106 (b)

Both the statements are true but the statements are made about two different kinds of antenna.

5.107 (b)

To resolve the range ambiguity we calculate the minimum range which is given by

Min range = 16.4 × Pulse Width
and pulse width is used to determine the pulse bandwidth.

5.108 (d)

Both shape and size depend on the frequency range used.

5.109 (a)

$$l = \frac{\lambda}{8}$$
$$\eta = 89.159\%$$
$$R_{rad} = 18.25\ \Omega \quad (\text{For } \lambda/4 = 36.5)$$
$$\eta = \frac{R_{rad}}{R_{rad} + R_L}$$
$$R_L = 1.5\ \Omega$$

5.110 (b)

$$f = 1\text{ MHz}$$
$$\lambda = \frac{c}{f} = \frac{3\times 10^8}{10^6} = 300\text{ m}$$
$$d = \frac{\lambda}{6} = 50\text{m}$$

For any small dipole $1/r$ terms are radiation fields and $1/r^2$ and $1/r^3$ terms are induction fields.

5.111 (d)

Given: $f = 300$ MHz ; $p_t = 2$ kW ; $G_t = G_r = 1$
$R = 1$ km

$$P_r = \frac{P_t G_t G_r}{4\pi R^2}.A_e = \frac{P_t}{4\pi R^2}.\frac{\lambda^2}{4\pi}$$
$$= \frac{2000\times\dfrac{3\times 10^8}{300\times 10^6}}{(4\pi\times 10^3)^2} = \frac{2000}{16\pi^2}\times 10^{-6}$$
$$= 12.7\ \mu\text{W}$$

5.112 (c)

Diameter of the conductor of the dipole effects only losses and loss resistance

$$R_l \propto \frac{1}{A}$$

Losses reduce when area increases and gain increases. Bandwidth is unaffected as the gain increase is due to loss aspects but not beam angle aspects.
Unlike an amplifier we cannot relate gain × bandwidth relationship in antennas.

5.113 (a)

$$\text{Gain} = \frac{4\pi A_e}{\lambda^2} = 4\pi\frac{\pi\times 5^2}{(3.16)^2} \approx 100$$
$$10\log 100 = 20\text{ dB}$$

5.114 (a)

Group pattern is given by array factor (AF)

$$AF = \frac{\sin\dfrac{N\psi}{2}}{\sin\dfrac{\psi}{2}} = \frac{\sin 4\dfrac{\psi}{2}}{\sin\dfrac{\psi}{2}}$$

where, $\psi = \beta d\cos\theta + \alpha$ $\quad \left|\ \beta = \frac{2\pi}{\lambda},\ d = \frac{\lambda}{4},\ \alpha = 0\right.$

$$\psi = \frac{\pi}{2}\cos\theta$$
$$AF = \frac{\sin\left[\dfrac{4.\frac{\pi}{2}\cos\theta}{2}\right]}{\sin\left[\dfrac{\frac{\pi}{2}\cos\theta}{2}\right]}$$
$$= \frac{\sin[\pi\cos\theta]}{\sin\left[\dfrac{\pi\cos\theta}{4}\right]} = \frac{2\sin\left[\frac{\pi}{2}\cos\theta\right]\cos\left[\frac{\pi}{2}\cos\theta\right]}{\sin\left[\dfrac{\pi\cos\theta}{4}\right]}$$
$$= 4\cos\left[\frac{\pi}{4}\cos\theta\right]\cos\left[\frac{\pi}{2}\cos\theta\right]$$

5.115 (c)

$$P_{rad} = 100\text{ W}$$
$$f = 100\text{ MHz}$$
$$l = 0.01\text{ m}$$
$$R_{rad} = 80\pi^2\left(\frac{l}{\lambda}\right)^2 = 80\pi^2\left(\frac{0.01}{3}\right)^2$$
$$P_{rad} = \frac{I_0^2 R_{rad}}{2}$$
$$I_0 = \sqrt{\frac{2P_{rad}}{R_{rad}}} = \sqrt{\frac{2\times 100}{80\pi^2\left(\dfrac{0.01}{3}\right)^2}}$$
$$\approx 151\text{ A}$$

5.116 (a)

$$f = 30 \text{ MHz}$$

$$\lambda = \frac{c}{f} = \frac{3\times10^8}{3\times10^7} = 10$$

$$\frac{\lambda}{2} = 5 \text{ m}$$

5.117 (d)

$$d = 1 \text{ m}$$
$$\eta = 0.7$$
$$f = 20 \text{ GHz}$$

$$G = 6\left(\frac{d}{\lambda}\right)^2 \quad \text{(for 70\% efficiency)}$$

$$= 6\left(\frac{1}{\frac{3\times10^8}{20\times10^9}}\right)^2 = 26666$$

$$10 \log G = 44 \approx 40 \text{ dB}$$

5.118 (b)

$$\theta_{HPBW} = 20^\circ$$
$$D = \text{Directivity}$$

$$= \frac{4\pi}{(\theta_{HPBW})^2} = \frac{41800^\circ}{(20)^2} = 103.13$$

Nearest option is (b).

5.121 (b)

Longer the size of antenna or array more the number of minor lobes.

5.122 (a)

By definition of endfire array, the maximum radiation is at 0° (or) 180° to the axis of the array.

5.123 (a)

For half-wave dipole, $R_{rad} = 73\ \Omega$

$$P_{rad} = \frac{1}{2} I_0^2 \cdot R_{rad}$$

$$\Rightarrow \quad I_0^2 = \frac{2P_{rad}}{R_{rad}} = \frac{2\times0.1\times10^3}{73} = 2.74$$

$$\Rightarrow \quad I_o = 1.65 \text{ A}$$

We have, $|E_o| = \frac{60 I_o}{r} = \frac{60\times1.65}{28\times10^3}$

$$= 3.5 \text{ mV/m}$$

5.125 (d)

$$V_p = 0.6\times3\times10^8 \text{ m/sec}$$
$$f = 10 \text{ MHz}$$

The distance (r) at which induction and radiation fields have same magnitude for EM wave is

$$\lambda = \frac{V_p}{f} = \frac{0.6\times3\times10^8}{10\times10^6} = 18 \text{ m}$$

$$r = \frac{\lambda}{6} = 3 \text{ m}$$

5.126 (a)

$$f = 3 \text{ GHz} \ ; \quad A_{et} = 2.5 \text{ m}^2$$
$$P_t = 22.5 \text{ Watt}$$
$$A_{er} = 0.5\text{m}^2 \ ; \ d = 15\times10^3 \text{ m find } P_r$$

Friis equation is $\quad P_r = G_t G_r \left(\frac{\lambda}{4\pi R}\right)^2 P_t$

$$P_r = \frac{(A_{et})(A_{er})}{\lambda^2 R^2} P_t$$

where, $\quad \lambda = \frac{c}{f} = \frac{3\times10^8}{3\times10^9} = 10^{-1}$

$$P_r = \frac{(2.5)(0.5)}{(10^{-1})^2 (15\times10^3)^2} 22.5$$

$$= 0.125\times10^{-4} \text{ W} = 12.5\ \mu\text{W}$$

5.127 (b)

$$R_{max} = \left[\frac{P_t G_d \sigma}{(4\pi) P_{r\,\text{density (min)}}}\right]^{1/4}$$

$$G_d = 16 \text{ dB} = 40$$

$$R_{max} = \left[\frac{100\times10^3\times40\times200}{(4\pi)^2\times10^{-9}}\right]^{1/4}$$

$$= 8.436 \text{ km} \simeq 8.4 \text{ km}$$

5.128 (b)

For an antenna array,

$$\Psi = \beta d \cos\phi + \alpha$$

$$\Psi = \frac{2\pi}{\lambda} d \cos\phi + \alpha$$

For broadside array progressive phase shift $\alpha = 0$

However, the question has an ambiguity as d is not given.

If $\quad d = \frac{\lambda}{2}$ then $\psi = \frac{2\pi}{\lambda}\frac{\lambda}{2}\cos\phi + 0$

$$\Psi = \pi\cos\phi$$

5.129 (b)

$$h < \frac{\lambda}{8\sin\gamma}\ ;\ \gamma = \text{depression angle}$$

So, $$h < \frac{23.5}{8(0.94)}\ \text{cm}$$

$$h < 3.125\ \text{cm}$$

5.130 (d)

Ground waves are vertically polarized waves using vertical antennas with a small height.

5.131 (c)

$$P_t = 25\ \text{W}$$
$$r = 3\ \text{m}$$

$$\text{Light intensity} = \frac{P_t}{4\pi r^2} = \frac{25}{4\pi(3)^2}\left(\frac{\text{W}}{\text{m}^2}\right)$$
$$= 221\ \text{mW/m}^2$$

5.132 (a)

Broadside array has $\theta_{max} = 90°/270°$ giving $\psi \to 0$

$$\psi = \alpha + \beta d\cos\theta$$
$$\Rightarrow \quad \alpha = 0$$

End fire array has $\theta_{max} = 0°/180°$ giving $\psi \to 0$

$$\Rightarrow \quad \psi = \alpha + \beta d\cos(0°/180°) = 0$$
$$\alpha = \pm\beta d$$

5.133 (b)

$$G = \pi^2\left(\frac{D}{\lambda}\right)^2 K\ ;\quad D = 2r = \text{Diameter}$$

K is aperture efficiency.

$$\lambda = \frac{c}{f}$$

$$G = \pi^2\left(\frac{D}{\lambda}\right)^2 K = \pi^2\left(\frac{2r}{\lambda}\right)^2 (0.7)$$
$$= 27.63\left(\frac{r}{\lambda}\right)^2$$

5.134 (a)

Antenna radiates isotropically over a half space ($2\pi r^2$).

Given, $E_{rms} = 50\times10^{-3}\left(\frac{V}{m}\right)$ at $r = 1$ km

Antenna terminal current $= I_{rms} = 3.5$ A

$$P_{rad} = I_{rms}^2 R_{rad} \quad \text{...(i)}$$

$$P_{rad} = \left|\vec{P}_{avg}\right|(\text{area})$$

$$P_{rad} = \frac{E_{rms}^2}{\eta_0}(2\pi r^2) \quad \text{...(ii)}$$

$$R_{rad} = \frac{P_{rad}}{I_{rms}^2} \quad \text{...(iii)}$$

By equations (ii) and (iii)

$$R_{rad} = \frac{E_{rms}^2(2\pi r^2)}{\eta_0 I_{rms}^2}$$
$$= \frac{(50\times10^{-3})^2 2\pi\times(10^3)^2}{120\pi\,(3.5)^2}$$
$$= \frac{50\times50}{60(3.5)^2} = 3.401\,\Omega$$

5.135 (a)

$$A_e = \frac{\lambda^2}{4\pi}G_d$$

$$G_d = \frac{4\pi A_e}{\lambda^2} = \frac{4\pi A_e}{\left(\frac{c}{f}\right)^2} = \frac{4\pi f^2 A_e}{c^2}$$

5.136 (a)

$$\text{SNR} = \frac{\lambda^2}{16\pi^2 r^2 k T_{sys} B}$$

5.137 (a)

At higher frequencies, the electric field becomes more confined to the region between the microstrip and ground plane. It is due to the frequency dispersion of the effective dielectric constant and the change in characteristic impedance with frequency.

5.138 (b)

Given: $H = 5\times10^{-6}$ A/m at $r = 2$ km and $\theta = \frac{\pi}{2}$ in air.

For a half-wave dipole,

$$H_\phi = \frac{jI_0 e^{-j\beta r}\cos\left(\frac{\pi}{2}\cos\theta\right)}{2\pi r\sin\theta}$$

$$\Rightarrow \quad |H_\phi| = \frac{I_0}{2\pi r}\times\frac{\cos\left(\frac{\pi}{2}\cos\theta\right)}{\sin\theta}$$

$$\Rightarrow \quad I_0 = \frac{|H_\phi|(2\pi r \sin\theta)}{\cos\left(\frac{\pi}{2}\cos\theta\right)}$$

Substituting the given values, we get,

$$I_0 = \frac{(5\times10^{-6})2\pi(2\times10^{3})\sin 90°}{\cos\left(\frac{\pi}{2}\cos 90°\right)}$$

$$I_0 = 20\pi \times 10^{-3}\text{ A}$$

The power radiated by the antenna is

$$P_{rad} = I_{rms}^2 R_{rad} = \left(\frac{I_0}{\sqrt{2}}\right)^2 R_{rad}$$

$$P_{rad} = \frac{1}{2}\times(20\pi)^2\times10^{-6}\times73$$

$$P_{rad} = 144096\times10^{-6}\text{ W}$$

$$P_{rad} \approx 144\text{ mW}$$

5.139 (d)

The horn antenna is an example of an aperture antenna that provides a smooth transition from a waveguide to a larger aperture that couples more effectively into space.

Horn Antennas are used directly as radiators aboard satellites to illuminate comparatively large areas of the earth, and they are also widely used as primary feeds for reflector type antennas both in transmitting and receiving modes.

5.140 (b)

The term "superdirectivity" or "supergain" broadly means a directivity that is higher than obtained with the same array length and elements uniformly excited (constant amplitude and constant phase). The Hansen-Woodyard array is a first example of super gain array.

5.141 (b)

Antenna can be considered as a transformation device converting currents from a circuit into electromagnetic photons radiating into free space. It can also act as a receiving device converting photons back into current towards the load.

Hence, both Statement (I) and Statement (II) are individually true, but Statement (II) is not the correct explanation of Statement (I).

5.142 (b)

We know that,

$$\text{Directivity, } D = G_d(\theta,\phi)\big|_{max} = \frac{4\pi U(\theta,\phi)\big|_{max}}{P_{rad}}$$

$$\text{Now, } \quad U(\theta,\phi) = r^2 P_{avg} = r^2\cdot\frac{E^2}{2\eta}$$

$$\Rightarrow \quad U(\theta,\phi) \propto E^2 \equiv F^2(\theta)$$

$$\Rightarrow \quad U(\theta,\phi) = \cos^8\theta$$

$$\therefore \quad U(\theta,\phi)\big|_{max} = 1$$

$$\text{Now, } \quad P_{rad} = \int U(\theta,\phi)\,d\Omega$$

$$= \int \cos^8\theta\cdot\sin\theta\,d\theta\,d\phi$$

$$= \int_{\theta=0}^{\pi/2}\cos^8\theta\sin\theta\,d\theta\int_{\phi=0}^{2\pi}d\phi$$

Let $\cos\theta = t \Rightarrow -\sin\theta\,d\theta = dt$

$\therefore$ at $\theta = 0°$, $t = 1$

at $\theta = \pi/2$, $t = 0$

$$\text{So, } \quad P_{rad} = -\int_1^0 t^8 dt\cdot 2\pi = \frac{t^9}{9}\Bigg|_0^1\cdot 2\pi = 0.698$$

$$\text{Hence, } \quad D = \frac{4\pi * 1}{0.698} \equiv 18$$

$$\therefore \quad D = 18$$

5.144 (a)

Radiation resistance of infinitesimal dipole of length $dl = \frac{\lambda}{50}$

$$R_{rad} = 80\pi^2\left[\frac{dl}{\lambda}\right]^2$$

$$= 80\pi^2\left[\frac{1}{50}\right]^2 = 0.316\ \Omega$$

5.145 (d)

The interchange of energy between the two antennas is known as mutual coupling which can be due to different mechanisms depending on radiation characteristics, relative separation between antennas and relative orientation of antennas. Hence, option (d) is the correct answer.

5.146 (b)

The disadvantages of different types of antennas are as matched below:

- **Cavity-backed printed antennas:** Narrow bandwidth
- **Triplate:** Coupling occurs inside structure
- **Waveguide:** Difficult to make conformal
- **Microstrip antennas:** Low efficiency

Hence, P $\rightarrow$ 3, Q $\rightarrow$ 4, R $\rightarrow$ 1, S $\rightarrow$ 2

5.147 (c)

A slot antenna of dual monopole feed can provide two separate impedance bandwidths of 124 MHz (about 5.1% centered at 2.45 GHz) and 1124 MHz (about 22.4% centered at 5.5 GHz), making it easy to cover the specification for WLAN operation in the 2.45-GHz band and 5.2/5.8-GHz bands.

5.148 (c)

- In Broadside array, the direction of the maximum radiation is perpendicular to the axis of array; while in the end fire array, the direction of the maximum radiation is along the axis of array.
- The pattern multiplication principle states that the radiation patterns of an array of N identical antennas is equal to the product of the element pattern $F_e(\theta)$ (pattern of one of the antennas) and the array pattern or group pattern $F_a(\theta)$, where $F_a(\theta)$ is the pattern obtained upon replacing all of the actual antennas with isotropic sources.

Hence, statements 2 and 3 are correct.

■■■

UNIT

IV

Computer Organization and Architecture

Syllabus

Basic architecture, CPU, I/O organisation, memory organisation, peripheral devices, trends; Hardware / software issues; Pipeline concepts & functions; Data representation& Programming; Operating systems-basics, processes, characteristics, applications; Memory management, virtual memory, file systems, protection & security; Data bases, different types, characteristics and design; Transactions and concurrency control; Elements of programming languages, typical examples.

Contents

Sl.	Topic	Page No.
1.	Basics of Computer Organisation and Architecture	314
2.	I/O Organisation and Pipelining	322
3.	Memory Organisation	330
4.	Data Representation and Programming	342
5.	Data Structures	354
6.	Operating Systems and Data Bases	362
7.	Miscellaneous	372

oooo

1 Basics of Computer Organisation and Architecture

1.1 **Assertion (A):** A processor can reference a memory stack without specifying an address.
Reason (R): The address is always available and automatically updated in the stack pointer.
(a) Both A and R are true and R is the correct explanation of A
(b) Both A and R are true but R is NOT the correct explanation of A
(c) A is true but R is false
(d) A is false but R is true [ESE-2001]

1.2 Effective address is calculated by adding or subtracting displacement value to
(a) immediate address
(b) relative address
(c) absolute address
(d) base address [ESE-2001]

1.3 Consider the following:
1. Input device
2. Arithmetic and logic unit
3. Control unit
4. Auxiliary memory
5. Main memory
6. Active hub

Which of these form part of CPU?
(a) 1, 4 and 6 (b) 2, 3 and 6
(c) 2, 4 and 5 (d) 2, 3 and 5
[ESE-2002]

1.4 A microprocessor has 24 address lines and 32 data lines. If it uses 10 bits of opcode, the size of its Memory Buffer Register is
(a) 22 bits (b) 24 bits
(c) 32 bits (d) 14 bits
[ESE-2002]

1.5 The following micro operations are part of interrupt cycle of a control unit:
1. MAR ← save-address
   PC ← routine-address
2. MBR ← (PC)
3. Memory ← (MBR)

Which of the following is the correct order of their occurrence?
(a) 1, 2 and 3 (b) 2, 3 and 1
(c) 2, 1 and 3 (d) 3, 1 and 2
[ESE-2002]

1.6 Which of the following are included in the architecture of computer?
1. Addressing modes, design of CPU.
2. Instruction set, data formats.
3. Secondary memory, operating system.

Select the correct answer using the codes given below:
(a) 1 and 2 (b) 2 and 3
(c) 1 and 3 (d) 1, 2 and 3
[ESE-2003]

1.7 In a microprocessor when a CPU is interrupted, it
(a) Stops execution of instructions
(b) Acknowledges interrupt and branches of subroutine
(c) Acknowledges interrupt and continues
(d) Acknowledges interrupt and waits for the next instruction from the interrupting device
[ESE-2003]

1.8 The content of which of the following determines the state of the CPU at the end of the execute cycle (when the interrupt is recognized)?
1. Program counter
2. Processor register
3. Certain status conditions

Select the correct answer using the codes given below:
(a) 1 and 2 (b) 2 and 3
(c) 1 and 3 (d) 1, 2 and 3
[ESE-2003]

1.9 Consider a hypothetical processor with largest instruction length being 32-bit and 16 registers $R_0 - R_{15}$. Processor supports only following instructions:

ADD Ri, Rj
SUB Ri, Rj
AND Ri, Rj
NOT Ri
MOV Ri, Rj
LOAD Address//Loads with register R_0
STORE Address//Stores the content of R_0
JUMP Address
JZ Address

What is the maximum number of address pins on this processor?
(a) 27 (b) 28
(c) 29 (d) 30 [ESE-2004]

1.10 Assertion (A): There is no overflow after an addition, if one number is positive and the other is negative.
Reason (R): Adding a positive number to a negative number always produces a result which is smaller than the larger of the two.
(a) Both A and R are true and R is the correct explanation of A
(b) Both A and R are true but R is NOT the correct explanation of A
(c) A is true but R is false
(d) A is false but R is true [ESE-2005]

1.11 Which of the following are included in the architecture of computer?
1. Addressing modes, design of CPU
2. Instruction set, data formats
3. Secondary memory, operating system

Which of the statements given above are correct?
(a) 1 and 2 only (b) 2 and 3 only
(c) 1 and 3 only (d) 1, 2 and 3 [ESE-2007]

1.12 Assertion (A): Neumann machines are called Control Flow Computers.
Reason (R): Instructions are executed sequentially as controlled by a program counter.
(a) Both A and R are true and R is the correct explanation of A
(b) Both A and R are true but R is NOT the correct explanation of A
(c) A is true but R is false
(d) A is false but R is true [ESE-2007]

1.13 Assertion (A): Data-flow computers exploit maximum parallelism.
Reason (R): Data-flow computers require no program counter.
(a) Both A and R are true and R is the correct explanation of A
(b) Both A and R are true but R is NOT the correct explanation of A
(c) A is true but R is false
(d) A is false but R is true [ESE-2007]

1.14 The following micro-operations are part of interrupt cycle of a control unit:
1. MAR $\leftarrow$ save-address
PC $\leftarrow$ routine address
2. MBR $\leftarrow$ (PC)
3. Memory $\leftarrow$ (MBR)

Which one of the following is the correct order of their occurrence?
(a) 1-2-3 (b) 2-3-1
(c) 2-1-3 (d) 3-1-2 [ESE-2007]

1.15 Which one of the following is correct?
A micro program
(a) is usually written in high level language
(b) is a program for micro computers
(c) is a program written in assembly language
(d) is a sequencing program for the control unit of any processor [ESE-2008]

1.16 Which one of the following is correct?
A micro program is
(a) any source program run on micro computers
(b) any set of instructions for primitive operation in a system
(c) a general name for 'macros' used in assemble language programming
(d) any program of a very small size [ESE-2008]

1.17 Which of the following is/are correct statements (s)?
1. Bus is a group of wires carrying information.
2. Bus is needed to achieve reasonable speed of operation.
3. Bus can carry data or address.
4. A bus can be shared by more than one device.

Select the correct answer from the codes given below :
(a) 1 only (b) 1 and 2 only
(c) 2, 3 and 4 only (d) 1, 2, 3 and 4 [ESE-2009]

1.18 Which of the following are included in the architecture of a computer ?
1. Addressing modes, CPU
2. Instruction set, data formats
3. Secondary memory, operating system

Select the correct answer from the codes given below :
(a) 1 and 2 (b) 1 and 3
(c) 2 and 3 (d) 3 only [ESE-2009]

1.19 Identification of highest priority interrupt can be achieved in a minimum time by which of the following schemes ?
(a) Hardwired polling
(b) Priority encoder circuit
(c) Software polling
(d) Subdivision of interrupt register into a number of sub-registers which are checked parallely
[ESE-2009]

1.20 On receiving an interrupt from an I/O device, the CPU
(a) Halts for a predetermined time.
(b) Branches off to the interrupt service routine after completion of the current instruction.
(c) Branches off to the interrupt service routine immediately.
(d) Hands over control of address bus and data bus to the interrupting device.
[ESE-2010]

1.21 A single bus structure is primarily found in
(a) Main frames
(b) Mini and micro computers
(c) Super computers
(d) High performance machines [ESE-2010]

1.22 Consider the following statements:
1. Compilers
2. Design
3. Evaluation
4. Instruction set architecture

Which of these are included in the present definition of computer architecture to design a full computer system?
(a) 1, 2 and 3 (b) 1, 3 and 4
(c) 2, 3 and 4 (d) 1, 2, 3 and 4
[ESE-2010]

1.23 Which of the following are the problems with using Millions Instructions Per Second (MIPS) as a measure for comparing computer performance?
1. It does not take into account the capabilities of the instructions.
2. MIPS can vary inversely with performance.
3. MIPS varies between programs on the same computer.

(a) 1 and 2 only (b) 2 and 3 only
(c) 1 and 3 only (d) 1, 2 and 3
[ESE-2011]

1.24 A bus organized processor consists of 15 registers. The number of selection lines in each multiplexer and in the destination decoder are respectively
(a) 2 and 4 (b) 4 and 2
(c) 4 and 4 (d) 4 and 8
[ESE-2011]

1.25 In writing the micro-program, there are two situations in which a field of the micro instruction can be kept blank when it
1. Controls a functional unit
2. Causes state to be written
3. Specifies the control of a multiplexer

Which of these statements are correct?
(a) 1 and 2 only (b) 2 and 3 only
(c) 1 and 3 only (d) 1, 2 and 3
[ESE-2011]

1.26 The incorrect match (when $n > 1$) is
(a) SISD model of computer : 1 control unit and 1 ALU
(b) SIMD model of computer : 1 control unit and n ALUs
(c) MISD model of computer : n control units and n ALUs
(d) MIMD model of computer : n control units and 1 ALU
[ESE-2012]

1.27 In case the code is position dependent, the most suitable addressing mode is
(a) Direct mode (b) Indirect mode
(c) Relative mode (d) Indexed mode
[ESE-2012]

1.28 **Statement (I):** The data path contains all the circuits to process data within the CPU with the help of which data is suitably transformed.
Statement (II): It is the responsibility of the control path to generate control and timing signals as required by the opcode.

(a) Both Statement (I) and Statement (II) are individually true and Statement (II) is the correct explanation of Statement (I).
(b) Both Statement (I) and Statement (II) are individually true but Statement (II) is not the correct explanation of Statement (I).
(c) Statement (I) is true but Statement (II) is false.
(d) Statement (I) is false but Statement (II) is true.
[ESE-2013]

1.29 The addressing mode that permits relocation, without any change whatsoever in the code, is
(a) Indirect addressing
(b) Base register addressing
(c) Indexed addressing
(d) PC relative addressing [ESE-2014]

1.30 Data transfer between the main memory and the CPU register takes place through two registers, namely,
(a) General purpose register and MDR
(b) Accumulator and Program counter
(c) MAR and MDR
(d) MAR and Accumulator [ESE-2016]

1.31 A processor has 32-bit architecture. Each instruction is 1 word long (32 bits). It has 64 registers. It supports 50 instructions, which have 2 register operands +1 immediate operand. Assuming that the immediate operand is an unsigned integer, what is its maximum value?
(a) 16383 (b) 32767
(c) 65536 (d) 1024 [ESE-2016]

1.32 Consider the following :
1. Operation code
2. Source operand reference
3. Result operand reference
4. Next instruction reference

Which of the above are typical elements of machine instructions?
(a) 1, 2 and 3 only (b) 1, 2 and 4 only
(c) 3 and 4 only (d) 1, 2, 3 and 4
[ESE-2016]

1.33 Which addressing mode helps to access table data in memory efficiently?
(a) Indirect mode
(b) Immediate mode
(c) Auto-decrement or Auto-increment mode
(d) Index mode [ESE-2016]

1.34 Consider the following statements:
1. A computer will have a multiply instruction.
2. Multiply instruction will be implemented by a special multiply unit.

Which of the following is correct?
(a) Both 1 and 2 are not architectural design issues.
(b) Both 1 and 2 are not organizational issues.
(c) 1 is an architectural design issue while 2 is an organizational issue.
(d) 1 is an organizational issue while 2 is an architectural design issue.
[ESE-2019]

1.35 A core of processor chip consists of
1. ALU
2. Instruction logic
3. Load/store logic
4. L3 cache
5. L1 data cache

(a) 1, 2, 3 and 4 only (b) 1, 2, 3 and 5 only
(c) 2, 3, 4 and 5 only (d) 1, 4 and 5 only
[ESE-2020]

1.36 Superscalar processor consists of
(a) Single pipeline for instruction execution
(b) Multiple-instruction pipelines for instruction execution
(c) No pipelines for instruction execution
(d) Multiple combination of hardware for execution
[ESE-2020]

1.37 Which one of the following characteristics is correct regarding RISC processor?
(a) Relatively very large addressing modes
(b) Multi-cycle instruction execution
(c) All operations are not done within the registers of the CPU
(d) Relatively few instructions
[ESE-2021]

1.38 Which one of the following instructions in a stack computer consists of an operation code only with no address field?
(a) PUSH X (b) POP X
(c) ADD (d) LOAD A
[ESE-2021]

1.39 Which of the following basic steps are needed in order to perform a memory read operation?

1. The address of the location from which the word is to be read is loaded into the memory address register (MAR).
2. A signal, called read (issued by the CPU indicating that the word whose address is in the MAR) is to be read into the MDR.
3. Corresponding to the memory delay in reading the specified word, the required word will be loaded by the memory into the MDR ready for use by the CPU.

Select the correct answer using the code given below.

(a) 1 and 2 only (b) 2 and 3 only
(c) 1 and 3 only (d) 1, 2 and 3

[ESE-2022]

1.40 Which of the following instructions are long and allow maximum parallelism since each bit controls a single control line?

(a) Horizontal microinstructions
(b) Vertical microinstructions
(c) Diagonal microinstructions
(d) Jumbled microinstructions [ESE-2022]

1.41 What is the overall number of Clock cycles Per Instruction (CPI) for a machine A for which the following performance measures were recorded when executing a set of benchmark programs? (Assume the clock rate of the CPU as 200 MHz and execution of 100 instructions)

Instruction category	Percentage of occurence	No. of cycles per instruction
ALU	38	1
Load and store	15	3
Branch	42	4
Others	5	5

(a) 2.76 (b) 4.76
(c) 6.76 (d) 8.76 [ESE-2023]

1.42 Which one of the following is a program that takes an object file generated and generates a file in a binary code called COM file or EXE file?

(a) Editor (b) Assembler
(c) Loader (d) Debugger

[ESE-2023]

1.43 Non-uniform memory access belongs to

(a) SISD (b) SIMD
(c) MISD (d) MIMD [ESE-2024]

■■■

Answers Basics of Computer Organisation and Architecture

1.1 (a)	1.2 (d)	1.3 (d)	1.4 (c)	1.5 (c)	1.6 (a)	1.7 (d)	1.8 (c)	1.9 (b)
1.10 (a)	1.11 (a)	1.12 (a)	1.13 (a)	1.14 (c)	1.15 (d)	1.16 (b)	1.17 (d)	1.18 (a)
1.19 (a)	1.20 (b)	1.21 (b)	1.22 (c)	1.23 (c)	1.24 (c)	1.25 (a)	1.26 (d)	1.27 (c)
1.28 (a)	1.29 (d)	1.30 (d)	1.31 (a)	1.32 (d)	1.33 (c)	1.34 (c)	1.35 (b)	1.36 (b)
1.37 (d)	1.38 (c)	1.39 (d)	1.40 (a)	1.41 (a)	1.42 (c)	1.43 (d)		

Explanations Basics of Computer Organisation and Architecture

1.1 (a)

The variable stack pointer always contains the address of the top of the stack.
Whenever there is an PUSH or POP operation on the stock the address value from stack pointer Ps accessed and updated upon the operations.

1.2 (d)

Effective address = base address ± displacement

1.3 (d)

ALU and control unit forms are the basic parts of a Central Processing Unit (CPU). Some authors also include main memory or primary memory also as a part of CPU. So, among the given options, (d) can be selected.

1.4 (c)

Size of MBR depends on the data pins.

1.5 (c)

2. MBR ← (PC)
1. MAR ← save-address
   PC ← routine-address
3. Memory ← (MBR)

2, 1, 3 is correct order.

1.6 (a)

- Addressing modes, designed of CPU, instructions set and data are particular to specific CPU.
- Secondary memory and operating system can work with different computer with different architecture hence, there are not part of computer architecture.

1.7 (d)

When a CPU is interrupted it acknowledges interrupt and waits for the next instruction from the interrupting device.

1.8 (c)

At the end of the execute cycle, the CPU checks for an interrupt, if interrupt has occurred then the address of Program Counter (PC) is moved to Memory Address Register and PC is updated with interrupt routine address. Meanwhile, certain status conditions (flags) may also determine state of CPU.

1.9 (b)

Instruction length is 32-bit.
Total 16 registers R_0 to R_{15}
Number of instructions given nine ($9 < 2^4$)
It need 4-bits to support the given instructions and remaining 28-bits of instruction can be used for addressing.
∴ 28-pins can present maximum for addressing.

1.10 (a)

Addition of positive number and negative number will not overflow because result is always less than the larger of two.

1.11 (a)

- Addressing modes, designed of CPU, instructions set and data are particular to specific CPU.
- Secondary memory and operating system can work with different computer with different architecture hence, there are not part of computer architecture.

1.12 (a)

In Von-Neumann machines, the next instruction to be executed by processor is decided by the program counter (PC), which is updated with sequential addresses.

1.14 (c)

2. MBR ← (PC)
1. MAR ← save-address
   PC ← routine-address
3. Memory ← (MBR)

2, 1, 3 is correct order.

1.15 (d)

Micro program is program for control unit of processor.

1.16 (b)

Micro program is the set of instructions for primitive operation in a system.

1.17 (d)

- Bus is group of wires (lines) carrying information.
- Bus is needed to achieve reasonable speed of operation.
- Bus can carry data or address.
- Bus can be shared by more than one device.

All statements are correct.

1.18 (a)

- Addressing modes, designed of CPU, instructions set and data are particular to specific CPU.
- Secondary memory and operating system can work with different computer with different architecture hence, there are not part of computer architecture.

1.19 (a)

Hardwired polling is best far identification of highest priority interrupt can be achieved in a minimum time.
Hardwired polling (Daisy chaining) in which each I/O device is connected to a common priority line. The advantage of this is that, when low priority interrupt is already generated, it can be overwritten by the highest priority signal before it reaches the processor.

1.20 (b)

On receiving an interrupt from an I/O device, the CPU branches off to the interrupt service routine after completion of the current instruction.

1.22 (c)

Compiler is never part of architecture and output is assembly code which can be executed on any architecture.

1.23 (c)

Architecture like RISC and CISC are different

CISC: Complex and more number of instructions.

RISC: Simple and reduced number of instructions are not compared using MIPS, so 1 is true.

Since CPU inserts variable delay cycles as per priorities hence MIPS varies between programs on the same computer.

As per previous example of RISC and CISC

Since they have different MIPS for executing a program (some time CISC takes less MIPS for a program and for same program RISC may take vice versa more MIPS and hence MIPS does not vary inversely with performance.

So statement 2 is not the problem.

1.24 (c)

$\because$ Number of registers = 15

$\therefore$ Selection lines in MUX will be

$$2^n \geq 15$$
$$n \geq \log_2 15$$
$$n_{min} = 4$$

This will be also number of input lines for decoders, the bus architecture will be like this:

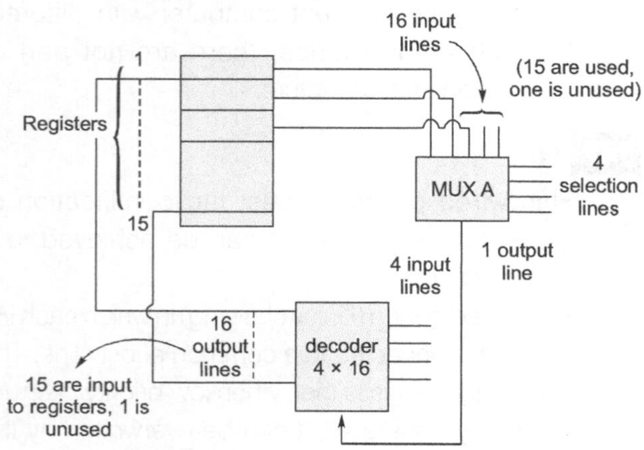

So, 4, 4 is the answer.

1.25 (a)

There are two situations in which micro instructions can be kept blank when it controls a functional unit, and causes state to be written.

1.26 (d)

MIMD model has n-control units and since in multiprocessing, it requires n ALU units also.

1.27 (c)

If the code is position dependent then suitable addressing mode is relative addressing mode. The effective address computed from X(PC), X bytes away from the address pointed by program counter.

Relative addressing mode uses program counter and indexed addressing mode uses some general purpose register.

1.28 (a)

The control unit tells the data path what to do, based on the instruction that is currently being executed.

The control unit sets the data path signals appropriately so that registers are read, ALU output is generated, data memory is read or written and branch target addresses are computed.

1.29 (d)

PC-relative addressing mode:

Effective PC address = Next instruction address + offset

Here, offset is a positive or negative constant which makes relocation without any change in the code.

Base register addressing, Indirect addressing and indexed addressing are dependent on the code, which can be used in assignment statement, array elements, etc.

1.31 (a)

Opcode	Reg	Reg	Immediate
$\log_2 50$	$\log_2 64$	$\log_2 64$	14 bit
↓ 6 bit	↓ 6 bit	↓ 6 bit	

Largest unsigned value

$$= (2^n - 1) = (2^{14} - 1) = 16383$$

1.37 (d)

RISC (Reduced Instruction Set Computer) is a type of microprocessor that has limited number of instructions. Hence, option (d) is correct.
In RISC, the instruction set contains simple and basic instructions and the most instructions are completed in one cycle. It supports simple addressing modes and contains more number of general purpose registers for faster execution of the instruction.

1.38 (c)

In the stack CPU, ALU operations are performed only on stack data. Hence, no need of an operand address. It is a zero address instruction.

1.39 (d)

Three basic steps are needed in order to perform a memory read operation.

1. The address of the location from which the word is to be read is loaded into the MAR (Memory Address Register).
2. A signal called read, is issued by the CPU indicating that the word whose address is in the MAR is to be read into the MDR (Memory Data Register).
3. After some time, corresponding to the memory delay in reading the specified word, the required word will be loaded by the memory into the MDR ready for use by the CPU.

Hence, all the given statements are correct.

1.40 (a)

There are two distinct ways of organising microinstructions:

(i) **Horizontal microinstruction:** It uses one bit for each control signal. Hence, the length of the microinstruction is high and allow maximum parallelism i.e. they control many hardware resources operating in parallel.

(ii) **Vertical microinstructions:** The encoded signals are represented in encoded binary format. Hence, it has low instruction length and low parallelism.

1.41 (a)

We know that,

$$\text{Average CPI} = \frac{\text{No.of Cycles/program}}{\text{No. of Instruction/program}};$$

$$= \frac{(38 \times 1) + (15 \times 3) + (42 \times 4) + (5 \times 5)}{100} = 2.76$$

1.42 (c)

Assembler O/P → Object file.

Loader:

1. Converts object file to hex file.
2. Finally dumps program to memory.

1.43 (d)

Non-uniform memory access, or NUMA, is a method of configuring a cluster of microprocessors in a multiprocessing system so they can share memory locally. In a NUMA setup, the individual processors in a computing system share local memory and can work together.
MIMD architectures allow multiple processors to execute different instructions on different data sets concurrently. NUMA systems belongs to MIMD architecture because they consist of multiple processors accessing the shared, with each capable of executing different instructions and accessing different data.

■■■

2 I/O Organisation and Pipelining

2.1 Consider the following statements:
The horizontal microinstruction has
1. longer control word than vertical microinstruction.
2. high degree of parallelism.
3. slower execution than vertical microinstruction.
Which of these statements is/are correct?
(a) 1 alone (b) 2 alone
(c) 1 and 2 (d) 2 and 3 [ESE-2000]

2.2 Which of the following is not a characteristic of transparent DMA mode of I/O operation?
(a) The external logic steals cycles from the CPU
(b) The normal rate of execution is slowed down
(c) Only one word can be transferred at a time
(d) Data is transferred to/from memory directly
[ESE-2002]

2.3 Which one of the following is not a characteristic of RISC processor design?
(a) One instruction per cycle
(b) Register-to-register operations only
(c) Simple address modes
(d) Register-to-memory operations only
[ESE-2002]

2.4 **Assertion (A):** Reduced Instruction Set Computers (RISC) use pipelined control unit.
Reason (R): Pipelining reduces memory requirements of programs.
(a) Both A and R are true and R is the correct explanation of A
(b) Both A and R are true but R is NOT the correct explanation of A
(c) A is true but R is false
(d) A is false but R is true
[ESE-2003]

2.5 A particular parallel program computation requires 100 seconds when executed on a single processor. If 40 percent of this computation is 'inherently sequential', then what are the theoretically best elapsed times for this program running with 2 and 4 processors, respectively?
(a) 20 and 10 seconds
(b) 30 and 15 seconds
(c) 50 and 25 seconds
(d) 70 and 55 seconds [ESE-2004]

2.6 Match **List-I** (Characteristic) with **List-II** (Processor Architecture) and select the correct answer using the code given below the lists:
List-I
A. Micro-code for several instructions
B. Lack of indirect addressing
C. Presence of on-chip cache
D. Simple optimizing compiler
List-II
1. Both RISC and CISC
2. CISC only
3. Neither RISC nor CISC
4. RISC only

Codes:

	A	B	C	D
(a)	2	4	1	3
(b)	1	3	2	4
(c)	2	3	1	4
(d)	1	4	2	3

[ESE-2004]

2.7 **Assertion (A):** The DMA technique is more efficient than the Interrupt-driven technique for high volume I/O data transfer.
Reason (R): The DMA technique does not make use of the Interrupt mechanism.
(a) Both A and R are true and R is the correct explanation of A
(b) Both A and R are true but R is NOT the correct explanation of A
(c) A is true but R is false
(d) A is false but R is true [ESE-2004]

2.8 Consider the following statements:
The advantage of cycle stealing in DMA is that
1. it increases the maximum I/O transfer rate.
2. it reduces the interference by the DMA controller in the CPU's memory access.

3. it is beneficially employed for I/O devices with shorter bursts of data transfer.

Which of the statements given above are correct?

(a) 1 and 2 (b) 1 and 3
(c) 2 and 3 (d) 1, 2 and 3

[ESE-2004]

2.9 For which of the following devices, is DMA the most suitable?

(a) Keyboard (b) Mouse
(c) Joy stick (d) Hard disk

[ESE-2006]

2.10 In a microcomputer, why are wait states used?

(a) To make the processor wait during a DMA operation
(b) To make the processor wait during an interrupt processing
(c) To make the processor wait during a power shutdown
(d) To interface slow peripherals to the processor

[ESE-2007]

2.11 **Assertion (A):** DMA is faster than either Interrupt initiated I/O or Polling based I/O for very large data transfers.

Reason (R): DMA takes control of the system buses and needs no processor intervention during the data transfer.

(a) Both A and R are true and R is the correct explanation of A
(b) Both A and R are true but R is NOT the correct explanation of A
(c) A is true but R is false
(d) A is false but R is true [ESE-2007]

2.12 The parallel computers are divided into which of the following architecture configurations?

1. Array processors
2. Data processors
3. Multi processor systems
4. Pipeline computers

Select the correct answer using the code given below:

(a) 1, 2 and 3 only (b) 2, 3 and 4 only
(c) 1, 3 and 4 only (d) 1, 2 and 4 only

[ESE-2008]

2.13 What is an interrupt in which the external device supplies its address as well as the interrupt request known as?

(a) Vectoral interrupt
(b) Maskable interrupt
(c) Non-maskable interrupt
(d) None of the above [ESE-2008]

2.14 What is the purpose of DMA facility in microprocessor based system?

(a) To increase the speed of data transfer between the microprocessor and I/O devices.
(b) To increase the speed of data transfer between the μP and the memory
(c) To increase the speed of data transfer between the memory and the I/O devices
(d) To improve the reliability of the system

[ESE-2009]

2.15 Match **List-I** with **List-II** and select the correct answer using the codes given below the lists:

List-I

A. Pipelined ALU
B. Simpler compiler
C. Separate data and instruction caches
D. Lesser cycles per instruction

List-II

1. RISC
2. CISC
3. Mixed
4. Mixed RISC-CISC

Codes:

	A	B	C	D
(a)	3	2	3	1
(b)	1	2	3	3
(c)	3	3	2	1
(d)	3	3	3	1

[ESE-2010]

2.16 The ability to halt the CPU temporarily and use this interval of time to send information on buses is called

(a) Cycle stealing
(b) Vectoring an interrupt
(c) Polling
(d) Direct memory access

[ESE-2010]

2.17 A 'DMA' transfer implies

(a) Direct transfer of data between memory and accumulator
(b) Direct transfer of data between memory and I/O devices without the use of microprocessor

(c) Transfer of data exclusively within microprocessor registers
(d) A fast transfer of data between microprocessor and I/O devices

[ESE-2010]

2.18 In microprocessor based systems DMA facility is required to increase the speed of data transfer between the
(a) Microprocessor and the I/O devices
(b) Microprocessor and the memory
(c) Memory and the I/O devices
(d) Memory and the reliability system

[ESE-2010]

2.19 The speed gained by an 'n' segment pipeline executing 'm' tasks is

(a) $\frac{(n+m-1)}{mn}$ (b) $\frac{mn}{(n+m-1)}$

(c) $\frac{n+m}{(mn-1)}$ (d) $\frac{n+m}{(mn+1)}$

[ESE-2011]

2.20 Match **List-I** with **List-II** and select the correct answer using the code given below the lists:

List-I	List-II
A. DMA I/O	**1.** High speed RAM
B. Cache	**2.** Disk
C. Interrupt I/O	**3.** Printer
D. Condition code registers	**4.** ALU

Code:

	A	B	C	D
(a)	4	1	3	2
(b)	2	1	3	4
(c)	4	3	1	2
(d)	2	3	1	4

[ESE-2011]

2.21 Which of the following instruction processing activity of the CPU can be pipelined?
1. Instruction encoding
2. Operand loading
3. Operand storing

(a) 1 and 2 only (b) 2 and 3 only
(c) 1 and 3 only (d) 1, 2 and 3

[ESE-2011]

2.22 In a microprocessor, the service routine for a certain interrupt starts from a fixed location of memory which cannot be externally set, but the interrupt can be delayed or rejected. Such an interrupt is
(a) Non-maskable and non-vectored
(b) Maskable and non-vectored
(c) Non-maskable and vectored
(d) Maskable and vectored [ESE-2012]

2.23 The following register holds the instruction before it goes to the decoder:
(a) Control register (b) Accumulator
(c) Address register (d) Data register

[ESE-2012]

2.24 **Statement (I):** The DMA technique is more efficient than the Interrupt-driven technique for high volume I/O data transfer.
Statement (II): The DMA technique does not make use of the Interrupt mechanism.
(a) Both Statement (I) and Statement (II) are individually true and Statement (II) is the correct explanation of Statement (I).
(b) Both Statement (I) and Statement (II) are individually true but Statement (II) is not the correct explanation of Statement (I).
(c) Statement (I) is true but Statement (II) is false.
(d) Statement (I) is false but Statement (II) is true.

[ESE-2012]

2.25 The type of device used to interface a parallel data format with external equipment's serial format is
(a) Key matrix (b) UART
(c) Memory chip (d) Serial-in, parallel-out

[ESE-2012]

2.26 During direct memory access (DMA), the INTR and INTA (interrupt and interrupt acknowledge) lines
(a) are never used
(b) are used by the DMA controller to acquire the data and address bus from the microprocessor
(c) are used by the IO device to invoke an error routine if a problem occurs during DMA transfer
(d) are used by the DMA controller to signal the microprocessor about the end of the current data transfer

[ESE-2013]

2.27 An interrupt in which an external device supplies its address and the interrupt request, is called
(a) poll interrupt
(b) maskable interrupt
(c) non-maskable interrupt
(d) vectored interrupt [ESE-2013]

2.28 The method used for resolving data dependency conflict by the compiler itself is
(a) Delayed load
(b) Operand forwarding
(c) Prefetch target instruction
(d) Loop buffer [ESE-2014]

2.29 Consider the following statements:
The advantage of cycle stealing in DMA is that
1. It increases the maximum I/O transfer rate.
2. It reduces the interference by the DMA controller in the CPU's memory access.
3. It is beneficially employed for I/O device with shorter bursts of data transfer.

Which of the above statements are correct?
(a) 1 and 2 only (b) 1 and 3 only
(c) 2 and 3 only (d) 1, 2 and 3
[ESE-2015]

2.30 The correct instruction execution sequence is
(a) Execute, Decode and Fetch
(b) Fetch, Execute and Decode
(c) Execute, Fetch and Decode
(d) Fetch, Decode and Execute [ESE-2015]

2.31 The serial connection of interrupt lines for establishing hardware priority is known as
(a) Daisy Chaining Priority
(b) Parallel Priority
(c) Polling
(d) Serial-line Priority [ESE-2016]

2.32 **Statement (I):** Isolated I/O method isolates memory and I/O addresses.
Statement (II): In isolated I/O method, memory and I/O interfaces have their own individual address space.
(a) Both Statement I and Statement II are individually true and Statement II is the correct explanation of Statement I
(b) Both Statement I and Statement II are individually true but Statement II is not the correct explanation of Statement I
(c) Statement I is true but Statement II is false
(d) Statement I is false but Statement II is true
[ESE-2018]

2.33 **Statement (I):** The internal interrupt is initiated by some exceptional condition caused by the program itself rather than by an external event.
Statement (II): External interrupt depends on external conditions that are independent of the program being executed at the time.
(a) Both Statement I and Statement II are individually true and Statement II is the correct explanation of Statement I
(b) Both Statement I and Statement II are individually true but Statement II is not the correct explanation of Statement I
(c) Statement I is true but Statement II is false
(d) Statement I is false but Statement II is true
[ESE-2018]

2.34 In which one of the following situations is the CPU often idle?
(a) The speeds of the mechanical I/O devices are intrinsically slower than those of electronic devices.
(b) The speeds of the electromechanical I/O devices are intrinsically faster than those of electronic devices.
(c) The speeds of the electrical I/O devices are intrinsically slower than those of electronic devices.
(d) The speeds of the electrical I/O devices are intrinsically faster than those of electronic devices.
[ESE-2021]

2.35 Which one of the following is a drawback of Programmed and Interrupt-Driven I/O?
(a) The processor is tied up in managing an I/O transfer; a number of instructions must be executed for each I/O transfer.
(b) A more efficient technique is to use a daisy chain, which provides, in effect, a hardware poll.
(c) When the processor detects an interrupt, it branches to an interrupt service routine whose job is to poll each I/O module.
(d) A more efficient technique is not to use a daisy chain, which provides, in effect, a hardware poll.
[ESE-2023]

2.36 **Statement (I) :** Pipeline processing cycle overlaps computer instruction cycle in execution for the performance improvement.
Statement (II) : Pipelining is a technique of decomposing a sequential process into sub-operations, with each sub-process being executed in a special dedicated segment that operates concurrently with all other segments.

(a) Both Statement (I) and Statement (II) is the correct explanation of Statement (I).
(b) Both Statement (I) and Statement (II) are individually true, but Statement (II) is not the correct explanation of Statement (I).
(c) Statement (I) is true, but Statement (II) is false.
(d) Statement (I) is false, but Statement (II) is true

[ESE-2023]

2.37 Which one of the following is not related to the CISC architecture?
(a) Typically 100 - 250 instructions
(b) 5-20 addressing modes
(c) Fixed length instruction formats
(d) Instructions that manipulate operands in memory

[ESE-2024]

2.38 Consider the following statements regarding the instruction pipelining :
1. It is used for increasing the efficiency of the advanced microprocessors.
2. Pipelining a processor means breaking down its instruction into a series of discrete pipeline stages which can be completed in sequence by specialized hardware.
3. The performance and programming simplification is achieved by pipelining operation with the elimination of pipeline interlocks and the control of pipeline is simplified.

Which of the above statements are correct?
(a) 1 and 2 only
(b) 1 and 3 only
(c) 2 and 3 only
(d) 1, 2 and 3

[ESE-2024]

■■■

Answers I/O Organisation and Pipelining

2.1 (c)	2.2 (b)	2.3 (d)	2.4 (a)	2.5 (d)	2.6 (b)	2.7 (a)	2.8 (d)	2.9 (d)
2.10 (d)	2.11 (a)	2.12 (c)	2.13 (a)	2.14 (c)	2.15 (a)	2.16 (d)	2.17 (b)	2.18 (c)
2.19 (b)	2.20 (b)	2.21 (a)	2.22 (d)	2.23 (a)	2.24 (a)	2.25 (b)	2.26 (a)	2.27 (b)
2.28 (a)	2.29 (b)	2.30 (d)	2.31 (a)	2.32 (a)	2.33 (b)	2.34 (a)	2.35 (a)	2.36 (a)
2.37 (c)	2.38 (d)							

Explanations I/O Organisation and Pipelining

2.1 (c)

Horizontal microstructure has:
(a) 1 bit/control signal (longer control word)
(b) little encoding (faster)
(c) high degree of parallelism (more than 1 control signal enabled at a time).

2.2 (b)

Transparent DMA mode of I/O operation:
- External logic may steel cycles from CPU.
- Only one word can be transferred at a time.
- Data is transferred to/from memory directly.

The normal rate of execution not affected by this mode.

2.3 (d)

Most RISC instructions use register-to register operations.

2.4 (a)

In pipe lining we use the concept overlapped execution of program hence there by reducing its memory requirements.

2.5 (d)

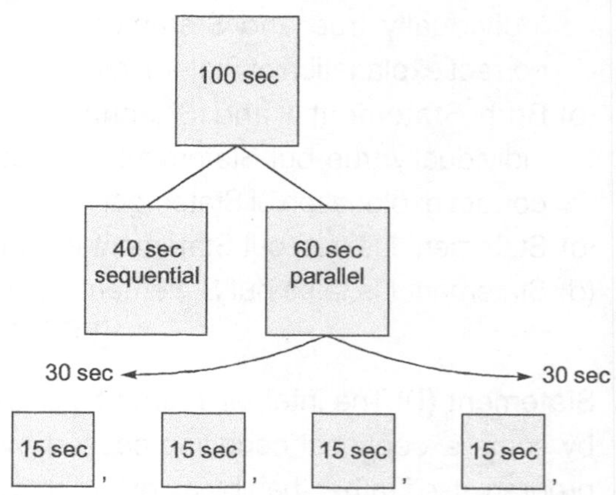

Take a single processor which requires 100 seconds for computation. It's 40 sec computation is done sequentially.

⇒ 60 sec computation is done parallel.

If 2 processors are used

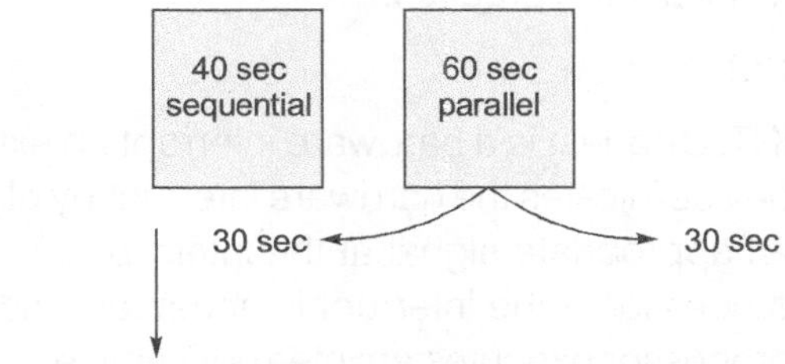

It leads to maximum i.e. (40 + 30) sec = 70 sec

If 4 processors are used

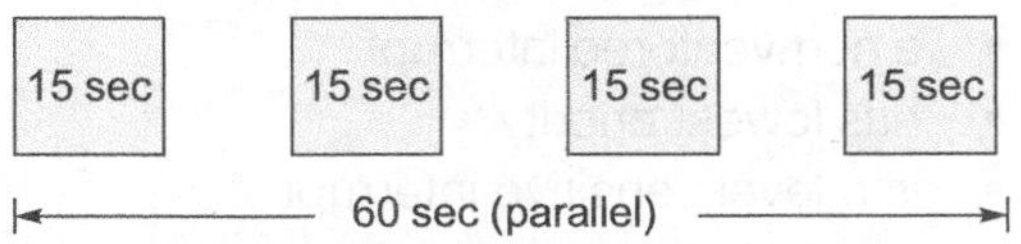

Here maximum elapsed time

$$= (40 + 15)\text{sec} = 55 \text{ sec.}$$

2.7 (a)

The DMA technique does not make use of the interrupt mechanism, that's why it is more efficient than the interrupt-driven technique for high volume I/O data transfer.

2.8 (d)

All are advantages of cycle stealing in DMA.

2.9 (d)

Keyboard, mouse and joy-stick interact with system to get the immediate response, but hard disk and memory transfers need DMA to interleave in burst mode or cycle stealing mode, without the processor-intervention.

2.10 (d)

Generally the processor is fast and I/O devices reading data from the memory takes lot of time. During this time we use the waiting state (or I/O buffer) so that the processor can do meaningful work in mean time.

2.11 (a)

(a) In diagram controlled I/O, the processor keeps polling till the device ready. Then the processor has to take care of transfer. This is too much overhead for processor.

(b) In interrupt driven I/O, the polling overhead is reduced but during transfer the processor does not do any useful work.

(c) In DMA there is no overhead on processor as DMA takes the control over the bus.

2.13 (a)

Vectored interrupt (RST n) is an interrupt which supplies it address as well as interrupt request. The processor multiplies n by 8 to know the address.

2.14 (c)

In DMA (Direct Memory Access) scheme, data are directly transferred from on I/O device to or from RAM to I/O device. The microprocessor has to relinquish the control of the address and data buses for DMA operation on the request of I/O device. Thus the speed increases in DMA operation.

2.16 (d)

The method used to transfer data from I/O units to memory by suspending the memory CPU data transfer for one memory cycle is called **Cycle stealing**.

2.17 (b)

DMA : Direct memory access

In microprocessor environment the DMA concept arised only to increase speed of access between memory and I/O devices. The is lot of difference between the speed of CPU, memory and I/O devices and DMA takes care of all those issues.

2.18 (c)

In microprocessor environment the DMA concept arised only to increase speed of access between memory and I/O devices. There is lot of difference between the speed of CPU, memory and I/O devices and DMA takes care of all those issues.

2.19 (b)

Tasks → m

Stages in pipeline = n

Without pipelining number of cycles required to execute m tasks = $n\,m$.

(∵ each task required n cycle)

When we pipeline the tasks for Ist task it requires

n cycles and for next $(m-1)$ 1 cycle for each $(m-1)$ tasks.
So total cycles required with pipelining

$$= n + (m-1) \times 1$$
$$= (n + m - 1)$$

∴ Speed gained by pipeline

$$= \frac{\text{Number of cycles without pipelining}}{\text{Number of cycles with pipeline}}$$

$$= \frac{nm}{(n+m-1)}$$

2.20 (b)

Cache is used along with main memory with speed up the execution → Cache → High speed RAM (1)
DMA I/O is used for data transfers between memories (disks) taking the control from the CPU using main memory.
DMA I/O → disk (2)

2.21 (a)

Pipelining basically refer to queuing of instructions such that the next instruction is being immediately followed after the previous instruction is being proceed. Hence pipelining is basically push and process.
Hence operand storing can not be done using pipeline.

2.22 (d)

Maskable interrupt is one which can be delayed or rejected due to non-urgency when the microprocessor is busy.
If the instruction execution program is already written at a particular fixed location, then it is called vectored instruction.

2.24 (a)

DMA controller requests for the data bus lines from the microprocessor through the HOLD pin. If the microprocessor acknowledges the request it sends the acknowledgment through the HLDA pin and thus the DMA controller can transfer the bulk data to the memory through these buses without interrupt mechanism.

2.25 (b)

UART (Universal Asynchronous Receiver Transmitter) is a device used to interface a parallel data format with external equipments to serial format in conjunction with communication standards such as EIA, RS-23C etc.

2.26 (a)

Direct memory access is used to transfer the data directly without interference from the CPU or microprocessor using the pin. HOLD and HLDA to initiate the process.

2.27 (b)

INTR request is a hardware interrupt an external device initiates the hardware interrupt by placing an appropriate signal at the interrupt pin of the processor. If the interrupt is accepted, then the processor executes an interrupt service routine. Thus, INTR is

- a maskable interrupt
- a non-vectored interrupt
- has lowest priority
- is a level sensitive interrupt

2.28 (a)

Compiler uses delayed load method to resolve the data depending conflict by inserting no operation instructions.

2.29 (b)

In cycle stealing mode, the DMA sends the data 1 byte at a time. This would increase the DMA controllers interference with CPU's memory access. Hence (2) is false.

2.30 (d)

Steps of executions are
(i) Opcode fetching
(ii) Decoding in decoder
(iii) Execution

2.31 (a)

In daisy chaining priority, each device is verified whether it has interrupt or not separately so more time is wasted.

2.34 (a)

In a computer, CPU and memory are electronic devices. I/O devices have components: mechanical components and electronic components.
If the speed of the mechanical I/O device is intrinsically slower than those of electronic devices, CPU will often remain idle awaiting response from I/O devices.

2.35 (a)

In programmed IO interrupt driven IO both of the modes, CPU is responsible to transfer the I/O data. CPU utilization is inefficient.

2.36 (a)

Pipelining allows overlapping execution. So, instruction level parallelism present in the system.
∴ System performance is improved using pipelining technique.

2.37 (c)

CISC architecture has a large variety of addressing modes and hence, variable length instructions, where instructions can vary in length depending on the number of operands involved. Fixed length instruction formats are more commonly associated with RISC architectures.

2.38 (d)

All the given statements are correct.

■■■

3 Memory Organisation

3.1 If a RAM has 34 bits in its MAR and 16 bits in its MDR, then its capacity will be
(a) 32 GB (b) 16 GB
(c) 32 MB (d) 16 MB **[ESE-2000]**

3.2 **Assertion (A):** LRU (Least Recently Used) replacement policy is not applicable to direct mapped caches.
Reason (R): A unique memory page is associated with every cache page in direct mapped caches.
(a) Both A and R are true and R is the correct explanation of A
(b) Both A and R are true but R is NOT the correct explanation of A
(c) A is true but R is false
(d) A is false but R is true **[ESE-2000]**

3.3 **Assertion (A):** Most personal computers use static RAMs for their main memory.
Reason (R): Static RAMs are much faster than dynamic RAMs.
(a) Both A and R are true and R is the correct explanation of A
(b) Both A and R are true but R is NOT the correct explanation of A
(c) A is true but R is false
(d) A is false but R is true
[ESE-2000]

3.4 The use of a cache in a computer system increases the
(a) available memory space for the program
(b) available memory space for data
(c) average speed of memory access
(d) addressing range of CPU
[ESE-2002]

3.5 Match **List-I** (Type of Memory) with **List-II** (Used as) and select the correct answer using the code given below the lists:

List-I	List-II
A. DRAM	**1.** Cache memory
B. SRAM	**2.** Main memory
C. Parallel Access	**3.** BIOS memory
D. ROM	**4.** CPU registers

Codes:

	A	B	C	D
(a)	1	2	3	4
(b)	2	1	4	3
(c)	1	2	4	3
(d)	2	1	3	4

[ESE-2003]

3.6 Given a 32-bit processor with 16-MB main memory, 32-kB 4-way set-associative on-chip cache and a cache block size (or line size) of 16 words. What is the total number of tag bits in the memory address format?
(a) 9 (b) 20
(c) 11 (d) 24 **[ESE-2004]**

3.7 Consider the following statements:
1. Cache memory is low cost and fast memory.
2. Cache memory is fast but costly memory.
3. Performance of cache during program execution is measured by hit ratio.

Which of the following statements given above are correct?
(a) 1 and 2 (b) 2 and 3
(c) 3 and 4 (d) 1 and 4 **[ESE-2005]**

3.8 Which one of the following correctly defines temporal locality?
(a) Adjacent instructions for current instruction may be needed soon
(b) Current instruction being fetched may be needed again soon
(c) Instructions temporarily residing in memory
(d) None of the above
[ESE-2005]

3.9 Which block replacement algorithm is not generally used in cache operation?
(a) LIFO (b) FIFO
(c) LRU (d) Random
[ESE-2006]

3.10 Three memory chips are of size 1 kB, 2 kB and 4 kB. Their address bus is 10 bits. What are the data bus sizes of the chips?
(a) 8 bits, 16 bits and 24 bits respectively
(b) 8 bits, 16 bits and 32 bits respectively
(c) 8 bits, 16 bits and 64 bits respectively
(d) 8 bits, 16 bits and 128 bits respectively
[ESE-2007]

3.11 In which unit is the performance of cache memory measured?
(a) Hz (b) Bits/s
(c) Cache constant (d) Hit ratio
[ESE-2007]

3.12 Match **List-I** (Type of Memory) with **List-II** (Used as) and select the correct answer using the code given below the lists:

	List-I		List-II
A.	DRAM	**1.**	Cache memory
B.	SRAM	**2.**	Main memory
C.	Parallel Access Registers	**3.**	BIOS memory
D.	ROM	**4.**	CPU registers

Codes:

	A	B	C	D
(a)	1	2	3	4
(b)	2	1	4	3
(c)	1	2	4	3
(d)	2	1	3	4

[ESE-2007]

3.13 Match **List-I** (Type of Memory) with **List-II** (Access time in μsec/nsec) and select the correct answer using the code given below the lists:

	List-I		List-II
A.	DRAM	**1.**	1
B.	SRAM	**2.**	10
C.	Hard Disk	**3.**	100000
D.	Magnetic Tape	**4.**	10000000

Codes:

	A	B	C	D
(a)	1	2	3	4
(b)	1	2	4	3
(c)	2	1	3	4
(d)	2	1	4	3

[ESE-2008]

3.14 Match **List-I** (Type of Memory) with **List-II** (Used as) and select the correct answer using the code given below the lists :

	List-I		List-II
A.	DRAM	**1.**	Cache memory
B.	SRAM	**2.**	Main memory
C.	Parallel Access Registers	**3.**	BIOS memory
D.	ROM	**4.**	CPU registers

Codes:

	A	B	C	D
(a)	1	4	2	3
(b)	3	4	2	1
(c)	1	2	4	3
(d)	3	2	4	1

[ESE-2009]

3.15 Which of the following are the memory performance parameters ?
1. Access time and latency
2. Block size and Block access time
3. Cycle time and Bandwidth

Select the correct answer from the codes given below :
(a) 1 only (b) 1 and 2 only
(c) 2 and 3 only (d) 1, 2 and 3
[ESE-2009]

3.16 A 3 × 8 decoder with two enable inputs is to be used to address 8 blocks of memory. What will be the size of each memory block when addressed from a sixteen bit bus with two MSBs used to enable the decoder ?
(a) 2 K (b) 4 K
(c) 16 K (d) 64 K
[ESE-2009]

3.17 Given that main memory access time is 1200 ns and Cache access time is 100 ns; the average memory access time is not to exceed 120 ns. The hit ratio for the Cache must be at least.
(a) 90% (b) 98%
(c) 80% (d) 75% [ESE-2010]

3.18 An I/O processor controls the flow of information between
(a) Cache memory and I/O device
(b) Main memory and I/O devices
(c) Two I/O devices
(d) Cache and main memory [ESE-2010]

3.19 **Assertion (A):** A memory module presents a specific memory interface to the processor or other unit that references memory.

Reason (R): Memory module contains buffer registers for the address and data.
(a) Both A and R are true and R is the correct explanation of A
(b) Both A and R are true but R is NOT the correct explanation of A
(c) A is true but R is false
(d) A is false but R is true [ESE-2011]

3.20 Among memory types, the abbreviation MPDRAM stands for
(a) Multiport Dynamic Random Access Memory
(b) Multipoint Dynamic Random Access Memory
(c) Multipoint Disk Random Access Memory
(d) Multiport Dimensional Random Access Memory
[ESE-2012]

3.21 The access time of a cache memory is 100 ns and that of main memory is 1 μs. 80% of the memory requests are for read and others are for write. Hit ratio for read only accesses is 0.9. A write through procedure is used. The average access time of the system for both read and write requests is
(a) 200 ns (b) 360 ns
(c) 720 ns (d) 1100 ns [ESE-2012]

3.22 A memory system has a total of 8 memory chips, each with 12 address lines and 4 data lines. The size of the memory system is
(a) 16 k bytes (b) 32 k bytes
(c) 48 k bytes (d) 64 k bytes
[ESE-2012]

3.23 A computer system has a 4 K word cache organized in block-set associative manner with 4 blocks per set, 64 words per block. The numbers of bits in the SET and WORD fields of the main memory address formula are respectively
(a) 15 and 4 (b) 6 and 4
(c) 7 and 2 (d) 4 and 6 [ESE-2013]

3.24 Statement (I): Associative memory is fast memory.
Statement (II): Associative memory searches by content and not by accessing of address.
(a) Both Statement (I) and Statement (II) are individually true and Statement (II) is the correct explanation of Statement (I).
(b) Both Statement (I) and Statement (II) are individually true but Statement (II) is not the correct explanation of Statement (I).
(c) Statement (I) is true but Statement (II) is false.
(d) Statement (I) is false but Statement (II) is true.
[ESE-2013]

3.25 The semiconductor random access memory of a computer has 65,536 words, each of 8-bits. It can perform two basic operations Read and Write. How many bits are there in the Address Register of this memory?
(a) 8 (b) 12
(c) 16 (d) 24 [ESE-2014]

3.26 The speed mismatch between processor and memory in a computer is alleviated by using a small fast memory as an intermediate buffer between memory and processor. This buffer memory is known as
(a) Volatile ROM (b) Non-volatile ROM
(c) Cache memory (d) EPROM [ESE-2014]

3.27 Statement (I): On chip Cache memory is used for temporary storage of commonly used code/data copied from the main memory.
Statement (II): Provision of Cache memory eliminates the need for the processor to go off the chip to access the main memory thus improving the processor performance.
(a) Both Statement (I) and Statement (II) are individually true and Statement (II) is the correct explanation of Statement (I).
(b) Both Statement (I) and Statement (II) are individually true but Statement (II) is not the correct explanation of Statement (I).
(c) Statement (I) is true but Statement (II) is false.
(d) Statement (I) is false but Statement (II) is true.
[ESE-2014]

3.28 Statement (I): In the main memory of a computer, RAM is used as a short term memory.
Statement (II): RAM is a volatile memory.
(a) Both Statement (I) and Statement (II) are individually true and Statement (II) is the correct explanation of Statement (I).
(b) Both Statement (I) and Statement (II) are individually true but Statement (II) is not the correct explanation of Statement (I).
(c) Statement (I) is true but Statement (II) is false.
(d) Statement (I) is false but Statement (II) is true.
[ESE-2014]

3.29 The semiconductor RAM of a digital computer has a word length of 16 bits and a capacity of 65,536 words. It has a cycle time of 80 ns. If the CPU is much faster than the memory, the time required to fill 1024 bytes of this memory with all 0's will be

(a) 40.96 μs (b) 81.92 μs
(c) 5.24 ms (d) 10.48 ms
[ESE-2014]

3.30 Which of the following statements are correct?
1. DRAM offers reduced power consumption.
2. An associative memory is cheaper than RAM.
3. The fastest and most flexible cache organization uses content addressable memory.
4. The address generated by a segmented program is called a physical address.

(a) 1 and 3 (b) 1 and 4
(c) 2 and 3 (d) 2 and 4
[ESE-2015]

3.31 Techniques that automatically move program and data blocks into the physical main memory when they are required for execution are called
(a) Main memory techniques
(b) Virtual memory techniques
(c) Cache memory techniques
(d) Associate mapping techniques
[ESE-2015]

3.32 Four memory chips of 16 × 4 sizes have their address buses connected together. This system will be of size
(a) 64 × 4 (b) 32 × 8
(c) 16 × 16 (d) 256 × 1 [ESE-2015]

3.33 Consider the following statements comparing static RAM with dynamic RAM :
1. In static RAM typical cell requires more number of transistors than the dynamic RAM.
2. Power consumption per bit of static RAM is less than that of dynamic RAM.
3. Dynamic RAM is less expensive than the static RAM.

Which of the above statements are correct?
(a) 1, 2 and 3 (b) 1 and 2 only
(c) 2 and 3 only (d) 1 and 3 only
[ESE-2016]

3.34 A computer system has a cache with access time 10 ns, a hit ratio of 80% and average memory access time is 20 ns. Then what is the access time for physical memory?
(a) 50 ns (b) 40 ns
(c) 30 ns (d) 20 ns
[ESE-2016]

3.35 A computer employs RAM chips of 256 bytes and ROM chips of 1024 bytes. If the computer system needs 1 kB of RAM and 1 kB of ROM, then how many address lines are required to access the memory?
(a) 10 (b) 11
(c) 12 (d) 13
[ESE-2016]

3.36 A cache line has 128 bytes. The main memory has latency 64 ns and bandwidth 1 GB/s. The time required to fetch the entire cache line is
(a) 32 ns (b) 64 ns
(c) 96 ns (d) 192 ns
[ESE-2017]

3.37 **Statement (I):** Stack is organized as 8-bit storage in the microprocessor.
Statement (II): Stack is a set of memory locations in R/W memory reserved for storing information temporarily during the execution of a program.
(a) Both Statement I and Statement II are individually true and Statement II is the correct explanation of Statement I
(b) Both Statement I and Statement II are individually true but Statement II is not the correct explanation of Statement I
(c) Statement I is true but Statement II is false
(d) Statement I is false but Statement II is true
[ESE-2018]

3.38 **Statement (I):** Static RAM memory devices retain data for as long as power is supplied.
Statement (II): SRAM is used when the size of read/write memory required is large.
(a) Both Statement I and Statement II are individually true and Statement II is the correct explanation of Statement I
(b) Both Statement I and Statement II are individually true but Statement II is not the correct explanation of Statement I
(c) Statement I is true but Statement II is false
(d) Statement I is false but Statement II is true
[ESE-2018]

3.39 Consider a disk with an average seek time of 4 ms, rotational delay of 2 ms, rotation speed of 15000 r.p.m. and 512 byte sectors with 500 sectors per track. A file occupies all of the sectors on 5 adjacent tracks. After reading the first track, if remaining tracks can be read with no seek time, then the time required in sequential organization to transfer the file will be nearly

(a) 0.01 second (b) 0.034 second
(c) 0.34 second (d) 3.4 second

[ESE-2019]

3.40 A soft error is a

(a) regular-nondestructive event
(b) random-nondestructive event
(c) random-destructive event
(d) regular-destructive event [ESE-2019]

3.41 **Statement (I):** SRAM is used for cache memory and DRAM is used for main memory.

Statement (II): SRAM is somewhat faster than DRAM.

(a) Both Statement (I) and Statement (II) are individually true and Statement (II) is the correct explanation of Statement (I)
(b) Both Statement (I) and Statement (II) are individually true but Statement (II) is .not the correct explanation of Statement (I)
(c) Statement (I) is true but Statement (II) is false
(d) Statement (I) is false but Statement (II) is true

[ESE-2019]

3.42 In a cache with 64-byte cache lines, how many bits are used to determine which byte within a cache line an address points to?

(a) 16 (b) 8
(c) 6 (d) 3 [ESE-2019]

3.43 The memory that communicates directly with CPU is called

(a) Auxiliary memory
(b) USB storage
(c) Main memory
(d) Micro-program memory [ESE-2020]

3.44 Virtual memory is normally implemented by

(a) Demand paging (b) Buses
(c) Device drivers (d) Bus matrix

[ESE-2020]

3.45 Which of the following are the computer memory performance parameters?

1. Access time (Latency)
2. Memory cycle time
3. Transfer rate

(a) 1 and 2 only (b) 1 and 3 only
(c) 2 and 3 only (d) 1, 2 and 3

[ESE-2020]

3.46 Which one of the following statements is **not** correct regarding a usage of virtual memory?

(a) To free user programs from the need to carryout storage allocation and to permit efficient sharing of the available memory space among different users
(b) To make program independent of the configuration and capacity of the physical memory for their execution
(c) To achieve higher CPU performance
(d) To achieve the very low access time and cost per bit that is possible with a memory hierarchy

[ESE-2020]

3.47 Which one of the following techniques inherits the simplicity of the direct mapping technique in terms of determining the target set?

(a) Set-associative-mapping technique
(b) Set-associative-direct mapping technique
(c) Direct mapping set technique
(d) Indirect mapping set technique

[ESE-2021]

3.48 Which one of the following systems provides a mechanism for translating program-generated addresses into correct main memory locations?

(a) Virtual memory system
(b) Main memory system
(c) Physical addresses system
(d) Memory space system [ESE-2021]

3.49 Which one of the following messaging systems attempts to avoid double copy operations by using virtual-memory management techniques?

(a) Mach message system
(b) Duplex message system
(c) Packet message system
(d) Data message system [ESE-2021]

3.50 Which one of the following storage devices used is when the operating system abstracts from the physical properties of its storage devices to define a logical storage unit?

(a) Volatile storage devices
(b) Non-volatile devices
(c) Flash storages
(d) Cache storage devices [ESE-2021]

3.51 The most personal computers (PCs)-compatible computer systems use a 20-bit address code to identify each of over 1 million memory locations. What is the 5-digit hexadecimal address of the 500th memory location?

(a) 001F3 (b) 001F4
(c) 001F5 (d) 001F6 [ESE-2022]

3.52 If the memory chip size is 1024 x 1, then the number of memory chips needed to design 8 Kbyte memory is
(a) 16 (b) 32
(c) 64 (d) 128 [ESE-2022]

3.53 What is the number of bits in the main memory address for a memory system having the following specification?
Size of the main memory is 4 K blocks, size of the cache is 128 blocks and the block size is 16 words
(Assume that the system uses set-associative mapping with four blocks per set)
(a) 18 (b) 20
(c) 24 (d) 16 [ESE-2023]

3.54 Consider the following reference string of a pages made by a processor:
4, 7, 5, 7, 6, 7, 10 4, 8, 5, 8, 6, 8, 11, 4, 9, 5, 9, 6, 9, 12, 4, 7, 5, 7
Assume that the number of page frames allocated in the main memory is four. What is the number of page faults generated using Least Recently Used (LRU) replacement technique?
(a) 15 (b) 17
(c) 18 (d) 16 [ESE-2023]

3.55 Which one of the following statements is correct with respect to bounded buffer in shared memory systems?
(a) The consumer may have to wait for new items, but the producer can always produce new items.
(b) The consumer must wait if the buffer is empty, and the producer must wait if the buffer is full.
(c) The producer and consumer must be synchronized, so that the consumer does not try to consume an item.
(d) Shared memory suffers from cache coherency issues, which arise because shared data migrate among the several caches.
[ESE-2023]

3.56 Assume we have a computer where the clocks per instruction (CPI) is 1.0 when all memory accesses hit in the cache. The only data accesses are loads and stores, and these total 50% of the instructions. If the miss penalty is 25 clock cycles and the miss rate is 2%, how much faster would the computer be if all instructions were cache hits?
(a) 2.75 times faster (b) 1.75 times faster
(c) 3.75 times faster (d) 4.75 times faster
[ESE-2024]

3.57 Which one of the following is not correct when one goes down a memory hierarchy?
(a) Decreasing cost per bit
(b) Increasing capacity
(c) Decreasing access time
(d) Decreasing frequency of access of the memory by the processor
[ESE-2024]

3.58 Which one of the following is not correct about Rambus DRAM?
(a) The special DRAM bus delivers address and control information using an asynchronous block oriented protocol.
(b) Data rate is 1.6 GBps.
(c) Its operation can be controlled by the explicit RAS, CAS, R/W & CE signals.
(d) It sends data to the controller synchronously to the clock to master, and the controller sends data to an RDRAM synchronously with the clock signal in opposite direction.
[ESE-2024]

3.59 A better measure of memory-hierarchy performance is the average time to access memory. This can be represented as
(a) Average memory access time = Hit time + 2(Miss rate x Miss penalty)
(b) Average memory access time = Hit time + Miss rate x Miss penalty
(c) Average memory access time = 2 Hit time + Miss rate x Miss penalty
(d) Average memory access time = Hit time + (Miss rate x Miss penalty)/2
[ESE-2024]

■■■

Answers Memory Organisation

3.1 (a)	3.2 (d)	3.3 (d)	3.4 (c)	3.5 (b)	3.6 (c)	3.7 (b)	3.8 (b)	3.9 (a)
3.10 (b)	3.11 (d)	3.12 (b)	3.13 (c)	3.14 (*)	3.15 (d)	3.16 (a)	3.17 (b)	3.18 (b)
3.19 (a)	3.20 (a)	3.21 (b)	3.22 (a)	3.23 (d)	3.24 (a)	3.25 (c)	3.26 (c)	3.27 (a)
3.28 (a)	3.29 (a)	3.30 (a)	3.31 (b)	3.32 (c)	3.33 (d)	3.34 (a)	3.35 (b)	3.36 (d)
3.37 (d)	3.38 (c)	3.39 (b)	3.40 (b)	3.41 (a)	3.42 (c)	3.43 (c)	3.44 (a)	3.45 (d)
3.46 (c)	3.47 (a)	3.48 (a)	3.49 (a)	3.50 (b)	3.51 (a)	3.52 (c)	3.53 (d)	3.54 (c)
3.55 (b)	3.56 (b)	3.57 (c)	3.58 (c)	3.59 (b)				

Explanations Memory Organisation

3.1 (a)

MAR stores address-bits

$\Rightarrow$ 34 bits of address bus.

MDR stores data

$\Rightarrow$ 16-bits of data,

Capacity of RAM = $2^{34} * 16$ bits = 32 GB

3.2 (d)

Direct mapped cache forces to use a specific block only, inspite other are free because of (KMODN) rule i.e. K^{th} main memory block in $(KMODN)^{th}$ cache location.

Where N = number of blocks in cache.

Thus, LRU replacement policy is not applicable and Reason (R) is also false.

Every main memory block has fixed location in cache i.e. the ith block of a page must be mapped to i^{th} block of cache.

3.3 (d)

Dynamic RAMs are used in the main memory, whereas Static RAMs are used in cache memory. Cache memory is faster than main memory because Static RAMs are much faster than Dynamic RAMs.

3.4 (c)

Cache memory is a high speed RAM that increases average speed of memory access by the locality of reference.

3.5 (b)

DRAM is used in main memory.

SRAM is used in cache memory.

Parallel access is done by CPU registers. System programs are written and stored in ROM which we can only read. BIOS is loaded in the main memory by the bootstrap loader to initialise the system.

3.6 (c)

24 bits

TAG	Set offset	Word offset
9 bits	$\log_2 2^{11}$ = 11 bits	$\log_2 16$ = 4 bits

Given: 32-bit processor,

Cache line size = 16 words $\Rightarrow 16 \times 4B = 2^6 B$

$$\text{\# blocks in cache} = \frac{2^{15}}{2^6} = 2^9$$

$$\text{\# sets} = \frac{2^9}{2^2} = 2^7$$

TAG bits = 24 – (7 + 6) = 11 bits

3.7 (b)

Cache is a fast memory and it is costly also as it uses SRAM in it's manufacture.

Performance of cache during program execution is measured by hit ratio.

3.8 (b)

Spatial locality $\rightarrow$ Adjacent instructions for current instruction may be needed soon.

Temporal locality $\rightarrow$ current instruction being fetched may be needed again soon.

3.9 (a)

Most frequently used block replacement algorithms in cache operation are:

Random and LRU
Other block replacement algorithms are:
FIFO, MRU, MFU etc.
LIFO is never used.

3.10 (b)

$$\text{Data bus size} = \frac{\text{Memory chip size}}{\text{Address bus size}}$$

For 1 kB memory chips,

$$\text{data bus size} = \frac{2^{10} \times 8}{2^{10}} = 8 \text{ bits}$$

For 2 kB memory chip,

$$\text{data bus size} = \frac{2 \times 2^{10} \times 8}{2^{10}} = 16 \text{ bits}$$

For 4 kB memory chip,

$$\text{data bus size} = \frac{4 \times 2^{10} \times 8}{2^{10}} = 32 \text{ bits}$$

3.11 (d)

The performance of cache memory is measured in terms of hit ratio.

3.12 (b)

DRAM is used in main memory.
SRAM is used in cache memory.
Parallel access is done by CPU registers system programs are written and stored in ROM which we can only read. BIOS is loaded in the main memory by the bootstrap loader to initialise the system.

3.13 (c)

DRAM takes 10 μsec.
SRAM takes 1 μsec.
Hard disk takes 100000 μsec.
Magnetic tape takes 10000000 μsec.

3.14 (*)

DRAM used in main memory.
SRAM used in cache memory.
Parallel access is done by CPU registers system programs are written and stored in ROM which we can only read. BIOS is loaded in the main memory by the bootstrap loader to initialise the system.

3.15 (d)

All are memory performance parameters.

3.16 (a)

$$\frac{2^{14}}{2^{3}} = 2^{11} = 2\text{ K}$$

as two MSBs used to enable the decoder logic.

3.17 (b)

Average memory access time = Hit ratio × access time in cache memory + (1 – Hit ratio) × access time in main memory

$\Rightarrow$ 120 = Hit ratio × 100 + (1 – Hit ratio) × 1200
$\Rightarrow$ 12 = (10 – 120) Hit ratio + 120
$\Rightarrow$ 110 × Hit ratio = 108

$$\therefore \text{ Hit ratio} = \frac{108}{110} = 0.98 \quad \text{or} \quad \text{Hit ratio} = 98\%$$

3.18 (b)

I/O processor is part of I/O interface between the main memory and I/O devices.
This interfaces are required because I/O either too slow or too fast.

3.21 (b)

Read access time
$= HT_c + (1 - H)(T_c + T_m)$
$= 0.9 \times 100 + 0.1 \times (100 + 1000)$nsec = 200 nsec.
Total access time
$= (0.8 \times 200 + 0.2 \times 1000)$nsec = 360 nsec.

3.22 (a)

Total size of memory
$= 2^A \times$ data lines × number of memory chips
$A \rightarrow$ number of address lines

$$= 2^{12} \times 4 \times 8 = 2^{10} \times 2^2 \times 2^2 \times 2^3$$

$$\begin{bmatrix} 2^{10} = 1\text{k} \\ 2^3 = 8 = 1\text{byte} \end{bmatrix}$$

$$= 2^4 \text{ kbytes} = 16 \text{ kbytes}$$

3.23 (d)

$$\text{\# blocks} = \frac{4K}{64} = 64$$

$$\text{\# sets} = \frac{64}{4} = 16$$

$\Rightarrow$ 4 bits for SET
Each block has 64 words
$\Rightarrow$ 6-bits required for word.

Tag	Set	Word
	(4)Bit	(6) Bit

3.24 (a)

Associative memory is any device that associates a set of predefined output patterns with specific input patterns. Associative memory can also be called as content addressable memory (CAM) and is accessed simultaneously and in parallel on the basis of data rather than by specific address or location.

3.25 (c)

$$2^n = 65536$$

where, n = number of bits address line

$\therefore$ $2^n = (2)^{16}$

$$n = 16$$

3.26 (c)

Cache memory is faster compared to the main memory. CPU is much faster than cache and main memory. Cache memory acts as buffer between CPU and main memory to avoid the speed mismatch.

3.27 (a)

Cache memory eliminates the need for the processor to access the main memory. Cache memory stores the temporary and language commonly used data which is copied from main memory.

Cache memory access time is less compared to main memory.

Hence, it improves the processor performance.

3.28 (a)

Both statements are correct, and statement (I) is correct explanation of statement (II).

Main memory is primary memory.

There are two types:

(i) RAM (Random Access Memory)

(ii) ROM (Read Only Memory)

RAM is volatile memory, Hence it is called as short-term memory.

ROM is non-volatile memory.

3.29 (a)

1 word = 16-bits

1024 bytes = 512 words

1 word $\rightarrow$ 80ns

512 words $\rightarrow$?

Required time = 512 × 80ns

= 40960ns = 40.96 μs

3.30 (a)

2. Associative memory is costlier than memory. Hence it is false.

4. The address generated by a segmented program is called logical address and not physical address.

$\therefore$ False

3.31 (b)

Virtual memory technique is used to move the program or data from secondary memory to main memory.

3.32 (c)

Address lines of memory chip = 16

data lines of 1 memory chip = 4

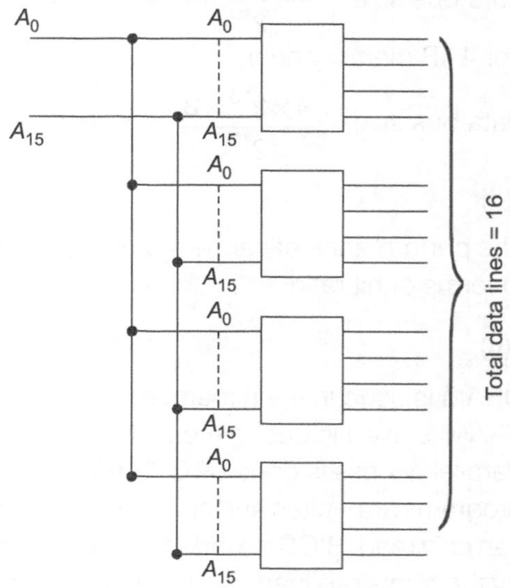

$\Rightarrow$ Size of memory = 16 × 16

3.33 (d)

Static RAM $\rightarrow$ More cost, power consumption

DRAM $\rightarrow$ low cost and power consumption

3.34 (a)

$$T_{avg} = HT_c + (1-H)(T_m + T_c)$$

$$20 = (0.8 \times 10) + (1-0.8)(x + 10)$$

$$20 = 8 + 0.2x + 2$$

$$0.2x = 10$$

$$x = 50 \text{ ns}$$

3.35 (b)

RAM $\rightarrow$ 256 B

ROM $\rightarrow$ 1024 B. i.e. 1 kB

1 kB of RAM and 1 kB of ROM total 2 kB

$$2^n = 2 \text{ kB} = \log_2 2^{11}$$

$\therefore$ $n = 11$ (bits are required)

3.36 (d)

Total time = Access time + transfer time

$$= 64\text{ ns} + \begin{Bmatrix} 1\text{GB} & - & 1\text{sec} \\ 128\text{ B} & - & ? \end{Bmatrix}$$

$$= 64\text{ ns} + 128\text{ ns} = 192\text{ ns}$$

3.38 (c)

Statements II is incorrect.
The correct statements is that DRAM is used when the size of read/write.

3.39 (b)

1. Seek time = 4 ms
2. 15000 revolutions — 1 min (60 sec)
   1 revolution — ?

$$\Rightarrow \frac{60}{15000}\text{ sec} = 4\text{ ms}$$

∴ Average rotational latency

$$= \frac{1}{2}\text{revolution time} = \frac{1}{2}\times 4\text{ ms} = 2\text{ ms}$$

3. Transfer time
   1 revolution time = 1 track data
   i.e., 4 ms/track
   File stored in 5 adjacent track which doesn't requires seek time for every track.
   ∴ Time required to access the file is
   $\Rightarrow 4\text{ ms} + (2\text{ ms} + 4\text{ ms})5 \Rightarrow 34\text{ ms}$
   $\Rightarrow 0.034\text{ sec}$

3.41 (a)

SRAM is faster than DRAM.

3.42 (c)

Word offset = $\log_2^{64} = 6\text{ bit}$

3.43 (c)

1. The memory unit that directly communicates with CPU is called main memory.
2. The devices that provide backup storage are called auxiliary memory.

3.44 (a)

Virtual memory system is an example of demand paging, in the demand paging, pages of data are only brought into the main memory when a program accesses them.

3.45 (d)

From a user's point of view, the two most important characteristics of memory are capacity and performance. Three performance parameters are used:

Access time (latency): For random-access memory, this is the time it takes to perform a read or write operation, that is, the time from the instant that an address is presented to the memory to the instant that data have been stored or made available for use. For non-random-access memory, access time is the time it takes to position the read-write mechanism at the desired location.

Memory cycle time: This concept is primarily applied to random-access-memory and consists of the access time pulse any additional time may be required for transients to die out on signal lines or to regenerate data if they are read destructively. Note that memory cycle time is concerned with the system bus, not the processor.

Transfer rate: This is the rate at which data can be transferred into or out of a memory unit. For random-access memory, it is equal to 1/(cycle time).

3.47 (a)

The set-associative mapping uses combination of direct mapping and fully-associative mapping. In set-associative mapping, the cache is divided into a number of sets of cache lines; each main memory block can be mapped into any line in a particular set as in fully-associative mapping. However, the technique inherits the simplicity of the direct mapping technique in terms of determining the target set.

3.48 (a)

A virtual memory system provides a mechanism for translating program generated addresses into correct main memory locations. This is done dynamically, while programs are being executed in the CPU. The translation or mapping is handled automatically by the hardware by means of a mapping table.

3.49 (a)

Mach messaging system design includes the use of memory mapping techniques, notably copy-on-write, to avoid copying data when, for example, messages are passed between tasks.

Copy-on-write (COW) sometimes referred to as implicit sharing or shadowing is a resource-management technique to efficiently implement a "duplicate" or "copy" operation on modifiable resources. If a resource is duplicated but not modified, it is not necessary to create a new resource, the resource can be shared between the copy and the original. Modifications must still create a copy, hence the technique: the copy operation is deferred until the first write.

3.50 (b)

On a hard drive, we see files and folders when you open it from your explorer. On a pen drive, you again see files and folders.
This is the abstraction the operating system provides, by hiding the details about how the data is actually read/written from the hard-drive or pen-drive which are non-volatile devices.
Application program is always stored in the secondary memory (Hard-disk) which is a non-volatile storage device. Later, OS transfer the program from the secondary memory to main memory to execute.

3.51 (a)

The addresses of the memory begin with 0000H.
Hence, first memory location → 0000H
500th memory location → $(500-1)_{10} = (499)_{10}$
Converting decimal address to hexadecimal address,

16	499	
16	31	3
16	1	F
	0	1

$\therefore \quad (499)_{10} = (1F3)_{16}$
Hence, the address of 500th memory location is 001F3H.

3.52 (c)

Given : Memory chip size = 1024 × 1 bits
Memory to be designed = 8 kB
∴ Number of memory chips needed to design 8 Kbyte memory

$$= \frac{\text{Memory to be designed}}{\text{Available Capacity}}$$

$$= \frac{8\text{ kB}}{1024 \times 1\text{bits}} = \frac{8(1024 \times 8)\text{ bits}}{1024 \times 1\text{bits}} = 64$$

3.53 (d)

We know that,
Number of Blocks in main memory = 4 K
Block size = 16 Words
Number of lines in cache memory = 128
∴ Main memory size = 4 K × 16 W

$$= 2^{12} \times 2^4 \text{ W} = 2^{16} \text{ W}$$

So, Main memory address = $\log_2 2^{16} = 16$ bit

3.54 (c)

From Least Recently Used (LRU) technique,

				6	7	10	4	8
		5	7	7	6	7	10	4
	7	7	5	5	5	6	7	10
4	4	4	4	4	4	5	6	7
4-m	7-m	5-m	7-Hit	6-Hit	7-Hit	10-m	4-m	8-m

5	8	6	8	11	4	9	5	9
8	5	8	6	8	11	4	9	5
4	4	5	5	6	8	11	4	4
10	10	4	4	5	6	8	11	11
5-m	8-Hit	6-m	8-Hit	11-m	4-m	9-m	5-m	9-Hit

6	9	12	4	7	5	7
9	6	9	12	4	7	5
5	5	6	9	12	4	4
4	4	5	6	9	12	12
6-m	9-Hit	12-m	4-m	7-m	5-m	7-Hit

Misses (Page faults) = 18

3.55 (b)

- The bounded-buffer problems (producer-consumer problem) is a classic example of concurrent access to a shared resource.
- Multiple producers and multiple consumers share a single buffer.
- Producer write data into the buffer and consumers read data from the buffer.
- Producers must block when the buffer is full.
- Consumer must block when the buffer is empty.

3.56 (b)

For the computer where all instructions (say N) were cache hits,
CPU execution time

$$= (\text{CPU clock cycles} + \text{Memory stall cycles}) \times \text{clock cycles}$$

$$= (N \times \text{CPI} + 0) \times \text{clock cycles}$$

$$= N \text{ clock cycles}$$

[∵ CPI = 1 for cache hit]

For the computer with real cache,
Memory stall cycles

$$= N \times \frac{\text{Memory accesses}}{\text{Instruction}}$$

$\times$ Miss rate $\times$ Miss penalty

$= N \times (1 + 0.5) \times 0.02 \times 25$

$= 0.75$ N clock cycles

where the term (1 + 0.5) represents one instruction access and 0.5 data accesses per instruction. The CPU execution time is, thus

CPU execution time$_{cache}$

$= (N \times 1 + N \times 0.75) \times$ clock cycles

$= 1.75$ N clock cycles

Therefore, the computer with no cache misses is faster by

$$\frac{\text{CPU execution time}_{cache}}{\text{CPU execution time}} = \frac{1.75 \text{ N clock cycles}}{\text{N clock cycles}}$$

$= 1.75$ times

3.57 (c)

The below figure represents the memory hierarchy:

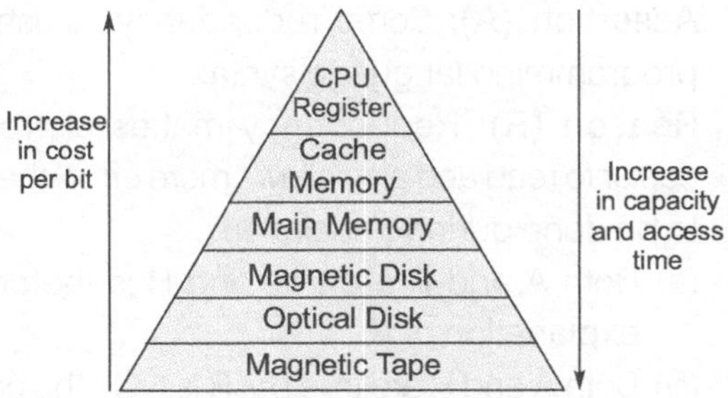

As we go down a memory hierarchy,

(i) Cost per bit decreases.

(ii) Capacity increases

(iii) access time increases

(iv) frequency of access of the memory by the processor is decreased as faster memories are accessed more.

Hence, option (c) is not correct.

3.58 (c)

- The RDRAM, developed by Rambus delivers address and control information using an asynchronous block-oriented protocol. After an initial 480 ns access time, this produces the 1.6 Gbps data rate.
- Rather than being controlled by the explicit RAS, CAS, R/W, and CE signals used in the conventional DRAMs, RDRAM gets a memory request over the high-speed bus. This request contains the desired address, the type of operation, and the number of bytes in the operation.
- RDRAM system consists of a controller and a number of RDRAM modules connected together via a common bus. The bus includes 18 data lines, separate 8 lines (RC) used for address and control signals. There is also a clock signal that starts at the far end from the controller propagates to the controller end and then loops back. A RDRAM module sends data to the controller synchronously to the clock to master, and the controller sends data to an RDRAM synchronously with the clock signal in opposite direction.

3.59 (b)

T_{avg} = Hit time + (Miss rate $\times$ Miss penalty)

OR

$T_{avg} = H_C T_C + (1 - H_C)(T_M + T_C)$

where H_C is the cache hit rate, T_C is the cache access time and T_M is the main memory access time required in case of a cache miss.

4 Data Representation and Programming

4.1 If the value of x in decimal number 3954, the value of y in decimal number computed by the given flow chart is

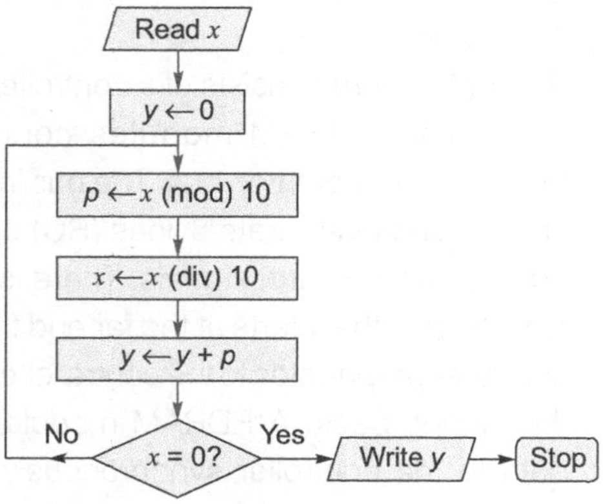

(a) 20 (b) 22
(c) 21 (d) 3954

[ESE-2000]

4.2 In *C* language, $f-=9$ is equivalent to
(a) $f = -9$ (b) $f = f - 9$
(c) $f = 9 - 1$ (d) $-f = 9$ [ESE-2000]

4.3 A primitive computer uses a single register. The following fragment of assembly code is written for the machine

```
LOAD X
MULT Y
STORE T1
MULT T1
STORE T1
LOAD Z
ADD Y
ADD T1
STORE R
```

Which one of the following expressions is evaluated?
(a) $R := (XY) + Y + Z$
(b) $R := (XY)^2 + Y + Z$
(c) $R := (XY)^2 \pm Y + Z$
(d) $R : XY^2 + (Y + Z)$ [ESE-2000]

4.4 In an assembler, which one of the following is required for variable names in symbol table?
(a) Addresses (b) Values
(c) Registers (d) Storage [ESE-2001]

4.5 The microprograms provided by a manufacturer to be used on his micro-programmed computer are generally called
(a) software (b) netware
(c) firmware (d) hardware [ESE-2001]

4.6 **Assertion (A):** Some redundancy is useful in programming language syntax.
Reason (R): Redundancy makes a program easier to read and also allows more error-checking to be done during translation.
(a) Both A and R are true and R is the correct explanation of A
(b) Both A and R are true but R is NOT the correct explanation of A
(c) A is true but R is false
(d) A is false but R is true [ESE-2001]

4.7 The computer program which converts statements written in high level language to object code is known as
(a) Assembler (b) Compiler
(c) Disassembler (d) Operating system

[ESE-2002]

4.8 Consider the following C structure and declaration:

```
Struct date {
int day;
int month;
int year;
};
Struct date * pd;
```

Which of the following is the correct method to refer to the year number?
(a) (* pd) . year (b) (* pd) * year
(c) (* pd) → year (d) pd → year

[ESE-2002]

4.9 What will be values of A and B, respectively, when printed for the given flow chart?

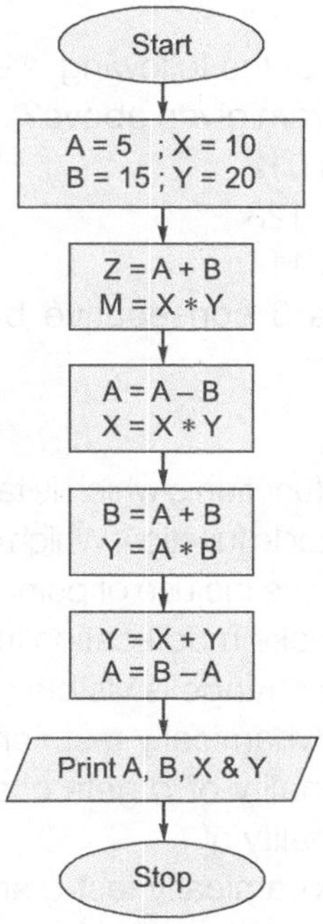

(a) 10 and 20 (b) 5 and 15
(c) 20 and 10 (d) 15 and 5

[ESE-2002]

4.10 A certain well-known computer family represents the exponents of its floating-point numbers as 'Excess-64' integers. Which one of the following numbers is represented by the exponent $e_6e_5e_4e_3e_2e_1e_0$?

(a) $e = -64 + \sum_{i=0}^{6} 2^i e_i$ (b) $e = -64 + \sum_{i=0}^{6} 2e_i$

(c) $e = 64 - \sum_{i=0}^{6} 2^i e_i$ (d) $e = 64 - \sum_{i=0}^{6} 2e_i$

[ESE-2003]

4.11 Assertion (A): In call-by-value parameter passing technique, function call overheads are less as compared to that of call-by-reference.

Reason (R): In call-by-reference parameter passing technique, address of actual parameter is pushed into the stack.

(a) Both A and R are true and R is the correct explanation of A
(b) Both A and R are true but R is NOT the correct explanation of A
(c) A is true but R is false
(d) A is false but R is true

[ESE-2003]

4.12 Which one of the following correctly defines a C-macro for computing the square?

(a) # defines sqr $(x)x^*x$
(b) # defines sqr (x) $(x * x)$
(c) # defines sqr (x) $((x) * (x))$
(d) # defines sqr (x) $(x) * (x)$

[ESE-2003]

4.13 Consider the following declaration of C:
int(*P) ( );
Which of the following is true for the above declaration?

(a) P is pointed to function returning integer
(b) P is pointer to an array of integers
(c) P is an array of pointers
(d) P is a function returning pointer to integer

[ESE-2003]

4.14 Consider the following Pascal program fragment:

```
var i, j: integer;
procedure Y(p, q : integer);
begin
    p : p – q;
    q : p + q;
    p : q – p
end;
    i : = 2;
    j : = 3;
    Y (i, j);
```

If both parameters to Y are passed by reference, what are the values of i and j at the end of the program fragment?

(a) $i = 0, j = 2$ (b) $i = 1, j = 5$
(c) $i = 2, j = 3$ (d) $i = 3, j = 2$

[ESE-2003]

4.15 Consider the following C program:

```
# include < stdio.h >
main ( )
{ float total = 0.0, q = 1.0, p = 2.0;
        while (p/q > 0.001)
        {
        q = q + q;
        total = total + p / q;
        print f ("% f\n", total);
        }
}
```

Which one of the following is the integer that best approximates the last number printed?

(a) 0 (b) 1
(c) 2 (d) 3

[ESE-2003]

4.16 Which of the following correctly declares a pointer to an array of integers in C?
(a) int * P [20];
(b) int *P ;
(c) int (*P) [20];
(d) int *(P[20]);
[ESE-2003]

4.17 Consider the following information:
An array A [1...m] is said to be p-ordered if
$A(i-p) \leq A[i] \leq A[i+p]$
for each i such that $p < i \leq m - p$. For example, the array 1 4 2 6 3 7 5 8 is 2-ordered.
In a 2-ordered array of 2N elements, what is the maximum number of positions that an element can be from its position if the array were 1-ordered?
(a) 1
(b) 2
(c) N/2
(d) N
[ESE-2003]

4.18 Given three integer variables a, b, c where each one will take positive value. Which one of the following expressions in C avoids overflow?
(a) $a + b - c$
(b) $b + a - c$
(c) $b - c + a$
(d) $c + a + b - c - c$
[ESE-2004]

4.19 It is given that two pointer variables p and q are of the same type and $p < q$. Which one of the following operations is logically not correct?
(a) $p - q$
(b) $p + q$
(c) $p + 5$
(d) p ++
[ESE-2004]

4.20 x: = 1; y: = 0;
while y < k do
begin
x: = 2 * x;
y: = y + 1
end;
For the above Pascal program fragment involving integers x, y, and k, which one of the following is a loop invariant; i.e. true at the beginning of each execution of the loop and at the completion of the loop?
(a) $x = 2^y$
(b) $x = y + 1$
(c) $x = (y + 1)^2$
(d) $x = (y+1)2^y$
[ESE-2004]

4.21 Consider the following 'C' program:
```
main ()
{
pri () ; pri (); pri ();
}
pri ()
{
static int k;
print ("%d", ++k);
}
```
Which one of the following is correct in respect of the program given above?
(a) It prints 012
(b) It prints 123
(c) It prints 111
(d) It prints 3 consecutive but unpredictable numbers
[ESE-2004]

4.22 Consider a function g which is taking a parameter f of type pointer to function. Which one of the following best describes the use of pointer to function?
(a) g can make modification in the definition of f and the change is visible after return from g
(b) g can dynamically test f and modify it
(c) Functionality of g gets customized through functionality of f
(d) f can dynamically test g and modify it
[ESE-2005]

4.23 Which one of the following is not associated with a Logic analyzer?
(a) Delayed state
(b) Delayed sweep
(c) Disassembler
(d) Pre-triggering
[ESE-2005]

4.24 Consider the following statements in respect of the expression: A + XYZ (&A):
1. Result of expression may depend upon order of evaluation of operands of the operator 'x'
2. Result of expression would never depend upon order of evaluation of operands of the operator 'x'
3. XYZ is not a valid name of function

Which of the statements correctly defines temporal locality?
(a) 1 only
(b) 2 only
(c) 1 and 3
(d) 1, 2 and 3
[ESE-2005]

4.25 Match **List-I** (Operator of C Language) with **List-II** (Characteristic of the Operator) and select the correct answer using the code given below the lists:

List-I	List-II
A. ^	1. Unary operator
B. &	2. Binary operator
C. mod	3. Ternary operator
D. ?	4. Invalid operator

Codes:

	A	B	C	D
(a)	2	1	4	3
(b)	4	3	2	1
(c)	2	3	4	1
(d)	4	1	2	3

[ESE-2005]

4.26 What is the output of the following program?

```
#include < Stdio. h>
main ( )
{
float f;
f = 10/3;
print f ("%f," f);
}
```

(a) 3.3 (b) 3.0
(c) 3 (d) 0.3 [ESE-2005]

4.27 Self-referential structures of 'C' programming language are very useful in applications that involve
(a) Graphs (b) Lists
(c) Queues (d) Stacks [ESE-2006]

4.28 **Assertion (A):** Binary search function can be easily implemented using recursion.
Reason (R): Recursion is based on number of elements in array to be searched.
(a) Both A and R are true and R is the correct explanation of A
(b) Both A and R are true but R is NOT the correct explanation of A
(c) A is true but R is false
(d) A is false but R is true [ESE-2006]

4.29 Which one of the following is correct for ASCII codes of all upper case English letters compared to all lower case English letters?
(a) Are all larger
(b) Are all smaller
(c) Are all equal
(d) Some are larger and some are smaller [ESE-2006]

4.30 Consider the following C statements:
1. $x = = 1; x++$;
2. #define max = 40 ;
3. $x = 1$; $x--$;
4. For (i = 0; i < = 1; i++)
{ printf ("i = %d\n",i);
}

Which of the C statements given above are not correct?
(a) 1 and 2 only (b) 1, 2 and 3 only
(c) 2 and 3 only (d) 3 and 4 only [ESE-2007]

4.31 For which one of the following problems, recursive solution exists but non-recursive solution does not?
(a) Tower of Hanoi
(b) Factorial computation
(c) Tree traversal
(d) No such problem exists [ESE-2007]

4.32 Division by zero in a program gives rise to which one of the following?
(a) Syntax error (b) Run-time error
(c) Logical error (d) Semantic error [ESE-2007]

4.33 Match **List-I** (Pascal Operator) with **List-II** (C operator) and select the correct answer using the code given below lists:

List-I	List-II
A. : =	**1.** ! =
B. <>	**2.** = =
C. =	**3.** =

Codes:

	A	B	C
(a)	1	2	3
(b)	3	1	2
(c)	2	1	3
(d)	2	3	1

[ESE-2008]

4.34 For interfacing of assembly language routines with some high level language, one must address which of the following?
1. How is the subroutines invoked ?
2. How are parameters passed ?
3. How are values returned?
4. How many parameters are passed?

Select the correct answer using the code given below:
(a) 1 and 3 only (b) 2 and 4 only
(c) 1, 2 and 3 only (d) 1, 2, 3 and 4 [ESE-2008]

4.35 Which one of the following is correct?
The convention is that the number 0 has the normal form representation as
(a) 0 (b) ·0
(c) 0·0 (d) $0 \cdot 10^0$ [ESE-2008]

4.36 Which are the characteristics of vertical micro instructions?
1. Considerable encoding of control information

2. Limited ability to express parallel micro operation
3. Long format

Select the correct answer using the code given below:
(a) 1 and 2 only (b) 2 and 3 only
(c) 1 and 3 only (d) 1, 2 and 3
[ESE-2008]

4.37 Consider the following statements :
1. Strictly speaking C supports 1-dimensional arrays only.
2. An array element may be an array by itself.
3. Array elements need not occupy contiguous memory locations.

Which of the above statements is/are correct ?
(a) 1 only (b) 2 only
(c) 1 and 2 (d) 2 and 3 [ESE-2009]

4.38 Consider the following statements :
1. Taking 2's complement is equivalent to sign change.
2. In the 2's complement representation the most significant bit (MSB) is zero for a positive number.
3. In a 4 bit binary representation of a binary number A, A + 1's complement of A = 2^4.

Which of the above statements are correct ?
(a) 1 and 2 only (b) 1 and 3 only
(c) 2 and 3 only (d) 1, 2 and 3
[ESE-2009]

4.39 What can be the maximum dimension of an array in C language program?
(a) 3
(b) 4
(c) 5
(d) It is compiler dependent [ESE-2009]

4.40 With reference to C programming language, which of the following statements are correct?
1. An identifier may start with an underscore.
2. An identifier may end with an underscore.
3. IF is a valid identifier.
4. The number of significant characters in an identifier is implementation dependent.

Select the correct answer from the codes given below :
(a) 1 and 2 only (b) 2 and 3 only
(c) 1, 2, 3 and 4 (d) 3 and 4 only
[ESE-2009]

4.41 Which of the following has a major role in implementation of function calls in C?
(a) Processor's registers
(b) Data segment
(c) The heap
(d) System stack [ESE-2009]

4.42 The default parameter-passing mechanism in C-programming language is:
(a) Call by reference (b) Call as random
(c) Call by value (d) Call by value result
[ESE-2010]

4.43 Which of the following data structure is used in compiler to manage information about variables and their attributes?
(a) Abstract syntax tree
(b) Linked list
(c) Parse table
(d) Symbol table
[ESE-2010]

4.44 The data type defined by user is:
(a) Built-in data type
(b) Abstract data type
(c) Homogeneous data type
(d) Real data type
[ESE-2010]

4.45 **Assertion (A):** Logic analyzer offers a 'delayed sweep'.
Reason (R): Because the logic analyzer 'sweep' is really a clock signal.
(a) Both A and R are true and R is the correct explanation of A
(b) Both A and R are true but R is NOT the correct explanation of A
(c) A is true but R is false
(d) A is false but R is true
[ESE-2010]

4.46 Which of the following is/are NOT the functions of assembly-language directions?
1. Define system parameters
2. Assign specific symbolic memory location
3. Control the output of the assembly process

Which of these statements are correct?
(a) 1 and 2 only (b) 2 and 3 only
(c) 1 and 3 only (d) 1, 2 and 3
[ESE-2011]

4.47 Which one of the following operators of high level language is used to eliminate the run-time cost of redundant address calculations?
(a) Arithmetic (b) Assignment
(c) Logical (d) Relational
[ESE-2011]

4.48 Assertion (A): Most high level programming languages include a notion of 'type' for expression.

Reason (R): Type provides implicit context for many operations and it limits the set of operation that may be performed in a semantically valid program.

(a) Both A and R are true and R is the correct explanation of A
(b) Both A and R are true but R is NOT the correct explanation of A
(c) A is true but R is false
(d) A is false but R is true **[ESE-2011]**

4.49 Assertion (A): The activity reading from or writing into one of the stand along register and into register file is same.

Reason (R): The register file has additional control and access overhead compared to the single stand alone register.

(a) Both A and R are true and R is the correct explanation of A
(b) Both A and R are true but R is NOT the correct explanation of A
(c) A is true but R is false
(d) A is false but R is true **[ESE-2011]**

4.50 Aliasing in the context of programming language refers to multiple variables having the same
(a) memory location (b) value
(c) identifier (d) none of these **[ESE-2013]**

4.51 The creation of the variable will automatically create a special variable associated with it, called as
(a) buffer variable (b) text variable
(c) allocated variable (d) floating variable **[ESE-2014]**

4.52 Expression $C = i$++ causes
(a) value of i to be assigned to C, and then i to be incremented by 1
(b) i to be incremented by 1, and then value of i to be assigned to C
(c) value of i to be assigned to C
(d) i to be incremented by 1 **[ESE-2014]**

4.53 The micro programs provided by a manufacturer to be used on his micro programmed computer are generally called
(a) Software (b) Netware
(c) Firmware (d) Hardware **[ESE-2014]**

4.54 Locality of reference concept will fail in which of the following cases?
Where there are
(a) many conditional jumps
(b) many unconditionally jumps
(c) many operands
(d) none of the above **[ESE-2014]**

4.55 If, $A = 60$ and $B = 3$, then using C-programming $A >> B$ results in
(a) 11100000 (b) 00000111
(c) 00111111 (d) 01111000 **[ESE-2016]**

4.56 Consider the following statements : Pointers in C-programming are useful to
1. Handle the data tables efficiently
2. Reduce the length of a program
3. Reduce the complexity of a program Which of the above statements are correct?

(a) 1 and 2 only (b) 1, 2 and 3
(c) 2 and 3 only (d) 1 and 3 only **[ESE-2016]**

4.57 It is awkward to employ signed-magnitude system in computer arithmetic, because
1. sign and magnitude have to be handled separately.
2. it has two representations for '0'.

Which of the above statements is/are correct?
(a) 1 only (b) 2 only
(c) Both 1 and 2 (d) Neither 1 nor 2 **[ESE-2016]**

4.58 A program structure that permits repeated operation of a particular sequence of instructions is known as
(a) subroutine (b) loop
(c) module (d) microprogramming **[ESE-2018]**

4.59 Statement (I): The break statement provides an early exit from for, while, do/while and switch structures, and the execution continues with the first statement after the structure.

Statement (II): The continue statement, used in while, for, do/while loop, skips the remaining statements in the body of that structure, and proceeds with the next iteration of the loop.

(a) Both Statement I and Statement II are individually true and Statement II is the correct explanation of Statement I

(b) Both Statement I and Statement II are individually true but Statement II is not the correct explanation of Statement I
(c) Statement I is true but Statement II is false
(d) Statement I is false but Statement II is true

[ESE-2018]

4.60 **Statement (I):** MS DOS and Windows products have always used implicit mounting when an attempt is first made to access the media.

Statement (II): Unix has traditionally used explicit mount command to access the removable medium.

(a) Both Statement I and Statement II are individually true and Statement II is the correct explanation of Statement I
(b) Both Statement I and Statement II are individually true but Statement II is not the correct explanation of Statement I
(c) Statement I is true but Statement II is false
(d) Statement I is false but Statement II is true

[ESE-2018]

4.61 Which one of the following is an advantage of assembly language over high-level language?

(a) Assembly language program runs faster.
(b) Writing of assembly language programming is easy.
(c) Assembly language program is portable
(d) Assembly language program contains less instruction.

[ESE-2019]

4.62 Which of the following statements are correct?

1. A pseudoinstruction is a machine instruction.
2. A pseudoinstruction is an instruction to the assembler.
3. The ORG (origin) is an example of pseudoinstruction.
4. It is not possible to use ORG more than once in a program.

Select the correct answer using the code given below.

(a) 1 and 3 (b) 2 and 3
(c) 1 and 4 (d) 2 and 4 [ESE-2019]

4.63 Booth algorithm is associated with

(a) Binary division
(b) Binary integer multiplication
(c) Sorting binary integers
(d) Searching of binary data

[ESE-2020]

4.64 Which one of the following statements is correct regarding arithmetic and logical operations?

(a) When two 3-bit unsigned numbers are multiplied, the result is an 8-bit product that must be stored in three memory words.
(b) When two 16-bit unsigned numbers are multiplied, the result is a 32-bit product that must be stored in two memory words.
(c) Operations that are implemented in a computer with one machine instruction are said to be implemented by software.
(d) Operations implemented by a set of instructions that constitute a program are said to be implemented by hardware.

[ESE-2021]

4.65 Which one of the following techniques is used to speed up the multiplication of two signed binary numbers in 2's complement notation?

(a) K-Map algorithm
(b) Booth's algorithm
(c) Pascal's algorithm
(d) De Morgan's algorithm

[ESE-2022]

4.66 A floating-point (FP) number is said to be normalized, if the leftmost bit of the mantissa is

(a) 1 (b) 0
(c) –1 (d) 2

[ESE-2022]

4.67 Which one of the following is the entity that can combine object modules that may have resulted from assembling multiple assembly modules separately?

(a) Compiler (b) Interpreter
(c) Linker (d) Loader

[ESE-2022]

4.68 What single-precision number does the following 32-bit word represent?

11000000101000000000000000000000

(a) –2 (b) –4
(c) –5 (d) –7

[ESE-2024]

■■■

Answers Data Representation and Programming

4.1 (c)	4.2 (b)	4.3 (b)	4.4 (c)	4.5 (c)	4.6 (d)	4.7 (b)	4.8 (a, d)	4.9 (d)
4.10 (a)	4.11 (a)	4.12 (c)	4.13 (a)	4.14 (d)	4.15 (c)	4.16 (c)	4.17 (c)	4.18 (c)
4.19 (a, b)	4.20 (a)	4.21 (b)	4.22 (a)	4.23 (b)	4.24 (a)	4.25 (a)	4.26 (b)	4.27 (a)
4.28 (c)	4.29 (b)	4.30 (a)	4.31 (d)	4.32 (b)	4.33 (b)	4.34 (c)	4.35 (c)	4.36 (a)
4.37 (b)	4.38 (a)	4.39 (d)	4.40 (c)	4.41 (d)	4.42 (c)	4.43 (d)	4.44 (b)	4.45 (a)
4.46 (a)	4.47 (b)	4.48 (a)	4.49 (d)	4.50 (a)	4.51 (c)	4.52 (a)	4.53 (c)	4.54 (c)
4.55 (b)	4.56 (b)	4.57 (a)	4.58 (a)	4.59 (b)	4.60 (b)	4.61 (a)	4.62 (b)	4.63 (b)
4.64 (b)	4.65 (b)	4.66 (a)	4.67 (c)	4.68 (b)				

Explanations Data Representation and Programming

4.1 (c)

$x = 3954$
$y = 0$
p = remainder of 3954/10 $\Rightarrow p = 4$
x = greatest integer of 3954/10
$\Rightarrow$ $x = 395$
$y = 0 + 4 = 4$
Since $x \neq 0$, loop will continue.
p = remainder of 395/10
= 5
x = greatest integer of 395/10
= 39
$y = 4 + 5 = 9$
Since, $x \neq 0$, loop will continue.
So p = remainder of 39/10
= 9
x = greatest integer of 39/10
= 3
$y = 9 + 9 = 18$
Since $x \neq 0$, loop will continue.
So p = remainder of 3/10
= 3
x = greatest integer of 3/10
= 0
$y = 18 + 3 = 21$
Since $x = 0$, y has final value.
$y = 21$

4.2 (b)

$f-=9;$	$f*=9;$	$f/=9;$
$\Downarrow$	$\Downarrow$	$\Downarrow$
$f=f-9;$	$f=f*9;$	$f=f/9;$

−= , *=, etc. are called as multi character operator.

4.3 (b)

Assembly code		Operation
LOAD	X	X
MULT	Y	XY
STORE	T1	XY
MULT	T1	$(XY)^2$
STORE	T1	$(XY)^2$
LOAD	Z	Z
ADD	Y	Z + Y
ADD	T1	$Z + Y + (XY)^2$
STORE	R	R: = $(XY)^2 + Y + Z$

4.5 (c)

Firmwares are the programs stored in ROMS or other devices which permanently keep their stored information.

4.6 (d)

Redundancy is eliminated in the programs to save the execution time of the program.
Redundancy may be easier to read and may allow more error checking to be done.

4.7 (b)

A program that translates a high-level language program into a machine language program is called a compiler.

4.8 (a, d)

Pd is a pointer to structure of type data

(a) Pd → year
(b) (∗ Pd) year
Both (a) and (d) are correct method of refer year.

4.9 (d)

$$A = 5 \quad ; \quad x = 10$$
$$B = 15 \quad ; \quad y = 20$$
$$Z = 5 + 15 = 20$$
$$M = 10 \times 20 = 200$$
$$A = 5 - 15 = -10$$
$$X = 10 \times 20 = 200$$
$$B = -10 + 15 = 5$$
$$Y = -10 \times 5 = -50$$
$$Y = 200 - 50 = 150$$
$$A = 5 - (-10) = 15$$

Therefore, $A = 15$, $B = 5$.

4.10 (a)

Excess-64 is also written as $E - 64$
Exponent $h\, e_6\, e_5\, e_4\, e_3\, e_2\, e_1\, e_0$ can be represented by

$$e = -64 + \sum_{i=0}^{6} 2^i e_i$$

$$\Rightarrow \quad e = -64 + (2^6 e_6 + 2^5 e_5 + 2^4 e_4 + 2^3 e_3 + 2^2 e_2 + 2^1 e_1 + 2^0 e_0)$$

4.11 (a)

A call by value parameter passing technique only passes the copy of variable.
In call by reference parameter passing technique address is passed.

4.12 (c)

define sqr (x) ((x) ∗ (x)).

4.13 (a)

P is a pointer to the function returning an integer.

4.14 (d)

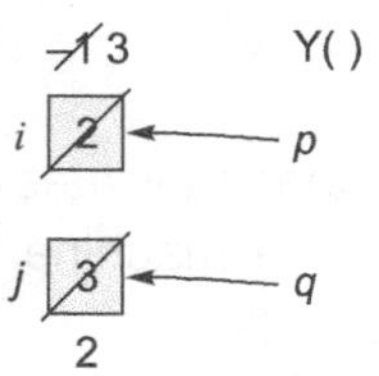

$i = 3, j = 2$

4.15 (c)

```
p = 2.0 ; q = 1.0 ;
While (p/q > 0.001)
{
  q = q+q;
  total = total + p/q;
  printf ("%f", total);
}
```

1st	2nd	3rd	4th
2 > 0.001 q = 2.0 total = 1.0 1.0	1 > 0.001 q = 4.0 total = 1.5 1.5	0.5 > 0.001 q = 8.0 total = 1.75 1.75	0.25 > 0.001 q = 16.0 total = 1.875 1.875

5th	6th	7th
0.125 > 0.001 q = 32.0 total = 1.9375 1.9375	0.0625 > 0.001 q = 64.0 total = 1.96875 1.96875	0.03125 > 0.001 q = 128 total = 1.984375 1.984375

After 1 interation loop will terminate total value ≅ 2.

4.16 (c)

int (∗*P*) [20];
A pointer to an array of integer
int ∗ *P*[20];
An array of pointers.

4.17 (c)

If array is 1-ordered then it is sorted.
So maximum number of positions that an element can from its position is *N*/2.

4.18 (c)

Since *a*, *b*, *c* are positive values $(b - c) + a$ avoids overflow.

4.19 (a, b)

p and *q* are two pointer variables of the same type and $p < q$.

(a) $p - q$ is logically not correct because this operation result in negative pointer. There is no such term called negative pointer.

(b) $p + q$ is logically not correct, two pointer variables cannot be added.

(c) $p + 5$ is logically correct, an integer value may be added or subtracted from a pointer variable.

(d) p++ is logically correct, a pointer variable can be pre-fixed or post-fixed with increment or decrement operators.

4.20 (a)

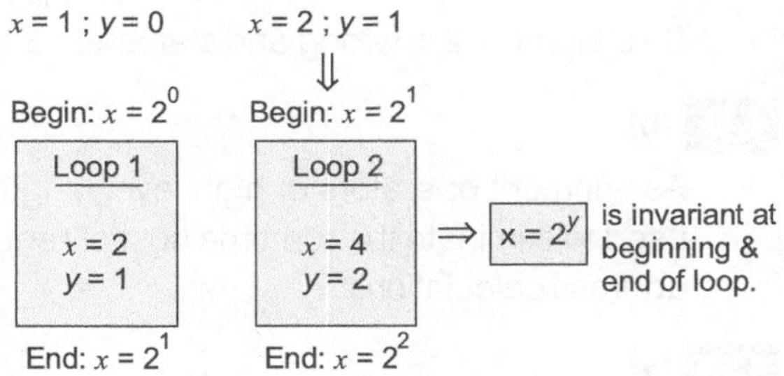

4.21 (b)

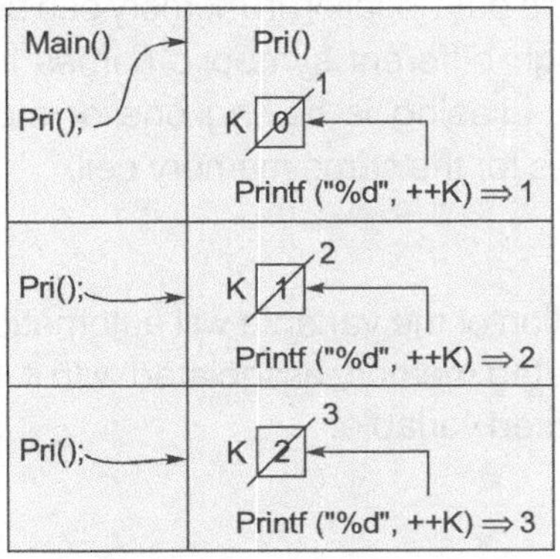

∴ It prints 123.

4.22 (a)

Passing a pointer to a function to g enables the g to make modification to function whose pointer is passed to the *g*.

4.24 (a)

Locality of reference tells us whether memory locations in a program are likely to be accessed in the near future. Result of expression may depend upon the order of evaluation of operands of the operator 'x', defines temporal locality.

4.25 (a)

^ is binary operator [BITWISE XOR]
& is unary operator [Address of]
& is binary operator [Bitwise AND]
mod is not an operator in *C*.
? is ternary operator.

A – 2
B – 1 or 2
C – 4
D – 3

∴ Option (a) is correct.

4.26 (b)

$$f = \frac{10}{3};$$

int/int produces integer value.
3.0 is printed as '*f*' value.

4.27 (a)

In graphs, self loops of nodes can be represented by self-referential structure.

4.28 (c)

Binary search can be implemented using recursion. Recursion is not depend on number of elements in the array.
So A is true and R is false.

4.29 (b)

ASCII codes of upper case English letters ranges from 65 to 90.
ASCII codes of lower case English letters ranges from 97 to 122.

4.30 (a)

Note: Question has typing error.

1. x == 1; is condition and can not be used as simple statement.
2. #define max = 40 ; is not a correct syntax.

∴ Statement 1 and 2 are incorrect.
Option (a) is correct.

4.31 (d)

There is no such problem because every recursive solution can be converted into similar non recursive function.

4.32 (b)

Divide by zero is run time error.

4.33 (b)

PASCAL		C
:=	⟷	=
<>	⟷	!=
=	⟷	==

4.35 (c)

[Representation of normalised form of a number = 0·Mantissa × 2^{exponent}]
Normal form of 0 is 0·0, mantissa = 0 and exponent = 0.

4.36 (a)

Vertical micro instruction.
→ Encoding
→ Low degree of parallelism
→ Log_2 N bits for N control signal (small format).
Only 1 and 2 are correct

4.37 (b)

C-supports 1-dimensional, 2-dimensioal, etc.
Array elements-occupy contiguous memory locations.

4.38 (a)

Taking 2's complement is equivalent to sign change. In the 2's complement representation the most significant bit (MSB) is zero for a positive number.

4.39 (d)

The maximum dimension of an array depends on the compiler.

4.40 (c)

An underscore can be used anywhere within an identifier.
"IF" is an identifies and "if" is a keyword.
All given statements are correct.

4.41 (d)

Function call uses the system stack by creating an activation record to store the function information by maintaining PC, locals, access link, control link, etc.

4.42 (c)

C-program uses call by value mechanism.

4.43 (d)

A compiler uses symbol table to manage information about variables and their attributes.

4.44 (b)

The data type defined by user is abstract data type.

4.46 (a)

Assembly languages directions define the machine code (by some α–bits). It implements a symbolic representation of the machine codes and other constants needed to program a given CPU architecture it does not implement specific symbolic memory location. It is based on mnemonics that symbolize processing steps (instructions), processor registers, and other language features. So control the output of assembly process.
So statement 2 is wrong and statement 3 is right.

4.47 (b)

Assignment operators of high level language is used to eliminate the run-time cost of redundant address calculations.

4.50 (a)

In computing aliasing describes a situation in which a data location in memory can be accessed through different symbolic names in program. Thus, aliasing is having one or more distinct names for the same memory cell.

4.51 (c)

Creation of the variable will automatically create allocated memory associated with it. Hence, it is allocated variable.

4.52 (a)

$$C = i{+}{+};$$

The above statement is equivalent to the following sequence of two statements.

$C = i$; /* Assignment of i to C*/
$i = i + 1$; /* Increment of i by 1*/

4.53 (c)

Firmware is a micro program which is embedded on a hardware device and provided by a manufacturer.

4.54 (c)

Locality of reference concept will fail if a program has many operands.

4.55 (b)

00111100
↓
Right shift by "3" times
↓
00000111

4.56 (b)

Using pointer we can,

1. Goto table any entry change contents accordingly.

2. If you want to pass array no need to pass total array pass only 1st element address. It will reduce the length of the program and it will reduce complexity.

4.57 (a)

The signed magnitude system both sign bit and magnitude bits should be processed separately.

4.61 (a)

Assembly language is a low-level language. Assembler can be used to convert into machine code directly so runs faster.

4.62 (b)

Pseudoinstruction is a hint given to assembler by programmer. They do not require any memory space.

Example: ORG, DB, DW.

4.64 (b)

When we multiply two n-bit numbers, the maximum result size can be of 2n bits.

For example: for a 3-bit unsigned number, maximum value = 7. The multiplication of 3-bit unsigned numbers would result in a maximum value of $7 \times 7 = 49$ which requires 6-bits to represent that can be stored in one memory word. Hence, option (a) is incorrect.

When two 16-bit unsigned numbers are multiplied, the result will be of 32-bits which can be stored in two memory words. Hence, option (b) is correct.

4.65 (b)

Booth's algorithm is a multiplication algorithm that multiplies two signed binary numbers in 2's complement notation. Booth used desk calculators that were faster at shifting than adding and created the algorithm to increase their speed.

4.66 (a)

A floating point number (FP) is said to be normalized if the integer part of the mantissa is 1.

4.67 (c)

Linker is a computer system program that takes one or more object files (generated by a compiler or an assembler) and combines them into a single object file.

4.68 (b)

The format of the single-precision floating point representation of a number as per the IEEE 754 standard is as below.

32 bits

Sign Bit (S)	Biased Exponent (BE)	Mantissa (M)
1 Bit	8 Bit	23 bits

The floating point number can be obtained by the formula:

$(-1)^s \times 1 \cdot M \times 2^{BE-127}$

From the given representation,

1	10000001	01000000000000000000000
S = 1	BE = 129	

$\therefore \quad N = (-1)^1 \times (1.0100...) \times 2^{129-127}$

$N = -1.01 \times 2^2 = -4.04$

5 Data Structures

Note: This topic is out of syllabus for ESE from 2016, it may be useful for PSUs.

5.1 A single edge is added to a tree without increasing the number of nodes. The number of cycles in the resulting graph is equal to

(a) zero (b) one
(c) two (d) indeterminate

[ESE-2000]

5.2 The prefix form of the expression $X + Y - Z$ is

(a) $- + XYZ$ (b) $+ - XYZ$
(c) $XYZ - +$ (d) $XYZ + -$

[ESE-2000]

5.3 Which of the following operations are performed on linear queues?

1. Testing a linear queue for underflow
2. Enqueue operation
3. Dequeue operation
4. Testing a linear queue for overflow

Select the correct answer using the codes given below:

(a) 1, 2 and 3 (b) 2, 3 and 4
(c) 1, 3 and 4 (d) 1, 2, 3 and 4

[ESE-2001]

5.4 Which one of the following is not a linear data structure?

(a) Array (b) Linked list
(c) Stack (d) Tree

[ESE-2002]

5.5 A data structure in which insertions and deletions are possible at either end, is called a

(a) queue (b) deque
(c) stack (d) enque [ESE-2002]

5.6 **Assertion (A):** Insertion and deletion in a sorted array can be time-consuming.

Reason (R): All the elements following the inserted or deleted element must be shifted approximately.

(a) Both A and R are true and R is the correct explanation of A
(b) Both A and R are true but R is NOT the correct explanation of A
(c) A is true but R is false
(d) A is false but R is true [ESE-2003]

5.7 Consider a complete graph with n vertices. What is the total number of spanning trees?

(a) $\frac{n(n-1)}{2}$ (b) $2n - 1$
(c) $n!$ (d) n^{n-2}

[ESE-2004]

5.8 Match **List-I** (Type of Data Structure) with **List-II** (Used in Application) and select the correct answer using the code given below the lists:

List-I	List II
A. Stack	1. Solving linear simultaneous equations
B. Tree	2. Subroutine linkage
C. Record	3. File processing
D. Array	4. Sorting

Codes:

	A	B	C	D
(a)	3	4	2	1
(b)	2	1	3	4
(c)	2	4	3	1
(d)	3	1	2	4

[ESE-2004]

5.9 What is the depth of a complete binary tree with '*n*' nodes?

(a) $\log_2(n + 1) - 1$ (b) $\log_2(n - 1) + 1$
(c) $\log_2(n) + 1$ (d) $\log_2(n) - 1$

[ESE-2004]

5.10 Consider the following statements:

1. Infix, Prefix and Postfix notations for expressing sum of *A* and *B* are *A*+*B*, +*AB*, and *AB*+, respectively.
2. AVL tree is a binary tree in which the difference in heights between the left and the right sub tree is not more than one for every node.

3. Stack data structure is used to save and retrieve information in reverse order.
4. Queue data structure is known as LIFO.

Which of the statements given above are correct?

(a) 1, 2 and 3 (b) 2, 3 and 4
(c) 1, 3 and 4 (d) 1, 2 and 4

[ESE-2005]

5.11 In a 500 × 500 matrix, 95% of the elements are zeros and these elements are randomly distributed. Which is an appropriate data structure to store this efficiently?

(a) An array (b) A tree
(c) A list (d) A stack

[ESE-2005]

5.12 If push and pop operation on a stack takes 1 unit time, how much time would it take to delete an element at n^{th} position from bottom?

(a) 1
(b) n
(c) n^2
(d) Cannot be determined from the given data

[ESE-2006]

5.13 Which one of the following is not correct about recursion?

(a) Depth of stack is proportional to the depth of recursion
(b) Some of the recursive function can not be written as non-recursive one
(c) For some of functions, writing recursive version is easier than non-recursive version
(d) Recursive functions have terminating condition which limits the depth of recursion

[ESE-2006]

5.14 Which one of the following is the most suitable definition of ARRAY?

(a) It is a collection of items which share a common name
(b) It is a collection of items which share a common name and occupy consecutive memory locations
(c) It is a collection of items of the same type and storage class which share a common name and occupy consecutive memory locations
(d) It is just a collection of unordered items

[ESE-2007]

5.15 Consider the following statements regarding binary tree:

1. A tree with n nodes has $(n-1)$ edges.
2. A labelled and rooted binary tree can be uniquely constructed given its post-order and pre-order traversal results.
3. The maximum number of nodes in a binary tree of height (depth) h is $(2^{h+1}-1)$.
4. A complete binary tree with n internal nodes has $(n+1)$ leaves.

Which of the statements given above are correct?

(a) 1, 2, 3 and 4 (b) 1, 2 and 3 only
(c) 1, 3 and 4 (d) 2 and 4 only

[ESE-2007]

5.16 If the mantissa in the floating point representation of a number is 37 bits long, then what is the accuracy of the digital computer?

(a) 37 decimal places
(b) 23 decimal places
(c) 11 decimal places
(d) 10 decimal places

[ESE-2008]

5.17 How many distinct binary tree can be constructed with three nodes ?

(a) 1 (b) 2
(c) 3 (d) 5

[ESE-2009]

5.18 Consider the following statements :

1. Internal sorting is used if the number of items to be sorted is very large.
2. External sorting is used if the number of items to be sorted is very large.
3. External sorting needs auxiliary storage.
4. Internal sorting needs auxiliary storage.

Which of the above statements are correct ?

(a) 1 and 2 (b) 2 and 3
(c) 3 and 4 (d) 2 and 4

[ESE-2009]

5.19 Which one of the following algorithms design techniques is used in quick sort algorithms?

(a) Dynamic programming
(b) Backtracking
(c) Divide and conquer
(d) Greedy

[ESE-2009]

5.20 There are four different algorithms A1, A2, A3 and A4 to solve a given problem with the complexity order log(n), log(log(n)), $n$log(n) and n/log(n) respectively. Which is the best algorithms

(a) A1 (b) A2
(c) A3 (d) A4 [ESE-2009]

5.21 Insertion of a record in a circularly single linked list involves the modification of

(a) 4 pointers (b) 3 pointers
(c) 2 pointers (d) 1 pointer [ESE-2010]

5.22 The number of edges in a regular graph of degree D and N vertices is equal to

(a) ND (b) $ND/2$
(c) $N+D$ (d) N^D [ESE-2010]

5.23 Sorting is useful for

1. Report generation
2. Making searching easier and efficient
3. Responding to queries easily
4. Minimizing the storage needed

(a) 1, 2 and 3 only (b) 1, 3 and 4 only
(c) 2, 3 and 4 only (d) 1, 2, 3 and 4 [ESE-2011]

5.24 Given below are some applications. Choosing from the options pick the one that allocates a suitable data structure for implementing these applications:

1. Representation of a sparse matrix
2. Fast access to any item from a set of data
3. Convert infix expression to postfix expression
4. Storing the terms of a long polynomial with arbitrary number of terms

(a) Linked list, array listed list and stack
(b) Stack, array and stack listed list
(c) Array array, tree and stack
(d) Linked list, array and stack listed list [ESE-2011]

5.25 How many passes does a Bubble sort algorithm require for sorting a given list of 'n' items?

(a) n^2 (b) $\sqrt{n}$
(c) $n+1$ (d) $n-1$ [ESE-2011]

5.26 The data structure needed to convert infix notation to prefix notation is

(a) Queue (b) Stack
(c) Tree (d) Graph [ESE-2012]

5.27 A binary tree T has n leaf nodes. The number of nodes of degree 2 in T is

(a) $\log_2 n$ (b) $n-1$
(c) n (d) 2^n [ESE-2012]

5.28 An array multiplier is used to find the product of a 3 bit number with a 4 bit number. How many 4 bits address are required to perform multiplication?

(a) 1 (b) 2
(c) 3 (d) 4 [ESE-2013]

5.29 Wrapping of data functions together in a class is known as

(a) Overloading (b) Data Abstraction
(c) Polymorphism (d) Encapsulation [ESE-2014]

5.30 Which of the following algorithm design techniques is used in the quick sort algorithm?

(a) Dynamic programming
(b) Backtracking
(c) Divide and conquer
(d) Greedy method [ESE-2014]

5.31 Which of the following sorting methods will be the best, if the number of swapping done, is the only measure of efficiency?

(a) Bubble sort (b) Quick sort
(c) Insertion sort (d) Selection sort [ESE-2014]

5.32 In a circularly single linked list organisation, insertion of a record involves the modification of

(a) No pointer (b) 1 pointer
(c) 2 pointers (d) 3 pointers [ESE-2014]

5.33 The average successful search time for sequential search 'n' items is

(a) $\frac{(n+1)}{2}$ (b) $\frac{n}{2}$
(c) $(n-1)$ (d) $\log(n)+1$ [ESE-2014]

5.34 What is maximum number of nodes in a binary tree that has N levels, if the root level is zero?

(a) 2^{2N} (b) $2^{N+1}-1$
(c) 2^N-1 (d) 2^N-2N [ESE-2014]

5.35 To arrange a binary search tree in ascending order, we need
(a) post order traversal only
(b) in order traversal only
(c) pre-order traversal only
(d) post order traversal and pre-order traversal
[ESE-2014]

5.36 In a binary tree, the number of internal nodes of degree 1 is 3, and the number of internal nodes of degree 2 is 6. The number of leaf nodes in the binary tree is
(a) 7 (b) 8
(c) 9 (d) 10 [ESE-2016]

5.37 A weighted complete graph with n vertices has weights $2(i-j)$ at edges (v_i, v_j). The weight of a minimum spanning tree is
(a) $\frac{n^2}{2}$ (b) $\frac{n}{2}$
(c) $2n-2$ (d) $n-1$ [ESE-2017]

5.38 Which one of the following is ***not*** correct when we define either a class that does not implement either a mathematical entity like a matrix or a complex number or a low-level type such as linked list?
(a) Don't use global data (use members)
(b) Don't use local functions
(c) Don't use public data members
(d) Don't use inline functions, except as a significant optimization
[ESE-2021]

5.39 In a program, the following declaration : float number1, number2, sum, average; is represented as
(a) variables (b) floating numbers
(c) input data (d) decimal numbers
[ESE-2024]

Answers Data Structures

5.1 (b)	5.2 (a)	5.3 (d)	5.4 (d)	5.5 (b)	5.6 (a)	5.7 (d)	5.8 (c)	5.9 (a)
5.10 (a)	5.11 (a)	5.12 (d)	5.13 (b)	5.14 (c)	5.15 (c)	5.16 (a)	5.17 (d)	5.18 (b)
5.19 (c)	5.20 (b)	5.21 (c)	5.22 (b)	5.23 (d)	5.24 (d)	5.25 (d)	5.26 (b)	5.27 (b)
5.28 (b)	5.29 (d)	5.30 (c)	5.31 (d)	5.32 (c)	5.33 (a)	5.34 (c)	5.35 (b)	5.36 (a)
5.37 (c)	5.38 (d)	5.39 (a)						

Explanations Data Structures

5.1 (b)

The diagram shows a complete binary tree.

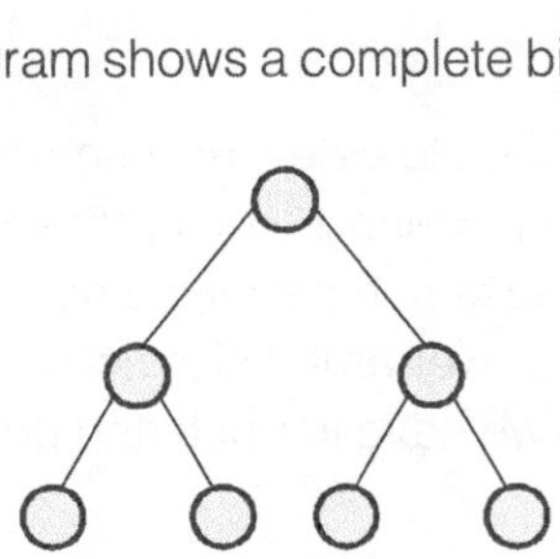

A tree is a graph with no cycles. If we add a new edge such that it forms a cycle as shown in figure, then we will have only one cycle.

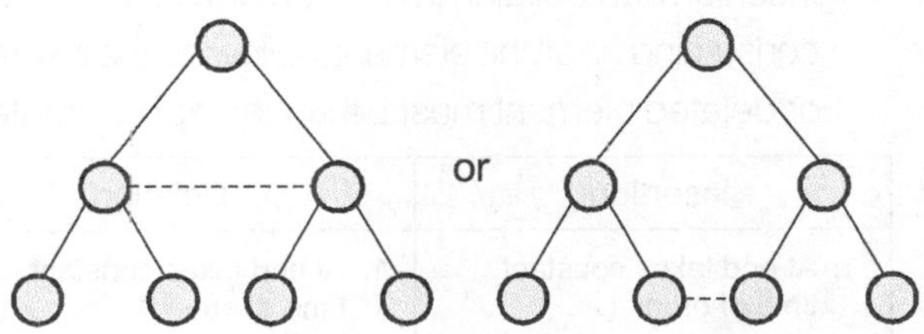

5.2 (a)

Infix : X + Y − Z

Prefix : X + Y − Z

⇒ +X Y − Z

⇒ −+X Y Z

5.3 (d)

All the four operations are performed on linear queues.

5.4 (d)

Array is represented as

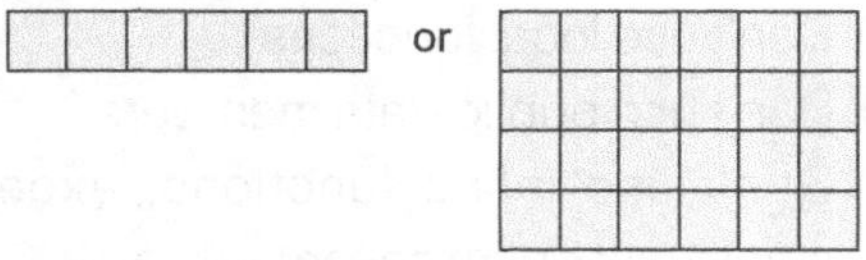

Linked list is represented as

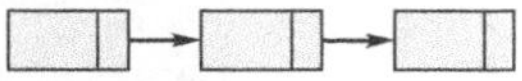

Stack list is represented as

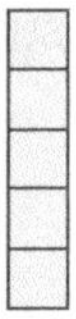

Tree is represented as

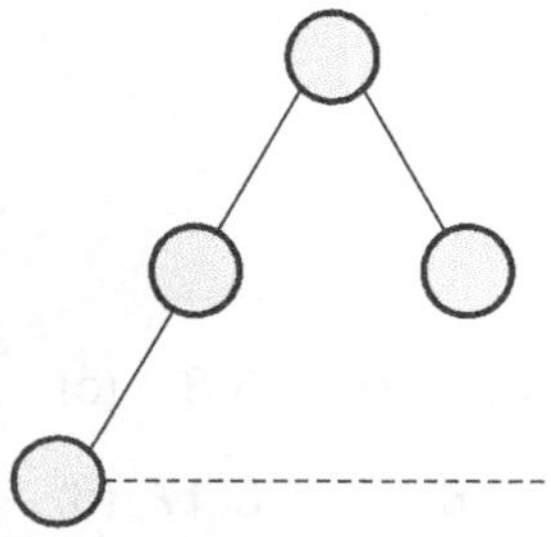

⇒ It is non linear data structure.

5.5 (b)

A data structure in which insertions and deletions are possible at either end is called deque.

5.6 (a)

Insertion and deletion in a sorted array can be time-consuming as all the elements following the inserted or deleted element must be shifted approximately.

Insertion	Deletion
1. At end takes constant time of order 1. 2. At beginning takes order of (n) worst case. 2. Average 1 2 3 4 . . . $n-1$ n $(n-1)$ $(n-(n-1))$ 0 $(n-2)$ Insertion operation for average case takes n(n – 1)/2n operations.	1. At end takes constant time of order 1. 2. At beginning takes order of (n) worst case. 2. Average 1 2 3 4 . . . $n-1$ n $(n-1)$ $(n-(n-1))$ 0 $(n-2)$ Insertion operation for average case takes $n(n-1)/2n$ operations.

5.7 (d)

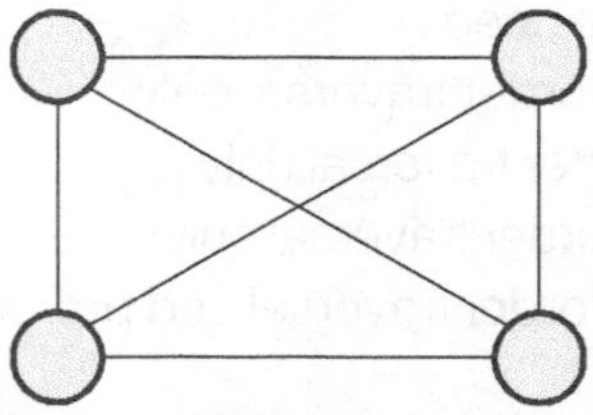

Complete graph with 4 vertices has 16 spanning trees.

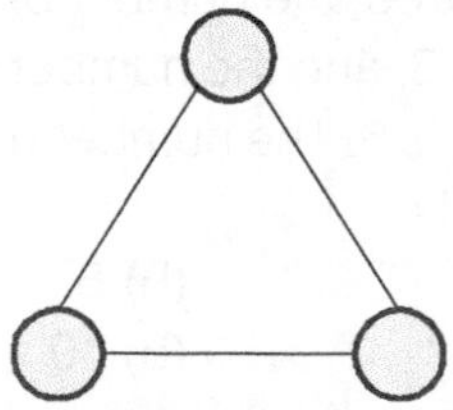

Complete graph with 3 vertices has 3 spanning trees.

Total # spanning trees = n^{n-2}.

5.9 (a)

Maximum number of nodes in a complete binary tree with depth (d) is given as:

$$n = 2^{d+1} - 1$$

$$\Rightarrow \quad n + 1 = 2^{d+1}$$

$$\Rightarrow \quad \log_2(n+1) = d + 1$$

$$\Rightarrow \quad \log_2(n+1) - 1 = d$$

5.10 (a)

Queue data structure is known as FIFO (First In First Out).

While

Stack is known as LIFO (Last In First Out).

Statements 1, 2 and 3 are correct.

5.11 (a)

An array is an appropriate data structure to store a 2-dimensional matrix.

5.12 (d)

Time required to delete an element at n^{th} position from bottom cannot be determined from the given data because number of elements i.e. n value is not given as well as it is also not mentioned, which elements we have to push and pop.

5.13 (b)

All recursive functions can be written non-recursive.

5.14 (c)

An array is a collection of items of the same type and storage class which share a common name and occupy consecutive memory locations.

int x[10] ;

x is an array of 10 integer elements located in contiguous/consecutive memory locations. Storage class of every element is same.

5.15 (c)

Only statement 2 is incorrect.

1. A tree with n nodes have $(n - 1)$ edges because a tree is not closed and it does not contain any cycle.
2. A labelled and rooted binary tree can be uniquely constructed given its inorder and pre/post order traversal results.
3. Maximum # nodes in a binary tree of height h $= 2^{h+1} - 1$

   where h = height of binary tree
4. A complete binary tree with n internal nodes have $(n + 1)$ leaves.

   # nodes with degree 0 (left nodes)

   = # nodes with degree 2 + 1

   $= n + 1$

5.17 (d)

Distinct binary trees constructed with 3 nodes.

$$= \frac{{}^{2n}C_n}{n+1} = \frac{{}^{6}C_3}{4} = 5$$

5.18 (b)

External sorting is used if the number of items to be sorted is very large and it needs auxiliary storage.

5.19 (c)

Quick sort algorithm works on the principle of divide and conquer.

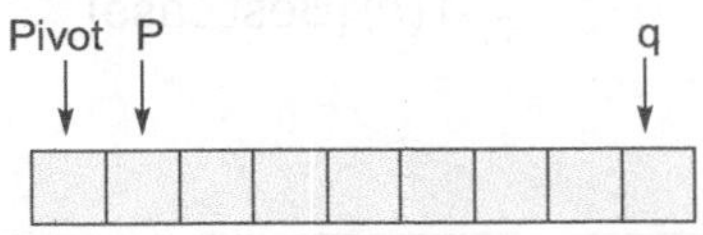

First element is chosen as pivot.

Two pointers p and q are also set as shown in the figure. Quick sort algorithm is run and after first pass, the list becomes:

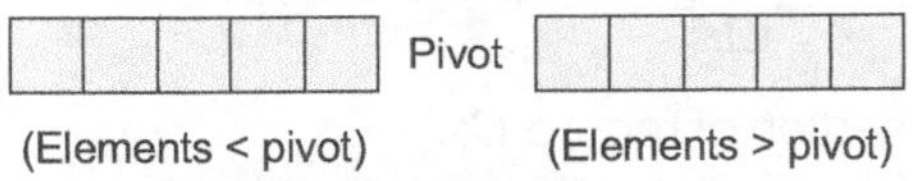

Here the list is divided into two lists.

Now, two lists are obtained, each one is sorted in the same manner.

5.20 (b)

Algorithm	A1	A2	A3	A4
Complexity order	log(n)	log(log(n))	$n$log(n)	n/log(n)

Increasing order of complexities are:

$1 < \log(\log(n)) < \log(n) < n/\log(n) < n <$ nlogn

$< n\log_n^2 < n^2 < n^3 < \cdots < 2n < \cdots < n!$

A2 is the best algorithm as it takes less time as compared to A1, A3 and 4.

5.21 (c)

Suppose we want to insert node A to which we have pointer p, after pointer q then we will have following pointer operations.

1. p->next = q->next;
2. q->next = p;

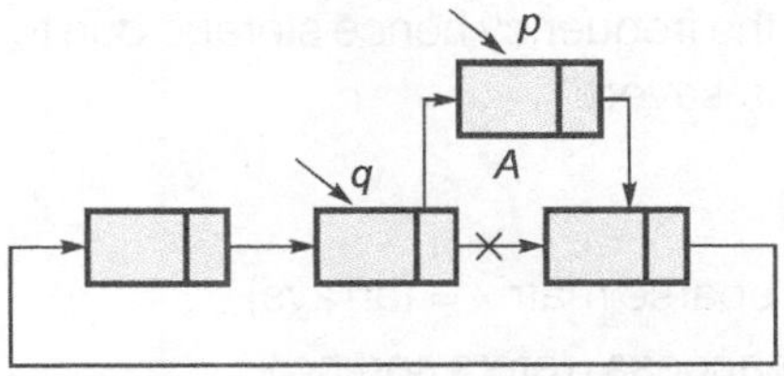

Circular single linked list involves modification of two (2) pointers.

5.22 (b)

A regular graph is a graph in which degree of every vertex is same.

For eg : K_3

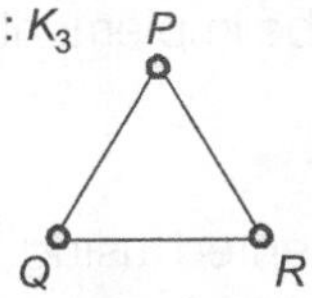

Degree of each vertex = 2

vertices = 3

$$\text{\# edges} = \frac{2 \times 3}{2} = 3$$

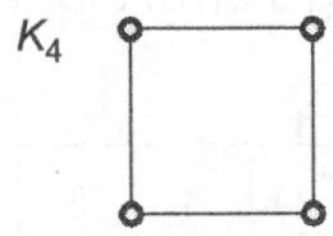

$$\text{\# edges} = \frac{4 \times 2}{2} = 4$$

or

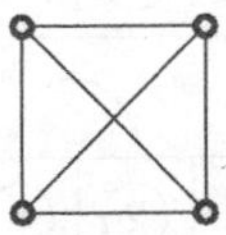

$$\text{\# edges} = \frac{4 \times 3}{2} = 6$$

$\therefore$ # edges = ND/2

5.23 (d)

Distribution sort is very useful for report generation. It refers to any sorting algorithm where data is distributed from its input to multiple intermediate structure which are then gathered and placed on the output.

- Making searches easier and efficient → because in sorting, data is arranged in a ordered manner as per our needs and by using algorithm such as quick sort and bucket sort sorting (searching and related option) becomes fast and efficient.
- Since our data is in preferred order manner which may include the priorities hence responding become easy.
- Minimizing the storage needed → sorting such as counting sort stores the ith bit along with the frequency hence storage due to repetition is saved.

5.24 (d)

For sparse matrix = {arrays}
fast access (refers sorting)
= {array}
Infix to postfix = {stack}
Sorting in terms of
long polynomial = {linked list, arrays}.
So by using linked, array and stack listed list the required functions can be implemented.

5.25 (d)

Bubble sort is implemented using sorting two element at a time → bubble is a simple sorting algorithm that works by repeatedly stepping through the list to be sorted, comparing each pair of adjacent items swapping them if they are in wrong order.

i.e.

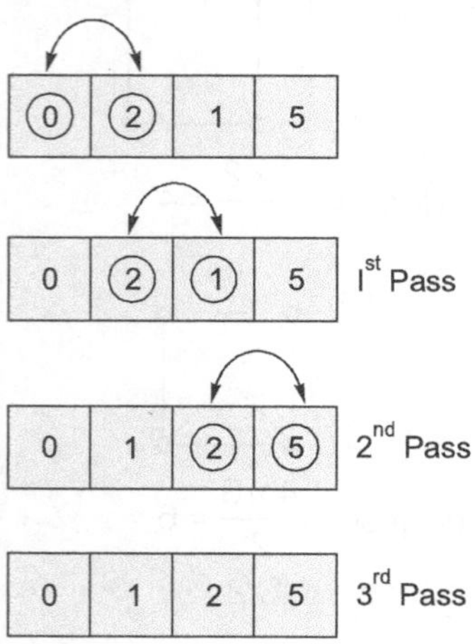

So for 4 element there 3 = (4 – 1) passes required similarly for n items (n – 1) passes are needed.

5.26 (b)

Stack is used in conversion of infix expression to prefix or postfix expression.

5.27 (b)

nodes of degree 0 = # nodes of degree 2 + 1
⇒ # nodes of degree 2 = n – 1

Note: As per RTI filed and reply of **UPSC** answer is option (a).

5.29 (d)

Wrapping of data and functions together in a class is called as Encapsulation.

5.30 (c)

Quick sort algorithm uses partition approach by choosing the pivot element. In every iteration of algorithm, the array is divided into two sublists by finding the pivot position.

It uses divide and conquer approach.

5.31 (d)

Efficiency is measured using only swaps. The best algorithm which does less number of swappings is selection sort.

In each iteration (pass) of algorithm the maximum number of swapping is 1.

Total number of swappings in entire algorithm $\leq n - 1$.

Selection sort (swappings)

= 0(n) [worst case]

= 1(n) [Best case]

5.32 (c)

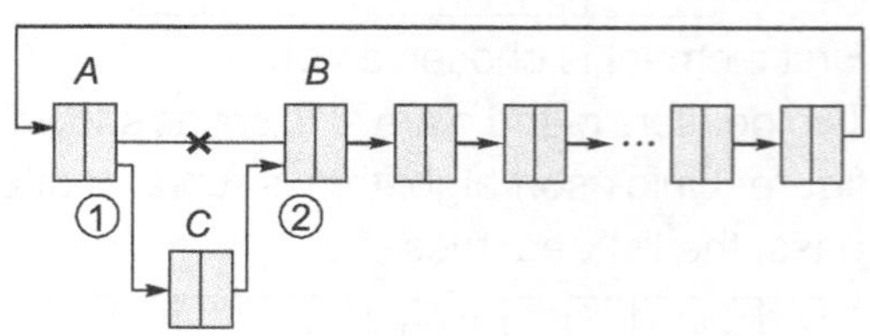

Insertion of record C;

(1) $A \rightarrow$ next = C

(2) $C \rightarrow$ next = B

∴ $A \rightarrow$ next and $C \rightarrow$ next are modified

(two pointers are modified)

5.33 (a)

1	2	3	4	....	$n-1$	n

Search 1 → 1 comparison
Search 2 → 2 comparison

•
•
•

Search n → n comparison

$$\text{Total comparison} = 1+2+\ldots+n = \frac{n(n+1)}{2}$$

Average successful comparisons

$$= \frac{n(n+1)}{2n} = \frac{(n+1)}{2}$$

5.34 (c)

Number of levels = N
Maximum number of nodes in a binary tree = ?
If $N = 1$ ⇒ 1 node in tree (only root)
$N = 2$ ⇒ 3 node (maximum) in tree

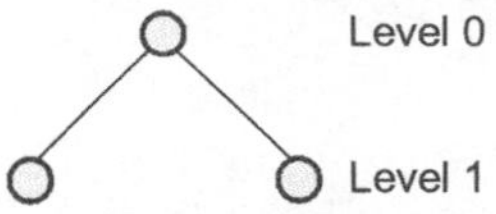

$N = 3$ ⇒ 7 node (maximum) in tree

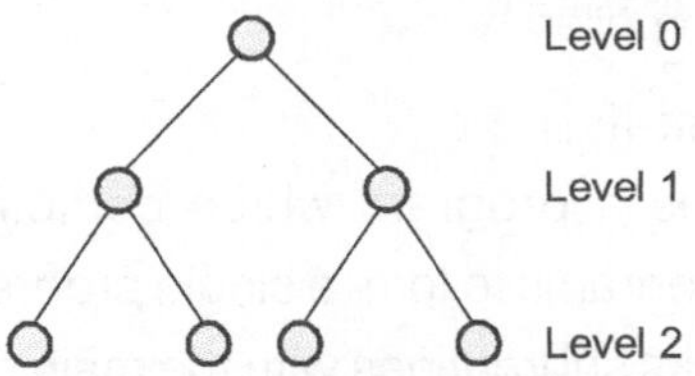

[Note: If number of levels is zero then no tree is possible]

∴ Maximum number of nodes in a binary tree $= (2^N - 1)$

5.35 (b)

Given "Binary Tree" no traversal order gives ascending order.
Binary tree has no restriction but binary search tree has restrictions at every node of the tree such that left has less than it and right has greater than it. So, in binary search tree, in order traverse gives the elements in ascending order.

5.36 (a)

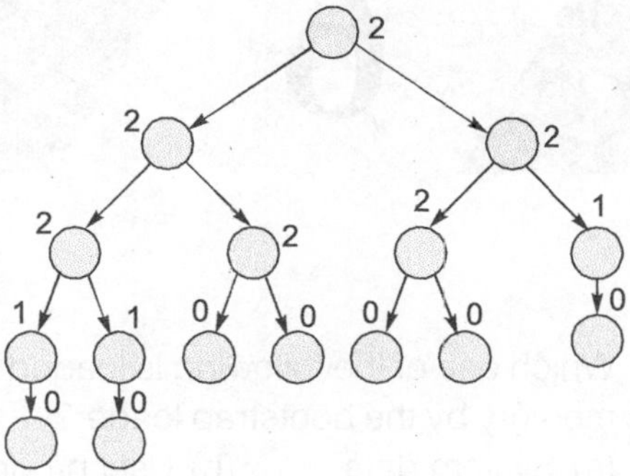

degree – 2 ⇒ 6
degree – 1 ⇒ 3
degree – 0 ⇒ 7 [leaf]

5.37 (c)

For graph with '3' vertices numbered (V_1, V_2, V_3)

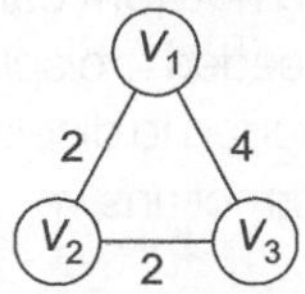

The minimum spanning tree of the above graph is

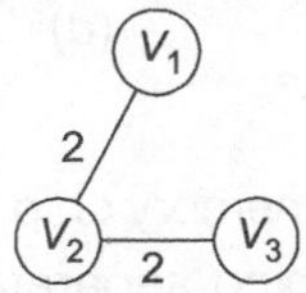

Which will gives the '4' as a minimum spanning tree value. By potting $n = 3$ in a given options, we get $\{2n - 2 = 4\}$.

5.38 (d)

Since, all the operations are best handled by local functions (member functions) as it enables abstraction and prevents external entities from handling private data. Hence, local functions must be used. However, global functions should not be used in such case.

5.39 (a)

The 'float' keyword specifies the data type of the variables with 'number 1', 'number 2', 'sum', 'average' as the names (identifiers) assigned to the variables. These variables are used to store floating-point numbers. Hence, the given declaration is used to define the variables of 'float' data type.

■■■

6 Operating Systems and Data Bases

6.1 Which one of the following is loaded in the main memory by the bootstrap loader?
(a) System data (b) User program
(c) BIOS (d) Parts of DOS
[ESE-2000]

6.2 Which of the following are required for a multimedia PC?
1. CD-ROM drive, speaker and sound card.
2. Modem and network card.
3. Hardware needed to display videos & animation.
4. Software needed to display videos & animation.
Selection the correct answer using the codes given below:
(a) 1, 2 and 3 (b) 1, 2, 3 and 4
(c) 1, 2 and 4 (d) 1, 3 and 4
[ESE-2001]

6.3 A Read/Write memory chip has a capacity of 64 kBytes. Assuming separate data and address lines and availability of chip enable signal, what is the minimum number of pins required in the IC chip?
(a) 28 (b) 26
(c) 24 (d) 22 [ESE-2002]

6.4 A disc drive has a rotational speed of 3600 rpm, an average seek time of 10 ms, 64 sectors per track and 512 bytes of data per sector. What is the average time to access the entire data of a 16 kbytes file stored sequentially on the disk?
(a) 18.85 ms (b) 10 ms
(c) 27.15 ms (d) 9 ms [ESE-2004]

6.5 A disc drive has a average seek time of 10 ms, 32 sectors on each track and 512 bytes per sector. If the average time to read 8 kbytes of continuously stored data is 20 ms, what is the rotational speed of the disc drive?
(a) 3600 rpm (b) 6000 rpm
(c) 3000 rpm (d) 2400 rpm
[ESE-2004]

6.6 A disc rotates at a speed of 7200 rpm. It has 4000 cylinders, 16 surfaces and 256 sectors per track. What is the average latency time of the disk?
(a) 8.33 ms (b) 4.166 ms
(c) 41.66 ms (d) 8.33 μs
[ESE-2005]

6.7 Match **List-I** (Program Required in System Software) with **List-II** (Definition) and select the correct answer using the code given below the lists:

List-I
A. Linker
B. Loader
C. Interpreter
D. Compiler

List-II
1. It is a program which combines smaller programs to form a single program; and also links subroutines with the main program
2. It is a program which loads machine codes of a program into the system memory
3. It is a program which translates a high-level program into machine code, executes it and reads one statement at a time, executes and then goes to the next statements of the program
4. It is a program which translates a high-level program into a machine language, reads the entire program and then executes it

Codes:

	A	B	C	D
(a)	1	2	3	4
(b)	3	4	1	2
(c)	1	4	3	2
(d)	3	2	1	4

[ESE-2007]

6.8 Why does an increase of the RAM of a computer typically improve performance?
(a) Virtual memory increases
(b) Larger RAMs are faster
(c) Fewer page faults occur
(d) Fewer segmentation faults occur
[ESE-2007]

6.9 A memory system of size 16 Kbytes is required to be designed using memory chips which have 12 address lines and 4 data lines each. What is the number of such chips required to design the memory system?
(a) 2 (b) 4
(c) 8 (d) 16 [ESE-2007]

6.10 A magnetic drum of 8 inch diameter has 100 tracks and storage density of 200 bits/inch. What is its storage capacity?
(a) 8402 bits (b) 202400 bits
(c) 502400 bits (d) 1004800 bits
[ESE-2008]

6.11 A memory system of size 32 kbytes is required to be designed using memory chips which have 12 address lines and 4 data lines each. What is the number of such chips required to design the memory system?
(a) 4 (b) 8
(c) 16 (d) 32 [ESE-2008]

6.12 An example of a spooled device is
(a) A graphical display device
(b) A line printer used to print the output of a number of jobs
(c) A secondary storage device in a virtual memory system
(d) A terminal used to enter input data to a running program
[ESE-2010]

6.13 **Assertion (A):** When you turn on your PC, a process called POST (Power-on-self-test) begins with an electrical signal.
Reason (R): The electrical signal restores left over data from the chip's internal memory register.
(a) Both A and R are true and R is the correct explanation of A
(b) Both A and R are true but R is NOT the correct explanation of A
(c) A is true but R is false
(d) A is false but R is true [ESE-2010]

6.14 A virtual memory system has an address space of 8 k words, a memory space of 4 k words and page and block sizes of 1 k words. The number of page faults using LRU policy, for following page references is
1 0 2 4 6 2 1 5 7 0 0
(a) 5 (b) 7
(c) 9 (d) 10 [ESE-2012]

6.15 The seek time of a disk is 30 msec. It rotates at the rate of 30 rotations per second. Each track has a capacity of 300 words. The access time is
(a) 47 msec (b) 50 msec
(c) 60 msec (d) 62 msec
[ESE-2013]

6.16 **Statement (I):** External fragmentation is NOT possible in paging.
Statement (II): Pages of any program are fixed sized and can be located anywhere in memory.
(a) Both Statement (I) and Statement (II) are individually true and Statement (II) is the correct explanation of Statement (I).
(b) Both Statement (I) and Statement (II) are individually true but Statement (II) is not the correct explanation of Statement (I).
(c) Statement (I) is true but Statement (II) is false.
(d) Statement (I) is false but Statement (II) is true.
[ESE-2013]

6.17 OS that permits multiple programs to be run simultaneously using a single processor is referred to as
(a) Multitasking (b) Multithreading
(c) Multi-user (d) Multi-processing
[ESE-2015]

6.18 Consider the following statements regarding the functions of an operating system in a computer.
1. It controls hardware access.
2. It manages files and folders.
3. It provides a user interface.
4. It manages application.

Which of the above statements are correct?
(a) 1, 2 and 3 only (b) 1, 2 and 4 only
(c) 3 and 4 only (d) 1, 2, 3 and 4
[ESE-2017]

6.19 Consider the following processes which arrived in the order P_1, P_2 and P_3.

Process	Burst time
P_1	24 ms
P_2	3 ms
P_3	3 ms

What is the average waiting time by FCFS scheduling?
(a) 17 ms (b) 19 ms
(c) 21 ms (d) 23 ms
[ESE-2017]

6.20 A disk unit has 24 recording surfaces. It has a total of 14000 cylinders. There is an average of 400 sectors per track. Each sector contains 512 bytes of data. What is the data transfer rate at a rotational speed of 7200 r.p.m?
(a) 68.80×10^6 bytes/s
(b) 24.58×10^6 bytes/s
(c) 68.80×10^3 bytes/s
(d) 24.58×10^3 bytes/s
[ESE-2017]

6.21 In the demand paging memory, a page table is held in registers. If it takes 1000 ms to service a page fault and if the memory access time is 10 ms, what is the effective access time for a page fault rate of 0.01?
(a) 19.9 ms (b) 10.9 ns
(c) 9.99 ms (d) 0.99 ms
[ESE-2017]

6.22 Consider the following statements regarding database normal forms:
1. Any relation with two attributes is BCNF?
2. Lossless, dependency-preserving decomposition into BCNF is always possible.
3. Lossless, dependency-preserving decomposition into 3NF is always possible.
4. BCNF is stricter than 3NF.

Which of the above statements are correct?
(a) 1, 2 and 3 (b) 1, 3 and 4
(c) 1, 2 and 4 (d) 2, 3 and 4
[ESE-2017]

6.23 Consider the following schedules for transactions T_1, T_2 and T_3 :

T_1	T_2	T_3
Read (X)		
	Read (Y)	
		Read (Y)
	Write (Y)	
Write (X)		
		Write (X)
	Read (X)	
	Write (X)	

The correct schedule of serialization will be
(a) $T_1 \to T_2 \to T_3$ (b) $T_2 \to T_3 \to T_1$
(c) $T_3 \to T_1 \to T_2$ (d) $T_1 \to T_3 \to T_2$
[ESE-2017]

6.24 **Statement (I):** Swapping is sometimes called 'roll-out/ roll-in'.
Statement (II): Swapping is utilized in systems designed to support time-sharing.
(a) Both Statement I and Statement II are individually true and Statement II is the correct explanation of Statement I
(b) Both Statement I and Statement II are individually true but Statement II is not the correct explanation of Statement I
(c) Statement I is true but Statement II is false
(d) Statement I is false but Statement II is true
[ESE-2018]

6.25 **Statement (I):** Virtual memory is designated virtual because there is no such memory inside the computer.
Statement (II): Virtual memory uses some space of the hard disk as an extension of the primary storage of the computer.
(a) Both Statement I and Statement II are individually true and Statement II is the correct explanation of Statement I
(b) Both Statement I and Statement II are individually true but Statement II is not the correct explanation of Statement I
(c) Statement I is true but Statement II is false
(d) Statement I is false but Statement II is true
[ESE-2018]

6.26 Consider the following set of processes:

Process	P_1	P_2	P_3	P_4	P_5
Burst time	10 ms	29 ms	3 ms	7 ms	12 ms

First Come First Serve (FCFS), non-preemptive Shortest Job First (SJF) and Round Robin (RR) (quantum = 10 ms) Scheduling Algorithms for this process set would imply which of the following features?

1. The SJF policy results in less than half of the average waiting time obtained with FCFS scheduling.
2. The RR algorithm gives an intermediate value for the average waiting time.
3. Deterministic modelling takes a particular predetermined workload, and designs the performance of each algorithm for that workload.

Select the correct answer using the code given below.

(a) 1, 2 and 3 (b) 1 and 2 only
(c) 1 and 3 only (d) 2 and 3 only

[ESE-2018]

6.27 Three processors with their respective process IDs given by P_1, P_2 and P_3, having estimated completion time of 8 ms, 4 ms and 2 ms, respectively, enter a ready queue together in the order P_1, P_2 and P_3. What is the average turn time in the Round Robin Scheduling Algorithm with time 2 ms?

(a) 10 ms (b) 15 ms
(c) 20 ms (d) 25 ms

[ESE-2018]

6.28 Consider that a level of the memory hierarchy has a hit rate of 80%. Memory requests take 10 ns to complete if they hit in the level, and memory requests that miss in the level take 100 ns to complete. The average access time of the level is

(a) 110 ns (b) 100 ns
(c) 80 ns (d) 28 ns [ESE-2018]

6.29 Which of the following statements is/are correct?

1. An address generated by the CPU is commonly referred to as a physical address.
2. An address seen by the memory unit is commonly referred to as a logical address.
3. The run-time mapping from virtual to physical address is done by the memory management unit (MMU).

Select the correct answer using the code given below.

(a) 1 only (b) 2 only
(c) 3 only (d) 1, 2 and 3

[ESE-2019]

6.30 A system has 64-bit virtual addresses and 43-bit physical addresses. If the pages are 8 kB in size, the number of bits required for VPN and PPN will be respectively

(a) 51 bits and 30 bits (b) 30 bits and 51 bits
(c) 51 bits and 13 bits (d) 30 bits and 13 bits

[ESE-2019]

6.31 A main memory can hold 3 page frames and initially all of them are vacant. Consider the following stream of page requests

2, 3, 2, 4, 6, 2, 5, 6, 1, 4, 6

If the stream uses FIFO replacement policy, the hit ratio *h* will be

(a) $\frac{11}{3}$ (b) $\frac{1}{11}$
(c) $\frac{3}{11}$ (d) $\frac{2}{11}$ [ESE-2019]

6.32 **Statement (I):** Dynamic loading gives better memory-space utilization.

Statement (II): In dynamic loading, an unused routine is never loaded

(a) Both Statement (I) and Statement (II) are individually true and Statement (II) is the correct explanation of Statement (I)
(b) Both Statement (I) and Statement (II) are individually true but Statement (II) is .not the correct explanation of Statement (I)
(c) Statement (I) is true but Statement (II) is false
(d) Statement (I) is false but Statement (II) is true

[ESE-2019]

6.33 **Statement (I):** In a multiuser system, each user is assigned a section of usable memory area and is not allowed to go out of the assigned memory area.

Statement (II): In multiuser system, there is a software mechanism to prevent unauthorized access of memory by different users.

(a) Both Statement (I) and Statement (II) are individually true and Statement (II) is the correct explanation of Statement (I)
(b) Both Statement (I) and Statement (II) are individually true but Statement (II) is .not the correct explanation of Statement (I)
(c) Statement (I) is true but Statement (II) is false
(d) Statement (I) is false but Statement (II) is true

[ESE-2019]

6.34 The transfer time T of the disk is

(a) $\frac{2b}{rN}$ (b) $\frac{rb}{N}$

(c) $\frac{rN}{b}$ (d) $\frac{b}{rN}$

[ESE-2020]

where: b = Number of bytes to be transferred
N = Number of bytes on a track
r = Rotation speed in rps

6.35 What are the components of a memory management unit?

1. A facility for dynamic storage relocation.
2. Provision for preventing users for sharing programs stored in memory by different users.
3. Protection of information against unauthorised access.
4. Provision for users for changing operating system functions.

(a) 1 and 3 only (b) 1 and 4 only
(c) 2 and 3 only (d) 2 and 4 only

[ESE-2020]

6.36 Which one of the following makes permanently recorded transaction in the database?

(a) View (b) Commit
(c) Roll back (d) Flash back

[ESE-2020]

6.37 The advantage of optimistic locking is

(a) The lock is obtained only after the transaction has processed
(b) The lock is obtained only before the transaction has processed
(c) The lock never needs to be obtained
(d) The lock transactions are best suited with a lot of activity

[ESE-2020]

6.38 The ability to query information from the database, insert, delete and modify the tuples is

(a) Data Definition Language (DDL)
(b) Data Manipulation Language (DML)
(c) Storage Definition Language (SDL)
(d) Relational Schema [ESE-2020]

6.39 Which one of the following functions is not correct for an operating system?

(a) Allocating storage and memory
(b) Handling basic input and output operations
(c) Providing protected sharing of the computer among multiple applications
(d) Quick execution of instructions [ESE-2022]

6.40 Which one of the following is correct with respect to short-term scheduling?

(a) The decision as to which available process will be executed by the processor.
(b) The decision as to which process's pending I/O request shall be handled by an available I/O device.
(c) The decision to add to the pool of processes to be executed.
(d) The decision to add to the number of processes that are partially or fully in main memory.

[ESE-2023]

6.41 Which one of the following is relevant to non-preemptive kernels?

(a) Kernel allows a process to be preempted while it is running in kernel mode.
(b) Kernel data structure maintains a list of all open files in the system.
(c) Kernel does not allow a process running in kernel mode to be 'preempted; a kernel-mode process will run until it exits kernel mode, blocks, yields control of the CPU.
(d) Prone to possible race conditions include structures for maintaining memory allocation, for maintaining process lists and for interrupt handling.

[ESE-2023]

6.42 Windows keeps much of its configuration information in internal databases called

(a) system restore point
(b) service trigger
(c) service control manager
(d) hives [ESE-2023]

6.43 Which one of the following methods requires saving the value of the CPU registers from the thread being switched out and restoring the new thread being scheduled?

(a) Context switching between kernel level threads.
(b) Scheduling switching
(c) Kernel dispatcher
(d) Multilevel queue scheduling [ESE-2023]

6.44 In deadlock condition, at least one resource must be held in a non-shareable mode; that is, only one process at a time can use the resource. If another process requests that resource, the requesting process must be delayed until the resource has been released. This is called

(a) Circular wait (b) Mutual exclusion
(c) Hold and wait (d) No preemption

[ESE-2024]

6.45 Which one of the following statements is not correct for a buffer?

(a) Buffering is to cope with a speed mismatch between the producer and consumer of a data stream.
(b) Buffering is to adapt between devices that have different data-transfer sizes.
(c) Buffering is to support copy semantics for application I/O.
(d) Buffering holds a copy on faster storage of an item that resides elsewhere.

[ESE-2024]

■■■

Answers Operating Systems and Data Bases

6.1 (c)	6.2 (d)	6.3 (b)	6.4 (c)	6.5 (b)	6.6 (b)	6.7 (a)	6.8 (c)	6.9 (c)
6.10 (c)	6.11 (c)	6.12 (b)	6.13 (b)	6.14 (c)	6.15 (a)	6.16 (a)	6.17 (a)	6.18 (d)
6.19 (a)	6.20 (b)	6.21 (a)	6.22 (b)	6.23 (d)	6.24 (a)	6.25 (a)	6.26 (a)	6.27 (a)
6.28 (d)	6.29 (c)	6.30 (a)	6.31 (d)	6.32 (a)	6.33 (c)	6.34 (d)	6.35 (a)	6.36 (b)
6.37 (a)	6.38 (b)	6.39 (d)	6.40 (a)	6.41 (c)	6.42 (d)	6.43 (a)	6.44 (b)	6.45 (d)

Explanations Operating Systems and Data Bases

6.1 (c)

BIOS is loaded in the main memory by the bootstrap loader.

6.2 (d)

Modem and network card are required only for internet access not for media.

6.3 (b)

$$64 \text{ kB} = 2^{16} * 8\text{-bits}$$

$$26\text{-bits} \Rightarrow \begin{cases} 16\text{-Address pins} \\ + \\ 8\text{-Data pins} \\ + \\ 1\text{-Chip Enable pin} \\ + \\ 1\text{-Read/Write} \end{cases}$$

∴ Minimum pins required = 26-bits.

6.6 (b)

7200 rotations — 1 minute

$$1 \text{ rotation} — \frac{60}{7200} \text{sec}$$

$$= \frac{60 \times 10^{-3} \times 10^{3}}{7200} \text{sec}$$

$$= \frac{60 \times 1000}{7200} \text{msec} = 8.33 \text{ msec}$$

$$\text{Rotational latency} = \frac{\text{Rotation time}}{2}$$

$$= \frac{8.333}{2} \text{msec} = 4.166 \text{ msec}$$

6.8 (c)

If RAM size is increased, fewer page faults will occur.

6.9 (c)

$$\text{Number of chips required} = \frac{16 \times 2^{10} \times 8}{2^{12} \times 4} = 8$$

6.10 (c)

Storage capacity

$= \pi \times \text{diameter} \times \text{tracks} \times \text{storage density}$

$= 3.14 \times 8 \times 100 \times 200 = 502400$ bits

6.11 (c)

$$\text{Number of chips required} = \frac{32 \times 1024 \times 8}{2^{12} \times 4} = 16$$

6.12 (b)

A spooled device is the one that quess the spool of printer jobs and takes one job at a time and prints the output.

6.14 (c)

Memory space = 4 K words

Page size = Frame size = 1 K words

$$\therefore \quad \# \text{frames} = \frac{4 \text{K words}}{1 \text{K words}} = 4$$

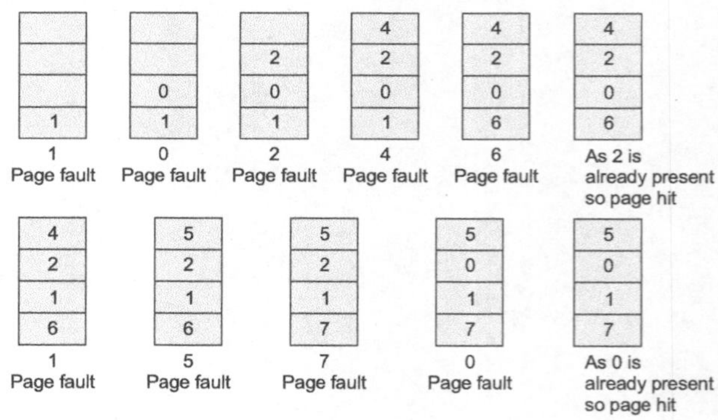

∴ # page faults = 9

6.15 (a)

$$\text{Rotational latency} = \frac{\text{Rotation time}}{2}$$

$$\text{Rotation time} = \frac{1}{30} \text{sec.}$$

$$\text{Avg. rotational latency} = \frac{1}{2} \times \frac{1}{30} \text{sec.}$$

$$= \frac{1}{60} \text{ sec} \cong 17 \text{ msec.}$$

$$\begin{aligned} \text{Access time} &= \text{Avg rotational latency} \\ &\quad + \text{seek time} \\ &= 17 \text{ msec} + 30 \text{ msec} \\ &= 47 \text{ msec.} \end{aligned}$$

6.16 (a)

Paging is a method to implement virtual memory. In paging virtual memory space is divided into equal, fixed-size pages and the physical address space is divided into equal-size frames and can be located any where in memory. Frames contains pages i.e. frame size = page size.

Paging suffers with internal fragmentation only in the last page and not with external fragmentation. The difference between two is that with external fragmentation, enough total memory space may exist to allocate to a process but this space is not contiguous, it exists as a large number of small unusable holes, so no process (might) fit into it.

6.17 (a)

Multitasking in operating system, is allowing a user to perform more than one computer task at a time. The operating system is able to keep track of where you are in there task and go from one to the other without loosing information.

6.19 (a)

If arrival time is not given then all are arrived at '0' ms.

Gantt chart:

P_1	P_2	P_3

0 24 27 30

P_{id}	AT	BT	CT	TAT	WT
P_1	0	24	24	24	0
P_2	0	3	27	27	24
P_3	0	3	30	30	27

$$WT = \frac{(0 + 24 + 27)}{3} = 17 \text{ ms}$$

6.20 (b)

7200 revolution — 60 sec

1 revolution — ?

$$\Rightarrow \frac{60}{7200}$$

$$\Rightarrow 8.33 \text{ ms}$$

1 revolution time — 1 track data (400 * 512 B) (# bytes)

1 sec — ?

$$\Rightarrow \frac{400 \times 512 \text{ B}}{8.33 \text{ mS}}$$

$$\Rightarrow 24585.83 \times 10^3 \text{ B/sec}$$

$$\Rightarrow 24.58 \times 10^6 \text{ B/sec}$$

6.21 (a)

Effective access time

$$\begin{aligned} &= (P \times S) + (1 - P)m \\ &= (0.01 \times 1000) + (1 - 0.01) \times 10 \\ &= 10 + 9.9 = 19.9 \text{ ms} \end{aligned}$$

6.22 (b)

It is not always necessary that BCNF relation is dependency preserving.

6.23 (d)

By using precedence graph

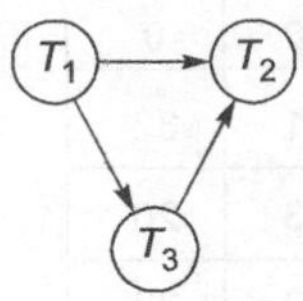

By applying topological shorting, it will give

$$T_1 \rightarrow T_3 \rightarrow T_2$$

6.26 (a)

FSFS

Chart AT = 0

P_1	P_2	P_3	P_4	P_5

0 10 39 42 49 61

	CT	TAT	WT
P_1:	10	10	0
P_2:	39	39	10
P_3:	42	42	39
P_4:	49	49	42
P_5:	61	61	49

$$\text{Avg. WT} = \frac{(0 + 10 + 39 + 42 + 49)}{5} = 28$$

Chart AT = 0

P_3	P_4	P_1	P_5	P_2

0 3 10 20 32 61

	CT	TAT (CT – AT)	WT (TAT – BT)
P_1 :	20	20	10
P_2 :	61	61	32
P_3 :	3	3	0
P_4 :	10	10	3
P_5 :	32	32	20

$$\text{Avg. WT} = \frac{(10+32+0+3+20)}{5} = 13$$

Rand Robin

Ready queue:

P_3	P_4	P_1	P_5	P_2

0 3 10 20 32 61

Chart :

P_1	P_2	P_3	P_4	P_5	P_2	P_5	P_2

0 10 20 23 30 40 50 52 61

	CT	TAT (CT – AT)	WT (TAT – BT)
P_1 :	10	10	0
P_2 :	61	61	32
P_3 :	23	23	20
P_4 :	30	30	23
P_5 :	52	52	40

$$\text{Avg. WT} = \frac{(0+32+20+23+40)}{5} = 23$$

6.27 (a)

Ready Queue :

P_1	P_2	P_3	P_4	P_2	P_1

Chart :

P_1	P_2	P_3	P_4	P_2	P_1

0 2 4 6 8 10 14

	CT	TAT
P_1:	14	14
P_2:	10	10
P_3:	6	6

Avg. TAT = 10

6.28 (d)

$$T_{avg} = HTC + (1-H)(T_m + T_c)$$

$(T_m + T_c) \Rightarrow$ Time required to service miss

$= (80\% * 10) + (1 - 80\%)100 = 28$ ns

6.30 (a)

$$\text{\#pages in SM} = \frac{\text{SM size}}{\text{page size}} = \frac{2^{64}}{8\text{ k}} = \frac{2^{64}}{2^{13}} = 2^{51}$$

64 bit virtual address

p	d
$\log_2^{2^{51}}$ ↓ 51 bit	$\log_2^{8\text{ k}}$ ↓ 13 bit

#pages (frames) in MM

$$= \frac{\text{MM size}}{\text{page size}} = \frac{2^{43}}{8\text{ k}} = \frac{2^{43}}{2^{13}} = 2^{30}$$

43 bit physical address

f	d
$\log_2^{2^{30}}$ ↓ 30 bit	$\log_2^{8\text{ k}}$ ↓ 13 bit

6.31 (d)

2 – M
3 – M
2 – H
4 – M
6 – M
2 – M
5 – M
6 – H
1 – M
4 – M
6 – M

MM

~~2~~ ~~6~~	1
~~3~~ ~~2~~	4
~~4~~ ~~5~~	6

$$H = \frac{\text{\#Hits}}{\text{Total\# Accesses}} = \frac{2}{11}$$

6.32 (a)

Dynamic loading is a mechanism by which a computer program can at run time, load a library into a memory, retrieve the addresses of functions and variables contained in the library execute those function (or) access those variables and unload the library from the memory.

6.33 (c)

To ensure memory protection the operating system provides memory addressing hardware that ensures that each process execute in its own address space.

6.34 (d)

Since time for "r" revolutions is 1 sec. So, time for 1 revolution will be $\frac{1}{r}$ sec.

Since "N" bytes are transferred in one rotation so "b" bytes will be transferred in

$$= \frac{b}{rN}$$

6.35 (a)

The basic components of a memory management unit are:

1. A facility for dynamic storage relocation that maps logical memory references into physical memory address.
2. A provision for sharing common programes stored in memory by different users.
3. Protection of information against unauthorized access between users and preventing users for changing operating system functions.

6.38 (b)

Data Definition Language : It includes statements that allows the database user to create and modify structure of database objects, such as tables, views and indexes.

Data Manipulation Language : It allows the data base user to add data to the database (in form of rows and tables), remove data from database, modify existing data in database.

INSERT, UPDATE, DELETE are considered part of DML.

6.39 (d)

Quick execution of instructions is not the functionality of operating system.

An operating system has three main functions:

(i) Manage the computer's resources such as the central processing unit, memory, disk drives and printers.
(ii) Establish a user interface.
(iii) Execute and provide services for applications software.

6.40 (a)

STS (Short term scheduler) is responsible to schedule the ready state processes into a running state.

6.41 (c)

In non-preemptive scheduling, processes are executed one after another. So, no process is suspended (pre-empted) in the middle of the execution.

6.42 (d)

- Windows Registry (Registry) is a centralized hierarchical data base to store system settings, H/W configurations and user preferences.
- Hives are the top most portions of the hierarchical data tree with each hive containing a certain category of information.

6.43 (a)

Context switching between the Kernal level threads are used to save the context of a process before suspend the execution (swapping).

6.44 (b)

A deadlock situation arises if the following four conditions hold simultaneously in a system:

Mutual Exclusion: Only one process can use a resource at any given time i.e. the resources are non-sharable.

Hold and Wait: A process must be holding at least one resource and waiting to acquire additional resources that are currently being held by other processes.

No Preemption: Resources cannot be preempted; that is, a resource can be released only voluntarily by the process holding it, after that process has completed its task.

Circular Wait: A set $\{P_0, P_1, P_2, ..., P_n\}$ of waiting processes must exist such that P_0 is waiting for a resource that is held by P_1, P_1 is waiting for a resource that is held by P_2, ..., P_{n-1} is waiting for a resource that is held by P_n, and P_n is waiting for a resource that is held by P_0.

6.45 (d)

The buffer is a temporary memory location situated in the main memory. The data transmitted from one device to another or the applications are temporarily stored in the buffer. Buffering is done to cope with device speed mismatch, device transfer size mismatch and to maintain copy semantics i.e. the version of data in the buffer is guaranteed to be the version of data at the time of system call, irrespective of any subsequent change to data in the buffer.

■■■

7 Miscellaneous

7.1 Which of the following conditional IF statements of Pascal are correct?

1. IF condition1 THEN statement1 ELSE IF condition2 THEN statement2 ELSE statement3
2. IF condition1 THEN IF condition2 THEN statement1 ELSE statement2 ELSE statement3
3. IF condition1 THEN IF condition2 THEN statement1 ELSE statement2

Select the correct answer using the codes given below:

(a) 1 and 2 (b) 2 and 3
(c) 1 and 3 (d) 1, 2 and 3

[ESE-2001]

7.2 The different classes of formal parameters used in PASCAL are

(a) value and variable parameters
(b) procedure and function parameters
(c) value, variable, procedure and function parameters
(d) variable, procedure and function parameters

[ESE-2002]

7.3 The technology used for display in PC note-books (laptop computers) is

(a) Light Emitting Diodes display
(b) Liquid Crystal display
(c) CRT display
(d) Plasma display [ESE-2002]

7.4 Consider the assembler directives:

ORG 8000
T : DW OFAOFH

Which one of the following is correct?

(a) The contents of the locations 8000 and 8001 get erased
(b) The contents of the locations 8000 and 8001 remain unchanged
(c) The least significant byte OF will be stored at location 8000 and the most significant byte FA will be stored at location 8001
(d) The least significant byte OF will be stored at location 8001 and the most significant byte FA will be stored at location 8000

[ESE-2003]

7.5 **Assertion (A):** The data which is keyed in can be viewed through a Visual Display Unit (VDU).
Reason (R): VDU is also called a terminal.

(a) Both A and R are true and R is the correct explanation of A.
(b) Both A and R are true but R is NOT the correct explanation of A
(c) A is true but R is false
(d) A is false but R is true [ESE-2003]

7.6 **Assertion (A):** Processor-level design is very much a heuristic process.
Reason (R): At this level of abstraction, components are very complex.

(a) Both A and R are true and R is the correct explanation of A
(b) Both A and R are true but R is NOT the correct explanation of A
(c) A is true but R is false
(d) A is false but R is true [ESE-2006]

7.7 What does the minimising of the memory requirement of a program mean?

(a) Maximizing number of macro calls
(b) Minimizing number of macro calls
(c) Maximizing depth of recursion
(d) Both (a) and (c) above

[ESE-2007]

7.8 **Assertion (A):** In a microprogrammed CPU, each machine instruction is executed by a real-time interpreter.
Reason (R): Real-time interpreter helps to achieve high degree parallelism in microprogrammed control.

(a) Both A and R are true and R is the correct explanation of A
(b) Both A and R are true but R is NOT the correct explanation of A
(c) A is true but R is false
(d) A is false but R is true

[ESE-2007]

7.9 A group of personal computers are configured to work together for speeding up the execution of a single program in
(a) simulated evaluation
(b) cluster computing
(c) network computing
(d) client server computing **[ESE-2008]**

7.10 Personal computer cannot be used for which one of the following ?
(a) Game playing (b) Weather forecasting
(c) Office Automation (d) Home computing
[ESE-2009]

7.11 Consider the following statements:
1. Wind tunnel simulation
2. Real-time video viewing
A computer is used for
(a) Neither 1 nor 2 (b) Both 1 and 2
(c) 1 only (d) 2 only
[ESE-2010]

7.12 **Assertion (A):** The low-level control of I/O device is easier at hardware level.
Reason (R): It requires managing a set of concurrent events.
(a) Both A and R are true and R is the correct explanation of A
(b) Both A and R are true but R is NOT the correct explanation of A
(c) A is true but R is false
(d) A is false but R is true **[ESE-2011]**

7.13 **Assertion (A):** Workstations are often used in engineering applications, especially for interactive design work.
Reason (R): Work stations with graphics I/O capability have a computational power that is significantly higher than that of personal computers.
(a) Both A and R are true and R is the correct explanation of A
(b) Both A and R are true but R is NOT the correct explanation of A
(c) A is true but R is false
(d) A is false but R is true **[ESE-2011]**

7.14 **Assertion (A):** Processor level design is heavily based on the use of prototype structures.
Reason (R): A prototype design is selected and modified to meet the given performance specifications.
(a) Both A and R are true and R is the correct explanation of A
(b) Both A and R are true but R is NOT the correct explanation of A
(c) A is true but R is false
(d) A is false but R is true **[ESE-2011]**

7.15 An algorithm is made up of 2 modules M_1 and M_2. If order of M_1 is $f(n)$ and that of M_2 is $g(n)$, then the order of the algorithm is
(a) $f(n) \times g(n)$ (b) $f(n) + g(n)$
(c) $\min(f(n), g(n))$ (d) $\max(f(n), g(n))$
[ESE-2014]

7.16 Which one of the following tables is generated in pass one and has an entry for every symbol in the program to perform its function?
(a) Symbol table
(b) Opcode table
(c) Pseudo-instruction table
(d) Addressing instruction table
[ESE-2022]

7.17 Consider the following statements.
1. Smoke test is a non-exhaustive test to ensure that the crucial requirements for the product are functioning.
2. Sanity testing is superficial testing performed to test the durability of the product.
3. Scalability test is used to test the performance aspects of the product after integration.

Which of the above statements are not correct?
(a) 1 and 2 only (b) 1 and 3 only
(c) 2 and 3 only (d) 1, 2 and 3
[ESE-2024]

■■■

Answers Miscellaneous

7.1 (c) 7.2 (c) 7.3 (b) 7.4 (c) 7.5 (b) 7.6 (d) 7.7 (b) 7.8 (c) 7.9 (b)
7.10 (b) 7.11 (c) 7.12 (a) 7.13 (a) 7.14 (a) 7.15 (d) 7.16 (a) 7.17 (c)

Explanations Miscellaneous

7.1 (c)

Pascal maintains/uses statements 1 and 3 conditional IF statements. Syntax of statement 2 is incorrect with respect to Pascal.

7.5 (b)

Assertion (A) is true.
VDU is also called terminal. But Reason (R) is not the correct explanation of Assertion (A).
VDU is an output device where we get output which can be viewed.

7.6 (d)

Processor level design is not heuristic process and based on the use of prototype structures.
At processors level of abstraction, components are very complex.

7.7 (b)

(a) Maximizing number of macro calls will maximize the requirement of memory.
(b) Minimizing number of macro calls will minimize the memory requirement.
(c) Maximizing depth of recursion will maximize the memory by storing the previous recursive calls on the stack space.

$\therefore$ Option (b) is correct.

7.8 (c)

In microprogrammed CPU, each machine instruction is executed by a real time interpreter but it will not helps to achieve the high degree parallelism.

7.10 (b)

Personal computer cannot be used for weather forecasting because it requires very huge calculations and storage of huge amounts of data and a high speed processor with maximum resources.
For these reasons PC can not be used for weather forecasting.

7.14 (a)

Prototype structures are used for processor level design. It is selected and modified to meet the given performance specifications.

7.15 (d)

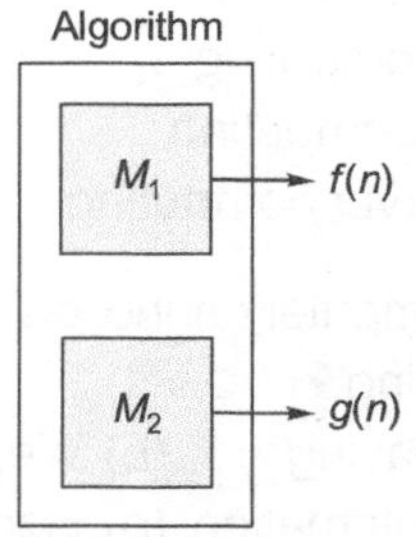

Order of Algorithm = Max ($f(n)$, $g(n)$)

7.16 (a)

Assemblers are programs that generate machine code instructions from a source code program written in assembly language. A simple assembler scans the entire assembly program twice, where each scan is called a pass.

(i) During the first pass, it generates a symbol table that includes all symbols and their binary values.

(ii) During the second pass, the assembler will use the symbol table and other tables to generate the object program.

7.17 (c)

- Smoke testing, also known as build verification testing or build acceptance testing, is a non-exhaustive testing type that ensures the proper function of basic and critical components of a single build. For example, smoke tests on Gmail only include testing the composing and sending email features - the core ones.
- Sanity Testing is performed after making changes to ensure the intended functionality is working as expected.
- A scalability test is a type of load testing that measures the application's ability to scale up or down as a reaction to an increase in the number of users. Integration testing is used to test the performance aspects of the product after integration.

Hence, statements 2 and 3 are incorrect.

UNIT

V

Microprocessors and Microcontrollers

Syllabus

Microprocessors & microcontrollers, basics, interrupts, DMA, instruction sets, interfacing; Controllers & uses; Embedded systems.

Contents

Sl.	Topic	Page No.
1.	8085 Microprocessors	376
2.	8086 Microprocessors	392
3.	Interfacing Devices and Microcontrollers	396

OOOO

1 8085 Microprocessors

1.1 The frequency of the driving network connected between pins 1 and 2 of a 8085 chip must be
(a) equal to the desired clock frequency
(b) twice the desired clock frequency
(c) four times the desired clock frequency
(d) eight times the desired clock frequency
[ESE-2001]

1.2 The logic operation that will selectively clear bits in register A in those positions where there are 1's in the bits of register B is given by
(a) $A \leftarrow A + B$ (b) $A \leftarrow \bar{A}B$
(c) $A \leftarrow \bar{A} + \bar{B}$ (d) $A \leftarrow A\bar{B}$
[ESE-2000]

1.3 The contents of accumulator in an 8085 microprocessor are altered after the execution of the instruction?
(a) CMP C (b) CPI 3A
(c) ANI 5C (d) ORA A
[ESE-2002]

1.4 An 8085 microprocessor after the execution of XRA A instruction
(a) the Carry flag is set
(b) the accumulator contains FF_H
(c) the content of accumulator is shifted by one
(d) the Zero flag is set
[ESE-2002]

1.5 Consider the following instructions of 8085 microprocessor:
1. MOV M, A 2. ADD C
3. MVI A, FF 4. CMP M
Which of these cause change in the status of flag(s)?
(a) 1 and 2 (b) 1, 2 and 3
(c) 3 and 4 (d) 2 and 4
[ESE-2002]

1.6 Consider the following:
1. Sign Flag 2. Zero Flag
3. Carry Flag 4. Parity Flag
Which of the above flags of 8085 get affected by the instruction SUB B?
(a) 1 and 2 (b) 1 and 3
(c) 3 and 4 (d) 1, 2, 3 and 4
[ESE-2002]

1.7 Consider the following registers:
1. Accumulator and B register
2. B and C registers
3. D and E registers
4. H and L registers
Which of these 8-bit registers of 8085 μP can be paired together to make a 16-bit register?
(a) 1, 3 and 4 (b) 2, 3 and 4
(c) 1 and 2 (d) 1, 2 and 3
[ESE-2002]

1.8 Consider the following features in an 8085 microprocessor system with memory mapped I/O:
1. I/O devices have 16-bit addresses.
2. I/O devices accessed using IN and OUT instructions.
3. There can be maximum of 256 input devices and 256 output devices.
4. Arithmetic and logic operations can be directly performed with the I/O data.
Select the correct answer using the codes given below:
(a) 1, 2 and 4 (b) 1, 3 and 4
(c) 2 and 3 (d) 1 and 4
[ESE-2003]

1.9 In 8085 microprocessor the value of the most significant bit of the result following the execution of any arithmetic or Boolean instruction is stored in
(a) The Carry-status flag
(b) The Auxiliary carry status flag
(c) The Sign status flag
(d) The Zero status flag [ESE-2003]

1.10 The first machine cycle of an instruction is always
(a) A memory read cycle
(b) A fetch cycle
(c) An I/O read cycle
(d) A memory write cycle

[ESE-2003]

1.11 Consider the following statements: The 8085 microprocessor will not enter into bus idle machine cycle whenever
1. INTR interrupts is recognized
2. RST x.5 is recognized
3. DAD rp instruction is executed

Which of the statements given above is/are correct?
(a) 1 only (b) 2 only
(c) 1 and 2 (d) 2 and 3

[ESE-2004]

1.12 Consider the following program:

```
        ORG 7000 H
BEGIN:  LXI H, 7000H
        MOV A, L
        ADD H
        JM END
        RST 0
END:    PCHL
        HLT
```

Which one of the following statements is correct?
(a) The program will halt the processor
(b) The program will be repeated infinitely
(c) The program will branch to 0007 H after JM END
(d) The program will branch to 0000 H after JM END

[ESE-2004]

1.13 Consider the following program intended to transfer a block of 5 bytes from A000H to 9000H:

```
START:  LXI B, 9000H
        LXI H, A000H
        MVI C, 05H
LOOP:   MOV A, M
        STAX B
        INX B
        INX H
        DCR C
        JNZ LOOP
        HLT
```

The above program will not work because
(a) C Register is used as counter
(b) DCR C instruction will not affect zero flag
(c) JNZ instruction is used instead of JZ
(d) The first two instructions in loop should have been LDAX D and MOV M, A

[ESE-2004]

1.14 Consider the following statements:
The SIM instruction outputs the contents of accumulator to define:
1. interrupt mask bit
2. interrupt pending bit
3. serial input data line
4. serial output data line

Which of the statements given above are correct?
(a) 1 and 2 (b) 2 and 3
(c) 3 and 4 (d) 1 and 4

[ESE-2005]

1.15 What are the number of machine cycles n, and the types of machine cycles carried out for PUSH B?
(a) $n = 2$, fetch and memory write
(b) $n = 3$, fetch and 2 memory write
(c) $n = 3$, fetch, memory write and read
(d) $n = 3$, fetch, and 2 memory read

[ESE-2005]

1.16 The following sequence of instructions is executed by an 8085 microprocessor:

```
1000    LXI SP, 27FF
1003    CALL 1006
1006    POP H
```

What are the contents of the stack pointer (SP) and the HL register pair after completion of execution of these instructions?
(a) SP = 27FF, HL = 1003
(b) SP = 27FD, HL = 1003
(c) SP = 27FF, HL = 1006
(d) SP = 27FD, HL = 1006

[ESE-2005]

1.17 Match **List-I** with **List-II** and select the correct answer using the code given below the lists:

List-I	List-II
A. Immediate addressing	**1.** LDA 308C
B. Implied addressing	**2.** MOV A, B
C. Register addressing	**3.** LXI H, 2050
D. Direct addressing	**4.** RRC

Codes:

	A	B	C	D
(a)	3	4	2	1
(b)	2	1	3	4
(c)	3	1	2	4
(d)	2	4	3	1

[ESE-2005]

1.18 Assertion (A): The development of a microprocessor based product requires the design of program and the hardware.

Reason (R): The design effort for an electronic product follows the same basic steps used in the development of software.

(a) Both A and R are true and R is the correct explanation of A
(b) Both A and R are true but R is NOT the correct explanation of A
(c) A is true but R is false
(d) A is false but R is true [ESE-2006]

1.19 If 8085 adds 87 H and 79 H, then
(a) both CARRY and ZERO flags will be set to 0
(b) CARRY flag will be set to 0, ZERO flag to 1
(c) CARRY flag will be set to 1, ZERO flag to 0
(d) both CARRY and ZERO flags will be set to 1

[ESE-2008]

1.20 On the 8085, which of the following machine cycles are not used in the CALL instruction?
1. Instruction Fetch 2. I/O
3. Memory Read 4. Memory Write

Select the correct answer using the code given below:
(a) 2 only (b) 1 and 4
(c) 2, 3 and 4 (d) None of these

[ESE-2008]

1.21 In 8085, if the clock frequency is 5 MHz, the time required to execute an instruction of 18 T-states is
(a) 3.0 μs (b) 3.6 μs
(c) 4.0 μs (d) 6.0 μs

[ESE-2008]

1.22 Match **List-I** with **List-II** and select the correct answer using the codes given below the lists:

List-I	List-II
A. Immediate addressing	1. LDA 30FF
B. Implied addressing	2. MOV A, B
C. Register addressing	3. LXI H, 2050
D. Direct addressing	4. RRC

Codes:

	A	B	C	D
(a)	3	4	2	1
(b)	2	1	3	4
(c)	3	1	2	4
(d)	2	4	3	1

[ESE-2008]

1.23 Which one of the following interrupts is both level and edge sensitive?
(a) RST 7.5 (b) RST 5.5
(c) TRAP (d) INTR [ESE-2008]

1.24 Which one of the following addressing modes is used in the instruction PUSH B ?
(a) Direct (b) Register
(c) Register indirect (d) Immediate

[ESE-2009]

1.25 Which of the following counters can be used to divide the clock frequency of a microprocessor by 5?
(a) 3 bit counter (b) 5 bit counter
(c) mod 3 counter (d) mod 5 counter

[ESE-2009]

1.26 The 8085 has two registers known as primary data pointers. These are registers
(a) B and C (b) D and E
(c) H and L (d) C and D

[ESE-2010]

1.27 READY signal in 8085 is useful when the CPU communicates with
(a) A slow peripherals device
(b) A fast peripheral device
(c) A DMA controller chip
(d) A PPI chip [ESE-2010]

1.28 A microprocessor based system can perform many different functions, because
(a) Its operation is controlled by software
(b) It is digital system
(c) It uses a RAM
(d) It can be controlled by input and output device

[ESE-2010]

1.29 The output data lines of microprocessors and memories are usually tristated, because
(a) More than one device can transmit information over the data bus by enabling only one device at a time

(b) More than one device can transmit information over the data bus at the same time
(c) The data lines can be multiplexed for both input and output
(d) It increases the speed of data transfers over the data bus

[ESE-2010]

1.30 Consider the following instructions of 8085 microprocessor
1. MOV B, C
2. STA address
3. ORI byte

The correct sequence in the decreasing order of their respective memory space requirement is
(a) 3, 2 and 1 (b) 1, 3 and 2
(c) 1, 2 and 3 (d) 2, 3 and 1

[ESE-2010]

1.31 If the CALL instruction of 8085 in the main program is conditional then RETURN instruction in the subroutine can be
(a) Conditional
(b) Conditional or unconditional
(c) Can be determined by DDA instruction
(d) Unconditional [ESE-2010]

1.32 In the following 8085 program how many times (decimal) is the DCR C executed?

```
LOOP:   MVI C, 78H
        DCR C
        JNZ LOOP
        HLT
```

(a) 119 (b) 120
(c) 78 (d) 77 [ESE-2010]

1.33 A single instruction to clear the lower four bits of the accumulator in 8085 assembly language is
(a) XRI 0F H (b) ANI F0 H
(c) XRI F0 H (d) ANI 0F H

[ESE-2010]

1.34 Which of the following instruction copies a byte of data into the accumulator from the memory address given in the instruction?
(a) LDA address (b) LDAX B
(c) LHLD address (d) STA address

[ESE-2010]

1.35 Consider a memory chip with 1024 bytes storage connected to a 8085 chip address lines (or any microprocessor with 16 address lines) as below. What is the range of memory address?

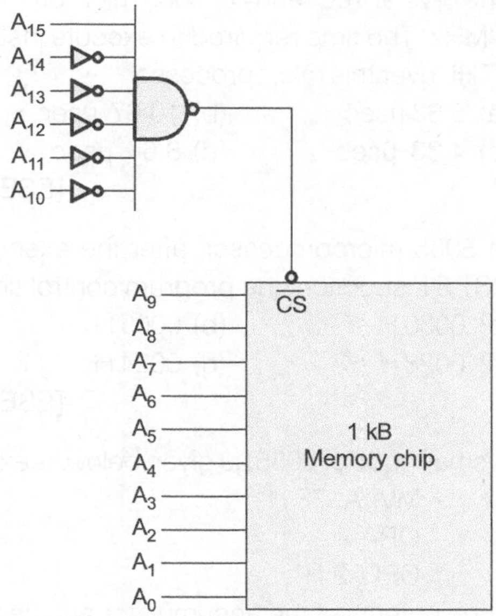

(a) 0000 H to 03FF H
(b) 8000 H to 83FF H
(c) F000 H to F3FF H
(d) 0000 H to FFFF H [ESE-2010]

1.36 If 8085 microprocessor adds 87 H and 79 H, the flags will be
(a) $S = 1, Z = 0, AC = 0$ and $Cy = 1$
(b) $S = 0, Z = 0, AC = 1$ and $Cy = 0$
(c) $S = 1, Z = 1, AC = 1$ and $Cy = 1$
(d) $S = 0, Z = 1, AC = 1$ and $Cy = 1$

[ESE-2011]

1.37 For the 8085 assembly language program given below, the content of the accumulator after the execution of the program is

```
3000  MVI A, 45 H
3002  MOV B, A
3003  STC
3004  CMC
3005  RAR
3006  XRA B
3007  HLT
```

(a) 00 H (b) 45 H
(c) 67 H (d) E7 H [ESE-2012]

1.38 The 8085 microprocessor instructions with required number of *T*-states are given below. Which pair is correctly matched?
(a) STAX : 8 *T*-states
(b) SPHL : 6 *T*-states
(c) SIM : 7 *T*-states
(d) STA : 12 *T*-states [ESE-2012]

1.39 The crystal frequency of 8085 microprocessor is 6 MHz. The time required to execute instruction XTHL over this microprocessor

(a) 5.33 μsec (b) 10.67 μsec
(c) 4.33 μsec (d) 8.67 μsec

[ESE-2012]

1.40 In 8085 microprocessor, after the execution of RST 5 instruction, the program control shifts to

(a) 0030 H (b) 0005 H
(c) 0028 H (d) 0024 H

[ESE-2012]

1.41 A small code of 8085 as given below, is executed

```
MVI A, 7F H
ORA A
CPI A2 H
```

The contents of the accumulator and flags after execution are

(a) $A = DD, S = 1, Z = 0, CY = 0$
(b) $A = 7F, S = 1, Z = 0, CY = 1$
(c) $A = DD, S = 0, Z = 1, CY = 0$
(d) $A = 7F, S = 0, Z = 1, CY = 1$ [ESE-2012]

1.42 Following program finds absolute value of N:

```
      MVI A, N
      ORA A
      JM ONE
      OUT 01 H
      HLT
ONE : (P)
      (Q)
      OUT 01 H
      HLT
```

The instruction of (P) and (Q) must be

(a) CMA and ADI 00 H
(b) CMC and ADI 00 H
(c) INR A and CMC
(d) INR A and CMA [ESE-2012]

1.43 An Intel 8085 processor is executing the program given below:

```
       MVI A, 10 H
       MVI B, 10 H
BACK : NOP
       ADD B
       RLC
       JNC BACK
       HLT
```

The number of times that the operation NOP will be executed is

(a) 1 (b) 2
(c) 3 (d) 4

[ESE-2012]

1.44 Match **List-I** with **List-II** and select the correct answer using the code given below the Lists:

List-I	List-II
A. Immediate addressing	1. LDA 30FF
B. Implicit addressing	2. MOV A, B
C. Register addressing	3. LXI H, 2050
D. Direct addressing	4. RRC

Codes:

	A	B	C	D
(a)	3	4	2	1
(b)	1	4	2	3
(c)	3	2	4	1
(d)	1	2	4	3

[ESE-2012]

1.45 While executing a program 8085 microprocessor completes fetching of instruction JMP 2050 stored at address 2057 H. The contents of the program counter after fetching the instruction would be

(a) 2050 H (b) 2057 H
(c) 205A H (d) 2051 H [ESE-2012]

1.46 For Opcode fetch operation in 8085 microprocessor

(a) $S_1 = 0, S_2 = 1, \overline{RD} = 0$, ALE high in T_1
(b) $S_1 = 1, S_2 = 1, \overline{RD} = 0$, ALE high in T_1
(c) $S_1 = 1, S_2 = 1, \overline{RD} = 0$, ALE high in T_2
(d) $S_1 = 0, S_2 = 1, \overline{RD} = 0$, ALE high in T_2

[ESE-2012]

1.47 Consider the following 8085 interrupts:

1. TRAP 2. INTR
3. RST 6 4. RST 7.5
5. RST 0

The software interrupts are

(a) 1 and 3 only (b) 2 and 5 only
(c) 3 and 5 only (d) 1, 2, 3, 4 and 5

[ESE-2012]

1.48 An 8085 microprocessor executes the following instructions:

The numbers are represented in signed 2's complement form as

$P = 11101101$ and $Q = 11100110$

If Q is subtracted from P, the value obtained in signed 2's complement form is
(a) 100000111 (b) 00000111
(c) 11111001 (d) 011111001
[ESE-2012]

1.49 TRAP is an interrupt in 8085. Which one of the following statements is true about the TRAP?
(a) It is level triggered
(b) It is negative edge triggered
(c) It is positive edge triggered
(d) It is both positive and negative edge triggered
[ESE-2013]

1.50 Which one of the following instructions uses implied addressing mode?
(a) CMA (b) JMP
(c) ADC (d) SBB
[ESE-2013]

1.51 In a microprocessor, the register which holds the address of the next instruction to be fetched is
(a) accumulator (b) program counter
(c) stack pointer (d) instructor register
[ESE-2013]

1.52 With reference to 8085 microprocessor, which of the following statements are correct?
1. The DAD instruction adds the contents of the specified register pair to the contents of the DE register pair.
2. The program counter is an instruction pointer.
3. Whenever an instruction uses the HL pointer, the addressing is called indirect addressing.
4. HL register pair is also called data pointer.
(a) 1 and 2 only (b) 1, 2 and 3
(c) 2, 3 and 4 (d) 2 and 4 only
[ESE-2013]

1.53 On the 8085, which of the following machine cycles are not used in the CALL instruction?
1. Instruction fetch
2. I/O
3. Memory Read
4. Memory Write
(a) 2 only (b) 1 and 4 only
(c) 2, 3 and 4 (d) 1, 2, 3 and 4
[ESE-2013]

1.54 The stack pointer will be affected by instructions:
1. PUSH PSW 2. CALL ADDR
3. XTHL 4. RST n
(a) 1, 2, 3 and 4 (b) 1 and 3 only
(c) 1, 2 and 4 only (d) 2, 3 and 4 only
[ESE-2013]

1.55 The program counter (PC) in a microprocessor
(a) counts the number of programs being executed by the microprocessor
(b) counts the number of instruction being executed by the microprocessor
(c) counts the number of interrupts handled by the microprocessor
(d) keeps the address of the next instruction to be fetched
[ESE-2013]

1.56 Consider the following statements regarding RESET instruction of 8085 microprocessor:
1. PC contents become 0000H.
2. All interrupts are enabled.
3. RESET OUT pin is at logic 0.
Which of the above statements is/are correct?
(a) 1 only (b) 2 only
(c) 1 and 2 (d) 2 and 3
[ESE-2014]

1.57 In a microprocessor, the register which holds address of the next instruction to be fetched is
(a) Accumulator (b) Program counter
(c) Stack pointer (d) Instruction register
[ESE-2014]

1.58 The following sequence of instructions is executed by an 8085 microprocessor:

```
1000    LXI    SP,    27FF
1003    CALL   1006
1006    POP    H
```

The contents of the stack pointer (SP) and the HL register pair on completion of execution of these instructions are
(a) SP = 27FF, HL = 1003
(b) SP = 27FD, HL = 1003
(c) SP = 27FF, HL = 1006
(d) SP = 27FD, HL = 1006 [ESE-2014]

1.59 What is the total memory range and memory map, if for a 16 bit address bus; $A_{15} = 1$, $A_{14} = 0$ and $A_{13} - A_{11}$ are connected to a 3 – 8 decoder input lines? A_{15} and A_{14} are connected to enable the decoder.

(a) 16 k, 8000 H – 8FFF H
(b) 2 k, 8000 H – BFFF H
(c) 16 k, 8000 H – BFFF H
(d) 2 k, 8000 H – 8 FFF H [ESE-2014]

1.60 In a microprocessor, WAIT states are used to
(a) Make the processor WAIT during a DMA operation
(b) Make the processor WAIT during an interrupt processing
(c) Make the processor WAIT during a power shut down
(d) Interface SLOW peripheral to the processor [ESE-2015]

1.61 A register of microprocessor which keeps track of the execution of a program and which contains the memory address of the next instruction to be executed is called
(a) Index register
(b) Memory address register
(c) Program counter
(d) Instruction register [ESE-2015]

1.62 In a microprocessor system with memory mapped I/O
(a) Devices have 8-bit addresses
(b) Devices are accessed using IN and OUT instructions
(c) There can be a maximum of 256 input devices and 256 output devices
(d) Arithmetic and logic operations can be directly performed with the I/O data [ESE-2015]

1.63 In a microprocessor when a CPU is interrupted, it
(a) Stops execution of instructions
(b) Acknowledges interrupt and branches of subroutine
(c) Acknowledges interrupt and continues
(d) Acknowledges interrupt and waits for the next instruction from the interrupting device [ESE-2015]

1.64 The statement label in a subroutine
(a) May be the same as in the main program
(b) Is always the same as in the main program
(c) Cannot be the same as in the main program
(d) Cannot be used [ESE-2015]

1.65 An addressing mode in which the location of the data is contained within the mnemonic, is known as
(a) Immediate addressing mode
(b) Implied addressing mode
(c) Register addressing mode
(d) Direct addressing mode [ESE-2016]

1.66 Consider the following loop:

```
    MOV CX, 8000h
L1: DEC CX
    JNZ L1
```

The processor is running at 14.7456/3 MHz and DEC CX requires 2 clock cycles and JNZ requires 16 clock cycles. The total time taken is nearly
(a) 0.01 s (b) 0.12 s
(c) 3.66 s (d) 4.19 s
[ESE-2017]

1.67 In the following 8085 assembly language program, assume that the carry flag is initially reset. What is the content of the accumulator after the execution of the program?

```
MVI  A,   04H
RRC
MOV  B, A
RLC
RLC
ADD  B
RRC
```

(a) 02H (b) 05H
(c) 15H (d) 25H [ESE-2018]

1.68 **Statement (I):** Stack works on the principle of LIFO.
Statement (II): Stack pointer contains address of the top of the stack.
(a) Both Statement I and Statement II are individually true and Statement II is the correct explanation of Statement I
(b) Both Statement I and Statement II are individually true but Statement II is not the correct explanation of Statement I
(c) Statement I is true but Statement II is false
(d) Statement I is false but Statement II is true
[ESE-2018]

1.69 **Statement (I):** I/O devices can be accessed using IN and OUT instructions.
Statement (II): Arithmetic and logic operations can be directly performed with I/O data.

(a) Both Statement I and Statement II are individually true and Statement II is the correct explanation of Statement I
(b) Both Statement I and Statement II are individually true but Statement II is not the correct explanation of Statement I
(c) Statement I is true but Statement II is false
(d) Statement I is false but Statement II is true

[ESE-2018]

1.70 **Statement (I):** On executing the HLT instruction, the microprocessor enters into a halt state and all the buses are tri-stated.

Statement (II): On executing the HLT instruction, the microprocessor is disconnected from the system bus till the reset is pressed.

(a) Both Statement I and Statement II are individually true and Statement II is the correct explanation of Statement I
(b) Both Statement I and Statement II are individually true but Statement II is not the correct explanation of Statement I
(c) Statement I is true but Statement II is false
(d) Statement I is false but Statement II is true

[ESE-2018]

1.71 Which one of the following types of instructions will be used to copy from the source to the destination location?
(a) Arithmetic instruction
(b) Data transfer instructions
(c) Logical instructions
(d) Machine control instructions [ESE-2019]

1.72 A 2-byte instruction which accepts the data from the input port specified in the second byte and loads into the accumulator is
(a) OUT <8-bit port address>
(b) IN <8-bit port address>
(c) OUT *R* <8-bit port address>
(d) IN *R* <8-bit port address> [ESE-2019]

1.73 Consider the following instruction :

```
EI
MVI A, 08H
SIM
```

It means
(a) disable all interrupts
(b) enable all interrupts
(c) disable RST 7.5 and 6.5
(d) enable RST 7.5 and 6.5

[ESE-2019]

1.74 The instruction BC 0×15 means
(a) jump 15 bytes relative to the program counter
(b) copy and load 15 words in reverse direction to the program counter
(c) move to a location by 15 bits to the program counter
(d) redirect (jump) to a location by 15 words relative to the program counter

[ESE-2019]

1.75 Which of the following will cause internal interrupt to CPU?
1. Stack overflow
2. Attempt to divide by zero
3. I/O device finished transfer of data
4. Power failure

(a) 1 and 2 only (b) 2 and 3 only
(c) 3 and 4 only (d) 1 and 4 only

[ESE-2020]

1.76 In an assembly language program END is a/an
(a) Machine instruction
(b) Pseudo instruction
(c) Micro instruction
(d) Interrupt [ESE-2020]

1.77 PUSH and POP operations are performed by
(a) Program counter register
(b) General purpose register
(c) Stack pointer register
(d) Link register [ESE-2020]

1.78 Consider the following statements regarding 8085 instruction set:
1. After execution of DAD instruction, if the result is larger than 16 bits, CY flag is set.
2. After execution of DCR instruction, S and CY flags will not be modified.
3. After execution of CMC instruction, only CY flag will be changed.

Which of the above statements is/are not correct?
(a) 1 only (b) 2 only
(c) 2 and 3 (d) 1 and 3

[ESE-2022]

1.79 Which of the following statements is/are correct for memory-mapped I/O?
1. The advantage of memory-mapped I/O is the ability to execute a number of memory-dedicated instructions on the registers in the I/O devices in addition to the elimination of the need for dedicated I/O instructions.

2. The disadvantage of memory-mapped I/O is the need to dedicate part of the memory address space for I/O devices.

Select the correct answer using the code given below.

(a) 1 only (b) 2 only
(c) Neither 1 nor 2 (d) Both 1 and 2

[ESE-2022]

1.80 The following sequence of instructions is executed by an 8085 microprocessor:

```
1000 H    LXI SP, 27FF H
1003 H    CALL 1006 H
1006 H    POP H
```

The contents of the stack pointer (SP) and the HL register pair on completion of execution of these instructions are

(a) SP = 27FF H and HL = 1003 H
(b) SP = 27FD H and HL = 1003 H
(c) SP = 27FF H and HL = 1006 H
(d) SP = 27FD H and HL = 1006 H

[ESE-2023]

1.81 The total number of memory accesses involved when an 8085 processor executes the instruction LDA 2003 H is

(a) 1 (b) 2
(c) 3 (d) 4

[ESE-2023]

1.82 The contents of register (B) and accumulator (A) of an 8085 microprocessor are 3C H and 89 H respectively. The contents of A and the status of carry flag (CY) and sign flag (S) after executing SUB B instructions are

(a) $A = C5$ H, $CY = 1$, $S = 1$
(b) $A = 5C$ H, $CY = 1$, $S = 1$
(c) $A = C5$ H, $CY = 0$, $S = 1$
(d) $A = 5C$ H, $CY = 0$, $S = 1$

[ESE-2023]

1.83 For an 8085 microprocessor, the following program is executed:

```
      MVI A, 05 H
      MVI B, 05 H
PTR:  ADD B
      DCR B
      JNZ PTR
      ADI 03 H
      HLT
```

At the end of the program, accumulator contains

(a) 17 H (b) 20 H
(c) 23 H (d) 05 H

[ESE-2023]

1.84 Which of the following opcodes is used if the contents of the accumulator are logically ANDed with the 8-bit data and the results are placed in the accumulator?

(a) CALL (b) POP
(c) ANI (d) ANA

[ESE-2023]

1.85 **Statement (I) :** Content in the flag register in 8085 microprocessor is read by PUSH PSW followed by POP instruction.

Statement (II) : Content in the flag register in 8085 microprocessor is not able to read and store to any general purpose register.

(a) Both Statement (I) and Statement (II) is the correct explanation of Statement (I).
(b) Both Statement (I) and Statement (II) are individually true, but Statement (II) is not the correct explanation of Statement (I).
(c) Statement (I) is true, but Statement (II) is false.
(d) Statement (I) is false, but Statement (II) is true

[ESE-2023]

1.86 Which of the following are used to make a decision that determines the program flow, based on ALU results or the contents of register?

(a) BCD codes (b) Grey codes
(c) Condition codes (d) Excess-3 codes

[ESE-2024]

■■■

Answers 8085 Microprocessors

1.1 (b)	1.2 (d)	1.3 (c)	1.4 (d)	1.5 (d)	1.6 (d)	1.7 (b)	1.8 (d)	1.9 (c)
1.10 (b)	1.11 (c)	1.12 (d)	1.13 (a)	1.14 (d)	1.15 (b)	1.16 (c)	1.17 (a)	1.18 (b)
1.19 (d)	1.20 (a)	1.21 (b)	1.22 (a)	1.23 (c)	1.24 (c)	1.25 (d)	1.26 (c)	1.27 (a)
1.28 (a)	1.29 (a)	1.30 (d)	1.31 (b)	1.32 (b)	1.33 (b)	1.34 (a)	1.35 (b)	1.36 (d)
1.37 (c)	1.38 (b)	1.39 (a)	1.40 (c)	1.41 (b)	1.42 (a)	1.43 (c)	1.44 (a)	1.45 (c)
1.46 (b)	1.47 (c)	1.48 (b)	1.49 (c)	1.50 (a)	1.51 (b)	1.52 (c)	1.53 (a)	1.54 (c)
1.55 (d)	1.56 (a)	1.57 (b)	1.58 (c)	1.59 (c)	1.60 (d)	1.61 (c)	1.62 (d)	1.63 (d)
1.64 (b)	1.65 (b)	1.66 (b)	1.67 (b)	1.68 (a)	1.69 (b)	1.70 (a)	1.71 (b)	1.72 (b)
1.73 (b)	1.74 (a, d)	1.75 (a)	1.76 (b)	1.77 (c)	1.78 (b)	1.79 (d)	1.80 (c)	1.81 (d)
1.82 (*)	1.83 (a)	1.84 (c)	1.85 (a)	1.86 (c)				

Explanations 8085 Microprocessors

1.1 (b)

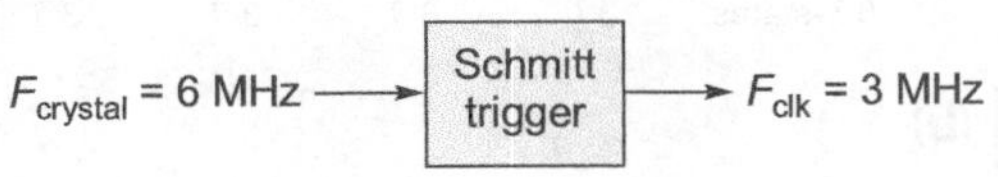

1.2 (d)

Register '*A*' bits will be cleared in those positions where there are 1's in the bits of Register '*B*'.

Example: $B = 1011$

$$A = a_1\, a_2\, a_3\, a_4$$

$$A \rightarrow A\bar{B}$$

$$\bar{B} = 0100$$

$$A = a_1\, a_2\, a_3\, a_4$$

$$A \leftarrow A\bar{B} \quad 0\, a_2\, 0\, 0$$

1.3 (c)

CMP, CPI, ORA instructions will not affect the accumulator.

ANI 5C instructions will changes the content of accumulator. (ANI is AND operation with 8-bit immediate data).

Accumulator AND 5C → Accumulator.

1.4 (d)

The content of accumulator A will be 0 after XRA A. So zero flag will set. In XRA operation carry and auxiliary flags are reset.

1.5 (d)

No flag changes in MOV and MVI instructions as they are data transfer instruction. All flags change in ADD and CMP as they are arithmetic instructions.

1.6 (d)

All flags get affected by SUB B.

1.7 (b)

Register pairs are BC, DE and HL,

1.8 (d)

(i) IN and OUT instruction are used in I/O mapped I/O scheme.

(ii) In memory mapped I/O scheme I/O devices limit can exceed 256.

1.9 (c)

The most significant bit represents the sign of the number. Therefore, it is stored in the sign status flag.

1.10 (b)

First cycle is always the fetch cycle. Read/write operations/cycle are with respect to microprocessor memory or I/O.

1.11 (c)

INTR and RST x·5 are interrupts. Bus idle machine cycle is executed when extra time is needed for internal operation of the processor.

DAD rp instruction may introduce bus idle cycle but when interrupt is recognized processor will not enter into bus idle machine cycles.

1.12 (d)

```
        ORG 4000H
BEGIN:  LXI H, 7000H    HL : 7000H
        MOV A, L          A : 00H
        ADD H             A : 70H
        JM END            Result is positive
        RST 0
END:    PCHL
        HLT
```

So program counter will be branched to RST 0 i.e. 0000H.

1.13 (a)

In this program, we are using BC register pair for memory address and register C is used as counter which is not be possible. That's why this program will not work.

1.14 (d)

SIM can mask RST 5.5, RST 6.5 and RST 7.5 and also can set SOD pin (serial output data line).

1.15 (b)

Machine cycles of PUSH B instruction are shown below:

Op-code fetch → 6 *T*-states

Memory write → 3 *T*-states

Memory write → 3 *T*-states

1.16 (c)

(i) LXI SP, 27FF

SP contain 27FF immediate data with 16-bit address.

(ii) CALL 1006

Current program counter (points to next address) value is pushed onto-stack and program counter sets to 1006.

(iii) POPH will POP from stack and stores in HL register with 1006.

SP = 27FF, HL = 1006

1.17 (a)

Immediate Addressing – LXI H, 2050

Implied Addressing – RRC

Register Addressing – MOV A, B

Direct Addressing – LDA 308C

1.18 (b)

Assertion (A): Microprocessor development requires the design of program and hardware.

Reason (R): Electronic product also follows the same basic steps used in software development.

Both A and R are correct but R is NOT the correct explanation of A.

1.19 (d)

87 H = 10000111

79 H = 01111001

$$\begin{array}{r} 1000\ 0111 \\ +\,0111\ 1001 \\ \hline \underbrace{1}_{\text{Carry}}\ \underbrace{00000000}_{\text{8-bit sum}} \end{array}$$

Therefore, both CARRY and ZERO flags will be set to 1.

1.20 (a)

Machine cycles of CALL instruction are given below:

$$\underbrace{\text{Fetch}}_{6\,T\text{-states}} + \underbrace{\text{Read}}_{3\,T} + \underbrace{\text{Read}}_{3\,T} + \underbrace{\text{Write}}_{3\,T} + \underbrace{\text{Write}}_{3\,T}$$

1.21 (b)

Clock frequency, $f = 5$ MHz

Time required to execute 1 T-state,

$$t = \frac{1}{f}$$

Time required to execute 18 T-states,

$$T = 18\,t$$

$$= \frac{18}{f} = \frac{18}{5 \times 10^6} = 3.6\ \mu\text{s}$$

1.22 (a)

Immediate Addressing – LXI H, 2050

Implied Addressing – RRC

Register Addressing – MOV A, B

Direct Addressing – LDA 308C

1.23 (c)

TRAP is level triggered as well as edge triggered interrupt.

1.24 (c)

Operands are present in BC pair.

1.25 (d)

(i) n-bit counter is used to divide the clock frequency by 2^n.

(ii) Mod-n counter is used to divide the clock frequency by *n*.

1.26 (c)

H and *L* registers are known as "primary data pointers".

1.27 (a)

READY signal introduces wait state and useful when CPU communicates with a slow peripheral device.

1.28 (a)

Microprocessor operation is controlled by software it can perform many different functions.

1.29 (a)

The output data lines of microprocessor and memories are tristate because more than one device can transmit information over the data bus by enabling only one device at a time.

1.30 (d)

Instruction	Size (in byte)
1. MOV B,C	1
2. STA address	3
3. ORI byte	2

So, the sequence of instructions in the decreasing order of their respective memory space requirement is 2, 3, 1 so the correct option is (d).

1.31 (b)

RETURN instruction can be conditional or unconditional in the subroutine.

1.32 (b)

Option (b) is correct if program is written as

```
      MVI C,  78H
LOOP  DCR C
      JNZ   LOOP
      HLT
```

Count register for loop is 'C' which has 78H.

So, $(78)_{16} = (120)_{10}$

1.33 (b)

To clear the lower 4-bits of an accumulator

ANI F0 H

$$A = a_7\ a_6\ a_5\ a_4\ \ a_3\ a_2\ a_1\ a_0$$

$$F0\ H = 1\ 1\ 1\ 1\ \ 0\ 0\ 0\ 0$$

$$ANI\quad a_7\ a_6\ a_5\ a_4\ \ 0\ 0\ 0\ 0$$

1.34 (a)

"LDA address" instruction copies a byte of data into the accumulator with the specified address".

1.35 (b)

To enable memory chip, $\overline{CS} \rightarrow$ low and is possible when output NAND gate is low.

When $A_{15} \rightarrow 1$

A_{14} to $A_{10} \rightarrow 0$

Thus output of NAND will be low.

The range of memory address will be as

A_{15}	A_{14}	A_{13}	A_{12}	A_{11}	A_{10}	A_9	A_8	A_7	A_6	A_5	A_4	A_3	A_2	A_1	A_0	
1	0	0	0	0	0	0	0	0	0	0	0	0	0	0	0	→ 8000 H
1	0	0	0	0	0	1	1	1	1	1	1	1	1	1	1	→ 83FF H

Hence range will be from 8000 H to 83FF H.

1.36 (d)

87 H → 1000 0111

79 H → 0111 1001

Add → 10000 0000

Flags:	*S*	*Z*	*AC*	*CY*
	0	1	1	1

1.37 (c)

```
MVI   A, 45 H   // A ← 45 H
MOV   B, A      // B ← A
STC   // Set the carry : CY = 1
CMC   // Complement the carry i.e. CY = 0
RAR   // Rotate accumulator right through carry
```

CY = 0

A → 0100 0101

A → 0010 0010 & CY = 1

XRA B // Perform EX-OR operation with B and store in A

A → 0010 0010

B → 0100 0101

0110 0111

A → 67 H

1.38 (b)

STAX	:	7 *T*-states
SPHL	:	6 *T*-states
SIM	:	4 *T*-states
STA	:	13 *T*-states

1.39 (a)

Microprocessor internally divides the crystal frequency by 2 with the help of flip-flop. Since, the crystal frequency is given as 6 MHz.

So, time period $= \frac{1}{3\times10^6\,\text{Hz}} = \frac{10^{-6}}{3}$ sec

Since instruction XTHL consumes 16T states to execute an instruction.

$$= \frac{16\times10^{-6}}{3} = 5.33\ \mu\text{sec}$$

1.40 (c)

RST 5 → Restart 5

Vector address = $(5\times8)_{10} = (0028)_H$

1.41 (b)

Flag register is of 8085 is

S	Z	X	AC	X	P	X	CY
↑	↑		↑		↑		↑
sign flag	zero flag		auxiliary flag		Parity flag		Carry flag

(i) MVI A, 7F H, A→ 7F H

(ii) ORAA → Perform OR operation will A

```
     0 1 1 1   1 1 1 1
     0 1 1 1   1 1 1 1
A ← 0 1 1 1   1 1 1 1
```

Flag

S	*Z*	*X*	*AC*	*X*	*P*	*X*	*CY*
0	0	0	0	0	0	0	0

(iii) CPI A2 H →Compare the immediate data with accumulator

This instruction performs three operation.

(i) Copy the contents of accumulator to the temporary register.

(ii) Subtract A2 H from the contents of accumulator and change the affected flags in flag register

(iii) Transfer the contents to the accumulator from the temporary register.

CPI instruction only affects the flag but does not alter the contents of accumulator.

SUB A2H

```
Accumulator    0 1 1 1   1 1 1 1
A2H         = −1 0 1 0   0 0 1 0
               1 1 0 1   1 1 0 1
```

flag is

S	*Z*	*X*	*AC*	*X*	*P*	*X*	*CY*
1	0	0	0	0	1	0	1

So A ← 7F H, S = 1, Z = 0, CY = 1

1.42 (a)

If *N* is negative as JM checks S → 1 then jump to ONE and find compliment otherwise display it is.

1.43 (c)

NOP

1st A ← 20 H

RLC A ← 40 H

2nd A ← 50 H

RLC A ← A0 H

3rd A ← B0 H

RLC A ← 61 H

1.44 (a)

Immediate Addressing	–	LXI H, 2050
Implied Addressing	–	RRC
Register Addressing	–	MOV A, B
Direct Addressing	–	LDA 308C

1.45 (c)

Program counter stores the address of the next instruction to be executed.

1.46 (b)

$S_1 = 1$, $S_2 = 1$, $\overline{RD} = 0$, ALE high in T_1.

1.47 (c)

Microprocessor 8085 has 5 hardware interrupts which are:

TRAP, RST 7.5, RST 6.5, RST 5.5, INTR

and rest interrupts are software interrupts.

So RST0 and RST6 are software interrupts.

1.48 (b)

$$P = \underline{1110}\ \underline{1101}$$
$$Q = \underline{1110}\ \underline{0110}$$

Take the 2's complement of *Q* and add it to *P*.

Using 2's complement addition *P* + (−*Q*)

```
     1 1 1 0 1 1 0 1
     0 0 0 1 1 0 0 1
=  ①0 0 0 0 0 1 1 0
```

Since CY is generated, add it to LSB (least significant bit) to get the correct subtraction result.

```
   0 0 0 0 0 1 1 0
+              1
   0 0 0 0 0 1 1 1
```

1.49 (c)

TRAP is a vectors interrupt having the maximum priority and it is both positive edge triggered and level triggered as

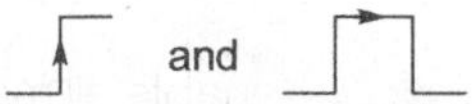

But the edge triggered is generally used for fast propagation of signals. So, the best option is (c).

1.50 (a)

Implied addressing mode are those in which neither data nor address is given in the instruction as they are implicit in mnemonic itself.
Eg : CMA

1.51 (b)

Program counter holds the address of next instruction to be fetched.

1.52 (c)

- DAD instruction means double addition in which contents of register pair is added with the contents of HL register pair and again the result is stored in HL
  DAD R_P
  HL ← $(H_L + R_P)$
- The program counter is an 16-bit register and works as an instruction pointer as it holds the address of the next instruction to be fetched.
- In indirect addressing, the address of the data is stored in HL register pair and pair is mentioned in the instruction as
  MOV C, M

Thus HL register pair also acts as data pointer.

1.53 (a)

CALL 16 bit address
Call a subroutine unconditionally
CALL ⟺ PUSH + JMP

So, it has 5 machine cycles and 18 T states

S, — Opcode fetch
W, W, — PC to store on stack
R, R — Read 2 bytes

1.54 (c)

Let us check each option

- **PUSH PSW** reduces the contents of stack pointer by 2.
- **CALL ADDR**
  as call instruction is a combination of PUSH + JMP. So, it also reduces the contents of stack pointer by 2.
- **XTHL**
  This exchanges HL pair with the top of stack but no effect on contents of stack pointer.

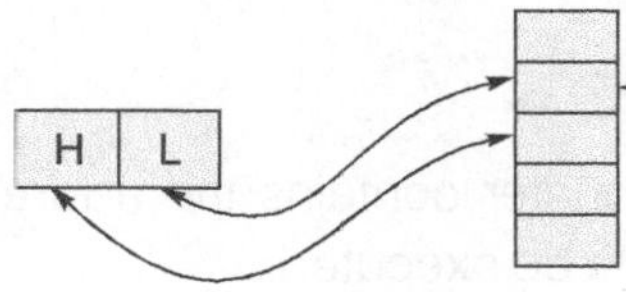

- **RSTn**
  Restart instruction is also a subroutine which jumps on the execution of instruction. So, changes the contents of stack pointer.

1.55 (d)

Program-counter is a 16-bit register used to store the address of next instruction to be fetched by the microprocessor.

1.56 (a)

The following actions are performed by 8085 RESET instruction.

- PC contents become 0000 H
- IR contents become 00 H
- All interrupts are disabled except TRAP.

1.57 (b)

Program counter is a register which holds address of the next instruction to be fetched.

1.58 (c)

1000 LXI SP, 27FF ; SP = 27FF
1003 CALL 1006; Decrements the top of stack by 2
→ SP = 27FD
1006 POP H; Stores 06 in L and 10 in H
→ HL = 1006 and SP is incremented by 2
and SP = 27FF
∴ SP = 27FF and HL = 1006

1.59 (c)

A_{15}	A_{14}	A_{13}	A_{12}	A_{11}	A_{10}	A_9	A_8	A_7	A_6	A_5	A_4	A_3	A_2	A_1	A_0
1	0	0	0	0	0	0	0	0	0	0	0	0	0	0	0
1	0	1	1	1	1	1	1	1	1	1	1	1	1	1	1

8000 – *BFFF*
Memory = $2^{10} \times 2^4$ = 16 k

1.60 (d)

When a slow peripheral is interfaced with the microprocessor, it has to wait until the peripheral is ready.

1.61 (c)

Program counter contains the address of next instruction to be executed.

1.62 (d)

In memory mapped I/O, I/O's are treated as memory locations and arithmetic and logic operation can be directly performed with the I/O data.

1.63 (d)

In a microprocessor when a CPU is interrupted, it acknowledges interrupt by INTA signal and then waits for the next instruction to be fetched by interrupting device.

1.64 (b)

Statement lable in a subroutine is always same as in main program.

1.65 (b)

Location of data within Mnemonic i.e. in CMA no operands are there but operation is done on Accumulator only.

1.66 (b)

```
    MOV CX, 8000h
L1: DEC CX
    JNZ L1
```

Loop run for 32768 times

So, Execution time = 32768 × 18*T*

$$T = \frac{1}{f_{clk}} = \frac{3}{14.7456}\,\mu s$$

$$\text{Execution time} = \frac{32768 \times 18 \times 3}{14.7456 \times 10^6} = 0.12 \text{ s}$$

1.71 (b)

Copy/data transfer instruction.

1.72 (b)

In 8-bit port to address to access data from 8-bit port address into accumulator.

1.73 (b)

EI	;	enable all maskable interrupts
MVI A, 08H	;	Accumulator has 08H.
SIM	:	set interrupt mask

D_7	D_6	D_5	D_4	D_3	D_2	D_1	D_0
SOD	SDE	xxx	R7.5	MSE	M7.5	M6.5	M5.5

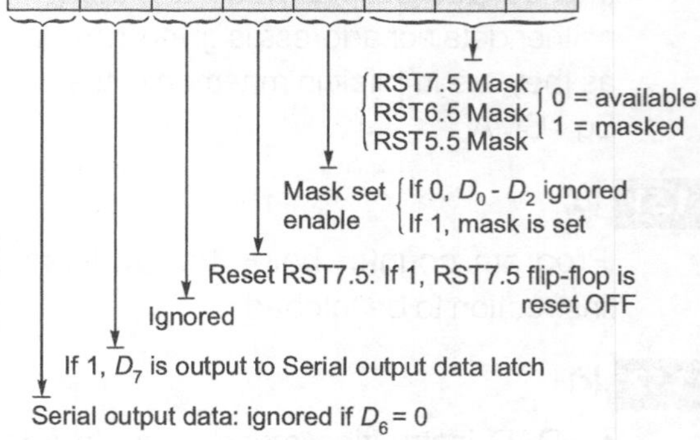

$D_7 \to 0$; $D_6 \to 0$; $D_5 \to 0$; $D_4 \to 0$
$D_3 \to 1$; $D_2 \to 0$; $D_1 \to 0$; $D_0 \to 0$

D_3 is 1 which implies D_2 to D_0 are significant.

D_2 to D_0 are 0 so all are available.

EI in also valid INTR.

1.74 (a, d)

This question is not with w.r.t to basic processors like 8085/8086. It is equivalent to jump if carry = 1, instead of 'JC' 'BC' i.e. branch if carry = 1 should be considered. This is used in PIC microcontrollers which can be 8-bit or 16-bit. As per UPSC answer, 15 words can be 15 × 1 byte or 15 × 2 bytes. So, option also suits if the controller is 8-bit.

1.75 (a)

I/O and power failure are related to external interrupts like DMA and TRAP.

1.76 (b)

END is an assembler directive/pseudo instruction.

1.77 (c)

PUSH and POP are used w.r.t. stack memory and the relevant register is stack pointer.

1.78 (b)

- DAD R_p instruction adds the content of register pair, R_p to HL register pair and the sum is stored in the HL register. If the result is larger than 16-bits, the CY (carry) flag is set. No other flags are affected.
- INR and DCR instructions affect all flags except carry flag. Hence, statement 2 is incorrect.

- CMC stands for "complement the carry flag". It performs complement operation on the carry flag and the result is stored back in the carry flag. Hence, after execution of CMC instruction, only CY flag will be changed.

1.79 (d)

- Memory mapped I/O is a scheme where an I/O device is mapped into the address space of the system, as if it were memory.
  Hence, memory related instructions are used for data transfer, thereby eliminating the need for dedicated I/O instructions.
- Since, the total memory space is shared between memory and I/O devices, so the actual memory space for instruction and data gets reduced.

Hence, both the given statements are correct.

1.80 (c)

LXI SP, 27FFH ; [SP] = 27FFH

CALL 1006H;

(1) Stack ← [PC] i.e., stack ← 1006 H

$\therefore$ [SP] ← [SP] – 2

(2) [PC] ← 1006 H i.e., subroutine address into 'PC'

POP H : Data @ Top of Stack memory i.e., 2 bytes accessed into 'HL' pair and [SP] ← [SP] + 2

$\therefore$ 'SP' content is unchanged and 'HL' pair contains 1006H.

1.81 (d)

We know that,

LDA 2003H → 3-byte Instruction

Load accumulator with the content @ 2003H

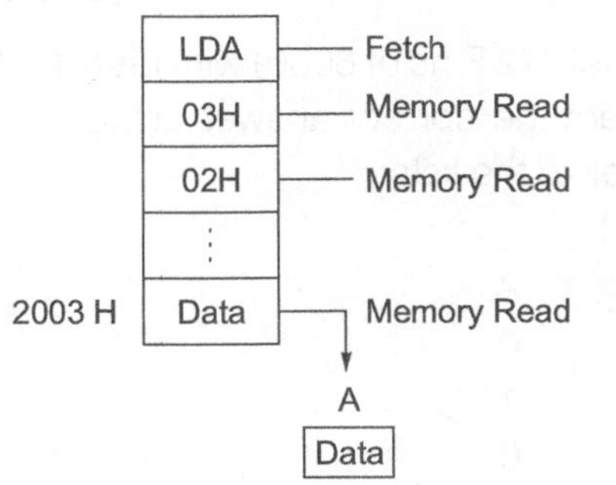

$\therefore$ FRRR → 4 Memory accesses

1.82 (*)

Given, content of accumulator and register B,

A: 89 H + B: 3C H   SUB B: [ACC] ← [ACC] – [B]

–3CH

4DH → 01001101 (Msb ...... Lsb)

S	Z	X	AC	X	P	X	CY
0	0	0	0	0	1	0	0

$Cy \rightarrow 0$; No borrow

$S \rightarrow 0$; sign flag

ACC = 4DH

A – B = A + (–B)

Microprocessor performs subtraction using 2's complement method.

1.83 (a)

The given program can be executed as,

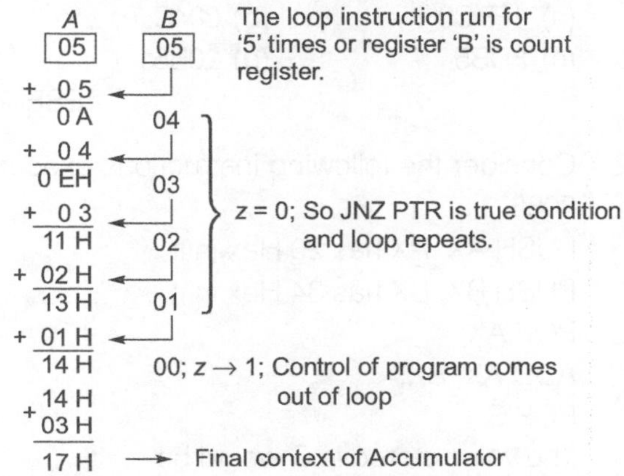

1.84 (c)

ANI 8-bit data.

'AND' immediate 8-bit data to accumulator.

1.85 (a)

In an 8085 microprocessor,

PSW → Program status word →

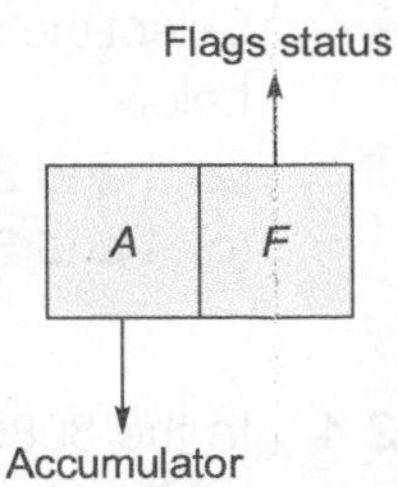

8085 doesn't has any separate instruction to read flag states into any general purpose register. So, PUSH PSW and 'POP' are used to know/read status of flags into some RP (Register Pair) using POP R_p. So, 'II' and 'I' are correct and II is right reason.

1.86 (c)

The condition codes, also known as flags are the status bits in the processor's status register that reflect the outcome of ALU operations and can be used to determine the program flow.

2 8086 Microprocessors

2.1 In 8086, if the content of the code segment register is 1FAB and the content of the IP register is 10A1, then the effective memory address is
(a) 1FBC0 (b) 304C
(c) FDB5 (d) 20B51
[ESE-2000]

2.2 Consider the following instructions executed in 8086.
PUSH AX; AX has 20 Hex in it
PUSH BX; BX has 34 Hex in it
POP AX;
ADD AX, BX;
POP G
The value stored in G would be
(a) 20 Hex (b) 34 Hex
(c) 54 Hex (d) 68 Hex [ESE-2000]

2.3 The 8086 arithmetic instructions work on
1. signed and unsigned numbers
2. ASCII data
3. unpacked BCD data.
Select the correct answer using the codes given below:
(a) 1 and 2 (b) 2 and 3
(c) 1 and 3 (d) 1, 2 and 3
[ESE-2001]

2.4 In the 8086 instruction ADD DX, [BX] [CI], the addressing mode of source operand is
(a) Register (b) Register Indirect
(c) Based Indexed (d) Direct
[ESE-2002]

2.5 Match **List-I** (8085 Register) with **List-II** (8086 Register) and select the correct answer using the code given below the lists:

List-I	List-II
A. A	1. CJ
B. H	2. AL
C. L	3. BL
D. B	4. BH

Codes:

	A	B	C	D
(a)	4	2	3	1
(b)	2	4	1	3
(c)	4	2	1	3
(d)	2	4	3	1

[ESE-2003]

2.6 Consider the Motorola 68008, 68010, 68012 and 68020 microprocessors. In systems that are severely constrained by the space available for the printed circuit board, it is better to use
(a) 68008 (b) 68010
(c) 68012 (d) 68020 [ESE-2003]

2.7
```
PUSH  a
PUSH  b
PUSH  c
POP   AX
POP   BX
SUB   AX, BX
POP   BX
ADD   AX, BX
```
The expression computed by the above program and stored in AX is
(a) a + b – c (b) c + b – a
(c) c – b + a (d) c – b – a
[ESE-2003]

2.8 Match **List-I** (2 Pins of 8086) with **List-II** (Status) and select the correct answer using the code given below the lists:

List-I

	$\overline{BHE}$	A_0
A.	0	0
B.	0	1
C.	1	0
D.	1	1

List-II

What is read/written?
1. 1 byte from/to odd address
2. 1 byte from/to even address
3. 1 16-bit word
4. NOP

Codes:

	A	B	C	D
(a)	4	2	1	3
(b)	3	1	2	4
(c)	4	1	2	3
(d)	3	2	1	4

[ESE-2005]

2.9 Which of the following instructions of an 8086 microprocessor uses the contents of a CX register as a counter?
1. LOCK 2. LOOP 3. ROTATE
Select the correct answer using the code given below:
(a) Only 1 and 2 (b) Only 1 and 3
(c) Only 2 and 3 (d) 1, 2 and 3
[ESE-2006]

2.10 What is the address space of 8086 CPU ?
(a) 1 megabytes (b) 256 kilobytes
(c) 1k megabytes (d) 64 kilobytes
[ESE-2009]

2.11 Which one of the following control bits of 8086 flag register is used to put the 8086 in single step mode?
(a) DF (b) IF
(c) TF (d) ZF [ESE-2011]

2.12 As compared to 16 bit microprocessor, 8 bit microprocessors are limited in
1. Speed
2. Directly addressable memory
3. Data handling capability
(a) 1 and 2 only (b) 2 and 3 only
(c) 1 and 3 only (d) 1, 2 and 3
[ESE-2011]

2.13 For 8086 microprocessor, the jump distance in bytes for short jump range is
(a) Forward 255 and Backward 256
(b) Forward 127 and Backward 128
(c) Forward 31 and Backward 32
(d) Forward 15 and Backward 16 [ESE-2012]

2.14 The action performed by the following instruction of 8086:
MOV [1234H], AX
(a) Move contents of memory location 1234 H to register AX
(b) Move the contents of register AX to memory offset 1234 H
(c) Add contents at 1234 H to register AX
(d) Add contents of 1234 H and AX and store the result in 1234 H
[ESE-2012]

2.15 What is the activity of the microprocessor 8086/8085 when the signals $\overline{SS_0}$, $DT/\overline{R}$ and $IO/\overline{M}$ are 1, 0 and 1 respectively in minimum mode?
(a) Read memory (b) Read I/O port
(c) Code Access (d) Write I/O port
[ESE-2015]

2.16 **Statement (I):** The direction flag D in 8086 selects increment or decrement mode for DI and/or SI registers.
Statement (II) : If $D = 0$, the registers are automatically decremented.
(a) Both Statement (I) and Statement (II) are individually true and Statement (II) is the correct explanation of Statement (I).
(b) Both Statement (I) and Statement (II) are individually true but Statement (II) is not the correct explanation of Statement (I).
(c) Statement (I) is true but Statement (II) is false.
(d) Statement (I) is false but Statement (II) is true.
[ESE-2016]

2.17 If the operating frequency of an 8086 microprocessor is 10 MHz and, if, for the given instruction, the machine cycle consists of 4 T-states, what will be the time taken by the machine cycle to complete the execution of the same instruction when three wait states are inserted?
(a) 0.4 μs (b) 0.7 μs
(c) 7 μs (d) 70 μs [ESE-2017]

2.18 Which one of the following is also called as pseudo instructions that are not directly translated into machine language instructions?
(a) Macro expansions (b) Assembly directives
(c) Micro expansions (d) Labels [ESE-2021]

2.19 Which one of the following addressing modes is used to compute the address of the operand by adding a constant value to the content of a register?
(a) Immediate addressing
(b) Indexed addressing
(c) Absolute addressing
(d) Direct addressing [ESE-2022]

2.20 Which one of the following addressing modes is similar to the register indirect addressing mode in the sense that the effective address of the operand is the content of a register, which is included in the instruction?
(a) Relative mode
(b) Autoincrement mode
(c) Autodecrement mode
(d) Immediate addressing mode [ESE-2022]

■■■

Answers 8086 Microprocessors

2.1 (d) 2.2 (a) 2.3 (c) 2.4 (c) 2.5 (d) 2.6 (d) 2.7 (c) 2.8 (d) 2.9 (c)

2.10 (a) 2.11 (c) 2.12 (d) 2.13 (b) 2.14 (b) 2.15 (b) 2.16 (c) 2.17 (b) 2.18 (b)

2.19 (b) 2.20 (a)

Explanations 8086 Microprocessors

2.1 (d)

Effective memory address

= 1 FAB0 + 10 A1 = 20 B51

2.2 (a)

PUSH AX ; AX has 20H and pushed onto stack.
PUSH BX ; BX has 34H and pushed onto stack.
POP AX ; 34 H stored in AX from top of stack.
ADD AX, BX ; AX contains sum of 34H and 34H.
POP G ; 20H stored in G from top of stack.
The value stored in G is 20H.

2.3 (c)

8086 arithmetic instructions work on signal/unisignal numbers and unpacked BCD data.

2.4 (c)

ADD DX, [BX] [CI]

BX store base address
CI stores index

∴ Source is based indexed.

2.5 (d)

A register of 8085 is similar to AL register of 8086.
H register of 8085 is similar to BH register of 8086.
L register of 8085 is similar to BL register of 8086.
B register of 8085 is similar to CJ register of 8086.

2.7 (c)

PUSH	a	⇒ Put a on the stack
PUSH	b	⇒ Put b on the stack
PUSH	c	⇒ Put c on the stack
POP	AX	⇒ Store c in AX
POP	BX	⇒ Store b in BX
SUB	AX, BX	⇒ Store c – b in AX
POP	BX	⇒ Store a in BX
ADD	AX, BX	⇒ Store c – b + a in AX

2.8 (d)

$\overline{BHE}$	A_0	Status
0	0	One 16-bit word
0	1	One byte from/to even address
1	0	One byte from/to odd address
1	1	NOP

2.9 (c)

Loop and rotate instructions uses the contents of a CX register as a counter in 8086.

2.10 (a)

Address space of 8086 CPU
= 2^{20} bytes = 1024 K byte = 1 Megabyte

2.11 (c)

TF → In 8086 if TRAP flag is set then microprocessor coil compile in single step mode.

2.12 (d)

In 16 bit microprocessor → clock frequency is 5 MHz high compare to 8 bit microprocessor 3 MHz so speed of 16 bit microprocessor will be high. So 1 – is right.

- Directly addressable memory in 16 bit microprocessor = 2^{16} byte
- Directly addressable memory in 8 bit microprocessor = 2^{8} byte

So, 16 bit microprocessor has higher number of addressable memory than 8 but microprocessor. So 16 bit is better than 8 bit.

16 bit microprocessor can handle 16 bit data at a time as having 16 data lines while 8 bit microprocessor can handle only 8 bit data at a time as having 8 data lines. So limited amp. are to 16 bit microprocessor.

So all are correct.

2.13 (b)

Jump distance for short jump range is:
127 bytes forward
128 bytes forward

2.14 (b)

MOV [1234 H], AX
It moves the content of register AX to the memory offset 1234H.

2.15 (b)

$$\overline{SS_0} = 1$$

$$DT/\overline{R} = 0$$

⇒ (Data is received by microprocessor)

$$IO/\overline{M} = 1 \quad \text{(I/O operation)}$$

⇒ Read I/O port.

2.16 (c)

D – Direction flag in to set direction in string manipulation instructions.

$D \rightarrow 0$; Auto increment mode
$D \rightarrow 1$; Auto decrement mode

2.17 (b)

$$\text{Execution time} = 4T + 3T$$

(∵ Three wait states are inserted)

$$= 7T$$

$$\Rightarrow \quad T = \frac{1}{10} \mu s$$

Execution time = 0.7 μsec

2.18 (b)

Assembler directives are the directions to the assembles to take some action or change a setting. Assembler directives do not represent instructions, and are not translated into machine code.

Example: END, DB, DW etc.

2.19 (b)

Immediate Addressing : The data is present in the address field of instruction. Example

MOV AL, 35H

Indexed Addressing : The contents of index register is added to address part of instruction to obtain effective address. The address field of instruction holds the base address. Example :

MOV AX, [SI + 05]

Absolute/Direct Addressing : The absolute addressing mode is also called direct addressing because the actual address of an operand is specified in the instruction. Example

LDA 2805H

Hence, option (b) is the correct answer i.e. indexed addressing mode is used to compute the address of operand by adding a constant value to the content of a register.

2.20 (a)

In relative addressing mode, the effective address of the operand is calculated by adding the content of a CPU register and the address part of instruction word.

Relative addressing mode can be classified into three types :

- PC (Program counter) Relative Addressing Mode.
- Index Register Relative Addressing Mode.
- Base Register Addressing Mode.

■■■

3 Interfacing Devices and Microcontrollers

3.1 The MODEM is used with a personal computer to do which of the following?
(a) Convert from serial to parallel and vice versa
(b) Convert signals between TTL ad RS 232 C standard and vice versa
(c) Convert from digital to analog signals and vice versa
(d) To convert the computer to a long distance communication link

[ESE-2002]

3.2 Which one of the following is used as the interface chip for data transmission between 8086 and a 16-bit ADC ?
(a) 8259 (b) 8255
(c) 8253 (d) 8251

[ESE-2009]

3.3 What is the purpose of a start bit in RS232 serial communication protocol ?
(a) To synchronise receiver for receiving every byte
(b) To synchronise receiver for receiving a sequence of byte
(c) Acts as a parity bit
(d) To synchronize receiver for receiving the last byte

[ESE-2009]

3.4 When compared with an RS232C serial port, the USB (Universal Serial Bus)
(a) Supports a lower range of peripherals
(b) Supports a faster transfer rate
(c) Does not support 'Hot plug-ability'
(d) Controller in PC can not detect the presence or absence of USB devices

[ESE-2010]

3.5 The purpose of a start bit in RS232 serial communication protocol is
(a) To synchronise receiver for receiving every byte
(b) To synchronise receiver for receiving a sequence of bytes
(c) As a parity bit
(d) To synchronise receiver for receiving the last byte

[ESE-2010]

3.6 The interface chip used for data transmission between 8086 and a 16-bit ADC is
(a) 8251 (b) 8253
(c) 8255 (d) 8259 [ESE-2010]

3.7 **Assertion (A):** The main difference between a microprocessor and a microcontroller is that the former does not have any on-chip main memory whereas latter has.
Reason (R): A microprocessor does not need memory to run program.
(a) Both A and R are true and R is the correct explanation of A
(b) Both A and R are true but R is NOT the correct explanation of A
(c) A is true but R is false
(d) A is false but R is true [ESE-2010]

3.8 To configure port A and port B as output ports, port C not being used in the 8155 programmable interface, the control word should have the value
(a) 03 H (b) 01 H
(c) 02 H (d) 60 H [ESE-2012]

3.9 The 8254 Programmable Interval Timer is set to work in MODE 5. The following would best describe its function:
(a) Software triggered strobe
(b) Hardware triggered strobe
(c) Square wave generator
(d) Interrupt on terminal count [ESE-2012]

3.10 With respect to the default IR7 routine in 8259 PIC, which of the following statements are correct?
1. Default CALL to the IR7 routine when an invalid interrupt occurs at IR pin.

2. A default IR7 call does not set the corresponding ISR bit.
3. A default IR7 routine is nothing but the same valid IR7 routine with lowest priority.

(a) 1, 2 and 3 (b) 1 and 2 only
(c) 1 and 3 only (d) 2 and 3 only

[ESE-2015]

3.11 In mode 0, Interrupt on terminal count of 8253, if the gate pin is made low while counter is decrementing, which one of the following operations will follow?

(a) Counter stops and cleared to '0' and starts decrementing when gate pin is made high
(b) Counter stops and thereafter it increments till gate pin = '1' high.
(c) Counter stops, the current contents are held and the decrement operation resumes only after gate pin is made high
(d) Counter stops, the current contents are held for one clock cycle and the decrement operation resumes

[ESE-2015]

3.12 Consider the following statements :
The 8259A Programmable Interrupt Controller can
1. manage eight interrupts
2. vector an interrupt request anywhere in the memory map
3. have 8-bit or 16-bit interval between interrupt vector locations
4. be initialized with operational command words

Which of the above statements are correct?
(a) 1, 2 and 3 only (b) 1, 2 and 4 only
(c) 3 and 4 only (d) 1, 2, 3 and 4

[ESE-2017]

3.13 What are the conditions which are necessary for using a parallel port?
1. Initializing by placing appropriate bits at the control register
2. Calling on interrupt whenever a status flag sets at the status register
3. Interrupting servicing (device driver) programming

Select the correct answer using code given below.
(a) 1 and 2 only (b) 1 and 3 only
(c) 1, 2 and 3 (d) 2 and 3 only

[ESE-2017]

3.14 An asynchronous link between two computers uses the start-stop scheme, with one start bit and one stop bit, and a transmission rate of 48.8 kbits/s. What is the effective transmission rate as seen by the two computers?
(a) 480 bytes/s (b) 488 bytes/s
(c) 4880 bytes/s (d) 4800 bytes/s

[ESE-2017]

3.15 **Statement (I):** Peripherals are electromechanical and/or electromagnetic devices.
Statement (II): All the peripherals are digital and parallel.
(a) Both Statement I and Statement II are individually true and Statement II is the correct explanation of Statement I
(b) Both Statement I and Statement II are individually true but Statement II is not the correct explanation of Statement I
(c) Statement I is true but Statement II is false
(d) Statement I is false but Statement II is true

[ESE-2018]

3.16 A controller that takes control of the buses and transfers data directly between source and destination bypassing the microprocessor is known as
(a) DMA controller
(b) read-write controller
(c) high-speed controller
(d) master-slave controller

[ESE-2019]

3.17 Which of the following constraints are to be considered by the designer while designing an embedded system?
1. Selecting the microcontroller as a controlling device
2. Selecting the language to write the software
3. Partitioning the tasks between hardware and software to optimize the cost

Select the correct answer using the code given below.
(a) 1, 2 and 3 (b) 1 and 2 only
(c) 1 and 3 only (d) 2 and 3 only

[ESE-2019]

3.18 In the 8051 microcontroller, direct addressing mode is used in
(a) Internal data memory
(b) External data memory
(c) Internal program memory
(d) External program memory

[ESE-2020]

3.19 Consider the following statements regarding the Read Only Memory (ROM):

1. The stored data is lost if the power is removed.
2. It consists of an address decoder with *n* input lines and a programmable OR array with *n* output lines.
3. The minterms are ORed through the programmable switches which can be made ON or OFF to select a particular minterm. The programmable switches can be implemented by either bipolar, CMOS, nMOS or pMOS technologies.
4. Mask-programed ROMs are used in the applications where the system requires data to be stored and to be changed during the operation.

Which of the above statements are ***not*** correct?

(a) 1 and 2 only (b) 1, 3 and 4 only
(c) 1, 2 and 4 only (d) 2, 3 and 4 only

[ESE-2021]

3.20 In 8051, the accumulator register contains 80H and B register contains 8FH. The content of the accumulator and status of the carry flag after the addition operation are respectively,

(a) 0FH, 1 (b) 10FH, 0
(c) FFH, 1 (d) 10FH, 1 **[ESE-2021]**

3.21 Which one of the following memories is primarily used to store machine microcode, desktop bootstrap loaders, and video game cartridges?

(a) Mask-programmed ROM
(b) Static-RAM
(c) Dynamic-RAM
(d) Non-Programmed ROM **[ESE-2021]**

3.22 Consider the following statements regarding memory:

1. Integrated circuit RAM chips are available in both static and dynamic modes.
2. The dynamic RAM stores the binary information in the form of electric charges that are applied to capacitors.
3. The static RAM is easier to use and has shorter read and write cycles.
4. RAM and ROM chips are available in a variety of sizes.

Which of the above statements are correct?

(a) 1 and 2 only (b) 1, 3 and 4 only
(c) 2, 3 and 4 only (d) 1, 2, 3 and 4

[ESE-2021]

3.23 Consider the following statements regarding PROM/EPROM:

1. The erasable programmable ROM using ultraviolet erasing is known as EPROM.
2. The ROM that makes use of the electrical voltage for erasing is known as electrically alterable ROM.
3. A PROM can be programmed many times after fabrication.

Which of the above statements are correct?

(a) 1 and 2 only (b) 1 and 3 only
(c) 1, 2 and 3 (d) 2 and 3 only

[ESE-2021]

3.24 Consider the following statements regarding programmable interrupt controller 8259A :

1. 8259A is specifically designed for use with the interrupt signals (INTR/INT) of 8085 microprocessor.
2. It can solve eight levels of interrupt priorities in a variety of modes.
3. With additional 8259A devices, the priority scheme can be expanded to 32 levels.

Which of the above statements is/are not correct?

(a) 1 and 3 (b) 2 only
(c) 3 only (d) 1 and 2

[ESE-2022]

3.25 Consider the following statements regarding 8051 microcontroller:

1. The size of RAM in 8051 is 128 bytes.
2. RAM locations 00-1FH are assigned to the register banks and stack.
3. RAM locations 20-2FH are available as a place to save byte-sized data.

Which of the above statements are correct?

(a) 1 and 2 only (b) 1 and 3 only
(c) 2 and 3 only (d) 1, 2 and 3

[ESE-2022]

3.26 Consider the following statements regarding nonvolatile RAM (NV-RAM) in 8051 :

1. When the power is turned off, the contents are lost.
2. NV-RAM uses extremely power-efficient SRAM cells built out of CMOS.
3. It uses an intelligent control circuitry.

Which of the above statements are correct?

(a) 1, 2 and 3 (b) 1 and 2 only
(c) 2 and 3 only (d) 1 and 3 only

[ESE-2022]

3.27 In order to keep track of the instruction locations, which one of the following contains the value of memory location assigned to the instruction or operand being processed?
(a) LD (b) STA
(c) END (d) ILC [ESE-2022]

3.28 Which one of the following is used to perform a transfer between two memory-mapped devices without the intervention of the CPU or the use of main memory?
(a) Direct virtual memory access
(b) Cycle stealing
(c) Direct memory access
(d) Programmed I/O [ESE-2023]

3.29 Which one of the following steps is not correct for an asynchronous data transfer?
(a) When the character is not being sent, the line is kept in the 0 state.
(b) The initiation of a character of a transmission is detected from the start bit which is always zero.
(c) The character bit always follows the start bit.
(d) After the last bit of the character is transmitted, the stop bit is detected.
[ESE-2024]

3.30 Which one of the following is the disadvantage of the software pole?
(a) Timing, and it can be rectified by daisy chain.
(b) Delay, and it can be rectified by vectored interrupt.
(c) Longest routing, and it can be rectified by multiple interrupt lines.
(d) Response time, and it can be rectified by interrupt controller.
[ESE-2024]

■■■

Answers Interfacing Devices and Microcontrollers

3.1 (c)	3.2 (b)	3.3 (b)	3.4 (b)	3.5 (b)	3.6 (c)	3.7 (b)	3.8 (a)	3.9 (b)
3.10 (d)	3.11 (c)	3.12 (b)	3.13 (c)	3.14 (c)	3.15 (c)	3.16 (a)	3.17 (a)	3.18 (a)
3.19 (c)	3.20 (a)	3.21 (a)	3.22 (d)	3.23 (a)	3.24 (a)	3.25 (d)	3.26 (c)	3.27 (d)
3.28 (c)	3.29 (a)	3.30 (d)						

Explanations Interfacing Devices and Microcontrollers

3.3 (b)

In RS 232 serial communication protocol, the frame consists of one start bit, seven or eight data bits, parity bits and stop bits.
Start bit is used to signal the beginning of a frame and stop bit is used to signal the end of a frame.

3.7 (b)

Microprocessor does not have any on-chip main memory and it does not need memory to run program. However, R is not correct explanation of A. A tells regarding difference between microprocessor and microcontroller.

3.8 (a)

8155 is a programmable I/O ports and timer control word of 8155 is given as

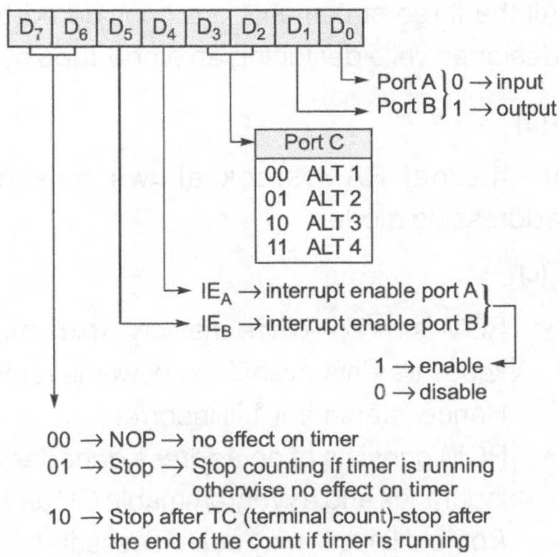

11 → Start timer if it is not running.
So, control word according to question is
0000 0011
03 H

3.9 (b)

8254 is a programmable interval/timer counter

D_7	D_6	D_5	D_4	D_3	D_2	D_1	D_0
SC_1	SC_0	RW_1	RW_0	M_2	M_1	M_0	BCD

Control word register.

8254 can operate in six different modes:

M_2	M_1	M_0		
0	0	0	Mode 0 :	Interrupt on terminal count
0	0	1	Mode 1 :	Hardware triggerable one-shot
X	1	0	Mode 2 :	Rate generator
X	1	1	Mode 3 :	Square wave generator
1	0	0	Mode 4 :	Software triggered strobe
1	0	1	Mode 5 :	Hardware triggered strobe

3.14 (c)

$$\text{Transmission rate} = 48.8 \text{ kbits/sec} = 48800 \text{ bits/sec}$$

In asynchronous transmission, each byte transmission requires 1 stop bit and 1 start bit. So total data transmission 10 bits.

$\therefore$ Effective transmission rate

$$= \frac{48800}{10} = 4880 \text{ bytes/sec}$$

3.16 (a)

DMA in between memory and I/O without the intervention of microprocessor.

3.17 (a)

All the three statements are considered by the designer while designing an embedded system.

3.18 (a)

In internal RAM stack allows only direct addressing mode.

3.19 (c)

- ROM is a non-volatile memory which means it keeps it's data even if the power is removed. Hence, statement 1 is incorrect.
- ROM consists of an address decoder with n input lines and a programmable OR array with k output lines. k may not necessarily be equal to n. Hence, statement 2 is incorrect.
- Mask-programmed ROMs is a type of ROM whose contents are programmed by the integrated circuit manufacturer rather than the user. Thus, it cannot be used in the applications where the system requires the data to be stored and changed during the operation.

$\therefore$ Option (c) is correct.

3.20 (a)

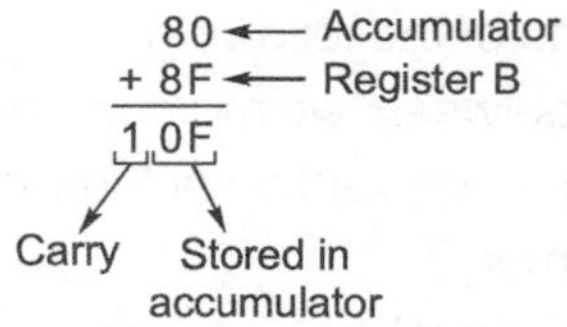

3.21 (a)

A Mask-programmed ROM is programmed at the manufacturer's site according to the specifications of the customer. Hence, Mask-programmed ROM is primarily used to store machine microcode, desktop bootstrap loaders and video game cartridges.

Static RAMs and Dynamic RAMs are volatile memory and therefore, can's be used for machine microcode, bootstrap loaders, etc.

3.22 (d)

- Integrated circuit RAM chips are available in both static and dynamic modes and are available in different sizes of 1 GB, 2 GB, 4 GB, 8 GB.
- Capacitor has high storing capability, so DRAM stores its binary information in the form of electric charges on capacitors.
- Static RAM is faster and therefore has shorter read and write cycles.

Therefore, all the given statements are correct.

3.23 (a)

A PROM is also called as OTPROM i.e. one time programmable read-only memory. So, statement 3 is wrong.

3.24 (a)

- 8259A is a programmable interrupt controller designed to work with Intel microprocessor 8080, 8085A, 8086, 8088.
- It manages 8 interrupt requests. It can vector an interrupt anywhere in memory without additional hardware. It can support 8 levels of interrupt priorities. The priority scheme can be extended to 64 levels using a hierarchy of 8259 devices.

Hence, statements 1 and 3 are not correct.

3.25 (d)

The 128 bytes of RAM inside the 8051 microcontroller are assigned the address 00 to 7FH. They can be accessed directly as memory locations and are divided into three different groups as follows :

- 32 bytes from 00H to 1FH locations are set aside for register banks and the stack.
- 16 bytes from 20H to 2FH locations are set aside for bit-addressable read/write memory.
- 80 bytes from 30H to 7FH locations are used for read and write storage.

3.26 (c)

NV-RAM (Non-volatile RAM) combines the best of RAM and ROM ; the read and write ability of RAM plus the non-volatility of ROM. To retain its contents, every NV-RAM chip internally is made of the following components:

1. It uses extremely power efficient SRAM cells built out of CMOS.
2. It uses an internal lithium battery as a backup energy source.
3. It uses an intelligent control circuitry. The main job of this control circuitry is to monitor the V_{CC} pin constantly to detect loss of the external power supply. If the power to V_{CC} pin falls below out-of-tolerance conditions, the control circuitry switches automatically to its internal power source, the lithium battery. The internal lithium power source is used to retain the NV-RAM contents only when the external power source is off. Hence, statement 2 is not correct.

3.27 (d)

- LD stands for load. It loads data from one place into another.
- STA instruction store the accumulator data into specified memory.
- END directive terminates an assembly language program.
- In order to keep track of the instruction locations, the assembler maintains a variable called Instruction Location Counter (ILC). The ILC contains the value of memory location assigned to the instruction or operand being processed. The ILC is initialized to 0 and is incremented after processing each instruction. The ILC is incremented by the length of the instruction being processed or the number of bytes allocated as a result of a data allocation pseudo instruction.

3.28 (c)

In Direct Memory Access, bulk amount of data will be transferred from the I/O to MM without involvement of a CPU.

3.29 (a)

The start and stop bits are used in asynchronous communication as a means of timing or synchronizing the data characters being transmitted. Another bit used to separate data characters during transmission is the mark (or idle) RS bit. A binary 1, is transmitted when the communication line is idle and no characters are being sent or received.
When a start bit (binary 0) is received by the system, a fixed number character bits, and even a parity bit follows the start bit. Then, a stop bit (binary 1) is received by the system after the last character bit is transmitted.

3.30 (d)

In software pole, the CPU has to continuously check the status of a device decreasing the responsiveness of the system. It is rectified using Interrupt controller that efficiently signal the CPU without the CPU constantly polling.

■■■

UNIT

VI

Analog and Digital Communication Systems

Syllabus

Random signals, noise, probability theory, information theory; Analog versus digital communication & applications: Systems- AM, FM, transmitters/receivers, theory/practice/ standards, SNR comparison; Digital communication basics: Sampling, quantizing, coding, PCM, DPCM, multiplexing-audio/video; Digital modulation: ASK, FSK, PSK. Multiple access : TDMA, FDMA, CDMA.

Contents

Sl.	Topic	Page No.
1.	Analog Communication Systems	403
2.	Random Variables, Random Process and Noise	425
3.	Digital Communication Systems	437
4.	Information Theory	460

OOOO

1 Analog Communication Systems

1.1 In an amplitude modulated system, if the total power is 600 W and the power in carrier is 400 W, then the modulation index is
(a) 0.5 (b) 0.75
(c) 0.9 (d) 1 **[ESE-2000]**

1.2 Which one of the following statements regarding the threshold effect in demodulators is correct?
(a) It is exhibited by all demodulators when the input signal to noise ratio is low
(b) It is the rapid fall in output signal to noise ratio when the input signal to noise ratio falls below a particular value
(c) It is the property exhibited by all AM suppressed carrier coherent demodulators
(d) It is the property exhibited by correlation receivers
[ESE-2000]

1.3 A band-pass signal has significant frequency components in the range of 1.5 MHz to 2 MHz. If the signal is to be reconstructed from its samples, the minimum sampling frequency will be
(a) 1 MHz (b) 2 MHz
(c) 3.5 MHz (d) 4 MHz
[ESE-2000]

1.4 A signal $m(t) = 5\cos(2\pi 100t)$ frequency modulates a carrier. The resulting FM signal is $10\cos\{(2\pi 10^5 t) + 15\sin(2\pi 100 t)\}$
The approximate bandwidth of the FM signal would be
(a) 0.1 kHz (b) 1 kHz
(c) 3.2 kHz (d) 100 kHz
[ESE-2000]

1.5 The essential blocks of a phase lock loop (PLL) are phase detector, amplifier,
(a) high-pass filter and crystal controlled oscillator
(b) low-pass filter and crystal controlled oscillator
(c) high-pass filter and voltage controlled oscillator
(d) low-pass filter and voltage controlled oscillator
[ESE-2000]

1.6 For an AM wave, the maximum voltage was found to be 10 V and the minimum voltage was found to be 5 V. The modulation index of the wave would be
(a) 0.33 (b) 0.52
(c) 0.40 (d) 0.1
[ESE-2001]

1.7 A signal $x(t) = 6\cos 10\pi t$ is sampled at the rate of 14 Hz. To recover the original signal, the cut-off frequency f_c of the ideal low-pass filter should be
(a) $5\text{ Hz} < f_c < 9\text{ Hz}$ (b) 9 Hz
(c) 10 Hz (d) 14 Hz
[ESE-2001]

1.8 The ratio $\dfrac{(S/N)_{WBFM}}{(S/N)_{AM}}$ for 100% amplitude modulation with identical total transmitted power (m_f is modulation index of FM) is
(a) $9/2\ m_f^2$ (b) $3/2\ m_f^2$
(c) $3/2\ m_f^3$ (d) $9/2\ m_f^3$
[ESE-2001]

1.9 A carrier of frequency 1 MHz is amplitude modulated by a signal of frequency of 1 kHz to a depth of 60%. This is passed through a filter of characteristics shown in the given Figure-I and Figure-II. If the filter output is fed to an envelope detector, the detector output will be

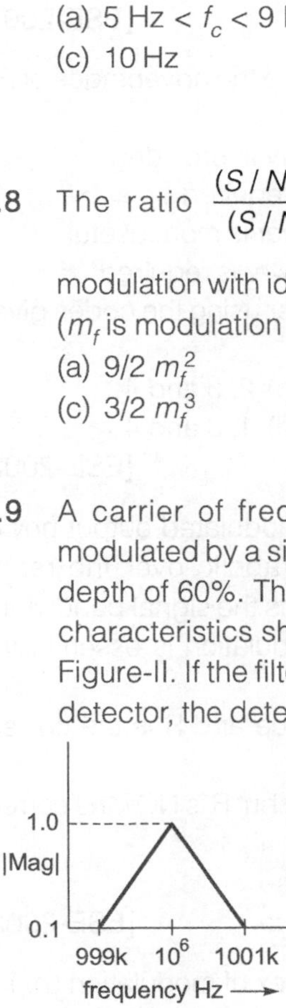

Figure-I

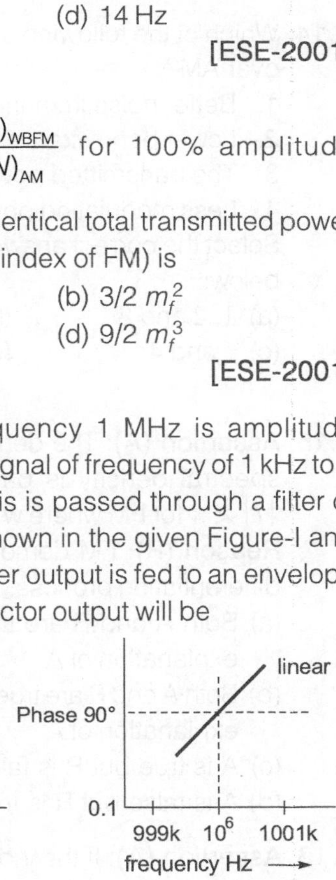

Figure-II

(a) zero (b) dc
(c) dc + 1 kHz signal (d) 1 kHz signal
[IES-2001]

1.10 If the radiated power of AM transmitter is 10 kW, the power in the carrier for modulation index of 0.6 is nearly
(a) 8.24 kW (b) 8.47 kW
(c) 9.26 kW (d) 9.6 kW [ESE-2001]

1.11 In a low-level AM system, the amplifier which follows the modulated stage must be the
(a) linear device (b) harmonic device
(c) class-C amplifier (d) non-linear device
[ESE-2001]

1.12 An angle-modulated signal is expressed by $f_a(t) = \cos(2 \times 10^8 \pi t + 75 \sin 2 \times 10^3 \pi t)$.
The peak frequency deviation of the carrier is then
(a) 1 kHz (b) 7.5 kHz
(c) 75 kHz (d) 100 MHz [ESE-2001]

1.13 MCVF telegraphy uses
(a) SSB techniques
(b) Pulse modulation
(c) out-of-band signalling
(d) FM [ESE-2001]

1.14 Which of the following are the advantages of FM over AM?
1. Better noise immunity is provided
2. Lower bandwidth is required
3. The transmitted power is more useful
4. Less modulating power is required
Select the correct answer using the codes given below:
(a) 1, 2 and 3 (b) 2, 3 and 4
(c) 2 and 4 (d) 1, 3 and 4
[ESE-2002]

1.15 **Assertion (A):** The demodulated output power spectral density is parabolic over the range $|f| < w$ for FM where w is the signal bandwidth.
Reason (R): FM demodulation is essentially a differentiation process.
(a) Both A and R are true and R is the correct explanation of A
(b) Both A and R are true but R is NOT the correct explanation of A
(c) A is true but R is false
(d) A is false but R is true [ESE-2002]

1.16 **Assertion (A):** If the index of modulation (m_f) of an FM signal gradually increases beyond 0.25, it will contribute to the generation of a progressively larger number of side bands separated by the modulation frequency (f_m).
Reason (R): A large modulation index will generate a finite number of side bands.
(a) Both A and R are true and R is the correct explanation of A
(b) Both A and R are true but R is NOT the correct explanation of A
(c) A is true but R is false
(d) A is false but R is true [ESE-2002]

1.17 **Assertion (A):** SSB transmission is not appropriate for rectangular pulse transmission.
Reason (R): Whenever SSB modulating signal has abrupt transitions, the resulting envelope $A(t)$ gives rise to envelope horns.
(a) Both A and R are true and R is the correct explanation of A
(b) Both A and R are true but R is NOT the correct explanation of A
(c) A is true but R is false
(d) A is false but R is true
[ESE-2002]

1.18 Consider the following statements:
1. Intermediate-frequency (IF) amplifier is designed for a bandwidth of a few kHz at a centre frequency of a few hundred kHz for voice-modulated signals.
2. IF amplifier is designed for a frequency of 6 MHz at a centre frequency of 40 MHz for TV receivers.
3. Tuned amplifiers are used for IF amplifier.
Which of these statements is/are correct?
(a) 1 only (b) 3 only
(c) 2 and 3 (d) 1, 2 and 3
[ESE-2003]

1.19 Consider the following:
1. Filter system
2. Phase-shift method
3. Third method
4. Balanced modulator
Which of these can be used to remove unwanted side-bands in SSB?
(a) 1, 2 and 4 (b) 1, 2 and 3
(c) 2, 3 and 4 (d) 1, 3 and 4
[ESE-2003]

1.20 In a modulator it is found that the amplitude spectrum consists of a component at f_c, the carrier frequency and one component each at $f_c - f_s$ and $f_c + f_s$, where f_s is the modulating signal frequency. Then the modulator used is

(a) AM
(b) AM and/or Narrow band FM with depth ≤ 0.5
(c) FM
(d) PAM [ESE-2003]

1.21 The modulation index of an AM wave is changed from 0 to 1. The transmitted power is
(a) Unchanged
(b) Halved
(c) Increased by 50%
(d) Quadrupled [ESE-2003]

1.22 In an SSB transmitter one is most likely to find
(a) Class-*C* audio amplifier
(b) Tuned modulator
(c) Class-*B RF* amplifier
(d) Class-*AB* power amplifier
[ESE-2003]

1.23 In the circuit shown below, the transformers are centre tapped and the diodes are connected as shown in a bridge. Between the terminals 1 and 2 an A.C voltage source of frequency 400 Hz is connected. Another A.C voltage of 1.0 MHz is connected between 3 and 4. The output between 5 and 6 contains components at

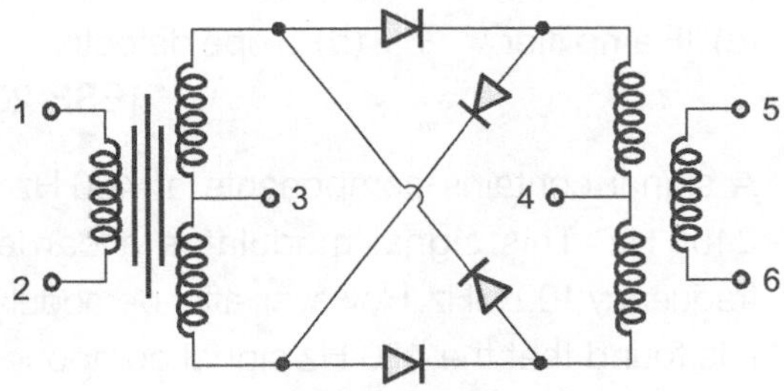

(a) 400 Hz, 1.0 MHz, 1000.4 kHz, 999.6 kHz
(b) 400 Hz, 1000.0 kHz, 999.6 kHz
(c) 1 MHz, 1000.4 kHz, 999.6 kHz
(d) 1000.4 kHz, 999.6 kHz
[ESE-2003]

1.24 A public broadcast system using amplitude modulation with double sideband should invariably have which one of the following features at the transmitter?
(a) Band limit the signal to 5 kHz
(b) Use a high frequency carrier oscillator
(c) use a stable crystal oscillator
(d) Use an all-round radiating transmitting antenna
[ESE-2003]

1.25 **Assertion (A):** Narrow band FM signal can be detected using an envelope detector.
Reason (R): Narrow band FM signal contains a carrier component and two sideband components.
(a) Both A and R are true and R is the correct explanation of A
(b) Both A and R are true but R is NOT the correct explanation of A
(c) A is true but R is false
(d) A is false but R is true [ESE-2003]

1.26 Which one of the following is represented by $v(t) = 5[\cos(10^6 \pi t) - \sin(10^3 \pi t) \times \sin(10^6 \pi t)]$?
(a) SSB under sideband signal
(b) DSB suppressed carrier signal
(c) AM signal
(d) Narrow band FM signal [ESE-2004]

1.27 Which one of the following statements is correct? In a ratio detector,
(a) linearity is worse than that of a phase discriminator
(b) stabilization is provided against signal strength variations
(c) the output is twice of that obtainable from a similar phase discriminator
(d) the circuit is same as that in a discriminator, except that the diode connections are reversed
[ESE-2004]

1.28 In a broadcast transmitter, the RF output is represented as $e(t) = 50[1 + 0.89 \cos 5000\, t + 0.30 \sin 9000\, t] \cos(6 \times 10^{10}\, t)$ volt.
What are the sidebands of the signals in radians?
(a) 5×10^3 and 9×10^3
(b) 5.991×10^6, 5.995×10^6, 6.005×10^6 and 6.009×10^6
(c) 4×10^3, 1.4×10^4
(d) 1×10^6, 1.1×10^7, 3×10^6 and 1.5×10^7
[ESE-2004]

1.29 **Assertion (A):** AM stereo broadcasting uses quadrature carrier multiplexing (QAM).
Reason (R): QAM involves more stringent synchronization than an FDM system.
(a) Both A and R are true and R is the correct explanation of A
(b) Both A and R are true but R is NOT the correct explanation of A
(c) A is true but R is false
(d) A is false but R is true [ESE-2004]

1.30 A composite signal $x_c(t)$ is expressed as:

$x_c(t) = A_c \cos \omega_c t - A_m \cos (\omega_c - \omega_m)t + A_m \cos (\omega_c + \omega_m)t$

Which of the following methods can be employed to retrieve the sinusoidal component at ω_m from $x_c(t)$?

(a) An envelop detector, square law detector
(b) Only a discriminator
(c) Only a square law detector
(d) Only an envelop detector [ESE-2005]

1.31 A circuit produces an output $y(t) = a + bx(t)^2$ where $x(t)$ is its input. This circuit can produce which one of the following?

(a) Rectified output
(b) Pulse modulation
(c) Amplitude modulation
(d) Frequency modulation [ESE-2005]

1.32 **Assertion (A):** The microprocessor based products used by the general consumer must comply FCC rules.

Reason (R): FCC regulates the use of radio transmission.

(a) Both A and R are true and R is the correct explanation of A
(b) Both A and R are true but R is NOT the correct explanation of A
(c) A is true but R is false
(d) A is false but R is true [ESE-2005]

1.33 Match **List-I** (Communication Service) with **List-II** (Bandwidth) and select the correct answer using the code given below the lists:

List-I	List-II
A. AM Broadcast	**1.** 10 kHz
B. Telephone	**2.** 4 kHz
C. Wideband FM	**3.** 200 kHz
D. TV	**4.** 7 MHz

Codes:

	A	B	C	D
(a)	1	2	3	4
(b)	3	4	1	2
(c)	1	4	3	2
(d)	3	2	1	4

[ESE-2006]

1.34 **Assertion (A):** The capture range of a PLL is always less than or equal to the lock range and is related to the cut-off frequency of the low-pass filter in the loop.

Reason (R): The capture range is range of frequencies, the VCO can produce.

(a) Both A and R are true and R is the correct explanation of A
(b) Both A and R are true but R is NOT the correct explanation of A
(c) A is true but R is false
(d) A is false but R is true [ESE-2006]

1.35 Match **List-I** (Modulation) with **List-II** (Power Input to Antenna in Watts) and select the correct answer using the code given below the lists:

List-I	List-II
A. 100% AM	**1.** 1.5
B. 50% AM	**2.** 1.125
C. 10% AM	**3.** 1.005
D. FM	**4.** 1.00

Codes:

	A	B	C	D
(a)	1	2	3	4
(b)	3	4	1	2
(c)	1	4	3	2
(d)	3	2	1	4

[ESE-2006]

1.36 Which one of the following blocks is not common in both AM and FM receivers?

(a) RF amplifier (b) Mixer
(c) IF amplifier (d) Slope detector

[ESE-2006]

1.37 A signal contains components at 400 Hz and 2400 Hz. This signal modulates a carrier of frequency 100 MHz. However, after demodulation it is found that the 400 Hz signal component is present. The channel BW is 15 kHz. What is the reason for the higher frequency signal not to be detected properly?

(a) Modulation used is FM and BW is insufficient
(b) Modulation used is AM and BW is insufficient
(c) Modulation used is FM but pre-emphasis is not used
(d) Modulation used is AM but detector is for FM

[ESE-2006]

1.38 In FDM systems used for telephone, which modulation scheme is adopted?

(a) AM (b) DSB-SC
(c) SSB-SC (d) FM

[ESE-2006]

1.39 FM broadcast standards specify a maximum deviation of frequency to be equal to 75 kHz and a maximum modulating frequency of 15 kHz. What is the modulation index for FM wave?

(a) 1/5 (b) 5
(c) 60 (d) 1125

[ESE-2007]

1.40 Why does an FM radio station perform better than an AM station radiating the same total power?

(a) FM is immune to noise
(b) AM has only two sidebands while FM has more
(c) FM uses larger bandwidth for large modulation depth
(d) Capture effect appears in FM

[ESE-2007]

1.41 On modulating a carrier of frequency f_c by an audio signal f_s, the following components have resulted:

$$f_c,\ f_c + f_s \text{ and } f_c - f_s$$

What is this type of modulation likely to be?

(a) Amplitude modulation DSB
(b) Single sideband modulation
(c) Frequency modulation only
(d) Amplitude modulation or frequency modulation

[ESE-2007]

1.42 A carrier signal at frequency 100 MHz is frequency modulated with modulation index 2 by a signal at 2 kHz. At what frequencies are the sidebands produced?

(a) 102 MHz and 98 MHz
(b) 100 MHz + 2 kHz, and 100 MHz – 2 kHz
(c) 100 MHz $\pm \Delta f_d$, where f_d is frequency deviation
(d) 100 MHz $\pm$ n 2 kHz, where n is 1, 2, ..., 5 integer

[ESE-2007]

1.43 The modulation index of an AM is changed from 0 to 1. How does the transmitted power change?

(a) Gets halved
(b) Gets doubled
(c) Gets increased by 50 percent
(d) Remains unchanged [ESE-2007]

1.44 Which one of the following is an indirect way of generating FM?

(a) Reactance FET modulator
(b) Varactor diode modulator
(c) Armstrong modulator
(d) Reactance tube modulator [ESE-2008]

1.45 Match **List-I** (Section) with **List-II** and select the correct answer using the code given below the lists:

List-I	List-II
A. IF amplifier	1. Demodulator
B. Mixer	2. Narrow band tuned amplifier
C. Detector	3. Variable gain amplifier
D. Automatic gain control	4. Frequency conversion

Codes:

	A	B	C	D
(a)	1	4	2	3
(b)	2	3	1	4
(c)	1	3	2	4
(d)	2	4	1	3

[ESE-2008]

1.46 A balanced modulator, is used in the generation of which of the following ?

(a) DSB-SC signal (b) FM signal
(c) PM signal (d) PAM signal

[ESE-2009]

1.47 An amplitude modulated signal occupies a frequency range from 395 kHz to 405 kHz. It can be demodulated by which of the following ?

(a) Using an envelope detector and filter
(b) Multiplying with a 395 kHz local signal
(c) Multiplying with a 405 kHz local signal
(d) Low pass filtering with cut off at 400 kHz

[ESE-2009]

1.48 Which one of the following statements is correct?

(a) AGC in radio receivers will keep the total signal output fairly constant but the noise component will be reduced; with the result the S/N ratio will improve.
(b) AGC in radio receivers is a linear operation with respect to both signals and noise
(c) Sudden changes in the output while tuning, which may cause damage to the components, is an avoidable feature of AGC
(d) AGC operation is independent of the filter components used in AGC circuit

[ESE-2009]

1.49 Consider the following statements with regard to a PLL

1. The 'capture range' of PLL primarily depends upon the slope of transition band edge of the low-pass filter.
2. The 'lock range' is not affected by the slope of transition band of the low-pass filter.

Which of these statements is/are correct?
(a) Both 1 and 2 (b) 1 only
(c) 2 only (d) Neither 1 nor 2
[ESE-2010]

1.50 Consider the following statements:
1. The lock range of PLL is the difference between the highest and lowest frequencies that the PLL can remain in lock onto.
2. The capture range of a PLL is the range of frequencies that the voltage controlled oscillator of a PLL can produce.
3. PLL can be used to synchronize the horizontal and vertical oscillators of TV receivers to incoming sync pulses.

Which of these statements is/are correct?
(a) 1 only (b) 3 only
(c) 1 and 3 only (d) 1, 2 and 3
[ESE-2010]

1.51 **Assertion (A):** Frequency modulation and phase modulation both produce different set of frequency bands for the same modulation depth.
Reason (R): Frequency modulation and phase modulation vary the carrier phase angle or its rate.
(a) Both A and R are true and R is the correct explanation of A
(b) Both A and R are true but R is NOT the correct explanation of A
(c) A is true but R is false
(d) A is false but R is true [ESE-2010]

1.52 **Assertion (A):** In amplitude modulation systems the value of modulation index should be around 1.
Reason (R): The power carried in the intelligence carrying sidebands increases with the modulation index.
(a) Both A and R are true and R is the correct explanation of A
(b) Both A and R are true but R is NOT the correct explanation of A
(c) A is true but R is false
(d) A is false but R is true [ESE-2010]

1.53 In a PLL
(a) Capture range = Lock range ≠ Free running frequency
(b) Capture range = Lock range = Free running frequency
(c) Capture range > Lock range
(d) Capture range < Lock range
[ESE-2011]

1.54 What bandwidth is needed for an FM signal that has a peak deviation of ±3 kHz and handles audio signals from 200 Hz to 5 kHz?
(a) 6 kHz (b) 16 kHz
(c) 10 kHz (d) 9.6 kHz [ESE-2011]

1.55 In a superheterodyne receiver arrange the following components sequentially from input to the output
1. Antenna
2. Mixer
3. IF amplifier
4. Audio amplifier

(a) 1, 2, 3 and 4 (b) 4, 2, 3 and 1
(c) 1, 3, 2 and 4 (d) 4, 3, 2 and 1
[ESE-2011]

1.56 **Statement (I):** FM spectrum consists of a carrier and an infinite number of side band components.
Statement (II): FM signals are used only in TV transmission to ensure better picture quality.
(a) Both Statement (I) and Statement (II) are individually true and Statement (II) is the correct explanation of Statement (I).
(b) Both Statement (I) and Statement (II) are individually true but Statement (II) is not the correct explanation of Statement (I).
(c) Statement (I) is true but Statement (II) is false.
(d) Statement (I) is false but Statement (II) is true.
[ESE-2012]

1.57 The threshold effect in demodulators is
(a) the rapid fall of output SNR when the input SNR falls below a particular value
(b) exhibited by all the demodulators when the input SNR is low
(c) exhibited by all AM suppressed carrier coherent demodulators
(d) exhibited by correlation receivers
[ESE-2012]

1.58 An amplitude modulated signal is $[A + 0.5\,A\cos\omega_m t]\cos\omega_c t$ where ω_m and ω_c are respectively, modulating and carrier frequencies. The power efficiency is
(a) 11.11% (b) 0.25%
(c) 4.32% (d) 50% [ESE-2012]

1.59 The signal $x(t) = \sin(200\pi t) + 2\sin(400\pi t)$ is modulated to produce a signal
$$g(t) = x(t)\sin(400\pi t).$$

This is passed through a low pass filter having a cut-off frequency of 400 π Hz and passband gain of 2. Then the output signal after the filter is
(a) 0 (b) 0.5 sin $(200\pi t)$
(c) sin $(200\pi t)$ (d) 2 sin $(200\pi t)$
[ESE-2012]

1.60 What is the ratio of modulating power to the total power at 100% AM modulation?
(a) 1 : 3 (b) 1 : 2
(c) 2 : 3 (d) 1 : 4
[ESE-2013]

1.61 A bandwidth of 10 kHz is required for AM system. If the lowest frequency component in the modulated signal is 555 kHz, carrier frequency in kHz is
(a) 525 (b) 550
(c) 560 (d) 565 [ESE-2013]

1.62 Determine the bandwidth occupied by a sinusoidal frequency modulated carrier for which the modulation index is 2.4.
(a) 4.8 f_m (b) 6.8 f_m
(c) 2.4 f_m (d) 3.8 f_m [ESE-2013]

1.63 An AM signal with a carrier of 1 kW has 200 W in each side band. The percentage of modulation is
(a) 20% (b) 89.4%
(c) 49.7% (d) 40% [ESE-2013]

1.64 An audio signal, 15sin$(2\pi\, 1500t)$ amplitude modulates 60sin$(2\pi\, 10^3 t)$. The modulation index will be
(a) 20% (b) 50%
(c) 25% (d) 100% [ESE-2013]

1.65 One of the main functions of the *RF* amplifiers in a super-heterodyne receiver is to
(a) provide improved tracking
(b) permit better adjacent channel rejection
(c) increase the tuning range of the receiver
(d) improve the rejection of the image frequency
[ESE-2014]

1.66 An FM signal has a carrier swing of 100 kHz when modulating signal has a frequency of 8 kHz. The modulation index is
(a) 12.5 (b) 7.5
(c) 6.25 (d) 15
[ESE-2014]

1.67 If a 400 Watt carrier is amplitude modulated to a depth of 75 percent, what is the total power in the modulated wave?
(a) 517.5 W (b) 463.3 W
(c) 448.5 W (d) 512.5 W [ESE-2014]

1.68 The signal $m(t)$ = sin$c(2 \times 10^4 t)$ is frequency modulated with $K = 10^3$ Hz/V. What is the maximum instantaneous frequency of the modulated signal when carrier frequency is 1 MHz?
(a) 0.999 MHz (b) 0.998 MHz
(c) 1.002 MHz (d) 1.001 MHz
[ESE-2014]

1.69 Amplitude modulation is used for broadcasting because
(a) it is more noise immune than other modulation systems.
(b) compared with other system it require less transmitting power.
(c) its use avoids receiver complexity
(d) no other modulation system can provide the necessary BW for high fidelity.
[ESE-2014]

1.70 In a superheterodyne receiver, if the intermediate frequency is 450 kHz, and the signal frequency is 1000 kHz, then the local oscillator frequency and image frequency respectively are
(a) 1450 kHz and 100 kHz
(b) 550 kHz and 1900 kHz
(c) 1450 kHz and 1900 kHz
(d) 550 kHz and 1450 kHz [ESE-2014]

1.71 What will be the total modulation index if a wave is amplitude modulated by three sine waves with modulation indices of 25%, 50% and 75%?
(a) $M_t = 1.5$ (b) $M_t = 0.93$
(c) $M_t = 1.22$ (d) $M_t = 1$ [ESE-2014]

1.72 Boosting of higher frequency at the transmitter is done by using
(a) De-emphasis (b) AGC circuit
(c) Pre-emphasis (d) Armstrong method
[ESE-2014]

1.73 The power contained in single sideband in amplitude modulation is
(a) $\dfrac{m^2 P_c}{2}$ (b) $2\, m^2 P_c$
(c) $\dfrac{m^2 P_c}{4}$ (d) $4\, m^2 P_c$ [ESE-2014]

1.74 The selectivity of tuned radio frequency (TRF) receiver is poor because
1. Q-factor requirement of tuned circuits in RF amplifier is not fixed.
2. Q-required for upper side of short wave is 2000.
3. Q-factor is zero.

Which of the above statements are correct?
(a) 1, 2 and 3 (b) 1 and 2 only
(c) 1 and 3 only (d) 3 only [ESE-2015]

1.75 For an AM signal, the bandwidth is 20 kHz and the highest frequency component present is 800 kHz. The carrier frequency used for this AM signal is
(a) 710 kHz (b) 705 kHz
(c) 700 kHz (d) 790 kHz
[ESE-2015]

1.76 The aerial current of an AM transmitter is 18 A when unmodulated, but increases to 20 A when modulated. The modulation index is
(a) 0.68 (b) 0.73
(c) 0.89 (d) 0.98 [ESE-2015]

1.77 A carrier wave is phase modulated with frequency deviation of 20 kHz by a single tone frequency of 2 kHz. If the single tone frequency is increased to 2 kHz, assuming that phase deviation remains unchanged, bandwidth of the PM signal is
(a) 20 kHz (b) 32 kHz
(c) 22 kHz (d) 44 kHz [ESE-2015]

1.78 A sinusoidal audio signal is given by $e_s = 15 \sin 2\pi(2000t)$. It modulates a sinusoidal carrier wave $e_c = 60 \sin 2\pi(10^5 t)$. The value of amplitude modulation index must be
(a) 0.5 (b) 4
(c) 0.6 (d) 0.25 [ESE-2015]

1.79 A signal of maximum frequency of 2 MHz modulated by 4 GHz carrier is DSB/SC. If the resultant signal is sampled ideally as low pass signal, the minimum sampling frequency should be
(a) 4 MHz (b) 8 MHz
(c) 4.004 GHz (d) 8.008 GHz
[ESE-2015]

1.80 For an angle modulated signal,

$$x(t) = 6 \cos [2\pi \times 10^6 t + 2 \sin (8000\ \pi t) + 4 \cos (8000\ \pi t)] \text{ V}.$$

The average power of $x(t)$ is
(a) 10 W (b) 18 W
(c) 20 W (d) 28 W [ESE-2015]

1.81 A single tone 4 kHz message signal is sampled with 10 kHz and 6 kHz. Aliasing effect will be seen in the reconstructed signal when the signal is sampled with
(a) 10 kHz
(b) 6 kHz
(c) Both 10 kHz and 6 kHz
(d) Neither 10 kHz nor 6 kHz [ESE-2015]

1.82 A modulated signal is given by
$s(t) = e^{-at}\cos[(\omega_c + \Delta\omega)t]u(t)$.
The complex envelop of $s(t)$ is
(a) $e^{-at}e^{[j(\omega_c+\Delta\omega)t]}u(t)$ (b) $e^{-at}e^{[j\Delta\omega t]}u(t)$
(c) $e^{[j\Delta\omega t]}u(t)$ (d) $e^{[j(\omega_c+\Delta\omega)t]}u(t)$
where a, ω_c and $\Delta\omega$ are positive constants and $\omega_c >> \Delta\omega$.
[ESE-2015]

1.83 **Statement (I):** The bandwidth of AM signal depends on the bandwidth of the modulating signal.
Statement (II): The amplitude-modulated signal contains f_c, $f_c \pm f_m$ frequency signals.
(a) Both Statement (I) and Statement (II) are individually true and Statement (II) is the correct explanation of Statement (I)
(b) Both Statement (I) and Statement (II) are individually true but Statement (II) is **NOT** the correct explanation of Statement (I)
(c) Statement (I) is true but Statement (II) is false
(d) Statement (I) is false but Statement (II) is true
[ESE-2015]

1.84 A carrier waveform 10 cos $\omega_c t$ and modulating signal 3 cos $\omega_m t$ have f_c = 100 kHz and f_m = 4 kHz. Given that sensitivity of FM is 4 kHz/V and FM spectra beyond J_6 is negligible, what are the channel bandwidth requirements for AM and FM, respectively?
(a) 12 kHz and 48 kHz
(b) 8 kHz and 48 kHz
(c) 12 kHz and 24 kHz
(d) 8 kHz and 24 kHz [ESE-2016]

1.85 When the modulating frequency is doubled, the modulation index is halved, and the modulating voltage remains constant. The modulation system is
1. Amplitude modulation
2. Phase modulation

3. Frequency modulation

Select the correct answer from the codes given below:

(a) 1 only (b) 2 only
(c) 3 only (d) 1, 2 and 3

[ESE-2016]

1.86 What is the modulation index of an FM signal having a carrier swing of 100 kHz and modulating frequency of 8 kHz?

(a) 4.75 (b) 5.50
(c) 6.25 (d) 7.50 [ESE-2016]

1.87 In the process of modulation

(a) Some characteristics of a high frequency sine wave are varied in accordance with the instantaneous value of a low frequency signal
(b) Parameters of carrier wave are held constant
(c) For proper and efficient radiation, the receiving antennas should have heights comparable to half-wavelength of the signal received
(d) The signal is converted first within the range of 10 Hz to 20 Hz

[ESE-2016]

1.88 In communication systems, modulation is the process of

(a) Improving frequency stability of transmitter
(b) Combining signal and radio frequency waves
(c) Generating constant frequency radio waves
(d) Reducing distortion in RF waves

[ESE-2016]

1.89 The Vestigial Side Band (VSB) modulation is preferred in TV systems because

1. It reduces the bandwidth requirement to half
2. It avoids phase distortion at low frequencies

Which of the above statement(s) is/are correct?

(a) 1 only (b) 2 only
(c) Both 1 and 2 (d) Neither 1 nor 2

[ESE-2016]

1.90 An FM broadcasting radio station transmits signals of frequency 100 MHz with a power of 10 kW. The bandwidth of the modulation signals is from 100 Hz to 1.5 kHz. If the maximum deviation set by the FCC, (δ), is 75 kHz, the range of the modulation index is

(a) 100 to 750 (b) 100 to 250
(c) 50 to 750 (d) 50 to 250

[ESE-2017]

1.91 An amplitude-modulated amplifier has a radio frequency output of 60 W at 100% modulation. The internal loss in the modulator is 6 W. What is the unmodulated carrier power?

(a) 33 W (b) 36 W
(c) 40 W (d) 44 W [ESE-2017]

1.92 The figure shows the block diagram of a frequency discriminator. What does the second block represent?

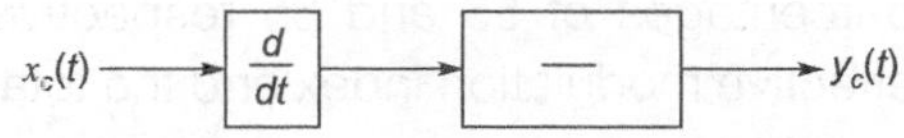

(a) Envelope detector (b) Low-pass filter
(c) Ratio detector (d) Band-reject filter

[ESE-2017]

1.93 A receiver tunes signals from 550 kHz to 1600 kHz with an IF of 455 kHz. The frequency tuning range ratio for the oscillator section of the receiver is nearly

(a) 2.90 (b) 2.05
(c) 1.65 (d) 1.30 [ESE-2017]

1.94 The phase detector circuit in the phase-locked loop demodulators recognizes

(a) voltage changes between the input and VCO signals
(b) frequency changes between the input and VCO signals
(c) impedance changes between the input and VCO signals
(d) resistance changes between the input and VCO signals

[ESE-2018]

1.95 A signal $m(t) = 10\cos(2\pi 100t)$ is frequency-modulated. The resulting FM signal is

$$x(t) = 20\cos\{2\pi 10^6 t + 15\sin(2\pi 100t)\}$$

The FM bandwidth is nearly

(a) 3.2 kHz (b) 9.6 kHz
(c) 32 kHz (d) 1000 kHz

[ESE-2018]

1.96 The minimum value of modulation index β for an FM system required to produce a noticeable improvement in SNR over a comparable AM system with $\mu = 1$ is

(a) 0.61 (b) 0.52
(c) 0.47 (d) 0.38 [ESE-2018]

1.97 An AM wave with modulation index 0.8 has total sideband power of 4.85 kW. The carrier power and the total power radiated will be nearly
(a) 12.2 kW and 20 kW
(b) 15.2 kW and 20 kW
(c) 12.2 kW and 25 kW
(d) 15.2 kW and 25 kW **[ESE-2019]**

1.98 A 360 W carrier is simultaneously modulated by two audio waves with modulation percentages of 55 and 65 respectively. The effective modulation index and the total power radiated are
(a) 0.85 and 490.5 W (b) 0.65 and 490.5 W
(c) 0.85 and 450.5 W (d) 0.65 and 450.5 W
[ESE-2019]

1.99 An amplitude modulated amplifier has a radio frequency output of 50 W at 100% modulation. The internal loss in the modulator is 10 W. The unmodulated carrier power is
(a) 40 W (b) 50 W
(c) 60 W (d) 80 W **[ESE-2019]**

1.100 For a PM modulator with a deviation sensitivity K = 2.5 rad/V and a modulating signal $v_m(t) = 2\cos(2\pi 2000t)$, the peak phase deviation m will be
(a) 1.25 rad (b) 2.5 rad
(c) 5.0 rad (d) 7.5 rad
[ESE-2019]

1.101 When unmodulated carrier alone is transmitted, the antenna current is 9A. When sinusoidal modulation is present, the antenna current is 11 A. The modulation index used will be nearly
(a) 0.994 (b) 0.764
(c) 0.546 (d) 0.326 **[ESE-2020]**

1.102 Frequency modulated signal with single-tone modulation-has a frequency deviation of 15 kHz and bandwidth of 50 kHz. The frequency of the modulating signal will be
(a) 05 kHz (b) 10 kHz
(c) 20 kHz (d) 30 kHz
[ESE-2020]

1.103 When the carrier and one of the sidebands are suppressed in an AM wave modulated to a depth of 50%, the power saving will be
(a) 84.4% (b) 88.6%
(c) 94.4% (d) 98.6% **[ESE-2020]**

1.104 An output of balanced modulator contains
(a) Carrier, LSB and USB
(b) Modulation frequency, carrier frequency and LSB
(c) Modulation frequency, carrier frequency and USB
(d) Modulation frequency, LSB and USB
[ESE-2020]

1.105 The antenna current of an AM transmitter is 8 A when only the carrier signal is transmitted. What is the antenna current when the carrier signal is modulated by sinusoidal signal $V(t) = 1.4 \sin(2\pi \times 500t)$ with modulation index 0.8?
(a) 3.2 A (b) 7.2 A
(c) 9.2 A (d) 11.2 A
[ESE-2021]

1.106 A 10 MHz carrier signal is frequency-modulated by analog-modulating signal. The maximum frequency deviation is 75 kHz. If the frequency of the modulating signal is 300 kHz, then the modulation index and the approximate transmission bandwidth of the FM signal are respectively,
(a) 0.25, 750 kHz (b) 0.25, 600 kHz
(c) 0.75, 750 kHz (d) 0.75, 450 kHz
[ESE-2021]

1.107 Which one of the following statements is **not** correct?
(a) The noise performance of a full AM receiver is always inferior to that of a DSB-SC receiver.
(b) The figure of merit of a DSB-SC receiver using coherent detection is always unity.
(c) The figure of merit of an SSB receiver using coherent detection is always unity.
(d) The figure of merit of an AM receiver using envelope detection is always greater than unity.
[ESE-2022]

1.108 The RF carrier range and IF bandwidth corresponding to AM radio receiver are respectively
(a) 0.535 MHz-1.605 MHz and 100 kHz
(b) 53.5 MHz-160.5 MHz and 10 kHz
(c) 0.535 MHz-1.605 MHz and 10 kHz
(d) 53.5 MHz-160.5 MHz and 100 kHz
[ESE-2022]

1.109 For the same average transmitted or modulated signal power and the same average noise power in the message bandwidth, a coherent SSB receiver will have exactly the same output signal-to-noise ratio as
(a) a coherent DSB-SC receiver
(b) an AM receiver
(c) a VSB receiver
(d) an SSB receiver [ESE-2022]

1.110 Identify the given circuit shown in the figure.

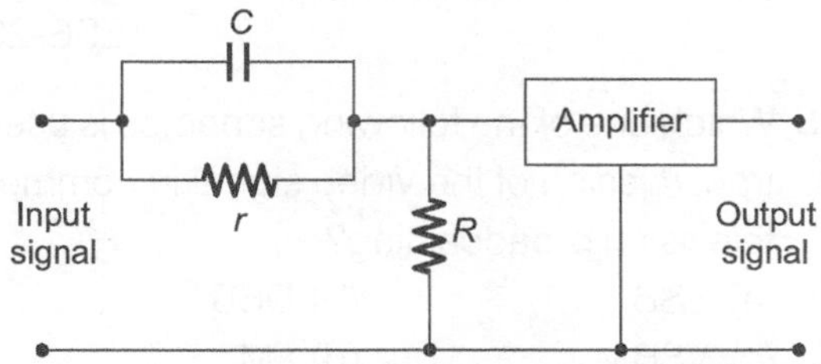

(a) Band-pass filter (b) De-emphasis filter
(c) Pre-emphasis filter (d) Band-reject filter
[ESE-2022]

1.111 The power transmitted by an SSB transmitter is 20 kW. It is required to be replaced by standard AM transmission having modulation index of 0.4 and same power. What is the transmission efficiency?
(a) 3.7% (b) 5.8%
(c) 7.4% (d) 21.6% [ESE-2023]

1.112 An angle modulated signal is given as

$x_c(t) = 20\cos\left[200\pi t + \frac{\pi}{4}\right]$. What is the instantaneous frequency?
(a) 50 Hz (b) 100 Hz
(c) 200 Hz (d) 400 Hz
[ESE-2023]

1.113 An FM modulator operates at carrier frequency of 250 kHz with frequency deviation sensitivity of 1.5 kHz/V. A PM modulator operates at carrier frequency of 500 kHz with phase deviation sensitivity of 1.5 rad/V. If both FM and PM modulators are modulated by the same modulating signal having peak amplitude of 5 V and modulating frequency of 5 kHz, then what is the relationship between frequency modulation index and phase modulation index?
(a) PM = FM (b) PM = 2FM
(c) PM = 4FM (d) PM = 5FM
[ESE-2023]

1.114 What is the relationship between the percentage efficiency saving when the carrier wave and one of the sidebands are suppressed in an AM wave modulated to a depth of 100% modulation index?
(a) $\eta_{DSB} = 2.5\eta_{AM}$ (b) $\eta_{DSB} = 4\eta_{AM}$
(c) $\eta_{DSB} = 5\eta_{AM}$ (d) $\eta_{DSB} = 2\eta_{AM}$
[ESE-2023]

1.115 An FM audio signal with single-tone modulation has a frequency deviation of 25 kHz and a bandwidth of 75 kHz. What is the frequency of the modulating signal using Carson's rule?
(a) 12.5 kHz (b) 25 kHz
(c) 50 kHz (d) 75 kHz
[ESE-2023]

1.116 An audio signal $s(t)$ is normalized, whose Fourier transform $S(f)$ is shown in the figure, so that $|s(t)| \leq 1$. This signal is to be transmitted using FM with a frequency deviation constant $k_f = 90$ kHz/V. What is the bandwidth required for transmission of the FM audio signal?

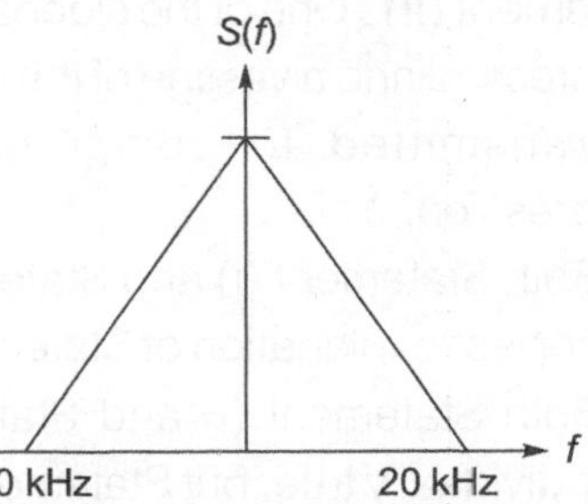

(a) 140 kHz (b) 180 kHz
(c) 220 kHz (d) 260 kHz
[ESE-2023]

1.117 A collector modulated class-C power amplifier is giving an amplitude modulated signal of 220 W average power at the output, while operating with a collector circuit efficiency of 40%. What is the power to be supplied by the modulating amplifier when the modulation index is 0.4?
(a) 16.3 W (b) 40.75 W
(c) 203.7 W (d) 220 W
[ESE-2023]

1.118 By considering standard notations, the normalized power of the AM signal is

(a) $S^2(t) = \frac{1}{2}A_c^2 + \frac{1}{2}A_c^2[m^2(t)]$

(b) $S^2(t) = A_c^2 + \frac{1}{2}A_c^2[m^2(t)]$

(c) $S^2(t) = \frac{1}{2}A_c^2 + A_c^2[m^2(t)]$

(d) $S^2(t) = \frac{1}{4}A_c^2 + \frac{1}{4}A_c^2[m^2(t)]$

[ESE-2023]

1.119 A certain AM transmitter is radiating 125 kW when a certain audio sine wave is modulating it to a depth of 70% and 144.5 kW when a second sinusoidal audio wave also modulates it simultaneously. What is the depth of the modulation for the second audio wave?

(a) $\sqrt{0.4}$ (b) $\sqrt{0.3}$

(c) $\sqrt{0.2}$ (d) $\sqrt{0.1}$

[ESE-2023]

1.120 **Statement (I) :** A popular method for generating a VSB modulated wave is to use the frequency discrimination method.

Statement (II) : One of the sidebands is partially suppressed and a vestige of the other sideband is transmitted to compensate for that suppression.

(a) Both Statement (I) and Statement (II) is the correct explanation of Statement (I).

(b) Both Statement (I) and Statement (II) are individually true, but Statement (II) is not the correct explanation of Statement (I).

(c) Statement (I) is true, but Statement (II) is false.

(d) Statement (I) is false, but Statement (II) is true

[ESE-2023]

1.121 Which one of the following mid-band frequency of IF section is used in FM radio?

(a) 100.7 MHz (b) 10.7 MHz

(c) 0.455 MHz (d) 455 MHz

[ESE-2024]

1.122 Consider the following statements

1. DSB-SC, SSB and VSB are examples of linear modulation
2. DSB-SC modulation is well suited for point-to-point communication.
3. SSB-SC modulation scheme is less affected by frequency selective fading

Which of the above statements is/are correct?

(a) 1 only (b) 2 and 3 only

(c) 1 and 3 only (d) 1, 2 and 3

[ESE-2024]

1.123 Which one of the following schemes is used for transmission of the video signal in commercial television broadcasting?

(a) SSB (b) DBB

(c) VSB (d) FM

[ESE-2024]

1.124 A carrier is frequency modulated with a sinusoidal signal of 2 kHz, resulting in a maximum frequency deviation of 5 kHz. The bandwidth of the modulated signal is

(a) 14 kHz (b) 24 kHz

(c) 10 kHz (d) 48 kHz

[ESE-2024]

1.125 In angle modulation, which of the following two factors limit the value of the modulation index β?

(a) Channel bandwidth and received power

(b) Pre-emphasis and de-emphasis

(c) Received noise power and signal power

(d) Pre-emphasis and received noise power

[ESE-2024]

■■■

Answers Analog Communication Systems

1.1 (d)	1.2 (b)	1.3 (a)	1.4 (c)	1.5 (d)	1.6 (a)	1.7 (a)	1.8 (a)	1.9 (b)
1.10 (b)	1.11 (a)	1.12 (c)	1.13 (d)	1.14 (d)	1.15 (b)	1.16 (c)	1.17 (a)	1.18 (d)
1.19 (b)	1.20 (b)	1.21 (c)	1.22 (c)	1.23 (d)	1.24 (a)	1.25 (d)	1.26 (d)	1.27 (b)
1.28 (b)	1.29 (b)	1.30 (b)	1.31 (c)	1.32 (d)	1.33 (a)	1.34 (c)	1.35 (a)	1.36 (d)
1.37 (c)	1.38 (c)	1.39 (b)	1.40 (a)	1.41 (d)	1.42 (d)	1.43 (c)	1.44 (c)	1.45 (d)
1.46 (a)	1.47 (a)	1.48 (c)	1.49 (a)	1.50 (c)	1.51 (d)	1.52 (a)	1.53 (d)	1.54 (b)
1.55 (a)	1.56 (c)	1.57 (a)	1.58 (a)	1.59 (*)	1.60 (a)	1.61 (c)	1.62 (b)	1.63 (b)
1.64 (c)	1.65 (d)	1.66 (c)	1.67 (d)	1.68 (d)	1.69 (c)	1.70 (c)	1.71 (b)	1.72 (c)
1.73 (c)	1.74 (b)	1.75 (d)	1.76 (a)	1.77 (d)	1.78 (d)	1.79 (b)	1.80 (b)	1.81 (b)
1.82 (b)	1.83 (a)	1.84 (b)	1.85 (c)	1.86 (c)	1.87 (a)	1.88 (b)	1.89 (a)	1.90 (c)
1.91 (d)	1.92 (a)	1.93 (b)	1.94 (b)	1.95 (a)	1.96 (c)	1.97 (b)	1.98 (a)	1.99 (a)
1.100 (c)	1.101 (a)	1.102 (b)	1.103 (c)	1.104 (d)	1.105 (c)	1.106 (b)	1.107 (d)	1.108 (c)
1.109 (a, d)	1.110 (c)	1.111 (c)	1.112 (b)	1.113 (d)	1.114 (d)	1.115 (a)	1.116 (c)	1.117 (b)
1.118 (a)	1.119 (a)	1.120 (d)	1.121 (b)	1.122 (d)	1.123 (c)	1.124 (a)	1.125 (a)	

Explanations Analog Communication Systems

1.1 (d)

$$P_t = P_c\left(1+\frac{m^2}{2}\right)$$

$$600 = 400\left(1+\frac{m^2}{2}\right)$$

$$\Rightarrow \quad \frac{m^2}{2} = \frac{6}{4}-1 = \frac{1}{2}$$

$$\Rightarrow \quad m = 1$$

1.2 (b)

The loss of a message in an envelope detector that operates at a low carrier-to-noise ratio is referred to as the threshold effect. By threshold we mean a value of carrier-to-noise ratio below which the noise performance of a detector deteriorates much more rapidly than proportionately to the CNR.

1.3 (a)

Minimum sampling frequency,

$$f_{s_{\min}} = \frac{2f_H}{k} \qquad \text{where, } k = \left\lfloor \frac{f_H}{f_H - f_L} \right\rfloor 1$$

where $\lfloor x \rfloor$ means the greatest integer lower than x

$$k = \left\lfloor \frac{2}{2-1.5} \right\rfloor = \left\lfloor \frac{2}{0.5} \right\rfloor = 4$$

$$f_{s_{\min}} = \frac{2\times 2}{4} = 1 \text{ MHz}$$

1.4 (c)

$$f_m = 100 \text{ Hz and } m_f = 15$$

$$\text{BW} = 2(m_f + 1)f_m = 2(15+1)\times 100 = 3.2 \text{ kHz}$$

1.5 (d)

The block diagram of a PLL is shown below:

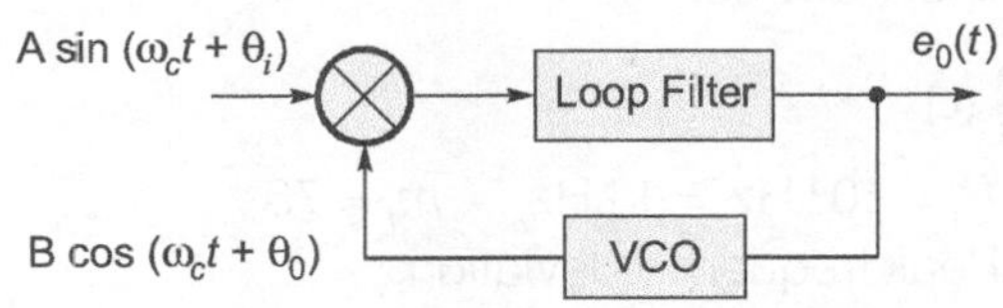

Loop filter is a low pass narrow band filter.

1.6 (a)

Modulation index,

$$m = \frac{V_{\max} - V_{\min}}{V_{\max} + V_{\min}} = \frac{10-5}{10+5} = \frac{5}{15} = 0.33$$

1.7 (a)

$$f_m = \frac{10\pi}{2\pi} = 5 \text{ Hz}$$

f_c of LPF should be $5 < f_c < 9$ Hz

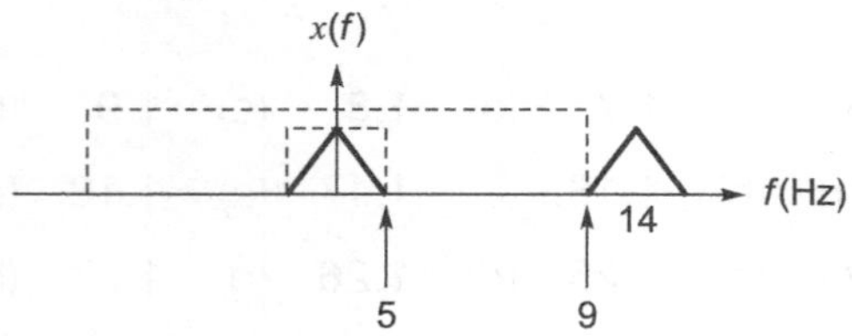

1.8 (a)

$(S/N)_{WBFM}$ at $m = 100\%$ modulation is

$$= \frac{3}{2}\beta^2 = \frac{3}{2}m_f^2$$

(m_f = modulation index of FM)

and $(S/N)_{AM}$ at 100% modulation = η = 1/3

$$\therefore \quad \frac{(S/N)_{WBFM}}{(S/N)_{AM}} = \frac{\frac{3}{2}m_f^2}{1/3} = \frac{9}{2}m_f^2$$

1.9 (b)

When the resultant AM signal passed through the given filter, the output of the filter will be the sinusoidal carrier signal with 90° phase shift. When this resultant sinusoidal carrier signal is passed through an envelope detector, the output of the envelope detector will be a DC signal.

1.10 (b)

$$P_T = P_c\left(1+\frac{m^2}{2}\right) \Rightarrow 10 = P_c\left(1+\frac{0.6^2}{2}\right)$$

$$\Rightarrow \quad P_C = 8.47 \text{ kW}$$

1.11 (a)

In a low-level AM system, the amplifier must be linear device.

1.12 (c)

$f_m = 10^3$ Hz = 1 kHz ; $m_f = 75$
Peak frequency deviation,

$$\delta = m_f \cdot f_m = 75 \text{ kHz}$$

1.13 (d)

The main common modulation technique used in telegraphy is Frequency Shift Keying (FSK). In telegraphy, the transmitting teletype writer produces a set of coded pulse when a given key is tapper. So, we can say, it is based on the Frequency Modulation (FM).

1.14 (d)

BW in FM is larger than that in AM.

1.18 (d)

Note:

(i) TV receivers in the VHF band (54 to 223 MHz) and in the UHF band (470 to 940 MHz) use an IF between 26 and 46 MHz, with approximately 36 and 46 MHz the two most popular values.

(ii) The IF amplifier is a fixed-frequency amplifier, with the very important function of rejecting adjacent unwanted frequencies.
Whereas FET and integrated circuit IF amplifiers generally are double-tuned at the input and at the output, BJT amplifiers often are single-tuned.

1.19 (b)

There are three main methods to remove unwanted sideband in SSB:

1. Filter method
2. Phase shift method
3. Third method

Note: Balanced modulator is used to suppress the carrier.

1.20 (b)

The similarity in AM and NBFM is that both have three frequency components: Carrier frequency component f_c and two sidebands at $f_c \pm f_s$.
Note: FM has infinite sidebands at $f_c \pm nf_m$, $n = 1, 2,...$

1.21 (c)

When $m = 0$, transmitted power is

$$P_{T_1} = P_C\left(1+\frac{0}{2}\right) = P_C$$

When $m = 1$, transmitted power is

$$P_{T_2} = P_c\left(1+\frac{1^2}{2}\right) = \frac{3P_C}{2}$$

% changed in transmitted power,

$$\%\Delta P_T = \frac{P_{T_2} - P_{T_1}}{P_{T_1}} \times 100 = \frac{3\frac{P_C}{2} - P_C}{P_C} \times 100 = 50\%$$

Thus, when the modulation index of AM wave is changed from 0 to 1, the transmitted power is increased by 50%.

1.23 (d)

The given circuit is Balanced Modulator. So carrier frequency will be suppressed and there will remain only two sidebands at $f_c + f_m$ and $f_c - f_m$.

Here, f_c = 1 MHz = 100 kHz

f_m = 400 Hz = 0.4 kHz

So, output contains 1000.4 kHz, 999.6 kHz

1.26 (d)

$v(t) = 5[\cos(10^6 \pi t) - \sin(10^3 \pi t) \cdot \sin(10^6 \pi t)]$

$v(t) = 5 \cos(10^6 \pi t) - 5/2 \quad [\cos(10^6 \pi t - 10^3 \pi t)]$

$+ 5/2 [\cos(10^6 \pi t + 10^3 \pi t)]$

which is an NBFM signal.

1.27 (b)

Ratio detector have worse linearity than discriminators.

1.28 (b)

Sidebands are $\omega_c \pm \omega_{m_1}$ and $\omega_c \pm \omega_{m_2}$

i.e., $6 \times 10^{10} \pm 5 \times 10^3$ and $6 \times 10^{10} \pm 9 \times 10^3$.

So, no answer is correct.

If $\omega_c = 6 \times 10^6$, then sidebands are

$6 \times 10^{10} \pm 5 \times 10^3$ and $6 \times 10^6 \pm 9 \times 10^3$ or

6.005×10^6, 5.995×10^6, 6.009×10^6 and 5.991×10^6.

1.30 (b)

It is an NBFM signal. NBFM signal can be demodulated by discriminator.

1.31 (c)

This is an example of nonlinear modulators to produce Amplitude Modulation.

1.35 (a)

For AM, $P_t = P_c\left(1 + \frac{m^2}{2}\right)$

For FM, $P_t = P_c$

(Here it is assumed that P_c = 1 W)

1.36 (d)

FM receivers contain slope detector but AM receivers do not contain that.

1.37 (c)

In FM, the amplitude of higher frequency components reduces significantly. So the high frequency components are lost if not boosted before transmission. This boosting of frequency signals is called pre-emphasis.

1.39 (b)

Modulation index,

$$m_f = \frac{\text{Maximum deviation of frequency } (\delta)}{\text{Maximum modulating frequency } (f_m)}$$

Given that δ = 75 kHz

and f_m = 15 kHz

$$\therefore \quad m_f = \frac{75}{15} = 5$$

1.40 (a)

FM receivers can be fitted with amplitude limiters to remove the amplitude variations caused by noise. This makes FM reception a good deal more immune to noise than AM reception.

1.41 (d)

The type of modulation is either amplitude modulation (DSB) or Narrow-band frequency modulation.

1.42 (d)

Expression for WBFM is

$$A_c \sum_{n=-\infty}^{\infty} J_n(\beta) \cos 2\pi(f_c \pm nf_m)t$$

$\therefore$ Side bands are present at $f_c \pm nf_m$

Given in question; Af_c = 100 MHz

f_m = 2 kHz

$\therefore$ 100 MHz + n2 kHz, where n = 1, 2, ⋯

1.46 (a)

Balanced modulator is used to suppress the carrier.

1.49 (a)

The LPF controls the capture range while VCO generates locking range.

1.52 (a)

$$\boxed{P_{SB} = \frac{m^2}{2} P_C}$$

1.53 (d)

The capture range is always less than or equal to the locking range.

1.54 (b)

Δf = 3 kHz

f_m = 5 kHz

So bandwidth of FM signal

$= 2(\Delta f + f_m) = 2(3 + 5) = 16$ kHz.

1.55 (a)

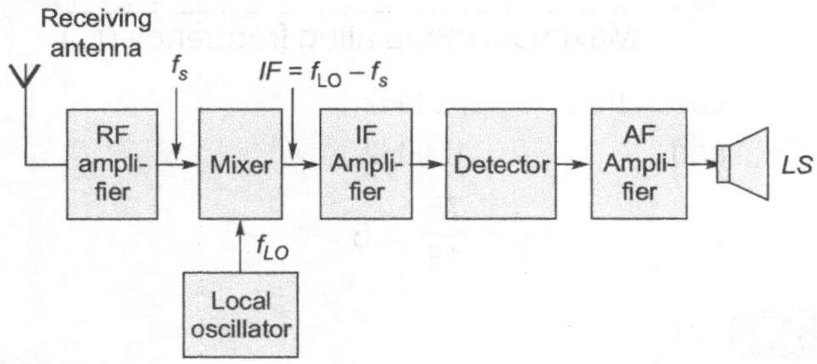

1.56 (c)

FM is used for audio transmission. VSB is for video transmission.

1.58 (a)

Amplitude modulated signal is

$$x_{AM}(t) = (A + 0.5A\cos\omega_m t)\cos\omega_c t$$
$$= A(1 + 0.5\cos\omega_m t)\cos\omega_c t$$

Now, compare this equation with the standard amplitude modulated signal equation i.e.

$$x_{AM}(t) = A_C(1 + m_a\cos\omega_m t)\cos\omega_c t.$$

$\therefore$ $m_a = 0.5 \rightarrow$ Modulation index.

$$\text{Power efficiency} = \frac{\text{Total side band power}}{\text{Total transmitted power}}$$

$$= \frac{\frac{P_c m^2}{2}}{P_c\left(\frac{1+m^2}{2}\right)} = \frac{m^2}{2+m^2} = \frac{(0.5)^2}{2+0.5^2} \times 100 = 11.11\%$$

1.59 (*)

$$x(t) = \sin(200\pi t) + 2\sin(400\pi t)$$
$$g(t) = x(t)\sin(400\pi t)$$
$$g(t) = [\sin(200\pi t) + 2\sin(400\pi t)]\sin(400\pi t)$$
$$= \frac{1}{2}[\cos 200\pi t - \cos 600\pi t] + 1 - \cos 800\pi t$$

Now this signal is passed through LPF having cut off frequency of 400π Hz = 1256.64 Hz and pass band gain of 2.

So, output signal = $2g(t)$

None of the options is matching.

1.60 (a)

$$\text{Total power, } P_T = P_c\left(1 + \frac{m^2}{2}\right)$$

Ratio of modulating power to total power

$$= \frac{P_c\frac{m^2}{2}}{P_c\left(1+\frac{m^2}{2}\right)} = \frac{m^2}{2+m^2}$$

at $m = 1$, Ratio = $\frac{1}{3}$

or 1 : 3

1.61 (c)

AM modulated signal in frequency domain.

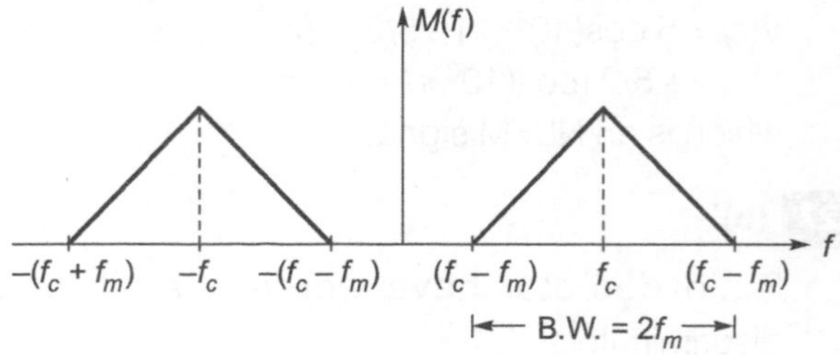

Lowest frequency component = $f_c - f_m$

$= 555$ kHz

Bandwidth = $2f_m = 10$ kHz

$\Rightarrow$ $f_m = 5$ kHz

Also, $f_c - f_m = 555$ kHz $\Rightarrow f_c = 560$ kHz

1.62 (b)

Bandwidth in frequency modulated signal is given by Carson's formula

$$= 2(\Delta f + f_m) = 2f_m(\beta + 1)$$

where, $\Delta f \rightarrow$ frequency deviation,

$f_m \rightarrow$ modulating signal frequency

$\beta \rightarrow$ modulation index

$\therefore$ B.W. $= 2f_m(2.4 + 1) = 6.8 f_m$

1.63 (b)

Carrier power, $P_c = 1000$ W

Power in each side band,

$$P_{USB} = P_{LSB} = \frac{P_c m^2}{4} = 200 \text{ W}$$

Where m is the modulation index

$$\therefore \quad P_{LSB} = P_{USB} = \frac{P_c m^2}{4}$$

$$200 = 1000\frac{m^2}{4}$$

$$m^2 = \frac{800}{1000} \Rightarrow m = 0.8944$$

% modulation = 89.44%

1.64 (c)

The modulating signal

$$m(t) = 15\sin(2\pi\, 1500t)$$

and carrier signal $c(t) = 60\sin(2\pi 10^3 t)$

Modulation index,

$$m = \frac{A_m}{A_c} = \frac{15}{60} \times 100 = 25\%$$

1.65 (d)

The main function of a RF amplifier is to increase selectivity, where selectivety is the ability to reject the unwanted frequency components.

1.66 (c)

Carrier swing = 100 kHz = $2 \times \Delta f$

$\therefore \quad \Delta f = 50$ kHz and $f_m = 8$ kHz

Thus, the modulation index

$$\beta = \frac{\Delta f}{f_m} = \frac{50}{8} = 6.25$$

1.67 (d)

$$P_C = 400 \text{ W}, \quad m = 75\%$$

$$\therefore \quad P_T = P_C\left(1+\frac{m^2}{2}\right)$$

$$= 400\left(1+\frac{(0.75)^2}{2}\right) = 512.5 \text{ W}$$

1.68 (d)

$$m(t) = \text{sinc}(2 \times 10^4 t)$$

$$\Rightarrow \quad f_m = 20 \text{ kHz}$$

$$K = 10^3 \text{ Hz/v}$$

$$f_c = 1 \text{ MHz}$$

$$(f_i)_{max} = f_c + K_f|m(t)|_{max}$$

$$= 1\text{ M} + 1 \text{ kHz} = 1.001 \text{ MHz}$$

1.69 (c)

In A.M. modulation a simple envelope detector can be used to detect the received signal thus making a very cost efficient and simple receiver system.

1.70 (c)

$$f_I = f_{LO} - f_S$$

$$\therefore \quad f_{LO} = f_I + f_S = 450 + 1000 = 1450 \text{ kHz}$$

$$f_{image} = f_S + 2f_I = 1000 + 900 = 1900 \text{ kHz}$$

1.71 (b)

$$M_t = \sqrt{M_1^2 + M_2^2 + M_3^2}$$

$$= \sqrt{(0.25)^2 + (0.5)^2 + (0.75)^2} = 0.93$$

1.72 (c)

For boosting of higher frequency components a pre-emphasis circuit is used.

1.73 (c)

$$\underbrace{P_T}_{\text{Total power}} = P_C\left(1+\frac{m^2}{2}\right) = \underbrace{P_C}_{\text{Power contained in carrier signal}} + \underbrace{P_C\frac{m^2}{2}}_{\text{Power contained in two sidebands}}$$

$\therefore$ The power contained in a single side bond is given by

$$P_{ssb} = \frac{P_C m^2}{4}$$

1.74 (b)

Selectivity of tuned radio frequency (TRF) receiver is poor because

- Q-factor requirement of tuned circuit varies with frequency.
- Q is very high for short waves and at upper side (say 20 MHz) it requires a Q-value of 2000.
- Q can't be zero.

1.75 (d)

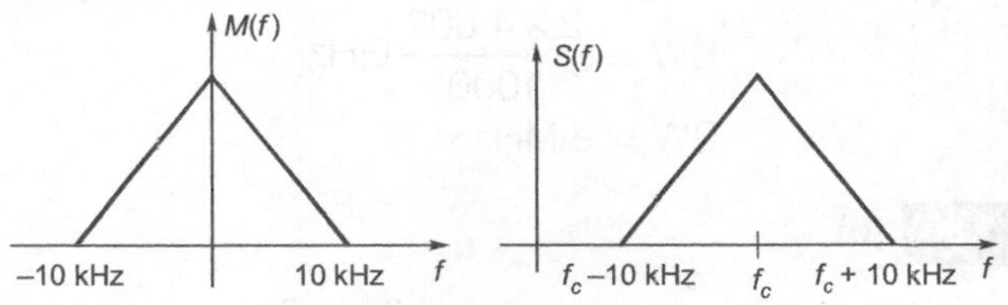

$$f_c + f_m = 800 \text{ kHz}$$

$$f_c = 800 - 10 = 790 \text{ kHz}$$

1.76 (a)

$$P_t = P_c\left(1+\frac{\mu^2}{2}\right)$$

$$P_t \propto I_t^2$$

and $\quad P_c \propto I_c^2$

$$\Rightarrow \quad I_t^2 = I_c^2\left(1+\frac{\mu^2}{2}\right)$$

$$(20)^2 = (18)^2\left(1+\frac{\mu^2}{2}\right)$$

$$\mu = 0.68$$

1.77 (d)

$$\Delta f = 20 \text{ kHz}$$

$$f_m = 2 \text{ kHz}$$

$$\text{BW} = 2(\Delta f + f_m) = 2(20 + 2) \text{ kHz}$$

$$= 44 \text{ kHz.}$$

1.78 (d)

Modulating signal $e_s = 15 \sin(2\pi \times 2000t)$

Carrier signal $e_c = 60 \sin(2\pi \times 10^5 t)$

$$A_c = 60 \quad A_m = 15$$

$$\mu = \frac{A_m}{A_c} = \frac{15}{60} = 0.25$$

1.79 (b)

$$\text{BW} = \frac{2f_H}{k}$$

$$k = \frac{f_H}{f_H - f_L} = \frac{4.002}{0.004} = 1000.5$$

$$k \approx 1000$$

$$\text{BW} = \frac{2 \times 4.002}{1000} \text{ GHz}$$

$$\text{BW} = 8 \text{ MHz}$$

1.80 (b)

$$\text{Average power of } x(t) = \frac{A^2}{2} = \frac{6^2}{2} = 18\text{W}$$

1.81 (b)

To avoid aliasing

$$f_s \geq 2 f_m$$

$$f_s \geq 2 \times 4$$

$$f_s \geq 8 \text{ Hz}$$

⇒ Aliasing effect is seen if f_s is 6 kHz.

1.82 (b)

$$s(t) = e^{-at} \cos[(\omega_c + \Delta\omega)t]u(t)$$

Complex signal = $s(t) + j\hat{s}(t)$

$= e^{-at} \cos[(\omega_c + \Delta\omega)t]u(t) + je^{-at} \cos[(\omega_c + \Delta\omega)t - 90°]\, u(t)$

$= e^{-at} \cos[(\omega_c + \Delta\omega)t + j\sin[(\omega_c + \Delta\omega)t]\, u(t)$

$= [e^{-at} . e^{j\Delta\omega t}\, u(t)]\, e^{j\omega_c t}$

Complex envelope = $e^{-at} . e^{[j\Delta\omega t]}\, u(t)$

1.83 (a)

Bandwidth of AM signal depends on bandwidth of the modulating signal

$$BW_{AM} = 2 f_m$$

Where f_m is bandwidth of modulating signal.

Amplitude modulated signal contains f_c, $f_c \pm f_m$ frequency signal.

highest frequency present = $f_c + f_m$

lowest frequency present = $f_c - f_m$

$$BW_{AM} = 2 f_m$$

1.84 (b)

$$\text{AM BW} = 2 f_m = 8 \text{ kHz}$$

$$\text{FM BW} = 12 f_m$$

(significant sidebands upto 6th order, so bandwidth = 12 f_m)

$$= 48 \text{ kHz}$$

1.85 (c)

For FM; $\beta = \dfrac{k_f A_m}{f_m}$

If modulating frequency is doubled, then modulation index is halved.

1.86 (c)

Carrier swing

$$2\Delta f = 100 \text{ kHz}$$

$$\Delta f = 50 \text{ kHz}$$

$$\beta = \frac{\Delta f}{f_m} = \frac{50}{8} = 6.25$$

1.88 (b)

Modulation is the process of transmitting low frequency message signal by combining with high frequency carrier signal.

1.90 (c)

$$\Delta f = 75 \text{ kHz}$$

$$\beta = \frac{\Delta f}{f_m}$$

for f_m = 100 Hz

$$\beta = \frac{75 \text{ kHz}}{100} = 750$$

for f_m = 1.5 kHz

$$\beta = \frac{75 \text{ kHz}}{1.5\text{k}} = 50$$

1.91 (d)

$$P_t = 60 \text{ W} + 6 \text{ W} = 66 \text{ W}$$

$$\mu = 1$$

$$P_t = P_c\left[1 + \frac{\mu^2}{2}\right]$$

$$66 = P_c\left[1+\frac{1}{2}\right]$$

$$P_c = 44\text{ W}$$

1.93 (b)

f_s varies from 550 kHz to 1600 kHz

$$IF = 455\text{ kHz}$$

$$f_I = f_s + IF$$

$$\frac{f_{I\max}}{f_{I\min}} = \frac{2055\text{ kHz}}{1005\text{ kHz}} = 2.044$$

1.95 (a)

$$BW = (\beta + 1)2f_m = [15 + 1]\,2 \times 100$$

$$= 3200\text{ Hz} = 3.2\text{ kHz}$$

1.96 (c)

$$(SNR)_{FM} > (SNR)_{AM}$$

$$\frac{3}{2}\beta^2 > \frac{\mu^2}{2+\mu^2}$$

Given $\mu = 1$

$$\frac{3}{2}\beta^2 > \frac{1}{3}$$

$$\beta > \sqrt{\frac{2}{9}}$$

$$\beta > 0.47$$

1.97 (b)

$$\mu = 0.8$$

$$P_{SB} = P_C\frac{\mu^2}{2}$$

$$4.85 = P_C\frac{(0.8)^2}{2}$$

$$9.7 = P_C \times (0.8)^2$$

$$P_C = 15.15\text{ kW}$$

$$P_t = P_C + P_{SB} = 20\text{ kW}$$

1.98 (a)

$$P_C = 360\text{ W}$$

$$\mu_1 = 0.55$$

$$\mu_2 = 0.65$$

$$\mu_t = \sqrt{\mu_1^2 + \mu_2^2} = 0.85$$

$$P_t = P_C\left[1+\frac{\mu_t^2}{2}\right] = 360\left[1+\frac{(0.85)^2}{2}\right]$$

$$= 490.5\text{ W}$$

1.99 (a)

Total AM power ⇒ 50 W + 10 W

$$P_t = 60\text{ W}$$

$$\mu = 1$$

$$P_t = P_C\left[1+\frac{\mu_t^2}{2}\right]$$

$$60 = P_C\left[1+\frac{1}{2}\right] = 40\text{ W}$$

1.100 (c)

Phase sensitivity, $K = 2.5$ rad/V

$$V_m(t) = 2\cos(2\pi \times 2000t)$$

Peak phase deviation $= K_pA_m = 2.5 \times 2 = 5$ rad

1.101 (a)

$$I_t = I_C\sqrt{1+\frac{\mu^2}{2}}$$

$$11 = 9\sqrt{1+\frac{\mu^2}{2}}$$

$$\mu \simeq 0.99$$

1.102 (b)

According to Carson's rule,

$$BW = 2[\Delta f + f_m]$$

$$50\text{ k} = 2[15k + f_m]$$

$$f_m = 10\text{ kHz}$$

1.103 (c)

$$\%\text{ of power saving} = \frac{4+\mu^2}{4+2\mu^2}$$

$$\mu = 0.5$$

% Power saving = 94.5%

1.105 (c)

We have, $I_t = I_C\sqrt{1+\frac{\mu^2}{2}}$

Given, $I_C = 8$A, $\mu = 0.8$

Substituting, we get,

$$I_t = 8\sqrt{1+\frac{(0.8)^2}{2}}$$

$$I_t = 9.2\text{ A}$$

1.106 (b)

Given $\Delta f = 75$ kHz, $f_m = 300$ kHz

The modulation index of FM signal is given as:

$$\beta = \frac{\Delta f}{f_m} = \frac{75}{300} = 0.25 < 1$$

Hence, it is a Narrow-Band FM signal. Therefore,

$$(BW)_{NBFM} = 2f_m = 600 \text{ kHz}$$

1.107 (d)

$$(FOM)_{AM} = \frac{\mu^2}{2+\mu^2}$$

For $\mu = 1$,

$$(FOM)_{AM} = \frac{1}{3}$$

and $(FOM)_{DSB\text{-}SC} = 1$

$(FOM)_{SSB\text{-}SC} = 1$

Since, $(FOM)_{AM} < 1$, therefore the noise performance of AM receiver is always inferior to that of a DSB-SC receiver.

For an AM receiver using envelop detection the figure of merit, is always less than 1. Hence, option (d) is the correct answer.

1.108 (c)

Typical frequency parameters of AM and FM radio receivers. :

	AM Radio	FM Radio
RF carrier range	0.535 to 1.605 MHz	80 to 108 MHz
Mid band frequency of IF section	0.455 MHz	10.7 MHz
IF bandwidth	10 kHz	200 kHz

1.109 (a, d)

$(FOM)_{DSB\text{-}SC} = (FOM)_{SSB\text{-}SC} = 1$

- For the same average transmitted or modulated signal power and the same average noise power in the message bandwidth, a coherent SSB receiver will have exactly the same output signal-to-noise ratio as a coherent DSB-SC receiver.
- In both cases, the noise performance of the receiver is exactly the same as that obtained by simply transmitting the message signal in the presence of the same channel noise. The only effect of the modulation process is to translate the message signal to a different frequency band to facilitate its transmission over a band-pass channel.

1.110 (c)

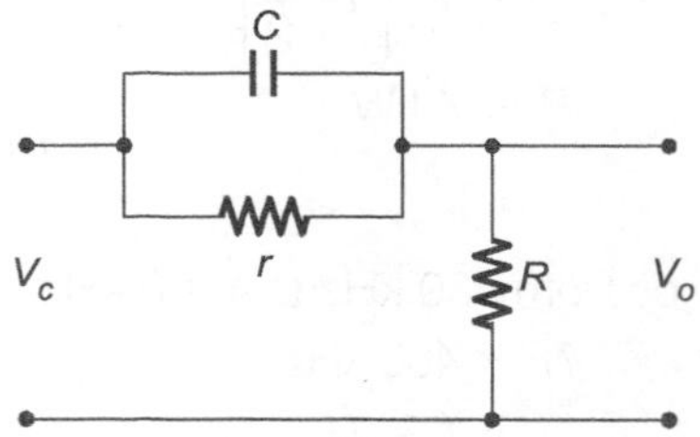

$$\frac{V_o(s)}{V_i(s)} = \frac{R}{R + \dfrac{\left(r \times \dfrac{1}{Cs}\right)}{r + \dfrac{1}{Cs}}}$$

$$\frac{V_o(s)}{V_i(s)} = \frac{1 + rCs}{1 + rCs + \dfrac{r}{R}}$$

$$\frac{V_o(s)}{V_i(s)} = \frac{1 + rCs}{1 + r/R + rCs}$$

The Bode-plot for the above transfer function can be obtained as below:

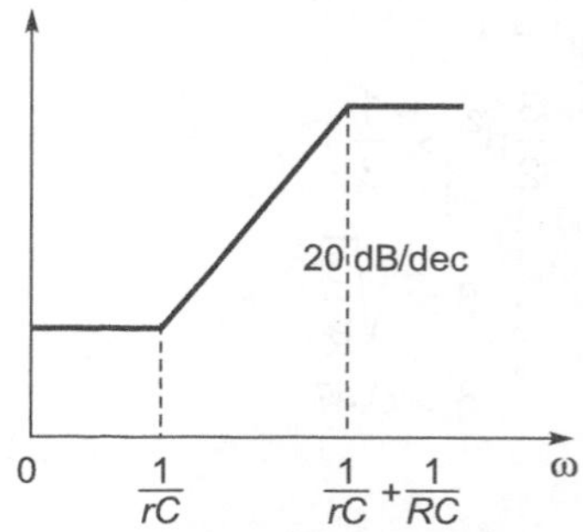

The given circuit boosts the higher frequency components of the signal. Hence, it is a pre-emphasis filter.

1.111 (c)

Given, power transmitted by an SSB transmitter,

$$P_{SSB} = 20 \text{ kW}$$

Modulation index,

$\mu = 0.4$ $\therefore$ AM Power = SSB Power; $P_t = 20$ kW

Transmission efficiency,

$$\eta = \frac{\mu^2}{2+\mu^2} = 7.4\%$$

1.112 (b)

Given, angle modulated signal,

$$x_c(t) = 20\cos\left[200\pi t + \frac{\pi}{4}\right] = A_c \cos[\theta_i(t)]$$

$$f_i = \frac{1}{2\pi}\frac{d}{dt}\theta_i(t) = \frac{1}{2\pi}[200\pi] = 100 \text{ Hz}$$

1.113 (d)

For FM:

Carrier frequency,

$$f_c = 250 \text{ kHz}$$

Frequency deviation sensitivity,

$$k_f = 1.5 \frac{\text{kHz}}{\text{volt}}$$

For PM: $f_c = 500$ kHz

Phase deviation sensitivity,

$$k_p = 1.5 \frac{\text{rad}}{\text{volt}}$$

$$A_m = 5 \text{ V}$$

$$f_m = 5 \text{ kHz}$$

Modulation index,

$$\beta_{FM} = \frac{k_f A_m}{f_m} = \frac{1.5 \times 1000 \times 5}{5000} = 1.5$$

$$\beta_{PM} = k_p A_m = 1.5 \times 5 = 5 \times \beta_{FM}$$

1.114 (d)

We know that, $$\frac{\eta_{DSB}}{\eta_{AM}} = \frac{P_c\mu^2/2}{P_c\mu^2/4} = 2$$

$$\eta_{DSB} = 2 \cdot \eta_{AM}$$

1.115 (a)

Frequency deviation,

$$\Delta_f = 25 \text{ kHz}$$

$$\text{FM BW} = 75 \text{ kHz}$$

$$\text{BW} = 2[\Delta f + f_m]$$

$$75 \text{ k} = 2\,[25\text{k} + f_m]$$

$$f_m = 12.5 \text{ kHz}$$

1.116 (c)

Given: $|s(t)| \le 1$

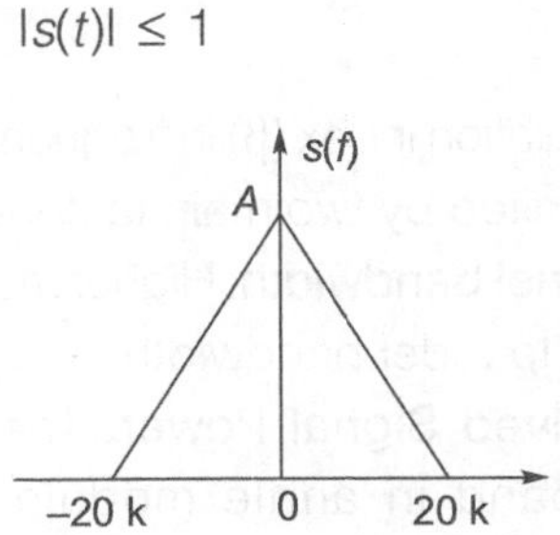

$$k_f = 90 \text{ kHz/volt}$$

$$\text{BW} = 2[\Delta f + f_{max}]$$

$$\Delta f = k_f \cdot |m(t)|_{max} = 90 \text{ kHz}$$

$$\because \quad |m(t)|_{max} = 1$$

$$\text{BW} = 2[90 \text{ k} + 20 \text{ k}] = 220 \text{ kHz}$$

1.117 (b)

Given, average output power of AM signal,

$$P_t = 220 \text{ W}$$

Modulation index,

$$\mu = 0.4$$

We know that,

$$P_t = P_c\left[1 + \frac{\mu^2}{2}\right]$$

$$P_c = 203.7 \text{ W}$$

Power supplied by modulating amplifier gets converted into output SB power.

Power to be supplied by class *C* power amplifier

$$= 220 - 203.7 = 16.29 \text{ W}$$

Given collector circuit efficiency = 40% = 0.4

DC power to be supplied by modulating amp

$$= \frac{16.29}{0.4} = 40.7 \text{ W}.$$

1.118 (a)

We know that, standard AM signal represented as

$$S_{AM}(t) = A_c[1 + k_a m(t)]\cos 2\pi f_c t$$

$$P_{AM} = \frac{A_c^2}{2} + \frac{A_c^2 k_a^2 m^2(t)}{2}$$

$$\because \quad k_a = \frac{1}{A_c}$$

Default, $k_a = 1$

$$P_{AM} = \frac{A_c^2}{2} + \frac{A_c^2 m^2(t)}{2}$$

1.119 (a)

Method 1: $P_{t1} = 125$ kW

$$\mu_1 = 0.7$$

$$P_{t1} = P_c\left[1 + \frac{\mu_1^2}{2}\right] \rightarrow P_c = \frac{125}{1.245} \text{kW}$$

$$P_t = 144.5 \text{ kW} \rightarrow \mu_2 = ?$$

$$\frac{P_t}{P_{t1}} = \frac{P_c\left[1 - \frac{\mu_t^2}{2}\right]}{P_c\left[1 + \frac{\mu_1^2}{2}\right]}$$

$$\frac{144.5K}{125K} = \frac{1 + \mu_1^2 + \mu_2^2}{1 + \frac{\mu_1^2}{2}}$$

$$1.156 = \frac{1+\frac{0.49+\mu_2^2}{2}}{1+\frac{0.49}{2}}$$

$$1.43922 = 1+0.245+\frac{\mu_2^2}{2}$$

$$\mu_2^2 = 0.388$$

$$\mu_2 = \sqrt{0.38}$$

Method 2:

$$P_t = P_c\left[1+\frac{\mu_t^2}{2}\right] = P_c\left[1+\frac{\mu_1^2}{2}\right]+\frac{P_c\mu_2^2}{2}$$

$$144.5\text{ K} = 125\text{ kW}+\frac{P_c\mu_2^2}{2}$$

$$19.5\text{ kW} = \frac{P_c\mu_2^2}{2}$$

$$39\text{ kW} = \frac{125}{1.245}\text{kW}\times\mu_2^2$$

$$\mu_2 = \sqrt{0.388}$$

1.121 (b)

Most commonly used intermediate frequencies for broadcast receivers are around 455 kHz for AM receivers and 10.7 MHz for FM receivers.

1.122 (d)

1. DSB-SC, SSB, and VSB are examples of linear modulation, in which the modulated signal is described by the canonical relation:
   $s(t) = s_i(t)\cos(2\pi f_c t) - s_o(t)\sin(2\pi f_c t)$
   The in-phase component $s_i(t)$ is a scaled version of the message signal $m(t)$. The quadrature component $s_o(t)$ is derived from $m(t)$ by some form of linear filtering. Accordingly, the principle of superposition can be used to calculate the modulator output $s(t)$ as the sum of the responses of the modulator to individual components of $m(t)$.
   Note: The ordinary AM modulation fails to meet the definition of a linear modulator with respect to the message signal.
2. DSB-SC requires much less power than standard AM to transmit the same message signal, however it is attained at the expense of increased receiver complexity. DSB-SC modulation is therefore well suited for point to point communication involving one transmitter and one receiver.
3. With only a single sideband transmitted in SSB modulation, it is the least affected by frequency-selective fading.

Hence, all the statements are correct.

1.123 (c)

In the video signal, very low frequency modulating components give rise to sidebands very close to the carrier frequency which are difficult to remove by physically realizable filters. The complete suppression of the lower sideband would result in objectionable phase distortion at these frequencies. Therefore, VSB modulation is used with only a part of the lower sideband and complete upper sideband being transmitted.

1.124 (a)

For a frequency modulated signal, it is given that
Message signal frequency,

$$f_m = 2\text{ kHz}$$

Maximum frequency deviation,

$$\Delta f = 5\text{ kHz}$$

Modulation Index,

$$\beta = \frac{\Delta f}{f_m} = \frac{5}{2} = 2.5$$

i.e., it is a WBFM signal
According to Carson's rule,

$$\text{Bandwidth} = 2(\beta + 1)f_m = 2(2.5 + 1) \times 2\text{ kHz} = 14\text{ kHz}$$

1.125 (a)

The modulation index (β) in frequency modulation (FM) is limited by two main factors:

1. **Channel bandwidth:** Higher modulation index leads to wider bandwidth.
2. **Received Signal Power:** The amplitude of sideband in angle modulated signal is proportional to $J_n(\beta)$ which decreases with increase in β. Hence, higher modulation index reduces the number of sidebands with significant amplitude.

■■■

2 Random Variables, Random Process and Noise

2.1 A linear system has the transfer function

$$H(j\omega) = \frac{1}{(j\omega + 1)}$$

When it is subjected to an input white noise process with a constant spectral density '*A*', the spectral density of the output will be

(a) $\frac{1}{(j\omega + 1)}$ (b) $\frac{A}{(j\omega + 1)^2}$

(c) $\frac{A}{(\omega^2 + 1)}$ (d) $\frac{A}{\sqrt{(\omega^2 + 1)}}$

[ESE-2000]

2.2 Which one of the following types of noise gains importance at high frequency?

(a) Shot noise (b) Random noise

(c) Impulse noise (d) Transit-time noise

[ESE-2001]

2.3 Two independent random signals *X* and *Y* are known to be Gaussian with mean values x_0 and y_0 and variance σ_x^2 and σ_y^2. A signal $Z = X - Y$ is obtained from them. The mean z_0 variance σ_z^2 and p.d.f. $p(z)$ of the signal *Z* are given by

(a) $x_0 \sim y_0$, $\sigma_x^2 \sim \sigma_y^2$, Gaussian

(b) $x_0 + y_0$, $\sigma_x^2 + \sigma_y^2$, Rayleigh

(c) $y_0 - x_0$, $\sigma_y^2 - \sigma_x^2$, Uniform

(d) $x_0 - y_0$, $\sigma_x^2 + \sigma_y^2$, Gaussian [ESE-2002]

2.4 **Assertion (A):** The noise generated by a resistor depends upon its operating temperature.

Reason (R): Average noise power generated in a resistor is given by $P_n = kT\Delta f$, where

k = Boltzmann's constant,

T = Temperature in degree Kelvin,

Δf = Bandwidth of interest

(a) Both A and R are true and R is the correct explanation of A

(b) Both A and R are true but R is NOT the correct explanation of A

(c) A is true but R is false

(d) A is false but R is true [ESE-2002]

2.5 If the cumulative distribution function is $F_X(x)$, then the probability density function $f_X(x)$ is given as

(a) $\int F_X(x)dx$ (b) $d/dx\, F_X(x)$

(c) $\int F_X(-x)dx$ (d) $d/dx\, F_X(-x)$

[ESE-2002]

2.6 Fading is

(a) change in polarization only at receiver end

(b) change in frequency only at receiver end

(c) fluctuation in signal strength at receiver

(d) change in phase only at receiver end

[ESE-2002]

2.7 The units of the spectrum obtained by Fourier transforming the covariance function of a stationary stochastic process is

(a) power per Hertz

(b) energy per Hertz

(c) power per second

(d) energy per second

[ESE-2002]

2.8 The covariance function of a band limited white noise is

(a) A Dirac delta function

(b) An exponentially decreasing function

(c) A sinc function

(d) A sinc^2 function

[ESE-2003]

2.9 In a satellite ground station, the received signal is directly amplified in a low noise parametric amplifier followed by a FET amplifier. The gains and effective noise temperatures of the amplifiers are 20 dB, 9.0 K and 10 dB, 200 K respectively. The effective noise temperature of the combination is

(a) 11.0 K (b) 20 K

(c) 29.0 K (d) 39 K

[ESE-2003]

2.10 Thermal noise is passed through an ideal low-pass filter having cut-off at $f_c = \omega$ Hz. The autocorrelation value of the noise at the output of the filter is given as
(a) A delta function at $t = 0$
(b) Gaussian over the range $-\infty \le t \le \infty$
(c) Sinc function over the range $-\infty < t < \infty$
(d) Triangular function over the range $-1/2\omega \le t \le 1/2\omega$

[ESE-2003]

2.11 A random process obeys Poisson's distribution. It is given that the mean of the process is 5. Then the variance of the process is
(a) 5 (b) 0.5
(c) 25 (d) 0 **[ESE-2003]**

2.12 Which one of the following gives the average value or expectation of the function $g(X)$ of the random variable X? {Given $f(X)$ is the probability density function}

(a) $E[g(X)] = \int_{-\infty}^{\infty} g(X)dX$

(b) $E[g(X)] = \int_{-\infty}^{\infty} g(X)f(X)dX$

(c) $E[g(X)] = \int_{-\infty}^{\infty} g^*(X)dX$

(d) $E[g(X)] = \int_{-\infty}^{\infty} \left[\frac{g(X)}{f(X)}\right]dX$ **[ESE-2004]**

2.13 The auto-correlation function $R_X(\tau)$ of a random process has the property that $R_X(0)$ is equal to
(a) The square of the mean value of the process
(b) The mean squared value of the process
(c) The smallest value of $R_X(\tau)$
(d) $1/2[R_X(\tau) + R_X(-\tau)]$ **[ESE-2005]**

2.14 Match **List-I** (Type of Random Process) with **List-II** (Property of the Random Process) and select the correct answer using the code given below the lists:

List-I
A. Stationary process
B. Ergodic process
C. Wide sense stationary process
D. Cyclostationary process

List-II
1. Statistical averages are periodic in time
2. Statistical averages are independent of time
3. Mean and autocorrelation are independent of time
4. Time averages equal corresponding ensemble average

Codes:

	A	B	C	D
(a)	3	1	2	4
(b)	2	4	3	1
(c)	3	4	2	1
(d)	2	1	3	4

[ESE-2005]

2.15 Match **List-I** (Type) with **List-II** (Application) and select the correct answer using the codes given below the lists:

List-I
A. Speech signal
B. Non-stationary
C. Random signal
D. Chaotic signal

List-II
1. The received signal of a radar system monitoring variation in prevalent weather condition
2. One dimensional signal where amplitude varies with time
3. Signals of coupled system of non-linear difference
4. Ensemble of unpredictable wave-forms

Codes:

	A	B	C	D
(a)	2	1	4	3
(b)	4	3	2	1
(c)	2	3	4	1
(d)	4	1	2	3

[ESE-2006]

2.16 For a random variable X having the probability density function (PDF) as shown in the below figure, what are the values of the mean and the variance, respectively?

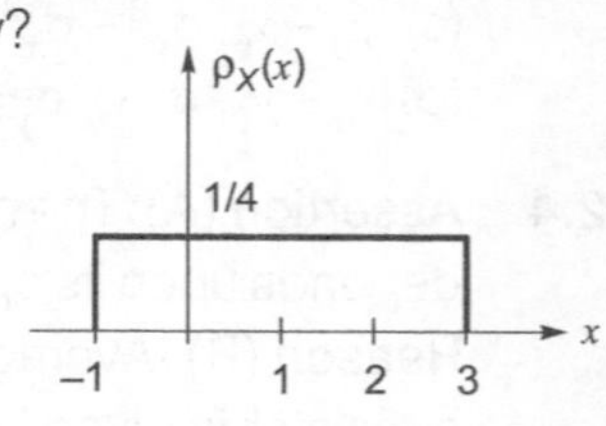

(a) 1/2 and 2/3
(b) 1 and 4/3
(c) 1 and 2/3
(d) 2 and 4/3

[ESE-2006]

2.17 If a linear time invariant system is excited by a true random signal like white noise, the output of the linear system will have which of the following properties?
(a) Output will be a white noise
(b) Output will be periodic
(c) Output will not be random
(d) Output will be correlated or coloured noise

[ESE-2007]

2.18 Which of the following is/are not a property/properties of a power spectral density function $S_X(\omega)$?
(a) $S_X(\omega)$ is a real function of ω
(b) $S_X(\omega)$ is an even function of ω
(c) $S_X(\omega)$ is a non-positive function of ω i.e. $S_X(\omega) \leq 0$ for all ω
(d) All of the above [ESE-2007]

2.19 A random variable X is defined by the double exponential distribution

$$\rho_x(x) = ae^{-b|x|}, -\infty < x < \infty$$

where a and b are +ve constants. What is the relation between a and b so that $\rho_x(x)$ is a probability density function ?
(a) $a = b/2$ (b) $b = a/2$
(c) $a = b$ (d) $a = 1/b$ [ESE-2008]

2.20 What is the spectral density of white noise?
(a) A constant
(b) $\delta(\omega)$
(c) $[\delta(\omega)]^2$
(d) A step function in ω [ESE-2008]

2.21 Let $x(n)$ be a real-valued sequence that is a sample sequence of a wide-sense stationary discrete-time random process. The power density function of this signal is
(a) real, odd and non -negative
(b) real, even and non-negative
(c) purely imaginary, even and negative
(d) purely imaginary, odd and negative
[ESE-2008]

2.22 Match **List-I** (Type of Noise) and **List-II** (Its property) and select the correct answer using the codes given below the lists:

List-I
A. Shot noise
B. Thermal noise
C. White noise
D. Narrow band noise

List-II
1. Noise generated in a resistor
2. Power spectral density is independent of frequency
3. Temperature-limited diode
4. Noise at the output of a filter

Codes:

	A	B	C	D
(a)	2	4	3	1
(b)	3	1	2	4
(c)	2	1	3	4
(d)	3	4	2	1

[ESE-2008]

2.23 Which one of the following is the correct statement? If the value of a resistor creating thermal noise is doubled, the noise generated is
(a) halved (b) doubled
(c) unchanged (d) slightly changed
[ESE-2008]

2.24 The outputs of two noise sources each producing uniformly distributed noise over the range $-a$ to $+a$ are added. What is the p.d.f. of the added noise?
(a) Uniformly distributed over the range $-2a$ to $+ 2a$
(b) Triangular over the range $-2a$ to $+ 2a$
(c) Gaussian over the range $-\infty$ to ∞
(d) None of the above
[ESE-2008]

2.25 If random process $X(t)$ and $Y(t)$ are orthogonal then
(a) $S_{XY}(f) = 0$
(b) $S_{XY}(f) = S_X(f) = S_Y(f)$
(c) $R_{XY}(\tau) = h(\tau)$
(d) $H(f) = 0$ [ESE-2010]

2.26 If a random process $X(t)$ is ergodic then, statistical averages
(a) and time averages are different
(b) and time averages are same
(c) are greater than time averages
(d) are smaller than time averages [ESE-2010]

2.27 A random variable is known to have a cumulative distribution function

$$F_x(x) = u(x)\left(1-\frac{x^2}{b}\right)$$

Its density function is
(a) $u(x)\dfrac{2x}{b}(1-e^{-x^2/b})$ (b) $u(x)\dfrac{2x}{b}e^{-x^2/b}$
(c) $u(x)\left(1-\dfrac{x^2}{b}\right)\delta(x)$ (d) $\left(1-\dfrac{x^2}{b}\right)\delta(x)+e^{-x^2/b}$
[ESE-2012]

2.28 The spectral density and autocorrelation function of white noise is
(a) Delta and uniform
(b) Uniform and delta
(c) Gaussian and uniform
(d) Gaussian and delta
[ESE-2012]

2.29 The probability cumulative distribution function must be monotone and

(a) increasing (b) decreasing
(c) non-increasing (d) non decreasing

[ESE-2012]

2.30 If the power spectral density is $\frac{\eta}{2}\frac{W}{Hz}$, and the auto correlation function is defined by

$$R(\tau) = \frac{\eta}{2}\int_{-\infty}^{\infty} e^{j\omega\tau} df.$$

The integral on the right represents the Fourier transform of

(a) Delta function (b) Step function
(c) Ramp function (d) Sinusoidal function

[ESE-2013]

2.31 Consider a discrete random variable assuming finitely many values. The cumulative distribution function of such a random variable is

(a) non-increasing function
(b) non-decreasing function with finitely many discontinuity and assuming values less than one
(c) non-decreasing function without discontinuity
(d) non-decreasing function assuming values larger than one

[ESE-2014]

2.32 A continuous random variable X has uncountable many values in the interval $[a, b]$. If C is a value in the interval $[a, b]$, then $P\{X = C\}$

(a) is zero
(b) is strictly non-zero
(c) depends on the limits $\{a, b\}$
(d) is less than one, but non-zero [ESE-2014]

2.33 In the case of a random variable dealing with non-deterministic signals.

(a) It is a function from space of outcomes to the real/complex numbers.
(b) It is a function with the probabilities of outcomes as random numbers.
(c) The values assumed by signals are always deterministic.
(d) Sometimes the events associated with random variable are deterministic.

[ESE-2014]

2.34 The correlation function of a wide sense stationary random process representing a non-deterministic signal is

(a) not a deterministic function
(b) deterministic, but not symmetric function
(c) sometimes non-deterministic function
(d) always deterministic and symmetric function

[ESE-2014]

2.35 An LTI system has a wide-sense stationary (WSS) input signal with zero mean. Its output is

(a) non-zero mean and non-WSS signal
(b) zero mean and WSS signal
(c) non-zero mean and WSS signal
(d) zero mean and non-WSS signal

[ESE-2014]

2.36 Consider an LTI system subjected to a wide-sense stationary input $\{x(n)\}$, which is a white noise sequence. The cross correlation $\Phi_{XY}[m]$ between input $x(n)$ and output $y(n)$ is

(a) $\sigma_x^2 h[m]$ (b) $\sigma_x h[m]$
(c) $\frac{\sigma_x^2}{2} h[m]$ (d) $\frac{\sigma_x}{2} h[m]$

where $\Phi_{xx}[m] = \sigma_x^2 \delta[m]$ and $h[\cdot]$ is impulse response

[ESE-2015]

2.37 For a random signal (continuous time) $x(t)$ defined for $t \geq 0$, its probability density function (pdf) at $t = t_0$ is such that

(a) It is non-negative and its integral equals 1
(b) Need not be non-negative, but integral equal 1
(c) It is non-negative, but integral is not 1
(d) None of the above

[ESE-2015]

2.38 If the variance σ_x^2 of $d(n) = X(n) - X(n-1)$ is one-tenth the variance σ_x^2 of a stationary zero-mean discrete-time signal $X(n)$, then the normalized autocorrelation function $R_{XY}(k)/\sigma_x^2$ at $k = 1$ is

(a) 0.95 (b) 0.90
(c) 0.10 (d) 0.05

[ESE-2015]

2.39 The response of a Gaussian random process applied to a stable linear system is
1. A Gaussian random process
2. Not a Gaussian random process
3. Completely specified by its mean and auto-covariance functions

Which of the above statements is/are correct?
(a) 1 only (b) 2 only
(c) 2 and 3 (d) 1 and 3 [ESE-2016]

2.40 Consider a discrete-random variable z assuming finitely many values. The cumulative distribution function, $F_z(z)$ has the following properties:
1. $\int_{-\infty}^{+\infty} F_z(z)\,dz = 1$
2. $F_z(z)$ is non-decreasing with finitely many jump-discontinuities
3. $F_z(z)$ is negative and non-decreasing

Which of the above properties is/are correct?
(a) 1 only (b) 2 only
(c) 3 only (d) 2 and 3
[ESE-2016]

2.41 Consider a random process given by:
$x(t)$ = A$\cos(2\pi f_c t + \theta)$, where A is a Rayleigh distributed random variable and θ is distributed in $[0, 2\pi]$. A and θ are independent. For any time t, the probability density function (PDF) of $x(t)$ is
(a) Gaussian (b) Rayleigh
(c) Rician (d) Uniform in $[-A, A]$
[ESE-2016]

2.42 The probability density function $F(x) = ae^{-b|x|}$, where x is a random variable whose allowable value range is from $x = -\infty$ to $x = +\infty$. The CDF for this function for $x \geq 0$ is
(a) $\frac{a}{b}e^{bx}$ (b) $\frac{a}{b}(2-e^{-bx})$
(c) $-\frac{a}{b}e^{bx}$ (d) $-\frac{a}{b}(2+e^{-bx})$
[ESE-2017]

2.43 The noise factor of an attenuator pad that has an insertion loss of 6 dB is
(a) 0.25 (b) 0.5
(c) 2 (d) 4 [ESE-2017]

2.44 The cumulative distribution function of a random variable x is the probability that X takes the value.
(a) less than or equal to x
(b) equal to x
(c) greater than x
(d) zero [ESE-2017]

2.45 A random process $X(t)$ is called 'white noise' if the power spectral density is equal to
(a) $\frac{\pi}{8}$ (b) $\frac{\pi}{2}$
(c) $\frac{3\pi}{4}$ (d) π
[ESE-2017]

2.46 For a random variable x having the PDF shown in the figure given below

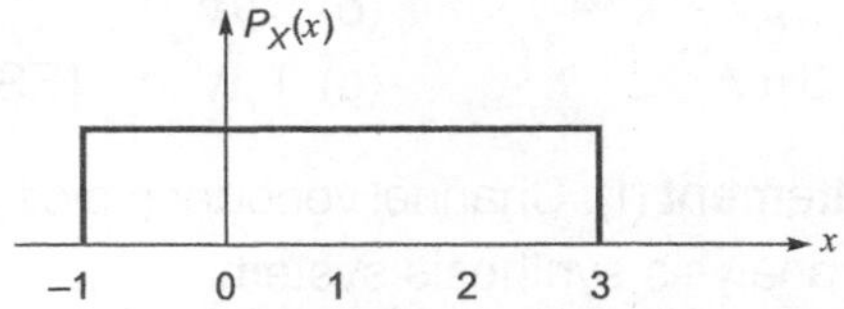

the mean and the variance are, respectively
(a) 0.5 and 0.66 (b) 2.0 and 1.33
(c) 1.0 and 0.66 (d) 1.0 and 1.33
[ESE-2017]

2.47 The power spectral density of the stationary noise process $N(t)$, having auto-correlation $R_{uu}(\tau) = Ke^{-3|\tau|}$, is
(a) $\frac{3K}{3+\omega^2}$ (b) $\frac{3K}{3-\omega^2}$
(c) $\frac{6K}{9+\omega^2}$ (d) $\frac{6K}{9-\omega^2}$
[ESE-2018]

2.48 A random process $X(t)$ is defined as
$$X(t) = 2\cos(2\pi t + Y)$$
where Y is a discrete random variable with $P(Y=0) = \frac{1}{2}$ and $P\left(Y = \frac{\pi}{2}\right) = \frac{1}{2}$. The mean $\mu_x(1)$ is
(a) $\frac{1}{4}$ (b) $\frac{1}{3}$
(c) $\frac{1}{2}$ (d) 1 [ESE-2019]

2.49 For an FM receiver with an input signal-to-noise ratio of 29 dB, a noise figure of 4 dB and an FM improvement factor of 16 dB, the pre-detection and post-detection signal-to-noise ratios are
(a) 25 dB and 41 dB (b) 30 dB and 49 dB
(c) 25 dB and 49 dB (d) 30 dB and 41 dB
[ESE-2019]

2.50 The mapping function that assigns a number to each outcome is called
(a) Sample space (b) Random variable
(c) Discrete variable (d) Event
[ESE-2020]

2.51 A device has 200 Ω equivalent noise resistance, 300 Ω input resistor and the bandwidth of the amplifier being 6 MHz. If the operating temperature of the amplifier is 290°K, the noise voltage at the input of a television RF amplifier will be nearly
(a) 7 μV (b) 5 μV
(c) 3 μV (d) 1 μV [ESE-2020]

2.52 **Statement (I):** Channel vocoder (voice coder) is an analysis synthesis system.
Statement (II): For voiced signal, the excitation is a white noise and for an unvoiced signal, the excitation is a periodic signal.
(a) Both Statement (I) and Statement (II) are individually true and Statement (II) is the correct explanation of Statement (I)
(b) Both Statement (I) and Statement (II) are individually true but Statement (II) is **NOT** the correct explanation of Statement (I)
(c) Statement (I) is true but Statement (II) is false
(d) Statement (I) is false but Statement (II) is true
[ESE-2020]

2.53 A white noise of magnitude 0.001 μW/Hz is applied to an RC low-pass filter of R = 1 kΩ and C = 0.1 μF. The output noise power of the RC low-pass filter is
(a) 0.5 μW (b) 1.5 μW
(c) 2.5 μW (d) 3.5 μW
[ESE-2021]

2.54 The two random variables X and Y are uncorrelated if and only if their covariance is
(a) 0 (b) 1
(c) –1 (d) infinity
[ESE-2021]

2.55 Consider the following statements regarding signals:
1. Deterministic signal is a signal about which there is no uncertainty with respect to its value at any time.
2. Each signal within the ensemble has a certain probability of occurrence and the ensemble of signals is referred to as a random process.
3. EEG signal is an example of random signal.

Which of the above statements are correct?
(a) 1 and 2 only (b) 1 and 3 only
(c) 2 and 3 only (d) 1, 2 and 3
[ESE-2022]

2.56 Which one of the following noises arises in electronic devices such as diodes and transistors because of the discrete nature of current flow?
(a) Shot noise (b) Thermal noise
(c) Gaussian noise (d) Random noise
[ESE-2022]

2.57 A mass of 10 kg is measured with an instrument and the readings are normally distributed with respect to the mean of 10 kg. Given that

$$\frac{2}{\sqrt{2\pi}}\int_0^{0.84} \exp\left(\frac{-\eta^2}{2}\right) d\eta = 0.6$$

and 60 percent of the recordings are found to be within 0.05 kg from the mean. The standard deviation of the data is
(a) 0.02 (b) 0.06
(c) 0.12 (d) 0.18
[ESE-2024]

2.58 If, in addition to being stationary, the random process is also ergodic in nature, then the relation between power content P_X and autocorrelation function is
(a) $P_X = R_X(0)$ (b) $P_X \neq R_X(0)$
(c) $P_X = 2R_X(0)$ (d) $P_X = \frac{1}{2}R_X(0)$
[ESE-2024]

2.59 Consider the following statements regarding a Gaussian process:
1. If a Gaussian process is wide sense stationary, then the process is not stationary in the strict sense.
2. If the Gaussian process $X(t)$ is applied to a stable linear filter, then the random process $Y(t)$ developed at the output of the filter is also Gaussian.
3. If the random variables $X(t_1)$, $X(t_2)$ ... $X(t_n)$, obtained by sampling a Gaussian process

$X(t)$ at times $t_1, t_2 \dots t_n$ are uncorrelated, then these random variables are not statistically independent.

Which of the above statements is/are correct?

(a) 1 and 2 (b) 1 and 3

(c) 1 only (d) 2 only

[ESE-2024]

■■■

Answers Random Variables, Random Process and Noise

2.1 (c)	2.2 (d)	2.3 (d)	2.4 (a)	2.5 (b)	2.6 (d)	2.7 (a)	2.8 (c)	2.9 (a)
2.10 (c)	2.11 (a)	2.12 (b)	2.13 (b)	2.14 (b)	2.15 (c)	2.16 (b)	2.17 (a)	2.18 (c)
2.19 (a)	2.20 (a)	2.21 (b)	2.22 (b)	2.23 (c)	2.24 (b)	2.25 (a)	2.26 (b)	2.27 (*)
2.28 (b)	2.29 (d)	2.30 (a)	2.31 (b)	2.32 (a)	2.33 (a)	2.34 (d)	2.35 (b)	2.36 (a)
2.37 (a)	2.38 (a)	2.39 (d)	2.40 (b)	2.41 (a)	2.42 (b)	2.43 (d)	2.44 (a)	2.45 (*)
2.46 (d)	2.47 (c)	2.48 (d)	2.49 (a)	2.50 (b)	2.51 (a)	2.52 (c)	2.53 (c)	2.54 (a)
2.55 (d)	2.56 (a)	2.57 (b)	2.58 (a)	2.59 (d)				

Explanations Random Variables, Random Process and Noise

2.1 (c)

$$S_0(\omega) = |H(\omega)|^2 S_i(\omega)$$

$$\Rightarrow S_0(\omega) = \left(\frac{1}{\omega^2+1}\right) A \quad \left(\because |H(\omega)| = \frac{1}{\sqrt{\omega^2+1}}\right)$$

$$\Rightarrow S_0(\omega) = \frac{A}{(\omega^2+1)}$$

2.2 (d)

At higher frequencies, the transit-time noise is the most important.

2.3 (d)

The mean value of the difference is equal to the difference of means.

So, $\overline{X-Y} = \bar{X} - \bar{Y} = x_0 - y_0$

variance of Z is

$$\sigma_Z^2 = \overline{(Z-\bar{Z})^2}$$

$$= \overline{[X-Y-(\bar{X}-\bar{Y})]^2}$$

$$= \overline{[(X-\bar{X})-(Y-\bar{Y})]^2}$$

$$= \overline{(X-\bar{X})^2} + \overline{(Y-\bar{Y})^2} - \overline{2(X-\bar{X})(Y-\bar{Y})}$$

$$= \sigma_X^2 + \sigma_Y^2 - \overline{2(X-\bar{X})(Y-\bar{Y})}$$

Since, X and y are independent RVs, $(X-\bar{X})$ and $(Y-\bar{Y})$ are also independent RVs.

Hence, $\overline{(X-\bar{X})(Y-\bar{Y})} = \overline{(X-\bar{X})}\,\overline{(Y-\bar{Y})}$

But $\overline{(X-\bar{X})} = \bar{X} - \bar{\bar{X}} = \bar{X} - \bar{X} = 0$

Similarly $\overline{(Y-\bar{Y})} = 0$

So $\sigma_Z^2 = \sigma_X^2 + \sigma_Y^2$

2.6 (d)

Fading is change in phase only at receiver end.

2.8 (c)

Fourier transform of the rectangular function is a sinc function

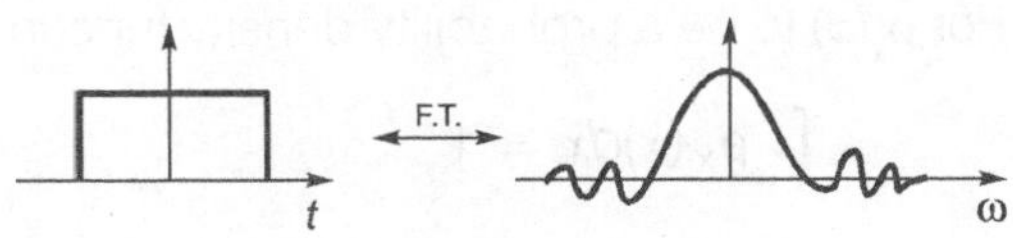

2.9 (a)

$$20 = 10\log_{10} G_1$$

$$\Rightarrow G_1 = (10)^2 = 100$$

$T_{e_1} = 9\text{ K}, T_{e_2} = 200\text{ K}$

Effective noise temperature

$$T_e = T_{e_1} + \frac{T_{e_2}}{G_1}$$

$$\Rightarrow \quad T_e = 9 + \frac{200}{100} = 11 \text{ K}$$

2.10 (c)

The auto-correlation of a rectangular signal is a triangular signal.
A triangular signal in frequency domain is a sinc function in time domain over the range $-\infty < t < \infty$.

2.11 (a)

We know that
Standard deviation = S.D

$$= \sqrt{\text{Mean}} = \sigma_X = \sqrt{\sigma_X^2} = \sqrt{\text{Variance}}$$

$\therefore$ Variance = mean = 5
For Poisson's distribution mean and variance are same.

2.16 (b)

Mean value, $\bar{X} = \int_{-\infty}^{\infty} x\rho_X(x)dx = \frac{1}{4}\int_{-1}^{3} X\,dx$

$$\Rightarrow \quad \bar{X} = \frac{1}{4}\cdot\frac{x^2}{2}\Big|_{-1}^{3} = \frac{1}{8}(9-1) = 1$$

$$\overline{X^2} = \int_{-\infty}^{\infty} x^2\rho_X(x)dx$$

$$= \frac{1}{4}\int_{-1}^{3} x^2 dx = \frac{1}{4\times 3}x^3\Big|_{-1}^{3}$$

$$= \frac{1}{12}(27+1) = \frac{28}{12} = \frac{7}{3}$$

$$\text{Variance} = \overline{X^2} - \bar{X}^2 = \frac{7}{3} - 1 = \frac{4}{3}$$

2.18 (c)

$$S_X(\omega) \geq 0 \text{ for all } \omega$$

2.19 (a)

For $\rho_x(x)$ to be a probability density function,

$$\int_{-\infty}^{\infty} \rho_X(x)dx = 1$$

$$\Rightarrow \quad \int_{-\infty}^{\infty} a e^{-b|x|} = 1$$

$$\Rightarrow \quad 2\int_0^{\infty} a e^{-bx}dx = 1$$

$$\Rightarrow \quad 2a\left[\frac{e^{-bx}}{-b}\right]_0^{\infty} = 1$$

$$\Rightarrow \quad \frac{2a}{b} = 1$$

$$\Rightarrow \quad a = b/2$$

2.20 (a)

Spectral density of white noise is a constant

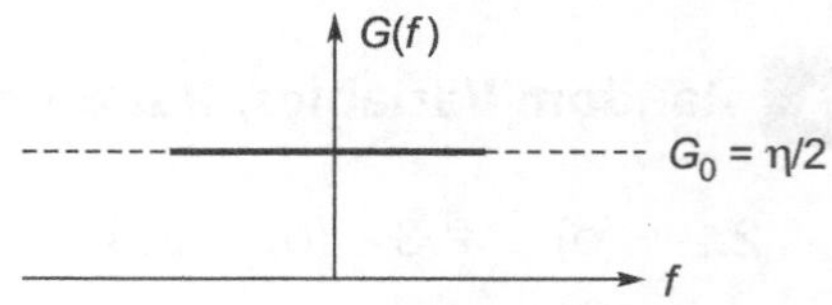

2.21 (b)

According to conjugate symmetry for real signals, the Fourier transform of a real valued sequence $x[n]$ is real and even.
Power density function
$G(\omega) = |X(e^{j\omega})|^2$
So, power density function is also real and even.

$\because \quad G(\omega) \geq 0 \ \forall \omega$

$\therefore$ Power density function is non-negative

2.23 (c)

Because, the resistor creating the normal noise is independent of the resistance.
Since, Noise Power = $P_N = kTB$

2.24 (b)

Addition of two random variable results in the convolution of their respective pdf's.

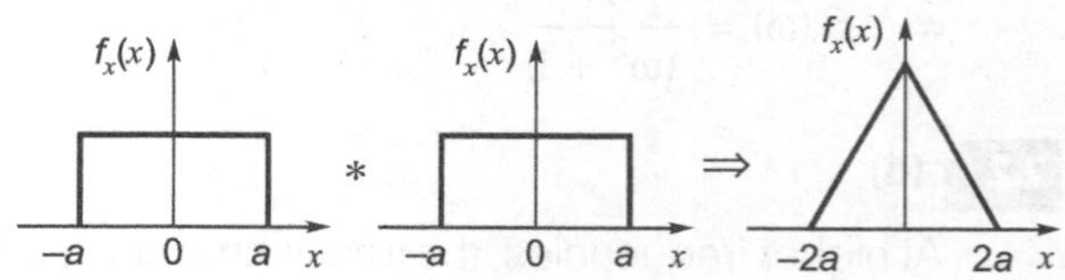

2.26 (b)

A random process is said to be ergodic:

1. When time average of a random process becomes equal to ensemble mean i.e.

$$\frac{1}{2T}\int_{-T}^{T} X(t)dt = m_x$$

2. When time average autocorrelation function becomes equal to ensemble autocorrelation

$$R_x(\tau) = E[X(t)\cdot X(t+\tau)] = \frac{1}{2}\int_{-T}^{T} X(t)\cdot X(t+\tau)dt$$

2.27 (*)

CDF function, $F_X(x) = u(x)\left(1 - \frac{x^2}{b}\right)$

Its density function

$$= \frac{d}{dx}F_X(x) = \frac{d}{dx}\left[u(x)\left(1-\frac{x^2}{b}\right)\right]$$

$$= \left[\frac{d}{dx}u(x)\right]\left(1-\frac{x^2}{b}\right) + u(x)\frac{d}{dx}\left(1-\frac{x^2}{b}\right)$$

$$= \delta(x)\left[1-\frac{x^2}{b}\right] + u(x)\left(-\frac{2x}{b}\right) = \delta(x) - \frac{2x}{b}u(x)$$

None of the given options are correct.

Note: As per RTI filed and reply of **UPSC** answer is option (b).

2.28 (b)

Spectral density of white noise is

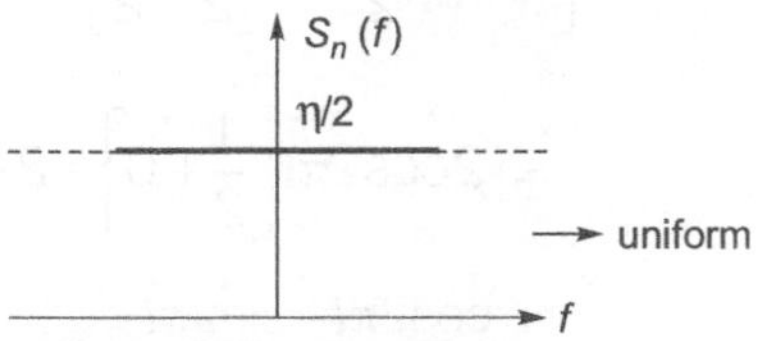

Auto correlation of white noise

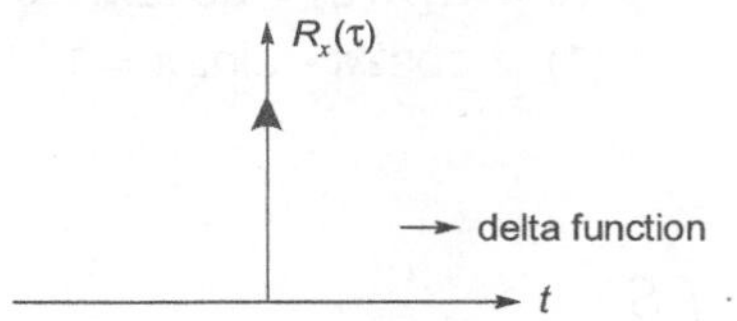

2.29 (d)

The cumulative distribution function must be non-decreasing function of x.

2.30 (a)

$$\text{Autocorrelation function} \xleftrightarrow{\text{F.T.}} \text{Power spectral density}$$

$$\Rightarrow \frac{\eta}{2}\int_{-\infty}^{\infty} e^{j\omega\tau} df \xleftrightarrow{\text{F.T.}} \frac{\eta}{2}$$

$$\Rightarrow \int_{-\infty}^{\infty} e^{j\omega\tau} df \xleftrightarrow{\text{F.T.}} 1$$

Hence, $\delta(t) = \int_{-\infty}^{\infty} e^{j\omega\tau} df = \text{delta function}$

2.31 (b)

C.D.F. of a discrete random variable has always values less than 1 and it is a non-decreasing function.

2.32 (a)

For a continuous random variable which has many values in a particular interval the value at a point is always zero.

$$P\{X = C\} = \int_{C}^{C} f_X(x)\,dx = 0$$

2.33 (a)

A random variable is nothing but the mapping of the outcome of the random experiment on the real or complex number line so that it is convenient for mathematical analysis. The outcomes of the random experiment consists of what is called as sample space

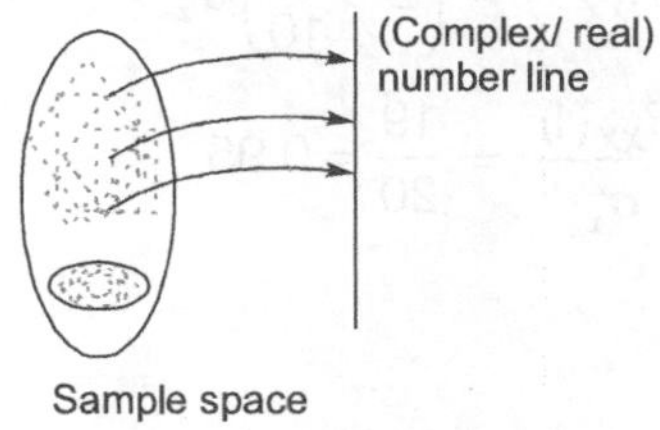

2.34 (d)

For a wide sense stationary random process correlation function depends upon time difference ($\tau = t_1 - t_2$) only. Thus, it will be always symmetric.

2.35 (b)

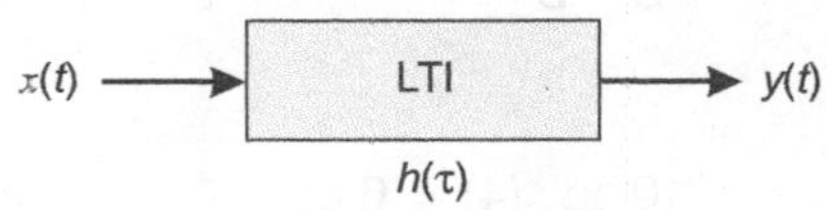

Mean $\quad E(Y) = \int_{-\infty}^{\infty} h(\tau)\cdot d\tau \cdot E[X]$

$\because \quad E(X) = 0$

$\therefore \quad E(Y) = 0$

Also, if input is WSS, then output will also be WSS.

2.36 (a)

Given, $\Phi_{xx}[m] = \sigma_x^2\,\delta[m]$

$h[n]$ is the impulse response of system

we know that $y[n] = h[n] * x[n]$

and $\Phi_{xx}[m] = \sum_{-\infty}^{\infty} x(k)x(m+k)$

then $\Phi_{xy}[m] = \sum_{-\infty}^{\infty} x(k)y(m+k)$

$$\Phi_{xy}[m] = \Phi_{xx}[m] * h(m)$$

$$= \sigma_x^2\,\delta[m] * h(m) = \sigma_x^2\,h[m]$$

2.37 (a)

Since probability of any event lies between 0 and 1, so probability density function at any point is non-negative and Area under probability density function (pdf) is equal to 1.

2.38 (a)

$$\text{Var } d(n) = \sigma(d^2(n))$$

$$d^2(n) = X^2(n) + X^2(n-1) - 2X(n)X(n-1)$$

$$\sigma(d^2(n)) = \sigma(X^2(n) + \sigma(X^2(n-1) - 2\sigma(X(n)X(n-1))$$

$$\frac{\sigma_x^2}{10} = \sigma_x^2 + \sigma_x^2 - 2R_{XX}(1)$$

$$2R_{XX}(1) = \left(2 - \frac{1}{10}\right)\sigma_x^2$$

$$\frac{R_{XX}(1)}{\sigma_x^2} = \frac{19}{20} = 0.95$$

2.42 (b)

$$CDF = \int_{-\infty}^{x} f_X(x)dx$$

$$= \int_{-\infty}^{0} ae^{bx}dx + \int_{0}^{x} ae^{-bx}dx$$

$$= \frac{a}{b} + \frac{a}{b}\left(1 - e^{-bx}\right) = \frac{a}{b}\left(2 - e^{-bx}\right)$$

2.43 (d)

$$10\log_{10}(4) = 6 \text{ dB}$$

For an alternator, Noise figure $(F) = A$

A = Attenuation = 4

So, $F = 4$

2.44 (a)

$$F_x(x) = P\{X \le x\}\ ; -\infty \le x \le \infty$$

2.45 (*)

- In this question options are not properly given.
- May be in the place of 'π', η should be given. If it is the intention of the paper setter, then the option (b) is correct.

2.46 (d)

$$E[X] = \int_{-\infty}^{\infty} x \cdot f_X(x)dx = \frac{1}{4}\int_{-1}^{3} x\,dx = 1$$

$$E[X^2] = m_x$$

$$= \int_{-\infty}^{\infty} x^2 \cdot f_X(x)dx = \frac{1}{4}\int_{-1}^{3} x^2\,dx = \frac{7}{3}$$

So, $\sigma_X^2 = E[X^2] - m_x^2 = \frac{7}{3} - 1 = \frac{4}{3} = 1.33$

2.47 (c)

$$\text{ACF} \rightarrow R(\tau) = Ke^{-3|\tau|}$$

$$PSD = \text{FT}[R(\tau)] = FT[Ke^{-3|\tau|}]$$

$$\text{PSD} = \frac{6K}{9 + \omega^2}$$

2.48 (d)

$$X(t) = 2\cos(2\pi t + y)$$

$$= 2\cos 2\pi t \cdot \cos y - 2\sin 2\pi t \cdot \sin y$$

$$E[X(t)] = 2\cos 2\pi t \cdot E[\cos y] - 2\sin 2\pi t \cdot E[\sin y]$$

$$= 2\cos 2\pi t\left[\cos 0 \times \frac{1}{2} + \cos\frac{\pi}{2} \times \frac{1}{2}\right]$$

$$-2\cos 2\pi t\left[\sin 0 \times \frac{1}{2} + \sin\frac{\pi}{2} \times \frac{1}{2}\right]$$

$$= 2\cos 2\pi t\left[\frac{1}{2} + 0\right] - 2\sin 2\pi t\left[\frac{1}{2}\right]$$

$$= \cos 2\pi t - \sin 2\pi t$$

$$\mu_X(t) = E[X(t)] = \cos 2\pi t - \sin 2\pi t$$

$$\mu_X(1) = \cos 2\pi - \sin 2\pi = 1$$

2.49 (a)

$$\left(\frac{S}{N}\right)_{i/p} = 29 \text{ dB}$$

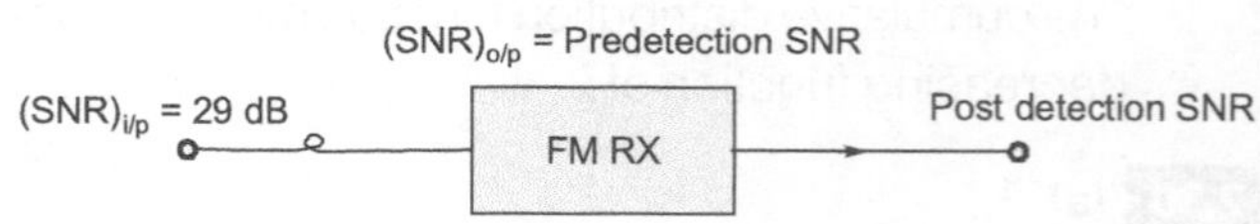

N.F = 4 dB

FM improvement factor = 16 dB

$$\text{Noise figure} = \frac{(\text{SNR})_{i/p}}{(\text{SNR})_{o/p}}$$

$$\Rightarrow (\text{N.F})_{dB} = (\text{SNR})_{i/p\,dB} - (\text{SNR})_{o/p\,dB}$$

$$(\text{SNR})_{o/p\,dB} = (\text{SNR})_{i/p\,dB} - (\text{N.F})_{dB}$$

Pre-detection SNR = 29 dB – 4 dB = 25 dB

Post-detection SNR

= Predetection SNR + FM improvement factor

= 25 dB + 16 dB = 41 dB

2.50 (b)

A random variable is a variable whose value is unknown or a function that assigns values to each of the experiments outcome.

2.51 (a)

$$V_n = \sqrt{4kTBR}$$
$$K = 1.38\times10^{-23}\ \frac{\text{Joules}}{\text{Kelvin}}$$
$$T = 290K$$
$$B = 6\text{ MHz}$$
$$R = 200 + 300 = 500\ \Omega$$
$$V_n = 7\ \mu V$$

2.52 (c)

A vocoder is an audio compressor that captures the characteristic elements of an audio signal and then using this characteristic signal to affect the other audio signal. The effect is called as vocoding and corresponding device is called as "talking vocoder". Voice speech is produced by taking impulse train as excitation. For unvoiced signal, a random white noise is used as the excitation.
Statement I is true and statement II is false.

2.53 (c)

The output noise power spectral density is given as:

$$S_0(f) = S_i(f)\cdot|H(f)|^2$$

where $H(f)$ = transfer function of RC low-pass filter

$$H(f) = \frac{1}{1+j2\pi fRC}$$

$$S_0(f) = \frac{N_0}{2}\times\frac{1}{(1+4\pi^2f^2R^2C^2)}$$

The output noise power is

$$P_{n0} = \frac{N_0}{2}\int_{-\infty}^{\infty}\frac{1}{1+4\pi^2f^2R^2C^2}\cdot df$$

$$P_{n0} = \frac{N_0}{2}\times\frac{1}{(2\pi RC)^2}\int_{-\infty}^{\infty}\frac{1}{\left(\frac{1}{2\pi RC}\right)^2+f^2}\cdot df$$

$$P_{n0} = \frac{N_0}{2}\times\frac{1}{(2\pi RC)^2}\times 2\pi RC\times\left[\tan^{-1}(2\pi fRC)\right]_{-\infty}^{\infty}$$

$$P_{n0} = \frac{N_0}{2}\times\frac{1}{2\pi RC}\times\pi$$

$$P_{n0} = \frac{N_0}{4RC}$$

Given, $N_0 = 10^{-9}$ W/Hz, $R = 1\text{ k}\Omega$ and $C = 0.1\ \mu F$.
Substituting, we get

$$P_{n0} = \frac{10^{-9}}{4\times1\times10^3\times0.1\times10^{-6}}$$

$$P_{n0} = 2.5\times10^{-6} = 2.5\ \mu W$$

2.54 (a)

The correlation coefficient between two random variables X and Y is defined as:

$$\rho_{XY} = \frac{COV(X,Y)}{E(X)\cdot E(Y)}$$

If X and Y are uncorrelated, $\rho_{XY} = 0$ which means

$COV(X, Y) = 0$

i.e., co-variance is zero.

2.55 (d)

- A signal is said to be deterministic if there is no uncertainty with respect to its value at any instant of time, or signals which can be defined exactly by a mathematical formula are known as deterministic signals.
- A random signal is a signal about which there is uncertainty before its actual occurrence. Such a signal may be viewed as belonging to an ensemble of signals with each signal within the ensemble having a certain probability of occurrence.
- EEG (Electroencephalogram) signal indicates the electrical activity of brain. They are highly random in nature.

Therefore, all the given statements are correct.

2.56 (a)

- Shot noise is a form of noise that arises because of the discrete nature of the charges carried by the charge carriers, electrons or holes.
- Thermal noise is the electronic noise generated by the thermal agitation of the electrons inside an electrical conductor at equilibrium.
- Gaussian noise denotes a kind of signal noise that has a probability density function (pdf) equal to that of the normal distribution.
- Random noise is a usually electric or acoustic signal that consists of equal amount of all frequencies.

2.57 (b)

The probability density function of a standard normal random variable X is given as,

$$f(x) = \frac{1}{\sqrt{2\pi}}\cdot e^{-x^2/2}; -\infty\le x\le\infty$$

The recording are distributed with mean, $\mu = 10$ kg and say, standard deviation, σ. If X denotes the random variable representing the measurement, then

$$X \sim N(\mu, \sigma^2)$$

$$\Rightarrow \quad Z = \frac{X-\mu}{\sigma} \sim N(0, 1)$$

i.e., normally distributed.

It is given that

$P[10 - 0.05 \le X \le 10 + 0.05] = 0.6$

$\Rightarrow \quad P[9.95 \le X \le 10.05] = 0.6$

$$\Rightarrow \quad P\left[\frac{9.95-10}{\sigma} \le \frac{X-\mu}{\sigma} \le \frac{10.05-10}{\sigma}\right] = 0.6$$

$$\Rightarrow \quad P\left[\frac{-0.05}{\sigma} \le z \le \frac{0.05}{\sigma}\right] = 0.6$$

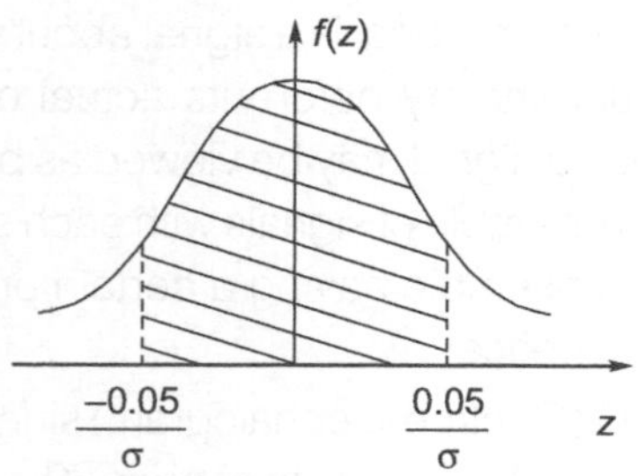

$$\Rightarrow \quad 2P\left[0 \le z \le \frac{0.05}{\sigma}\right] = 0.6$$

$$\Rightarrow \quad 2\int_0^{0.05/\sigma} \frac{1}{\sqrt{2\pi}} \cdot e^{-z^2/2} \cdot dz = 0.6$$

$$\Rightarrow \quad \frac{2}{\sqrt{2\pi}} \int_0^{0.05/\sigma} e^{-z^2/2} \cdot dz = 0.6$$

Comparing with the given expression, we get

$$\frac{0.05}{\sigma} = 0.84$$

$$\Rightarrow \quad \sigma = \frac{0.05}{0.84} = 0.06$$

2.58 (a)

For an ergodic random process,

$$\langle x(t) \cdot x(t+\tau) \rangle = E[X(t) \cdot X(t+\tau)]$$

$$\Rightarrow \quad \langle x(t) \cdot x(t+\tau) \rangle = R_X(\tau)$$

Put $\tau = 0$, we get

$$\langle x^2(t) \rangle = R_X(0)$$

where $\langle x^2(t) \rangle$ represents the power of the random process. Hence,

$$P_X = R_X(0)$$

2.59 (d)

1. A random process is called wide-sense stationary (WSS) if its mean and autocorrelation function do not change by shifts in time. The WSS processes are stationary in the strict sense if the probability density function (PDF) is independent of time. Since the PDF of the Gaussian process has only two parameters, mean and variance, which are independent of time shifts for WSS Gaussian process. Hence, if a Gaussian process is wide-sense stationary, then the process is also stationary in the strict sense.
2. If a Gaussian process $X(t)$ is applied to a stable linear filter, then the random process $Y(t)$ developed at the output of the filter is also Gaussian.
3. If the random variables $X(t_1)$, $X(t_2)$, ..., $X(t_n)$ obtained by sampling a Gaussian process $X(t)$ at time t_1, t_2, ..., t_n are uncorrelated, that is,

   $$E\left[(X(t_k) - m_{X(t_k)})\,(X(t_{0_1}) - m_{X(t_i)})\right] = 0; \; i \ne k$$

   then these random variables are also statistically independent. Hence, only the statement 2 is correct.

■■■

3 Digital Communication Systems

3.1 The bandwidth of a 'N' bit binary coded PCM signal for modulating a signal having bandwidth of 'f' Hz is

(a) $\frac{f}{N}$ Hz (b) $\frac{f}{N^2}$ Hz
(c) Nf Hz (d) N^2f Hz

[ESE-2000]

3.2 Time division multiplexing requires
(a) constant data transmission
(b) transmission of data samples
(c) transmission of data at random
(d) transmission of data of only one measurand

[ESE-2000]

3.3 A telephone channel has bandwidth B of 3 kHz and SNR ($S/\eta B$) of 30 dB. It is connected to a teletype machine having 32 different symbols. The symbol rate required for errorless transmission is nearly
(a) 1800 symbols/s (b) 3000 symbols/s
(c) 5000 symbols/s (d) 6000 symbols/s

[ESE-2000]

3.4 Quadrature multiplexing is
(a) same as FDM
(b) same as TDM
(c) a combination of FDM and TDM
(d) the scheme where same carrier frequency is used for two different signals

[ESE-2000]

3.5 A 8 kHz communication channel has an SNR of 30 dB. If the channel bandwidth is doubled, keeping the signal power constant, the SNR for the modified channel will be
(a) 27 dB (b) 30 dB
(c) 33 dB (d) 60 dB [ESE-2000]

3.6 The ramp signal $m(t)$ = at is applied to a delta modulator with sampling period T_s and step size δ. Slope overload distortion would occur if
(a) $\delta < a$ (b) $\delta > a$
(c) $\delta < aT_s$ (d) $\delta > aT_s$ [ESE-2000]

3.7 In a PCM system each quantisation level is encoded into 8 bits. The signal to quantisation noise ratio is equal to
(a) 24 dB (b) 48 dB
(c) 64 dB (d) 256 dB [ESE-2000]

3.8 If binary PSK modulation is used for transmission, the required minimum bandwidth is 9600 Hz. To reduce the transmission bandwidth to 2400 Hz, the modulation scheme to be adopted should be
(a) quadrature phase-shift keying
(b) minimum shift keying
(c) 16-ary quadrature amplitude modulation
(d) 8-ary PSK

[ESE-2000]

3.9 Four signals each band-limited to 5 kHz are sampled at twice the Nyquist rate. The resulting PAM samples are transmitted over a single channel after time division multiplexing. The theoretical minimum transmission bandwidth of the channel should be equal to
(a) 5 kHz (b) 20 kHz
(c) 40 kHz (d) 80 kHz

[ESE-2000]

3.10 Generally, a transversal type equaliser with 5 taps can take care of distorted signal due to intersymbol interference in the received signal at
(a) 4 sampling instants
(b) 5 sampling instants
(c) 9 sampling instants
(d) 10 sampling instants

[ESE-2000]

3.11 Match **List-I** (Operations) with **List-II** (Functions) and select the correct answer using the codes given below the lists:

List-I
A. Companding
B. Squelch
C. Preemphasis
D. Double conversion

List-II
1. Improving image rejection
2. Variation of step size in quantisation
3. Muting the receiver
4. Boosting of higher modulating frequencies at the transmitter

Codes:

	A	B	C	D
(a)	2	3	4	1
(b)	2	1	4	3
(c)	4	3	2	1
(d)	4	1	2	3

[ESE-2000]

3.12 For 10-bit PCM system, the signal to quantisation noise ratio is 62 dB. If the number of bits is increased by 2, then the signal to quantisation noise ratio will
(a) increase by 6 dB
(b) increase by 12 dB
(c) decrease by 6 dB
(d) decrease by 12 dB [ESE-2001]

3.13 For a given data rate, the bandwidth β_p of a BPSK signal and the bandwidth β_0 of the OQPSK signal are related as
(a) $\beta_p = \beta_0/2$ (b) $\beta_p = 2\beta_0$
(c) $\beta_p = \beta_0/4$ (d) $\beta_p = \beta_0$
[ESE-2001]

3.14 In a certain '12 channel TDM' system, it is found that channel number 3 and channel number 8 are connected to the same input signal. The technique
(a) wastes the channel capacity
(b) takes care of different sampling rates
(c) is required when different bandwidth signals are to be transmitted
(d) reduces noise [ESE-2001]

3.15 Match **List-I** with **List-II** and select the correct answer using the codes given below the lists:

List-I	List-II
A. Television	1. Either AM or FM used
B. Radio	2. Both AM & FM are used
C. Radar	3. PCM is used
D. Data communication	4. Digital system

Codes:

	A	B	C	D
(a)	4	3	1	2
(b)	2	1	3	4
(c)	4	1	3	2
(d)	2	3	1	4

[ESE-2002]

3.16 Which of the following pulse modulations is analog?
(a) PCM (b) Differential PCM
(c) PWM (d) Delta
[ESE-2002]

3.17 In asynchronous TDM, for *n* signal sources, each frame contains *m* slots, where *m* is usually
(a) less than *n* (b) 2*n*
(c) *n* (d) greater than 2*n*
[ESE-2002]

3.18 Consider a noisy binary channel with bit error probability $p_e = 5 \times 10^{-5}$. Assume 10000 bits are being transmitted over such a channel. The probability that there will be 2 or less number of bits in error is
(a) 10^{-4} (b) ~0.5
(c) ~1 (d) 10^4 [ESE-2003]

3.19 Consider the following:
1. Pulse-position modulation
2. Pulse-code modulation
3. Pulse-width modulation

Which of these communications are not digital?
(a) 1 and 2 (b) 2 and 3
(c) 1 and 3 (d) 1, 2 and 3
[ESE-2003]

3.20 The number of signalling bits per channel per frame in T1 multiplexer following CCITT hierarchy is
(a) 64000 (b) 128
(c) 4 (d) 400 [ESE-2003]

3.21 In a BPSK signal detector, the local oscillator has a fixed phase error of 20°. By what factor does this phase error deteriorate the SNR at the output?
(a) $\cos^2 20°$ (b) cos 20°
(c) cos 40° (d) cos 70°
[ESE-2004]

3.22 Consider minimum shift keying (MSK) also known as fast FSK with frequency spacing $2f_d$ between the state frequencies. r_b is the data rate. Which one of the following correctly relates r_b and f_d?
(a) $f_d = r_b/4$ (b) $f_d = r_b/2$
(c) $f_d = 3r_b/4$ (d) $f_d = r_b$ [ESE-2004]

3.23 What is the SNR improvement with FSK over ASK in most types of noise environment?
(a) 200 to 300 dB (b) 3 to 4dB
(c) 10 to 12 dB (d) 0 dB [ESE-2004]

3.24 The sum of two signals $e_1 = 3\sin(4\pi \times 10^3\, t)$ and $e_2 = 5\sin(2\pi \times 256\, t)$ is sampled at 1024 Hz. The sampled signal is passed through a low pass filter with cut off at 2048 Hz. The output of this filter will contain components at
(a) 256 Hz and 10^3 Hz
(b) 256 Hz and 1024 Hz
(c) 256 Hz only
(d) 1024 Hz only [ESE-2004]

3.25 Which one of the following statements is correct? Digital modulation techniques are used in satellite communication systems since
(a) they are easier to handle
(b) large bandwidth utilization is possible
(c) they have a higher spectral efficiency
(d) they are less prone to interference
[ESE-2004]

3.26 Assertion (A): Juxtaposition or bit-stuffing is encountered in asynchronous multiplexing.
Reason (R): The data rate at the output of a coaxial line change because of the change in the velocity of the propagating wave.
(a) Both A and R are true and R is the correct explanation of A
(b) Both A and R are true but R is NOT the correct explanation of A
(c) A is true but R is false
(d) A is false but R is true [ESE-2004]

3.27 A continuous signal has voltage range –2V to +2V. If this is quantized to 8 bits, what does the resulting signal have?
(a) 255 levels of step size 4/255
(b) 25 levels of step size 4/8
(c) 8 bits plus 2 levels at – 2 and + 2 volts
(d) 8 bits per sample if properly sampled
[ESE-2005]

3.28 Signal $x(t) = 3 \sin 2\pi 10^3\, t + 2 \sin 2\pi\, 660\, t$. At what sampling frequency should this signal be sampled to avoid aliasing?
(a) 2×660 Hz (b) 2×1000 Hz
(c) $2[1000 + 660]$ Hz (d) $2[1000 - 660]$ Hz
[ESE-2005]

3.29 An analog signal has significant spectral components from 1 kHz to 5 kHz. What is the Nyquist sampling rate for this signal?
(a) 5 k samples/s (b) 4 k samples/s
(c) 8 k samples/s (d) 10 k samples/s
[ESE-2005]

3.30 In delta modulation, when is the slope overloading noise absent?
(h is the pulse height and sampled every T_s seconds, $f(t)$ is the input signal, h and T_s are step-size of integration and sampling period, respectively:
(a) $\frac{df(t)}{dt} \le (h/T_s)$ (b) $\frac{df(t)}{dt} > (h/T_s)$
(c) $\frac{df(t)}{dt} \le (h \cdot T_s)$ (d) $\frac{df(t)}{dt} > (h \cdot T_s)$
[ESE-2005]

3.31 The constellation diagram of a modem is shown in the figure given below. What does the modem use?
(a) QAM
(b) PSK
(c) Both QAM and PSK
(d) ASK [ESE-2005]

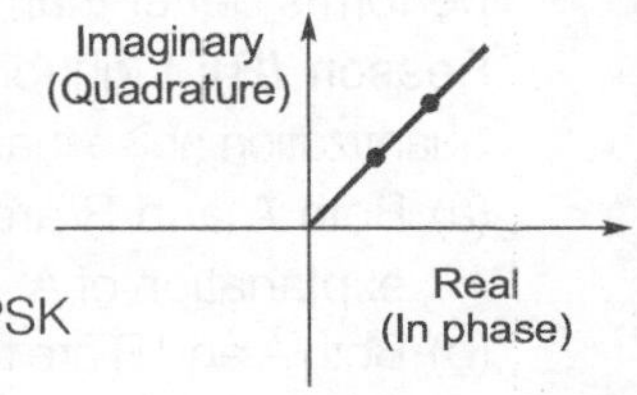

3.32 Four signals $g_1(t)$, $g_2(t)$, $g_3(t)$ and $g_4(t)$ are to be multiplexed and transmitted. $g_1(t)$ and $g_4(t)$ have a bandwidth of 4 kHz, and the remaining two signals have bandwidth of 8 kHz. Each sample requires 8 bit for encoding. What is the minimum transmission bit rate of the system?
(a) 512 kbps (b) 16 kbps
(c) 192 kbps (d) 384 kbps
[ESE-2005]

3.33 A 12 channel TDM system where each channel signal is 4 kHz is sampled at 8 kHz. What is the bandwidth requirement?
(a) 12 kHz (b) 12×4 kHz
(c) 12×8 kHz (d) $12 \times 8 \times 4$ kHz
[ESE-2005]

3.34 Which one of the following statements is correct?
(a) Equalisation is required in telephone lines but not in wireless communication
(b) Equalisation is required both in telephone lines and wireless communication
(c) Equalisation is required in wireless communication whereas this is not required in telephone lines

(d) Equalisation is not required in wireless communication since phase equalisation is required only when lines are used

[ESE-2005]

3.35 **Assertion (A):** TDM can be employed to transmit channels having unequal bandwidths.

Reason (R): If sampling theorem is strictly followed, any analog signal can be reconstructed from its samples.

(a) Both A and R are true and R is the correct explanation of A
(b) Both A and R are true but R is NOT the correct explanation of A
(c) A is true but R is false
(d) A is false but R is true [ESE-2006]

3.36 **Assertion (A):** An eight-bit PCM system performs better than a six-bit PCM system.

Reason (R): Eight-bit system produces smaller quantization noise than a six-bit system.

(a) Both A and R are true and R is the correct explanation of A
(b) Both A and R are true but R is NOT the correct explanation of A
(c) A is true but R is false
(d) A is false but R is true [ESE-2006]

3.37 Quantization process used in analog to digital conversion process is a

(a) One-to-many mapping which causes loss of information
(b) Many-to-one mapping. It causes a loss of information at the transmitter which can be recovered at the receiver
(c) One-to-one mapping
(d) Many-to-one mapping. The loss of information at the transmitter can not be recovered at the receiver

[ESE-2006]

3.38 How much is contributed by each bit in the code word of a PCM system under uniform and symmetric quantization to the signal to noise ratio?

(a) 3 dB (b) 4 dB
(c) 5 dB (d) 6 dB

[ESE-2006]

3.39 Which is the most important sub-system for recovering and reconstructing signals in a TDM system?

(a) Envelope detector followed by a low pass filter
(b) Synchronization circuit for proper timing
(c) Band pass filters to segregate channels
(d) Coherent detector to ensure frequency and phase connection

[ESE-2006]

3.40 Equalizers used in telephone systems to correct frequency distortion must have

(a) Flat-frequency response
(b) Similar frequency response as the medium to which they are connected
(c) Complementary response to that of the medium
(d) A band pass filter type of response

[ESE-2006]

3.41 What is the typical bandwidth of a 2-wire telephone line?

(a) 300 Hz to 3.4 kHz
(b) 20 Hz to 4 kHz
(c) d.c. to 4 kHz
(d) 20 Hz to 20 kHz [ESE-2007]

3.42 The constellation diagram of a modem has four states as shown in the figure given below. What is the bit rate of this modem?

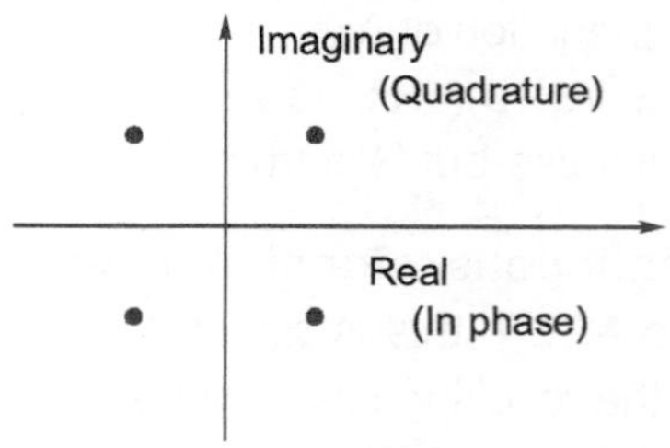

(a) Two times its baud rate
(b) Four times its baud rate
(c) Same as baud rate
(d) Sufficient information is not given to give the answer

[ESE-2007]

3.43 In a PCM system, if we increase the quantization levels from 2 to 8, how do the relative bandwidth requirements vary?

(a) Get doubled
(b) Get tripled
(c) Remain same
(d) Become eight times

[ESE-2007]

3.44 How can the quantizing noise be reduced?
(a) By using de-emphasis circuit
(b) By using RF amplifier in the receiver
(c) By increasing the number of quantization levels
(d) By sending sloping pulses **[ESE-2007]**

3.45 Which of the following pulse modulation are digital?
1. PCM
2. Differential PCM
3. PWM

Select the correct answer using the code given below:
(a) 1 and 2 only (b) 2 and 3 only
(c) 1 and 3 only (d) 1, 2 and 3
[ESE-2008]

3.46 Multiplexing is possible if signals are sampled. Two signals have bandwidths A = 0 to 4 kHz and B = 0 to 8 kHz respectively. The sampling frequency chosen is 12 kHz. Which one of the following is correct?
This choice of the sampling frequency
(a) is correct since A and B have an integral relationship of 2
(b) will not lead to aliasing
(c) does not obey sampling theorem
(d) can never lead to multiplexing
[ESE-2008]

3.47 Which one of the following is correct? In a TDM system each signal is allotted in a frame a unique and fixed
(a) frequency slot (b) time slot
(c) amplitude slot (d) phase slot
[ESE-2008]

3.48 As practical signals are time limited, they have infinite frequency bandwidth. Sampling of such signals lead to aliasing. To avoid aliasing, an anti-aliasing filter is used
(a) before the signal is reconstructed
(b) after the signal is reconstructed
(c) before the signal is sampled
(d) after the signal is sampled
[ESE-2008]

3.49 Which of the following statements about the matched filter in a communication receiver are correct?
1. Its impulse response depends on the signal shape.
2. It maximizes the SNR at the detection instant.
3. It produces ISI.
4. It may produce phase error if synchro-nization is improper.

Select the correct answer from the codes given below:
(a) 1 and 4 only (b) 1 and 2 only
(c) 2, 3 and 4 (d) 1, 2 and 4
[ESE-2009]

3.50 An audio signal is band limited to 4 kHz. It is sampled at 8 kHz. What will be the spectrum of the sampled signal?
(a) –4 kHz to 4 kHz
(b) –8 kHz to 8 kHz
(c) every 4n kHz and repeating
(d) every ±8n kHz and repeating as well as at zero (n integer)
[ESE-2009]

3.51 A signal occupies a band 5 kHz to 10 kHz. For proper error free reconstruction at what rate it should be sampled?
(a) 10 kHz (b) 20 kHz
(c) 5 kHz (d) (10 + 5) × 2 kHz
[ESE-2009]

3.52 The spectral range of a band pass signal extends from 10 MHz to 10.4 MHz. What is the minimum sampling frequency required for reconstruction?
(a) 20 MHz (b) 20.8 MHz
(c) 20.4 MHz (d) 0.8 MHz
[ESE-2009]

3.53 An audio signal is to be transmitted digitally. Which is the system best suited for good fidelity?
(a) 8 bit PCM
(b) 13 bit PCM
(c) 32 bit PCM
(d) PCM system with non-uniform quantizer
[ESE-2009]

3.54 A 13 bit PCM system performance is better than an 8 bit PCM system because:
(a) Noise is lower and is proportional to reciprocal of bandwidth
(b) Bandwidth is larger and detection is easier
(c) Quantization noise is lesser other things being equal
(d) Quantization noise is nearly zero since 2^{13} levels are available
[ESE-2010]

3.55 The spectrum of a band pass signal spans from 20 kHz to 30 kHz. The signal can be recovered ideally from the sampled values when the sampling rate is at least:

(a) 20 kHz (b) 60 kHz
(c) 50 kHz (d) 40 kHz **[ESE-2010]**

3.56 The Nyquist rate for the signal
$x(t) = 2\cos(2000\,\pi t)\cos(5000\,\pi t)$ is

(a) 7 kHz (b) 5 kHz
(c) 14 kHz (d) 10 kHz **[ESE-2010]**

3.57 The quantization noise of a PCM system depends on

(a) Number of quantization levels
(b) Step-size
(c) Both step-size and number of quantization levels
(d) Sampling rate **[ESE-2010]**

3.58 A TDM system is to be designed to multiplex the following two signals
$x_1 = 5\cos(2000\,\pi t)$
$x_2 = 2\cos(2000\,\pi t)\cos(3000\,\pi t)$
The minimum sampling rate is

(a) 4 kHz (b) 5 kHz
(c) 10 kHz (d) 6 kHz **[ESE-2010]**

3.59 Encoder

(a) Assigns quantized values
(b) Changes quantized values to binary values
(c) Changes quantized values to numerical values
(d) Changes numerical values to binary values **[ESE-2010]**

3.60 Assertion (A): When coding signals like speech signals A law or μ-law quantizers are used.
Reason (R): A-law and μ-law quantizers occupy smaller bandwidth than uniform quantizers.

(a) Both A and R are true and R is the correct explanation of A
(b) Both A and R are true but R is NOT the correct explanation of A
(c) A is true but R is false
(d) A is false but R is true **[ESE-2010]**

3.61 Assertion (A): PCM/FM systems transmit PCM pulses by modulating a high frequency carrier and hence occupy large bandwidth and eliminate distortion.
Reason (R): Large bandwidth ensures SNR trade off and hence distortionless transmission is ensured.

(a) Both A and R are true and R is the correct explanation of A
(b) Both A and R are true but R is NOT the correct explanation of A
(c) A is true but R is false
(d) A is false but R is true **[ESE-2010]**

3.62 The main factor that determines the accuracy of a reconstructed PCM signal is the

(a) signal bandwidth
(b) pulse repetition rate
(c) pulse amplitude
(d) number of bits used for quantization **[ESE-2011]**

3.63 Match **List-I** with **List-II** and select the correct answer using the code given below the lists:

List-I	List-II
A. Pilot carrier	**1.** Delta modulation
B. Tuned circuit	**2.** Frequency modulation
C. Slope overload	**3.** PCM
D. A to D converter	**4.** Single sideband AM

Code:

	A	B	C	D
(a)	3	2	1	4
(b)	4	2	1	3
(c)	3	1	2	4
(d)	4	1	2	3

[ESE-2011]

3.64 A high fidelity audio amplifier (MPEG) had frequency response as shown in below figure.

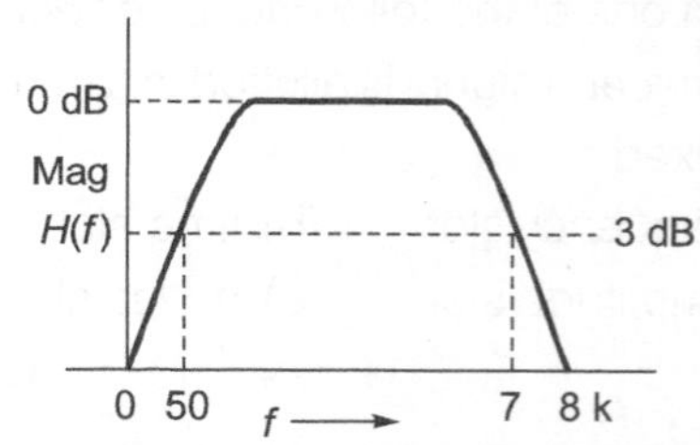

This response can be improved by which equalizer shown with frequency response $E(f)$ below?

(a) 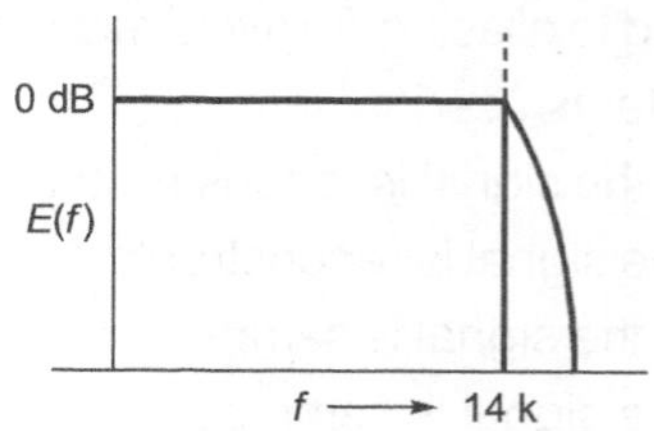

(b) 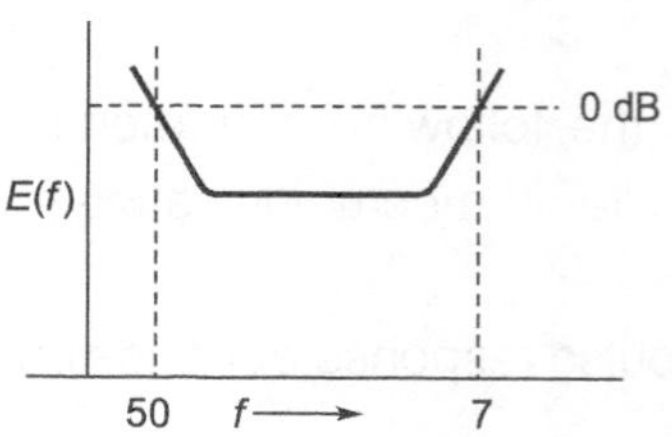

(c) $E(f)$

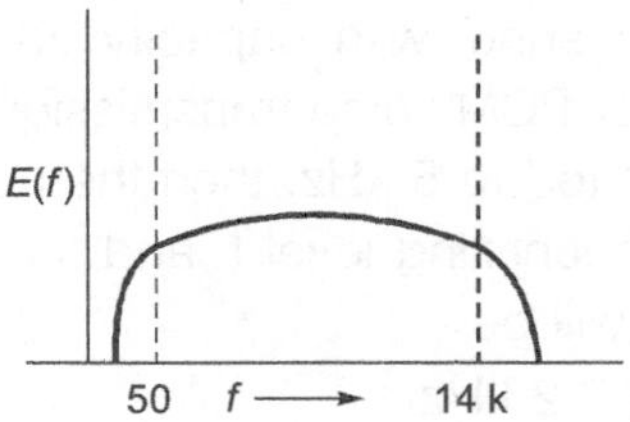

(d) $E(f)$

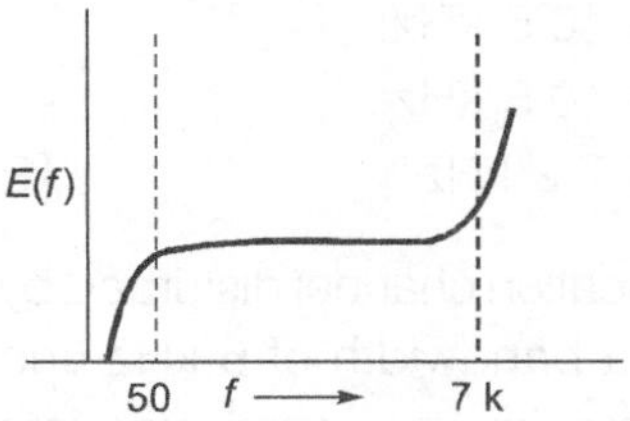

[ESE-2011]

3.65 Figure shows a block diagram of a system to recover a sampled signal shown as input.

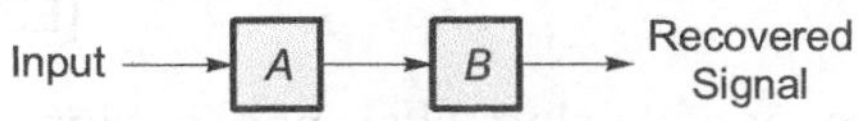

Blocks A and B can be respectively
(a) Zero order hold and low pass filter
(b) Multiplier and high pass filter
(c) Envelop detector and sampler
(d) Tuned circuit and mixer [ESE-2011]

3.66 Which one of the following scheme is a digital modulation technique?
(a) Pulse code modulation
(b) On-off keying
(c) Pulse width modulation
(d) Delta modulation [ESE-2011]

3.67 **Assertion (A):** The basic group in an FDM system occupies a band 60 kHz to 108 kHz in frequency.

Reason (R): The voice channels in FDM are band limited to 4 kHz and carrier frequencies.
$f_c = 60 + 4 \times n$ kHz are used for 12 channels in the basic group.
(a) Both A and R are true and R is the correct explanation of A
(b) Both A and R are true but R is NOT the correct explanation of A
(c) A is true but R is false
(d) A is false but R is true [ESE-2011]

3.68 In communication systems, noise due to quantization error is
(a) Linear and signal dependent
(b) Non-linear and signal dependent
(c) Linear and signal independent at low frequencies only
(d) Non-linear and signal dependent at low frequencies only

[ESE-2012]

3.69 In a PCM system, the amplitude levels are transmitted as 7 bit code-words. The sampling is done at 10 kHz. The bandwidth of the system is
(a) 10 kHz (b) 20 kHz
(c) 35 kHz (d) 70 kHz

[ESE-2013]

3.70 What multiplexing method do the CODECs use to transmit two PCM signals in opposite directions over the same transmission line?
(a) TDM
(b) FDM
(c) Amplitude-separation Multiplexing
(d) Code Division Multiplexing

[ESE-2013]

3.71 **Statement (I):** Baud Rate and symbols per second are often used interchangeably.
Statement (II): Baud Rate refers to the rate of change of signal on the transmission medium after encoding and modulation have occurred.
(a) Both Statement (I) and Statement (II) are individually true and Statement (II) is the correct explanation of Statement (I).
(b) Both Statement (I) and Statement (II) are individually true but Statement (II) is not the correct explanation of Statement (I).
(c) Statement (I) is true but Statement (II) is false.
(d) Statement (I) is false but Statement (II) is true.

[ESE-2013]

3.72 **Statement (I):** Most TDM systems use PCM to transmit analog signals.
Statement (II): PCM uses A/D conversion technique to translate analog signals into binary form for serial transmission.
(a) Both Statement (I) and Statement (II) are individually true and Statement (II) is the correct explanation of Statement (I).
(b) Both Statement (I) and Statement (II) are individually true but Statement (II) is not the correct explanation of Statement (I).
(c) Statement (I) is true but Statement (II) is false.
(d) Statement (I) is false but Statement (II) is true.

[ESE-2013]

3.73 An analog signal has a bit rate of 8000 bps and a baud rate of 1000 baud. The data elements carried by each signal and signal elements needed are respectively

(a) 8 and 256 (b) 4 and 16
(c) 2 and 4 (d) 7 and 128

[ESE-2013]

3.74 In a digital communication system employing Frequency Shift Keying (FSK), the 0 and 1 bit are represented by sine waves of 10 kHz and 25 kHz respectively. These waveforms will be orthogonal for a bit interval of

(a) 250 μsec (b) 200 μsec
(c) 50 μsec (d) 45 μsec

[ESE-2014]

3.75 In flat-top sampling a hold circuit is sometimes required. This hold circuit can be designed as a sampler followed by

(a) a shunt capacitor
(b) an envelope detector
(c) parallel RC circuit
(d) a series resistance along with parallel *RC* circuit in shunt

[ESE-2014]

3.76 As compared to A-law compander, the μ-law compander produces

(a) more companding at low amplitudes
(b) more companding at high amplitudes
(c) less companding for low amplitudes
(d) less companding for high amplitudes

[ESE-2014]

3.77 A carrier is modulated by a digital bit stream having one of the possible phase of 0°, 90°, 180° and 270°. Then the modulation is termed as

(a) BPSK (b) QPSK
(c) QAM (d) MSK

[ESE-2014]

3.78 Consider the following:

1. Pulse position modulation
2. Pulse code modulation
3. Pulse width modulation

Which of the above communication are not digital?

(a) 1 and 2 only (b) 2 and 3 only
(c) 1 and 3 only (d) 1, 2 and 3

[ESE-2014]

3.79 A binary channel with capacity 63 kbits/s is available for PCM voice transmission. If signal is band limited to 5 kHz, then the appropriate values of quantizing level L and the sampling frequency will be

(a) 64 and 7.2 kHz
(b) 32 and 10.5 kHz
(c) 64 and 10.5 kHz
(d) 32 and 7.2 kHz

[ESE-2015]

3.80 A communication channel disturbed by Gaussian noise has a bandwidth of 6 kHz and S/N ratio of 15. The maximum transmission rate that such a channel can support is

(a) 2.4 kbits/sec (b) 24 kbits/sec
(c) 32 kbits/sec (d) 48 kbits/sec

[ESE-2015]

3.81 The anti-aliasing filter has a higher cut-off frequency f_m. Its output is sampled periodically every T_S seconds. For proper recovery of the signal at the receiver, T_S must be

(a) $T_S \geq \frac{1}{(2f_m)}$ (b) $T_S \leq \frac{1}{(2f_m)}$

(c) $T_S \geq \frac{2}{f_m}$ (d) $T_S \leq \frac{2}{f_m}$

[ESE-2015]

3.82 A binary channel with capacity 36 kbits/s is available for PCM voice transmission. If signal is band limited to 3.2 kHz, then the appropriate values of quantizing level *L* and sampling frequency will be

(a) 64 and 7.2 kHz (b) 32 and 7.2 kHz
(c) 64 and 3.6 kHz (d) 32 and 3.6 kHz

[ESE-2015]

3.83 An analog signal is sampled at 36 kHz and quantized into 256 levels. The time duration of a bit of the binary coded signal is

(a) 7.43 μs (b) 6.43 μs
(c) 3.47 μs (d) 1.47 μs [ESE-2015]

3.84 Minimum sampling frequency required to reconstruct

$$y(t) = 0.5\left[\frac{\sin 2\pi 1000t}{\pi t}\right]^3 + 0.7\left[\frac{\sin 2\pi 100t}{\pi t}\right]^2$$ is

(a) 1×10^3 Hz (b) 2×10^3 Hz
(c) 4×10^3 Hz (d) 6×10^3 Hz

[ESE-2015]

3.85 In a Pulse Code Modulated system, the number of bits is increased from 7 to 8 bits. The improvement in signal to quantization noise ratio will be
(a) 2 dB (b) 4 dB
(c) 6 dB (d) 8 dB **[ESE-2016]**

3.86 If the sampling is carried out at a rate higher than twice the highest frequency of the original signal (f_{max}), then it is possible to receive the original signal from the sampled signal by passing it through
(a) A high-pass filter with the cut-off frequency equal to f_{max}
(b) A low-pass filter with the cut-off frequency equal to f_{max}
(c) A high-pass filter with the cut-off frequency greater than f_{max}
(d) A low-pass filter with the cut-off frequency greater than f_{max}
[ESE-2016]

3.87 The minimum bandwidth of the link needed for a guard band of 10 kHz frequency to prevent interference between six channels, each with 100 kHz frequency, is
(a) 425 kHz (b) 575 kHz
(c) 650 kHz (d) 725 kHz
[ESE-2016]

3.88 Which one of the following statements is correct?
(a) Sampling and quantization operate in amplitude domain.
(b) Sampling and quantization operate in time domain.
(c) Sampling operates in time domain and quantization operates in amplitude domain.
(d) Sampling operates in amplitude domain and quantization operates in time domain.
[ESE-2016]

3.89 Converting an analog signal to a digital signal is done by Sampling and
(a) Companding (b) Mixing
(c) Quantizing (d) Pre-emphasis
[ESE-2016]

3.90 In a synchronous TDM, there are four inputs and data rate of each input connection is 3 kbps. If 1 bit at a time is multiplexed, what is the duration of each frame?
(a) 0.02 ms (b) 0.03 ms
(c) 0.33 ms (d) 0.22 ms **[ESE-2016]**

3.91 **Statement (I) :** PAM can be demodulated using a suitable integrator.
Statement (II) : A suitable integrator practically acts as an envelope detector.
(a) Both Statement (I) and Statement (II) are individually true and Statement (II) is the correct explanation of Statement (I).
(b) Both Statement (I) and Statement (II) are individually true but Statement (II) is not the correct explanation of Statement (I).
(c) Statement (I) is true but Statement (II) is false.
(d) Statement (I) is false but Statement (II) is true.
[ESE-2016]

3.92 What is the minimum $\frac{E_b}{N_0}$ required to achieve a spectral efficiency of 6 bps/Hz?
(a) 5.2 (b) 5.3
(c) 10.5 (d) 15.8 **[ESE-2017]**

3.93 What is the required bandwidth of a PCM system for 256 quantization levels when 48 telephone channels, each band-limited to 4 kHz, are to be time-division multiplexed by the PCM?
(a) 6.246 MHz (b) 3.464 MHz
(c) 3.072 MHz (d) 1.544 MHz
[ESE-2017]

3.94 The basic motivation behind the development of digital modulation techniques is
(a) to develop a digital communication field
(b) to institute methods for translating digital message from baseband to passband
(c) to develop digitized versions of analog modulation schemes
(d) to improve upon pulse modulation schemes
[ESE-2017]

3.95 The received signal level for a particular digital system is −151 dBW and the effective noise temperature of the receiver system is 1500 K.

The value of $\frac{E_b}{N_0}$ required for a link transmitting 2400 bps is
(a) −12 dB (b) −1.2 dB
(c) +1.2 dB (d) +12 dB **[ESE-2017]**

3.96 Consider the following statements regarding Time-Division Multiplexing (TDM):
1. The information from different measuring points is transmitted serially on the same communication channel.

2. It involves transmission of data samples rather than continuous data transmission.
3. It is especially useful when telemetering fast-changing, high bandwidth data.

Which of the above statements are valid in respect to TDM?

(a) 1, 2 and 3 (b) 1 and 3 only
(c) 1 and 2 only (c) 2 and 3 only

[ESE-2017]

3.97 A signal is band-limited to 3.6 kHz and three other signals are band-limited to 1.2 kHz each. These signals are to be transmitted by means of time-division multiplexing. If each signal is sampled at its Nyquist rate, then the speed of the commutator (by assuming 6 samples per rotation) is

(a) 864000 r.p.m. (b) 144000 r.p.m.
(c) 86400 r.p.m. (d) 14400 r.p.m.

[ESE-2018]

3.98 1 Mbps BPSK receiver detects waveform $s_1(t) = A\cos\omega_0 t$ or $s_2(t) = -A\cos\omega_0 t$ with a matched filter. If A = 1 mV, then the average bit error probability assuming single-sided noise power expected density $N_0 = 10^{-11}$ W/Hz is nearly

(a) $Q(0.63)$ (b) $Q(0.16)$
(c) $Q(\sqrt{0.1})$ (d) $Q(\sqrt{0.3})$

[ESE-2018]

3.99 The outputs of 18 numbers of 20 Hz low-pass filters are sampled, multiplexed and A/D converted. If the sampling is at the Nyquist rate of 40 samples/s, corresponding to signal of 20 Hz bandwidth, and if we use 3 bits/sample to represent each voltage sample, the bit rate is

(a) 1.9×10^3 bits/s (b) 19×10^3 bits/s
(c) 2.16×10^3 bits/s (d) 21.6×10^3 bits/s

[ESE-2018]

3.100 The minimum double-sided Nyquist bandwidth for a QPSK modulator with an input data rate equal to 10 Mbps and a carrier frequency of 70 MHz is

(a) 72.5 MHz (b) 67.5 MHz
(c) 25.0 MHz (d) 5.0 MHz

[ESE-2018]

3.101 For a binary FSK signal with a mark frequency of 49 kHz, a space frequency of 51 kHz and an input bit rate of 2 kbps, the peak frequency deviation will be

(a) 0.5 kHz (b) 1.0 kHz
(c) 2.0 kHz (d) 4.0 kHz [ESE-2019]

3.102 In a PCM system, non-uniform quantization leads to

(a) increased quantizer noise
(b) simplification of the quantization process
(c) higher average SNR
(d) increased bandwidth [ESE-2019]

3.103 The bandwidth required in DPCM is less than that of PCM because

(a) the number of bits per code is reduced resulting in a reduced bit rate
(b) the difference signal is larger in amplitude than actual signal
(c) more quantization levels are needed
(d) the successive samples of signal often differ in amplitude

[ESE-2019]

3.104 The temperature of a particular place varies between 14°C and 34°C. For the purpose of transmitting the temperature record of that place using PCM the record is sampled at an appropriate sampling rate and the samples are quantized. If the error in the representation of the samples due to quantization is not to exceed ± 1% of the dynamic range, the minimum number of quantization levels that can be used will be

(a) 40 (b) 50
(c) 60 (d) 70 [ESE-2020]

3.105 A telephone signal band limited to 4 kHz is to be transmitted by PCM. If the signal to quantization noise is to be at least of 40 dB, the number of levels into which the signal is to be encoded will be

(a) 32 (b) 64
(c) 81 (d) 128 [ESE-2020]

3.106 To avoid slope overload error in delta modulation, the maximum amplitude of the input signal is

(a) $A \le 2\pi f_m$ (b) $A \le \sin 2\pi f_m$
(c) $A \le \dfrac{2\pi f_m}{\Delta f_s}$ (d) $A \le \dfrac{\Delta f_s}{2\pi f_m}$

[ESE-2020]

3.107 If bandwidth is of primary concern, which one of the following scheme is generally **not** considered?

(a) PSK (b) ASK
(c) DPSK (d) FSK [ESE-2020]

3.108 The process of assigning to each one of the sample values of the message signal, a discrete value from a prescribed set of a finite number of such discrete values is called
(a) Filtering (b) Noise removal
(c) Decoding (d) Quantization
[ESE-2020]

3.109 A coherent binary .phase shift keyed (BPSK) transmitter operates at a bit rate of 20 Mbps. For a probability of error $P(e)$ of 10^{-4}, the ratio $\frac{C}{N} = 8.8$ dB, the minimum theoretical $\frac{E_b}{N_0}$ ratio for a receiver bandwidth equal to the minimum double-sided Nyquist bandwidth will be
(a) 4.8 dB (b) 6.4 dB
(c) 8.8 dB (d) 10.4 dB
[ESE-2020]

3.110 Statement (I): PCM requires a very complex encoding and quantization circuitry.
Statement (II): PCM requires a less bandwidth compared to analog systems.
(a) Both Statement (I) and Statement (II) are individually true and Statement (II) is the correct explanation of Statement (I)
(b) Both Statement (I) and Statement (II) are individually true but Statement (II) is **NOT** the correct explanation of Statement (I)
(c) Statement (I) is true but Statement (II) is false
(d) Statement (I) is false but Statement (II) is true
[ESE-2020]

3.111 Which one of the following digital modulation schemes has the bit error rate as $\frac{1}{2}\text{erfc}\left(\sqrt{\frac{E_b}{N_0}}\right)$?
(a) Coherent binary PSK
(b) Coherent binary FSK
(c) DPSK
(d) Noncoherent binary FSK [ESE-2021]

3.112 Consider the following statements for baseband transmission model:
1. Channel signal-to-noise ratio is defined as the ratio of the average power of the modulated signal to the average power of noise in the message bandwidth, both measured at the receiver input.
2. Channel signal-to-noise ratio is defined as the ratio of the average power of the modulated signal to the average power of noise in the message bandwidth, both measured at the receiver output.
3. Figure of merit = $\frac{(\text{SNR})_o}{(\text{SNR})_c}$, where o stands for output and c stands for channel.
4. Figure of merit = $\frac{(\text{SNR})_c}{(\text{SNR})_o}$, where o stands for output and c stands for channel.

Which of the above statements are correct?
(a) 1 and 3 only (b) 2 and 3 only
(c) 1 and 4 only (d) 2 and 4 only
[ESE-2021]

3.113 Discrete samples of an analog signal are uniformly quantized to PCM. If the maximum value of analog sample is to be represented within 0.1% accuracy, then the minimum number of binary digits required per sample is
(a) 4 (b) 8
(c) 10 (d) 12
[ESE-2021]

3.114 Six analog information signals, each band-limited to 4 kHz, are required to be time-division multiplexed and transmitted by a TDM system. The minimum transmission bandwidth and the signalling rate of the PAM/TDM channel are respectively,
(a) 24 kHz and 48 kbps
(b) 24 kHz and 8 kbps
(c) 48 kHz and 48 kbps
(d) 48 kHz and 16 kbps [ESE-2021]

3.115 A 2000 bps binary information data signal is required to be transmitted in half-duplex mode using BFSK digital modulation technique. If the separation between two carrier frequencies is 4000 Hz, then the minimum bandwidth of the BFSK signal is
(a) 4 kHz (b) 6 kHz
(c) 8 kHz (d) 12 kHz [ESE-2021]

3.116 The temperature at a particular place varies between 14°C and 34°C. For the purpose of transmitting the temperature record of that place using PCM, the record is sampled at an appropriate sampling rate and the samples are quantized. If the error in representation of the samples due to quantization is not to exceed ± 1% of the dynamic range, what is the minimum number of quantization levels that can be used?

(a) 100 (b) 50
(c) 30 (d) 15 [ESE-2021]

3.117 Statements (I): M-ary PSK can be used to transmit digital data over a non-linear bandpass channel, whereas M-ary QAM requires the use of a linear channel.

Statement (II): M-ary PSK and M-ary QAM are examples of non-linear modulation.

(a) Both Statement (I) and Statement (II) are individually true and Statement (II) is the correct explanation of Statement (I).
(b) Both Statement (I) and Statement (II) are individually true but Statement (II) is **NOT** the correct explanation of Statement (I).
(c) Statement (I) is true, but Statement (II) is false.
(d) Statement (I) is false, but Statement (II) is true.

[ESE-2021]

3.118 Which one of the following represents the output signal-to-noise ratio of a uniform quantizer? (where P denotes average power of the message signal $m(t)$ and R denotes number of bits per sample)

(a) $\left(\frac{3P}{m_{max}^2}\right)2^{2R}$ (b) $\left(\frac{2P}{m_{max}^3}\right)2^{3R}$

(c) $\left(\frac{3P}{m_{max}^2}\right)2^{R}$ (d) $\left(\frac{3P}{m_{max}^2}\right)$

[ESE-2022]

3.119 Removing the redundant information before encoding is the basic idea behind

(a) quantization
(b) delta modulation
(c) pulse-code modulation
(d) differential pulse-code modulation

[ESE-2022]

3.120 In which one of the following techniques, RF binary signals are transmitted as given?

$0 : \sqrt{2}p'(t)\cos[\omega_c - (\Delta\omega/2)]t$

$1 : \sqrt{2}p'(t)\cos[\omega_c + (\Delta\omega/2)]t$

(a) ASK (b) FSK
(c) PSK (d) MPSK

[ESE-2022]

3.121 An audio signal $s(t) = 5\cos(2000\pi t)$ is quantized using 10-bit PCM. What is the signal-to-quantization noise ratio?

(a) 3.57×10^6 (b) 2.57×10^6
(c) 1.57×10^6 (d) 0.57×10^6

[ESE-2023]

3.122 An audio signal comprising of a single sinusoidal term $s(t) = 3\cos(2\pi 1000t)$ is quantized using DM. What is the signal-to-quantization noise ratio?

(a) 120 (b) 170
(c) 107 (d) 100

[ESE-2023]

3.123 The number of quantization levels is increased from 4 to 64. The bandwidth required for the transmission of a PCM signal increases by a factor of

(a) $\frac{1}{3}$ (b) $\frac{1}{4}$

(c) $\frac{1}{5}$ (d) $\frac{1}{6}$

[ESE-2023]

3.124 Statement (I) : In the sampling and quantizing operations, errors are introduced into the digital signal. These errors are reversible and it is possible to produce an exact replica of the original analog signal from its digital representation.

Statement (II) : The use of digital communication offers flexibility and compatibility in that the adoption of a common digital format makes it possible for a transmission system to sustain many different sources of information in a flexible manner.

(a) Both Statement (I) and Statement (II) is the correct explanation of Statement (I).
(b) Both Statement (I) and Statement (II) are individually true, but Statement (II) is not the correct explanation of Statement (I).
(c) Statement (I) is true, but Statement (II) is false.
(d) Statement (I) is false, but Statement (II) is true

[ESE-2023]

3.125 In which one of the following modulation schemes is the quantizer a 1-bit (two-level) quantizer with magnitudes $\pm\Delta$?

(a) PCM (b) DPCM
(c) DM (d) ADM

[ESE-2024]

3.126 The average probability of bit error in each channel of the coherent QPSK system is (where E and N_0 are the signal energy and noise spectral density respectively)

(a) $\frac{1}{2}\text{erfc}\left(\sqrt{\frac{N_0}{2}}\right)$ (b) $\frac{1}{2}\text{erfc}\left(\sqrt{\frac{E/2}{N_0}}\right)$

(c) $\frac{1}{2}\text{erfc}\left(\sqrt{\frac{N_0/2}{E}}\right)$ (d) $\text{erfc}\left(\sqrt{\frac{E}{N_0}}\right)$

[ESE-2024]

3.127 Consider the following statements regarding the communication system:

1. The baseband signal contains frequencies in the audio frequency range, some form of frequency-band shifting must be employed for the radio communication system to operate properly.
2. Digital repeaters reconstruct the received digital data by minimizing the effect of signal attenuation and distortion. Thus channel noise is not cumulative.
3. A digital signal can be propagated through a limited distance only, otherwise higher signal attenuation can introduce unacceptable errors.

Which of the above statements are correct?

(a) 1 and 2 only (b) 1 and 3 only
(c) 2 and 3 only (d) 1, 2 and 3

[ESE-2024]

Answers Digital Communication Systems

3.1 (c)	3.2 (b)	3.3 (d)	3.4 (d)	3.5 (a)	3.6 (c)	3.7 (b)	3.8 (c)	3.9 (c)
3.10 (b)	3.11 (a)	3.12 (b)	3.13 (b)	3.14 (b)	3.15 (b)	3.16 (c)	3.17 (a)	3.18 (c)
3.19 (c)	3.20 (c)	3.21 (a)	3.22 (a)	3.23 (b)	3.24 (*)	3.25 (d)	3.26 (c)	3.27 (a, d)
3.28 (b)	3.29 (d)	3.30 (a)	3.31 (d)	3.32 (d)	3.33 (b)	3.34 (b)	3.35 (a)	3.36 (a)
3.37 (d)	3.38 (d)	3.39 (b)	3.40 (c)	3.41 (c)	3.42 (a)	3.43 (b)	3.44 (c)	3.45 (a)
3.46 (c)	3.47 (b)	3.48 (c)	3.49 (b)	3.50 (d)	3.51 (a)	3.52 (d)	3.53 (d)	3.54 (d)
3.55 (a)	3.56 (a)	3.57 (c)	3.58 (b)	3.59 (b)	3.60 (c)	3.61 (c)	3.62 (d)	3.63 (b)
3.64 (b)	3.65 (a)	3.66 (*)	3.67 (a)	3.68 (b)	3.69 (c)	3.70 (a)	3.71 (a)	3.72 (a)
3.73 (a)	3.74 (b)	3.75 (a)	3.76 (a)	3.77 (b)	3.78 (c)	3.79 (c)	3.80 (b)	3.81 (b)
3.82 (b)	3.83 (c)	3.84 (d)	3.85 (c)	3.86 (b)	3.87 (c)	3.88 (c)	3.89 (c)	3.90 (c)
3.91 (a)	3.92 (c)	3.93 (d)	3.94 (b)	3.95 (d)	3.96 (c)	3.97 (b)	3.98 (c)	3.99 (c)
3.100 (d)	3.101 (b)	3.102 (c)	3.103 (a)	3.104 (b)	3.105 (d)	3.106 (d)	3.107 (d)	3.108 (d)
3.109 (c)	3.110 (c)	3.111 (a)	3.112 (a)	3.113 (c)	3.114 (a)	3.115 (b)	3.116 (b)	3.117 (c)
3.118 (a)	3.119 (d)	3.120 (b)	3.121 (c)	3.122 (*)	3.123 (a)	3.124 (d)	3.125 (c)	3.126 (b)
3.127 (d)								

Explanations Digital Communication Systems

3.1 (c)

$$\text{Bandwidth} = \frac{R_b}{2} = \frac{Wf_s}{2}$$

where, R_b = Bit rate (bits/sec)

f_s = Sampling frequency

and $f_s = 2f$

where, f = Signal bandwidth

$$\therefore \quad \text{B.W.} = \frac{N2f}{2} = Nf \text{ Hz}$$

3.2 (b)

TDM requires transmission of data samples.

3.3 (d)

$$10\log_{10}(\text{SNR}) = 30 \text{ dB}$$

$$\text{SNR} = 10^3 = 1000$$

The source is producing 32 symbols. By assuming equiprobable symbols case (or) fixed length coding,

Bit rate, $R_{b(\max)} = R_s \log_2(32) = 5R_s$

R_s = Symbol rate

For error-free transmission,

$$R_b \le C = B\log_2(1 + \text{SNR})$$

So, $5R_s = 3000 \log_2(1001)$ symbols/sec

$R_s = 5980 \simeq 6000$ symbols/sec

3.4 (d)

Quadrature multiplexing is a scheme where same carrier frequency is used in phase quadrature for two different signals both having the same bandwidths.

3.5 (a)

$B_2 = 2B_1$ and $(\text{SNR})_1 = 30$ dB

$$B_1(\text{SNR})_1 = B_2(\text{SNR})_2$$

$$\Rightarrow \quad (\text{SNR})_2 = \frac{B_1}{B_2} \times (\text{SNR})_1 = \frac{1}{2} \times (\text{SNR})_1$$

$\Rightarrow 10 \log (\text{SNR})_2 = -10 \log 2 + 10 \log (\text{SNR})_1$

$\Rightarrow (\text{SNR})_2 \text{ (dB)} = -3 + (\text{SNR})_1 \text{ (dB)}$

$\Rightarrow (\text{SNR})_2 \text{ (dB)} = -3 + 30 = 27$ dB

3.6 (c)

Slope overload distortion would occur if

$$a > \frac{\delta}{T_s} \Rightarrow \delta < aT_s$$

3.7 (b)

$$\frac{S}{N_q}(\text{dB}) \approx 6n$$

where $n \rightarrow$ number of bits

$$\frac{S}{N_q}(\text{dB}) \approx 6 \times 8 \approx 48 \text{ dB}$$

3.8 (c)

$$\text{BW}|_{\text{BPSK}} = 2R_b$$

$$\text{BW}|_{\text{QPSK}} = R_b$$

$$\text{BW}|_{\text{16-QAM}} = \frac{R_b}{2}$$

Bandwidth of 16-ary QAM modulation is one-fourth of the bandwidth required for BPSK.

3.9 (c)

$$\text{Transmission bandwidth} = \text{BW} = \frac{nmf_s}{2}$$

where, $n \rightarrow$ number of bits in a signal

$m \rightarrow$ number of signals

$f_s \rightarrow$ sampling frequency

$m = 4$, $f_s = 2f_{s\min} = 4f_m$ (Assuming $n = 1$)

$$\Rightarrow \quad \text{BW} = \frac{4 \times 4 \times 5 \text{ kHz}}{2} = 40 \text{ kHz}$$

3.10 (b)

Number of taps = Number of ISI check instances.

3.11 (a)

Companding $\rightarrow$ Variation of step size in quantisation.

Squelch $\rightarrow$ Muting the receiver.

Preemphasis $\rightarrow$ Boosting of higher modulating frequencies at the transmitter.

Double conversion $\rightarrow$ Improving image rejection.

3.12 (b)

$$\frac{S}{N_q} \cong 6n$$

In the number of bits is increased by 2, then the S/N_q will be increased by $6 \times 2 = 12$ dB.

3.13 (b)

For BPSK,

$$BW = \beta_P = 2f_b$$

For OQPSK,

$$BW = \beta_O = f_b$$

$$\Rightarrow \quad \beta_P = 2\beta_O$$

3.14 (b)

The input connected twice will have different sampling rate.

3.15 (b)

Television → Both AM and FM are used
Radio → Either AM or FM used
Radar → PCM is used
Data communication → Digital system

3.16 (c)

Only PWM in the given modulations is analog.

3.17 (a)

In asynchronous TDM, number of slots in each frame is less than that of number of sources.

3.18 (c)

The probability of occurring r events out of n is $P = {}^nC_r(p)^r(1-p)^{n-r}$ where p is the probability of occurring of event.

So, probability that there will be 2 or less number of bits in error is

$$P = {}^{10000}C_0(5\times10^{-5})^0(1-5\times10^{-5})^{10000}$$
$$+\ {}^{10000}C_1(5\times10^{-5})^1(1-5\times10^{-5})^{9999}$$
$$+\ {}^{10000}C_2(5\times10^{-5})^2(1-5\times10^{-5})^{9998}$$
$$P = 0.98$$
$$\Rightarrow \quad P \approx 1$$

3.19 (c)

PPM and PWM are analog communications. PCM is digital communication.

3.21 (a)

In BPSK signal detector, the phase error ϕ deteriorates the SNR at output by a factor of $\cos^2\phi$.

3.22 (a)

In the MSK system, given that,

$$f_H - f_L = 2f_d \quad \text{...(i)}$$

$$r_b = \text{data rate} = \frac{1}{T_b} = f_b$$

Now we have also,

$$f_H - f_L = \frac{f_b}{2} = \frac{r_b}{2} \quad \text{...(ii)}$$

From equation (i) and (ii),

$$2f_d = \frac{r_b}{2}$$

$$\Rightarrow \quad f_d = \frac{r_b}{4}$$

3.23 (b)

The probability of error of BASK and BFSK systems are,

$$P_{e(BASK)} = Q\left(\sqrt{\frac{E_1}{2N_0}}\right)$$

$$P_{e(BASK)} = Q\left(\sqrt{\frac{E_1}{N_0}}\right)$$

For same R_b, E_1 and P_e,

$$\frac{(SNR)_{BFSK}}{(SNR)_{BASK}} = 2 \Rightarrow 3.01 \text{ dB}$$

3.24 (*)

$$f_{m_1} = 2000 \text{ Hz} \quad ; \quad f_{m_2} = 256 \text{ Hz}$$
$$f_s = 1024 \text{ Hz} \quad ; \quad f_c = 2048 \text{ Hz}$$

The output of the filter will contain components f_{m1}, f_{m2}, $nf_s \pm f_{m1}$, $nf_s \pm f_{m2}$ provided, it would not be greater than f_c.

256 Hz, 2000 Hz, $n(1024) \pm (256)$, $n(1024) \pm (2000)$.
256 Hz, 2000 Hz, 768 Hz, 976 Hz, 1792 Hz, 48 Hz, 2816 Hz, 1072 Hz, 3840 Hz, 2096 Hz ...

Output of the filter contents components at, 48 Hz, 256 Hz, 768 Hz, 976 Hz, 1072 Hz, 1792 Hz, 2000 Hz.

None of the given options is matching with the correct answer.

3.25 (d)

Digital modulations technique are used in satellite communication system because they are less prone to interference.

3.26 (c)

The data rate depends on the bit duration but not on the velocity of the propagating wave.

Due to change in the velocity of propagation, the propagation delay of each bit change uniformly and there will not be any change in the bit duration or the difference between the adjacent pulses.

3.28 (b)

$$f_m = 1000 \text{ Hz}$$

To avoid aliasing, $f_s \geq 2f_m$.
So, signal should be sampled at 2 × 1000 Hz.
Actually, it should be given to determine the minimum sampling rate in the question; then only option (b) can be chosen, otherwise option (c) also satisfies the condition.
These type of ambiguities generally presents in UPSC question papers.

3.29 (d)

$$f_s = \frac{2f_H}{K}\ ;\quad \text{where, } K = \frac{f_H}{f_H - f_L} = \frac{5}{5-1} = 1.25$$

Greatest integer lower than 1.25 is 1.0.

$$\therefore \quad f_s = \frac{2\times 5}{1} = 10 \text{ k samples/sec}$$

3.30 (a)

Slope overloading noise is absent when the slope of input signal is less than or equal to that of approximated signal.
So, the condition required is

$$\frac{df(t)}{dt} \leq \frac{h}{T_s}$$

3.32 (d)

Bit rate, $r_b = nf_s$ and $f_{s_{\min}} = N \times 2f_m$

$$f_{s_{\min}} = 2\times 2\times 4 + 2\times 2\times 8$$

$$f_{s_{\min}} = 48 \text{ k samples/sec.}$$

$$r_{b_{\min}} = 8 \times 48 \text{ k} = 384 \text{ kbps}$$

3.33 (b)

Number of channels, $m = 12$

$$f_m = 4 \text{ kHz}$$

Number of bits in a signal = n

$$f_s = 8 \text{ kHz}$$

$$\text{Bit rate} = r_b = nmf_s$$

$$r_b = 1 \times 12 \times 8$$

(Assuming each signal has 1-bit)

$$\text{Bandwidth} = \frac{r_b}{2} = 12 \times 4 \text{ kHz}$$

3.36 (a)

$$(\text{SNR})_{dB} = 6n$$

Thus increasing number of bits, the performance will be better.
Number of bits ↑ → step size ↓ → quantization noise ↓.

3.38 (d)

$\left(\frac{S}{N_q}\right)_{dB} \cong 6n$ where n is number of bits. So, each bit contributes 6 dB.

3.39 (b)

The most important and most necessary thing is synchronization in TDM system.

3.41 (c)

Typical bandwidth of a 2-wire telephone line is from d.c. to 4 kHz.

3.42 (a)

For the constellation diagram shown in the figure each baud carries 2 bits.

3.43 (b)

In PCM, the BW required = nf_m
where, n = number of bits
Levels are increased from 2 to 8 means.

$$m = 2^n \text{ So, } n = 1 \text{ to } 3$$

Initially, when

$$m = 2 \text{ then } n = 1,$$

$$\text{B.W} = 1 \cdot f_m = f_m$$

$$m = 8, \text{ then } n = 3,$$

$$\text{B.W.} = 3 \cdot f_m$$

B.W. get tripled.

3.46 (c)

Here $f_m = 8$ kHz
Sampling frequency $f_s = 12$ kHz
Since $f_s < 2f_m$, therefore, the choice of sampling frequency does not obey sampling theorem.

3.48 (c)

Anti aliasing restricts the original signal in the 2 f_m bandwidth before we sample it.

3.50 (d)

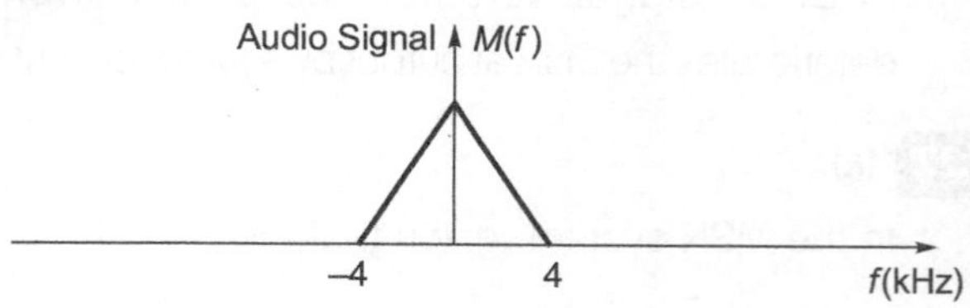

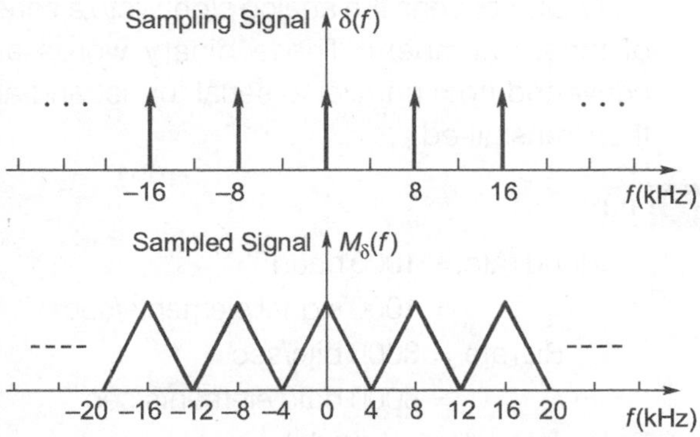

The spectrum of the sampled signal is at every $8n$ kHz (n = 0, ± 1, ± 2, ...)

3.51 (a)

$$f_s \geq \frac{2f_H}{K}$$

Where k = greatest integer $\left(\frac{f_H}{f_H - f_L}\right)$

Given that f_L= 5 kHz, f_H = 10 kHz

$$k = \text{greatest integer}\left(\frac{10}{10-5}\right) = 2$$

$$f_s = \frac{2\times 10}{2} = 10 \text{ kHz}$$

3.52 (d)

$$f_s = \frac{2f_H}{K}$$

where, K = Greatest integer lower than $\left[\frac{f_H}{f_H - f_L}\right]$

$$= \frac{10.4}{10.4-10} = 26$$

$$\therefore \quad f_s = \frac{2\times 10.4}{26} = 0.8 \text{ MHz}$$

3.53 (d)

Nonuniform quantizer improves the fidelity.

3.54 (d)

Quantization level = $M = 2^n$

where n is number of bits.

3.55 (a)

$$\text{Sampling rate} = \frac{2f_H}{K}$$

where, $K = \text{int}\left[\frac{f_H}{f_H - f_L}\right] = \text{int}\left[\frac{30}{10}\right] = 3$

$$\Rightarrow \text{Sampling rate} = \frac{2\times 30 \text{ kHz}}{3}$$

$$\Rightarrow \text{Sampling rate} = 20 \text{ kHz}$$

3.56 (a)

$$x(t) = 2\cos(2000\pi t)\cos(5000\pi t)$$

$$x(t) = x_1(t)\cdot x_2(t)\cdot x_3(t)....$$

Then, Nyquist rate

$$f_s = 2[f_{m1} + f_{m2} + f_{m3} + ...]$$

Here, f_{m1} = 1000 Hz;

f_{m2} = 2500 Hz

$\therefore \quad f_s = 2[1000 + 2500] = 7000$ Hz

$\Rightarrow \quad f_s = 7$ kHz

3.57 (c)

Quantization noise power

$$P = \frac{\Delta^2}{12}$$

$\therefore$ RMS value of noise $= \sqrt{\frac{\Delta^2}{12}} = \frac{\Delta}{2\sqrt{3}}$

as $\quad \Delta = \frac{V_H - V_L}{M}$

where, M = Number of quantization level.

and $V_H - V_L$ = Difference of highest and lowest level of voltage.

Hence quantization noise depends on both step-size and no. of quantization levels.

3.58 (b)

$$x_1 = 5\cos(2000\pi t)$$

$\Rightarrow \quad f_{m1}$ = 1000 Hz

$$x_2 = 2\cos(2000\pi t)\cos(3000\pi t)$$

$\Rightarrow \quad f_{m2}$ = 1000 + 1500 = 2500 Hz

minimum sampling rate is

$$f_{smin} = 2\max(f_{m1}, f_{m2}) = 2\times 2500 = 5 \text{ kHz}$$

3.59 (b)

Encoder assigns binary bits to the quantized data.

3.62 (d)

In PCM signal quantization noise depends upon only number of bits used for quantization.

3.63 (b)

The Correct match is:

Pilot carrier — Single sideband AM

Tuned circuit — Frequency modulation

Slope overload — Delta modulation

A to D converter — PCM

3.64 (b)

Frequency response of equalizer is given as

$$E(f) = \frac{1}{H(f)}$$

or $E(f)\,H(f) = 1$

3.65 (a)

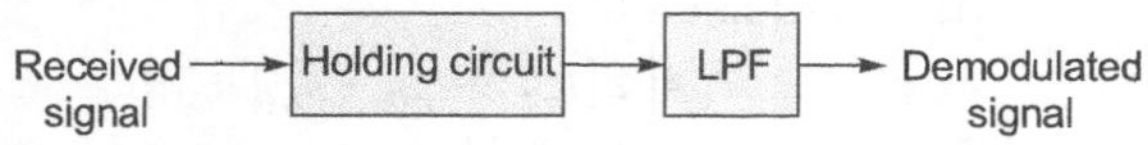

3.66 (*)

Pulse code modulation, on-off keying and delta modulation; all of these are digital modulation techniques.

3.67 (a)

Voice Channels	Bandwidth	Spectrum	Terminology
12	48 kHz	60-108 kHz	Group
60	240 kHz	312-552 kHz	Super group
300	1.232 MHz	812-2044 kHz	CCITT Master-group
600	2.520 MHz	564-3084 kHz	Master-group
900	3.872 MHz	8.516-12.383 MHz	CCITT Supermaster -group
3600	16.984 MHz	0.564-17.548 MHz	Jumbo-group
10800	57.442 MHz	3.124-60.566 MHz	

3.69 (c)

Minimum bandwidth can be taken as default bandwidth.

$$\text{BW} = \frac{nf_s}{2} = \frac{7\times 10}{2} = 35 \text{ kHz}$$

3.70 (a)

Codecs are used to digitize voice, music and video. TDM sequences analog data using codec, which samples the data into a stream. At the receiving end, a codec reassembles the data from the TDM slices.

3.72 (a)

Pulse code modulation is a form of digital modulation in which the code refers to a binary word that represents digital data. Multiple channels of serial digital data are transmitted with TDM by allowing each channel a time-slot in which to transmit one binary word of data.

The ADC converts the analog signal into a series of binary numbers. These binary words are converted from parallel to serial format and are then transmitted.

3.73 (a)

$$\text{Baud rate} = 1000 \text{ baud}$$
$$= 1000 \text{ signal elements/sec}$$
$$\text{Bit rate} = 8000 \text{ bits/sec}$$
$$= 8000 \text{ data elements/sec}$$

Data element per signal elements

$$= \frac{8000}{1000}$$

$$\boxed{N = 8 \text{ bits}}$$

So, signal elements needed

$$= 2^N = 2^8 = 256$$

3.74 (b)

$$(f_1 - f_2) = \frac{n}{2T_b}$$

$$(25 - 10)\times 10^3 = \frac{n}{T_b}\times\frac{1}{2}$$

or $$T_b = \frac{n}{30\times 10^3}$$

for $$n = 6$$
$$T_b = 200\ \mu\text{sec}$$

3.75 (a)

Hold circuit is basically a capacitor.

3.77 (b)

For *M*-array signaling,
Phase shift between the symbols.

$$\theta = \frac{2\pi}{m}$$

where, m = number of symbols = 2^n
for QPSK,

$$m = 4$$
$$\because \quad n = 2$$
$$\therefore \quad Q = 90°$$

3.78 (c)

PCM is a digital modulation technique.

3.79 (c)

$$f_s \geq 2 f_m$$
$$f_s \geq 2\times 5 \text{ kHz}$$

$$f_s \geq 10 \text{ kHz}$$

$$\frac{r_b}{f_s} = n$$

If $f_s = 10.5 \text{ kHz}$

$$n = \frac{63\text{ k}}{10.5\text{ k}} = 6$$

$$L = 2^6 = 64$$

3.80 (b)

Channel capacity $= B\log_2(1 + S/N)$

$$= 6\text{ k} \times \log_2(1 + 15)$$

$$= 6\text{ k} \times \log_2(16)$$

$$= 6\text{ k} \times 4 = 24\text{ k bits/sec.}$$

3.81 (b)

For an anti aliasing filter

$$f_s \geq 2f_m$$

$$\frac{1}{T_s} \geq 2f_m$$

$$T_s \leq \frac{1}{2f_m}$$

3.82 (b)

$$f_s \geq 2f_m$$

$$\geq 2 \times 3.2\text{ k} = 6.4 \text{ kHz}$$

$$f_s = 7.2 \text{ kHz (as given)}$$

$$r_b = nf_s$$

$$n = \frac{36k}{7.2k} = 5$$

$$L = 2^5 = 32$$

3.83 (c)

$$f_s = 36 \text{ kHz}$$

$$L = 256$$

$$n = \log_2 256$$

$$R_b = nf_s$$

$$T_b = \frac{1}{nf_s} = \frac{1}{8 \times 36\text{ k}} = 3.47\ \mu\text{sec.}$$

3.84 (d)

Maximum frequency present = 3 × 1000 Hz
= 3 kHz
Sampling frequency required = 3 × 2 = 6 kHz

3.85 (c)

If k bits increased, then SQNR is increased by '6 k' dB.

if one bit is increased, then SQNR is increased by '6 dB'

3.86 (b)

In sampling process, if sampling frequency is greater than nyquist frequency, for reconstruction of original signal, LPF is used.

3.87 (c)

6 signals multiplexed. So 5 guard bands will exist

BW = 6 × 100 kHz + 5 × 10 kHz = 650 kHz

3.88 (c)

Sampling is done in time domain and quantization is done in amplitude domain.

3.89 (c)

Analog signal can be converted to digital by performing sampling quantization and coding.

3.90 (c)

Number of inputs = 4
Data rate of each input = 3 kbps
Overall data rate = 12 kbps
Duration of each bit = (1/12) ms
Bits/frame = 4
Duration of each frame = (4/12) ms = 0.33 ms

3.91 (a)

PAM can be demodulated by a process of envelope detection, where suitable integrator will acts as envelope detector.

3.92 (c)

Spectral efficiency

$$= \frac{R_b}{B} = 6\frac{\text{bits/s}}{\text{Hz}}$$

$$C = R = B\log_2\left(1 + \frac{S}{N}\right)$$

$$\frac{R}{B} = \log_2\left(1 + \frac{S}{N}\right)$$

$$6 = \log_2\left(1 + \frac{S}{N}\right)$$

$$1 + \frac{S}{N} = 2^6$$

$$\frac{S}{N} = 2^6 - 1$$

$$\frac{E_b R_b}{N_0 B} = 2^6 - 1$$

$$6 \times \frac{E_b}{N_0} = 2^6 - 1$$

$$\frac{E_b}{N_0} = \frac{1}{6}[2^6 - 1] = 10.5$$

3.93 (d)

$$BW = \frac{R_b}{2} = \frac{Nnf_s}{2}$$

$$= \frac{48 \times 8 \times (2 \times 4)}{2} \text{ kHz} = 1.536 \text{ MHz}$$

3.95 (d)

$$E_b = P_r \times T_b = \frac{P_r}{R_b}$$

$$P_r = -15 \text{ dBW} = 10^{-15.1}$$

$$= 7.94 \times 10^{-16} \text{ W}$$

$$N_0 = kT_e = 1.38 \times 10^{-23} \times 1500$$

$$\frac{E_b}{N_0} = \frac{7.94 \times 10^{-16}}{(1.38 \times 10^{-23} \times 1500)(2400)}$$

$$\text{in dB} \rightarrow 10\log_{10}\frac{E_b}{N_0} = +12 \text{ dB}$$

3.96 (c)

TDM is not suitable for fast changing, high bandwidth signals.

3.97 (b)

Speed of the commutator = N.R. of minimum frequency

$$f_s = 2 \times f_m = 2 \times 1.2 \text{ kHz}$$

$$f_s = 2.4 \text{ kHz}$$

Speed of the commutator is 2400 rations/sec
= 2400 × 60 rotations/min = 144000 rpm

3.98 (c)

Average bit error probability of BPSK

$$= Q\left(\sqrt{\frac{A_c^2 T_b}{N_o}}\right)$$

Given, $A_c = 1$ Mbps

$R_b = 1$ Mbps – $T_b = 10^{-6}$ sec

$N_o = 10^{-11}$ W/Hz

$$\text{Bit error probability} = Q\left(\sqrt{\frac{10^{-12}}{10^{-11}}}\right) = Q(\sqrt{0.1})$$

3.99 (c)

$$N = 18$$

$$n = 3 \text{ bits/s}$$

$$f_s = 40 \text{ Hz}$$

$$R_b = Nnf_s = 18 \times 3 \times 40$$

$$= 2160 \text{ bits/s} = 2.16 \times 10^3 \text{ bits/sec}$$

3.100 (d)

$$\text{Min. BW} = \frac{R_b}{\log_2 M}$$

$$M = 4$$

$$= \frac{R_b}{2} = \frac{10 \text{ Mbps}}{2} = 5 \text{ Mbps}$$

3.101 (b)

$$f_L = 49 \text{ kHz}$$

$$R_b = 2 \text{ kbps}$$

$$f_H = 51 \text{ kHz}$$

Peak frequency deviation

$$\Rightarrow \quad 2\Delta f = f_H - f_L$$

$$2\Delta f = 51 \text{ k} - 49 \text{ k}$$

$$\Rightarrow \quad \Delta f = 1 \text{ kHz}$$

3.104 (b)

Given that quantization error,

$(Q_c)_{max} \leq 1\% \times$ Dynamic range (V_{p-p})

$\because$ Maximum quantization error $= \frac{\Delta}{2}$

$$\frac{\Delta}{2} \leq \frac{1}{100} \times V_{p-p}$$

$$\frac{1}{2} \times \frac{V_{p-p}}{L} \leq \frac{1}{100} \times V_{p-p} \qquad \left[\Delta = \frac{V_{p-p}}{L}\right]$$

$$\frac{1}{L} \leq \frac{1}{50}$$

$$L \geq 50$$

$$L_{min} = 50$$

3.105 (d)

Signal to quantization error (dB) for a PCM system
= 1.8 + 6.02 n

$$(SQNR)_{dB} \geq 40 \text{ dB}$$

$$1.8 + 6n \geq 40$$

$$n \geq 6.36$$

$$n_{min} = 7$$

$$\because \quad L = 2^n$$

$$L_{min} = 128$$

3.106 (d)

To overcome slope over load error

$$\Rightarrow \quad \frac{\Delta}{T_s} \geq 2\pi f_m A_m$$

$$A_m \leq \frac{\Delta}{T_s \times 2\pi f_m}$$

$$A_m \leq \frac{\Delta f_s}{2\pi f_m}$$

3.107 (d)

Bandwidth of FSK is greater than ASK, PSK and DPSK.

3.109 (c)

$$E_b = P \times T_b = C \times T_b$$

where, C = Carrier power

$$\frac{E_b}{N_o} = \frac{C \times T_b}{N/B} = \frac{C \times B}{N \times R_b}$$

$$\because \quad N = N_o B$$

$$N_o = \frac{N}{B}$$

$\because$ For sinc (or) Nyquist pulses

BW of BPSK $= B = R_b$

$$\frac{E_b}{N_o} = \frac{C \times R_b}{N \times R_b}$$

$$\frac{E_b}{N_o} = \frac{C}{N}$$

$$\frac{E_b}{N_o} = 8.8 \text{ dB}$$

3.110 (c)

Compared to DM, PCM is complex and compared to analog systems, PCM requires high bandwidth.

So, statement I is true and statement II is false.

3.111 (a)

$$\text{BER} = \frac{1}{2} erfc\left(\sqrt{\frac{E_b}{N_0}}\right)$$

We have,

$$Q(x) = \frac{1}{2}\,\text{erfc}\left(\frac{x}{\sqrt{2}}\right) \text{ Hence,}$$

$$\text{BER} = Q\left(\sqrt{\frac{2E_b}{N_0}}\right) = Q\left(\sqrt{\frac{d_{\min}^2}{2N_0}}\right)$$

We get, $d_{\min}^2 = 4E_b$

$$\Rightarrow \quad d_{\min} = 2\sqrt{E_b}$$

For coherent binary PSK, constellation diagram is as below:

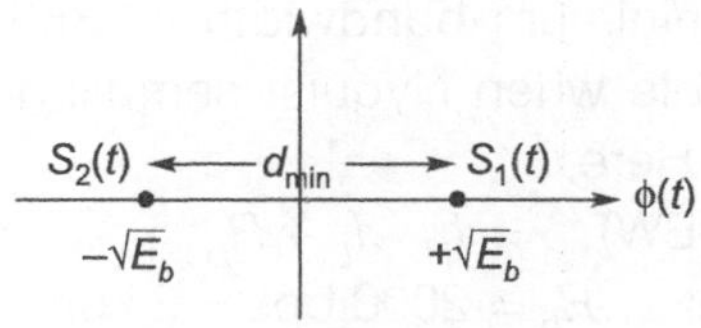

$$d_{\min} = 2\sqrt{E_b}$$

Hence, option (a) is correct.

3.112 (a)

$$(S/N)_C \longrightarrow \boxed{R_X} \longrightarrow (S/N)_0$$

Channel signal to noise ratio,

$$\left(\frac{S}{N}\right)_C = \frac{\text{Signal power}}{\text{Noise power in message bandwidth}}$$

is measured at receiver input

$$\text{FOM} = \frac{(S/N)_0}{(S/N)_C}$$

3.113 (c)

It is given that

$$|Q_e|_{\max} \leq 0.1\% \text{ of } A_m$$

where A_m is the peak amplitude of the signal. The maximum quantization error is given as

$$|Q_e|_{\max} = \frac{\Delta}{2} = \left[\frac{2A_m}{2^n}\right]\cdot\frac{1}{2}$$

where Δ = step-size

Hence, we get

$$\left[\frac{2A_m}{2^n}\right]\cdot\frac{1}{2} \leq \frac{0.1}{100} \times A_m$$

$$\Rightarrow \quad 2^n \geq 1000$$

$$\Rightarrow \quad n \geq 10$$

Therefore, the minimum number of binary digits required per sample is

$$n_{\min} = 10$$

3.114 (a)

Given : $N = 6$ and $f_m = 4$ kHz

$$\text{Signalling Rate} = \frac{R_b}{n} = Nf_s$$

where f_s = Nyquist Rate

$= 2f_m = 8$ kHz

∴ Signalling Rate = 6 × 8 k = 48 kbps

Minimum transmission bandwidth required

$$B_T = Nf_m = 6 \times 4\text{ k} = 24\text{ kHz}$$

3.115 (b)

The minimum bandwidth of BFSK signal is possible when Nyquist sampling pulses are used. Here,

$$(BW)_{min} = f_H - f_L + R_b$$

Given: $R_b = 2000$ bps = 2 kbps

$f_H - f_L = 4000$ Hz = 4 kHz

$(BW)_{min} = 4 + 2 = 6$ kHz

3.116 (b)

Given: $|Q_e|_{max} \le 1\%$ of Dynamic range

Dynamic range = Peak-to-peak amplitude of the signal = V_{p-p}

$$\frac{\Delta}{2} \le \frac{1}{100} \times V_{p-p}$$

$$\frac{1}{2}\left[\frac{V_{p-p}}{L}\right] \le \frac{1}{100} \times V_{p-p}$$

$$L \ge 50$$

The minimum number of quantization levels that can be used is 50.

3.117 (c)

M-ary PSK and M-ary QAM are linear modulation schemes. An M-ary PSK signal has a constant envelope, whereas an M-ary QAM signal involves changes in the carrier amplitude. Accordingly, M-ary PSK can be used to transmit digital data over a non-linear band-pass channel, whereas M-ary QAM requires the use of a linear channel. Hence, statement (I) is true but statement (II) is false.

3.118 (a)

$$\text{SNR} = \frac{\text{Signal power}}{\text{Quantization noise power}}$$

$$\text{SNR} = \frac{P}{(\Delta^2/12)}$$

where, Δ represents the step size of a uniform quantizer given by

$$\Delta = \frac{2m_{max}}{L}$$

where, L = Number of quantization levels

$L = 2^R$

where, R = Number of bits per sample

$$\therefore \quad \text{SNR} = \frac{P \times 12}{\left(\frac{2m_{max}}{2^R}\right)^2} = \frac{P \times 12}{4m_{max}^2} \times 2^{2R}$$

$$\therefore \quad \text{SNR} = \left(\frac{3P}{m_{max}^2}\right) 2^{2R}$$

Hence, option (a) is the correct answer.

3.119 (d)

The differential pulse code modulation works on the principle of prediction. The value of the present sample is predicted from the past samples, thereby eliminating the redundant information before encoding the sample value.

3.120 (b)

Binary '1' is transmitted at a frequency $(\omega_c + \Delta\omega/2)$ and binary '0' is transmitted at a frequency $(\omega_c - \Delta\omega/2)$. As the frequency of the carrier signal is changing for binary bits 0 and 1.

Hence, the given RF binary signals are transmitted using FSK technique.

3.121 (c)

Given, audio signal,

$$S(t) = 5\cos(2000\pi t)$$

$$n = 10$$

$$\frac{S}{N_Q} = \frac{3}{2} \times 2^{2n}$$

Signal to quantization noise ratio,

$$= \frac{3}{2} \times 2^{2 \times 10}$$

$$= 1.5 \times 2^{20} = 1.57 \times 10^6$$

3.122 (*)

For delta modulation,

Signal to quantization ratio,

$$\frac{S}{N_Q} = \frac{3}{8\pi^2}\left(\frac{f_s}{f_m}\right)^3 = \frac{3}{8\pi^2}\left(\frac{2f_m}{f_m}\right)^3$$

$$= \frac{3}{8\pi^2} \times 8 = \frac{3}{\pi^2} = 0.303$$

3.123 (a)

Given, number of quantization levels increased from,

$$L\uparrow \rightarrow 4 \text{ to } 64$$

Number of bits, $n\uparrow \rightarrow 2$ to 6

$$BW = nf_s$$

$$BW \propto n$$

$$\frac{BW_1}{BW_2} = \frac{n_1}{n_2} = \frac{2}{6} = \frac{1}{3}$$

3.125 (c)

Delta modulation quantizes the difference between successive samples of the signal rather than the absolute value of each sample using a 1-bit quantizer with output $+\Delta$ or $-\Delta$ depending on whether the sample is greater or less than the previous sample.

3.126 (b)

The QPSK signal is given by

$$S_i(t) = \begin{cases} \sqrt{\frac{2E}{T}}\cos\left[2\pi f_c t + \left((2i-1)\frac{\pi}{4}\right)\right], & 0 \le t \le T \\ 0 & \text{, otherwise} \end{cases}$$

We can write,

$$S_i(t) = \sqrt{\frac{2E}{T}}\cdot\cos\left[(2i-1)\frac{\pi}{4}\right]\cos(2\pi f_c t) - \sqrt{\frac{2E}{T}}\sin\left[(2i-1)\frac{\pi}{4}\right]\sin(2\pi f_c t)$$

The signal-space diagram of coherent QPSK system can be drawn as below:

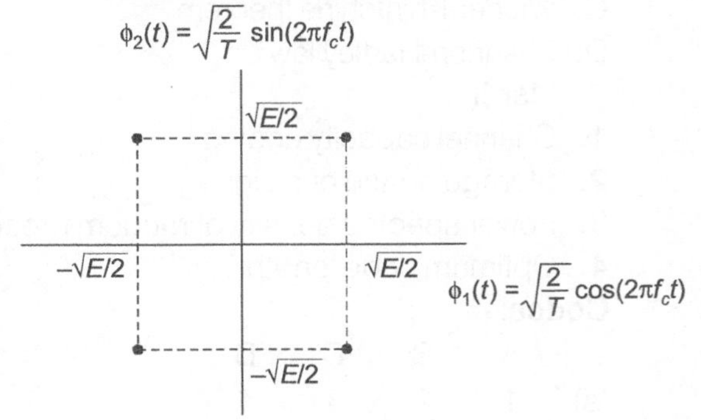

The in-phase channel output and the quadrature channel output may be viewed as the individual outputs of the two coherent binary PSK systems with

- Signal energy per bit as $E/2$
- Noise spectral density as $N_0/2$.

The average probability of bit error in each channel of the coherent QPSK system is given by:

$$P_e = \frac{1}{2}erfc\left(\sqrt{\frac{E_b}{N_0}}\right) = \frac{1}{2}erfc\left(\sqrt{\frac{E/2}{N_0}}\right)$$

3.127 (d)

- The baseband signal spectrum in the low frequency range is unsuitable for direct transmission as they require a large antenna. Therefore, it is modulated by a high frequency carrier, thus shifting the frequency spectrum to higher band which allows for practical size of antenna and helps to avoid interference with other signals.
- Digital repeaters are designed to regenerate or reconstruct digital signals to compensate for signal attenuation and distortion that occurs during transmission over long distances. Thus, channel noise is not cumulative and removed by the repeaters.
- At larger distances, the signal to noise ratio at the receiver decreases due to attenuation and noise which cause the signal to become distorted and more susceptible to errors beyond a certain distance. Hence, a digital signal can be propagated through a limited distance only. The digital repeaters can be used to extend the distance over which the digital signal can be transmitted.

Hence, all the given statements are correct.

4 Information Theory

4.1 In a single error correcting Hamming code, the number of message bits in a block is 26. The number of check bits in the block would be
(a) 3 (b) 4
(c) 5 (d) 7 [ESE-2000]

4.2 To permit the selection of 1 out of 16 equi-probable events, the number of bits required is
(a) 2 (b) $\log_{10} 16$
(c) 8 (d) 4
[ESE-2001]

4.3 When the channel is noisy, producing a conditional probability of error $p = 0.5$; the channel capacity and entropy function would be, respectively,
(a) 1 and 1 (b) 1 and 0.5
(c) 0.5 and 1 (d) zero and 1
[ESE-2001]

4.4 **Assertion (A):** Though raw binary bits can be transmitted over communication channels, often coding is done to transmit the data in a more error-free and reliable manner.
Reason (R): Coding assigns uneven distribution of binary digits to various symbols of the signal depending upon probability.
(a) Both A and R are true and R is the correct explanation of A
(b) Both A and R are true but R is NOT the correct explanation of A
(c) A is true but R is false
(d) A is false but R is true
[ESE-2003]

4.5 A ternary source produces alphabets *A*, *B* and *C* with probabilities $P_A = P_B =$ p and P_C. Which one of the following gives the correct values for the maximum value of the entropy of the source: and the corresponding value of p and the range of *p*?
(a) 1.58, 0.33, (0, 0.5) (b) 1.0, 0.5, (0, 1)
(c) 3.0, 0.67, (0, 0.5) (d) 2.0, 4.2, (0, 0.3)
[ESE-2004]

4.6 Which one of the following statements is correct? The Shannon's channel capacity formula indicates that in theory
(a) by using proper channel codes, we can get an error free transmission on a noisy channel
(b) it is not possible to get an error-free transmission on a noisy channel, since there will always be some error in the detected signal for finite noise on any channel
(c) it is true only for some wired channels and not wireless channels
(d) it works only for analog signals and not for digital signals on any channel
[ESE-2005]

4.7 Match **List-I** (Theorem/Law) with **List-II** (Specified Quantity) and select the correct answer using the codes given below the lists:
List-I
A. Shannon source theorem
B. Dimensionality theorem
C. Wiener-Khintchine theorem
D. Shannon-Hartley law
List-II
1. Channel capacity coding
2. Storage space of a signal
3. Power spectral density of random process
4. Optimum code length
Codes:

	A	B	C	D
(a)	1	2	3	4
(b)	4	3	2	1
(c)	1	3	2	4
(d)	4	2	3	1

[ESE-2005]

4.8 Match **List-I** with **List-II** and select the correct answer using the code given below the lists:

List-I	List-II
A. Entropy coding	1. McMillan's Rule
B. Channel capacity	2. Redundancy
C. Minimum length code	3. Shannon Fano
D. Equivocation	4. Shannon law

Codes:

	A	B	C	D
(a)	1	2	3	4
(b)	3	4	1	2
(c)	1	4	3	2
(d)	3	2	1	4

[ESE-2005]

4.9 A communication channel has a bandwidth of 100 MHz. The channel is extremely noisy such that the signal power is very much below the noise power. What is the capacity of this channel?
(a) 100 Mbps (b) 50 Mbps
(c) 2400 bps (d) Nearly 0 bps
[ESE-2006]

4.10 A source generates four messages with probability 1/8, 1/8, 1/4 and 1/2. What is the entropy of the source (bits/message)?
(a) 1 (b) 1.75
(c) 2 (d) 4 [ESE-2006]

4.11 A source produces 26 symbols with equal probability. What is the average information produced by this source?
(a) < 4 bits/symbol
(b) 6 bits/symbol
(c) 8 bits/symbol
(d) Between 4 and 6 bits/symbol [ESE-2006]

4.12 In order to permit the selection of 1 out of 16 equiprobable events, what is the number of bits required?
(a) 8 (b) 4
(c) $\log_{10} 16$ (d) 2 [ESE-2007]

4.13 Which one of the following is the code that is very close to 'trellis coded modulation'?
(a) Combines analog and digital modulations
(b) Combines modulation and encoding
(c) Encodes following trellis diagram
(d) Combines amplitude and frequency modulation
[ESE-2007]

4.14 A discrete source produces 8 symbols and is memory-less. Its entropy is
(a) 1 bit/symbol only
(b) 2 bits/symbol only
(c) 3 bits/symbol only
(d) ≤ 3 bits/symbol [ESE-2008]

4.15 Source S_1 produces 4 discrete symbols with equal probability. Source S_2 produces 6 discretes symbols with equal probability. If H_1 and H_2 are the entropies of sources S_1 and S_2 respectively, then which one of the following is correct?
(a) H_1 is always less than H_2
(b) H_1 is always greater than H_2
(c) H_1 is always equal to H_2
(d) H_2 is 1.5 times H_1 only [ESE-2008]

4.16 The entropy of a digital source is 2.7 bits/symbol. It is producing 100 different symbols per second. The source is likely to be which one of the following
(a) A binary source
(b) A quaternary source
(c) An octal source
(d) A hexadecimal source [ESE-2008]

4.17 A good line code should have which of the following?
1. Favourable psd
2. Low intersymbol interference
3. Adequate timing content
4. Transparency
Select the correct answer using the code given below:
(a) 1, 3 and 4 (b) 1, 2 and 4
(c) 2, 3 and 4 (d) 1, 2 and 3
[ESE-2008]

4.18 **Assertion (A):** Entropy of a binary source is maximum if the probabilities of occurrence of both the events are equal.
Reason (R): The average amount of information per source symbol is called entropy for a memoryless source.
(a) Both A and R are individually true and R is the correct explanation of A
(b) Both A and R are individually true but R is not the correct explanation of A
(c) A is true but R is false
(d) A is false but R is true [ESE-2008]

4.19 **Assertion (A):** Source produces two symbols A and B with probability 3/4 and 1/4 respectively. For error free transmission this source should be coded using Shannon-Fano code.
Reason (R): For better transmission efficiency, source and channel must be matched.
(a) Both A and R are true and R is the correct explanation of A
(b) Both A and R are true but R is NOT the correct explanation of A
(c) A is true but R is false
(d) A is false but R is true

[ESE-2010]

4.20 Discrete source S_1 has 4 equiprobable symbols while discrete source S_2 has 16 equiprobable symbols. When the entropy of these two sources is compared entropy of
(a) S_1 is greater than S_2
(b) S_1 is less than S_2
(c) S_1 is equal to S_2
(d) Depends on rate of symbols/second
[ESE-2011]

4.21 Consider the following codes:
1. Hamming code 2. Huffman code
3. Shannon-Fano code 4. Convolution code
Which of these are source codes?
(a) 1 and 2 only (b) 2 and 3 only
(c) 3 and 4 only (d) 1, 2, 3 and 4
[ESE-2011]

4.22 The average information associated with an extremely unlikely message is zero. What is the average information associated with an extremely likely message?
(a) Zero
(b) Infinity
(c) Depends on total number of messages
(d) Depends on speed of transmission of the message
[ESE-2012]

4.23 A coherent binary phase-shift keyed (BPSK) transmitter operates at a bit rate of 20 Mbps. For a probability of error $P(e)$ of 10^{-4} and given carrier-to-noise (C/N) density ratio of 8.8 dB, determine energy of bit-to-noise (E_b/N_o) density ratio for a receiver bandwidth equal to the minimum double-sided Nyquist bandwidth.
(a) 23 dB (b) 2.3 dB
(c) 8.8 dB (d) 0.88 dB
[ESE-2012]

4.24 An event has two possible outcomes with probability $P_1 = \frac{1}{2}$, and $P_2 = \frac{1}{64}$. The rate of information with 16 outcomes per second is
(a) $\frac{38}{4}$ bits/sec (b) $\frac{38}{64}$ bits/sec
(c) $\frac{38}{2}$ bits/sec (d) $\frac{38}{32}$ bits/sec
[ESE-2013]

4.25 A radio channel has bandwidth of 10 kHz and an S/N ratio of 15 dB. The maximum data rate that can be transmitted is
(a) 16.1 kb/sec (b) 24.2 kb/sec
(c) 32.3 kb/sec (d) 50.3 kb/sec
[ESE-2013]

4.26 Conveniently information $I(x)$ is expressed as
(a) (–) $\log_{10} P(x)$ Hartleys
(b) $\log_{10} P(x)$ Hartleys
(c) $\log_2 P(x)$ Hartleys
(d) $2\log_{10} \frac{1}{P(x)}$ Hartleys
[ESE-2013]

4.27 A source produces 4 symbols with probabilities $\frac{1}{2}, \frac{1}{4}, \frac{1}{8}$ and $\frac{1}{8}$. For this source, a practical coding scheme has an average code word length of 2 bits/symbol. The efficiency of the code is
(a) 1 (b) $\frac{7}{8}$
(c) $\frac{1}{2}$ (d) $\frac{1}{4}$
[ESE-2013]

4.28 An analog signal is band limited to 4 kHz. It is sampled at the Nyquist rate and samples are quantized into 4 levels. The quantization levels have probabilities $\frac{1}{8}, \frac{1}{8}, \frac{3}{8}$ and $\frac{3}{8}$. The information rate of the source is
(a) 15400 bps (b) 4000 bps
(c) 14400 bps (d) 16000 bps
[ESE-2013]

4.29 A source produces 26 symbols with equal probabilities. What is the average information produced by the source?
(a) Less than 4 bits/symbol
(b) 6 bits/symbol
(c) 7 bits/symbol
(d) Between 4 bits/symbol and 6 bits/symbol
[ESE-2015]

4.30 Consider a discrete memoryless source with source alphabet $S = \{s_0, s_1, s_2\}$ with probabilities

$$P(s_0) = \frac{1}{4},\ P(s_1) = \frac{1}{4} \text{ and } P(s_2) = \frac{1}{2}$$

The entropy of the source is

(a) $\frac{1}{2}$ bit (b) $\frac{2}{3}$ bit

(c) $\frac{3}{2}$ bits (d) $\frac{1}{3}$ bit [ESE-2017]

4.31 A source produces three symbols A, B and C with probabilities $P(A) = \frac{1}{2}$, $P(B) = \frac{1}{4}$ and $P(C) = \frac{1}{4}$. The source entropy is

(a) $\frac{1}{2}$ bit/symbol (b) 1 bit/symbol

(c) $1\frac{1}{4}$ bits/symbol (d) $1\frac{1}{2}$ bits/symbol

[ESE-2019]

4.32 For Gaussian and White channel noise, the capacity of a low-pass channel with a usable bandwidth of 3000 Hz and $\frac{S}{N} = 10^3$ at the channel output will be

(a) 15000 bits/s (b) 20000 bits/s

(c) 25000 bits/s (d) 30000 bits/s

[ESE-2019]

4.33 What is the entropy of a communication system that consists of six messages with probabilities 1/8, 1/8, 1/8, 1/8, 1/4 and 1/4 respectively?

(a) 1 bit/message (b) 2.5 bits/message

(c) 3 bits/message (d) 4.5 bits/message

[ESE-2021]

4.34 Over a certain binary communication channel, the symbol 0 is transmitted with probability 0.4 and 1 is transmitted with probability 0.6. It is given that $p(\varepsilon/0) = 10^{-6}$ and $p(\varepsilon/1) = 10^{-4}$, where $p(\varepsilon/x_i)$ is the probability of detecting the error given that x_i is transmitted. What is the error probability of the channel?

(a) 0.604×10^{-3} (b) 0.604×10^{-5}

(c) 0.604×10^{-6} (d) 0.604×10^{-4}

[ESE-2022]

4.35 **Statement (I) :** Shannon has shown that it is possible to achieve error-free communication by adding sufficient redundancy.

Statement (II) : The addition of an extra check digit increases redundancy.

(a) Both Statement (I) and Statement (II) are individually true and Statement (II) is the correct explanation of Statement (I).

(b) Both Statement (I) and Statement (II) are individually true, but Statement (II) is not the correct explanation of Statement (I).

(c) Statement (I) is true, but Statement (II) is false.

(d) Statement (I) is false, but Statement (II) is true.

[ESE-2022]

4.36 A high resolution black and white TV picture consists of about 2×10^6 picture elements and 16 different brightness levels. Pictures are repeated at the rate of 32 per second. All picture elements are assumed to be independent and all levels have equal likelihood of occurrence. The average rate of information conveyed by this TV picture source is

(a) 4 Mb/s (b) 256 Mb/s

(c) 64 Mb/s (d) 16 Mb/s

[ESE-2024]

4.37 Consider the following statements regarding information theory:

1. The information capacity of a continuous channel of bandwidth B Hertz, perturbed by additive white Gaussian noise of power spectral density $N_0/2$ and limited in bandwidth to B, is $C = B\log_2\left(1+\frac{P}{N_0 B}\right)$ bits per second, where P is the average transmitted power.
2. The mutual information is always non-negative.
3. The mutual information of a channel is symmetric
4. The mutual information of the channel is related to the joint entropy of the channel input and channel output by $I(x, y) = H(x) - H(y) + H(x, y)$.

Which of the above statements are correct?

(a) 1, 2 and 3 only (b) 2 and 3 only

(c) 2, 3 and 4 only (d) 1, 2, 3 and 4

[ESE-2024]

■■■

Answers Information Theory

4.1 (c)	4.2 (d)	4.3 (d)	4.4 (b)	4.5 (a)	4.6 (a)	4.7 (d)	4.8 (b)	4.9 (d)
4.10 (b)	4.11 (d)	4.12 (b)	4.13 (b)	4.14 (c)	4.15 (a)	4.16 (c)	4.17 (d)	4.18 (b)
4.19 (d)	4.20 (b)	4.21 (b)	4.22 (c)	4.23 (c)	4.24 (a)	4.25 (d)	4.26 (a)	4.27 (b)
4.28 (c)	4.29 (d)	4.30 (c)	4.31 (d)	4.32 (d)	4.33 (b)	4.34 (d)	4.35 (a)	4.36 (b)
4.37 (a)								

Explanations Information Theory

4.1 (c)

In a Hamming code, for k check bits and n message bits,

$$2^k - 1 - k \geq n$$

$$\Rightarrow \quad 2^k - 1 - k \geq 26$$

$$\Rightarrow \quad k = 5$$

4.2 (d)

Number of bits required,

$$n = \log_2 16 = 4$$

4.3 (d)

Channel capacity, $C = BW \log_2\left(1 + \frac{S}{N}\right)$

But $\frac{S}{N} \to 0$ ... for noisy channel

$$\Rightarrow \quad C \cong BW \cdot \log_2 1$$

$$\Rightarrow \quad C \approx 0 \text{ bps}$$

Probability of error, $P_e = 0.5 = 1/2$

Entropy for a noisy channel is

$$H = \left[P_e \log_2 \frac{1}{P_e} + (1 - P_e)\log_2 \frac{1}{1 - P_e}\right]$$

$$H = \left[\frac{1}{2}\log_2 2 + \frac{1}{2}\log_2 2\right] = 1$$

4.5 (a)

Maximum value of entropy is achieved only when all three alphabets are equiprobable, i.e.,

$$M = 3$$

$$H(x)\big|_{max} = \log_2 M$$

$$= \log_2 3 = 1.58 \text{ bits/symbol}$$

4.7 (d)

A. Shannon source coding theorem
→ Optimum code length

B. Dimensionality theorem
→ Storage space of a signal

C. Wiener-Khintchine theorem
→ PSD of random process

D. Shannon-Hartley law
→ Channel capacity

4.8 (b)

Entropy Coding → Shannon Fano
Channel capacity → Shanon law
Minimum Length code → McMillan's rule
Equivocation → Redundancy

4.9 (d)

Channel capacity, $C = BW \log_2\left(1 + \frac{S}{N}\right)$

But $\frac{S}{N} \to 0$ ... for extremely noisy channel

$$\Rightarrow \quad C \cong BW \cdot \log_2 1$$

$$\Rightarrow \quad C \approx 0 \text{ bps}$$

4.10 (b)

Entropy, $H(x) = \sum_{i=1}^{4} P(i)\log_2 \frac{1}{P(i)}$

$$= \frac{1}{8}\log_2 8 + \frac{1}{8}\log_2 8 + \frac{1}{4}\log_2 4 + \frac{1}{2}\log_2 2$$

= 1.75 bits/message

4.11 (d)

For equiprobable symbols, the average information

$$H(x) = \log_2 M$$

where, $M \to$ Number of symbols

$$H(x) = \log_2 26 = 4.7 \text{ bits/symbol}$$

So, the answer is 'between 4 and 6 bits/symbol'.

4.12 (b)

Number of bits required, $n = \log_2 m$
where, m = number of equiprobable events

$\therefore \quad n = \log_2 16 = 4$

4.13 (b)

Trellis coded modulation is a combination of coding and modulation.

4.14 (c)

When all symbols are equiprobable, then the entropy will be maximum.

$H(X)_{max.} = \log_2 8 = 3$ bits/symbol

Since the symbols are not mentioned equiprobable, therefore, entropy

$H(X) \leq 3$ bits/symbol

4.15 (a)

Entropy $H_1 = \log 4$
Entropy $H_2 = \log 6$
So, H_1 is always less than H_2.

4.16 (c)

The maximum entropy of a binary source is 1 bits/symbol.
The maximum entropy of a quaternary source is 2 bits/symbol.
The maximum entropy of a octal source is 3 bits/symbol.
The maximum entropy of a hexadecimal source is 4 bits/symbol.
The entropy of the given digital source is close to 3 bits/symbol. So, the source is most likely to be a quaternary source.
Note that the symbol rate has no role in the context of this problem.

4.20 (b)

Entropy of source $S_1 = \log_2 4$
$= 2$ symbols/sec.
Entropy of source $S_2 = \log_2 16$
$= 4$ symbols/sec

4.21 (b)

Hamming and convolution codes are error correcting codes.

4.22 (c)

If M messages are equally likely then average information or entropy $= \log_2$ M.
So entropy depends on number of messages.

4.23 (c)

Given that, $\left[\frac{S}{N}\right]_{dB} = \left[\frac{C}{N}\right]_{dB} = 8.8$ dB

Double sided Nyquist bandwidth of BPSK is, $B = R_b$.

$$\frac{S}{N} = \frac{E_b R_b}{N_o B} = \frac{E_b R_b}{N_o R_b} = \frac{E_b}{N_o}$$

So, $\left[\frac{E_b}{N_o}\right]_{dB} = \left[\frac{S}{N}\right]_{dB} = 8.8$ dB

4.24 (a)

$$P_1 = \frac{1}{2},\ P_2 = \frac{1}{64}$$

Entropy, $H = \sum_{i=1}^{M} P_i \log_2 \frac{1}{P_i}$

$$= \frac{1}{2}\log_2 2 + \frac{1}{64}\log_2 64$$

$$= \frac{19}{32} \text{ bits/outcomes}$$

$$r_s = 16 \text{ outcomes/sec.}$$

$\therefore$ Information rate

$\Rightarrow \quad R_s = r_s H = 16 \times \frac{19}{32}$

$$= \frac{19}{2} \text{ or } \frac{38}{4} \text{ bits/sec.}$$

4.25 (d)

Bandwidth, $B = 10$ kHz

$(S/N)_{dB} = 15 \text{ dB} = 10\log(S/N)$

$(S/N) = 31.62$

According to Shannon Hartley theorem, capacity of a channel having bandwidth B and additive Gaussian band limited white noise is

$$C = B\log_2\left(1 + \frac{S}{N}\right) = 10^4 \log_2(1 + 31.62)$$

$$C = 50.3 \text{ kb/sec.}$$

Capacity of a channel is defined as maximum data rate that can be supported by channel with minimum probability of error.

4.26 (a)

Information content in a signal will be high if its probability of occurrence will be less.
If the probability of occurrence of message is P, then its information content I will be

$$I = \log_2 \frac{1}{P} \text{ bits}$$

or $I = \log_2 \frac{1}{P}$ Hartley or decit

$\therefore$ $I = -\log_{10} P(x)$ Hartleys

4.27 (b)

$$P_1 = \frac{1}{2},\quad P_2 = \frac{1}{4},\quad P_3 = \frac{1}{8},\quad P_4 = \frac{1}{8}$$

Efficiency of the code word is given as $\frac{H}{L}$

$$H_i \to \text{Entropy} = \sum_{i=1}^{M} P_i \log_2 \frac{1}{P_i}$$

$$= \frac{1}{2}\log_2 2 + \frac{1}{4}\log_2 4 + \frac{1}{8}\log_2 8 + \frac{1}{8}\log_2 8$$

$$= \frac{7}{4}\text{ bits/symbol}$$

Average code word length,

$$L = 2 \text{ bits/symbol}$$

$$\text{Efficiency, } \eta = \frac{7/4}{2} = \frac{7}{8}$$

4.28 (c)

Analog signal is bandlimited to 4 kHz and signal is sampled at the Nyquist rate

$$\therefore \quad f_s = 2f_m$$
$$= 8 \text{ kHz or } 8 \text{ k samples/sec.}$$

Since, every sample generates one message signal, so, message per sec,

$$r_s = 8 \times 10^3 \text{ messages/sec.}$$

Also, entropy,

$$H = \sum_{i=1}^{M} P_1 \log_2 \frac{1}{P}$$

$$= \frac{1}{8}\log_2 8 + \frac{1}{8}\log_2 8 + \frac{3}{8}\log_2 \frac{8}{3} + \frac{3}{8}\log_2 \frac{8}{3}$$

$$= \frac{6}{8} + \frac{6}{8}\log_2 \frac{8}{3} = 1.811 \text{ bits/sample}$$

Information rate,

$$R = r_s H = 8 \times 1.811 \times 10^3$$
$$= 14.490 \approx 14400 \text{ bps.}$$

4.29 (d)

Average information $= \log_2(26)$

$$\log_2 16 = 4$$
$$\log_2 64 = 6$$

So, average information is between 4 bit/symbol and 6 bits/ symbol.

4.30 (c)

$$H = \sum_i P_i \log_2 \frac{1}{P_i} = \frac{1}{4}\log_2 4 + \frac{1}{4}\log_2 4 + \frac{1}{2}\log_2 2$$

$$= \frac{3}{2}\text{ bits/symbol}$$

4.31 (d)

$$P(A) = \frac{1}{2}$$
$$P(B) = \frac{1}{4}$$
$$P(C) = \frac{1}{4}$$

$$H = \Sigma P_i \log \frac{1}{P_i}$$

$$= \frac{1}{2}\log_2 2 + \frac{1}{4}\log_2 4 + \frac{1}{4}\log_2 4$$

$$= \frac{1}{2} + \frac{1}{2} + \frac{1}{2} = \frac{3}{2}$$

$$= 1\frac{1}{2}\text{ bits/symbol}$$

4.32 (d)

$$B = 3000 \text{ Hz}$$

$$\frac{S}{N} = 10^3$$

$$C = B\log_2\left(1 + \frac{S}{N}\right) = 3k \cdot \log_2(1 + 10^3)$$

$$= 29.9 \text{ kbps} \approx 30000 \text{ bps}$$

4.33 (b)

The entropy is given as

$$H = \sum_{i=1}^{6} P_i \log_2\left(\frac{1}{P_i}\right)$$

$$H = -\sum_{i=1}^{6} P_i \log_2(P_i)$$

$$H = -\left[4 \times \frac{1}{8}\log_2\left(\frac{1}{8}\right) + 2 \times \frac{1}{4}\log_2\left(\frac{1}{4}\right)\right]$$

$$H = \frac{4 \times 3}{8} + \frac{2 \times 2}{4}$$

$$H = 2.5 \text{ bits/message}$$

4.34 (d)

Given: $P(0) = 0.4$; $P(1) = 0.6$;
$P(\in/0) = 10^{-6}$; $P(\in/1) = 10^{-4}$

The error probability of the channel is given as

$$P(\epsilon) = P(0)\cdot P\left(\frac{\epsilon}{0}\right) + P(1)\cdot P\left(\frac{\epsilon}{1}\right)$$

$$P(\epsilon) = 0.4 \times 10^{-6} + 0.6 \times 10^{-4}$$

$$P(\epsilon) = 0.604 \times 10^{-4}$$

4.35 (a)

The addition of an extra check digits increases redundancy and allow detection or correction of errors to achieve error-free communication.

Hence, both statement (I) and statement (II) are individually true and Statement (II) is the correct explanation of Statement (I).

4.36 (b)

Number of picture elements transmitted per second

$$= 32 \times 2 \times 10^6$$

$\Rightarrow$ $r = 64 \times 10^6$ picture elements/sec

Given 16 different brightness levels having equal probability. Hence,

$$H = \log_2 16 = 4 \text{ bits/brightness level}$$

Each picture element exhibits one of the brightness level. Therefore,

$$H = 4 \text{ bits/picture element}$$

Information rate, $R = rH = 64 \times 10^6 \times 4$

$$= 256 \text{ Mbps}$$

4.37 (a)

- Shannon-Hartley Theorem:

$$C = B\log_2\left(1+\frac{S}{N}\right)$$

where, C = capacity of channel (bits/s), B = Bandwidth of channel, S = Signal power and N = Noise power.

For AWGN noise with power spectral density $N_0/2$, $N = N_0B$. Hence,

$$C = B\log_2\left(1+\frac{P}{N_0B}\right)$$

where, P is the average signal power.

- Properties of Mutual Information:

(i) $I(X; Y) \geq 0$ i.e., Mutual information is always non-negative.

(ii) $I(X; Y) = I(Y; X)$ i.e., Mutual information of a channel is symmetric.

(iii) $I(X; Y) = H(Y) - H(Y/X)$

$= H(X) + H(Y) - H(X, Y)$

Hence, only statements 1, 2 and 3 are correct.

■■■

UNIT

VII

Advanced Communication Topics

Syllabus

Communication networks: Principles /practices /technologies /uses /OSI model/security; Basic packet multiplexed streams/scheduling; Cellular networks, types, analysis, protocols (TCP/TCPIP); Microwave & satellite communication: Terrestrial/space type LOS systems, block schematics link calculations, system design; Communication satellites, orbits, characteristics, systems, uses; Fibre-optic communication systems: fibre optics, theory, practice/standards, block schematics, link calculations, system design.

Contents

Sl.	Topic	Page No.
1.	Communication and Cellular Networks	469
2.	Microwave and Satellite Communication Systems	479
3.	Fibre Optic Communication Systems	490

OOOO

1 Communication and Cellular Networks

1.1 Assuming that packet transmissions and retransmission can both be described as Poisson process, calculate the probability that a data packet transmission in an S-ALOHA system will experience a collision with another user. Assume the total traffic rate λ_t = 10 packets/sec and the packet duration τ = 10 msec.

(a) 0.9 (b) 0.09
(c) 0.009 (d) 9.0 [ESE-2013]

1.2 Consider a point-to-point communication network represented by a graph. In terms of the graph parameters, the maximum delay (quality of service) experienced by a packet employing Bellman-Ford routing algorithm is /are

1. diameter of the graph
2. shortest path on the graph
3. sum of all edge weights in the graph

Select the correct answer using the code given below.

(a) 1 only (b) 2 only
(c) 3 only (d) 1, 2 and 3

[ESE-2017]

1.3 Let RSA prime numbers be p = 3 and q = 11. If the corresponding public key e = 3, what is the private key?

(a) 4 (b) 5
(c) 6 (d) 7

[ESE-2017]

1.4 Consider the following statements regarding TCP:

1. It enables two hosts to establish a connection and exchange streams of data.
2. It guarantees delivery of data in the same order in which they are sent.
3. TCP segmentation offload is used to reduce the cpu overhead of TCP/IP on fast networks.

Which of the above statements are correct?

(a) 1 and 2 only (b) 1 and 3 only
(c) 2 and 3 only (d) 1, 2 and 3

[ESE-2017]

1.5 Consider a packet switched network based on a virtual circuit mode of switching. The delay jitter for the packets of a session from the source node to the destination node is/are

1. always zero
2. non-zero
3. for some networks, zero

Select the correct answer using the code given below:

(a) 1 only (b) 2 only
(c) 3 only (d) 2 and 3

[ESE-2017]

1.6 The modulation scheme used in GSM is

(a) frequency shift keying
(b) phase shift keying
(c) Gaussian minimum shift keying
(d) amplitude shift keying [ESE-2017]

1.7 The technique for using one set of addresses inside a network and remapping those addresses to a different set of addresses that are seen outside the local network on the Internet is called

(a) network address translation
(b) address resolution
(c) network address mapping
(d) virtual LAN [ESE-2018]

1.8 What type of network is the Internet?

(a) Circuit-switched network
(b) Message-switched network
(c) Packet-switched network
(d) Cell-switched network [ESE-2018]

1.9 Which of the following statements is correct in respect of TCP and UDP protocols?

(a) TCP is connection-oriented, whereas UDP is connectionless.
(b) TCP is connectionless, whereas UDP is connection-oriented.
(c) Both are connectionless.
(d) Both are connection-oriented.

[ESE-2018]

1.10 Which one of the following is the correct combination for a layer providing a service by means of primitives in an open systems interconnection?
(a) Request, Indication, Response and Confirm
(b) Request, Inform, Response and Service
(c) Request, Command, Response and Action
(d) Request, Confirm, Indication and Action
[ESE-2019]

1.11 A network uses a fully interconnected mesh topology to connect 10 nodes together. The number of links required will be
(a) 35 (b) 40
(c) 45 (d) 50
[ESE-2019]

1.12 Which of the following are the advantages of packet switching?
1. Greater link efficiency than circuit switching
2. Connections are not blocked when traffic congestion occurs
3. Direct channel established between transmitter and receiver
4. No time is taken to establish connection

Select the correct answer using the code given below.
(a) 1 and 3 (b) 1 and 2
(c) 2 and 3 (d) 3 and 4
[ESE-2019]

1.13 A message consisting of 2400 bits is to be passed over an internet. The message is passed to the transport layer which appends a 150-bit header, followed by the network layer which uses a 120-bit header. Network layer packets are transmitted via two networks, each of which uses a 26-bit header. The destination network only accepts up to 900 bits long. The number of bits, including headers delivered to the destination network, is
(a) 2706 bits (b) 2634 bits
(c) 2554 bits (d) 2476 bits
[ESE-2019]

1.14 In a communication network, 4 $T1$ streams are multiplexed to form 1 $T2$ stream and 7 T1 streams are multiplexed to form 1 T3 stream. Further 6 T3 streams are multiplexed to form 1 T4 stream. If each T1 stream is of 1.544 Mbps, the data rate of 1 T4 stream should be
(a) 211.8 Mbps (b) 232.6 Mbps
(c) 243.4 Mbps (d) 274.2 Mbps
[ESE-2019]

1.15 Which of the following statements are correct regarding CDMA?
1. It is similar to GSM.
2. It allows each station to transmit over the entire frequency spectrum all the time.
3. It assumes that multiple signals add linearly.

Select the correct answer using the code given below.
(a) 1 and 2 only (b) 1 and 3 only
(c) 2 and 3 only (d) 1, 2 and 3
[ESE-2019]

1.16 Which of the following regarding cellular systems with small cells are correct?
1. Higher capacity and robustness
2. Need less transmission power and have to deal with local interference only
3. Frequency planning and infrastructure needed
4. These require both circuit switching and packet switching

Select the correct answer using the code given below
(a) 1, 2 and 4 (b) 1, 3 and 4
(c) 1, 2 and 3 (d) 2, 3 and 4
[ESE-2019]

1.17 Network Interface Card (NIC) has a unique six-byte permanent address as
(a) IP address (b) MAC address
(c) DNS address (d) Local address
[ESE-2020]

1.18 The data-link layer is responsible for
(a) Incoming bit stream and simply repeats to other devices connected
(b) An error free communication across the physical link connecting primary and secondary stations within a network
(c) An end-to-end integrity of data message propagated through the network between two devices
(d) Logical connection at application layer
[ESE-2020]

1.19 The transmit carrier frequency f_t and receive carrier frequency f_r for AMPS channels-3 are nearly
(a) 875 MHz and 870 MHz
(b) 825 MHz and 870 MHz
(c) 875 MHz and 830 MHz
(d) 825 MHz and 830 MHz
[ESE-2020]

1.20 Which one of the following mode is called a two-way simultaneous, communication between two stations?
(a) Simplex (SX)
(b) Half duplex (HDX)
(c) Full duplex (FDX)
(d) Full/Full duplex (F/FDX)
[ESE-2020]

1.21 Blocked calls may be handled in one of two ways. First blocked calls can be put in a queue awaiting a free channel. This is termed as
(a) Lost Calls Cleared (LCQ
(b) Lost Calls Delayed (LCD)
(c) Lost Calls Held (LCH)
(d) Lost Calls Hand off
[ESE-2020]

1.22 A combination of direct sequence and frequency hopping is called
(a) Direct sequence hopping
(b) Hybrid direct frequency hopping
(c) Direct sequence frequency hopping
(d) Hybrid direct sequence frequency hopping
[ESE-2020]

1.23 Which one of the following traffics can adjust, over wide ranges, to changes in delay and throughput across an internet and still meet the needs of its applications?
(a) Elastic traffic (b) Inelastic traffic
(c) Internet traffic (d) Service traffic
[ESE-2021]

1.24 A data message of 10 ms duration having 4800 bits crosses 9 nodes (10 hops) to reach its destination. The data rate and total delay for circuit switched connection (assuming node delays as 1 ms) are respectively,
(a) 240 kbps, 20 ms (b) 240 kbps, 21 ms
(c) 120 kbps, 21 ms (d) 480 kbps, 20 ms
[ESE-2021]

1.25 Consider the following statements regarding the OSI model:
1. Application layer provides the control structure for communication between applications; establishes, manages and terminates connections between cooperating applications.
2. Data link layer provide for the reliable transfer of information across the physical link.
3. Transport layer provides end-to-end error recovery and flow control of data.

Which of the above statements are correct?
(a) 1 and 2 only (b) 1 and 3 only
(c) 1, 2 and 3 (d) 2 and 3 only
[ESE-2021]

1.26 Consider the following statements regarding the TCP/IP protocol :
1. TELNET is an application of TCP protocol.
2. SMTP provides a basic electronic mail facility and makes use of TCP to send message to an SMTP module on another host.
3. The Internet Protocol is used at internet layer to provide the routing function across multiple networks.

Which of the above statements are correct?
(a) 1 and 2 only (b) 1 and 3 only
(c) 1, 2 and 3 (d) 2 and 3 only
[ESE-2021]

1.27 If a cellular system has p number of co-channel interfering cells, S is the desired signal power from serving base station and I_P is interference power from p^{th} interfering co-channel cell base station, then what is the signal to interference ratio for a mobile receiver in the functioning cell?
(a) $\dfrac{S}{\sum_{i=1}^{P} I_i}$ (b) $\dfrac{2\times S}{\sum_{i=1}^{P} I_i}$
(c) $\dfrac{\sum_{i=1}^{P} I_i}{S}$ (d) $\dfrac{S}{I_P}$
[ESE-2021]

1.28 Which one of the following provides a reliable connection for the transfer of data between applications?
(a) TCP (b) UDP
(c) FTP (d) SMTP
[ESE-2021]

1.29 Consider the following statements for public circuit-switching network :

1. Private branch exchange (PBX) is an application of circuit switching.
2. A switching centre that directly supports subscribers is known as digital PBX.
3. The link between the subscriber and the network, is also referred to as the local loop.

Which of the above statements are correct?

(a) 1 and 2 only (b) 1 and 3 only
(c) 1, 2 and 3 (d) 2 and 3 only

[ESE-2021]

1.30 Consider the following statements regarding the cellular system/network :

1. Cellular systems use open-loop power control and closed-loop power control.
2. For FDMA system, the capacity of a cell is more as compared to the number of frequency channels allocated to it.
3. A cell has *L* potential subscribers and is able to handle *N* simultaneous users. If $L > N$, then the system is referred to as nonblocking.

Which of the above statements are ***not*** correct?

(a) 1 and 2 only (b) 1 and 3 only
(c) 1, 2 and 3 (d) 2 and 3 only

[ESE-2021]

1.31 If voice activity interference reduction factor is 2.5, antenna sectorization gain factor is 2.5 and interference increase factor is 1.6, then the performance improvement factor in CDMA digital cellular system is

(a) 1.2 (b) 2.5
(c) 3.1 (d) 3.9 [ESE-2021]

1.32 The internet protocol (IP) (RFC 791) is the heart of the TCP/IP protocol suite. IP corresponds to the network layer in the OSI reference model and provides.

(a) a connectionless service to the application layer which requires a virtual circuit
(b) high reliability in packet delivery
(c) a connectionless best effort delivery service to the transport layer
(d) reliability functions within a higher layer protocol

[ESE-2022]

1.33 With frequency reuse, several cells with no space in a given coverage area use the same set of frequencies. Two cells using the same set of frequencies are called

(a) hexagonal shape cells
(b) adjacent channel cells
(c) cluster neighbour cells
(d) co-channel cells

[ESE-2022]

1.34 Which one of the following statements is **not** correct regarding the features of CDMA?

(a) Multipath fading may be substantially increased because the signal is spread over a large spectrum.
(b) Channel data rates are very high in CDMA systems.
(c) The near-far problem occurs at a CDMA receiver if an undesired user has a high detected power as compared to the desired user.
(d) Many users of a CDMA system share the same frequency.

[ESE-2022]

1.35 How many 6 MHz wide TV channels can be multiplexed on 800 MHz coaxial cable?

(a) 96 (b) 266
(c) 133 (d) 48 [ESE-2022]

1.36 In an advance mobile phone system (AMPS), which of the following separate channels in a link is/are used?

(a) TDMA only (b) FDMA only
(c) SDMA only (d) Both TDMA and FDMA

[ESE-2023]

1.37 Which one of the following threats is used to facilitate the designer of a program or system which might leave a hole in the software that only he/she is capable of using?

(a) Spyware (b) Trap Door
(c) Trojan Horse (d) Logic Bomb

[ESE-2023]

1.38 Match the following lists :

List-I

P. Format control
Q. Transmission control
R. Information separator
S. Miscellaneous

List-II
1. NAK, ETB
2. RS, US
3. EM, DC1, DC2, DC3, DC4
4. FF, VT

Select the correct answer using the code given below :

	P	Q	R	S
(a)	2	1	3	4
(b)	4	3	1	2
(c)	3	1	4	2
(d)	4	1	2	3

[ESE-2024]

1.39 Common error control techniques are not based on
(a) error detection
(b) positive acknowledgement and retransmission
(c) retransmission after time out
(d) negative acknowledgement [ESE-2024]

1.40 Match the following lists :

List-I	List-II
P. SCMA	1. Interleavers are used as the only means for user separation
Q. MC-CDMA	2. Applies a new set of spreading codes that exhibit interference free window to separate users in the wireless channel.
R. IDMA	3. Enables two or more users, associated with the same base station, to use the same time and frequency and code resources on the grounds of their physical location or spatial separation
S. LAS-CDMA	4. Spreads each user symbol in the frequency domain

Select the correct answer using the code given below:

	P	Q	R	S
(a)	2	4	3	1
(b)	3	4	1	2
(c)	1	2	4	3
(d)	4	3	2	1

[ESE-2024]

1.41 Which one of the following is **not** the property of WiseMAC?
(a) The preamble duration is determined by the potential clock drift since last synchronization.
(b) It uses additional contents of ACK packets.
(c) Preamble T_P is min($L\theta$, $4T_W$).
(d) The packets in WiseMAC also contain a "more" bit.

[ESE-2024]

1.42 Which one of the following is **not** the property of flooding?
(a) An effective routing approach when the information in the routing tables is not available.
(b) It may easily swamp the network as one packet creates multiple packets that in turn create multiples of multiple packets, generating an exponential growth rate.
(c) TTL is used to reduce the resource consumption in the network.
(d) It requires the network to provide multiple paths for each source—destination pair.

[ESE-2024]

1.43 Which one of the following is a disadvantage of CDMA?
(a) It is simple to operate.
(b) It does not require any transmission synchronization between stations. The only synchronization required is that of the receiver to the sequence of the received carrier.
(c) It offers sufficient protection against interference from other stations and due to multiple paths.
(d) It has low throughput. [ESE-2024]

1.44 Which one of the following statements is correct for ST connector?
(a) It uses a ceramic ferrule and a rugged metal housing. It is latched in place by twisting.
(b) It has a shockproof plastic housing with a retractable shroud to protect the ceramic ferrules from damage.
(c) It is designed for multimode data communication applications and it is used for instrumentation connections.
(d) It allows high-density connections between network equipment in telecommunication rooms.

[ESE-2024]

1.45 Network and switching subsystem does **not** have which one of the following switches and databases?
(a) Mobile services switching center
(b) Home location register
(c) Visitor location register
(d) Operation and maintenance center
[ESE-2024]

1.46 Which one of the following is **not** an advantage of a WLAN?
(a) Only wireless ad-hoc networks allow for communication without previous planning.
(b) Several Government and non-Government institutions worldwide regulate the operation and restrict frequencies to minimize interference.
(c) Wireless networks allow for the design of small, independent devices.
(d) Wireless networks are robust.
[ESE-2024]

■■■

Answers Communication and Cellular Networks

1.1 (b)	1.2 (a)	1.3 (d)	1.4 (d)	1.5 (a)	1.6 (c)	1.7 (a)	1.8 (c)	1.9 (a)
1.10 (a)	1.11 (c)	1.12 (b)	1.13 (a)	1.14 (d)	1.15 (c)	1.16 (c)	1.17 (b)	1.18 (b)
1.19 (b)	1.20 (c)	1.21 (b)	1.22 (d)	1.23 (a)	1.24 (d)	1.25 (d)	1.26 (c)	1.27 (a)
1.28 (a)	1.29 (b)	1.30 (d)	1.31 (d)	1.32 (c)	1.33 (d)	1.34 (a)	1.35 (c)	1.36 (b)
1.37 (b)	1.38 (d)	1.39 (b)	1.40 (b)	1.41 (c)	1.42 (d)	1.43 (d)	1.44 (a)	1.45 (d)
1.46 (b)								

Explanations Communication and Cellular Networks

1.1 (b)

$$P(K=1) = \left.\frac{(\tau\lambda_t)^K e^{-\tau\lambda_t}}{K!}\right|_{K=1}$$

$$= \frac{(10\times 0.01)^1 e^{-(10\times 0.01)}}{1!}$$

$$= 0.1\, e^{-0.1} = 0.09$$

1.3 (d)

$p = 3,\ q = 11,\ e = 3$

$$z = (p-1)(q-1)$$
$$= (3-1)(11-1) = 20$$
$$(e\times d) \bmod z = 1$$
$$(3\times d) \bmod z = 1$$
$$(3\times d) \bmod 20 = 1$$
$$3d = \text{multiple of } 20 + 1$$
$$3d = 20\times 1 + 1$$
$$d = 7$$

1.5 (a)

Delay jitter is a situation where different packets transmitted from source to destination experience different amounts of transmission delay.

In a virtual circuit switching network, once the path is established between the source and destination, all the packets follow that path. Since all the packets are following the same path, all the packets experience the same delay and delay jitter will be zero.

1.6 (c)

The modulation scheme used in GSM is gaussian minimum shift keying.

1.10 (a)

In OSI model there is layered architecture and the service primitives are used to define client server model.

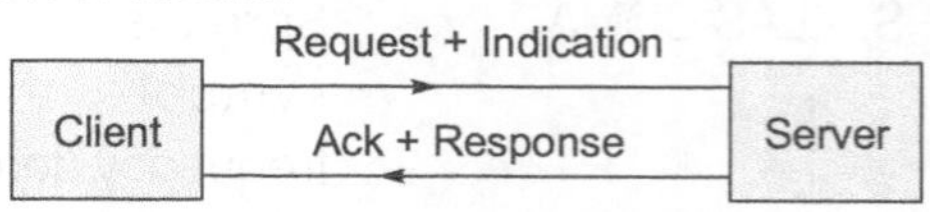

1.11 (c)

For a fully connected mesh set each node is connected to another node by a link, hence number of links

$$= {}^nC_2 = \frac{n(n-1)}{2} = \frac{(10)(10-1)}{2} = 45$$

1.12 (b)

The link is shared between number of users and no dedicated connection is there. It also supports first come first serve basis hence the blocking is addressed.

1.13 (a)

$$\text{Number of message bits} = 2400 \text{ bits}$$
$$\text{TL segment} = \text{Data} + \text{Header} = 150 + 2400 = 2550 \text{ bits}$$
$$\text{Data supported by network} = 900 - 26 \text{ bits} = 874 \text{ bits}$$
$$\text{Packet 1} = 874 \text{ bits}$$
$$\text{Packet 2} = 874 \text{ bits}$$
$$\text{Packet 3} = 802 \text{ bits}$$
$$\text{Total} = 2550 \text{ bits}$$

Total number of bits including header as well as travelling through two different network. Hence, the number of bits delivered to the destination network is equal to,

$$(26 + 26 + 874) + (26 + 26 + 874) + (26 + 26 + 802) = 2706 \text{ bits}$$

1.14 (d)

$$4T_1 = 1T_2$$
$$7T_2 = 1T_3$$
$$6T_3 = 1T_4$$
$$T_1 = 1.544 \text{ Mbps}$$
$$T_4 = ?$$
$$4T_1 = T_2 = 6.312 \text{ Mbps}$$
$$7T_2 = 28\text{T}_1 = 44.736 \text{ Mbps } (T_3)$$
$$6T_3 = 1T_4 = 274.2 \text{ Mbps}$$

Note: There is a printing error in the given question, the statement should be like:

"In a communication network, 4 T1 streams are multiplexed to form 1 T2 stream and 7 T2 streams are multiplexed to form 1 T3 stream. Further 6 T3 streams are multiplexed to form 1 T4 stream."

1.15 (c)

In CDMA there is no concept of frequency reuse and it is more secure by using codes. Hence unlike GSM, each station can transmit over entire frequency spectrum as each signal is having different codes. With the help of rake receiver multiple signals can add linearly.

1.16 (c)

In cellular system if cell size is small than the population density is high hence number of users are more and cellular system is decentralised hence robust.

Size ↓ d ↓ hence less radiated power to coverage of cell.

Small cells requires more frequency planning and infrastructure to avoid adjacent channel interference and co-channel interference is constant because Q/R is constant so need to address local interference.

1.17 (b)

MAC address defined at data link layer to indicate device i.e. NIC or ethernet address.

1.18 (b)

DLL provides flow control and error control through retransmission and CRC.

1.23 (a)

- Elastic traffic has the ability of making adjustment of wide-ranging changes in delay and throughput across the internet and still meets the needs of the applications. It adjusts its throughput between end hosts in response to network condition. Network load or congestion may cause packet loss.
- It the data transfer does not lose data, no matter how fast or slow it is, the data will sooner or later get transferred to the destination device. We say that, from the point of view of the network, this source of traffic is elastic. Many store-and-forward services (with the exception of media streaming services) are elastic e.g. file transfer, WWW download, e-mail.
- Inelastic traffic does not easily adapt to changes in delay and throughput. Real-time multimedia (audio streaming, video, VoIP) is an example of inelastic traffic.
- Service traffic is various non-TCP/IP traffic, like DNS requests, PINGs, etc.

1.24 (d)

The data message of 10 ms duration have 4800 bits to be transmitted. Therefore,

$$\text{Data Rate} = \frac{4800 \text{ bits}}{10 \times 10^{-3} \text{ sec}} = 480 \text{ kbps}$$

Since, there is a propagation delay of 1 ms per hop. Therefore,
Total propagation delay = 1 ms × 10 hops = 10 ms
The first packet is received at 10 ms. Since, the data message is of 10 ms duration, therefore,

Total delay = 10 ms + 10 ms = 20 ms

1.25 (d)

- The session layer controls the dialogues (connections) between applications. It provides mechanism for opening, closing and managing a session between end-user application processes. Therefore, statement (1) is incorrect.
- The data link layer provides reliable transmission of data over the physical layer by framing (i.e. breaking the packets into frames), error-detection, retransmission and flow-control. Therefore, statement (2) is correct.
- Transport layer is responsible for end-to-end error recovery and flow control of data between hosts. Therefore, statement (3) is correct.

1.26 (c)

- TELNET is an application protocol that allows a user at one site to access another user at a remote system. Telnet uses TCP to transmit data. Hence, statement (1) is correct.
- SMTP is an internet standard communication protocol for electronic mail transmission over a TCP connection. Hence, statement (2) is correct.
- The network layer is responsible for routing messages between different local networks. Hence, statement (3) is correct.

1.27 (a)

The signal-to-cochannel interference ratio (S/I) at the desired mobile receiver which monitors the forward channel is defined by

$$\frac{S}{I} = \frac{S}{I_1 + I_2 + \ldots + I_P} = \frac{S}{\sum_{i=1}^{P} I_i}$$

where,

S = Signal power from desired base station.

I_i = Interference power caused by the ith interfering co-channel cell base station.

P = No. of co-channel interfering cells.

1.28 (a)

- TCP is a reliable, connection-oriented protocol. It provides reliability with a mechanism called Positive Acknowledgment with Re-transmission (PAR).
- UDP is an unreliable, connection-less protocol. It doesn't include handshaking dialogues to provide reliability, ordering or data integrity. Hence, UDP's service is unreliable and packets may arrive out of order or even lost.
- FTP and SMTP are application layer protocols that uses TCP for reliable transfer of data.

1.29 (b)

- Private Branch Exchange (PBX) uses circuit switching to interconnect telephones within a building or office.
- A switching centre that directly supports subscribers is known as end-office. Hence, statement (2) is incorrect.
- In telephony, the local loop is the physical link or circuit that connects from the demarcation point of the customer premises to the edge of the common carrier or telecommunications service provider's network.

1.30 (d)

- For an FDMA system, the capacity of a cell is equal to the number of channels allocated to it.
- If a cell has L potential subscribers and is able to handle N simultaneous users, then
  (i) If $L < N$, the system is referred to as non-blocking and all calls can be handled all the time.
  (ii) If $L > N$, the system is blocking, a subscriber may attempt a call and find the capacity fully in use and therefore, be blocked.

Hence, statements 2 and 3 are incorrect.

1.31 (d)

The performance improvement factor P_f in CDMA digital cellular system is given by

$$P_f = \frac{G_V \cdot G_A}{\rho}$$

where G_V = Voice-activity interference reduction factor = 2.5

G_A = Cell sectorization gain factor = 2.5

ρ = Multi-user interference factor = 1.6

Substituting the values, we get

$$P_f = \frac{2.5 \times 2.5}{1.6} = 3.9$$

1.32 (c)

Internet Protocol (IP) is a unreliable protocol because it does not guarantee the delivery of a datagram to the destination. The reliability must be provided by the upper layer protocols like TCP. We can say that IP provides a connectionless best effort delivery service to the transport layer. Hence, option (c) is correct.

1.33 (d)

The cells that use the same set of frequencies in a coverage area (frequency reuse concept) are called co-channel cells.

1.34 (a)

Multipath fading may be substantially reduced because the signal is spread over a large spectrum. If the speed spectrum bandwidth is greater than the coherence bandwidth of the channel, the inherent frequency diversity will mitigate the effects of small-scale fading.

1.35 (c)

Since, the bandwidth of coaxial cable is given as 800 MHz, the number of 6 MHz wide TV channels that can be multiplexed on 800 MHz coaxial cable is given by

$$N = \frac{\text{Channel Bandwidth}}{\text{Signal Bandwidth}}$$

$$= \frac{800\,\text{MHz}}{6\,\text{MHz}} \simeq 133$$

1.36 (b)

We know that,

AMPS is a analog cellular phone system, using FDMA only.

D-AMPS is a digital cellular phone system using TDMA and FDMA.

GSM is a digital cellular phone system using TDMA and FDMA.

1.37 (b)

A trap door is kind of a **secret entry point** into a program that allows anyone to gain access to any system without going through the usual security access procedures.

1.38 (d)

ASCII code has different generic classes of control characters as well as a number of individual characters:

1. **Transmission Controls (TC1 - TC10):** Eg: *NAK* (Negative acknowledge character, a definite flag for, indicating a problem in reception), *ETB* (End of transmission block character to indicate the end of a block of data), *SOH*, *STX*, *ETX*, *EOT*, etc.
2. **Information Separators:** Eg: *Record Separator* **(RS)** used to position the text cursor to the upper left character position in the screen, *Unit Separator* **(US)** character is used to reset the code environment to a known initial state.
3. **Format Effectors:** Eg: *Vertical Tab* **(VT)** is used to position the text cursor a distance equal to the dimension of the character field lying perpendicular to the character path in a direction perpendicular to the character path, *Form Feed* **(FF)** is used to position the text cursor to the upper left character position on the screen, *Backspace* **(BS)**, *Horizontal Tab* **(HT)**, etc.
4. **Device Controls:** DC1 - DC4 are used to control device parameters. Typically, the characters DC1 and DC3 are used as stop/ start flow control codes in many transmission protocols.

Hence, P → 4, Q → 1, R → 2, S → 3

1.39 (b)

The error control technique is based on retransmission on negative acknowledgement. Hence, option (b) is correct.

1.40 (b)

Non-Orthogonal Multiple Access (NOMA) allows multiple users within the same spectrum, thus improving spectral usage efficiency. The various NOMA schemes are:

1. **Code Division Multiple Access (CDMA):** Several transmitters can send information simultaneously over a single communication channel. CDMA employs spread-spectrum technology and a special coding scheme.
2. **Large area synchronous CDMA (LAS-CDMA):** Applies a new set of spreading codes that exhibit interference free window to separate users in the wireless channel.
3. **Multi-Carrier CDMA (MC-CDMA):** Spreads each user symbol in the frequency domain. That is, each user symbol is carried over multiple orthogonal subcarriers.
4. **Interleave Division Multiple Access (IDMA):** Interleavers are used as the only means for user separation. IDMA inherits many advantages from CDMA, such as diversity against fading and mitigation of the worst-case other-cell user interference problem.
5. **Low Density Signature CDMA (LDS-CDMA):** Uses spreading sequences of low density for spreading the data symbols in time domain. This technique benefits from a less complex MUD compared to conventional CDMA with optimum MUD.
6. **Space Division Multiple Access (SDMA):** Enables two or more users, associated with the same base station, to use the same time and frequency and code resources on the grounds of their physical location or spatial separation.

Hence, P → 3, Q → 4, R → 1, S → 2

1.41 (c)

The long preamble that the transmitter needs to send causes throughput reduction and energy wastage for both transmitter and receiver. The WiseMAC protocol build upon preamble sampling to correct this deficiency. Using additional contents of ACK packets, each node learns the periodic sampling times of its neighboring nodes, and uses this information to send a shorter wake-up preamble at just the right time.

The preamble duration is determined by the potential clock drift since last synchronization. Let T_w be the receiver sampling period, θ the clock drift and L the interval between communications, then the duration of the preamble T_p need only be:

$$T_p = \min(4L\theta, T_w)$$

The packets in WiseMAC also contain a "more" bit which the transmitter uses to signal to the receiver if it needs to stay awake a little longer in order to receive additional packets intended for it.

1.42 (d)

Flooding involves broadcasting packets to all neighbors, and while having multiple paths can be beneficial, it's not a strict requirement. Flooding can work even in networks where only a single path exists between the source and destination.

1.43 (d)

CDMA provides increased capacity, security and efficient utilization. It has typically lower throughput compared to other media access techniques like TDMA and FDMA.

1.44 (a)

1. ST connector uses a ceramic ferrule and a rugged metal housing. It is latched in place by twisting.
2. ESCON connector has a shock proof plastic housing with a retractable shroud to protect the ceramic ferrules inside the connector.
3. SC connectors are available in both single-mode and multimode versions, accommodating different fiber types and applications.
4. LC connectors are ideal for high-density applications where space is limited, such as data centers and telecommunications rooms.

1.45 (d)

Network and switching subsystem (NSS) is the core network of GSM, which carries out call and mobility management functions for mobile phones present in the network. The key elements of NSS are MSC, HLR and VLR. The Operations and Maintenance Center (OMC) serves as a central hub for monitoring, managing, and maintaining the network's operational status and typically, is not a part of NSS.

1.46 (b)

Several Government and non-Government institutions worldwide regulate the operation and restrict frequencies to minimize interference. This is not an advantage of WLAN.

■■■

2 Microwave and Satellite Communication Systems

2.1 Antenna elevation angle at the ground station for satellite communication is always kept above 5° to
(a) minimise the sky noise temperature
(b) reduce the effect of oxygen and water vapour absorption on the antenna noise temperature
(c) minimise the slant range
(d) increase the visibility of the satellite
[ESE-2000]

2.2 Random satellite moves in
(a) random paths (b) polar orbits
(c) geostationary orbits (d) equatorial plane
[ESE-2000]

2.3 **Assertion (A):** Several satellites have been designed to operate in the Ku-band in spite of higher atmospheric absorption than the *C*-band.
Reason (R): Use of smaller, narrower beam antennas in the Ku-band results in lower interference.
(a) Both A and R are true and R is the correct explanation of A
(b) Both A and R are true but R is NOT the correct explanation of A
(c) A is true but R is false
(d) A is false but R is true
[ESE-2000]

2.4 An earth station employs a 1 kW high power amplifier (HPA) and a 20 m Cassegrain antenna whose transmitted gain is 65 dB at a free space wavelength of 2.1 cm. If the loss of the wavelength that connects HPA to the feed is 1 dBm, then the earth station EIRP is
(a) 29 dB m (b) 59 dB m
(c) 94 dB m (d) 124 dB m
[ESE-2000]

2.5 Which of the following statements on communication by geosynchronous satellites are correct?
1. The propagation loss over each direction (uplink and downlink) is about 200 dB at 5 GHz.
2. Radius of the geosynchronous orbit is 42,250 km.
3. An arc of about 17° must be covered by the satellite antenna in the equatorial plane.
4. The polar regions are well covered by the satellite antenna.

Select the correct answer using the code given below:
(a) 1, 2 and 4 (b) 2 and 3
(c) 1 and 4 (d) 1, 2 and 3
[ESE-2001]

2.6 A TV station works at 300 MHz and radiates 100 kW of power. This signal received at a distance of 200 km will be
(a) a few μV/m due to attenuation by atmosphere
(b) 1–5 mV due to ionospheric reflection
(c) nil since propagation is line of sight
(d) nil because signal fades completely at 300 MHz
[ESE-2001]

2.7 In free space line of sight propagation case, the transmission losses between transmitter and receiver increase with frequency (f) as
(a) f (b) f^2
(c) f^4 (d) $f^{1/2}$
[ESE-2002]

2.8 A TV signal transmitted in New Delhi can be received in Nagpur which is about 1000 km away if and only if
(a) Transmitting antenna is at a height of 5000 m
(b) A cable TV system is employed
(c) Power of the order of 100 kW is radiated
(d) A satellite is employed for this
[ESE-2003]

2.9 Which one of the following techniques is not suitable for automatic satellite tracking?
(a) Monopulse (b) Step-track
(c) Conical scanning (d) Lobe switching
[ESE-2004]

2.10 What should be the frequencies used for communication inside the coal mines, from the viewpoint of minimizing the propagation path loss?

(a) In the range from 30 MHz to 400 MHz
(b) In the range from 1 GHz to 4 GHz
(c) In the range from 500 MHz to 1 GHz
(d) In the range from 1 GHz to 2 GHz

[ESE-2004]

2.11 Which one of the following statements is not correct?
(a) A geosynchronous satellite remains practically stationary relative to earth antennas
(b) A geosynchronous satellite means the same thing as geostationary satellite
(c) There is a trade-off between the cost of a communication satellite and cost of its earth stations
(d) Three geosynchronous satellites cannot give 100% global coverage

[ESE-2004]

2.12 If a satellite is launched at an orbital radius of twice that of a geostationary satellite, how much time will the launched satellite take to travel around the earth?
(a) 48 hrs (b) 12 hrs
(c) 96 hrs (d) 6 hrs [ESE-2005]

2.13 A geostationary satellite located at about 35000 km from earth can cover
(a) Complete surface of the earth
(b) One hemisphere in one pass
(c) One side of the earth
(d) An area depending on antenna used

[ESE-2005]

2.14 What is the reason for using frequencies of the order of a few GHz in satellite communication?
(a) Antenna sizes are small and ionosphere does not reflect frequencies
(b) Helical antennas can be used at these frequencies
(c) Frequencies can pass through ionosphere without attenuation
(d) Easy availability of components at these frequencies. [ESE-2006]

2.15 Communication satellites are allotted bandwidth of 500 MHz. This can be increased by using
(a) frequency and polarization re-use
(b) time division multiplexing
(c) frequency division multiplexing
(d) triple modular redundancy

[ESE-2008]

2.16 Which one of the following is correct?
In satellite communication links, generally
(a) the uplink carrier frequencies are greater than downlink carrier frequencies
(b) the uplink carrier frequencies are lesser than downlink carrier frequencies
(c) both uplink and downlink carrier frequencies are same
(d) it is not necessary to use carrier at all

[ESE-2008]

2.17 During power measurement the out power measured was –90 dBm. What is the measured power in W?
(a) 1 mW (b) 1 pW
(c) 10 W (d) 1 W

[ESE-2008]

2.18 What was the first commercial geostationary communication satellite ?
(a) INTELSAT 1 (b) ECHO
(c) INSAT-1A (d) SPUTNIK

[ESE-2009]

2.19 What does a link budget for satellite communication include?
(a) Total cost of satellite
(b) Cost of satellite plus launch vehicle
(c) Signal and noise levels in dB
(d) Margins of error permitted [ESE-2009]

2.20 In a typical satellite communication system, which of the following could be the up-link and down-link frequencies respectively?
(a) 40 GHz and 60 GHz
(b) 60 GHz and 40 GHz
(c) 6 GHz and 4 GHz
(d) 4 GHz and 6 GHz [ESE-2010]

2.21 The output stage of a transponder onboard a satellite has a maximum power output of 10 watts. However, it is not operated at the maximum power output in order to
(a) Conserve the available limited battery power
(b) Reduce noise due to device
(c) Avoid intermodulation distortion
(d) Avoid heating gup of the satellite beyond a per set value

[ESE-2010]

2.22 **Assertion (A):** It is not necessary to incorporate a very low noise amplifier in a communication satellite.
Reason (R): The noise temperature of the satellite antenna is usually high, since the beam is covered by the earth.
(a) Both A and R are true and R is the correct explanation of A
(b) Both A and R are true but R is NOT the correct explanation of A
(c) A is true but R is false
(d) A is false but R is true [ESE-2010]

2.23 **Assertion (A):** In satellite communication technique, frequency reuse effectively doubles the bandwidth and information capacity of a satellite.
Reason (R): Electromagnetic waves radiated from a transmitting antenna could be received by a distant receiving antenna by tuning the receiver in terms of frequency and polarization.
(a) Both A and R are true and R is the correct explanation of A
(b) Both A and R are true but R is NOT the correct explanation of A
(c) A is true but R is false
(d) A is false but R is true [ESE-2010]

2.24 For communication from satellite to the earth station, microwave frequencies are used because
(a) loss is minimum
(b) noise added to signal is low in this window
(c) these do not get reflected back by ionosphere
(d) many channels can be used [ESE-2011]

2.25 A Geostationary orbit is chosen for communication satellites because
(a) it is stationary at one point in space
(b) with respect to a spot on earth it looks stationary
(c) this orbit provides earth's coverage of more than 50% using a single satellite
(d) the length of 4700 km is convenient for launching [ESE-2011]

2.26 A low earth orbit satellite can provide large signal strength at an earth station because
(a) Path loss is low
(b) These orbits are immune to noise
(c) Large solar power can be generated at these orbits
(d) Lower microwave frequencies in S-band can be used

[ESE-2011]

2.27 If 'r' is the radius of circular orbit then the orbital period of a satellite is directly proportional to
(a) $r^{3/2}$ (b) $r^{1/3}$
(c) $r^{1/2}$ (d) $r^{2/3}$

[ESE-2011]

2.28 **Assertion (A):** A passive satellite only reflects back signals.
Reason (R): Communication satellite is a repeater between many transmitting stations and many receiving stations.
(a) Both A and R are true and R is the correct explanation of A
(b) Both A and R are true but R is NOT the correct explanation of A
(c) A is true but R is false
(d) A is false but R is true [ESE-2011]

2.29 **Assertion (A):** A geostationary orbit is same as a geosynchronous orbit.
Reason (R): A geostationary orbit does not necessarily lie in equatorial plane.
(a) Both A and R are true and R is the correct explanation of A
(b) Both A and R are true but R is NOT the correct explanation of A
(c) A is true but R is false
(d) A is false but R is true [ESE-2011]

2.30 Consider the following statements about the maximum usable frequency (MUF) for radio communication between two specified points using an ionospheric layer
1. MUF is equal to critical frequency
2. MUF is more than the critical frequency
3. MUF depends upon the height of the ionospheric layer
4. MUF depends upon the distance between the two points

Which of these statements are correct?
(a) 1, 2, 3 and 4 (b) 2 and 3 only
(c) 3 and 4 only (d) 2 and 4 only

[ESE-2011]

2.31 **Statement (I):** In the two body system consisting of the earth and a satellite, centre of mass always coincides with centre of the earth.
Statement (II): Mass of earth is much greater than mass of satellite.
(a) Both Statement (I) and Statement (II) are individually true and Statement (II) is the correct explanation of Statement (I).

(b) Both Statement (I) and Statement (II) are individually true but Statement (II) is not the correct explanation of Statement (I).
(c) Statement (I) is true but Statement (II) is false.
(d) Statement (I) is false but Statement (II) is true.
[ESE-2012]

2.32 For an earth station transmitter with an antenna output power of 40 dBW (10,000 W), a back-off loss of 3 dB, a total branching and feeder loss of 3 dB and transmit antenna gain of 4 dB, the effective isotropic radiated power (EIRP) will be
(a) 38 dBW (b) 40 dBW
(c) 36 dBW (d) 47 dBW [ESE-2012]

2.33 In a satellite communication, the uplink power level is generally high and power level of the received downlink signal is low because:
1. High signal attenuation due to free space loss.
2. Limited available downlink power.
3. Finite satellite down link antenna gain, which is dictated by the required coverage area.

(a) 1 and 2 only (b) 1 and 3 only
(c) 2 and 3 only (d) 1, 2 and 3
[ESE-2013]

2.34 A radio cab company with its antenna at a height of 15 m communicates with a cab having its antenna 1.5 m. The maximum communication distance without obstacles is roughly
(a) 10 km (b) 20 km
(c) 28 km (d) 36 km [ESE-2013]

2.35 In microwave relay communication the repeater is usually an amplifier for the amplification of
(a) Carrier signal
(b) Baseband signal
(c) Amplitude modulated IF signal
(d) Frequency modulated IF signal
[ESE-2014]

2.36 In a geostationary satellite communication system, a message signal is transmitted from an earth station via an uplink to a satellite, amplified in a transponder on board the satellite and then transmitted via a downlink to another earth station. The most popular frequency band for satellite communication is
(a) 16 MHz for the uplink and 14 MHz for the downlink
(b) 4 GHz for the uplink and 6 GHz for the downlink
(c) 6 GHz for the uplink and 4 GHz for the downlink
(d) 10 GHz for the downlink and 8 GHz for the uplink
[ESE-2014]

2.37 As per WARC-1979 allocation, commercial communication satellites use bandwidth of
(a) 4 MHz (b) 40 MHz
(c) 200 MHz (d) 500 MHz [ESE-2014]

2.38 In satellite communication, Faraday rotation is caused by
(a) Plasma frequency
(b) Earth's magnetic field
(c) Non-Gaussian nature of uplink noise when received in the downlink channel
(d) Ionospheric reflections that occur multiple times
[ESE-2014]

2.39 A satellite link uses different frequencies for receiving and transmitting in order to
(a) avoid interference from terrestrial microwave link
(b) minimize free space losses
(c) maximize antenna gain
(d) avoid interference between its powerful transmitted signal and weak incoming signal
[ESE-2014]

2.40 **Statement (I):** Space wave is used for propagation of FM broadcast system.
Statement (II): Several independent interference free transmitters can be operated on the same frequency because of line-of-sight propagation.
(a) Both Statement (I) and Statement (II) are individually true and Statement (II) is the correct explanation of Statement (I).
(b) Both Statement (I) and Statement (II) are individually true but Statement (II) is not the correct explanation of Statement (I).
(c) Statement (I) is true but Statement (II) is false.
(d) Statement (I) is false but Statement (II) is true.
[ESE-2014]

2.41 The frequency range of very high frequency (VHF) is
(a) 300 MHz – 3000 MHz
(b) 30 MHz – 300 MHz
(c) 3 MHz – 30 MHz
(d) 30 THz – 3000 Hz [ESE-2014]

2.42 Maximum frequency reflected from ionosphere is 9 MHz on a particular day. The maximum ion density/metre^3 is
(a) 10^8 (b) 10^6
(c) 10^{12} (d) 10^9 [ESE-2015]

2.43 The frequency range for satellite communication is
(a) 1 kHz – 100 kHz
(b) 100 kHz – 1 MHz
(c) 10 MHz – 30 MHz
(d) 1 GHz – 30 GHz **[ESE-2015]**

2.44 Consider the following statements for multiple access system in a satellite earth station:
1. Access to same repeater sub-systems and same RF channel is possible.
2. Frequency division multiple access is used.
3. Several carriers are not amplified by same TWT.

Which of the above statements are correct?
(a) 2 and 3 only (b) 1 and 3 only
(c) 1 and 2 only (d) 1, 2 and 3
[ESE-2016]

2.45 The different access methods which permit many satellite users to operate in parallel through a single transponder without interfering with each other are
1. Frequency Division Multiple Access (FDMA)
2. Time Division Multiple Access (TDMA)
3. Code Division Multiple Access (CDMA)

Which of the above are correct?
(a) 1 and 2 only (b) 1 and 3 only
(c) 2 and 3 only (d) 1, 2 and 3
[ESE-2016]

2.46 The transmission path loss for a geostationary satellite signal for uplink frequency of 6 GHz is
(a) 60 dB (b) 92 dB
(c) 184 dB (d) 200 dB **[ESE-2017]**

2.47 What is the maximum signal propagation time for a geosynchronous satellite transmission system?
(a) 140 ms (b) 220 ms
(c) 280 ms (d) 560 ms **[ESE-2017]**

2.48 A microwave communication link employs two antennas for transmission and reception elevated at 200 m and 80 m, respectively. Considering obliqueness of the Earth, the maximum possible link distance is
(a) 46 km (b) 64 km
(c) 96 km (d) 102 km **[ESE-2017]**

2.49 A receiving antenna with a gain of 40 dB looks at a sky with a noise temperature of 15 K. The loss between the antenna and LNA input due to the feed horn is 0.4 dB, and the LNA has a noise temperature of 40 K. The $\frac{G}{T}$ value is
(a) 11.2 dB/K (b) 13.4 dB/K
(c) 20.6 dB/K (d) 39.0 dB/K
[ESE-2018]

2.50 A company operating taxicabs communicates from its central office with the help of an antenna at the top of a 16 m tower. The antenna on the taxicabs is 1.44 m above the ground. The communication distance between the central office and a taxicab ***cannot*** exceed
(a) 18.6 km (b) 24.0 km
(c) 50.0 km (d) 62.3 km
[ESE-2018]

2.51 The linear velocity of a satellite, when in a circular orbit, is
(a) directly proportional to its mass
(b) directly proportional to the square root of its mass
(c) directly proportional to the square of its mass
(d) independent of its mass
[ESE-2018]

2.52 The orbital period of a satellite in a circular orbit of 500 km above the Earth's surface, taking mean radius of the Earth as 6400 km and Kepler's constant μ as 4×10^5 km^3/s^2 is nearly
(a) 1.6 hours (b) 2.4 hours
(c) 3.2 hours (d) 6.4 hours
[ESE-2018]

2.53 A satellite is orbiting in the equatorial plane with a period from perigee to perigee of 12 h. If the eccentricity = 0.002, $i = 0°$, $K_1 = 66063.17$ km^2, $\mu = 3.99 \times 10^{14}$ m^3/s^2 and the earth's equatorial radius = 6378.14 km, the semi-major axis will be
(a) 34232 km (b) 30424 km
(c) 26612 km (d) 22804 km
[ESE-2019]

2.54 An earth station at sea level communicates at an elevation angle of 35° with GEO satellite. The vertical height of the stratiform rain is 3 km. The physical path length L through the rain will be nearly
(a) 6.3 km (b) 5.2 km
(c) 4.1 km (d) 3.0 km **[ESE-2019]**

2.55 Satellite communication among stations in different areas can be achieved if the satellite has the ability to switch time slots from one beam to another. This is known as satellite switched
(a) TDMA (b) TSMA
(c) FAMA (d) SCPC **[ESE-2020]**

2.56 For a total transmit power (P_t) of 1000 W and for a transmission rate of 50 Mbps, the energy per bit (E_b) will be

(a) 10 μJ (b) 20 μJ
(c) 30 μJ (d) 40 μJ

[ESE-2020]

2.57 Each earth station's transmission is encoded with a unique binary word called

(a) Station code (b) Chip code
(c) Access code (d) Gold code

[ESE-2020]

2.58 For a GEO satellite, what is the free space loss (L_{db}) at the equator in terms of carrier wavelength (λ)?

(a) $20\log(\lambda)+21.98$ (b) $-20\log(\lambda)+173.07$
(c) $20\log(\lambda)-173.07$ (d) $-20\log(\lambda)+21.98$

[ESE-2021]

2.59 If R = Earth's radius, h = orbit height, β = coverage angle, and θ = minimum elevation angle, then which one of the following relations is correct?

(a) $\frac{R}{R+h}=\frac{\cos(\beta+\theta)}{\cos(\theta)}$

(b) $\frac{R}{h}=\frac{\cos(\beta)}{\cos(\theta)}$

(c) $\frac{h}{R}=\frac{\cos(\beta+\theta)}{\cos(\beta)}$

(d) $\frac{R+h}{h}=\cos(\beta+\theta)-\cos(\theta)$ [ESE-2021]

2.60 The orbital satellites are

(a) symmetrical (b) asymmetrical
(c) synchronous (d) nonsynchronous

[ESE-2022]

2.61 The requirement for reliability, long life, stability, high efficiency and suitability for space environment are met by the use of

(a) IF amplifier
(b) RF amplifier
(c) travelling wave tube amplifier
(d) ultrasonic amplifier [ESE-2022]

2.62 The highest frequency that is returned to the earth at a given distance is called

(a) maximum available frequency
(b) maximum communication frequency
(c) maximum bandwidth frequency
(d) maximum usable frequency [ESE-2022]

2.63 The mean sun does move at a uniform speed but otherwise requires the same time as the real sun to complete one orbit of the earth, this time being

(a) the tropical year
(b) the leap year
(c) the Julian calendar year
(d) the Gregorian calendar year [ESE-2022]

2.64 The component in the satellite that takes an uplink signal and converts into a downlink signal is called

(a) uplink device (b) transponder
(c) downlink device (d) transmitter

[ESE-2022]

2.65 An advantage of passive satellites is that they do not require sophisticated electronic equipment on board, although they are not necessary void of power. Some passive satellites require which one of the following for tracking and ranging purposes?

(a) Radio beacon transmitter
(b) Defense launched courier
(c) Radio beacon receiver
(d) Satellite beacon launcher [ESE-2022]

2.66 In antennas, if S_A is the input signal generated by the source (by the receiving antenna) at temperature T_A, T_E is the effective noise temperature of the receiver network (referred to the input terminal), B is bandwidth and k is Boltzmann's constant, then the output signal-to-noise ratio is

(a) $\frac{S}{N}=\frac{S_A}{(T_A+T_E)kB}$ (b) $\frac{S}{N}=\frac{S_AT_A}{kBT_E}$

(c) $\frac{S}{N}=\frac{S_A(1+T_A)}{kB(1+T_E)}$ (d) $\frac{S}{N}=\frac{S_AT_A}{kB(1+T_E)}$

[ESE-2022]

2.67 The frequency range for the broadcast satellite service is

(a) 2 GHz to 4 GHz
(b) 4 GHz to 8 GHz
(c) 8 GHz to 12.5 GHz
(d) 12.5 GHz to 26.5 GHz

[ESE-2023]

2.68 A cell-site transmitter generates +15 dBm RF signal and is connected to an antenna using an RF coaxial cable that induces a 3 dB loss. The cable has two connectors at its either end that induce a loss of 2 dB each. What is the signal level at the input of the antenna?
(a) +5 dBm (b) +6 dBm
(c) +7 dBm (d) +8 dBm
[ESE-2024]

2.69 Which one of the following statements is **not** correct about the D-channel contention-resolution algorithm?
(a) When a subscriber device has an LAPD frame to transmit, it transmits a series of binary ones on the D-channel using the pseudoternary encoding scheme.
(b) The NT, on receipt of a D-channel bit, reflects back the binary value as a D-channel echo bit.
(c) When a terminal is ready to transmit an LAPD frame, it listens to the stream of incoming D-channel echo bits.
(d) To overcome collision, a transmitting TE monitors the E bits and compares them to its transmitted D bits. If a discrepancy is detected, the terminal ceases to transmit and returns to a listen state.
[ESE-2024]

■■■

Answers Microwave and Satellite Communication Systems

2.1 (d)	2.2 (b)	2.3 (a)	2.4 (d)	2.5 (a)	2.6 (c)	2.7 (b)	2.8 (d)	2.9 (b)
2.10 (a)	2.11 (d)	2.12 (*)	2.13 (d)	2.14 (a)	2.15 (a)	2.16 (a)	2.17 (b)	2.18 (a)
2.19 (c)	2.20 (c)	2.21 (a)	2.22 (d)	2.23 (a)	2.24 (c)	2.25 (b)	2.26 (a)	2.27 (a)
2.28 (b)	2.29 (c)	2.30 (d)	2.31 (a)	2.32 (a)	2.33 (d)	2.34 (b)	2.35 (d)	2.36 (c)
2.37 (d)	2.38 (b)	2.39 (d)	2.40 (a)	2.41 (b)	2.42 (c)	2.43 (d)	2.44 (c)	2.45 (d)
2.46 (d)	2.47 (c)	2.48 (c)	2.49 (c)	2.50 (a)	2.51 (d)	2.52 (a)	2.53 (c)	2.54 (b)
2.55 (a)	2.56 (b)	2.57 (b)	2.58 (b)	2.59 (a)	2.60 (d)	2.61 (c)	2.62 (d)	2.63 (a)
2.64 (b)	2.65 (a)	2.66 (a)	2.67 (d)	2.68 (d)	2.69 (a)			

Explanations Microwave and Satellite Communication Systems

2.4 (d)

$P_t = 10^3$ W ; $G_t = 65$ dB

$$= (10)^{65/10} = 3162277.66$$

$$\text{EIRP} = P_t G_t = 10^3 \times 3162277.66$$

$$= 3162277660 \text{ W}$$

$$(\text{EIRP})_{\text{dBm}} = 10\log\frac{3162277660}{10^{-3}} = 125 \text{ dBm}$$

Taking the loss into consideration,

$$\text{EIRP} = 125 - 1 = 124 \text{ dBm}$$

2.5 (a)

Geosynchronous satellites are placed above or below the equator. So, these satellite can offer services to the polar region that might be beyond the sight of geostationary satellites.

2.7 (b)

Free space losses $\propto f^2$

Free space losses $\propto d^2$

(Free space losses)$_{\text{dB}}$

$$= 32.45 + 20\log_{10}d + 20\log_{10}f$$

where d in km and f in MHz.

2.9 (b)

Step-track technique is not suitable for automatic satellite tracking.

2.10 (a)

$$\text{Path loss} = \left(\frac{4\pi d}{\lambda}\right)^2 = \left(\frac{4\pi d}{C}\right)^2 \cdot f^2$$

Since path loss ∝ f^2, so frequencies used for communication inside the coal mines should be low to minimize the path loss.

2.11 (d)

Three geosynchronous satellites give 100% global coverage.

2.12 (*)

$T^2 \propto r^3$ by Kepler's law.

$$\left(\frac{T}{24}\right)^2 = (2)^3 \left[\because \text{time period of geostationary satellite is 24 hrs.}\right]$$

$$\Rightarrow \quad T^2 = (24)^2 (8)$$

$$\Rightarrow \quad T = 24\sqrt{8} = 48\sqrt{2} \simeq 67.88 \text{ hrs}$$

2.13 (d)

The coverage area of a satellite depends on its orbital height and the antenna used.

2.14 (a)

$$\text{Size of antenna} \propto \frac{1}{\text{frequency}}$$

For satellite communication, the frequencies used should be such that it must penetrate the ionosphere.

2.16 (a)

The receiver antenna gain

$$G = \frac{4\pi A}{\lambda^2} = \frac{4\pi A f^2}{c^2}$$

where, A = Cross sectional area of antenna
f = Uplink frequency
c = Velocity of light

Since the size of receiver antenna in satellite is very small in comparison to that in earth station, therefore, to achieve higher gain uplink carrier frequency is greater than downlink carrier frequency.

2.17 (b)

$$-90 \text{ dBm} = 10 \log \frac{P(\text{Watts})}{1 \times 10^{-3}}$$

$$\frac{P(\text{watts})}{10^{-3}} = 10^{-9}$$

$$P = 10^{-12} \text{ watts} = 1 \text{ pW}$$

2.21 (c)

A link budget involves a relatively simple addition and substraction of gains and losses in dB within an RF link.

2.25 (b)

A satellite in geostationary orbit looks stationary with respect a spot on earth.

2.24 (a)

$$\text{Path loss} = \left(\frac{4\pi R}{\lambda}\right)^2$$

If R is less then path loss will be low.

2.27 (a)

Satellite orbit period is

$$T^2 = \frac{4\pi^2 r^3}{GM}$$

or $$T \propto r^{3/2}$$

2.28 (b)

An active satellite first amplifies the signal and then reflects back. Satellite works as a repeater whereas passive one only reflects back the signal.

2.29 (c)

A geostationary orbit must be in equatorial plane.

2.30 (d)

$$\text{MUF} = f_c \sec \phi$$

where, f_c = critical frequency
ϕ = incidence angle

So MUF > f_c

Also angle of incidence is governed by the distance between the two points.

2.31 (a)

Let mass of earth = M
Mass of satellite = m
Separation between M and m = L

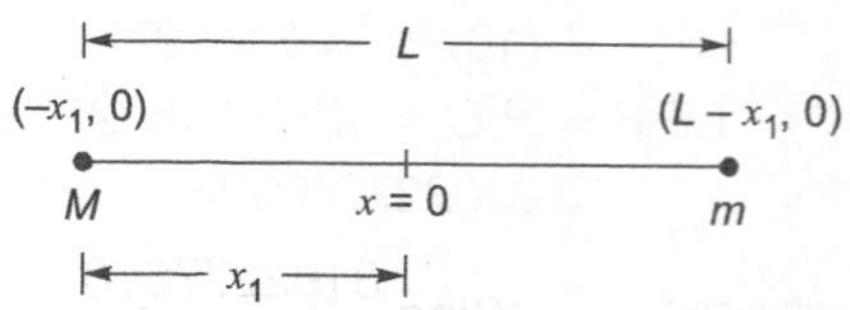

∴ Center of mass is at $x = 0$

$$\therefore \quad \text{CM} = 0 = \frac{M(-x_1) + m(L - x_1)}{M + m}$$

$$Mx_1 = mL - mx_1$$

$$x_1 = \frac{mL}{m + M} = \frac{\frac{m}{M}L}{1 + \frac{m}{M}}$$

If M is very large

$$x_1 \approx 0$$

∴ Centre of mass of satellite coincides with the centre of earth when M is very large.

2.32 (a)

Output power, P_{out} = 40 dBW

Back-off loss = 3 dB

Total branching and feeder less = 3 dB

So, $P_T = (40 - 3 - 3)\text{dB} = 34$ dB

Antenna gain = 4 dB

$$\text{EIRP} = P_T(\text{dB}) + G_T(\text{dB}) = 38 \text{ dB}$$

2.33 (d)

Path loss in satellite system is given as,

$$L_p = \left(\frac{4\pi R}{\lambda}\right)^2$$

So, path loss depends upon the frequency of the operation. Higher the frequency, more will be the path loss.

Also Gain $G = \left(\frac{4\pi A_e}{\lambda^2}\right)$

Power required to transmit a signal is proportional to the frequency of operation. More power requirement leads to more weight of the satellite. The satellite gets power from solar cell. So, power available at satellite is limited. On the other hand, the ground station can have much higher power. As the frequency is increased, the amount of interference or the path loss increases, so to overcome it, high power transmitters are required.

2.34 (b)

Maximum compensation distance,

$$= 4(\sqrt{d_1} + \sqrt{d_2}) = 4(\sqrt{15} + \sqrt{1.5}) \text{ kms} \approx 20 \text{ km}$$

2.36 (c)

In satellite communication generally the communication takes place in C-band (4-8 GHz) and as a standard two different frequencies of 6 GHz and 4 GHz are chosen for the process to avoid interference in the uplink and downlink.

2.37 (d)

As per WARC - 1979 a band of 500 MHz was assigned for commercial communication satellites. The bandwidth of the band allotted could however be increased by using different types of polarization but for single polarised wave the bandwidth will be equal to the band allotted i.e. 500 MHz.

2.41 (b)

Frequency range:

VLF = 3 – 30 kHz

LF = 30 – 300 kHz

MF = 0.3 – 3 MHz

HF = 3 – 30 MHz

VHF = 30 – 300 MHz

UHF = 300 – 3 GHz

SHF = 3 – 30 GHz

EHF = 30 – 300 GHz

2.42 (c)

$$f = 9\sqrt{N_{max}}$$

$$9 \times 10^6 = 9 \times \sqrt{N_{max}}$$

$$N_{max} = (10^6)^2$$

$$N_{max} = 10^{12} / \text{m}^3$$

2.43 (d)

The frequency range for satellite communication is in GHz. so (d) option is correct.

2.46 (d)

$$\text{FSL} = 32.45 + 20\log d_{km} + 20\log f_{MHz}$$

$$= 32.45 + 20\log 36000 + 20\log 6000 \text{ MHz}$$

$$= 199.2 \text{ dB} \cong 200 \text{ dB}$$

2.47 (c)

Round trip propagation delay,

$$t = \frac{2d}{c}$$

$$d \simeq 36000 \text{ km}$$

So, $t = \dfrac{2 \times 36000}{3 \times 10^5} = 240$ msec

The maximum delay also includes the processing time along with propagation delay. So, option (c) can be selected.

2.48 (c)

$$h_t = 200 \text{ m}$$

$$h_r = 80 \text{ m}$$

$$d = 4.12\left[\sqrt{h_t} + \sqrt{h_r}\right] \text{km}$$

$$= 4.12\left[\sqrt{200} + \sqrt{80}\right] \text{km}$$

$$= 95.11 \text{ km} \simeq 96 \text{ km}$$

2.49 (c)

$$T = T_{ant} + T_{LNA}$$

$$= 15 \text{ K} + 40 \text{ K}$$

$$= 55\text{ K}$$

$$10\log T = 10\log 55 = 17.4$$

$$\text{Gain} = 40 - 0.4 = 39.6\text{ dB}$$

$$\frac{G}{T} = G_{dB} - 10\log T$$

$$= 39.6 - 17.4 = 22.2\text{ dB/K}$$

Nearest answer 20.6 dB/K

2.50 (a)

For radio horizon consideration,

$$\text{Range} = 4.12\left[\sqrt{h_{t_m}} + \sqrt{h_{r_m}}\right]\text{km}$$

$$= 4.12\left[\sqrt{16} + \sqrt{1.44}\right]\text{km}$$

$$= 21.424\text{ km} \quad \text{...(i)}$$

For optical horizon consideration,

$$\text{Range} = 3.57\left[\sqrt{h_{t_m}} + \sqrt{h_{r_m}}\right]\text{km}$$

$$= 3.57\left[\sqrt{16} + \sqrt{1.44}\right]\text{km}$$

$$= 18.564\text{ km} \quad \text{...(ii)}$$

From equations (i) and (ii), option (a) can be selected as the answer.

2.52 (a)

$$T = \frac{2\pi(r_{th})^{3/2}}{\sqrt{GM}} = \frac{2\pi(500 + 6400)^{3/2}}{\sqrt{GM}}$$

$$= \frac{2\times 3.14\,(6900)^{3/2}}{\sqrt{4\times 10^5}} = 1.58\text{ hours}$$

$$\approx 1.6\text{ hours}$$

2.53 (c)

For an elliptical orbit according to Kepler's law

$$T^2 = \frac{4\pi^2}{GM}a^3$$

Here, a = Perigee distance hence = a_p

$T = 12h = 43200$ sec

$$\therefore \quad a_p^3 = \frac{T^2 GM}{4\pi^2}$$

$$= \frac{43200\times 43200\times 4\times 10^{14}}{4\times(3.14)^2}$$

$$a_p = 26525\times 10^3\text{m} = 26525\text{ km}$$

Also, $a_p = a(1-e)$

Hence, a = Semi-major axis

$$= \frac{a_p}{(1-e)} = \frac{26525}{0.998} \quad (\because e = 0.002)$$

$$a = 26578\text{ km}$$

Nearest possible answer is option (c).

2.54 (b)

Angle of elevation = 35°

Stratiform rain height = 3 km

$$\text{Physical path length} = \frac{R_{SRH}}{\sin\theta_e} = \frac{3\text{ km}}{\sin 35°}$$

$$= 5.19\text{ km} \simeq 5.2\text{ km}$$

2.56 (b)

$$E_b = P\times T_b = \frac{P}{R_b} = \frac{1000}{50\times 10^6} = 20\,\mu\text{J}$$

2.58 (b)

The free-space path loss is given by

$$L_s = \left(\frac{4\pi d}{\lambda}\right)^2$$

In decibels,

$$L_{db} = 10\log\left(\frac{4\pi d}{\lambda}\right)^2 = 20\log\left(\frac{4\pi d}{\lambda}\right)$$

A GEO satellite is located at 35,786 kms above Earth equator. Hence,

$$d = 35{,}786\times 10^3\text{ m}$$

$$L_{db} = 20\log\left(\frac{4\pi\times 35786\times 10^3}{\lambda}\right)$$

$$L_{db} = 173.07 - 20\log(\lambda)$$

2.59 (a)

Here,

θ = Minimum elevation angle is the angle of the horizontal of the earth surface to the center line of the satellite transmission beam.

β = Coverage angle is a measure of the portion of the earth surface visible to a satellite taking the minimum elevation angle into account.

h = Satellite orbit height

R = Radius of Earth

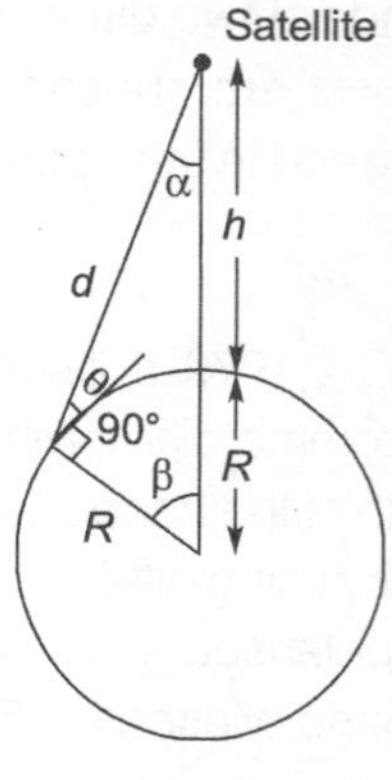

From the above figure, applying Sine's law, we get

$$\frac{\sin(\alpha)}{\sin\left(\theta+\frac{\pi}{2}\right)} = \frac{R}{R+h}$$

$$\Rightarrow \quad \frac{R}{R+h} = \frac{\sin\left(\frac{\pi}{2}-\beta-\theta\right)}{\sin\left(\theta+\frac{\pi}{2}\right)} = \frac{\cos(\beta+\theta)}{\cos(\theta)}$$

2.60 (d)

- A satellite in synchronous (geostationary) orbit is always positioned over the same spot on earth.
- A satellite in asynchronous orbit flies at a lower altitude of approximately 644 km passing over the planet during different times.
- Most of the orbital satellites are asynchronous (non synchronous).

2.61 (c)

In space communication, we use microwave frequencies. The requirement for reliability, long life, stability, high efficiency and suitability for space environment are met by the use of Travelling Wave Tube Amplifier (TWTA).

2.62 (d)

Maximum Usable Frequency (MUF) is the highest frequency that can be used for transmission between two points via reflection from the ionosphere. Beyond maximum usable frequency (MUF), frequency doesn't get reflected by ionosphere.

2.63 (a)

Mean Sun is a fictitious Sun used for time keeping that moves at a uniform speed but otherwise requires the same time as the real Sun to complete one orbit of the earth, this time being the tropical year.

2.64 (b)

Satellite transponder refers to a transmitter-receiver subsystem on board the satellite that processes, amplifies and retransmits the signal received (uplink signal) to earth station as downlink signal.

2.65 (a)

Passive satellites require radio beacon transmitters for tracking and ranging purposes. A beacon is a continuously transmitted unmodulated carrier that an earth station can lock on to and use to determine the exact location of a satellite so the earth station can align its antennas.

2.66 (a)

The equivalent noise temperature,

$$T_n = T_E + T_A$$

Hence, the noise power is

$$N = k(T_E + T_A)\,B$$

If S_A is the input signal generated by the source, the output signal to noise ratio is given as

$$\frac{S}{N} = \frac{S_A}{k(T_A + T_E)B}$$

2.67 (d)

For mobile satellite service we use L band (1-2 GHz).

For fixed satellite service we use C band (4-8 GHz)

For broadcast satellite service we use Ku and K band

Ku – 12.5 – 18 GHz

K – 18 – 26.5 GHz

2.68 (d)

Given that,

$$L_{cable} = 3 \text{ dB}$$

$$L_{connector} = 2 \times 2 \text{ dB} = 4 \text{ dB}$$

(due to 2 connectors)

If $P_{antenna}$ is the signal level at the input of the antenna, we have

$$P_{antenna} = P_{in} - L_{cable} - L_{connector}$$

$$P_{antenna} = 15 \text{ dBm} - 3 \text{ dB} - 4 \text{ dB}$$

$$P_{antenna} = 8 \text{ dBm}$$

2.69 (a)

When a subscriber device has no LAPD frames to transmit, it transmits a series of binary ones on the D-channel. Using the pseudoternary encoding scheme, this corresponds to the absence of line signal.

■■■■

3 Fibre Optic Communication Systems

3.1 **Assertion (A):** Optical fibres have broader bandwidth compared to conventional copper cables.
Reason (R): The information carrying capacity of optical fibres is limited by Rayleigh scattering loss.
(a) Both A and R are true and R is the correct explanation of A
(b) Both A and R are true but R is NOT the correct explanation of A
(c) A is true but R is false
(d) A is false but R is true [ESE-2000]

3.2 Consider the following statements:
In optical communications, the losses in optical fibres can be caused by
1. impurities
2. microbending
3. attenuation in glass
4. stepped index operations

Which of these statements are correct?
(a) 1, 2 and 3 (b) 1, 3 and 4
(c) 1, 2 and 4 (d) 2, 3 and 4
[ESE-2001]

3.3 **Assertion (A):** Optical fibres are highly desirable for communication links for lasers.
Reason (R): Active nature of optical fibres provides high spectral purity of the signal.
(a) Both A and R are true and R is the correct explanation of A
(b) Both A and R are true but R is NOT the correct explanation of A
(c) A is true but R is false
(d) A is false but R is true [ESE-2001]

3.4 A certain optical fibre has refractive index of clad (n_1) = 1.40 and that of core (n_2) = 1.05. Its numerical aperture will be
(a) 0.8575 (b) 0.9260
(c) 0.3500 (d) 0.1585
[ESE-2002]

3.5 Velocity of light travelling in an optical fibre is
(a) Equal to c
(b) Greater than c by a few per cent
(c) Less than c by a few per cent
(d) Much greater than c, approaching the magnitude of c^2
[ESE-2003]

3.6 What is the responsivity of a photodiode having quantum efficiency of 65% with photons of energy 1.5×10^{-19} J incident upon it?
(a) 0.832 AW^{-1} (b) 0.714 AW^{-1}
(c) 0.694 AW^{-1} (d) 0.452 AW^{-1}
[ESE-2004]

3.7 The normalized frequency of a step index fibre is 28 at 1300 nm wavelength. What is the total number (approximately) of guided modes that can be supported by the fibre?
(a) 50 (b) 200
(c) 400 (d) 800
[ESE-2004]

3.8 Which one of the following is not a correct reason for biasing the LASER near but below threshold in the off state for a laser drive circuit?
(a) It reduces the switch-on delay
(b) It allows easy compensation for changes in ambient
(c) It eliminates the spontaneous emission of light in the off state
(d) It reduces the junction heating caused by the digital drive current
[ESE-2004]

3.9 Which one of the following statements is correct? The modal noise in multimode fibres cannot be reduced by
(a) use of a broad spectrum source
(b) using fibres having large numerical apertures
(c) phase correlation between the modes
(d) use of a single mode fibre
[ESE-2004]

3.10 Match **List -I** (Different Fibre Generation) with **List-II** (Optical Components Used) and select the correct answer using the code given below the lists:

List-I

A. First
B. Second
C. Third
D. Fourth

List-II

1. Graded index multimode fibre and 1310 nm LD
2. Graded index multimode fibre and 850 nm LD
3. Step index monomode fibre and 1550 nm LD
3. Step index monomode fibre and 1310 nm LD

Codes:

	A	B	C	D
(a)	1	2	3	4
(b)	2	1	3	4
(c)	1	2	4	3
(d)	2	1	4	3

[ESE-2005]

3.11 **Assertion (A):** In a step index fibre, for single mode operation, the fibre parameters should be tailored such that the *V*-parameter is less than 2.4048.

Reason (R): Modal fields of step index fibre are represented by Bessel functions $J_n(x)$ in the core and modified Bessel functions $K_n(x)$ in the cladding. The first zero for $J_1(x)$ falls at $x = 2.4048$.

(a) Both A and R are true and R is the correct explanation of A
(b) Both A and R are true but R is NOT the correct explanation of A
(c) A is true but R is false
(d) A is false but R is true

[ESE-2005]

3.12 **Assertion (A):** For optical communication, three communication wavelength windows are decided and they are located near 850 nm, near 1300 nm and near 1550 nm.

Reason (R): At 850 nm the first semiconductor laser for optical communi-cation was fabricated, at 1300 nm the fibre dispersion is minimum and at 1550 nm wavelength multiplexing is possible due to optical amplifiers.

(a) Both A and R are true and R is the correct explanation of A
(b) Both A and R are true but R is NOT the correct explanation of A
(c) A is true but R is false
(d) A is false but R is true

[ESE-2005]

3.13 Which of the following are the advantages of a fibre optic link over a conventional copper wire link?

1. A fibre optic link has greater bandwidth.
2. A fibre optic link has lower cost.
3. A fibre optic link is immune to cross link.
4. A fibre optic link is easy to split.

Select the correct answer using the code given below:

(a) 1 and 2 (b) 1 and 3
(c) 2 and 3 (d) 1 and 4 [ESE-2005]

3.14 Light travels along the optical fibres by which mechanism?

(a) Refraction
(b) Reflection
(c) Scattering
(d) Total internal reflection

[ESE-2006]

3.15 Which one of the following is the correct statement?

The graded index optical fibre cable can be considered to have

(a) smaller numerical aperture than step index cable
(b) several layers of clad and each layer having a smaller index number relative to its distance from the core
(c) several layers of clad and each layer having a higher index number relative to its distance from the core
(d) large beam spread

[ESE-2007]

3.16 In an optical fiber, the light beam propagates due to which one of the following?

(a) Simple reflection of light at a boundary between two media
(b) Refraction of light in the medium
(c) Total internal reflection at the boundary of the fibre
(d) Scattering of light in the medium

[ESE-2008]

3.17 Dispersion in an optical fibre used in a communication link is of which type?
(a) Angular dispersion
(b) Modal dispersion
(c) Chromatic dispersion
(d) Dispersion arising due to structural irregularities in the fibre
[ESE-2008]

3.18 In optical communication, wavelength 1550 nm is used nowadays. What is/are the possible reason(s)?
1. Dispersion is very low
2. Loss is very low
3. WDM and optical amplification are feasible
Select the correct answer from the codes given below:
(a) 1, 2 and 3 (b) 2 only
(c) 1 only (d) 1 and 3 only
[ESE-2008]

3.19 **Assertion (A):** Pulse broadening is a phenomenon associated with multimode fibres.
Reason (R): Multimode fibres have core diameter much larger compared with single mode fibres.
(a) Both A and R are individually true and R is the correct explanation of A
(b) Both A and R are individually true but R is not the correct explanation of A
(c) A is true but R is false
(d) A is false but R is true [ESE-2008]

3.20 Why are semiconductor lasers important for communication engineering?
1. They possess large bandwidth
2. They operate at low voltage
3. They are suitable for coupling to fibers at wavelengths, where the attenuation is very small
4. Their integration with other components is easy
Select the correct answer using the code given below:
(a) 1 only (b) 1 and 2 only
(c) 2 and 3 only (d) 1, 2, 3 and 4
[ESE-2008]

3.21 A Ruby Maser is considered to be superior to an Ammonia Maser because of
(a) higher power output
(b) low signal to noise ratio
(c) greater bandwidth
(d) low amplification factor and low noise levels
[ESE-2008]

3.22 Losses in optical fibres can be caused by which of the following?
1. Impurities
2. Microbending
3. Stepped index operation
Select the correct answer using the code given below:
(a) 1 and 3 (b) 2 and 3
(c) 1 and 2 (d) 3 only [ESE-2008]

3.23 A single mode fibre does not suffer from which type of dispersion?
(a) Waveguide dispersion
(b) Material dispersion
(c) Intermodal dispersion
(d) Polarization mode dispersion [ESE-2009]

3.24 Which of the following introduces mode partition noise?
(a) Coaxial line
(b) Wave guide
(c) Fibre transmission line
(d) Both coaxial line and wave guide
[ESE-2009]

3.25 Which of the following does not cause losses in optical fibre cables ?
(a) Impurities
(b) Microbending
(c) Attenuation in glass
(d) Stepped index operation [ESE-2009]

3.26 On which bands, do the optical fibres operate ?
1. Ultra violet band
2. Ultra high frequency band
3. Visible light band
4. Infra red band
Select the correct answer from the codes given below:
(a) 1 only (b) 1 and 2 only
(c) 1, 2 and 3 (d) 1, 3 and 4
[ESE-2009]

3.27 Which one of the following photo-detector does not provide gain?
(a) Photo-transistor
(b) Photo conductor
(c) Avalanche photodiode
(d) *p-i-n* photodiode [ESE-2009]

3.28 Population inversion in semiconductor laser diode is achieved by
(a) Lightly doping *p* and *n* sides
(b) Introducing trap centres on *p* and *n* sides
(c) Heavily doping *p* and *n* sides
(d) Reverse biasing the junction [ESE-2010]

3.29 Consider the following advantages of optical fiber-cables:

1. Small diameter
2. Immunity to cross talk and electromagnetic interference
3. Laser and LED modulation methods lend themselves ideally to digital operation

Which of these advantages are correct?

(a) 1 and 2 only (b) 2 and 3 only
(c) 3 and 1 only (d) 1, 2 and 3

[ESE-2011]

3.30 Polarization mode dispersion (PMD) is mainly observed in

(a) Multiple step-index fiber
(b) Single mode fiber
(c) Multimode graded-index fiber
(d) Plastic fiber [ESE-2011]

3.31 An optical fibre has refractive index core

(a) high and low refractive index cladding
(b) low and high refractive index cladding
(c) uniform surrounded by variable index cladding
(d) variable with refractive index increases from low at the centre to high at the junction with cladding

[ESE-2013]

3.32 **Statement (I):** Single mode step indexed fibers are widely used for wide band communications.
Statement (II): With only path of light along the axis, minimum refraction and no pulse spreading permits high pulse repetition rates.

(a) Both Statement (I) and Statement (II) are individually true and Statement (II) is the correct explanation of Statement (I).
(b) Both Statement (I) and Statement (II) are individually true but Statement (II) is not the correct explanation of Statement (I).
(c) Statement (I) is true but Statement (II) is false.
(d) Statement (I) is false but Statement (II) is true.

[ESE-2013]

3.33 Optical fibers are preferred as communication links for Lasers because they

(a) prevent interference by other Lasers
(b) ensure that the beam does not spread
(c) prevent atmospheric interference
(d) ensure amplification of the signal

[ESE-2013]

3.34 The cladding which surrounds the fibre core

(a) is used to protect the fibre
(b) is used to reduce optical interference
(c) helps to guide the light in the core
(d) ensures that refractive index remains unaltered

[ESE-2014]

3.35 If n_1 and n_2 are refractive indices of the core and cladding respectively, the maximum acceptance angle at the air-core interface should be

(a) $\tan^{-1}\frac{n_2}{n_1}$ (b) $\sin^{-1}\sqrt{n_2^2 - n_1^2}$
(c) $\sin^{-1}\sqrt{n_1^2 - n_2^2}$ (d) $\tan^{-1}\frac{n_1}{n_2}$

[ESE-2015]

3.36 A fibre-optic cable has the indices of refraction of core of 1.6 and of cladding of 1.4. For an angle of incidence 70°, the angle of return light ray will be

(a) 35° (b) 61°
(c) 70° (d) 90° [ESE-2015]

3.37 In an optical communication system, having an operating wavelength λ in metres, only X% of its source frequency can be used as its channel bandwidth. The system is to be used for transmitting TV signal requiring a bandwidth of f Hz. The number of channels transmitted by this system simultaneously is (c = speed of light)

(a) $\frac{100Xc}{\lambda f}$ (b) $\frac{100\lambda f}{Xc}$
(c) $\frac{Xc}{\lambda f}$ (d) $\frac{Xc}{100\lambda f}$ [ESE-2015]

3.38 An optical fibre communication system works on a wavelength of 1.3 μm. The number of subscribers it can feed, if a channel requires 20 kHz, is

(a) 2.3×10^{10} (b) 1×10^5
(c) 1.15×10^{10} (d) 10^{10} [ESE-2015]

3.39 In an optical fibre, the pulse dispersion effect is minimized by

1. Using a high frequency light source
2. Using plastic cladding
3. Minimizing the core diameter

Which of the above statements is/are correct?

(a) 1 only (b) 2 only
(c) 3 only (d) 1, 2 and 3

[ESE-2016]

3.40 **Statement (I) :** Multimode fibres are now used for long distance communication.

Statement (II): Multimode fibres have larger core-radius than single-mode fibres.

(a) Both Statement (I) and Statement (II) are individually true and Statement (II) is the correct explanation of Statement (I).

(b) Both Statement (I) and Statement (II) are individually true but Statement (II) is not the correct explanation of Statement (I).

(c) Statement (I) is true but Statement (II) is false.

(d) Statement (I) is false but Statement (II) is true.

[ESE-2016]

3.41 A ray of light is incident on a glass slab (of refractive index 1.5) with an angle $\frac{\pi}{4}$, then the value of sine of angle of refraction is

(a) $\frac{1}{\sqrt{2}}$ (b) $\frac{3}{\sqrt{2}}$

(c) $\frac{\sqrt{2}}{3}$ (d) $\sqrt{2}$

[ESE-2016]

3.42 LASER beam of light essentially finds its application in transmission of a signal in the optical fibre communication systems due to

(a) Incredible speed of signal communication

(b) Low loss transmission of the signal

(c) Inexpensive installation cost

(d) Bulk availability of LASER sources

[ESE-2016]

3.43 In graded index multimode optical fiber, the refractive index of the core is

(a) uniform across its radial distance, except for the cladding

(b) maximum at the fiber axis and decreases stepwise towards the cladding

(c) maximum at the fiber axis and decreases gradually towards the cladding

(d) maximum at the fiber axis and increases stepwise towards the cladding

[ESE-2017]

3.44 The cut-off wavelength for an optical fibre with 5 μm core diameter and core cladding indices of refraction 1.484 and 1.402 is nearly

(a) 0.5 μm (b) 1 μm

(c) 3 μm (d) 6 μm [ESE-2018]

3.45 What is/are the advantage(s) of step-index mono-mode fibre optical cable?

1. Manufacturing process is simple.
2. Bandwidth of several GHz-km is possible.
3. Splicing is easier.

Select the correct answer using the code given below.

(a) 1 only (b) 2 only

(c) 3 only (d) 1, 2 and 3

[ESE-2018]

3.46 In optical communication, the maximum angle in which external light rays may strike the air/glass interface and still propagate down the fiber is called as

(a) critical angle (b) numerical aperture

(c) angle of refraction (d) acceptance angle

[ESE-2019]

3.47 Which one of the following specifications does **not** fit for a single-mode fiber?

(a) The bandwidth of 1 GHz/km.

(b) The digital communication rate is excess of 2000 Mbytes/s.

(c) More than 100000 voice channels are available

(d) The mode field diameter (MFD; spot size) is larger than the core diameter.

[ESE-2019]

3.48 A single-mode optical fiber has a beat length of 8 cm at 1300 nm. The value of birefringence B_f will be nearly

(a) 1.6×10^{-5} (b) 2.7×10^{-5}

(c) 3.2×10^{-5} (d) 4.9×10^{-5}

[ESE-2019]

3.49 Which one of the following instruments is useful while measuring the optical power as a function of wavelength?

(a) Optical power attenuator

(b) Optical power meter

(c) Optical spectrum analyzer

(d) Optical return loss tester

[ESE-2019]

3.50 The optical performance monitoring involves
(a) transport layer monitoring, optical signal monitoring and protocol performance monitoring
(b) physical layer, network layer and application layer monitoring
(c) data-link layer, presentation layer and session layer monitoring
(d) transport layer, session layer and application layer monitoring

[ESE-2019]

3.51 Which one of the following types of fiber suffers with modal dispersion?
(a) Single-mode step-index fiber
(b) Multimode graded-index fiber
(c) Multimode step-index fiber
(d) Single-mode graded-index fiber

[ESE-2020]

3.52 For a 300 m optical fibre cable with BLP of 600 MHz-km, the bandwidth will be
(a) 8 GHz (b) 6 GHz
(c) 4 GHz (d) 2 GHz

[ESE-2020]

3.53 Numerical aperture (*NA*) in optical fibre transmission is used to describe
(a) Light spreading ability
(b) Light gathering or light collecting ability
(c) Light output from external shield
(d) Light leakage ability [ESE-2020]

3.54 The fundamental parameter of a single-mode fiber is
(a) the core diameter and cladding diameter.
(b) the mode-field diameter.
(c) the cladding diameter.
(d) the buffer coating diameter.

[ESE-2021]

3.55 Consider the following statements regarding the advantages of optical fiber communication:
1. Enormous potential bandwidth.
2. Electrical isolation.
3. Immunity to interference and crosstalk.
4. System reliability and ease of maintenance.

Which of the above statements are correct?
(a) 1 and 4 only (b) 2, 3 and 4 only
(c) 1, 2 and 3 only (d) 1, 2, 3 and 4

[ESE-2021]

3.56 A multimode step index fiber with a core diameter of 80 μm and a relative index difference of 1.5% is operating at a wavelength of 0.85 μm. If the core refractive index is 1.48, then the normalized frequency for the fiber is
(a) 37.9 (b) 75.8
(c) 151.6 (d) 303.2 [ESE-2021]

3.57 The total efficiency of an injection laser with a GaAs active region is 18%. The voltage applied to the device is 2.5 V and the bandgap energy for GaAs is 1.43 eV. The external power efficiency of the device is
(a) 5% (b) 10%
(c) 15% (d) 20% [ESE-2021]

3.58 Which one of the following statements is not correct regarding SONET networks?
(a) A point-to-point network is normally made of an STS multiplexer.
(b) A linear synchronous optical network can be point-to-point or multipoint.
(c) The signal flow can be unidirectional or bidirectional.
(d) A multipoint network uses STS multiplexers to allow the communications between several terminals.

[ESE-2022]

3.59 A digital optical fiber communication system operating at a wavelength of 1 μm requires a maximum bit error rate of 10^{-9}. What is the theoretical quantum limit at the receiver in terms of the quantum efficiency of the detector and the energy of an incident photon?
(a) $\frac{20.7hf}{\eta}$ (b) $\frac{10.7hf}{\eta}$
(c) $\frac{20.7\eta}{hf}$ (d) $\frac{10.7\eta}{hf}$ [ESE-2022]

3.60 Some of the diffracted light continues down the fiber and some of it escapes through the cladding. The light rays that escape represent a loss in light power which is called
(a) Rayleigh scattering loss
(b) chromic distortion loss
(c) chromatic distortion loss
(d) predominant fiber loss

[ESE-2022]

3.61 Numerical aperture is closely related to acceptance angle and is the figure of merit commonly used to measure
(a) the sine of the maximum angle
(b) the magnitude of the acceptance angle
(c) the maximum angle of light ray
(d) the intensity of light ray entering the cable

[ESE-2022]

3.62 Which one of the following fibers has the distinct advantage of low intermodal dispersion?
(a) Multimode step-index fiber
(b) Single-mode step-index fiber
(c) Graded-index multimode fiber
(d) Spatially incoherent multimode fiber

[ESE-2022]

3.63 Which one of the following is **not** applicable to Channel guides?
(a) It cannot provide confinement in the x dimension.
(b) It is confinement of the light within the film plane, i.e. the y-z plane.
(c) It is used in many active and passive devices of integrated optics, including lasers, modulators, switches and directional couplers.
(d) It is used in the design of single-mode structures that are compatible with single-mode fiber guides.

[ESE-2024]

■■■

Answers Fibre Optic Communication Systems

3.1 (b)	3.2 (a)	3.3 (a)	3.4 (b)	3.5 (c)	3.6 (c)	3.7 (c)	3.8 (c)	3.9 (c)
3.10 (d)	3.11 (a)	3.12 (a)	3.13 (b)	3.14 (d)	3.15 (b)	3.16 (c)	3.17 (b)	3.18 (a)
3.19 (b)	3.20 (d)	3.21 (c)	3.22 (c)	3.23 (c)	3.24 (c)	3.25 (d)	3.26 (d)	3.27 (d)
3.28 (c)	3.29 (d)	3.30 (b)	3.31 (a)	3.32 (a)	3.33 (b)	3.34 (c)	3.35 (c)	3.36 (c)
3.37 (d)	3.38 (c)	3.39 (c)	3.40 (d)	3.41 (c)	3.42 (a)	3.43 (c)	3.44 (c)	3.45 (b)
3.46 (d)	3.47 (a)	3.48 (a)	3.49 (c)	3.50 (a)	3.51 (c)	3.52 (d)	3.53 (b)	3.54 (b)
3.55 (d)	3.56 (b)	3.57 (b)	3.58 (d)	3.59 (a)	3.60 (a)	3.61 (b)	3.62 (b)	3.63 (a)

Explanations Fibre Optic Communication Systems

3.2 (a)

Losses in OFC is caused due to

1. Impurities
2. Microbending
3. Attenuation in glass

3.4 (b)

$$n_1 = 1.40$$
$$n_2 = 1.05$$

Numerical aperture,

$$NA = (n_1^2 - n_2^2)^{1/2}$$
$$\Rightarrow \quad NA = [(1.40)^2 - (1.05)^2]^{1/2}$$
$$\Rightarrow \quad NA = 0.9260$$

3.6 (c)

Responsivity,

$$R = \frac{\text{Output photocurrent}}{\text{Incident optical power}} = \frac{I_P}{P_o}$$
$$= \frac{\eta q}{P_o}$$
$$\Rightarrow \quad R = \frac{0.65 \times 1.6 \times 10^{-19}}{1.5 \times 10^{-19}}$$
$$\Rightarrow \quad R = 0.694 \text{ AW}^{-1}$$

3.7 (c)

For a step index fiber,

$$M_S \cong \frac{V^2}{2}$$

where, $M_S \rightarrow$ Total number of guided modes

$V \rightarrow$ Normalized frequency of fiber

$$M_S \cong \frac{(28)^2}{2} = 392 \simeq 400$$

3.9 (c)

Phase correlation between modes is a condition to rise modal noise.

3.14 (d)

The principle of Optical Fiber communication is 'total internal reflection'.

3.23 (c)

Intermodal dispersion arises in a multimode fibre.

3.24 (c)

Mode partition noise exist in fiber line. Mode partition noise is intensity fluctuation in longitudinal modes of LASER.

3.29 (d)

All the given statements are advantages of optical fibre-cables.

3.30 (b)

PMD is a form of dispersion that affects high speed signals in longer range systems. It is a major type of dispersion that becomes significant in a system that has already minimized other forms of dispersion and is operating at Gigabit rate. PMD arises from the fact that even a single mode can have two polarization states. A typical PMD values is 0.5 psec/km. It is significant in case of single mode fiber.

3.31 (a)

In an optical fibre refractive index of core is high in comparison to refractive index of cladding so that total internal reflection can take place.

3.32 (a)

Single-mode step index fibers are the most widely used in today's wideband communication area. With this fiber a light ray can travel on only one path; therefore modal dispersion is zero. The extra cladding thickness tends to set an overall fibre size standard and makes the fiber less fragile. Some specifications are:

(i) The bandwidth is from 50 to 100 GHz/km.

(ii) The digital communication rate is in excess of 2000 Mbytes/sec.

3.35 (c)

Reflective index of core = n_1

refractive index of cladding = n_2

Numerical aperture $NA = \sqrt{n_1^2 - n_2^2}$

Acceptance angle $\phi_a = \sin^{-1}$ (NA)

$$= \sin^{-1}\left(\sqrt{n_1^2 - n_2^2}\right)$$

3.36 (c)

$$\theta_c = \sin^{-1}\left(\frac{1.4}{1.6}\right) = 61°$$

Since incident angle is greater than 61°, ray will be totally internal reflected.

3.37 (d)

$$\text{Frequency of source} = \frac{c}{\lambda}$$

$$\text{Total BW of the system} = \left(\frac{X}{100}\right)\left(\frac{c}{\lambda}\right)$$

BW of each TV channel = f Hz

Total number of TV channels that can be transmitted simultaneously is,

$$N = \frac{\left(\frac{X}{100}\right)\left(\frac{c}{\lambda}\right)}{f} = \frac{Xc}{100\lambda f}$$

3.38 (c)

$$f_m = \frac{c}{\lambda} = \frac{3\times 10^8}{1.3\times 10^{-6}} = 2.30 \times 10^{14} \text{ Hz}$$

Number of subscribers

$$= \frac{2.30\times 10^{14}}{20\times 10^3} = 1.15\times 10^{10}$$

3.41 (c)

$$n_1 \sin\theta_1 = n_2 \sin\theta_2$$
$$\sin 45° = 1.5 \sin\theta_2$$
$$\frac{1}{\sqrt{2}} = 1.5 \times \sin\theta_2$$
$$\sin\theta_2 = \frac{\sqrt{2}}{3}$$

3.44 (c)

$$V = \frac{2\pi a\, N.A}{\lambda} = \frac{2\pi a\sqrt{\eta_1^2 - \eta_2^2}}{\lambda}$$

$$0 \le V \le 2.405$$

To find the minimum wavelength that the fibre allows (cut-off wavelength),

Put, $V = 2.405$

$$\lambda = \frac{2\pi\times 2.5\times 10^{-6}\sqrt{1.484^2 - 1.402^2}}{2.405}$$

$$\lambda = 3.17\ \mu\text{m}$$

$$\lambda \simeq 3\ \mu\text{m}$$

3.47 (a)

For a single-mode fibre, the bandwidth ranges from 50 to 100 GHz/km.

3.48 (a)

Beat length,

$$L = \frac{2\pi}{\beta_x - \beta_y} \quad \beta \to \text{Propagation constant}$$

$$L = \frac{2\pi}{\delta B_f} = \frac{\lambda_c}{B_f}$$

$$B_f = \frac{2\pi}{L} = \frac{\lambda}{L} = \frac{1300}{8}\times 10^{-9} = 1.6\times 10^{-5}$$

3.49 (c)

Optical spectrum analyzer measures optical power as a function of wavelength.

3.51 (c)

Modal dispersion is most prominent multimode step index fibres.

One of the ways to reduce modal dispersion is graded index fiber. With graded index fibre, the light follows a more curved path. The higher order modes spend most of the time travelling in lower index cladding layer near the outside of the fibre. Modal dispersion is completely eliminated by using single mode fibre.

3.52 (d)

$$BW \times L = 600 \text{ MHz-km} \quad \Rightarrow L = 0.3 \text{ km}$$

$$BW = \frac{600 \text{ MHz-km}}{0.3 \text{ km}} = 2 \text{ GHz}$$

3.54 (b)

For single-mode fiber, only the fundamental mode can exist. Many properties of the fundamental mode are determined by the radial extent of its electromagnetic field including losses at launching and jointing, microbend losses, waveguide dispersion and the width of the radiation pattern. Therefore, the mode-field diameter (MFD) is an important parameter for characterizing single-mode fiber properties which takes into account the wavelength-dependent field penetration into the fiber cladding.

Mode field diameter is a measure of light intensity in a single-mode fiber cross-section.

3.55 (d)

The following are the advantages of optical fiber communication:

(i) **Large Potential Bandwidth:** As the optical carrier frequency is in the range of 10^{12} to 10^{15} Hz, it yields a far greater potential transmission bandwidth than metallic cable systems.

(ii) **Electrical Isolation:** Optical fibers which are fabricated from glass or sometimes a plastic polymer are electrical insulator, they do not exhibit earth loop and interface problems.

(iii) **Immunity to interference and cross-talk:** Optical fibers form a dielectric waveguide and are, therefore, free from electromagnetic interference (EMI), radio frequency (RF), etc. Hence, the operation of an optical fiber communication system is unaffected by transmission through an electrically noisy environment and fiber cable requires no shielding from EMI.

(iv) **System reliability and ease of maintenance:** The low loss property of optical fiber cable reduces the requirement of intermediate repeaters or line amplifiers to boost the transmitted signal strength. Hence, with fewer repeaters, the system reliability is greatly enhanced. High reliability reduces it's expenditure on maintenance.

3.56 (b)

The normalized frequency for the fiber is given as

$$V = \frac{2\pi a}{\lambda} \cdot NA = \frac{2\pi_a}{\lambda} \cdot n\sqrt{2\Delta}$$

where a = core-radius = 40 μm

n = core refractive index = 1.48

D = relative index difference = 0.015

λ = Wavelength = 0.85 μm

Substituting the values, we get

$$V = \frac{2\pi \times 40}{0.85} \times 1.48 \times \sqrt{2 \times 0.015}$$

$$V = 75.8$$

3.57 (b)

The external power efficiency is defined as

$$\eta = \eta_T \left(\frac{E_g}{V}\right) \times 100\%$$

$$\eta = 0.18 \times \left(\frac{1.43}{2.5}\right) \times 100\%$$

$$\eta = 10.296\% \approx 10\%$$

3.58 (d)

- In SONET network, a point-to-point network is normally made of an STS multiplexer, STS demultiplexer and zero or more repeaters as shown below :

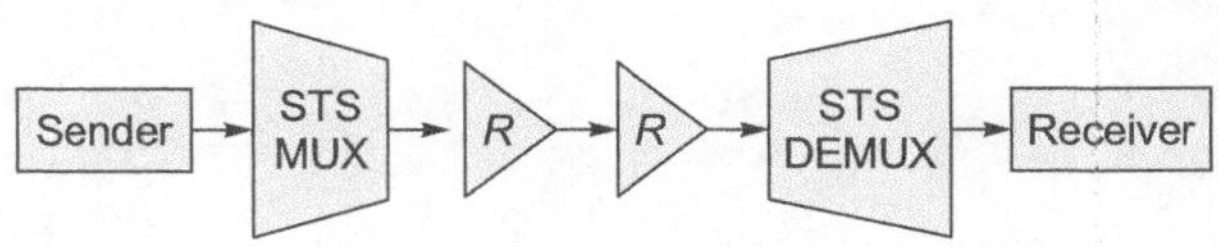

- The signal flow can be unidirectional or bidirectional.
- A linear SONET network can be point-to-point or multipoint. A multipoint network uses Add-Drop Multiplexers (ADMs) to allow the communications between several terminals.

3.59 (a)

The minimum pulse energy, E_{min} required to maintain a given bit-error rate (BER) is known as the quantum limit.

The bit-error is given as

$$P(e) = \exp(-Z_m) \quad \text{...(i)}$$

where Z_m corresponds to average number of photons detected in time period τ for a BER of 10^{-9}.

$$Z_m = \frac{\eta \cdot P_o \tau}{hf} \quad \text{...(ii)}$$

Using equation (i),

$$10^{-9} = \exp(-Z_m)$$

$$\Rightarrow \quad Z_m = \frac{\eta \cdot P_o \tau}{hf} = 20.7$$

Hence, the minimum pulse energy or quantum limit,

$$E_{min} = P_o \tau = \frac{20.7hf}{\eta}$$

3.60 (a)

- When light rays propagating down a fiber strike the impurities, they are diffracted. Diffraction causes the light to disperse or spread out in many directions. Some of the diffracted light continues down the fiber and some of it escapes through the cladding. The light rays that escape represent a loss in light power. This is called Rayleigh scattering loss.
- Chromatic distortion or wavelength dispersion is due to the wavelengths within the composite light signal travelling at a different velocity when propagating through fiber.

3.61 (b)

Numerical Aperture (NA) is closely related to acceptance angle and is the figure of merit commonly used to measure the magnitude of the acceptance angle given as

$$\theta_a = \sin^{-1}(\text{NA})$$

3.62 (b)

Intermodal dispersion is due to the optical path differences between different modes. Since, a single mode step-index fiber has only a single mode, it has the distinct advantage of low intermodal dispersion compared to multimode fiber.

3.63 (a)

In planar optical guides, the confinement takes place in one dimension only. Optical channel guides can provide the additional confinement in one more dimension i.e. in a plane. Hence, option (a) is not applicable to Channel guides.

■■■

UNIT

VIII

Advanced Electronics Topics

Syllabus

Basics of Integrated Circuits (ICs); Bipolar, MOS and CMOS ICs; VLSI technology: Processing, lithography, interconnects, packaging, testing; VLSI design: Principles, Design for testability, examples.

Contents

Sl.	Topic	Page No.
1.	VLSI Technology	502
2.	Digital Design	510

○○○○

1 VLSI Technology

1.1 Almost all resistors are made in a monolithic integrated circuit
(a) during the emitter diffusion
(b) while growing the epitaxial layer
(c) during the base diffusion
(d) during the collector diffusion [ESE-2000]

1.2 In the fabrication of a buried layer *n-p-n* transistor, the processes involved are
1. diffusion 2. oxidation
3. epitaxy 4. lithography
The correct sequence in which these processes are to be carried out, is
(a) 2, 4, 3, 1 (b) 4, 2, 1, 3
(c) 2, 4, 1, 3 (d) 4, 2, 3, 1
[ESE-2001]

1.3 Moor's law relates to
(a) speed of operation of bipolar devices
(b) speed of operation of MOS devices
(c) power rating of MOS devices
(d) level of integration of MOS devices
[ESE-2002]

1.4 In fabricating silicon BJT in ICs by the epitaxial process, the number of diffusions used is usually
(a) 2 (b) 3
(c) 4 (d) 6
[ESE-2003]

1.5 In the fabrication of *n-p-n* transistor in an IC, the buried layer on the *p*-type substrate is
(a) p^+-doped
(b) n^+-doped
(c) Used to reduce the parasitic capacitance
(d) Located in the emitter region
[ESE-2003]

1.6 The scaling factor of a MOS device is α. Using constant voltage scaling model, the gate area of the device will be scaled as
(a) $1/\alpha$ (b) $1/\alpha^2$
(c) $1/\alpha^3$ (d) $1/\alpha^4$ [ESE-2003]

1.7 Which one of the following statements is correct? In the context of IC fabrication, metallisation means
(a) connecting metallic wires
(b) formation of interconnecting conduction pattern and bonding pads
(c) doping SiO_2 layer
(d) covering with a metallic cap
[ESE-2004]

1.8 Consider the following:
1. Cheaper method
2. Low residual oxygen content in the grown crystal is possible
3. Uniform doping is possible
4. Larger diameter crystals can be grown
Which of the above are the advantages of the float zone (FZ) method over the Czochralski (CZ) method in single crystal growth of silicon?
(a) 1 and 2 (b) 2 and 3
(c) 3 and 4 (d) 1 and 4
[ESE-2004]

1.9 Diffusion of impurities in a semiconductor is carried out in a furnace through which a steady stream of impurity atoms is passed during the entire diffusion process. What would be the type of the profile of the impurity atoms inside the semiconductor?
(a) Linear
(b) Gaussian
(c) Complementary error function
(d) Exponential
[ESE-2005]

1.10 Match **List-I** (Isolation Technique in IC) with **List-II** (Related Characteristic) and select the correct answer using the codes given below the lists:
List-I
A. Reverse bias *p-n* junction isolation
B. Resistive isolation using the bulk resistivity of the layer

C. Native oxide isolation
D. Oxide (other than native isolation)

List-II

1. Requires large area of the wafer, thereby increasing the IC size
2. Best choice for silicon ICs with low parasitic capacitance
3. Suitable for ICs of III-V semiconductors
4. Introduces bias-dependent parasitic capacitance

Codes:

	A	B	C	D
(a)	4	1	2	3
(b)	2	3	4	1
(c)	4	3	2	1
(d)	2	1	4	3

[ESE-2005]

1.11 Consider the following statements:
X-rays are used for lithography in IC technology because
1. high resolution is achievable.
2. scattering effects are small.
3. they can be focussed easily
4. they can be deflected easily.

Which of the statements given above are correct?
(a) 1 and 2 (b) 1 and 3
(c) 1 and 4 (d) 2 and 3

[ESE-2005]

1.12 The basic function of buried n^+ layer in an *n-p-n* transistor in IC is to
(a) Reduce the magnitude of the base spreading resistance
(b) Reduce the collector series resistance
(c) Reduce the base width of the transistor
(d) Increase the gain of the transistor

[ESE-2005]

1.13 Which one of the statements concerning IC fabrication is not correct?
(a) A typical wafer of doped Si may be 400 μm thick, of diameter 5 - 15 cm. The purity of the wafer does not matter and can be even polycrystalline in nature.
(b) Resistors are obtained by utilizing the bulk resistivity of one of the regions; for example, the DS channel of a MOSFET can serve as a resistor.
(c) Semiconductors lack magnetic properties, so they cannot exhibit inductance. However, the inductors can be realized by combination of active and passive components.
(d) In a reverse biased *p-n* junction, the positive and negative ions exist on opposite sides of the *p-n* junction; because of that *p-n* junction behaves like a parallel plate capacitor.

[ESE-2005]

1.14 **Assertion (A):** An IC is more reliable as compared to its discrete circuit equivalent.
Reason (R): All the joints in an IC are soldered simultaneously.
(a) Both A and R are true and R is the correct explanation of A
(b) Both A and R are true but R is NOT the correct explanation of A
(c) A is true but R is false
(d) A is false but R is true

[ESE-2006]

1.15 In integrated circuits, the design of electronic circuits is based on the approach of use of
(a) maximum number of resistors in the circuit
(b) large sized capacitor
(c) minimum chip area irrespective of the type of components in the design
(d) use of only bipolar transistors

[ESE-2006]

1.16 **Assertion (A):** The resistors and capacitors fabricated using IC technology have poor tolerances with respect to their absolute values.
Reason (R): As all the components of the IC are fabricated simultaneously, their ratio of tolerances is very low.
(a) Both (A) and (R) are individually true and (R) is the correct explanation of (A).
(b) Both (A) and (R) are individually true but (R) is not the correct explanation of (A).
(c) (A) is true but (R) is false.
(d) (A) is false but (R) is true.

[ESE-2007]

1.17 Which of the following capacitors are made use of widely for a capacitance application in monolithic ICs.
1. MOS capacitor
2. Collector Substrate capacitor
3. Collector-Base capacitor
4. Base -Emitter capacitor

Select the correct answer using the code given below:
(a) 1 and 2 only (b) 2 and 3 only
(c) 3 and 4 only (d) 1 and 4 only

[ESE-2008]

1.18 Why is silicon dioxide (SiO_2) layer used in ICs?
(a) To protect the surface of the chip from external contaminants and to allow for selective formation of the *n* and *p* regions by diffusion
(b) Because it facilitates the penetration of the desired impurity by diffusion
(c) To control the concentration of the diffused impurities
(d) Because of its high heat conduction
[ESE-2008]

1.19 Why is the term 'planer technology' for fabrication of devices in ICs used?
(a) The variety of manufacturing processes by which devices are fabricated, takes place through a single plane
(b) The aluminium contacts to the collector, base and emitter regions of the tran-sistors in the ICs are laid in the same plane
(c) The collector, base and emitter regions of the transistors in ICs are laid in the same plane
(d) The device looks like a thin plane wafer
[ESE-2008]

1.20 The process of extension of a single-crystal surface by growing a film in such a way that the added atoms form a continuation of the single-crystal structure is called
(a) Ion implantation
(b) Chemical vapour deposition
(c) Electroplating
(d) Epitaxy [ESE-2009]

1.21 The maximum concentration of the element which can be dissolved in solid silicon at a given temperature is termed as
(a) Solid solubility
(b) Dissolution coefficient
(c) Solidification index
(d) Concentration index [ESE-2009]

1.22 **Assertion (A):** Si is mainly used for making ICs and not Ge.
Reason (R): In Si, SiO_2 layer which acts as an insulator can be formed for isolation purposes. Corresponding oxide layer cannot be formed in Ge.
(a) Both A and R are true and R is the correct explanation of A
(b) Both A and R are true but R is NOT the correct explanation of A
(c) A is true but R is false
(d) A is false but R is true [ESE-2010]

1.23 The *p*-type epitaxial layer grown over an *n*-type substrate for fabricating a bipolar transistor will function as
(a) The collector of a *p-n-p* transistor
(b) The base of an *n-p-n* transistor
(c) The emitter of a *p-n-p* transistor
(d) The collector contact for a *p-n-p* transistor
[ESE-2011]

1.24 The biasing of an IC BJT is done by the following biasing scheme:
(a) Potential-divider biasing scheme
(b) Fixed biasing scheme
(c) Current mirror biasing scheme
(d) Collector to base feedback biasing scheme
[ESE-2011]

1.25 **Statement (I):** Concentration of acceptor atoms in the region between isolation islands will be much higher p^+ than in the *p*-type substrate in an integrated circuit.
Statement (II): This is to prevent the depletion region of the reverse-biased isolation to substrate junction from extending into p^+-type material.
(a) Both Statement (I) and Statement (II) are individually true and Statement (II) is the correct explanation of Statement (I).
(b) Both Statement (I) and Statement (II) are individually true but Statement (II) is not the correct explanation of Statement (I).
(c) Statement (I) is true but Statement (II) is false.
(d) Statement (I) is false but Statement (II) is true.
[ESE-2011]

1.26 **Statement (I):** Concentration of acceptor atoms in the region between isolation islands in a monolithic integrated circuit will be much higher than in the *p*-type substrate.
Statement (II): The higher density is provided to prevent the depletion region of the reverse-biased isolation to substrate junction from extending into the p^+-type material.
(a) Both Statement (I) and Statement (II) are individually true and Statement (II) is the correct explanation of Statement (I).
(b) Both Statement (I) and Statement (II) are individually true but Statement (II) is not the correct explanation of Statement (I).
(c) Statement (I) is true but Statement (II) is false.
(d) Statement (I) is false but Statement (II) is true.
[ESE-2012]

1.27 When the photo resist coating (during IC fabrication) is exposed to ultraviolet light the photo resist becomes
(a) oxidized (b) ionized
(c) polymerized (d) brittle [ESE-2013]

1.28 **Statement (I):** All passive components can be fabricated in a single chip.
Statement (II): As opposed to discrete circuits where all components are separately inserted and connected, in an integrated circuit, they are simultaneously created on a chip of semiconductor material during manufacturing.
(a) Both Statement (I) and Statement (II) are individually true and Statement (II) is the correct explanation of Statement (I)
(b) Both Statement (I) and Statement (II) are individually true but Statement (II) is **NOT** the correct explanation of Statement (I)
(c) Statement (I) is true but Statement (II) is false
(d) Statement (I) is false but Statement (II) is true
[ESE-2015]

1.29 In VLSI *n*-MOS process, the thinox mask
(a) patterns the ion implantation within the thinox region
(b) deposits polysilicon all over the thinox region
(c) patterns thickox regions to expose silicon where source, drain or gate areas are required
(d) grows thickox over thinox regions in gate areas
[ESE-2017]

1.30 Which of the following are the advantages of Silicon over Insulator (SOI)?
1. Lower diffusion capacitance
2. Smaller parasitic delay and lower dynamic power consumption
3. Lower threshold voltages
Select the correct answer using the code given below.
(a) 1, 2 and 3 (b) 1 and 2 only
(c) 1 and 3 only (d) 2 and 3 only
[ESE-2019]

1.31 In EPROMs, applying a high voltage to the upper gate causes electrons to jump through the thin oxide onto the floating gate through the process known as
(a) mask programming
(b) one-time programming
(c) avalanche injection or Fowler-Nordheim tunneling
(d) erasing [ESE-2019]

1.32 Silicon dioxide (SiO_2) is used in *IC*s, because it
(a) Facilitates the penetration of diffusants
(b) Has high heat conduction
(c) Prevents diffusion of impurities
(d) Controls the concentration of diffusants
[ESE-2020]

1.33 Consider the following statements for Poly-Si deposition :
1. Poly-Si layer is used for gate electrode of MOSFET because it has similar lattice constants with SiO_2.
2. Poly-Si layer used for gate electrode of MOSFET for the better mechanical stability due to different thermal expansion coefficients.
3. In VLSI circuits, interconnects can be completed in one or two metal levels.
4. Poly-Si is used for short interconnects.
Which of the above statements are correct?
(a) 1 and 2 only (b) 2 and 3 only
(c) 1 and 4 only (d) 2, 3 and 4 only
[ESE-2021]

1.34 If each core in a 16-core processor has a yield of 90% and nothing else on the chip fails, what is the yield of the chip?
(a) $(0.9)^8$ (b) $(0.9)^{16}$
(c) $(0.1)^8$ (d) $(0.1)^{16}$ [ESE-2021]

1.35 Consider the following statements regarding Epitaxial Growth :
1. Thin layers are grown on a substrate wafer, this technique is known as epitaxial growth.
2. Physical vapour deposition is also called vapour phase epitaxy.
3. OMCVD is a technique to grow epitaxial layers from metalorganic compounds.
4. High throughout and slow deposition rate are the disadvantages of the CVD technique.
Which of the above statements are correct?
(a) 2 and 3 only (b) 1 and 3 only
(c) 1, 2 and 3 only (d) 1, 3 and 4 only
[ESE-2021]

1.36 Which one of the following statements is correct regarding integrated circuit fabrication?
(a) IC offers increased reliability, improved performance, high speed and lower power consumption.

(b) IC is a miniature, low cost electronic circuit fabricated on a multi crystal chip of silicon.
(c) IC is a miniature, high cost electronic circuit fabricated on a multi crystal chip of silicon.
(d) IC offers decreased reliability, improved performance, low speed and higher power consumption.

[ESE-2021]

1.37 Which one of the following technologies consumes less power?
(a) Surface-mount technology
(b) CMOS
(c) NMOS
(d) PMOS [ESE-2022]

1.38 Which one of the following requires additional process steps in their fabrication?
(a) Thin film resistor (b) Epitaxial resistor
(c) Pinched resistor (d) Junction resistor

[ESE-2022]

1.39 In monolithic integrated circuits, the concentration of acceptor atoms in the region between isolation islands will be
(a) much higher than in the *p*-type substrate
(b) much lesser than in the *p*-type substrate
(c) equal to the p-type substrate
(d) not equal to the p-type substrate

[ESE-2022]

1.40 Consider the following statements regarding MOS circuit design process :
1. MOS circuits are formed on four basic layers: *n*-diffusion, p-diffusion, polysilicon and metal, which are isolated from one another by thick or thin silicon dioxide insulating layers.
2. Thin oxide (thinox) mask region includes n-diffusion, p-diffusion and transistor channels.
3. Polysilicon and thinox regions interact so that a transistor is formed where they cross one another.

Which of the above statements are correct?
(a) 1 and 2 only (b) 1 and 3 only
(c) 2 and 3 only (d) 1, 2 and 3

[ESE-2022]

1.41 The ratio of total number of transistors on the chip to the number of transistor circuits that must be designed in detail is known as
(a) capacity (b) regularity
(c) design efficiency (d) design density

[ESE-2022]

1.42 The power dissipation under constant field after scaling on MOS device characteristics is
(a) $\frac{P}{S}$ (b) $\frac{2P}{S^2}$
(c) $\frac{P}{S^2}$ (d) $\frac{P}{2S^2}$ [ESE-2023]

1.43 Source/Drain regions doping concentration value used for analysis and simulation of short-channel SOI MESFET is
(a) 10^{10} cm^{-3} (b) 10^{20} cm^{-3}
(c) 10^{15} cm^{-3} (d) 10^{25} cm^{-3}

[ESE-2023]

1.44 In which one of the following is it used that the particular operating wavelength can be selected at the time of device fabrication by an appropriate choice of the reflector spacing?
(a) Distributed feedback laser
(b) Fabry-Perot laser
(c) Tunable lasers
(d) A vertical cavity surface emitting laser

[ESE-2024]

1.45 Consider the following statements for an edge-emitting laser:
1. For small threshold current, Γ should be small to make the best use of the material gain.
2. A thick low band gap active region. (large Γ) leads to increased losses by internal absorption α_i.
3. For large material gain, the quantum well thickness should be large.
4. Absorption losses are often caused by absorption in the high band gap outer barriers forming the waveguide.

Which of the above statements are correct?
(a) 1 and 4 only (b) 1, 2, 3 and 4
(c) 1 and 3 only (d) 2 and 4 only

[ESE-2024]

1.46 Consider the following statements :
1. The VLSI circuit (all the devices and their interconnections) is fabricated on a die. In a single wafer, there are always two dies. The complete one VLSI circuit is equally distributed (fabricated) on each die.

2. Modern technologies use a diamond saw to separate the dies from the wafer.
3. A suitable compound material is used to package the die to provide protection to the die from environmental damage, to prevent the interface to the outside world, to provide the cooling of heat generated due to power dissipation.

Which of the above statements is/are correct ?

(a) 1 and 2 (b) 2 and 3
(c) 2 only (d) 3 only

[ESE-2024]

1.47 Consider the following statements regarding the VLSI design:

1. Parallelism and pipelining have been effective ways to reduce power consumption.
2. Replacing a single functional unit with an N-stage pipelined unit reduces the amount of logic in a clock cycle at the expense of more registers.
3. When leakage is unimportant, parallelism offers a slight edge because the multiplexer has less overhead than the pipeline registers.

Which of the above statements are correct?

(a) 1 and 2 only (b) 2 and 3 only
(c) 1 and 3 only (d) 1, 2 and 3

[ESE-2024]

Answers VLSI Technology

1.1 (c)	1.2 (b)	1.3 (d)	1.4 (a)	1.5 (b)	1.6 (b)	1.7 (b)	1.8 (b)	1.9 (c)
1.10 (a)	1.11 (a)	1.12 (b)	1.13 (a)	1.14 (c)	1.15 (c)	1.16 (a)	1.17 (a)	1.18 (a)
1.19 (a)	1.20 (d)	1.21 (a)	1.22 (a)	1.23 (a)	1.24 (c)	1.25 (a)	1.26 (a)	1.27 (c)
1.28 (d)	1.29 (c)	1.30 (a)	1.31 (c)	1.32 (c)	1.33 (c)	1.34 (b)	1.35 (b)	1.36 (a)
1.37 (b)	1.38 (a)	1.39 (a)	1.40 (d)	1.41 (b)	1.42 (c)	1.43 (b)	1.44 (a)	1.45 (d)
1.46 (b)	1.47 (d)							

Explanations VLSI Technology

1.4 (a)

For fabricating BJT in IC by epitaxial process, Number of diffusion required = 2

1.5 (b)

The heavily doped n^+ buried layer on the p-type substrate is the common method to reduce the collector series resistance.

1.6 (b)

Scaling factor for surface dimensions is $1/\alpha$. Therefore, the scaling factor for the gate area is $1/\alpha \cdot 1/\alpha$, i.e. $1/\alpha^2$.

1.14 (c)

It is the advantage of IC technology over discrete components that it is highly reliable.

1.21 (a)

The solid solubility is defined as the maximum concentration N_0 of the element which can be dissolved in the solid silicon at a given temperature.

1.23 (a)

The p-type epitaxial layer grown over an n-type substrate functions as the collector of p-n-p transistor.

1.24 (c)

To reduce fabrication space current mirror biasing scheme is used in an IC BJT.

1.25 (a)

The depletion width in a region decreases as the doping concentration in that region increases. Since concentration of acceptor atoms is higher, depletion region will not extend more into the p^+ substrate.

1.26 (a)

The depletion width in a region decreases as the doping concentration in that region increases. Since concentration of acceptor atoms is higher, depletion region will not extend more into the p^+ substrate.

1.27 (c)

During IC fabrication when photo resist coating is exposed to UV light then it will become polymerized.

1.28 (d)

All passive components except inductor can be fabricated on single chip because practical values of inductance cannot be realized.
So, statement (I) is false.
In IC, all components are fabricated simultaneously on a single chip.
Hence, statement (II) is true.

1.32 (c)

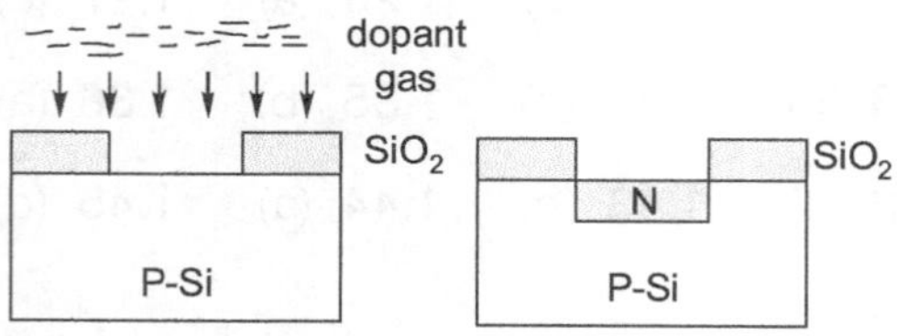

SiO_2 is mask against diffusion.

1.33 (c)

- Polysilicon has similar lattice constants with SiO_2.
- Polysilicon is used for short interconnect only for long interconnect metal is used.
- Metallization (or) interconnection is completed in multiple levels of metals.

Hence, statement 1 and 4 are correct.

1.34 (b)

If a system is built from N components, each of which must work, then the yield of the system Y_t is the product of the yield Y_c of the components

$$Y_t = Y_c^N$$

For a 16-core processor, if each core has yield of 0.9, then

Yield of chip $= (0.9)^{16}$

1.35 (b)

- Epitaxy refers to a type of crystal growth or material deposition in which new crystalline layers are formed with one or more well-defined orientations with respect to the crystalline substrate. Hence, statement 1 is correct.
- Chemical vapour deposition is also called vapour phase epitaxy. Hence, statement 2 is incorrect.
- Organometallic chemical vapor deposition (OMCVD) technique is used for epitaxial deposition. Hence, statement 3 is correct.
- CVD has fast deposition rate. Hence, statement 4 is incorrect.

1.36 (a)

- IC is fabricated on single crystal chip of silicon. Therefore, statements 2 and 3 are incorrect.
- IC offers increased reliability, improved performance, high speed and low power consumption. Therefore, statement 1 is correct.

1.37 (b)

CMOS logic consumes less power. It dissipates less power than NMOS and PMOS logic circuits because CMOS dissipates power only when switching ("dynamic power").

1.38 (a)

In thin film resistor, the fabrication requires additional processing steps. A very thin metallic film of thickness less than 1 μm is deposited on the silicon dioxide layer by vapour deposition techniques. Desired geometry is then achieved using masked etching processes to obtain suitable value of resistors.

1.39 (a)

The concentration of the acceptor atoms in the region between isolation islands is usually kept much higher (p^+) than the p-type substrate. This prevents the depletion region of the reverse biased diode from penetrating more into p^+ region and possibly connecting the isolation islands.

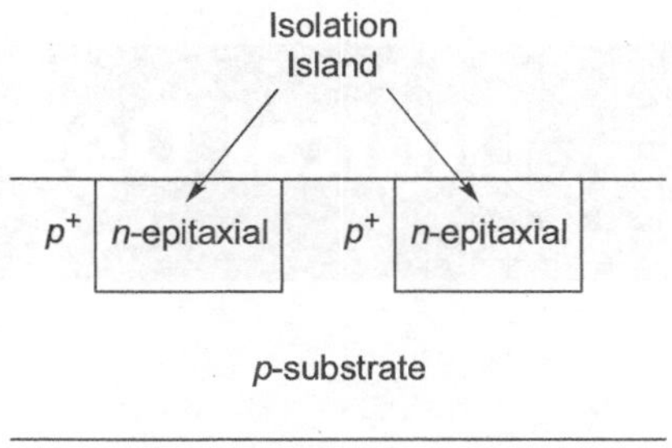

1.40 (d)

All the given statements are correct.

1.41 (b)

The regularity of an integrated circuit design, also called the chip regularity factor is defined as the ratio of the total number of transistors on the chip to the drawn transistors i.e. the number of transistor circuits that must be designed in detail. Thus, a design that requires a unique layout for a circuit element and then uses this circuit element n times without change would exhibit a regularity factor of n. At the other extreme, a design with unique layout for each circuit component would exhibit a regularity factor of only 1.

1.42 (c)

The power dissipation under constant field after scaling on MOS device is $\frac{P}{S^2}$.

1.44 (a)

In distributed feedback laser, a series of closely spaced reflectors provides light feedback in a distributed fashion throughout the semiconductor cavity for enabling lasing. Through a suitable design of these reflectors, the device can be made to oscillate in only a single mode with very narrow linewidth. The particular operating wavelength can be selected at the time of device fabrication by an appropriate choice of the reflector spacing.

1.45 (d)

The design considerations for the edge-emitting laser diodes include:

1. For small threshold current, confinement factor Γ should be large to make the best use of the material gain. The confinement factor describes the overlap of the electric field of the light (photons) with the gain material in the quantum well.
2. However, for large material gain, the quantum well thickness should be small. The active semiconductor material should be very thin (quantum well) in order to increase the local carrier density.
3. A thick low band gap active region (large Γ) leads to increased losses by internal absorption α_i.
4. Absorption losses are often caused by absorption in the high band gap outer barriers forming the waveguide.

1.46 (b)

- Many dies are manufactured on a single wafer in the IC fabrication process.
- Wafer sawing is a common technique used to separate a wafer into individual dies often performed with a diamond blade saw.
- Packaging involves encapsulating the die in a protective material (such as plastic or ceramic), providing electrical connections for interfacing with external devices. It also helps dissipate heat generated by the integrated circuit.

Hence, statements 2 and 3 are correct.

1.47 (d)

All the given statements are correct.

■■■

2 Digital Design

2.1 Consider the following circuit of a scrambler:

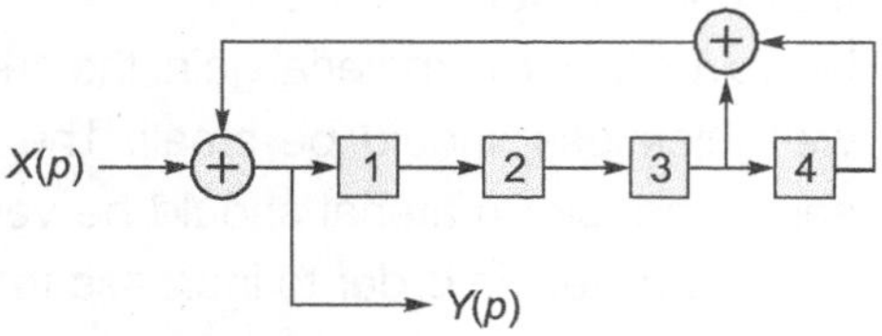

$X(p) \triangleq$ input data stream polynomial

$Y(p) \triangleq$ output data stream polynomial

$\square$'s $\triangleq$ shift-register stages

$\oplus$'s $\triangleq$ XOR gates

Which one of the following relates $X(p)$ and $Y(p)$?

(a) $Y(p) = X(p) / [p^4 + p^3 + 1]$
(b) $Y(p) = p^4X(p) / [p^4 + p^3 + 1]$
(c) $Y(p) = X(p) / [p^4 + p + 1]$
(d) $Y(p) = p^4X(p) / [p^4 + p + 1]$

[ESE-2004]

2.2 In a 3-input CMOS NAND gate, the substrate terminals of NMOS transistors are grounded (lowest potential available in the circuit) and the substrate terminals of PMOS transistors are connected to V_{DD} (maximum positive potential available in the circuit). Which of the following transistors may suffer in this circuit from body bias effect?

(a) 2 NMOS transistors
(b) 2 PMOS transistors
(c) 1 NMOS transistor
(d) 1 PMOS transistor

[ESE-2018]

2.3 The finite state machine in which

1. the output is a function of the current state and inputs.
2. the output is a function of only the current state.

Which of the following machines are respectively correct for these styles?

(a) Mealy machine and Moore machine
(b) Moore machine and Mealy machine
(c) State machine and Mealy machine
(d) State machine and State machine

[ESE-2019]

2.4 Which of the following steps are followed by HIS during synthesis ?

1. Data model generation
2. Data flow analysis
3. Scheduling and allocation
4. Data path optimization
5. Control optimization

(a) 1, 3 and 5 only (b) 2, 4 and 5 only
(c) 1, 2, 3 and 4 only (d) 1, 2, 3, 4 and 5

[ESE-2020]

2.5 Pipelining

1. Reduces clock period of long combinational operations
2. Allows large combinational functions to be broken down into pieces whose delays are balanced with the rest of the system components

(a) 1 only (b) 2 only
(c) Both 1 and 2 (d) Neither 1 nor 2

[ESE-2020]

2.6 **Statement (I):** Control logic in CMOS is constructed using two-level SOP logic and multi-level logic.

Statement (II): Typical PLA uses multi-level logic.

(a) Both Statement (I) and Statement (II) are individually true and Statement (II) is the correct explanation of Statement (I)
(b) Both Statement (I) and Statement (II) are individually true but Statement (II) is **NOT** the correct explanation of Statement (I)
(c) Statement (I) is true but Statement (II) is false
(d) Statement (I) is false but Statement (II) is true

[ESE-2020]

2.7 Consider the following statements regarding the Moore and Mealy models :

1. In the Mealy circuit, the final output depends only on the present state of memory elements.
2. In the Moore circuit, output can change in between the clock edges if the external inputs change.
3. The implementation of a logic function in Mealy circuit needs more number of states than Moore circuit.

Which of the above statements are **not** correct?

(a) 1 and 2 only (b) 1 and 3 only
(c) 1, 2 and 3 (d) 2 and 3 only

[ESE-2021]

2.8 Consider the following statements regarding the sequential circuit :

1. The Moore model sequential circuit is called transition-assigned circuit.
2. In the Mealy model sequential circuit, the output is associated with the state transitions, that is, the arcs in the state diagram.
3. In the Moore model sequential circuit, outputs are the functions of both, the inputs and the present state.

Which of the above statements are not correct?

(a) 1 and 2 only (b) 1 and 3 only
(c) 2 and 3 only (d) 1, 2 and 3

[ESE-2024]

Answers Digital Design

2.1 (d) 2.2 (a) 2.3 (a) 2.4 (d) 2.5 (c) 2.6 (c) 2.7 (c) 2.8 (b)

Explanations Digital Design

2.4 (d)

The main steps followed by industrial high level synthesis systems during synthesis include:

1. **Data modal generation :** Control flow and data flow graphs are generated from the input high level description.
2. **Data flow analysis :** Variable lifetimes are determined explicit clocking constraints are analysed.
3. **Scheduling and allocation :** The operations in the behaviour are scheduled. Register and multiplexers are allocated. Control signals are generated.
4. **Data path optimisation :** Data path optimizations try to efficiently share resources.
5. **Control Optimisations :** State assignment.

2.7 (c)

- In the Mealy machine, the output depends on the present state as well as the present input.
- In the Moore machine, the output depends only on the present state and hence, the output cannot change in between the clock edges if the external inputs change.
- Mealy machine requires fewer states than Moore machine.

Hence, all the statements are incorrect.

2.8 (b)

- The Mealy model represents a transition-assigned circuit, in which the circuit's output is associated with the state transition.
- In the Moore machine, the outputs are a function of the present state only whereas in the Mealy machine, the output depends on the present state as well as the present input.

Hence, statements 1 and 3 are incorrect.